Essentials of Computational Chemistry

Second Edition

Essentials of Computational Chemistry

Theories and Models

Second Edition

Christopher J. Cramer

*Department of Chemistry and Supercomputing Institute,
University of Minnesota, USA*

John Wiley & Sons, Ltd

Copyright © 2004   John Wiley & Sons Ltd, The Atrium, Southern Gate, Chichester,
West Sussex PO19 8SQ, England

Telephone (+44) 1243 779777

Email (for orders and customer service enquiries): cs-books@wiley.co.uk
Visit our Home Page on www.wileyeurope.com or www.wiley.com

Reprinted with corrections March 2006

This publication is designed to provide accurate and authoritative information in regard to the subject matter
covered. It is sold on the understanding that the Publisher is not engaged in rendering professional services. If
professional advice or other expert assistance is required, the services of a competent professional should be
sought.

Other Wiley Editorial Offices

John Wiley & Sons Inc., 111 River Street, Hoboken, NJ 07030, USA

Jossey-Bass, 989 Market Street, San Francisco, CA 94103-1741, USA

Wiley-VCH Verlag GmbH, Boschstr. 12, D-69469 Weinheim, Germany

John Wiley & Sons Australia Ltd, 33 Park Road, Milton, Queensland 4064, Australia

John Wiley & Sons (Asia) Pte Ltd, 2 Clementi Loop #02-01, Jin Xing Distripark, Singapore 129809

John Wiley & Sons Canada Ltd, 22 Worcester Road, Etobicoke, Ontario, Canada M9W 1L1

Wiley also publishes its books in a variety of electronic formats. Some content that appears
in print may not be available in electronic books.

Library of Congress Cataloging-in-Publication Data

Cramer, Christopher J., 1961–
   Essentials of computational chemistry : theories and models /
Christopher J. Cramer. – 2nd ed.
        p. cm.
   Includes bibliographical references and index.
   ISBN 0-470-09181-9 (cloth : alk. paper) – ISBN 0-470-09182-7 (pbk.
: alk. paper)
 1.   Chemistry, Physical and theoretical – Data processing. 2.
Chemistry, Physical and theoretical – Mathematical models.   I. Title.
   QD455.3.E4C73 2004
   541′.0285 – dc22

                                              2004015537

British Library Cataloguing in Publication Data

A catalogue record for this book is available from the British Library

ISBN 0-470-09181-9 (cased)
ISBN 0-470-09182-7 (pbk)

Typeset in 10/12pt Times by Laserwords Private Limited, Chennai, India
Printed and bound in Great Britain by Antony Rowe Ltd, Chippenham, Wiltshire
This book is printed on acid-free paper responsibly manufactured from sustainable forestry
in which at least two trees are planted for each one used for paper production.

For Katherine

Contents

Preface to the First Edition   **xv**

Preface to the Second Edition   **xix**

Acknowledgments   **xxi**

1   What are Theory, Computation, and Modeling?   **1**
1.1   Definition of Terms   1
1.2   Quantum Mechanics   4
1.3   Computable Quantities   5
     1.3.1   Structure   5
     1.3.2   Potential Energy Surfaces   6
     1.3.3   Chemical Properties   10
1.4   Cost and Efficiency   11
     1.4.1   Intrinsic Value   11
     1.4.2   Hardware and Software   12
     1.4.3   Algorithms   14
1.5   Note on Units   15
     Bibliography and Suggested Additional Reading   15
     References   16

2   Molecular Mechanics   **17**
2.1   History and Fundamental Assumptions   17
2.2   Potential Energy Functional Forms   19
     2.2.1   Bond Stretching   19
     2.2.2   Valence Angle Bending   21
     2.2.3   Torsions   22
     2.2.4   van der Waals Interactions   27
     2.2.5   Electrostatic Interactions   30
     2.2.6   Cross Terms and Additional Non-bonded Terms   34
     2.2.7   Parameterization Strategies   36
2.3   Force-field Energies and Thermodynamics   39
2.4   Geometry Optimization   40
     2.4.1   Optimization Algorithms   41
     2.4.2   Optimization Aspects Specific to Force Fields   46

	2.5	Menagerie of Modern Force Fields	50
		2.5.1 Available Force Fields	50
		2.5.2 Validation	59
	2.6	Force Fields and Docking	62
	2.7	Case Study: (2R*,4S*)-1-Hydroxy-2,4-dimethylhex-5-ene	64
		Bibliography and Suggested Additional Reading	66
		References	67

3   Simulations of Molecular Ensembles   69

	3.1	Relationship Between MM Optima and Real Systems	69
	3.2	Phase Space and Trajectories	70
		3.2.1 Properties as Ensemble Averages	70
		3.2.2 Properties as Time Averages of Trajectories	71
	3.3	Molecular Dynamics	72
		3.3.1 Harmonic Oscillator Trajectories	72
		3.3.2 Non-analytical Systems	74
		3.3.3 Practical Issues in Propagation	77
		3.3.4 Stochastic Dynamics	79
	3.4	Monte Carlo	80
		3.4.1 Manipulation of Phase-space Integrals	80
		3.4.2 Metropolis Sampling	81
	3.5	Ensemble and Dynamical Property Examples	82
	3.6	Key Details in Formalism	88
		3.6.1 Cutoffs and Boundary Conditions	88
		3.6.2 Polarization	90
		3.6.3 Control of System Variables	91
		3.6.4 Simulation Convergence	93
		3.6.5 The Multiple Minima Problem	96
	3.7	Force Field Performance in Simulations	98
	3.8	Case Study: Silica Sodalite	99
		Bibliography and Suggested Additional Reading	101
		References	102

4   Foundations of Molecular Orbital Theory   105

	4.1	Quantum Mechanics and the Wave Function	105
	4.2	The Hamiltonian Operator	106
		4.2.1 General Features	106
		4.2.2 The Variational Principle	108
		4.2.3 The Born–Oppenheimer Approximation	110
	4.3	Construction of Trial Wave Functions	111
		4.3.1 The LCAO Basis Set Approach	111
		4.3.2 The Secular Equation	113
	4.4	Hückel Theory	115
		4.4.1 Fundamental Principles	115
		4.4.2 Application to the Allyl System	116
	4.5	Many-electron Wave Functions	119
		4.5.1 Hartree-product Wave Functions	120
		4.5.2 The Hartree Hamiltonian	121
		4.5.3 Electron Spin and Antisymmetry	122
		4.5.4 Slater Determinants	124
		4.5.5 The Hartree-Fock Self-consistent Field Method	126
		Bibliography and Suggested Additional Reading	129
		References	130

5   Semiempirical Implementations of Molecular Orbital Theory   131

   5.1   Semiempirical Philosophy   131
       5.1.1   Chemically Virtuous Approximations   131
       5.1.2   Analytic Derivatives   133
   5.2   Extended Hückel Theory   134
   5.3   CNDO Formalism   136
   5.4   INDO Formalism   139
       5.4.1   INDO and INDO/S   139
       5.4.2   MINDO/3 and SINDO1   141
   5.5   Basic NDDO Formalism   143
       5.5.1   MNDO   143
       5.5.2   AM1   145
       5.5.3   PM3   146
   5.6   General Performance Overview of Basic NDDO Models   147
       5.6.1   Energetics   147
       5.6.2   Geometries   150
       5.6.3   Charge Distributions   151
   5.7   Ongoing Developments in Semiempirical MO Theory   152
       5.7.1   Use of Semiempirical Properties in SAR   152
       5.7.2   d Orbitals in NDDO Models   153
       5.7.3   SRP Models   155
       5.7.4   Linear Scaling   157
       5.7.5   Other Changes in Functional Form   157
   5.8   Case Study: Asymmetric Alkylation of Benzaldehyde   159
       Bibliography and Suggested Additional Reading   162
       References   163

**6   *Ab Initio* Implementations of Hartree–Fock Molecular Orbital
    Theory   165**

   6.1   *Ab Initio* Philosophy   165
   6.2   Basis Sets   166
       6.2.1   Functional Forms   167
       6.2.2   Contracted Gaussian Functions   168
       6.2.3   Single-ζ, Multiple-ζ, and Split-Valence   170
       6.2.4   Polarization Functions   173
       6.2.5   Diffuse Functions   176
       6.2.6   The HF Limit   176
       6.2.7   Effective Core Potentials   178
       6.2.8   Sources   180
   6.3   Key Technical and Practical Points of Hartree–Fock Theory   180
       6.3.1   SCF Convergence   181
       6.3.2   Symmetry   182
       6.3.3   Open-shell Systems   188
       6.3.4   Efficiency of Implementation and Use   190
   6.4   General Performance Overview of *Ab Initio* HF Theory   192
       6.4.1   Energetics   192
       6.4.2   Geometries   196
       6.4.3   Charge Distributions   198
   6.5   Case Study: Polymerization of 4-Substituted Aromatic Enynes   199
       Bibliography and Suggested Additional Reading   201
       References   201

7   Including Electron Correlation in Molecular Orbital Theory   203

7.1   Dynamical vs. Non-dynamical Electron Correlation   203
7.2   Multiconfiguration Self-Consistent Field Theory   205
       7.2.1   Conceptual Basis   205
       7.2.2   Active Space Specification   207
       7.2.3   Full Configuration Interaction   211
7.3   Configuration Interaction   211
       7.3.1   Single-determinant Reference   211
       7.3.2   Multireference   216
7.4   Perturbation Theory   216
       7.4.1   General Principles   216
       7.4.2   Single-reference   219
       7.4.3   Multireference   223
       7.4.4   First-order Perturbation Theory for Some Relativistic Effects   223
7.5   Coupled-cluster Theory   224
7.6   Practical Issues in Application   227
       7.6.1   Basis Set Convergence   227
       7.6.2   Sensitivity to Reference Wave Function   230
       7.6.3   Price/Performance Summary   235
7.7   Parameterized Methods   237
       7.7.1   Scaling Correlation Energies   238
       7.7.2   Extrapolation   239
       7.7.3   Multilevel Methods   239
7.8   Case Study: Ethylenedione Radical Anion   244
       Bibliography and Suggested Additional Reading   246
       References   247

8   Density Functional Theory   249

8.1   Theoretical Motivation   249
       8.1.1   Philosophy   249
       8.1.2   Early Approximations   250
8.2   Rigorous Foundation   252
       8.2.1   The Hohenberg–Kohn Existence Theorem   252
       8.2.2   The Hohenberg–Kohn Variational Theorem   254
8.3   Kohn–Sham Self-consistent Field Methodology   255
8.4   Exchange-correlation Functionals   257
       8.4.1   Local Density Approximation   258
       8.4.2   Density Gradient and Kinetic Energy Density Corrections   263
       8.4.3   Adiabatic Connection Methods   264
       8.4.4   Semiempirical DFT   268
8.5   Advantages and Disadvantages of DFT Compared to MO Theory   271
       8.5.1   Densities vs. Wave Functions   271
       8.5.2   Computational Efficiency   273
       8.5.3   Limitations of the KS Formalism   274
       8.5.4   Systematic Improvability   278
       8.5.5   Worst-case Scenarios   278
8.6   General Performance Overview of DFT   280
       8.6.1   Energetics   280
       8.6.2   Geometries   291
       8.6.3   Charge Distributions   294
8.7   Case Study: Transition-Metal Catalyzed Carbonylation of Methanol   299
       Bibliography and Suggested Additional Reading   300
       References   301

9 Charge Distribution and Spectroscopic Properties 305

9.1 Properties Related to Charge Distribution 305
    9.1.1 Electric Multipole Moments 305
    9.1.2 Molecular Electrostatic Potential 308
    9.1.3 Partial Atomic Charges 309
    9.1.4 Total Spin 324
    9.1.5 Polarizability and Hyperpolarizability 325
    9.1.6 ESR Hyperfine Coupling Constants 327
9.2 Ionization Potentials and Electron Affinities 330
9.3 Spectroscopy of Nuclear Motion 331
    9.3.1 Rotational 332
    9.3.2 Vibrational 334
9.4 NMR Spectral Properties 344
    9.4.1 Technical Issues 344
    9.4.2 Chemical Shifts and Spin–spin Coupling Constants 345
9.5 Case Study: Matrix Isolation of Perfluorinated p-Benzyne 349
Bibliography and Suggested Additional Reading 351
References 351

10 Thermodynamic Properties 355

10.1 Microscopic–macroscopic Connection 355
10.2 Zero-point Vibrational Energy 356
10.3 Ensemble Properties and Basic Statistical Mechanics 357
    10.3.1 Ideal Gas Assumption 358
    10.3.2 Separability of Energy Components 359
    10.3.3 Molecular Electronic Partition Function 360
    10.3.4 Molecular Translational Partition Function 361
    10.3.5 Molecular Rotational Partition Function 362
    10.3.6 Molecular Vibrational Partition Function 364
10.4 Standard-state Heats and Free Energies of Formation and Reaction 366
    10.4.1 Direct Computation 367
    10.4.2 Parametric Improvement 370
    10.4.3 Isodesmic Equations 372
10.5 Technical Caveats 375
    10.5.1 Semiempirical Heats of Formation 375
    10.5.2 Low-frequency Motions 375
    10.5.3 Equilibrium Populations over Multiple Minima 377
    10.5.4 Standard-state Conversions 378
    10.5.5 Standard-state Free Energies, Equilibrium Constants, and Concentrations 379
10.6 Case Study: Heat of Formation of H_2NOH 381
Bibliography and Suggested Additional Reading 383
References 383

11 Implicit Models for Condensed Phases 385

11.1 Condensed-phase Effects on Structure and Reactivity 385
    11.1.1 Free Energy of Transfer and Its Physical Components 386
    11.1.2 Solvation as It Affects Potential Energy Surfaces 389
11.2 Electrostatic Interactions with a Continuum 393
    11.2.1 The Poisson Equation 394
    11.2.2 Generalized Born 402
    11.2.3 Conductor-like Screening Model 404
11.3 Continuum Models for Non-electrostatic Interactions 406
    11.3.1 Specific Component Models 406
    11.3.2 Atomic Surface Tensions 407

11.4 Strengths and Weaknesses of Continuum Solvation Models   410
      11.4.1 General Performance for Solvation Free Energies   410
      11.4.2 Partitioning   416
      11.4.3 Non-isotropic Media   416
      11.4.4 Potentials of Mean Force and Solvent Structure   419
      11.4.5 Molecular Dynamics with Implicit Solvent   420
      11.4.6 Equilibrium vs. Non-equilibrium Solvation   421
11.5 Case Study: Aqueous Reductive Dechlorination of Hexachloroethane   422
      Bibliography and Suggested Additional Reading   424
      References   425

12 Explicit Models for Condensed Phases   **429**
12.1 Motivation   429
12.2 Computing Free-energy Differences   429
      12.2.1 Raw Differences   430
      12.2.2 Free-energy Perturbation   432
      12.2.3 Slow Growth and Thermodynamic Integration   435
      12.2.4 Free-energy Cycles   437
      12.2.5 Potentials of Mean Force   439
      12.2.6 Technical Issues and Error Analysis   443
12.3 Other Thermodynamic Properties   444
12.4 Solvent Models   445
      12.4.1 Classical Models   445
      12.4.2 Quantal Models   447
12.5 Relative Merits of Explicit and Implicit Solvent Models   448
      12.5.1 Analysis of Solvation Shell Structure and Energetics   448
      12.5.2 Speed/Efficiency   450
      12.5.3 Non-equilibrium Solvation   450
      12.5.4 Mixed Explicit/Implicit Models   451
12.6 Case Study: Binding of Biotin Analogs to Avidin   452
      Bibliography and Suggested Additional Reading   454
      References   455

13 Hybrid Quantal/Classical Models   **457**
13.1 Motivation   457
13.2 Boundaries Through Space   458
      13.2.1 Unpolarized Interactions   459
      13.2.2 Polarized QM/Unpolarized MM   461
      13.2.3 Fully Polarized Interactions   466
13.3 Boundaries Through Bonds   467
      13.3.1 Linear Combinations of Model Compounds   467
      13.3.2 Link Atoms   473
      13.3.3 Frozen Orbitals   475
13.4 Empirical Valence Bond Methods   477
      13.4.1 Potential Energy Surfaces   478
      13.4.2 Following Reaction Paths   480
      13.4.3 Generalization to QM/MM   481
13.5 Case Study: Catalytic Mechanism of Yeast Enolase   482
      Bibliography and Suggested Additional Reading   484
      References   485

14 Excited Electronic States   **487**
14.1 Determinantal/Configurational Representation of Excited States   487

14.2	Singly Excited States	492
	14.2.1 SCF Applicability	493
	14.2.2 CI Singles	496
	14.2.3 Rydberg States	498
14.3	General Excited State Methods	499
	14.3.1 Higher Roots in MCSCF and CI Calculations	499
	14.3.2 Propagator Methods and Time-dependent DFT	501
14.4	Sum and Projection Methods	504
14.5	Transition Probabilities	507
14.6	Solvatochromism	511
14.7	Case Study: Organic Light Emitting Diode Alq3	513
	Bibliography and Suggested Additional Reading	515
	References	516

15 Adiabatic Reaction Dynamics — **519**

15.1	Reaction Kinetics and Rate Constants	519
	15.1.1 Unimolecular Reactions	520
	15.1.2 Bimolecular Reactions	521
15.2	Reaction Paths and Transition States	522
15.3	Transition-state Theory	524
	15.3.1 Canonical Equation	524
	15.3.2 Variational Transition-state Theory	531
	15.3.3 Quantum Effects on the Rate Constant	533
15.4	Condensed-phase Dynamics	538
15.5	Non-adiabatic Dynamics	539
	15.5.1 General Surface Crossings	539
	15.5.2 Marcus Theory	541
15.6	Case Study: Isomerization of Propylene Oxide	544
	Bibliography and Suggested Additional Reading	546
	References	546

Appendix A Acronym Glossary — **549**

Appendix B Symmetry and Group Theory — **557**

B.1	Symmetry Elements	557
B.2	Molecular Point Groups and Irreducible Representations	559
B.3	Assigning Electronic State Symmetries	561
B.4	Symmetry in the Evaluation of Integrals and Partition Functions	562

Appendix C Spin Algebra — **565**

C.1	Spin Operators	565
C.2	Pure- and Mixed-spin Wave Functions	566
C.3	UHF Wave Functions	571
C.4	Spin Projection/Annihilation	571
	Reference	574

Appendix D Orbital Localization — **575**

D.1	Orbitals as Empirical Constructs	575
D.2	Natural Bond Orbital Analysis	578
	References	579

Index — **581**

Preface to the First Edition

Computational chemistry, alternatively sometimes called theoretical chemistry or molecular modeling (reflecting a certain factionalization amongst practitioners), is a field that can be said to be both old and young. It is old in the sense that its foundation was laid with the development of quantum mechanics in the early part of the twentieth century. It is young, however, insofar as arguably no technology in human history has developed at the pace that digital computers have over the last 35 years or so. The digital computer being the 'instrument' of the computational chemist, workers in the field have taken advantage of this progress to develop and apply new theoretical methodologies at a similarly astonishing pace.

The evidence of this progress and its impact on Chemistry in general can be assessed in various ways. Boyd and Lipkowitz, in their book series *Reviews in Computational Chemistry*, have periodically examined such quantifiable indicators as numbers of computational papers published, citations to computational chemistry software packages, and citation rankings of computational chemists. While such metrics need not necessarily be correlated with 'importance', the exponential growth rates they document are noteworthy. My own personal (and somewhat more whimsical) metric is the staggering increase in the percentage of exposition floor space occupied by computational chemistry software vendors at various chemistry meetings worldwide – *someone* must be buying those products!

Importantly, the need for at least a cursory understanding of theory/computation/modeling is by no means restricted to practitioners of the art. Because of the broad array of theoretical tools now available, it is a rare problem of interest that does not occupy the attention of both experimental *and* theoretical chemists. Indeed, the synergy between theory and experiment has vastly accelerated progress in any number of areas (as one example, it is hard to imagine a modern paper on the matrix isolation of a reactive intermediate and its identification by infrared spectroscopy not making a comparison of the experimental spectrum to one obtained from theory/calculation). To take advantage of readily accessible theoretical tools, and to understand the results reported by theoretical collaborators (or competitors), even the wettest of wet chemists can benefit from some familiarity with theoretical chemistry. My objective in this book is to provide a survey of computational chemistry – its underpinnings, its jargon, its strengths and weaknesses – that will be accessible to both the experimental and theoretical communities. The level of the presentation assumes exposure to quantum and statistical mechanics; particular topics/examples span the range of inorganic, organic, and biological chemistry. As such, this text could be used in a course populated by senior undergraduates and/or beginning graduate students without regard to specialization.

The scope of theoretical methodologies presented in the text reflects my judgment of the degree to which these methodologies impact on a broad range of chemical problems, i.e., the degree to which a practicing chemist may expect to encounter them repeatedly in the literature and thus should understand their applicability (or lack thereof). In some instances, methodologies that do not find much modern use are discussed because they help to illustrate in an intuitive fashion how more contemporary models developed their current form. Indeed, one of my central goals in this book is to render less opaque the fundamental natures of the various theoretical models. By understanding the assumptions implicit in a theoretical model, and the concomitant limitations imposed by those assumptions, one can make informed judgments about the trustworthiness of theoretical results (and economically sound choices of models to apply, if one is about to embark on a computational project).

With no wish to be divisive, it must be acknowledged: there are some chemists who are not fond of advanced mathematics. Unfortunately, it is simply not possible to describe computational chemistry without resort to a fairly hefty number of equations, and, particularly for modern electronic-structure theories, some of those equations are fantastically daunting in the absence of a detailed knowledge of the field. That being said, I offer a promise to present no equation without an effort to provide an intuitive explanation for its form and the various terms within it. In those instances where I don't think such an explanation *can* be offered (of which there are, admittedly, a few), I will provide a qualitative discussion of the area and point to some useful references for those inclined to learn more.

In terms of layout, it might be preferable from a historic sense to start with quantum theories and then develop classical theories as an approximation to the more rigorous formulation. However, I think it is more pedagogically straightforward (and far easier on the student) to begin with classical models, which are in the widest use by experimentalists and tend to feel very intuitive to the modern chemist, and move from there to increasingly more complex theories. In that same vein, early emphasis will be on single-molecule (gas-phase) calculations followed by a discussion of extensions to include condensed-phase effects. While the book focuses primarily on the calculation of equilibrium properties, excited states and reaction dynamics are dealt with as advanced subjects in later chapters.

The quality of a theory is necessarily judged by its comparison to (accurate) physical measurements. Thus, careful attention is paid to offering comparisons between theory and experiment for a broad array of physical observables (the first chapter is devoted in part to enumerating these). In addition, there *is* some utility in the computation of things which cannot be observed (e.g., partial atomic charges), and these will also be discussed with respect to the performance of different levels of theory. However, the best way to develop a feeling for the scope and utility of various theories is to apply them, and instructors are encouraged to develop computational problem sets for their students. To assist in that regard, case studies appear at the end of most chapters illustrating the employ of one or more of the models most recently presented. The studies are drawn from the chemical literature;

depending on the level of instruction, reading and discussing the original papers as part of the class may well be worthwhile, since any synopsis necessarily does away with some of the original content.

Perversely, perhaps, I do not include in this book specific problems. Indeed, I provide almost no discussion of such nuts and bolts issues as, for example, how to enter a molecular geometry into a given program. The reason I eschew these undertakings is not that I think them unimportant, but that computational chemistry software is not particularly well standardized, and I would like neither to tie the book to a particular code or codes nor to recapitulate material found in users' manuals. Furthermore, the hardware and software available in different venues varies widely, so individual instructors are best equipped to handle technical issues themselves. With respect to illustrative problems for students, there *are* reasonably good archives of such exercises provided either by software vendors as part of their particular package or developed for computational chemistry courses around the world. Chemistry 8021 at the University of Minnesota, for example, has several years worth of problem sets (with answers) available at pollux.chem.umn.edu/8021. Given the pace of computational chemistry development and of modern publishing, such archives are expected to offer a more timely range of challenges in any case.

A brief summary of the mathematical notation adopted throughout this text is in order. Scalar quantities, whether constants or variables, are represented by italic characters. Vectors and matrices are represented by boldface characters (individual matrix *elements* are scalar, however, and thus are represented by italic characters that are indexed by subscript(s) identifying the particular element). Quantum mechanical operators are represented by italic characters if they have scalar expectation values and boldface characters if their expectation values are vectors or matrices (or if they are typically *constructed* as matrices for computational purposes). The only deliberate exception to the above rules is that quantities represented by Greek characters typically are made neither italic nor boldface, irrespective of their scalar or vector/matrix nature.

Finally, as with most textbooks, the total content encompassed herein is such that only the most masochistic of classes would attempt to go through this book cover to cover in the context of a typical, semester-long course. My intent in coverage is not to act as a firehose, but to offer a reasonable degree of flexibility to the instructor in terms of optional topics. Thus, for instance, Chapters 3 and 11–13 could readily be skipped in courses whose focus is primarily on the modeling of small- and medium-sized molecular systems. Similarly, courses with a focus on macromolecular modeling could easily choose to ignore the more advanced levels of quantum mechanical modeling. And, clearly, time constraints in a typical course are unlikely to allow the inclusion of more than one of the last two chapters. These practical points having been made, one can always hope that the eager student, riveted by the content, will take time to read the rest of the book him- or herself!

Christopher J. Cramer
September 2001

Preface to the Second Edition

Since publication of the first edition I have become increasingly, painfully aware of just how short the half-life of certain 'Essentials' can be in a field growing as quickly as is computational chemistry. While I utterly disavow any hubris on my part and indeed blithely assign all blame for this text's title to my editor, that does not detract from my satisfaction at having brought the text up from the ancient history of 2001 to the present of 2004. Hopefully, readers too will be satisfied with what's new and improved.

So, what *is* new and improved? In a nutshell, *new* material includes discussion of docking, principal components analysis, force field validation in dynamics simulations, first-order perturbation theory for relativistic effects, tight-binding density functional theory, electronegativity equalization charge models, standard-state equilibrium constants, computation of pK_a values and redox potentials, molecular dynamics with implicit solvent, and direct dynamics. With respect to *improved* material, the menagerie of modern force fields has been restocked to account for the latest in new and ongoing developments and a *new* menagerie of density functionals has been assembled to help the computational innocent navigate the forest of acronyms (in this last regard, the acronym glossary of Appendix A has also been expanded with an additional 64 entries). In addition, newly developed basis sets for electronic structure calculations are discussed, as are methods to scale various theories to infinite-basis-set limits, and new thermochemical methods. The performances of various more recent methods for the prediction of nuclear magnetic resonance chemical shifts are summarized, and discussion of the generation of condensed-phase potentials of mean force from simulation is expanded.

As developments in semiempirical molecular orbital theory, density functional theory, and continuum solvation models have proceeded at a particularly breakneck pace over the last three years, Chapters 5, 8, and 11 have been substantially reworked and contain much fresh material. In addition, I have tried wherever possible to update discussions and, while so doing, to add the most modern references available so as to improve the text's connection with the primary literature. This effort poses something of a challenge, as I definitely do not want to cross the line from writing a text to writing instead an outrageously lengthy review article – I leave it to the reader to assess my success in that regard. Lastly, the few remaining errors, typographical and otherwise, left over from the second printing of the first edition have been corrected – I accept full responsibility for all of them (with particular apologies to any descendants of Leopold Kronecker) and I thank those readers who called some of them to my attention.

As for important things that have *not* changed, with the exception of Chapter 10 I have chosen to continue to use all of the existing case studies. I consider them still to be sufficiently illustrative of modern application that they remain useful as a basis for thought/discussion, and instructors will inevitably have their own particular favorites that they may discuss 'off-text' in any case. The thorough nature of the index has also, hopefully, not changed, nor I hope the deliberate and careful explanation of all equations, tables, and figures.

Finally, in spite of the somewhat greater corpulence of the second edition compared to the first, I have done my best to maintain the text's liveliness – at least to the extent that a scientific tome can be said to possess that quality. After all, to what end science without humor?

Christopher J. Cramer
July 2004

Acknowledgments

It is a pleasure to recognize the extent to which conversations with my computationally minded colleagues at the University of Minnesota – Jiali Gao, Steven Kass, Ilja Siepmann, Don Truhlar, Darrin York, and the late Jan Almlöf – contributed to this project. As a long-time friend and collaborator, Don in particular has been an invaluable source of knowledge and inspiration. So, too, this book, and particularly the second edition, has been improved based on the input of graduate students either in my research group or taking Computational Chemistry as part of their coursework. Of these, Ed Sherer deserves special mention for having offered detailed and helpful comments on the book when it was in manuscript form. In addition, my colleague Bill Tolman provided inspirational assistance in the preparation of the cover art, and I am grateful to Sheryl Frankel for exceedingly efficient executive assistance. Finally, the editorial staff at Wiley have been consummate professionals with whom it has been a pleasure to work.

Most of the first edition of this book was written during a sabbatical year spent working with Modesto Orozco and Javier Luque at the University of Barcelona. Two more gracious hosts are unlikely to exist anywhere (particularly with respect to ignoring the vast amounts of time the moonlighting author spent writing a book). Support for that sabbatical year derived from the John Simon Guggenheim Foundation, the Spanish Ministry of Education and Culture, the Foundation BBV, and the University of Minnesota, and the generosity of those agencies is gratefully acknowledged. The writing of the second edition was shoehorned into whatever free moments presented themselves, and I thank the members of my research group for not complaining about my assiduous efforts to hide myself from them over the course of a long Minnesota winter.

Finally, if it were not for the heroic efforts of my wife Katherine and the (relative) patience of my children William, Matthew, and Allison – all of whom allowed me to spend a ridiculous number of hours hunched over a keyboard in a non-communicative trance – I most certainly could never have accomplished anything.

1

What are Theory, Computation, and Modeling?

1.1 Definition of Terms

A clear definition of terms is critical to the success of all communication. Particularly in the area of computational chemistry, there is a need to be careful in the nomenclature used to describe predictive tools, since this often helps clarify what approximations have been made in the course of a modeling 'experiment'. For the purposes of this textbook, we will adopt a specific convention for what distinguishes theory, computation, and modeling.

In general, 'theory' is a word with which most scientists are entirely comfortable. A theory is one or more rules that are postulated to govern the behavior of physical systems. Often, in science at least, such rules are quantitative in nature and expressed in the form of a mathematical equation. Thus, for example, one has the theory of Einstein that the energy of a particle, E, is equal to its relativistic mass, m, times the speed of light in a vacuum, c, squared,

$$E = mc^2 \tag{1.1}$$

The quantitative nature of scientific theories allows them to be tested by experiment. This testing is the means by which the applicable range of a theory is elucidated. Thus, for instance, many theories of classical mechanics prove applicable to macroscopic systems but break down for very small systems, where one must instead resort to quantum mechanics. The observation that a theory has limits in its applicability might, at first glance, seem a sufficient flaw to warrant discarding it. However, if a sufficiently large number of 'interesting' systems falls within the range of the theory, practical reasons tend to motivate its continued use. Of course, such a situation tends to inspire efforts to find a more *general* theory that is not subject to the limitations of the original. Thus, for example, classical mechanics can be viewed as a special case of the more general quantum mechanics in which the presence of macroscopic masses and velocities leads to a simplification of the governing equations (and concepts).

Such simplifications of general theories under special circumstances can be key to getting anything useful done! One would certainly *not* want to design the pendulum for a mechanical

Essentials of Computational Chemistry, 2nd Edition Christopher J. Cramer
© 2004 John Wiley & Sons, Ltd   ISBNs: 0-470-09181-9 (cased); 0-470-09182-7 (pbk)

clock using the fairly complicated mathematics of quantal theories, for instance, although the process would ultimately lead to the same result as that obtained from the simpler equations of the more restricted classical theories. Furthermore, at least at the start of the twenty-first century, a generalized 'theory of everything' does not yet exist. For instance, efforts to link theories of quantum electromagnetics and theories of gravity continue to be pursued.

Occasionally, a theory has proven so robust over time, even if only within a limited range of applicability, that it is called a 'law'. For instance, Coulomb's law specifies that the energy of interaction (in arbitrary units) between two point charges is given by

$$E = \frac{q_1 q_2}{\varepsilon r_{12}} \tag{1.2}$$

where q is a charge, ε is the dielectric constant of a homogeneous medium (possibly vacuum) in which the charges are embedded, and r_{12} is the distance between them. However, the term 'law' is best regarded as honorific – indeed, one might regard it as hubris to imply that experimentalists *can* discern the laws of the universe within a finite span of time.

Theory behind us, let us now move on to 'model'. The difference between a theory and a model tends to be rather subtle, and largely a matter of intent. Thus, the goal of a theory tends to be to achieve as great a generality as possible, irrespective of the practical consequences. Quantum theory, for instance, has breathtaking generality, but the practical consequence is that the equations that govern quantum theory are intractable for all but the most ideal of systems. A model, on the other hand, typically involves the deliberate introduction of simplifying approximations into a more general theory so as to extend its practical utility. Indeed, the approximations sometimes go to the extreme of rendering the model deliberately qualitative. Thus, one can regard the valence-shell-electron-pair repulsion (VSEPR; an acronym glossary is provided as Appendix A of this text) model familiar to most students of inorganic chemistry as a drastic simplification of quantum mechanics to permit discrete choices for preferred conformations of inorganic complexes. (While serious theoreticians may shudder at the empiricism that often governs such drastic simplifications, and mutter gloomily about lack of 'rigor', the value of a model is not in its intrinsic beauty, of course, but in its ability to solve practical problems; for a delightful cartoon capturing the hubris of theoretical dogmatism, see Ghosh 2003.)

Another feature sometimes characteristic of a *quantitative* 'model' is that it incorporates certain constants that are derived wholly from experimental data, i.e., they are empirically determined. Again, the degree to which this distinguishes a model from a theory can be subtle. The speed of light and the charge of the electron are fundamental constants of the universe that appear either explicitly or implicitly in Eqs. (1.1) and (1.2), and we know these values only through experimental measurement. So, again, the issue tends to be intent. A model is often designed to apply specifically to a restricted volume of what we might call chemical space. For instance, we might imagine developing a model that would predict the free energy of activation for the hydrolysis of substituted β-lactams in water. Our motivation, obviously, would be the therapeutic utility of these species as antibiotics. Because we are limiting ourselves to consideration of only very specific kinds of bond-making and bond-breaking, we may be able to construct a model that takes advantage of a few experimentally known free energies of activation and correlates them with some other measured or predicted

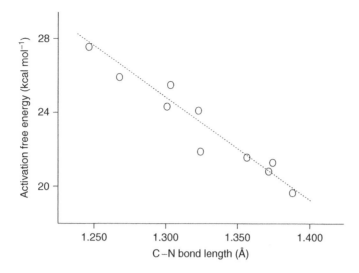

Figure 1.1   Correlation between activation free energy for aqueous hydrolysis of β-lactams and lactam C–N bond lengths as determined from X-ray crystallography (data entirely fictitious)

quantity. For example, we might find from comparison with X-ray crystallography that there is a linear correlation between the aqueous free energy of activation, $\Delta G^{\ddagger}$, and the length of the lactam C–N bond in the crystal, r_{CN} (Figure 1.1). Our 'model' would then be

$$\Delta G^{\ddagger} = ar_{\mathrm{CN}} + b \tag{1.3}$$

where a would be the slope (in units of energy per length) and b the intercept (in units of energy) for the empirically determined correlation.

Equation (1.3) represents a very simple model, and that simplicity derives, presumably, from the small volume of chemical space over which it appears to hold. As it is hard to imagine deriving Eq. (1.3) from the fundamental equations of quantum mechanics, it might be more descriptive to refer to it as a 'relationship' rather than a 'model'. That is, we make some attempt to distinguish between correlation and causality. For the moment, we will not parse the terms too closely.

An interesting question that arises with respect to Eq. (1.3) is whether it may be more broadly applicable. For instance, might the model be useful for predicting the free energies of activation for the hydrolysis of γ-lactams? What about amides in general? What about imides? In a statistical sense, these chemical questions are analogous to asking about the degree to which a correlation may be trusted for extrapolation vs. interpolation. One might say that we have derived a correlation involving two axes of multi-dimensional chemical space, activation free energy for β-lactam hydrolysis and β-lactam C–N bond length. Like any correlation, our model is expected to be most robust when used in an interpolative sense, i.e., when applied to newly measured β-lactam C–N bonds with lengths that fall within the range of the data used to derive the correlation. Increasingly less certain will be application of Eq. (1.3) to β-lactam bond lengths that are *outside* the range used to derive the correlation, or assumption that other chemical axes, albeit qualitatively similar (like γ-lactam C–N bond lengths), will be coincident with the abscissa.

Thus, a key question in one's mind when evaluating any application of a theoretical model should be, 'How similar is the system being studied to systems that were employed in the development of the model?' The generality of a given model can only be established by comparison to experiment for a wider and wider variety of systems. This point will be emphasized repeatedly throughout this text.

Finally, there is the definition of 'computation'. While theories and models like those represented by Eqs. (1.1), (1.2), and (1.3), are not particularly taxing in terms of their mathematics, many others can only be efficiently put to use with the assistance of a digital computer. Indeed, there is a certain synergy between the development of chemical theories and the development of computational hardware, software, etc. If a theory cannot be tested, say because solution of the relevant equations lies outside the scope of practical possibility, then its utility cannot be determined. Similarly, advances in computational technology can permit existing theories to be applied to increasingly complex systems to better gauge the degree to which they are robust. These points are expanded upon in Section 1.4. Here we simply close with the concise statement that 'computation' is the use of digital technology to solve the mathematical equations defining a particular theory or model.

With all these definitions in hand, we may return to a point raised in the preface, namely, what is the difference between 'Theory', 'Molecular Modeling', and 'Computational Chemistry'? To the extent members of the community make distinctions, 'theorists' tend to have as their greatest goal the development of new theories and/or models that have improved performance or generality over existing ones. Researchers involved in 'molecular modeling' tend to focus on target systems having particular chemical relevance (e.g., for economic reasons) and to be willing to sacrifice a certain amount of theoretical rigor in favor of getting the right answer in an efficient manner. Finally, 'computational chemists' may devote themselves not to chemical aspects of the problem, *per se*, but to computer-related aspects, e.g., writing improved algorithms for solving particularly difficult equations, or developing new ways to encode or visualize data, either as input to or output from a model. As with any classification scheme, there are no distinct boundaries recognized either by observers or by individual researchers, and certainly a given research endeavor may involve significant efforts undertaken within all three of the areas noted above. In the spirit of inclusiveness, we will treat the terms as essentially interchangeable.

1.2   Quantum Mechanics

The postulates and theorems of quantum mechanics form the rigorous foundation for the prediction of observable chemical properties from first principles. Expressed somewhat loosely, the fundamental postulates of quantum mechanics assert that microscopic systems are described by 'wave functions' that completely characterize all of the physical properties of the system. In particular, there are quantum mechanical 'operators' corresponding to each physical observable that, when applied to the wave function, allow one to predict the probability of finding the system to exhibit a particular value or range of values (scalar, vector, etc.) for that observable. This text assumes prior exposure to quantum mechanics and some familiarity with operator and matrix formalisms and notation.

However, many successful chemical models exist that do not necessarily have obvious connections with quantum mechanics. Typically, these models were developed based on intuitive concepts, i.e., their forms were determined inductively. In principle, any successful model *must* ultimately find its basis in quantum mechanics, and indeed *a posteriori* derivations have illustrated this point in select instances, but often the form of a good model is more readily grasped when rationalized on the basis of intuitive chemical concepts rather than on the basis of quantum mechanics (the latter being desperately non-intuitive at first blush).

Thus, we shall leave quantum mechanics largely unreviewed in the next two chapters of this text, focusing instead on the intuitive basis for classical models falling under the heading of 'molecular mechanics'. Later in the text, we shall see how some of the fundamental approximations used in molecular mechanics can be justified in terms of well-defined approximations to more complete quantum mechanical theories.

1.3 Computable Quantities

What predictions can be made by the computational chemist? In principle, if one can measure it, one can predict it. In practice, some properties are more amenable to accurate computation than others. There is thus some utility in categorizing the various properties most typically studied by computational chemists.

1.3.1 Structure

Let us begin by focusing on isolated molecules, as they are the fundamental unit from which pure substances are constructed. The minimum information required to specify a molecule is its molecular formula, i.e., the atoms of which it is composed, and the manner in which those atoms are connected. Actually, the latter point should be put more generally. What is required is simply to know the relative positions of all of the atoms in space. Connectivity, or 'bonding', is itself a property that is open to determination. Indeed, the determination of the 'best' structure from a chemically reasonable (or unreasonable) guess is a very common undertaking of computational chemistry. In this case 'best' is defined as having the lowest possible energy given an overall connectivity roughly dictated by the starting positions of the atoms as chosen by the theoretician (the process of structure optimization is described in more detail in subsequent chapters).

This sounds relatively simple because we are talking about the modeling of an isolated, single molecule. In the laboratory, however, we are much more typically dealing with an equilibrium mixture of a very large number of molecules at some non-zero temperature. In that case, *measured* properties reflect thermal averaging, possibly over multiple discrete stereoisomers, tautomers, etc., that are structurally quite different from the idealized model system, and great care must be taken in making comparisons between theory and experiment in such instances.

1.3.2   Potential Energy Surfaces

The first step to making the theory more closely mimic the experiment is to consider not just one structure for a given chemical formula, but all possible structures. That is, we fully characterize the potential energy surface (PES) for a given chemical formula (this requires invocation of the Born–Oppenheimer approximation, as discussed in more detail in Chapters 4 and 15). The PES is a hypersurface defined by the potential energy of a collection of atoms over all possible atomic arrangements; the PES has $3N - 6$ coordinate dimensions, where N is the number of atoms ≥ 3. This dimensionality derives from the three-dimensional nature of Cartesian space. Thus each structure, which is a point on the PES, can be defined by a vector $\mathbf{X}$ where

$$\mathbf{X} \equiv (x_1, y_1, z_1, x_2, y_2, z_2, \ldots, x_N, y_N, z_N) \tag{1.4}$$

and x_i, y_i, and z_i are the Cartesian coordinates of atom i. However, this expression of $\mathbf{X}$ does not *uniquely* define the structure because it involves an arbitrary origin. We can reduce the dimensionality without affecting the structure by removing the three dimensions associated with translation of the structure in the x, y, and z directions (e.g., by insisting that the molecular center of mass be at the origin) and removing the three dimensions associated with rotation about the x, y, and z axes (e.g., by requiring that the principal moments of inertia align along those axes in increasing order).

A different way to appreciate this reduced dimensionality is to imagine constructing a structure vector atom by atom (Figure 1.2), in which case it is most convenient to imagine the dimensions of the PES being internal coordinates (i.e., bond lengths, valence angles, etc.). Thus, choice of the first atom involves no degrees of geometric freedom – the atom defines the origin. The position of the second atom is specified by its distance from the first. So, a two-atom system has a single degree of freedom, the bond length; this corresponds to $3N - 5$ degrees of freedom, as should be the case for a linear molecule. The third atom must be specified either by its distances to each of the preceding atoms, or by a distance to one and an angle between the two bonds thus far defined to a common atom. The three-atom system, if collinearity is not enforced, has 3 total degrees of freedom, as it should. Each additional atom requires three coordinates to describe its position. There are several ways to envision describing those coordinates. As in Figure 1.2, they can either be a bond length, a valence angle, and a dihedral angle, or they can be a bond length and two valence angles. Or, one can imagine that the first three atoms have been used to create a fixed Cartesian reference frame, with atom 1 defining the origin, atom 2 defining the direction of the positive x axis, and atom 3 defining the upper half of the xy plane. The choice in a given calculation is a matter of computational convenience. Note, however, that the *shapes* of particular surfaces necessarily depend on the choice of their coordinate systems, although they will map to one another in a one-to-one fashion.

Particularly interesting points on PESs include local minima, which correspond to optimal molecular structures, and saddle points (i.e., points characterized by having no slope in any direction, downward curvature for a single coordinate, and upward curvature for all of the other coordinates). Simple calculus dictates that saddle points are lowest energy barriers

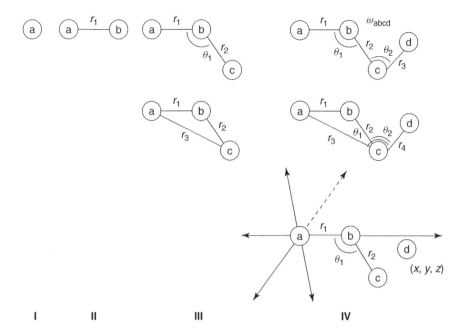

Figure 1.2 Different means for specifying molecular geometries. In frame **I**, there are no degrees of freedom as only the nature of atom 'a' has been specified. In frame **II**, there is a single degree of freedom, namely the bond length. In frame **III**, location of atom 'c' requires two additional degrees of freedom, either two bond lengths or a bond length and a valence angle. Frame **IV** illustrates various ways to specify the location of atom 'd'; note that in every case, three new degrees of freedom must be specified, either in internal or Cartesian coordinates

on paths connecting minima, and thus they can be related to the chemical concept of a transition state. So, a complete PES provides, for a given collection of atoms, complete information about all possible chemical structures and all isomerization pathways interconnecting them.

Unfortunately, complete PESs for polyatomic molecules are very hard to visualize, since they involve a large number of dimensions. Typically, we take slices through potential energy surfaces that involve only a single coordinate (e.g., a bond length) or perhaps two coordinates, and show the relevant reduced-dimensionality energy curves or surfaces (Figure 1.3). Note that some care must be taken to describe the nature of the slice with respect to the *other* coordinates. For instance, was the slice a hyperplane, implying that all of the non-visualized coordinates have fixed values, or was it a more general hypersurface? A typical example of the latter choice is one where the non-visualized coordinates take on values that minimize the potential energy given the value of the visualized coordinate(s). Thus, in the case of a single visualized dimension, the curve attempts to illustrate the minimum energy path associated with varying the visualized coordinate. [We must say 'attempts' here, because an actual continuous path connecting any two structures on a PES may involve any number of structures all of which have the same value for a single internal coordinate. When that

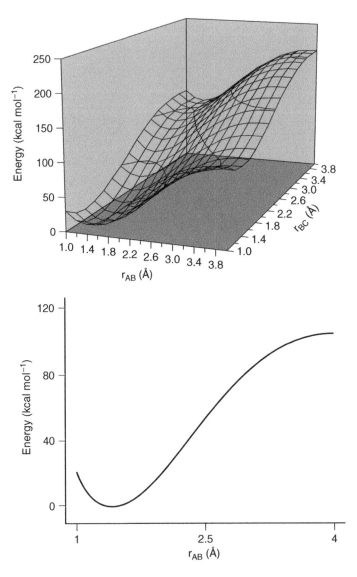

Figure 1.3 The full PES for the hypothetical molecule ABC requires four dimensions to display ($3N - 6 = 3$ coordinate degrees of freedom plus one dimension for energy). The three-dimensional plot (top) represents a hyperslice through the full PES showing the energy as a function of two coordinate dimensions, the AB and BC bond lengths, while taking a fixed value for the angle ABC (a typical choice might be the value characterizing the global minimum on the full PES). A further slice of this surface (bottom) now gives the energy as a function of a single dimension, the AB bond length, where the BC bond length is now also treated as frozen (again at the equilibrium value for the global minimum)

path is projected onto the dimension defined by that single coordinate (or any reduced number of dimensions including it) the resulting curve is a non-single-valued function of the dimension. When we arbitrarily choose to use the lowest energy point for each value of the varied coordinate, we may introduce discontinuities in the actual structures, even though the curve may appear to be smooth (Figure 1.4). Thus, the generation and interpretation of such 'partially relaxed' potential energy curves should involve a check of the individual structures to ensure that such a situation has not arisen.]

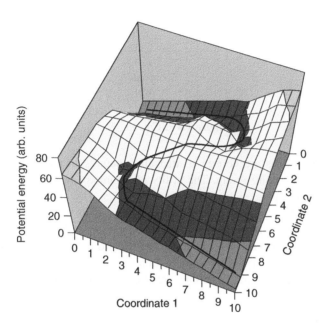

Figure 1.4 The bold line in (a) traces out a lowest-energy path connecting two minima of energy 0, located at coordinates (0,1) and (10,9), on a hypothetical three-dimensional PES – shaded regions correspond to contour levels spanning 20 energy units. Following the path starting from point (0,1) in the upper left, coordinate 1 initially smoothly increases to a value of about 7.5 while coordinate 2 undergoes little change. Then, however, because of the coupling between the two coordinates, coordinate 1 begins *decreasing* while coordinate 2 changes. The 'transition state structure' (saddle point) is reached at coordinates (5,5) and has energy 50. On this PES, the path downward is the symmetric reverse of the path up. If the full path is projected so as to remove coordinate 2, the two-dimensional potential energy diagram (b) is generated. The solid curve is what would result if we only considered lowest energy structures having a given value of coordinate 1. Of course, the solid curve is discontinuous in coordinate 2, since approaches to the 'barrier' in the solid curve from the left and right correspond to structures having values for coordinate 2 of about 1 and 9, respectively. The dashed curve represents the higher energy structures that appear on the smooth, continuous, three-dimensional path. If the lower potential energy diagram were to be generated by driving coordinate 1, and care were not taken to note the discontinuity in coordinate 2, the barrier for interconversion of the two minima would be underestimated by a factor of 2 in this hypothetical example. (For an actual example of this phenomenon, see Cramer *et al.* 1994.)

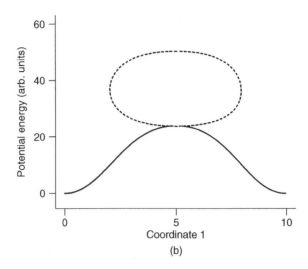

Figure 1.4 (*Continued*)

Finally, sometimes slices are chosen so that all structures in the slicing surface belong to a particular symmetry point group. The utility of symmetry will be illustrated in various situations throughout the text.

With the complete PES in hand (or, more typically, with the region of the PES that would be expected to be chemically accessible under the conditions of the experimental system being modeled), one can take advantage of standard precepts of statistical mechanics (see Chapter 10) to estimate equilibrium populations for situations involving multiple stable molecular structures and compute ensemble averages for physical observables.

1.3.3 Chemical Properties

One can arbitrarily divide the properties one might wish to estimate by computation into three classes. The first is 'single-molecule' properties, that is, properties that could in principle be measured from a single molecule, even though, in practice, use of a statistical ensemble may be required for practical reasons. Typical examples of such properties are spectral quantities. Thus, theory finds considerable modern application to predicting nuclear magnetic resonance (NMR) chemical shifts and coupling constants, electron paramagnetic resonance (EPR) hyperfine coupling constants, absorption maxima for rotational, vibrational, and electronic spectra (typically in the microwave, infrared, and ultraviolet/visible regions of the spectrum, respectively), and electron affinities and ionization potentials (see Chapter 9).

With respect to molecular energetics, one can, in principle, measure the total energy of a molecule (i.e., the energy required to separate it into its constituent nuclei and electrons all infinitely separated from one another and at rest). More typically, however, laboratory measurements focus on thermodynamic quantities such as enthalpy, free energy, etc., and this is the second category into which predicted quantities fall. Theory is extensively used to estimate equilibrium constants, which are derived from free energy differences between minima on a PES, and rate constants, which, with certain assumptions (see Chapter 15), are derived from free energy differences between minima on a PES and connected transition-state structures. Thus, theory may be used to predict reaction thermochemistries, heats of formation and combustion, kinetic isotope effects, complexation energies (key to molecular recognition), acidity and basicity (e.g., pK_a values), 'stability', and hydrogen bond strengths, to name a few properties of special interest. With a sufficiently large collection of molecules being modeled, theory can also, in principle, compute bulk thermodynamic phenomena such as solvation effects, phase transitions, etc., although the complexity of the system may render such computations quite challenging.

Finally, there are computable 'properties' that do not correspond to physical observables. One may legitimately ask about the utility of such ontologically indefensible constructs! However, one should note that unmeasurable properties long predate computational chemistry – some examples include bond order, aromaticity, reaction concertedness, and isoelectronic, -steric, and -lobal behavior. These properties involve *conceptual* models that have proven sufficiently useful in furthering chemical understanding that they have overcome objections to their not being uniquely defined.

In cases where such models take measurable quantities as input (e.g., aromaticity models that consider heats of hydrogenation or bond-length alternation), clearly those measurable quantities are also computable. There are additional non-observables, however, that are unique to modeling, usually being tied to some aspect of the computational algorithm. A good example is atomic partial charge (see Chapter 9), which can be a very useful chemical concept for understanding molecular reactivity.

1.4   Cost and Efficiency

1.4.1   Intrinsic Value

Why has the practice of computational chemistry skyrocketed in the last few years? Try taking this short quiz: Chemical waste disposal and computational technology – which of these two keeps getting more and more expensive and which less and less? From an economic perspective, at least, theory is enormously attractive as a tool to reduce the costs of doing experiments.

Chemistry's impact on modern society is most readily perceived in the creation of materials, be they foods, textiles, circuit boards, fuels, drugs, packaging, etc. Thus, even the most ardent theoretician would be unlikely to suggest that theory could ever *supplant* experiment. Rather, most would opine that opportunities exist for *combining* theory with experiment so as to take advantage of synergies between them.

With that in mind, one can categorize efficient combinations of theory and experiment into three classes. In the first category, theory is applied *post facto* to a situation where some ambiguity exists in the interpretation of existing experimental results. For example, photolysis of a compound in an inert matrix may lead to a single product species as analyzed by spectroscopy. However, the identity of this unique product may not be obvious given a number of plausible alternatives. A calculation of the energies and spectra for *all* of the postulated products provides an opportunity for comparison and may prove to be definitive.

In the second category, theory may be employed in a simultaneous fashion to optimize the design and progress of an experimental program. Continuing the above analogy, *a priori* calculation of spectra for plausible products may assist in choosing experimental parameters to permit the observation of minor components which might otherwise be missed in a complicated mixture (e.g., theory may allow the experimental instrument to be tuned properly to observe a signal whose location would not otherwise be predictable).

Finally, theory may be used to predict properties which might be especially difficult or dangerous (i.e., costly) to measure experimentally. In the difficult category are such data as rate constants for the reactions of trace, upper-atmospheric constituents that might play an important role in the ozone cycle. For sufficiently small systems, levels of quantum mechanical theory can now be brought to bear that have accuracies comparable to the best modern experimental techniques, and computationally derived rate constants may find use in complex kinetic models until such time as experimental data are available. As for dangerous experiments, theoretical pre-screening of a series of toxic or explosive compounds for desirable (or undesirable) properties may assist in prioritizing the order in which they are prepared, thereby increasing the probability that an acceptable product will be arrived at in a maximally efficient manner.

1.4.2 Hardware and Software

All of these points being made, even computational chemistry is not without cost. In general, the more sophisticated the computational model, the more expensive in terms of computational resources. The talent of the well-trained computational chemist is knowing how to maximize the accuracy of a prediction while minimizing the investment of such resources. A primary goal of this text is to render more clear the relationship between accuracy and cost for various levels of theory so that even relatively inexperienced users can make informed assessments of the likely utility (before the fact) or credibility (after the fact) of a given calculation.

To be more specific about computational resources, we may, without going into a great deal of engineering detail, identify three features of a modern digital computer that impact upon its utility as a platform for molecular modeling. The first feature is the speed with which it carries out mathematical operations. Various metrics are used when comparing the speed of 'chips', which are the fundamental processing units. One particularly useful one is the number of floating-point operations per second (FLOPS) that the chip can accomplish. That is, how many mathematical manipulations of decimally represented numbers can be carried out (the equivalent measure for integers is IPS). Various benchmark computer codes are available for comparing one chip to another, and one should always bear in mind that measured processor speeds are dependent on which code or set of codes was used. Different kinds of mathematical operations or different orderings of operations can have effects as large as an order of magnitude on individual machine speeds because of the way the processors are designed and because of the way they interact with other features of the computational hardware.

The second feature affecting performance is memory. In order to carry out a floating-point operation, there must be floating-point numbers on which to operate. Numbers (or characters) to be processed are stored in a magnetic medium referred to as memory. In a practical sense, the size of the memory associated with a given processor sets the limit on the total amount of information to which it has 'instant' access. In modern multiprocessor machines, this definition has grown more fuzzy, as there tend to be multiple memory locations, and the speed with which a given processor can access a given memory site varies depending upon their physical locations with respect to one another. The somewhat unsurprising bottom line is that more memory and shorter access times tend to lead to improved computational performance.

The last feature is storage, typically referred to as disk since that has been the read/write storage medium of choice for the last several years. Storage is exactly like memory, in the sense that it holds number or character data, but it is accessible to the processing unit at a much slower rate than is memory. It makes up for this by being much cheaper and being, in principle, limitless and permanent. Calculations which need to read and/or write data to a disk necessarily proceed more slowly than do calculations that can take place entirely in memory. The difference is sufficiently large that there are situations where, rather than storing on disk data that will be needed later, it is better to throw them away (because memory limits require you to overwrite the locations in which they are stored), as subsequent recomputation of the needed data is faster than reading it back from disk storage. Such a protocol is usually called a 'direct' method (see Almlöf, Faegri, and Korsell 1982).

Processors, memory, and storage media are components of a computer referred to as 'hardware'. However, the efficiency of a given computational task depends also on the nature of the instructions informing the processor how to go about implementing that task. Those instructions are encoded in what is known as 'software'. In terms of computational chemistry, the most obvious piece of software is the individual program or suite of programs with which the chemist interfaces in order to carry out a computation. However, that is by no means the only software involved. Most computational chemistry software consists of a large set of instructions written in a 'high-level' programming language (e.g., FORTRAN or C++), and choices of the user dictate which sets of instructions are followed in which order. The collection of all such instructions is usually called a 'code' (listings of various computational chemistry codes can be found at websites such as http://cmm.info.nih.gov/modeling/software.html). But the language of the code cannot be interpreted directly by the processor. Instead, a series of other pieces of software (compilers, assemblers, etc.) translate the high-level language instructions into the step-by-step operations that are carried out by the processing unit. Understanding how to write code (in whatever language) that takes the best advantage of the total hardware/software environment on a particular computer is a key aspect to the creation of an efficient software package.

1.4.3   Algorithms

In a related sense, the manner in which mathematical equations are turned into computer instructions is also key to efficient software development. Operations like addition and subtraction do not allow for much in the way of innovation, needless to say, but operations like matrix diagonalization, numerical integration, etc., are sufficiently complicated that different algorithms leading to the same (correct) result can vary markedly in computational performance. A great deal of productive effort in the last decade has gone into the development of so-called 'linear-scaling' algorithms for various levels of theory. Such an algorithm is one that permits the cost of a computation to scale roughly linearly with the size of the system studied. At first, this may not sound terribly demanding, but a quick glance back at Coulomb's law [Eq. (1.2)] will help to set this in context. Coulomb's law states that the potential energy from the interaction of charged particles depends on the pairwise interaction of all such particles. Thus, one might expect any calculation of this quantity to scale as the *square* of the size of the system (there are $n(n-1)/2$ such interactions where n is the number of particles). However, for sufficiently large systems, sophisticated mathematical 'tricks' permit the scaling to be brought down to linear.

In this text, we will not be particularly concerned with algorithms – not because they are not important but because such concerns are more properly addressed in advanced textbooks aimed at future practitioners of the art. Our focus will be primarily on the conceptual aspects of particular computational models, and not necessarily on the most efficient means for implementing them.

We close this section with one more note on careful nomenclature. A 'code' renders a 'model' into a set of instructions that can be understood by a digital computer. Thus, if one applies a particular model, let us say the molecular mechanics model called MM3 (which will be described in the next chapter) to a particular problem, say the energy of chair cyclohexane, the results should be completely independent of which code one employs to carry out the calculation. If two pieces of software (let's call them MYPROG and YOURPROG) differ by more than the numerical noise that can arise because of different round-off conventions with different computer chips (or having set different tolerances for what constitutes a converged calculation) then one (or both!) of those pieces of software is *incorrect*. In colloquial terms, there is a 'bug' in the incorrect code(s).

Furthermore, it is never correct to refer to the results of a calculation as deriving from the code, e.g., to talk about one's 'MYPROG structure'. Rather, the results derive from the model, and the structure is an 'MM3 structure'. It is not simply incorrect to refer to the results of the calculation by the name of the code, it is confusing: MYPROG may well contain code for several *different* molecular mechanics models, not just MM3, so simply naming the program is insufficiently descriptive.

It is regrettable, but must be acknowledged, that certain models found in the chemical literature are themselves not terribly well defined. This tends to happen when features or parameters of a model are updated without any change in the name of the model as assigned by the original authors. When this happens, codes implementing older versions of the model will disagree with codes implementing newer versions even though each uses the same name for the model. Obviously, developers should scrupulously avoid ever allowing this situation

Table 1.1 Useful quantities in atomic and other units

Physical quantity (unit name)	Symbol	Value in a.u.	Value in SI units	Value(s) in other units
Angular momentum	$\hbar$	1	1.055×10^{-34} J s	2.521×10^{-35} cal s
Mass	m_e	1	9.109×10^{-31} kg	
Charge	e	1	1.602×10^{-19} C	1.519×10^{-14} statC
Vacuum permittivity	$4\pi\varepsilon_0$	1	1.113×10^{-10} C^2 J^{-1} m^{-1}	2.660×10^{-21} C^2 cal^{-1} Å^{-1}
Length (bohr)	a_0	1	5.292×10^{-11} m	0.529 Å
				52.9 pm
Energy (hartree)	E_h	1	4.360×10^{-18} J	627.51 kcal mol^{-1}
				2.626×10^3 kJ mol^{-1}
				27.211 eV
				2.195×10^5 cm^{-1}
Electric dipole moment	ea_0	1	8.478×10^{-30} C m	2.542 D
Electric polarizability	$e^2 a_0^2 E_h^{-1}$	1	1.649×10^{-41} C^2 m^2 J^{-1}	
Planck's constant	h	2π	6.626×10^{-34} J s	
Speed of light	c	1.370×10^2	2.998×10^8 m s^{-1}	
Bohr magneton	μ_B	0.5	9.274×10^{-24} J T^{-1}	
Nuclear magneton	μ_N	2.723×10^{-4}	5.051×10^{-27} J T^{-1}	

to arise. To be safe, scientific publishing that includes computational results should always state what code or codes were used, *to include version numbers*, in obtaining particular model results (clearly version control of computer codes is thus just as critical as it is for models).

1.5 Note on Units

In describing a computational model, a clear equation can be worth 1000 words. One way to render equations more clear is to work in atomic (or theorist's) units. In a.u., the charge on the proton, e, the mass of the electron, m_e, and $\hbar$ (i.e., Planck's constant divided by 2π) are all defined to have magnitude 1. When converting equations expressed in SI units (as opposed to Gaussian units), $4\pi\varepsilon_0$, where ε_0 is the permittivity of the vacuum, is also defined to have magnitude 1. As the magnitude of these quantities is unity, they are dropped from relevant equations, thereby simplifying the notation. Other atomic units having magnitudes of unity can be derived from these three by dimensional analysis. For instance, $\hbar^2/m_e e^2$ has units of distance and is defined as 1 a.u.; this atomic unit of distance is also called the 'bohr' and symbolized by a_0. Similarly, e^2/a_0 has units of energy, and defines 1 a.u. for this quantity, also called 1 hartree and symbolized by E_h. Table 1.1 provides notation and values for several useful quantities in a.u. and also equivalent values in other commonly used units. Greater precision and additional data are available at http://www.physics.nist.gov/PhysRefData/.

Bibliography and Suggested Additional Reading

Cramer, C. J., Famini, G. R., and Lowrey, A. 1993. 'Use of Quantum Chemical Properties as Analogs for Solvatochromic Parameters in Structure–Activity Relationships', *Acc. Chem. Res.*, **26**, 599.

Irikura, K. K., Frurip, D. J., Eds. 1998. *Computational Thermochemistry*, American Chemical Society Symposium Series, Vol. **677**, American Chemical Society: Washington, DC.

Jensen, F. 1999. *Introduction to Computational Chemistry*, Wiley: Chichester.

Jorgensen, W. L. 2004. 'The Many Roles of Computation in Drug Discovery', *Science*, **303**, 1813.

Leach, A. R. 2001. *Molecular Modelling*, 2nd Edn., Prentice Hall: London.

Levine, I. N. 2000. *Quantum Chemistry*, 5th Edn., Prentice Hall: New York.

Truhlar, D. G. 2000. 'Perspective on "Principles for a direct SCF approach to LCAO-MO *ab initio* calculations"' *Theor. Chem. Acc.*, **103**, 349.

References

Almlöf, J., Faegri, K., Jr., and Korsell, K. 1982. *J. Comput. Chem.*, **3**, 385.

Cramer, C. J., Denmark, S. E., Miller, P. C., Dorow, R. L., Swiss, K. A., and Wilson, S. R. 1994. *J. Am. Chem. Soc.*, **116**, 2437.

Ghosh, A. 2003. *Curr. Opin. Chem. Biol.*, **7**, 110.

2

Molecular Mechanics

2.1 History and Fundamental Assumptions

Let us return to the concept of the PES as described in Chapter 1. To a computational chemist, the PES is a surface that can be generated point by point by use of some computational method which determines a molecular energy for each point's structure. However, the concept of the PES predates any serious efforts to "compute" such surfaces. The first PESs (or slices thereof) were constructed by molecular spectroscopists.

A heterodiatomic molecule represents the simplest case for study by vibrational spectroscopy, and it also represents the simplest PES, since there is only the single degree of freedom, the bond length. Vibrational spectroscopy measures the energy separations between different vibrational levels, which are quantized. Most chemistry students are familiar with the simplest kind of vibrational spectroscopy, where allowed transitions from the vibrational ground state ($v = 0$) to the first vibrationally excited state ($v = 1$) are monitored by absorption spectroscopy; the typical photon energy for the excitation falls in the infrared region of the optical spectrum. More sensitive experimental apparati are capable of observing other allowed absorptions (or emissions) between more highly excited vibrational states, and/or forbidden transitions between states differing by more than 1 vibrational quantum number. Isotopic substitution perturbs the vibrational energy levels by changing the reduced mass of the molecule, so the number of vibrational transitions that can be observed is arithmetically related to the number of different isotopomers that can be studied. Taking all of these data together, spectroscopists are able to construct an extensive ladder of vibrational energy levels to a very high degree of accuracy (tenths of a wavenumber in favorable cases), as illustrated in Figure 2.1.

The spacings between the various vibrational energy levels depend on the potential energy associated with bond stretching (see Section 9.3.2). The data from the spectroscopic experiments thus permit the derivation of that potential energy function in a straightforward way.

Let us consider for the moment the potential energy function in an abstract form. A useful potential energy function for a bond between atoms A and B should have an analytic form. Moreover, it should be continuously differentiable. Finally, assuming the dissociation energy for the bond to be positive, we will define the minimum of the function to have a potential energy of zero; we will call the bond length at the minimum r_{eq}. We can determine the value

Essentials of Computational Chemistry, 2nd Edition Christopher J. Cramer
© 2004 John Wiley & Sons, Ltd   ISBNs: 0-470-09181-9 (cased); 0-470-09182-7 (pbk)

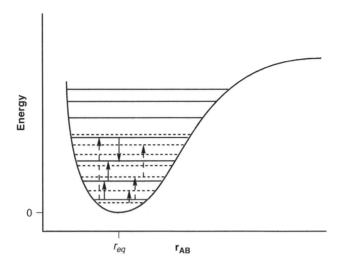

Figure 2.1 The first seven vibrational energy levels for a lighter (solid horizontal lines) and heavier (horizontal dashed lines) isotopomer of diatomic AB. Allowed vibrational transitions are indicated by solid vertical arrows, forbidden transitions are indicated by dashed vertical arrows

of the potential energy at an arbitrary point by taking a Taylor expansion about r_{eq}

$$U(r) = U(r_{eq}) + \left.\frac{dU}{dr}\right|_{r=r_{eq}} (r - r_{eq}) + \frac{1}{2!} \left.\frac{d^2U}{dr^2}\right|_{r=r_{eq}} (r - r_{eq})^2$$

$$+ \frac{1}{3!} \left.\frac{d^3U}{dr^3}\right|_{r=r_{eq}} (r - r_{eq})^3 + \cdots \tag{2.1}$$

Note that the first two terms on the r.h.s. of Eq. (2.1) are zero, the first by arbitrary choice, the second by virtue of r_{eq} being the minimum. If we truncate after the first non-zero term, we have the simplest possible expression for the vibrational potential energy

$$U(r_{AB}) = \tfrac{1}{2} k_{AB}(r_{AB} - r_{AB,eq})^2 \tag{2.2}$$

where we have replaced the second derivative of U by the symbol k. Equation (2.2) is Hooke's law for a spring, where k is the 'force constant' for the spring; the same term is used for k in spectroscopy and molecular mechanics. Subscripts have been added to emphasize that force constants and equilibrium bond lengths may vary from one pair of atoms to another.

Indeed, one might expect that force constants and equilibrium lengths might vary substantially even when A and B remain constant, but the bond itself is embedded in different molecular frameworks (i.e., surroundings). However, as more and more spectroscopic data became available in the early 20th century, particularly in the area of organic chemistry, where hundreds or thousands of molecules having similar bonds (e.g., C–C single bonds) could be characterized, it became empirically evident that the force constants and equilibrium bond lengths were largely the same from one molecule to the next. This phenomenon came to be called 'transferability'.

Concomitant with these developments in spectroscopy, thermochemists were finding that, to a reasonable approximation, molecular enthalpies could be determined as a sum of bond enthalpies. Thus, assuming transferability, if two different molecules were to be composed of identical bonds (i.e., they were to be isomers of one kind or another), the sum of the differences in the 'strains' of those bonds from one molecule to the other (which would arise from different bond lengths in the two molecules – the definition of strain in this instance is the positive deviation from the zero of energy) would allow one to predict the difference in enthalpies. Such prediction was a major goal of the emerging area of organic conformational analysis.

One might ask why any classical mechanical bond would deviate from its equilibrium bond length, insofar as that represents the zero of energy. The answer is that in polyatomic molecules, other energies of interaction must also be considered. For instance, repulsive van der Waals interactions between nearby groups may force some bonds connecting them to lengthen. The same argument can be applied to bond angles, which also have transferable force constants and optimal values (vide infra). Energetically unfavorable non-bonded, non-angle-bending interactions have come to be called 'steric effects' following the terminology suggested by Hill (1946), who proposed that a minimization of overall steric energy could be used to predict optimal structures. The first truly successful reduction to practice of this general idea was accomplished by Westheimer and Mayer (1946), who used potential energy functions to compute energy differences between twisted and planar substituted biphenyls and were able to rationalize racemization rates in these molecules.

The rest of this chapter examines the various components of the molecular energy and the force-field approaches taken for their computation. The discussion is, for the most part, general. At the end of the chapter, a comprehensive listing of reported/available force fields is provided with some description of their form and intended applicability.

2.2 Potential Energy Functional Forms

2.2.1 Bond Stretching

Before we go on to consider functional forms for all of the components of a molecule's total steric energy, let us consider the limitations of Eq. (2.2) for bond stretching. Like any truncated Taylor expansion, it works best in regions near its reference point, in this case r_{eq}. Thus, if we are interested primarily in molecular structures where no bond is terribly distorted from its optimal value, we may expect Eq. (2.2) to have reasonable utility. However, as the bond is stretched to longer and longer r, Eq. (2.2) predicts the energy to become infinitely positive, which is certainly not chemically realistic. The practical solution to such inaccuracy is to include additional terms in the Taylor expansion. Inclusion of the cubic term provides a potential energy function of the form

$$U(r_{AB}) = \tfrac{1}{2}[k_{AB} + k_{AB}^{(3)}(r_{AB} - r_{AB,eq})](r_{AB} - r_{AB,eq})^2 \qquad (2.3)$$

where we have added the superscript '(3)' to the cubic force constant (also called the 'anharmonic' force constant) to emphasize that it is different from the quadratic one. The cubic force constant is negative, since its function is to reduce the overly high stretching energies predicted by Eq. (2.2). This leads to an unintended complication, however; Eq. (2.3) diverges to *negative* infinity with increasing bond length. Thus, the lowest possible energy for a molecule whose bond energies are described by functions having the form of Eq. (2.3) corresponds to all bonds being dissociated, and this can play havoc with automated minimization procedures.

Again, the simple, practical solution is to include the next term in the Taylor expansion, namely the quartic term, leading to an expression of the form

$$U(r_{AB}) = \tfrac{1}{2}[k_{AB} + k_{AB}^{(3)}(r_{AB} - r_{AB,eq}) + k_{AB}^{(4)}(r_{AB} - r_{AB,eq})^2](r_{AB} - r_{AB,eq})^2 \qquad (2.4)$$

Such quartic functional forms are used in the general organic force field, MM3 (a large taxonomy of existing force fields appears at the end of the chapter). Many force fields that are designed to be used in reduced regions of chemical space (e.g., for specific biopolymers), however, use quadratic bond stretching potentials because of their greater computational simplicity.

The alert reader may wonder, at this point, why there has been no discussion of the Morse function

$$U(r_{AB}) = D_{AB}[1 - e^{-\alpha_{AB}(r_{AB} - r_{AB,eq})}]^2 \qquad (2.5)$$

where D_{AB} is the dissociation energy of the bond and α_{AB} is a fitting constant. The hypothetical potential energy curve shown in Figure 2.1 can be reproduced over a much wider range of r by a Morse potential than by a quartic potential. Most force fields decline to use the Morse potential because it is computationally much less efficient to evaluate the exponential function than to evaluate a polynomial function (vide infra). Moreover, most force fields are designed to study the energetics of molecules whose various degrees of freedom are all reasonably close to their equilibrium values, say within 10 kcal/mol. Over such a range, the deviation between the Morse function and a quartic function is usually negligible.

Even in these instances, however, there is some utility to considering the Morse function. If we approximate the exponential in Eq. (2.5) as its infinite series expansion truncated at the cubic term, we have

$$U(r_{AB}) = D_{AB} \left\{ 1 - \left[1 - \alpha_{AB}(r_{AB} - r_{AB,eq}) + \tfrac{1}{2}\alpha_{AB}^2(r_{AB} - r_{AB,eq})^2 \right. \right.$$
$$\left. \left. - \tfrac{1}{6}\alpha_{AB}^3(r_{AB} - r_{AB,eq})^3 \right] \right\}^2 \qquad (2.6)$$

Squaring the quantity in braces and keeping only terms through quartic gives

$$U(r_{AB}) = D_{AB} \left[\alpha_{AB}^2 - \alpha_{AB}^3(r_{AB} - r_{AB,eq}) + \frac{7}{12}\alpha_{AB}^4(r_{AB} - r_{AB,eq})^2 \right] (r_{AB} - r_{AB,eq})^2 \qquad (2.7)$$

where comparison of Eqs. (2.4) and (2.7) makes clear the relationship between the various force constants and the parameters D and α of the Morse potential. In particular,

$$k_{AB} = 2\alpha_{AB}^2 D_{AB} \tag{2.8}$$

Typically, the simplest parameters to determine from experiment are k_{AB} and D_{AB}. With these two parameters available, α_{AB} can be determined from Eq. (2.8), and thus the cubic and quartic force constants can also be determined from Eqs. (2.4) and (2.7). Direct measurement of cubic and quartic force constants requires more spectral data than are available for many kinds of bonds, so this derivation facilitates parameterization. We will discuss parameterization in more detail later in the chapter, but turn now to consideration of other components of the total molecular energy.

2.2.2 Valence Angle Bending

Vibrational spectroscopy reveals that, for small displacements from equilibrium, energy variations associated with bond angle deformation are as well modeled by polynomial expansions as are variations associated with bond stretching. Thus, the typical force field function for angle strain energy is

$$U(\theta_{ABC}) = \tfrac{1}{2}[k_{ABC} + k_{ABC}^{(3)}(\theta_{ABC} - \theta_{ABC,eq}) + k_{ABC}^{(4)}(\theta_{ABC} - \theta_{ABC,eq})^2 + \cdots]$$
$$(\theta_{ABC} - \theta_{ABC,eq})^2 \tag{2.9}$$

where θ is the valence angle between bonds AB and BC (note that in a force field, a bond is *defined* to be a vector connecting two atoms, so there is no ambiguity about what is meant by an angle between two bonds), and the force constants are now subscripted ABC to emphasize that they are dependent on three atoms. Whether Eq. (2.9) is truncated at the quadratic term or whether more terms are included in the expansion depends entirely on the balance between computational simplicity and generality that any given force field chooses to strike. Thus, to note two specific examples, the general organic force field MM3 continues the expansion through to the sextic term for some ABC combinations, while the biomolecular force field of Cornell *et al.* (see Table 2.1, first row) limits itself to a quadratic expression in all instances. (Original references to all the force fields discussed in this chapter will be found in Table 2.1.)

While the above prescription for angle bending seems useful, certain issues do arise. First, note that no power expansion having the form of Eq. (2.9) will show the appropriate chemical behavior as the bond angle becomes linear, i.e., at $\theta = \pi$. Another flaw with Eq. (2.9) is that, particularly in inorganic systems, it is possible to have *multiple* equilibrium values; for instance, in the trigonal bipyramidal system PCl_5 there are stable Cl–P–Cl angles of $\pi/2$, $\pi/3$, and π for axial/equatorial, equatorial/equatorial, and axial/axial combinations of chlorine atoms, respectively. Finally, there is another kind of angle bending that is sometimes discussed in molecular systems, namely 'out-of-plane' bending. Prior to addressing these various issues, it is instructive to consider the manner in which force fields typically handle potential energy variations associated with torsional motion.

2.2.3   Torsions

If we consider four atoms connected in sequence, ABCD, Figure 1.2 shows that a convenient means to describe the location of atom D is by means of a CD bond length, a BCD valence angle, and the torsional angle (or dihedral angle) associated with the ABCD linkage. As depicted in Figure 2.2, the torsional angle is defined as the angle between bonds AB and CD when they are projected into the plane bisecting the BC bond. The convention is to define the angle as positive if one must rotate the bond in front of the bisecting plane in a clockwise fashion to eclipse the bond behind the bisecting plane. By construction, the torsion angle is periodic. An obvious convention would be to use only the positive angle, in which case the torsion period would run from 0 to 2π radians (0 to 360°). However, the minimum energy for many torsions is for the antiperiplanar arrangement, i.e., $\omega = \pi$. Thus, the convention that $-\pi < \omega \le \pi (-180° \le \omega \le 180°)$ also sees considerable use.

Since the torsion itself is periodic, so too must be the torsional potential energy. As such, it makes sense to model the potential energy function as an expansion of periodic functions, e.g., a Fourier series. In a general form, typical force fields use

$$U(\omega_{\text{ABCD}}) = \tfrac{1}{2} \sum_{\{j\}_{\text{ABCD}}} V_{j,\text{ABCD}}[1 + (-1)^{j+1} \cos(j\omega_{\text{ABCD}} + \psi_{j,\text{ABCD}})] \qquad (2.10)$$

where the values of the signed term amplitudes V_j and the set of periodicities $\{j\}$ included in the sum are specific to the torsional linkage ABCD (note that deleting a particular value of j from the evaluated set is equivalent to setting the term amplitude for that value of j

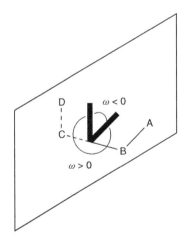

Figure 2.2   Definition and sign convention for dihedral angle ω. The bold lines are the projections of the AB and CD bonds into the bisecting plane. Note that the sign of ω is independent of whether one chooses to view the bisecting plane from the AB side or the CD side

equal to zero). Other features of Eq. (2.10) meriting note are the factor of 1/2 on the r.h.s., which is included so that the term amplitude V_j is equal to the maximum the particular term can contribute to U. The factor of $(-1)^{j+1}$ is included so that the function in brackets within the sum is zero for all j when $\omega = \pi$, if the phase angles ψ are all set to 0. This choice is motivated by the empirical observation that most (but not all) torsional energies are minimized for antiperiplanar geometries; the zero of energy for U in Eq. (2.10) thus occurs at $\omega = \pi$. Choice of phase angles ψ other than 0 permits a fine tuning of the torsional coordinate, which can be particularly useful for describing torsions in systems exhibiting large stereoelectronic effects, like the anomeric linkages in sugars (see, for instance, Woods 1996).

While the mathematical utility of Eq. (2.10) is clear, it is also well founded in a chemical sense, because the various terms can be associated with particular physical interactions when all phase angles ψ are taken equal to 0. Indeed, the magnitudes of the terms appearing in an individual fit can be informative in illuminating the degree to which those terms influence the overall rotational profile. We consider as an example the rotation about the C–O bond in fluoromethanol, the analysis of which was first described in detail by Wolfe *et al.* (1971) and Radom, Hehre and Pople (1971). Figure 2.3 shows the three-term Fourier decomposition of the complete torsional potential energy curve. Fluoromethanol is somewhat unusual insofar as the antiperiplanar structure is *not* the global minimum, although it is a local minimum. It is instructive to note the extent to which each Fourier term contributes to the overall torsional profile, and also to consider the physical factors implicit in each term.

One physical effect that would be expected to be onefold periodic in the case of fluoromethanol is the dipole–dipole interaction between the C–F bond and the O–H bond. Because of differences in electronegativity between C and F and O and H, the bond dipoles

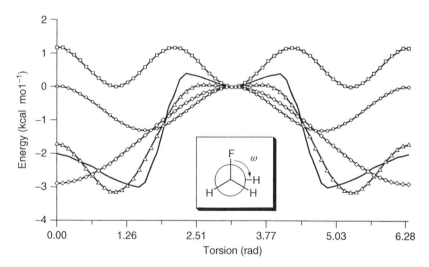

Figure 2.3 Fourier decomposition of the torsional energy for rotation about the C–O bond of fluoromethanol (bold black curve, energetics approximate). The Fourier sum ($\triangle$) is composed of the onefold ($\diamond$), twofold ($\circ$), and threefold ($\square$) periodic terms, respectively. In the Newman projection of the molecule, the oxygen atom lies behind the carbon atom at center

for these bonds point from C to F and from H to O, respectively. Thus, at $\omega = 0$, the dipoles are antiparallel (most energetically favorable) while at $\omega = \pi$ they are parallel (least energetically favorable). Thus, we would expect the V_1 term to be a minimum at $\omega = 0$, implying V_1 should be negative, and that is indeed the case. This term makes the largest contribution to the full rotational profile, having a magnitude roughly double either of the other two terms.

Twofold periodicity is associated with hyperconjugative effects. Hyperconjugation is the favorable interaction of a filled or partially filled orbital, typically a σ orbital, with a nearby empty orbital (hyperconjugation is discussed in more detail in Appendix D within the context of natural bond orbital (NBO) analysis). In the case of fluoromethanol, the filled orbital that is highest in energy is an oxygen lone pair orbital, and the empty orbital lowest in energy (and thus best able to interact in a resonance fashion with the oxygen lone pair) is the C–F σ^* antibonding orbital. Resonance between these orbitals, which is sometimes called negative hyperconjugation to distinguish it from resonance involving filled σ orbitals as donors, is favored by maximum overlap; this takes place for torsion angles of roughly $\pm \pi/2$. The contribution of this V_2 term to the overall torsional potential of fluoromethanol is roughly half that of the V_1 term, and of the expected sign.

The remaining V_3 term is associated with unfavorable bond–bond eclipsing interactions, which, for a torsion involving sp^3-hybridized carbon atoms, would be expected to show three-fold periodicity. To be precise, true threefold periodicity would only be expected were each carbon atom to bear all identical substituents. Experiments suggest that fluorine and hydrogen have similar steric behavior, so we will ignore this point for the moment. As expected, the sign of the V_3 term is positive, and it has roughly equal weight to the hyperconjugative term.

[Note that, following the terminology introduced earlier, we refer to the unfavorable eclipsing of chemical bonds as a steric interaction. Since molecular mechanics in essence treats molecules as classical atomic balls (possibly charged balls, as discussed in more detail below) connected together by springs, this terminology is certainly acceptable. It should be borne in mind, however, that real atoms are most certainly not billiard balls bumping into one another with hard shells. Rather, the unfavorable steric interaction derives from exchange-repulsion between filled molecular orbitals as they come closer to one another, i.e., the effect is electronic in nature. Thus, the bromide that all energetic issues in chemistry can be analyzed as a combination of electronic and steric effects is perhaps overly complex... *all* energetic effects in chemistry, at least if we ignore nuclear chemistry, are *exclusively* electronic/electrical in nature.]

While this analysis of fluoromethanol is instructive, it must be pointed out that a number of critical issues have been either finessed or ignored. First, as can be seen in Figure 2.3, the actual rotational profile of fluoromethanol cannot be perfectly fit by restricting the Fourier decomposition to only three terms. This may sound like quibbling, since the 'perfect' fitting of an arbitrary periodic curve takes an infinite number of Fourier terms, but the poorness of the fit is actually rather severe from a chemical standpoint. This may be most readily appreciated by considering simply the four symmetry-unique stationary points – two minima and two rotational barriers. We are trying to fit their energies, but we also want their nature as stationary points to be correct, implying that we are trying to fit their first derivatives as well (making the first derivative equal to zero defines them as stationary points). Thus, we are trying to fit eight constraints using only three variables (namely, the term amplitudes). By construction, we are actually guaranteed that 0 and π will have correct first derivatives, and that the energy value for π will be correct (since it is required to be the relative zero), but that still leaves five constraints on three variables. If we add non-zero phase angles ψ, we can do a better (but still not perfect) job.

Another major difficulty is that we have biased the system so that we can focus on a single dihedral interaction (FCOH) as being dominant, i.e., we ignored the HCOH interactions, and we picked a system where one end of the rotating bond had only a single substituent. To illustrate the complexities introduced by more substitution, consider the relatively simple case of n-butane (Figure 2.4). In this case, the three-term Fourier fit is in very good agreement with the full rotational profile, and certain aspects continue to make very good chemical sense. For instance, the twofold periodic term is essentially negligible, as would be expected since there are no particularly good donors or acceptors to interact in a hyperconjugative fashion. The onefold term, on the other hand, makes a very significant contribution, and this clearly cannot be assigned to some sort of dipole–dipole interaction, since the magnitude of a methylene–methyl bond dipole is very near zero. Rather, the magnitudes of the one- and threefold symmetric terms provide information about the relative steric strains associated with the two possible eclipsed structures, the lower energy of which has one H/H and two H/CH_3 eclipsing interactions, while the higher energy structure has two H/H and one CH_3/CH_3 interactions. While one might be tempted to try to derive some sort of linear combination rule for this still highly symmetric case, it should be clear that by the time one tries to analyze the torsion about a C–C bond bearing six different substituents, one's ability

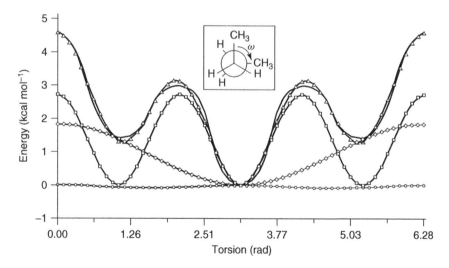

Figure 2.4 Fourier decomposition of the torsional energy for rotation about the C–C bond of n-butane (bold black curve, energetics approximate). The Fourier sum ($\triangle$) has a close overlap, and is composed of the onefold ($\diamond$), twofold ($\circ$), and threefold ($\square$) periodic terms, respectively

to provide a physically meaningful interpretation of the many different term amplitudes is quite limited.

Moreover, as discussed in more detail later, force field parameters are not statistically orthogonal, so optimized values can be skewed by coupling with other parameters. With all of these caveats in mind, however, there are still instances where valuable physical insights derive from a term-by-term analysis of the torsional coordinate.

Let us return now to a question raised above, namely, how to handle the valence angle bending term in a system where multiple equilibrium angles are present. Such a case is clearly analogous to the torsional energy, which also presents multiple minima. Thus, the inorganic SHAPES force field uses the following equations to compute angle bending energy

$$U(\theta_{ABC}) = \sum_{\{j\}_{ABC}} k_{j,ABC}^{Fourier}[1 + \cos(j\theta_{ABC} + \psi)] \tag{2.11}$$

$$k_{j,ABC}^{Fourier} = \frac{2k_{ABC}^{harmonic}}{j^2} \tag{2.12}$$

where ψ is a phase angle. Note that this functional form can also be used to ensure appropriate behavior in regions of bond angle inversion, i.e., where $\theta = \pi$. [As a digression, in metal coordination force fields an alternative formulation designed to handle multiple ligand–metal–ligand angles is simply to remove the angle term altogether. It is replaced by a non-bonded term specific to 1,3-interactions (a so-called 'Urey–Bradley' term) which tends to be repulsive. Thus, a given number of ligands attached to a central atom will tend to organize themselves so as to maximize the separation between any two. This 'points-on-a-sphere' (POS) approach is reminiscent of the VSEPR model of coordination chemistry.]

A separate situation, also mentioned in the angle bending discussion, arises in the case of four-atom systems where a central atom is bonded to three otherwise unconnected atoms, e.g., formaldehyde. Such systems are good examples of the second case of step IV of Figure 1.2, i.e., systems where a fourth atom is more naturally defined by a bond length to the central atom and its two bond angles to the other two atoms. However, as Figure 2.5 makes clear, one *could* define the final atom's position using the first case of step IV of Figure 1.2, i.e., by assigning a length to the central atom, an angle to a third atom, *and then a dihedral angle to the fourth atom even though atoms three and four are not defined as connected.* Such an

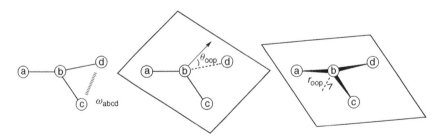

Figure 2.5   Alternative molecular coordinates that can be used to compute the energetics of distortions from planarity about a triply substituted central atom

assignment makes perfect sense from a geometric standpoint, even though it may seem odd from a chemical standpoint. Torsion angles defined in this manner are typically referred to as 'improper torsions'. In a system like formaldehyde, an improper torsion like OCHH would have a value of π radians (180°) in the planar, minimum energy structure. Increasing or decreasing this value would have the effect of moving the oxygen atom out of the plane defined by the remaining three atoms. Many force fields treat such improper torsions like any other torsion, i.e., they use Eq. (2.10). However, as Figure 2.5 indicates, the torsional description for this motion is only one of several equally reasonable coordinates that one might choose. One alternative is to quantify deviations from planarity by the angle $\theta_{o.o.p.}$ that one substituent makes with the plane defined by the other three (o.o.p. = 'out of plane'). Another is to quantify the elevation $r_{o.o.p.}$ of the central atom above/below the plane defined by the three atoms to which it is attached. Both of these latter modes have obvious connections to angle bending and bond stretching, respectively, and typically Eqs. (2.9) and (2.4), respectively, are used to model the energetics of their motion.

Let us return to the case of the butane rotational potential. As noted previously, the barriers in this potential are primarily associated with steric interactions between eclipsing atoms/groups. Anyone who has ever built a space-filling model of a sterically congested molecule is familiar with the phenomenon of steric congestion – some atomic balls in the space-filling model push against one another, creating strain (leading to the apocryphal 'drop test' metric of molecular stability: from how great a height can the model be dropped and remain intact?) Thus, in cases where dipole–dipole and hyperconjugative interactions are small about a rotating bond, one might question whether there is a need to parameterize a torsional function at all. Instead, one could represent atoms as balls, each having a characteristic radius, and develop a functional form quantifying the energetics of ball–ball interactions. Such a prescription provides an intuitive model for more distant 'non-bonded' interactions, which we now examine.

2.2.4 van der Waals Interactions

Consider the mutual approach of two noble gas atoms. At infinite separation, there is no interaction between them, and this defines the zero of potential energy. The isolated atoms are spherically symmetric, lacking any electric multipole moments. In a classical world (ignoring the chemically irrelevant gravitational interaction) there is no attractive force between them as they approach one another. When there are no dissipative forces, the relationship between force F in a given coordinate direction q and potential energy U is

$$F_q = -\frac{\partial U}{\partial q} \tag{2.13}$$

In this one-dimensional problem, saying that there is no force is equivalent to saying that the slope of the energy curve with respect to the 'bond length' coordinate is zero, so the potential energy remains zero as the two atoms approach one another. Associating non-zero size with our classical noble gas atoms, we might assign them hard-sphere radii r_{vdw}. In that case, when the bond length reaches twice the radius, the two cannot approach one another more

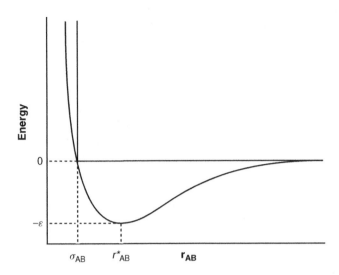

Figure 2.6 Non-attractive hard-sphere potential (straight lines) and Lennard–Jones potential (curve). Key points on the energy and bond length axes are labeled

closely, which is to say the potential energy discontinuously becomes infinite for $r < 2r_{vdw}$. This potential energy curve is illustrated in Figure 2.6.

One of the more profound manifestations of quantum mechanics is that this curve does *not* accurately describe reality. Instead, because the 'motions' of electrons are correlated (more properly, the electronic wave functions are correlated), the two atoms simultaneously develop electrical moments that are oriented so as to be mutually attractive. The force associated with this interaction is referred to variously as 'dispersion', the 'London' force, or the 'attractive van der Waals' force. In the absence of a permanent charge, the strongest such interaction is a dipole–dipole interaction, usually referred to as an 'induced dipole–induced dipole' interaction, since the moments in question are not permanent. Such an interaction has an inverse sixth power dependence on the distance between the two atoms. Thus, the potential energy becomes increasingly negative as the two noble gas atoms approach one another from infinity.

Dispersion is a fascinating phenomenon. It is sufficiently strong that even the dimer of He is found to have one bound vibrational state (Luo *et al.* 1993; with a vibrationally averaged bond length of 55 Å it is a remarkable member of the molecular bestiary). Even for molecules with fairly large *permanent* electric moments in the gas phase, dispersion is the dominant force favoring condensation to the liquid state at favorable temperatures and pressures (Reichardt 1990).

However, as the two atoms continue to approach one another, their surrounding electron densities ultimately begin to interpenetrate. In the absence of opportunities for bonding interactions, Pauli repulsion (or 'exchange repulsion') causes the energy of the system to rise rapidly with decreasing bond length. The sum of these two effects is depicted in Figure 2.6;

the contrasts with the classical hard-sphere model are that (i) an attractive region of the potential energy curve exists and (ii) the repulsive wall is not infinitely steep. [Note that at $r = 0$ the potential energy is that for an isolated atom having an atomic number equal to the sum of the atomic numbers for the two separated atoms; this can be of interest in certain formal and even certain practical situations, but we do no modeling of nuclear chemistry here.]

The simplest functional form that tends to be used in force fields to represent the combination of the dispersion and repulsion energies is

$$U(r_{AB}) = \frac{a_{AB}}{r_{AB}^{12}} - \frac{b_{AB}}{r_{AB}^{6}} \tag{2.14}$$

where a and b are constants specific to atoms A and B. Equation (2.14) defines a so-called 'Lennard–Jones' potential.

The inverse 12th power dependence of the repulsive term on interatomic separation has no theoretical justification – instead, this term offers a glimpse into the nuts and bolts of the algorithmic implementation of computational chemistry. Formally, one can more convincingly argue that the repulsive term in the non-bonded potential should have an exponential dependence on interatomic distance. However, the evaluation of the exponential function (and the log, square root, and trigonometric functions, *inter alia*) is roughly a factor of five times more costly in terms of central processing unit (cpu) time than the evaluation of the simple mathematical functions of addition, subtraction, or multiplication. Thus, the evaluation of r^{12} requires only that the theoretically justified r^6 term be multiplied by itself, which is a very cheap operation. Note moreover the happy coincidence that all terms in r involve *even* powers of r. The relationship between the internal coordinate r and Cartesian coordinates, which are typically used to specify atomic positions (see Section 2.4), is defined by

$$r_{AB} = \sqrt{(x_A - x_B)^2 + (y_A - y_B)^2 + (z_A - z_B)^2} \tag{2.15}$$

If only even powers of r are required, one avoids having to compute a square root. While quibbling over relative factors of five with respect to an operation that takes a tiny fraction of a second in absolute time may seem like overkill, one should keep in mind how many times the function in question may have to be evaluated in a given calculation. In a formal analysis, the number of non-bonded interactions that must be evaluated scales as N^2, where N is the number of atoms. In the process of optimizing a geometry, or of searching for many energy minima for a complex molecule, hundreds or thousands of energy evaluations may need to be performed for interim structures. Thus, seemingly small savings in time can be multiplied so that they are of practical importance in code development.

The form of the Lennard–Jones potential is more typically written as

$$U(r_{AB}) = 4\varepsilon_{AB} \left[\left(\frac{\sigma_{AB}}{r_{AB}} \right)^{12} - \left(\frac{\sigma_{AB}}{r_{AB}} \right)^{6} \right] \tag{2.16}$$

where the constants a and b of Eq. (2.14) are here replaced by the constants ε and σ. Inspection of Eq. (2.16) indicates that σ has units of length, and is the interatomic separation at which repulsive and attractive forces exactly balance, so that $U = 0$. If we differentiate Eq. (2.16) with respect to r_{AB}, we obtain

$$\frac{dU(r_{AB})}{dr_{AB}} = \frac{4\varepsilon_{AB}}{r_{AB}} \left[-12 \left(\frac{\sigma_{AB}}{r_{AB}} \right)^{12} + 6 \left(\frac{\sigma_{AB}}{r_{AB}} \right)^{6} \right] \qquad (2.17)$$

Setting the derivative equal to zero in order to find the minimum in the Lennard–Jones potential gives, after rearrangement

$$r_{AB}^* = 2^{1/6} \sigma_{AB} \qquad (2.18)$$

where r^* is the bond length at the minimum. If we use this value for the bond length in Eq. (2.16), we obtain $U = -\varepsilon_{AB}$, indicating that the parameter ε is the Lennard–Jones well depth (Figure 2.6).

The Lennard–Jones potential continues to be used in many force fields, particularly those targeted for use in large systems, e.g., biomolecular force fields. In more general force fields targeted at molecules of small to medium size, slightly more complicated functional forms, arguably having more physical justification, tend to be used (computational times for small molecules are so short that the efficiency of the Lennard–Jones potential is of little consequence). Such forms include the Morse potential [Eq. (2.5)] and the 'Hill' potential

$$U(r_{AB}) = \varepsilon_{AB} \left[\frac{6}{\beta_{AB} - 6} \exp \left(\beta_{AB} \frac{1 - r_{AB}}{r_{AB}^*} \right) - \frac{\beta_{AB}}{\beta_{AB} - 6} \left(\frac{r_{AB}^*}{r_{AB}} \right)^6 \right] \qquad (2.19)$$

where β is a new parameter and all other terms have the same meanings as in previous equations.

Irrespective of the functional form of the van der Waals interaction, some force fields reduce the energy computed for 1,4-related atoms (i.e., torsionally related) by a constant scale factor.

Our discussion of non-bonded interactions began with the example of two noble gas atoms having no permanent electrical moments. We now turn to a consideration of non-bonded interactions between atoms, bonds, or groups characterized by non-zero local electrical moments.

2.2.5   Electrostatic Interactions

Consider the case of two molecules A and B interacting at a reasonably large distance, each characterized by classical, non-polarizable, permanent electric moments. Classical electrostatics asserts the energy of interaction for the system to be

$$U_{AB} = \mathbf{M}^{(A)} \mathbf{V}^{(B)} \qquad (2.20)$$

where $\mathbf{M}^{(A)}$ is an ordered vector of the multipole moments of A, e.g., charge (zeroth moment), x, y, and z components of the dipole moment, then the nine components of the quadrupole moment, etc., and $\mathbf{V}^{(B)}$ is a similarly ordered row vector of the electrical potentials deriving from the multipole moments of B. Both expansions are about single centers, e.g., the centers of mass of the molecules. At long distances, one can truncate the moment expansions at reasonably low order and obtain useful interaction energies.

Equation (2.20) can be used to model the behavior of a large collection of individual molecules efficiently because the electrostatic interaction energy is pairwise additive. That is, we may write

$$U = \sum_A \sum_{B>A} \mathbf{M}^{(A)} \mathbf{V}^{(B)} \tag{2.21}$$

However, Eq. (2.21) is not very convenient in the context of *intra*molecular electrostatic interactions. In a protein, for instance, how can one derive the electrostatic interactions between spatially adjacent amide groups (which have large local electrical moments)? In principle, one could attempt to define moment expansions for functional groups that recur with high frequency in molecules, but such an approach poses several difficulties. First, there is no good experimental way in which to measure (or even define) such local moments, making parameterization difficult at best. Furthermore, such an approach would be computationally quite intensive, as evaluation of the moment potentials is tedious. Finally, the convergence of Eq. (2.20) at short distances can be quite slow with respect to the point of truncation in the electrical moments.

Let us pause for a moment to consider the fundamental constructs we have used thus far to define a force field. We have introduced van der Waals balls we call atoms, and we have defined bonds, angles, and torsional linkages between them. What would be convenient would be to describe electrostatic interactions in some manner that is based on these available entities (this convenience derives in part from our desire to be able to optimize molecular geometries efficiently, as described in more detail below). The simplest approach is to assign to each van der Waals atom a partial charge, in which case the interaction energy between atoms A and B is simply

$$U_{AB} = \frac{q_A q_B}{\varepsilon_{AB} r_{AB}} \tag{2.22}$$

This assignment tends to follow one of three formalisms, depending on the intent of the modeling endeavor. In the simplest case, the charges are 'permanent', in the sense that all atoms of a given type are defined to carry that charge in all situations. Thus, the atomic charge is a fixed parameter.

Alternatively, the charge can be determined from a scheme that depends on the electronegativity of the atom in question, and also on the electronegativities of those atoms to which it is defined to be connected Thus, the atomic electronegativity becomes a parameter and some functional form is adopted in which it plays a role as a variable. In a force field with a reduced number of atomic 'types' (see below for more discussion of atomic types) this preserves flexibility in the recognition of different chemical environments. Such flexibility is critical for the charge because the electrostatic energy can be so large compared to other components of the force field: Eq. (2.22) is written in a.u.; the conversion to energy units of kilocalories per mole and distance units of ångstorms involves multiplication of the r.h.s. by a factor of 332. Thus, even at 100 Å separation, the interaction energy between two unit charges in a vacuum would be more than 3 kcal/mol, which is of the same order of energy we expect for distortion of an individual stretching, bending, or torsional coordinate.

Finally, in cases where the force field is designed to study a particular molecule (i.e., generality is not an issue), the partial charges are often chosen to accurately reproduce some experimental or computed electrostatic observable of the molecule. Various schemes in common use are described in Chapter 9.

If, instead of the atom, we define charge polarization for the chemical bonds, the most convenient bond moment is the dipole moment. In this case, the interaction energy is defined between bonds AB and CD as

$$U_{AB/CD} = \frac{\mu_{AB}\mu_{CD}}{\varepsilon_{AB/CD} r_{AB/CD}^3} (\cos \chi_{AB/CD} - 3 \cos \alpha_{AB} \cos \alpha_{CD}) \tag{2.23}$$

where the bond moment vectors having magnitude μ are centered midway along the bonds and are collinear with them. The orientation vectors χ and α are defined in Figure 2.7.

Note that in Eqs. (2.22) and (2.23) the dielectric constant ε is subscripted. Although one might expect the best dielectric constant to be that for the permittivity of free space, such an assumption is not necessarily consistent with the approximations introduced by the use of atomic point charges. Instead, the dielectric constant must be viewed as a parameter of the model, and it is moreover a parameter that can take on multiple values. For use in Eq. (2.22),

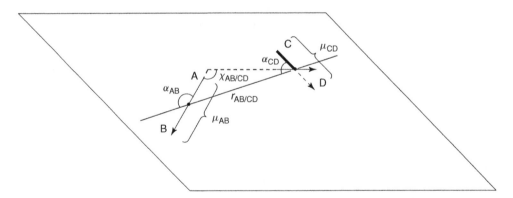

Figure 2.7 Prescription for evaluating the interaction energy between two dipoles. Each angle α is defined as the angle between the positive end of its respective dipole and the line passing through the two dipole centroids. The length of the line segment connecting the two centroids is r. To determine χ, the AB dipole and the centroid of the CD dipole are used to define a plane, and the CD dipole is projected into this plane. If the AB dipole and the projected CD dipole are parallel, χ is defined to be 0; if they are not parallel, they are extended as rays until they intersect. If the extension is from the same signed end of both dipoles, χ is the interior angle of the intersection (as illustrated), otherwise it is the exterior angle of the intersection

a plausible choice might be

$$\varepsilon_{AB} = \begin{cases} \infty & \text{if A and B are 1,2- or 1,3-related} \\ 3.0 & \text{if A and B are 1,4-related} \\ 1.5 & \text{otherwise} \end{cases} \qquad (2.24)$$

which dictates that electrostatic interactions between bonded atoms or between atoms sharing a common bonded atom are not evaluated, interactions between torsionally related atoms *are* evaluated, but are reduced in magnitude by a factor of 2 relative to all other interactions, which are evaluated with a dielectric constant of 1.5. Dielectric constants can also be defined so as to have a continuous dependence on the distance between the atoms. Although one might expect the use of high dielectric constants to mimic to some extent the influence of a surrounding medium characterized by that dielectric (e.g., a solvent), this is rarely successful – more accurate approaches for including condensed-phase effects are discussed in Chapters 3, 11, and 12.

Bonds between heteroatoms and hydrogen atoms are amongst the most polar found in non-ionic systems. This polarity is largely responsible for the well-known phenomenon of hydrogen bonding, which is a favorable interaction (usually ranging from 3 to 10 kcal/mol) between a hydrogen and a heteroatom to which it is *not* formally bonded. Most force fields account for hydrogen bonding implicitly in the non-bonded terms, van der Waals and electrostatic. In some instances an additional non-bonded interaction term, in the form of a 10–12 potential, is added

$$U(r_{XH}) = \frac{a'_{XH}}{r_{XH}^{12}} - \frac{b'_{XH}}{r_{XH}^{10}} \qquad (2.25)$$

where X is a heteroatom to which H is not bound. This term is analogous to a Lennard–Jones potential, but has a much more rapid decay of the attractive region with increasing bond length. Indeed, the potential well is so steep and narrow that one may regard this term as effectively forcing a hydrogen bond to deviate only very slightly from its equilibrium value.

Up to now, we have considered the interactions of *static* electric moments, but actual molecules have their electric moments *perturbed* under the influence of an electrical field (such as that deriving from the electrical moments of another molecule). That is to say, molecules are polarizable. To extend a force field to include polarizability is conceptually straightforward. Each atom is assigned a polarizability tensor. In the presence of the permanent electric field of the molecule (i.e., the field derived from the atomic charges or the bond–dipole moments), a dipole moment will be induced on each atom. Following this, however, the total electric field is the *sum* of the permanent electric field and that created by the induced dipoles, so the determination of the 'final' induced dipoles is an iterative process that must be carried out to convergence (which may be difficult to achieve). The total electrostatic energy can then be determined from the pairwise interaction of all moments and moment potentials (although the energy is determined in a pairwise fashion, note that many-body effects are incorporated by the iterative determination of the induced dipole moments). As a rough rule, computing the electrostatic interaction energy for a polarizable force field is about an order of magnitude more costly than it is for a static force field. Moreover, except for the most accurate work in very large systems, the benefits derived from polarization appear to be small. Thus, with the possible exception of solvent molecules in condensed-phase models (see Section 12.4.1), most force fields tend to avoid including polarization.

2.2.6   Cross Terms and Additional Non-bonded Terms

Bonds, angles, and torsions are not isolated molecular coordinates: they couple with one another. To appreciate this from a chemical point of view, consider BeH_2. In its preferred, linear geometry, one describes the Be hybridization as sp, i.e., each Be hybrid orbital used to bond with hydrogen has 50% 2s character and 50% 2p character. If we now decrease the bond angle, the p contribution increases until we stop at, say, a bond angle of $\pi/3$, which is the value corresponding to sp^2 hybridization. With more p character in the Be bonding hybrids, the bonds should grow longer. While this argument relies on rather basic molecular orbital theory, even from a *mechanical* standpoint, one would expect that as a bond angle is compressed, the bond lengths to the central atom will lengthen to decrease the non-bonded interactions between the terminal atoms in the sequence.

We can put this on a somewhat clearer mathematical footing by expanding the full molecular potential energy in a multi-dimensional Taylor expansion, which is a generalization of the one-dimensional case presented as Eq. (2.1). Thus

$$
U(\mathbf{q}) = U(\mathbf{q}_{eq}) + \sum_{i=1}^{3N-6} (q_i - q_{i,eq}) \frac{\partial U}{\partial q_i}\bigg|_{\mathbf{q}=\mathbf{q}_{eq}}
$$

$$
+ \frac{1}{2!} \sum_{i=1}^{3N-6} \sum_{j=1}^{3N-6} (q_i - q_{i,eq})(q_j - q_{j,eq}) \frac{\partial^2 U}{\partial q_i \partial q_j}\bigg|_{\mathbf{q}=\mathbf{q}_{eq}}
$$

$$
+ \frac{1}{3!} \sum_{i=1}^{3N-6} \sum_{j=1}^{3N-6} \sum_{k=1}^{3N-6} (q_i - q_{i,eq})(q_j - q_{j,eq})(q_k - q_{k,eq}) \frac{\partial^3 U}{\partial q_i \partial q_j \partial q_k}\bigg|_{\mathbf{q}=\mathbf{q}_{eq}} + \cdots
$$

$$(2.26)$$

where $\mathbf{q}$ is a molecular geometry vector of $3N - 6$ internal coordinates and the expansion is taken about an equilibrium structure. Again, the first two terms on the r.h.s. are zero by definition of U for $\mathbf{q}_{eq}$ and by virtue of all of the first derivatives being zero for an equilibrium structure. Up to this point, we have primarily discussed the 'diagonal' terms of the remaining summations, i.e., those terms for which all of the summation indices are equal to one another. However, if we imagine that index 1 of the double summation corresponds to a bond stretching coordinate, and index 2 to an angle bending coordinate, it is clear that our force field will be more 'complete' if we include energy terms like

$$
U(r_{AB}, \theta_{ABC}) = \frac{1}{2} k_{AB,ABC}(r_{AB} - r_{AB,eq})(\theta_{ABC} - \theta_{ABC,eq}) \tag{2.27}
$$

where $k_{AB,ABC}$ is the mixed partial derivative appearing in Eq. (2.26). Typically, the mixed partial derivative will be negligible for degrees of freedom that do not share common atoms.

In general force fields, stretch–stretch terms can be useful in modeling systems characterized by π conjugation. In amides, for instance, the coupling force constant between CO and CN stretching has been found to be roughly 15% as large as the respective diagonal bond-stretch force constants (Fogarasi and Balázs, 1985). Stretch–bend coupling terms tend to be most useful in highly strained systems, and for the computation of vibrational frequencies (see Chapter 9). Stretch–torsion coupling can be useful in systems where eclipsing interactions lead to high degrees of strain. The coupling has the form

$$U(r_{BC}, \omega_{ABCD}) = \tfrac{1}{2}k_{BC,ABCD}(r_{BC} - r_{BC,eq})[1 + \cos(j\omega + \psi)] \tag{2.28}$$

where j is the periodicity of the torsional term and ψ is a phase angle. Thus, if the term were designed to capture extra strain involving eclipsing interactions in a substituted ethane, the periodicity would require $j = 3$ and the phase angle would be 0. Note that the stretching bond, BC, is the *central* bond in the torsional linkage.

Other useful coupling terms include stretch–stretch coupling (typically between two adjacent bonds) and bend–bend coupling (typically between two angles sharing a common central atom). In force fields that aim for spectroscopic accuracy, i.e., the reproduction of vibrational spectra, still higher order coupling terms are often included. However, for purposes of general molecular modeling, they are typically not used.

In the case of non-bonded interactions, the discussion in prior sections focused on atom–atom type interactions. However, for larger molecules, and particularly for biopolymers, it is often possible to adopt a more coarse-grained description of the overall structure by focusing on elements of secondary structure, i.e., structural motifs that recur frequently, like α-helices in proteins or base-pairing or -stacking arrangements in polynucleotides. When such structural motifs are highly transferable, it is sometimes possible to describe an entire fragment (e.g., an entire amino acid in a protein) using a number of interaction sites and potential energy functions that is very much reduced compared to what would be required in an atomistic description. Such reduced models sacrifice atomic detail in structural analysis, but, owing to their simplicity, significantly expand the speed with which energy evaluations may be accomplished. Such efficiency can prove decisive in the simulation of biomolecules over long time scales, as discussed in Chapter 3. Many research groups are now using such coarse-grained models to study, *inter alia*, the process whereby proteins fold from denatured states into their native forms (see, for example, Hassinen and Peräkylä 2001).

As a separate example, Harvey *et al.* (2003) have derived expressions for pseudobonds and pseudoangles in DNA and RNA modeling that are designed to predict base-pairing and -stacking interactions when rigid bases are employed. While this model is coarse-grained, it is worth noting that even when a fully atomistic force field is being used, it may sometimes be helpful to add such additional interaction sites so as better to enforce elements of secondary structure like those found in biopolymers.

Finally, for particular biomolecules, experiment sometimes provides insight into elements of secondary structure that can be used in conjunction with a standard force field to more accurately determine a complete molecular structure. The most typical example of this approach is the imposition of atom–atom distance restraints based on nuclear Overhauser effect (nOe) data determined from NMR experiments. For each nOe, a pseudobond between the two atoms involved is defined, and a potential energy 'penalty' function depending on their interatomic distance is added to the overall force field energy. The most typical form for these penalty functions is a flat-bottomed linearized parabola. That is, there is no penalty over a certain range of bond distances, but outside that range the energy increases quadratically up to a certain point and then linearly thereafter. When the structure of a particular biomolecule is referred to as an 'NMR structure', what is meant is that the structure was determined from a force-field minimization incorporating experimental NMR restraints. Typically, a *set* of NMR structures is generated and deposited in the relevant database(s), each member of which satisfied the experimental restraints to within a certain level of tolerance. The quality of any NMR structure depends on the number of restraints that were available experimentally–the more (and the more widely distributed throughout the molecule) the better.

2.2.7   Parameterization Strategies

At this stage, it is worth emphasizing the possibly obvious point that a force field is nothing but a (possibly very large) collection of functional forms and associated constants. With that collection in hand, the energy of a given molecule (whose atomic connectivity must in general be specified) can be evaluated by computing the energy associated with every defined type of interaction occurring in the molecule. Because there are typically a rather large number of such interactions, the process is facilitated by the use of a digital computer, but the mathematics is really extraordinarily simple and straightforward.

Thus, we have detailed how to construct a molecular PES as a sum of energies from chemically intuitive functional forms that depend on internal coordinates and on atomic (and possibly bond-specific) properties. However, we have not paid much attention to the individual parameters appearing in those functional forms (force constants, equilibrium coordinate values, phase angles, etc.) other than pointing out the relationship of many of them to certain spectroscopically measurable quantities. Let us now look more closely at the 'Art and Science' of the parameterization process.

In an abstract sense, parameterization can be a very well-defined process. The goal is to develop a model that reproduces experimental measurements to as high a degree as possible. Thus, step 1 of parameterization is to assemble the experimental data. For molecular mechanics, these data consist of structural data, energetic data, and, possibly, data on molecular electric moments. We will discuss the issues associated with each kind of datum further below, but for the moment let us proceed abstractly. We next need to define a 'penalty function', that is, a function that provides a measure of how much deviation there is between our predicted values and our experimental values. Our goal will then be to select force-field parameters that minimize the penalty function. Choice of a penalty function is necessarily completely arbitrary. One example of such a function is

$$Z = \left[\sum_i^{\text{Observables}} \sum_j^{\text{Occurrences}} \frac{(\text{calc}_{i,j} - \text{expt}_{i,j})^2}{w_i^2} \right]^{1/2} \tag{2.29}$$

where observables might include bond lengths, bond angles, torsion angles, heats of formation, neutral molecular dipole moments, etc., and the weighting factors w carry units (so as to make Z dimensionless) and take into account not only possibly different numbers of data for different observables, but also the degree of tolerance the penalty function will have for the deviation of calculation from experiment for those observables. Thus, for instance, one might choose the weights so as to tolerate equally 0.01 Å deviations in bond lengths, 1° deviations in bond angles, 5° deviations in dihedral angles, 2 kcal/mol deviations in heats of formation, and 0.3 D deviations in dipole moment. Note that Z is evaluated using optimized geometries for all molecules; geometry optimization is discussed in Section 2.4. Minimization of Z is a typical problem in applied mathematics, and any number of statistical or quasi-statistical techniques can be used (see, for example, Schlick 1992). The minimization approach taken, however, is rarely able to remove the chemist and his or her intuition from the process.

To elaborate on this point, first consider the challenge for a force field designed to be general over the periodic table – or, for ease of discussion, over the first 100 elements. The number of unique bonds that can be formed from any two elements is 5050. If we were to operate under the assumption that bond-stretch force constants depend only on the atomic numbers of the bonded atoms (e.g., to make no distinction between so-called single, double, triple, etc. bonds), we would require 5050 force constants and 5050 equilibrium bond lengths to complete our force field. Similarly, we would require 100 partial atomic charges, and 5050 each values of σ and ε if we use Coulomb's law for electrostatics and a Lennard–Jones formalism for van der Waals interactions. If we carry out the same sort of analysis for bond angles, we need on the order of 10^6 parameters to complete the force field. Finally, in the case of torsions, somewhere on the order of 10^8 different terms are needed. If we include coupling terms, yet more constants are introduced.

Since one is unlikely to have access to $100\,000\,000+$ relevant experimental data, minimization of Z is an underdetermined process, and in such a case there will be many different combinations of parameter values that give similar Z values. What combination is optimal? Chemical knowledge can facilitate the process of settling on a single set of parameters. For instance, a set of parameters that involved fluorine atoms being assigned a partial positive charge would seem chemically unreasonable. Similarly, a quick glance at many force constants and equilibrium coordinate values would rapidly eliminate cases with abnormally large or small values. Another approach that introduces the chemist is making the optimization process stepwise. One optimizes some parameters over a smaller data set, then holds those parameters frozen while optimizing others over a larger data set, and this process goes on until all parameters have been chosen. The process of choosing which parameters to optimize in which order is as arbitrary as the choice of a penalty function, but may be justified with chemical reasoning.

Now, one might argue that no one would be foolish enough to attempt to design a force field that would be completely general over the first 100 elements. Perhaps if we were to restrict ourselves to organic molecules composed of {H, C, N, O, F, Si, P, Cl, Br, and I} – which certainly encompasses a large range of interesting molecules – then we could ameliorate the data sparsity problem. In principle, this is true, but in practice, the results are not very satisfactory. When large quantities of data are in hand, it becomes quite clear that atomic 'types' cannot be defined by atomic number alone. Thus, for instance, bonds involving two C atoms fall into at least four classes, each one characterized by its own particular stretching force constant and equilibrium distance (e.g., single, aromatic, double, and triple). A similar situation obtains for any pair of atoms when multiple bonding is an option. Different atomic hybridizations give rise to different angle bending equilibrium values. The same is true for torsional terms. If one wants to include metals, usually different oxidation states give rise to differences in structural and energetic properties (indeed, this segregation of compounds based on similar, discrete properties is what inorganic chemists sometimes use to *assign* oxidation state).

Thus, in order to improve accuracy, a given force field may have a very large number of atom types, even though it includes only a relatively modest number of nuclei. The primarily organic force fields MM3 and MMFF have 153 and 99 atom types, respectively. The two general biomolecular force fields (proteins, nucleic acids, carbohydrates) OPLS (optimized potentials for liquid simulations) and that of Cornell *et al.* have 41 atoms types each. The completely general (i.e., most of the periodic table) universal force field (UFF) has 126 atom types. So, again, the chemist typically faces an underdetermined optimization of parameter values in finalizing the force field.

So, what steps can be taken to decrease the scope of the problem? One approach is to make certain parameters that depend on more than one atom themselves functions of single-atom-specific parameters. For instance, for use in Eq. (2.16), one usually defines

$$\sigma_{AB} = \sigma_A + \sigma_B \tag{2.30}$$

and

$$\varepsilon_{AB} = (\varepsilon_A \varepsilon_B)^{1/2} \tag{2.31}$$

thereby reducing in each case the need for $N(N + 1)/2$ diatomic parameters to only N atomic parameters. [Indeed, truly general force fields, like DREIDING, UFF, and VALBOND attempt to reduce almost all parameters to being derivable from a fairly small set of atomic parameters. In practice, these force fields are not very robust, but as their limitations continue to be addressed, they have good long-range potential for broad, general utility.]

Another approach that is conceptually similar is to make certain constants depend on bond order or bond hybridization. Thus, for instance, in the VALBOND force field, angle bending energies at metal atoms are computed from orbital properties of the metal–ligand bonds; in the MM2 and MM3 force fields, stretching force constants, equilibrium bond lengths, and two-fold torsional terms depend on computed π bond orders between atoms. Such additions to the force field somewhat strain the limits of a 'classical' model, since references to orbitals or computed bond orders necessarily introduce quantum mechanical aspects to the calculation. There is, of course, nothing wrong with moving the model in this direction – aesthetics and accuracy are orthogonal concepts – but such QM enhancements add to model complexity and increase the computational cost.

Yet another way to minimize the number of parameters required is to adopt a so-called 'united-atom' (UA) model. That is, instead of defining only atoms as the fundamental units of the force field, one also defines certain functional groups, usually hydrocarbon groups, e.g., methyl, methylene, aryl CH, etc. The group has its own single set of non-bonded and other parameters – effectively, this reduces the total number of atoms by one less than the total number incorporated into the united atom group.

Even with the various simplifications one may envision to reduce the number of parameters needed, a vast number remain for which experimental data may be too sparse to permit reliable parameterization (thus, for example, the MMFF94 force field has about 9000 defined parameters). How does one find the best parameter values? There are three typical responses to this problem.

The most common response nowadays is to supplement the experimental data with the highest quality *ab initio* data that can be had (either from molecular orbital or density functional calculations). A pleasant feature of using theoretical data is that one can compare regions on a PES that are far from equilibrium structures by direct computation rather than by trying to interpret vibrational spectra. Furthermore, one can attempt to make force-field energy derivatives correspond to those computed *ab initio*. The only limitation to this approach is the computational resources that are required to ensure that the *ab initio* data are sufficiently accurate.

The next most sensible response is to do nothing, and accept that there will be some molecules whose connectivity places them outside the range of chemical space to which the force field can be applied. While this can be very frustrating for the general user (typically the software package delivers a message to the effect that one or more parameters are lacking and then quits), if the situation merits, the necessary new parameters can be determined in relatively short order. Far more objectionable, when not well described, is the third response, which is to estimate missing parameter values and then carry on. The estimation process can be highly suspect, and unwary users can be returned nonsense results with no indication that some parameters were guessed at. If one suspects that a particular linkage or linkages in one's molecule may be outside the well-parameterized bounds of the force field, it is always wise to run a few test calculations on structures having small to moderate distortions of those linkages so as to evaluate the quality of the force constants employed.

It is worth noting that sometimes parameter estimation takes place 'on-the-fly'. That is, the program is designed to guess without human intervention parameters that were not explicitly coded. This is a somewhat pernicious aspect of so-called graphical user interfaces (GUIs): while they make the submission of a calculation blissfully simple – all one has to do is draw the structure – one is rather far removed from knowing what is taking place in the process of the calculation. Ideally, prominent warnings from the software should accompany any results derived from such calculations.

2.3 Force-field Energies and Thermodynamics

We have alluded above that one measure of the accuracy of a force field can be its ability to predict heats of formation. A careful inspection of all of the formulas presented thus far, however, should make it clear that we have not yet established any kind of connection between the force-field energy and any kind of thermodynamic quantity.

Let us review again the sense of Eqs. (2.4) and (2.9). In both instances, the minimum value for the energy is zero (assuming positive force constants and sensible behavior for odd power terms). An energy of zero is obtained when the bond length or angle adopts its equilibrium value. Thus, a 'strain-free' molecule is one in which every coordinate adopts its equilibrium value. Although we accepted a negative torsional term in our fluoromethanol example above, because it provided some chemical insight, by proper choice of phase angles in Eq. (2.10) we could also require this energy to have zero as a minimum (although not necessarily for the dihedral angle $\omega = \pi$). So, neglecting non-bonded terms for the moment, we see that the raw force-field energy can be called the 'strain energy', since it represents the positive deviation from a hypothetical strain-free system.

The key point that must be noted here is that strain energies for two different molecules *cannot be meaningfully compared unless the zero of energy is identical.* This is probably best illustrated with a chemical example. Consider a comparison of the molecules ethanol and dimethyl ether using the MM2(91) force field. Both have the chemical formula C_2H_6O. However, while ethanol is defined by the force field to be composed of two sp^3 carbon atoms, one sp^3 oxygen atom, five carbon-bound hydrogen atoms, and one alcohol hydrogen atom, dimethyl ether differs in that all six of its hydrogen atoms are of the carbon-bound type. Each strain energy will thus be computed relative to a different hypothetical reference system, and there is no *a priori* reason that the two hypothetical systems should be thermodynamically equivalent.

What is necessary to compute a heat of formation, then, is to define the heat of formation of each hypothetical, unstrained atom type. The molecular heat of formation can then be computed as the sum of the heats of formation of all of the atom types plus the strain energy. Assigning atom-type heats of formation can be accomplished using additivity methods originally developed for organic functional groups (Cohen and Benson 1993). The process is typically iterative in conjunction with parameter determination.

Since the assignment of the atomic heats of formation is really just an aspect of parameterization, it should be clear that the possibility of a negative force-field energy, which could derive from addition of net negative non-bonded interaction energies to small non-negative strain energies, is not a complication. Thus, a typical force-field energy calculation will report any or all of (i) a strain energy, which is the energetic consequence of the deviation of the internal molecular coordinates from their equilibrium values, (ii) a force-field energy, which is the sum of the strain energy and the non-bonded interaction energies, and (iii) a heat of formation, which is the sum of the force-field energy and the reference heats of formation for the constituent atom types (Figure 2.8).

For some atom types, thermodynamic data may be lacking to assign a reference heat of formation. When a molecule contains one or more of these atom types, the force field cannot compute a molecular heat of formation, *and energetic comparisons are necessarily limited to conformers, or other isomers that can be formed without any change in atom types.*

2.4 Geometry Optimization

One of the key motivations in early force-field design was the development of an energy functional that would permit facile optimization of molecular geometries. While the energy

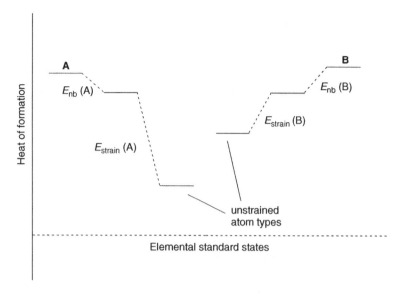

Figure 2.8 Molecules **A** and **B** are chemical isomers but are composed of different atomic types (atomic typomers?). Thus, the sums of the heats of formation of their respective unstrained atom types, which serve as their zeroes of force-field energy, are different. To each zero, strain energy and non-bonded energy (the sum of which are force-field energy) are added to determine heat of formation. In this example, note that **A** is predicted to have a lower heat of formation than **B** even though it has a substantially larger strain energy (and force-field energy); this difference is more than offset by the difference in the reference zeroes

of an *arbitrary* structure can be interesting, real molecules vibrate thermally about their equilibrium structures, so finding minimum energy structures is key to describing equilibrium constants, comparing to experiment, etc. Thus, as emphasized above, one priority in force-field development is to adopt reasonably simple functional forms so as to facilitate geometry optimization. We now examine the optimization process in order to see how the functional forms enter into the problem.

2.4.1 Optimization Algorithms

Note that, in principle, geometry optimization could be a separate chapter of this text. In its essence, geometry optimization is a problem in applied mathematics. How does one find a minimum in an arbitrary function of many variables? [Indeed, we have already discussed that problem once, in the context of parameter optimization. In the case of parameter optimization, however, it is not necessarily obvious how the penalty function being minimized *depends* on any given variable, and moreover the problem is highly underdetermined. In the case of geometry optimization, we are working with far fewer variables (the geometric degrees of freedom) and have, at least with a force field, analytic expressions for how the energy depends on the variables. The mathematical approach can thus be quite different.] As the problem is general, so, too, many of the details presented below will be general to any energy functional. However, certain special considerations associated with force-field calculations merit discussion, and so we will proceed first with an overview of geometry optimization, and then examine force-field specific aspects.

Because this text is designed primarily to illuminate the conceptual aspects of computational chemistry, and not to provide detailed descriptions of algorithms, we will examine only the most basic procedures. Much more detailed treatises of more sophisticated algorithms are available (see, for instance, Jensen 1999).

For pedagogical purposes, let us begin by considering a case where we do not know how our energy depends on the geometric coordinates of our molecule. To optimize the geometry, all we can do is keep trying different geometries until we are reasonably sure that we have found the one with the lowest possible energy (while this situation is atypical with force fields, there are still many sophisticated electronic structure methods for which it is indeed the only way to optimize the structure). How can one most efficiently survey different geometries?

It is easiest to proceed by considering a one-dimensional case, i.e., a diatomic with only the bond length as a geometric degree of freedom. One selects a bond length, and computes the energy. One then changes the bond length, let us say by shortening it 0.2 Å, and again computes the energy. If the energy goes down, we want to continue moving the bond length in that direction, and we should take another step (which need not necessarily be of the same length). If the energy goes up, on the other hand, we are moving in the wrong direction, and we should take a step in the opposite direction. Ultimately, the process will provide three adjacent points where the one in the center is lower in energy than the other two. Three non-collinear points uniquely define a parabola, and in this case the parabola must have a minimum (since the central point was lower in energy than the other two). We next calculate the energy for the bond length corresponding to the parabolic minimum (the degree to which the computed energy agrees with that from the parabolic equation will be an indication of how nearly harmonic the local bond stretching coordinate is). We again step left and right on the bond stretching coordinate, this time with smaller steps (perhaps an order of magnitude smaller) and repeat the parabolic fitting process. This procedure can be repeated until we are satisfied that our step size falls below some arbitrary threshold we have established as defining convergence of the geometry. Note that one can certainly envision variations on this theme. One could use more than three points in order to fit to higher order polynomial equations, step sizes could be adjusted based on knowledge of previous points, etc.

In the multi-dimensional case, the simplest generalization of this procedure is to carry out the process iteratively. Thus, for LiOH, for example, we might first find a parabolic minimum for the OH bond, then for the LiO bond, then for the LiOH bond angle (in each case holding the other two degrees of freedom fixed), and then repeat the process to convergence. Of course, if there is strong coupling between the various degrees of freedom, this process will converge rather slowly.

What we really want to do at any given point in the multi-dimensional case is move not in the direction of a *single* coordinate, but rather in the direction of the greatest downward slope in the energy with respect to *all* coordinates. This direction is the opposite of the gradient vector, **g**, which is defined as

$$
\mathbf{g}(\mathbf{q}) =
\begin{bmatrix}
\dfrac{\partial U}{\partial q_1} \\[2mm]
\dfrac{\partial U}{\partial q_2} \\[2mm]
\dfrac{\partial U}{\partial q_3} \\[2mm]
\vdots \\[2mm]
\dfrac{\partial U}{\partial q_n}
\end{bmatrix}
\tag{2.32}
$$

where **q** is an n-dimensional coordinate vector ($n = 3N - 6$ where N is the number of atoms if we are working in internal coordinates, $n = 3N$ if we are working in Cartesian coordinates, etc.) If we cannot compute the partial derivatives that make up **g** analytically, we can do so numerically. However, that numerical evaluation requires at least one additional energy calculation for each degree of freedom. Thus, we would increase (or decrease) every degree of freedom by some step size, compute the slope of the resulting line derived from the energies of our initial structure and the perturbed structure, and use this slope as an estimate for the partial derivative. Such a 'forward difference' estimation is typically not very accurate, and it would be better to take an additional point in the opposite direction for each degree of freedom, and then compute the 'central difference' slope from the corresponding parabola. It should be obvious that, as the number of degrees of freedom increases, it can be particularly valuable to have an energy function for which the first derivative is known *analytically*.

Let us examine this point a bit more closely for the force-field case. For this example, we will work in Cartesian coordinates, in which case $\mathbf{q} = \mathbf{X}$ of Eq. (1.4). To compute, say, the partial derivative of the energy with respect to the x coordinate of atom A, we will need to evaluate the changes in energy for the various terms contributing to the full force-field energy as a function of moving atom A in the x direction. For simplicity, let us consider only the bond stretching terms. Clearly, only the energy of those bonds that have A at one terminus will be affected by A's movement. We may then use the chain rule to write

$$
\frac{\partial U}{\partial x_A} = \sum_{\substack{i \text{ bonded} \\ \text{to A}}} \frac{\partial U}{\partial r_{Ai}} \frac{\partial r_{Ai}}{\partial x_A}
\tag{2.33}
$$

Differentiation of E with respect to r_{Ai} for Eq. (2.4) gives

$$
\frac{\partial U}{\partial r_{Ai}} = \tfrac{1}{2}[2k_{Ai} + 3k_{Ai}^{(3)}(r_{Ai} - r_{Ai,eq}) + 4k_{Ai}^{(4)}(r_{Ai} - r_{Ai,eq})^2](r_{Ai} - r_{Ai,eq})
\tag{2.34}
$$

The bond length r_{Ai} was defined in Eq. 2.15, and its partial derivative with respect to x_A is

$$\frac{\partial r_{Ai}}{\partial x_A} = \frac{(x_A - x_i)}{\sqrt{(x_A - x_i)^2 + (y_A - y_i)^2 + (z_A - z_i)^2}} \qquad (2.35)$$

Thus, we may quickly assemble the bond stretching contributions to this particular component of the gradient. Contributions from the other terms in the force field can be somewhat more tedious to derive, but are nevertheless available analytically. This makes force fields highly efficient for the optimization of geometries of very large systems.

With $\mathbf{g}$ in hand, we can proceed in a fashion analogous to the one-dimensional case outlined above. We step along the direction defined by $-\mathbf{g}$ until we locate a minimum in the energy for this process; since we are taking points in a linear fashion, this movement is called a 'line search' (even though we may identify our minimum by fitting our points to a polynomial curve). Then, we recompute $\mathbf{g}$ at the located minimum and repeat the process. Our new search direction is necessarily orthogonal to our last one, since we minimized E in the last direction. This particular feature of a steepest descent curve can lead to *very* slow convergence in unfavorable cases.

A more robust method is the Newton–Raphson procedure. In Eq. (2.26), we expressed the full force-field energy as a multidimensional Taylor expansion in arbitrary coordinates. If we rewrite this expression in matrix notation, and truncate at second order, we have

$$U(\mathbf{q}^{(k+1)}) = U(\mathbf{q}^{(k)}) + (\mathbf{q}^{(k+1)} - \mathbf{q}^{(k)})\mathbf{g}^{(k)} + \frac{1}{2}(\mathbf{q}^{(k+1)} - \mathbf{q}^{(k)})^{\dagger}\mathbf{H}^{(k)}(\mathbf{q}^{(k+1)} - \mathbf{q}^{(k)}) \quad (2.36)$$

where the reference point is $\mathbf{q}^{(k)}$, $\mathbf{g}^{(k)}$ is the gradient vector for the reference point as defined by Eq. (2.32), and $\mathbf{H}^{(k)}$ is the 'Hessian' matrix for the reference point, whose elements are defined by

$$H_{ij}^{(k)} = \left. \frac{\partial^2 U}{\partial q_i \partial q_j} \right|_{\mathbf{q}=\mathbf{q}^{(k)}} \qquad (2.37)$$

If we differentiate Eq. (2.36) term by term with respect to the ith coordinate of $\mathbf{q}^{(k+1)}$, noting that no term associated with point k has any dependence on a coordinate of point $k + 1$ (and hence the relevant partial derivative will be 0), we obtain

$$\frac{\partial U(\mathbf{q}^{(k+1)})}{\partial q_i^{k+1}} = \frac{\partial \mathbf{q}^{(k+1)}}{\partial q_i^{k+1}}\mathbf{g}^{(k)} + \frac{1}{2}\frac{\partial \mathbf{q}^{(k+1)\dagger}}{\partial q_i^{k+1}}\mathbf{H}^{(k)}(\mathbf{q}^{(k+1)} - \mathbf{q}^{(k)})$$

$$+ \frac{1}{2}(\mathbf{q}^{(k+1)} - \mathbf{q}^{(k)})^{\dagger}\mathbf{H}^{(k)}\frac{\partial \mathbf{q}^{(k+1)}}{\partial q_i^{k+1}} \qquad (2.38)$$

The l.h.s. of Eq. (2.38) is the ith element of the vector $\mathbf{g}^{(k+1)}$. On the r.h.s. of Eq. (2.38), since the partial derivative of $\mathbf{q}$ with respect to its ith coordinate is simply the unit vector in the ith coordinate direction, the various matrix multiplications simply produce the ith element of the multiplied vectors. Because mixed partial derivative values are independent of the order of differentiation, the Hessian matrix is Hermitian, and we may simplify

Eq. (2.38) as

$$g_i^{(k+1)} = g_i^{(k)} + [\mathbf{H}^{(k)}(\mathbf{q}^{(k+1)} - \mathbf{q}^{(k)})]_i \tag{2.39}$$

where the notation $[\,]_i$ indicates the ith element of the product column matrix. The condition for a stationary point is that the l.h.s. of Eq. (2.39) be 0 for *all* coordinates, or

$$\mathbf{0} = \mathbf{g}^{(k)} + \mathbf{H}^{(k)}(\mathbf{q}^{(k+1)} - \mathbf{q}^{(k)}) \tag{2.40}$$

which may be rearranged to

$$\mathbf{q}^{(k+1)} = \mathbf{q}^{(k)} - (\mathbf{H}^{(k)})^{-1}\mathbf{g}^{(k)} \tag{2.41}$$

This equation provides a prescription for the location of stationary points. In principle, starting from an arbitrary structure having coordinates $\mathbf{q}^{(k)}$, one would compute its gradient vector $\mathbf{g}$ and its Hessian matrix $\mathbf{H}$, and then select a new geometry $\mathbf{q}^{(k+1)}$ according to Eq. (2.41). Equation (2.40) shows that the gradient vector for this new structure will be the $\mathbf{0}$ vector, so we will have a stationary point.

Recall, however, that our derivation involved a truncation of the full Taylor expansion at second order. Thus, Eq. (2.40) is only approximate, and $\mathbf{g}^{(k+1)}$ will not necessarily be $\mathbf{0}$. However, it will probably be smaller than $\mathbf{g}^{(k)}$, so we can repeat the whole process to pick a point $k + 2$. After a sufficient number of iterations, the gradient will hopefully become so small that structures $k + n$ and $k + n + 1$ differ by a chemically insignificant amount, and we declare our geometry to be converged.

There are a few points with respect to this procedure that merit discussion. First, there is the Hessian matrix. With n^2 elements, where n is the number of coordinates in the molecular geometry vector, it can grow somewhat expensive to construct this matrix at every step even for functions, like those used in most force fields, that have fairly simple analytical expressions for their second derivatives. Moreover, the matrix must be *inverted* at every step, and matrix inversion formally scales as n^3, where n is the dimensionality of the matrix. Thus, for purposes of efficiency (or in cases where analytic second derivatives are simply not available) approximate Hessian matrices are often used in the optimization process – after all, the truncation of the Taylor expansion renders the Newton–Raphson method *intrinsically* approximate. As an optimization progresses, second derivatives can be estimated reasonably well from finite differences in the analytic first derivatives over the last few steps. For the first step, however, this is not an option, and one typically either accepts the cost of computing an initial Hessian analytically for the level of theory in use, or one employs a Hessian obtained at a less expensive level of theory, when such levels are available (which is typically *not* the case for force fields). To speed up slowly convergent optimizations, it is often helpful to compute an analytic Hessian every few steps and replace the approximate one in use up to that point. For *really* tricky cases (e.g., where the PES is fairly flat in many directions) one is occasionally forced to compute an analytic Hessian for *every* step.

Another key issue to note is that Eq. (2.41) provides a prescription to get to what is usually the *nearest* stationary point, but there is no guarantee that that point will be a *minimum*. The condition for a minimum is that all coordinate second derivatives (i.e., all diagonal elements of the Hessian matrix) be positive, but Eq. (2.41) places no constraints on the second derivatives. Thus, if one starts with a geometry that is very near a transition state (TS) structure, the Newton–Raphson procedure is likely to converge to that structure. This can be a pleasant feature, if one is looking for the TS in question, or an annoying one, if one is not. To verify the nature of a located stationary point, it is necessary to compute an accurate Hessian matrix and inspect its eigenvalues, as discussed in more detail in Chapter 9. With force fields, it is often cheaper and equally effective simply to 'kick' the structure, which is to say, by hand one moves one or a few atoms to reasonably distorted locations and then reoptimizes to verify that the original structure is again found as the lowest energy structure nearby.

Because of the importance of TS structures, a large number of more sophisticated methods exist to locate them. Many of these methods require that two minima be specified that the TS structure should 'connect', i.e., the TS structure intervenes in some reaction path that connects them. Within a given choice of coordinates, intermediate structures are evaluated and, hopefully, the relevant stationary point is located. Other methods allow the specification of a particular coordinate with respect to which the energy is to be maximized while minimizing it with respect to all other coordinates. When this coordinate is one of the normal modes of the molecule, this defines a TS structure. The bottom line for all TS structure location methods is that they work best when the chemist can provide a reasonably good initial guess for the structure, and they tend to be considerably more sensitive to the availability of a good Hessian matrix, since finding the TS essentially amounts to distinguishing between different local curvatures on the PES.

Most modern computational chemistry software packages provide some discussion of the relative merits of the various optimizers that they make available, at least on the level of providing practical advice (particularly where the user can set certain variables in the optimization algorithm with respect to step size between structures, tolerances, use of redundant internal coordinates, etc.), so we will not try to cover all possible tricks and tweaks here. We will simply note that it is usually a good idea to visualize the structures in an optimization as it progresses, as every algorithm can sometimes take a pathological bad step, and it is usually better to restart the calculation with an improved guess than it is to wait and hope that the optimization ultimately returns to normalcy.

A final point to be made is that most optimizers are rather good at getting you to the *nearest* minimum, but an individual researcher may be interested in finding the *global* minimum (i.e., the minimum having the lowest energy of all minima). Again, this is a problem in applied mathematics for which no one solution is optimal (see, for instance, Leach 1991). Most methods involve a systematic or random sampling of alternative conformations, and this subject will be discussed further in the next chapter.

2.4.2   Optimization Aspects Specific to Force Fields

Because of their utility for very large systems, where their relative speed proves advantageous, force fields present several specific issues with respect to practical geometry optimization that merit discussion. Most of these issues revolve around the scaling behavior that the speed of a force-field calculation exhibits with respect to increasing system size. Although we raise the issues here in the context of geometry optimization, they are equally important in force-field simulations, which are discussed in more detail in the next chapter.

If we look at the scaling behavior of the various terms in a typical force field, we see that the internal coordinates have very favorable scaling – the number of internal coordinates is $3N - 6$, which is linear in N. The non-bonded terms, on the other hand, are computed from pairwise interactions, and therefore scale as N^2. However, this scaling assumes the evaluation of *all* pairwise terms. If we consider the Lennard–Jones potential, its long-range behavior decays proportional to r^{-6}. The total number of interactions should grow at most as r^2 (i.e., proportional to the surface area of a surrounding sphere), so the net energetic contribution should decay with an r^{-4} dependence. This quickly becomes negligible (particularly from a gradient standpoint) so force fields usually employ a 'cut-off' range for the evaluation of van der Waals energies – a typical choice is 10 Å. Thus, part of the calculation involves the periodic updating of a 'pair list', which is a list of all atoms for which the Lennard–Jones interaction needs to be calculated (Petrella *et al.* 2003). The update usually occurs only once every several steps, since, of course, evaluation of interatomic distances *also* formally scales as N^2.

In practice, even though the use of a cut-off introduces only small disparities in the energy, the discontinuity of these disparities can cause problems for optimizers. A more stable approach is to use a 'switching function' which multiplies the van der Waals interaction and causes it (and possibly its first and second derivatives) to go smoothly to zero at some cut-off distance. This function must, of course, be equal to 1 at short distances.

The electrostatic interaction is more problematic. For point charges, the interaction energy decays as r^{-1}. As already noted, the number of interactions increases by up to r^2, so the total energy in an infinite system might be expected to diverge! Such formal divergence is avoided in most real cases, however, because in systems that are electrically neutral there are as many positive interactions as negative, and thus there are large cancellation effects. If we imagine a system composed entirely of neutral groups (e.g., functional groups of a single molecule or individual molecules of a condensed phase), the long-range interaction between groups is a dipole–dipole interaction, which decays as r^{-3}, and the total energy contribution should decay as r^{-1}. Again, the actual situation is more favorable because of positive and negative cancellation effects, but the much slower decay of the electrostatic interaction makes it significantly harder to deal with. Cut-off distances (again, ideally implemented with smooth switching functions) must be quite large to avoid structural artifacts (e.g., atoms having large partial charges of like sign anomalously segregating at interatomic distances just in excess of the cut-off).

In infinite periodic systems, an attractive alternative to the use of a cut-off distance is the Ewald sum technique, first described for chemical systems by York, Darden and Pedersen (1993). By using a reciprocal-space technique to evaluate long-range contributions, the total electrostatic interaction can be calculated to a pre-selected level of accuracy (i.e., the Ewald sum limit is exact) with a scaling that, in the most favorable case (called 'Particle-mesh Ewald', or PME), is $N\log N$. Prior to the introduction of Ewald sums, the modeling of polyelectrolytes (e.g., DNA) was rarely successful because of the instabilities introduced by cut-offs in systems having such a high degree of localized charges (see, for instance, Beveridge and McConnell 2000).

In aperiodic systems, another important contribution has been the development of the so-called 'Fast Multipole Moment' (FMM) method (Greengard and Rokhlin 1987). In essence, this approach takes advantage of the significant cancellations in charge–charge interactions between widely separated regions in space, and the increasing degree to which those interactions can be approximated by highly truncated multipole–multipole interactions. In the most favorable case, FMM methods scale linearly with system size.

It should be remembered, of course, that scaling behavior is informative of the relative time one system takes compared to another of different size, and says *nothing* about the *absolute* time required for the calculation. Thus, FMM methods scale linearly, but the initial overhead can be quite large, so that it requires a very large system before it outperforms PME for the same level of accuracy. Nevertheless, the availability of the FMM method renders conceivable the molecular modeling of extraordinarily large systems, and refinements of the method, for example the use of multiple grids (Skeel, Tezcan, and Hardy 2002), are likely to continue to be forthcoming.

An interesting question that arises with respect to force fields is the degree to which they can be used to study reactive processes, i.e., processes whereby one minimum-energy compound is converted into another with the intermediacy of some transition state. As noted at the beginning of this chapter, one of the first applications of force-field methodology was to study the racemization of substituted biphenyls. And, for such 'conformational reactions', there seems to be no reason to believe force fields would not be perfectly appropriate modeling tools. Unless the conformational change in question were to involve an enormous amount of strain in the TS structure, there is little reason to believe that any of the internal coordinates would be so significantly displaced from their equilibrium values that the force-field functional forms would no longer be accurate.

However, when it comes to reactions where bonds are being made and/or broken, it is clear that, at least for the vast majority of force fields that use polynomial expressions for the bond stretching energy, the 'normal' model is inapplicable. Nevertheless, substantial application of molecular mechanics to such TS structures has been reported, with essentially three different approaches having been adopted.

One approach, when sufficient data are available, is to define new atom types and associated parameters for those atoms involved in the bond-making/bond-breaking coordinate(s). This is rather tricky since, while there may be solid experimental data for activation energies, there are unlikely to be any TS structural data. Instead, one might choose to use structures computed from some QM level of theory for one or more members of the molecular data set. Then, if one assumes the reaction coordinate is highly transferable from one molecule to the next (i.e., this methodology is necessarily restricted to the study of a single reaction amongst a reasonably closely related set of compounds), one can define a force field where TS structures are treated as 'minima' – minima in quotes because the equilibrium distances and force constants for the reactive coordinate(s) have values characteristic of the transition state.

This methodology has two chief drawbacks. A philosophical drawback is that movement along the reaction coordinate raises the force-field energy instead of lowering it, which is opposite to the real chemical system. A practical drawback is that it tends to be data limited – one may need to define a fairly large number of parameters using only a rather limited number of activation energies and perhaps some QM data. As noted in Section 2.2.7, this creates a tension between chemical intuition and statistical rigor. Two papers applying this technique to model the acid-catalyzed lactonization of organic hydroxy-acids illustrate the competing extremes to which such optimizations may be taken (Dorigo and Houk 1987; Menger 1990).

An alternative approach is one that is valence-bond like in its formulation. A possible TS structure is one whose molecular geometry is computed to have the same energy *irrespective of whether the atomic connectivity is that of the reactant or that of the product* (Olsen and Jensen 2003). Consider the example in Figure 2.9 for a hypothetical hydride transfer from an alkoxide carbon to a carbonyl. When the C–H bond is stretched from the reactant structure, the energy of the reactant-bonded structure goes up, while the energy of the product-bonded structure goes down because that structure's C–H bond is coming closer to its equilibrium value (from which it is initially very highly displaced). The simplest way to view this process is to envision two PESs, one defined for the reactant and one for the product. These two surfaces will intersect along a 'seam', and this seam is where the energy is independent of which connectivity is employed. The TS structure is then defined as the *minimum* on the seam. This approach is only valid when the reactant and product energies are computed

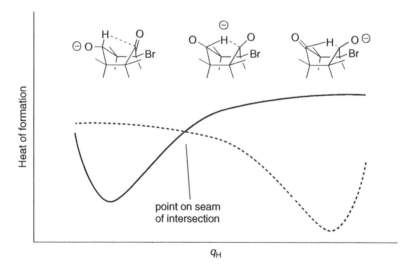

Figure 2.9   Slice through two intersecting enthalpy 'surfaces' along an arbitrary coordinate describing the location of a transferring H atom. The solid curve corresponds to bond stretching of the solid bond from carbon to the H atom being transferred. The dashed curve corresponds analogously to the dashed bond. At the point of intersection, the structure has the same energy irrespective of which bonding scheme is chosen. [For chemical clarity, the negative charge is shown shifting from one oxygen to the other, but for the method to be valid the two oxygen atom types could not change along either reaction coordinate. Note also that the bromine atom lifts the symmetry that would otherwise be present in this reaction.]

relative to a common zero (e.g., heats of formation are used; see Section 2.3), but one of its chief advantages is that it should properly reflect movement of the TS structure as a function of reaction thermicity. Because the seam of intersection involves structures having highly stretched bonds, care must be taken to use bond stretching functional forms that are accurate over larger ranges than are otherwise typical. When the VB formalism goes beyond the seam approach, and is adopted in full, a new ground-state potential energy surface can be generated about a true TS structure; such an approach is sometimes referred to as multiconfiguration molecular mechanics (MCMM) and is described in detail in Section 13.4.

The third approach to finding TS structures involves either adopting bond making/breaking functional forms that are accurate at all distances (making evaluation of bond energies a rather unpleasant N^2 process), or mixing the force-field representation of the bulk of the molecule with a QM representation of the reacting region. Mixed QM/MM models are described in detail in Chapter 13.

2.5   Menagerie of Modern Force Fields

2.5.1   Available Force Fields

Table 2.1 contains an alphabetic listing of force fields which for the most part continue to be in use today. Nomenclature of force fields can be rather puzzling because developers rarely change the name of the force field as development progresses. This is not necessarily a major issue when new development extends a force field to functionality that had not previously been addressed, but can be singularly confusing if pre-existing parameters or functional forms are changed from one version to the next without an accompanying name change. Many developers have tried to solve this problem by adding to the force field name the last two digits of the year of the most recent change to the force field. Thus, one can have MM3(92) and MM3(96), which are characterized by, *inter alia*, different hydrogen bonding parameters. Similarly, one has consistent force field (CFF) and Merck molecular force field (MMFF) versions identified by trailing year numbers. Regrettably, the year appearing in a version number does not necessarily correspond to the year in which the modifications were published in the open literature. Moreover, even when the developers themselves exercise adequate care, there is a tendency for the user community to be rather sloppy in referring to the force field, so that the literature is replete with calculations inadequately described to ensure reproducibility.

Further confusing the situation, certain existing force fields have been used as starting points for development by new teams of researchers, and the name of the resulting product has not necessarily been well distinguished from the original (which may itself be in ongoing development by its original designers!). Thus, for instance, one has the MM2* and MM3* force fields that appear in the commercial program MACROMODEL and that are based on early versions of the unstarred force fields of the same name (the * indicates the use of point charges to evaluate the electrostatics instead of bond dipoles, the use of a non-directional 10–12 potential for hydrogen bonding in place of an MM3 Buckingham potential, and a different formalism for handling conjugated systems). The commercial program Chem3D

Table 2.1 Force fields

Name (if any)	Range	Comments	Refs	$\Sigma(error)^a$
—	Biomolecules (2nd generation includes organics)	Sometimes referred to as AMBER force fields; new versions are first coded in software of that name. All-atom (AA) and united-atom (UA) versions exist.	Original: Weiner, S. J., Kollman, P. A., Nguyen, D. T., and Case, D. A. 1986. *J. Comput. Chem.*, **7**, 230. Latest generation: Duan, Y., Wu, C., Chowdhury, S., Lee, M. C., Xiong, G. M., Zhang, W., Yang, R., Cieplak, P., Luo, R., Lee, T., Caldwell, J., Wang, J. M., and Kollman, P. A. 2003. *J. Comput. Chem.*, **24**, 1999.; Ryjacek, F., Kubar, T., and Hobza. P. 2003. *J. Comput. Chem.*, **24**, 1891. See also amber.scripps.edu	7 (AMBER*)
—	Organics and biomolecules	The program MACROMODEL contains many modified versions of other force fields, e.g., AMBER*, MM2*, MM3*, OPLSA*.	Mohamadi, F., Richards, N. J. G., Guida, W. C., Liskamp, R., Lipton, M., Caufield, C., Chang, G., Hendrickson, T., and Still, W. C. 1990. *J. Comput. Chem.* **11**, 440. Recent extension: Senderowitz, H. and Still, W. C. 1997. *J. Org. Chem.*, **62**, 1427. See also www.schrodinger.com	4 (MM2*) 5 (MM3*)
BMS	Nucleic Acids		Langley, D. R. 1998. *J. Biomol. Struct. Dyn.*, **16**, 487.	

(continued overleaf)

Table 2.1   (*continued*)

Name (if any)	Range	Comments	Refs	$\Sigma(\text{error})^a$
CHARMM	Biomolecules	Many versions of force field parameters exist, distinguished by ordinal number. All-atom and united-atom versions exist.	Original: Brooks, B. R., Bruccoleri, R. E., Olafson, B. D., States, D. J., Swaminathan, S., and Karplus, M. 1983. *J. Comput. Chem.,* **4**, 187; Nilsson, L. and Karplus, M. 1986. *J. Comput. Chem.* **7**, 591. Latest generation: MacKerell, A. D., Bashford, D., Bellott, M., Dunbrack, R. L., Evanseck, J. D., Field, M. J., Gao, J., Guo, H., Ha, S., Joseph-McCarthy, D., Kuchnir, L., Kuczera, K., Lau, T. F. K., Mattos, C., Michnick, S., Nago, T., Nguyen, D. T., Prodhom, B., Reiher, W. E., Roux, B., Schlenkrich, M., Smith, J. C., Stote, R., Straub, J., Watanabe, M., Wiórkievicz-Kuczera, J., Yin, D., and Karplus, M. 1998. *J. Phys. Chem. B,* **102**, 3586; MacKerell, A. D. and Banavali, N. 2000. *J. Comput. Chem.,* **21**, 105; Patel, S. and Brooks, C. L. 2004. *J. Comput. Chem.,* **25**, 1. See also yuri.harvard.edu	
CHARMm	Biomolecules and organics	Version of CHARMM somewhat extended and made available in Accelrys software products.	Momany, F. A. and Rone, R. 1992. *J. Comput. Chem.,* **13**, 888. See also www.accelrys.com	

Chem-X	Organics	Available in Chemical Design Ltd. software.	Davies, E. K. and Murrall, N. W. 1989. *J. Comput. Chem.*, **13**, 149.	12
CFF/CVFF	Organics and biomolecules	CVFF is the original; CFF versions are identified by trailing year digits. Bond stretching can be modeled with a Morse potential. Primarily available in Accelrys software.	CVFF: Lifson, S., Hagler, A. T., and Stockfisch, J. P. 1979. *J. Am. Chem. Soc.*, **101**, 5111, 5122, 5131. CFF: Hwang, M.-J., Stockfisch, T. P., and Hagler, A. T. 1994. *J. Am. Chem. Soc.*, **116**, 2515; Maple, J. R., Hwang, M.-J., Stockfisch, T. P., Dinur, U., Waldman, M., Ewig, C. S., and Hagler, A. T. 1994. *J. Comput. Chem.*, **15**, 162; Maple. J. R., Hwang, M.-J., Jalkanen, K. J., Stockfisch, T. P., and Hagler, A. T. 1998. *J. Comput. Chem.*, **19**, 430; Ewig, C. S., Berry, R., Dinur, U., Hill, J.-R., Hwang, M.-J., Li, C., Maple, J., Peng, Z., Stockfisch, T. P., Thacher, T. S., Yan, L., Ni, X., and Hagler, A. T. 2001. *J. Comput. Chem.*, **22**, 1782. See also www.accelrys.com	13 (CVFF) 7 (CFF91)
DREIDING	Main-group organics and inorganics	Bond stretching can be modeled with a Morse potential.	Mayo, S. L., Olafson, B. D., and Goddard, W. A., III, 1990. *J. Phys. Chem.* **94**, 8897.	10

(continued overleaf)

Table 2.1 (*continued*)

Name (if any)	Range	Comments	Refs	$\Sigma(\text{error})^a$
ECEPP	Proteins	Computes only non-bonded interactions for fixed structures. Versions identified by /(ordinal number) after name.	Original: Némethy, G., Pottle, M. S., and Scheraga, H. A. 1983. *J. Phys. Chem.*, **87**, 1883. Latest generation: Kang, Y. K., No, K. T., and Scheraga, H. A. 1996. *J. Phys. Chem.*, **100**, 15588.	
ESFF	General	Bond stretching is modeled with a Morse potential. Partial atomic charges from electronegativity equalization.	Original: Barlow, S., Rohl, A. L., Shi, S., Freeman, C. M., and O'Hare, D. 1996. *J. Am. Chem. Soc.*, **118**, 7578. Latest generation: Shi, S., Yan, L., Yang, Y., Fisher-Shaulsky, J., and Thacher, T. 2003. *J. Comput. Chem.*, **24**, 1059.	
GROMOS	Biomolecules	Coded primarily in the software having the same name.	Daura, X., Mark, A. E., and van Gunsteren, W. F. 1998. *J. Comput. Chem.*, **19**, 535.; Schuler, L. D., Daura, X., and van Gunsteren, W. F. 2001. *J. Comput. Chem.*, **22**, 1205. See also igc.ethz.ch/gromos	

MM2	Organics	Superseded by MM3 but still widely available in many modified forms.	Comprehensive: Burkert, U. and Allinger, N. L. 1982. *Molecular Mechanics*, ACS Monograph 177, American Chemical Society: Washington, DC.	5 (MM2(85), MM2(91), Chem-3D)
MM3	Organics and biomolecules	Widely available in many modified forms.	Original: Allinger, N. L., Yuh, Y. H., and Lii, J.-H. 1989. *J. Am. Chem. Soc.*, **111**, 8551. MM3(94): Allinger, N. L., Zhou, X., and Bergsma, J. 1994. *J. Mol. Struct. (Theochem)*, **312**, 69. Recent extension: Stewart, E. L., Nevins, N., Allinger, N. L., and Bowen, J. P. 1999. *J. Org. Chem.* **64**, 5350.	5 (MM3(92))
MM4	Hydrocarbons, alcohols, ethers, and carbohydrates		Allinger, N. L., Chen, K. S., and Lii, J. H. 1996. *J. Comput. Chem.*, **17**, 642; Nevins, N., Chen, K. S., and Allinger, N. L. 1996. *J. Comput. Chem.*, **17**, 669; Nevins, N., Lii, J. H., and Allinger, N. L. 1996. *J. Comput. Chem.*, **17**, 695; Nevins, N. and Allinger, N. L. 1996. *J. Comput. Chem.*, **17**, 730. Recent extension: Lii, J. H., Chen, K. H., and Allinger, N. L. 2004. *J. Phys. Chem. A*, **108**, 3006.	

(continued overleaf)

Table 2.1 (*continued*)

Name (if any)	Range	Comments	Refs	Σ(error)[a]
MMFF	Organics and biomolecules	Widely available in relatively stable form.	Halgren, T. A. 1996. *J. Comput. Chem.*, **17**, 490, 520, 553, 616; Halgren, T. A., and Nachbar, R. B. 1996. *J. Comput. Chem.*, **17**, 587. See also www.schrodinger.com	4 (MMFF93)
MMX	Organics, biomolecules, and inorganics	Based on MM2.	See www.serenasoft.com	5
MOMEC	Transition metal compounds		Original: Bernhardt, P. V. and Comba, P. 1992. *Inorg. Chem.*, **31**, 2638. Latest generation: Comba, P. and Gyr, T. 1999. *Eur. J. Inorg. Chem.* 1787 See also www.uni-heidelberg.de/institute/fak12/AC/comba/molmod_momec.html	
OPLS	Biomolecules, some organics	Organic parameters are primarily for solvents. All-atom and united-atom versions exist.	Proteins: Jorgensen, W. L., and Tirado-Rives, J. 1988. *J. Am. Chem. Soc.*, **110**, 1657; Kaminski, G. A., Friesner, R. A., Tirado-Rives, J., and Jorgensen, W. L. 2001. *J. Phys. Chem. B*, **105**, 6474.	

			Nucleic acids: Pranata, J., Wierschke, S. G., and Jorgensen, W. L. 1991. *J. Phys. Chem. B*, **113**, 2810. Sugars: Damm, W., Frontera, A., Tirado-Rives, J., and Jorgensen, W. L. 1997. *J. Comput. Chem.*, **18**, 1955. Recent extensions: Rizzo, R. C., Jorgensen, W. L. 1999. *J. Am. Chem. Soc.*, **121**, 4827. Carbohydrates: Kony, D., Damm, W., Stoll, S., and van Gunsteren, W. F. 2002. *J. Comput. Chem.*, **2**, 1416.
PEF95SAC	Carbohydrates	Based on CFF form.	Fabricius, J., Engelsen, S. B., and Rasmussen, K. 1997. *J. Carbohydr. Chem.*, **16**, 751.
PFF	Proteins	Polarizable electrostatics	Kaminski, G. A., Stern, H. A., Berne, B. J., Friesner, R. A., Cao, Y. X., Murphy, R. B., Zhou, R., and Halgren, T. A. 2002. *J. Comput. Chem.*, **23**, 1515.

(continued overleaf)

Table 2.1   (*continued*)

Name (if any)	Range	Comments	Refs	Σ(error)[a]
SHAPES	Transition metal compounds		Allured, V. S., Kelly, C., and Landis, C. R. 1991. *J. Am. Chem. Soc.*, **113**, 1.	
SYBYL/Tripos	Organics and proteins	Available in Tripos and some other software.	Clark, M., Cramer, R. D., III, and van Opdenbosch, N. 1989. *J. Comput. Chem.*, **10**, 982. See also www.tripos.com and www.scivision.com	8–12
TraPPE	Organic	Primarily for computing liquid/vapor/supercritical fluid phase equilibria	Original: Martin, M. G. and Siepmann, J. I. 1998. *J. Phys. Chem. B*, **102**, 2569. Latest Generation: Chen, B., Potoff, J. J., and Siepmann, J. I. 2001. *J. Phys. Chem. B*, **105**, 3093.	
UFF	General	Bond stretching can be modeled with a Morse potential.	Rappé, A. K., Casewit, C. J., Colwell, K. S., Goddard, W. A., III, and Skiff, W. M. 1992. *J. Am. Chem. Soc.*, **114**, 10024, 10035, 10046.	21
VALBOND	Transition metal compounds	Atomic-orbital-dependent energy expressions.	Root, D. M., Landis, C. R., and Cleveland, T. 1993. *J. Am. Chem. Soc.*, **115**, 4201.	

[a]Kcal mol^{-1}. From Gundertofte *et al.* (1991, 1996); see text.

also has force fields based on MM2 and MM3, and makes no modification to the names of the originals.

As a final point of ambiguity, some force fields have not been given names, *per se*, but have come to be called by the names of the software packages in which they first became widely available. Thus, the force fields developed by the Kollman group (see Table 2.1) have tended to be referred to generically as AMBER force fields, because this software package is where they were originally coded. Kollman preferred that they be referred to by the names of the authors on the relevant paper describing their development, e.g., 'the force field of Cornell *et al.*' This is certainly more informative, since at this point the AMBER program includes within it *many* different force fields, so reference to the 'AMBER force field' conveys no information.

Because of the above ambiguities, and because it is scientifically unacceptable to publish data without an adequate description of how independent researchers might reproduce those data, many respected journals in the chemistry field now have requirements that papers reporting force-field calculations include as supplementary material a complete listing of all force field parameters (and functional forms, if they too cannot be adequately described otherwise) required to carry out the calculations described. This also facilitates the dissemination of information to those researchers wishing to develop their own codes for specific purposes.

Table 2.1 also includes a general description of the chemical space over which the force field has been designed to be effective; in cases where multiple subspaces are addressed, the order roughly reflects the priority given to these spaces during development. Force fields which have undergone many years worth of refinements tend to have generated a rather large number of publications, and the table does not try to be exhaustive, but effort is made to provide key references. The table also includes comments deemed to be particularly pertinent with respect to software implementing the force fields. For an exhaustive listing, by force field, of individual papers in which parameters for specific functional groups, metals, etc., were developed, readers are referred to Jalaie and Lipkowitz (2000).

2.5.2 Validation

The vast majority of potential users of molecular mechanics have two primary, related questions: 'How do I pick the best force field for my problem?' and, 'How will I know whether I can trust the results?' The process of testing the utility of a force field for molecules other than those over which it was parameterized is known as 'validation'.

The answer to the first question is obvious, if not necessarily trivial: one should pick the force field that has previously been shown to be most effective for the most closely related problem one can find. That demonstration of effectiveness may have taken place within the process of parameterization (i.e., if one is interested in conformational properties of proteins, one is more likely to be successful with a force field specifically parameterized to model proteins than with one which has not been) or by post-development validation. Periodically in the literature, papers appear comparing a wide variety of force fields for some well-defined problem, and the results can be quite useful in guiding the choices of subsequent researchers (see also, Bays 1992). Gundertofte *et al.* (1991, 1996) studied the accuracy of 17 different force fields with respect to predicting 38 experimental conformational energy differences or rotational barriers in organic molecules. These data were grouped into eight separate categories (conjugated compounds, halocyclohexanes, haloalkanes, cyclohexanes, nitrogen-containing compounds, oxygen-containing compounds, hydrocarbons, and rotational barriers). A summary of these results appears for relevant force fields in Table 2.1, where the number cited represents the sum of the mean errors over all eight categories. In some cases a range is cited because different versions of the same force field and/or different software packages were compared. In general, the best performances are exhibited by the MM2 and MM3 force fields and those other force fields based upon them. In addition, MMFF93 had similar accuracy. Not surprisingly, the most general force fields do rather badly, with UFF faring quite poorly in every category other than hydrocarbons.

Broad comparisons have also appeared for small biomolecules. Barrows *et al.* (1998) compared 10 different force fields against well-converged quantum mechanical calculations for predicting the relative conformational energies of 11 different conformers of D-glucose. GROMOS, MM3(96), and the force field of Weiner *et al.* were found to have average errors of 1.5 to 2.1 kcal/mol in relative energy, CHARMM and MMFF had average errors of from 0.9 to 1.5, and AMBER*, Chem-X, OPLS, and an unpublished force field of Simmerling and Kollman had average errors from 0.6 to 0.8 kcal/mol, which compared quite well with vastly more expensive *ab initio* methods. Shi *et al.* (2003) compared the performance of the very general force fields ESFF, CFF91, and CVFF over five of these glucose conformers and found average errors of 1.2, 0.6, and 1.9 kcal/mol, respectively; a more recent comparison by Heramingsen *et al.* (2004) of 20 carbohydrate force fields over a larger test of sugars and sugar–water complexes did not indicate any single force field to be clearly superior to the others. Beachy *et al.* (1997) carried out a similar comparison for a large number of polypeptide conformations and found OPLS, MMFF, and the force field of Cornell *et al.* to be generally the most robust. Price and Brooks (2002) compared protein dynamical properties, as opposed to polypeptide energetics, and found that the force fields of Cornell *et al.*, CHARM22, and OPLS-AA all provided similarly good predictions for radii of gyration, backbone order parameters, and other properties for three different proteins.

Of course, in looking for an optimal force field there is no guarantee that *any* system sufficiently similar to the one an individual researcher is interested in has *ever* been studied, in which case it is hard to make a confident assessment of force-field utility. In that instance, assuming some experimental data are available, it is best to do a survey of several force fields to gauge their reliability. When experimental data are *not* available, recourse to well-converged quantum mechanical calculations for a few examples is a possibility, assuming the computational cost is not prohibitive. QM values would then take the place of experimental data. Absent any of these alternatives, any force field calculations will simply carry with them a high degree of uncertainty and the results should be used with caution.

Inorganic chemists may be frustrated to have reached this point having received relatively little guidance on what force fields are best suited to *their* problems. Regrettably, the current state of the art does not provide any single force field that is both robust and accurate over a large range of inorganic molecules (particularly metal coordination compounds). As noted above, parameter transferability tends to be low, i.e., the number of atom types potentially requiring parameterization for a single metal atom, together with the associated very large number of geometric and non-bonded constants, tends to significantly exceed available data. Instead, individual problems tend to be best solved with highly tailored force fields, when they are available (see for example, Comba and Remenyi 2002), or by combining QM and MM methods (see Chapter 13), or by accepting that the use of available highly generalized force fields increases the risk for significant errors and thus focusing primarily on structural perturbations over a related series of compounds rather than absolute structures or energetics is advised (see also Hay 1993; Norrby and Brandt 2001).

A last point that should be raised with regard to validation is that any comparison between theory and experiment must proceed in a consistent fashion. Consider molecular geometries. Chemists typically visualize molecules as having 'structure'. Thus, for example, single-crystal X-ray diffractometry can be used to determine a molecular structure, and at the end of a molecular mechanics minimization one has a molecular structure, but is it strictly valid to compare them?

It is best to consider this question in a series of steps. First, recall that the goal of a MM minimization is to find a local minimum on the PES. That local minimum has a unique structure and each molecular coordinate has a precise value. What about the structure from experiment? Since most experimental techniques for assigning structure sample an ensemble of molecules (or one molecule many, many times), the experimental measurement is properly referred to as an expectation value, which is denoted by angle brackets about the measured variable. Real molecules vibrate, even at temperatures arbitrarily close to absolute zero, so measured structural parameters are actually expectation values over the molecular vibrations. Consider, for example, the length of the bond between atoms A and B in its ground vibrational state. For a quantum mechanical harmonic oscillator, $\langle r_{AB} \rangle = r_{AB,eq}$, but real bond stretching coordinates are anharmonic, and this inevitably leads to $\langle r_{AB} \rangle > r_{AB,eq}$ (see Section 9.3.2). In the case of He_2, mentioned above, the effect of vibrational averaging is rather extreme, leading to a difference between $\langle r_{AB} \rangle$ and $r_{AB,eq}$ of more than 50 Å! Obviously, one should not judge the quality of the calculated $r_{AB,eq}$ value based on comparison to the experimental $\langle r_{AB} \rangle$ value. Note that discrepancies between $\langle r_{AB} \rangle$ and $r_{AB,eq}$ will increase if the experimental sample includes molecules in excited vibrational states. To be rigorous in comparison, either the calculation should be extended to compute $\langle r_{AB} \rangle$ (by computation of the vibrational wave function(s) and appropriate averaging) or the experiment must be analyzed to determine $r_{AB,eq}$, e.g., as described in Figure 2.1.

Moreover, the above discussion assumes that the experimental technique measures exactly what the computational technique does, namely, the separation between the nuclear centroids defining a bond. X-ray crystallography, however, measures maxima in scattering amplitudes, and X-rays scatter not off nuclei but off electrons. Thus, if electron density maxima do not correspond to nuclear positions, there is no reason to expect agreement between theory and experiment (for heavy atoms this is not much of an issue, but for very light ones it can be). Furthermore, the conditions of the calculation typically correspond to an isolated molecule acting as an ideal gas (i.e., experiencing no intermolecular interactions), while a technique like X-ray crystallography obviously probes molecular structure in a condensed phase where crystal packing and dielectric effects may have significant impact on the determined structure (see, for example, Jacobson *et al.* 2002).

The above example illustrates some of the caveats in comparing theory to experiment for a structural datum (see also Allinger, Zhou and Bergsma 1994). Care must also be taken in assessing energetic data. Force-field calculations typically compute potential energy, whereas equilibrium distributions of molecules are dictated by free energies (see Chapter 10). Thus, the force-field energies of two conformers should not necessarily be expected to reproduce an experimental equilibrium constant between them. The situation can become still more confused for transition states, since experimental data typically are either activation free energies or Arrhenius activation energies, neither of which corresponds directly with the difference in potential energy between a reactant and a TS structure (see Chapter 15). Even in those cases where the force field makes possible the computation of heats of formation and the experimental data are available as enthalpies, it must be remembered that the effect of zero-point vibrational energy is accounted for in an entirely average way when atom-type reference heats of formation are parameterized, so some caution in comparison is warranted.

Finally, any experimental measurement carries with it some error, and obviously a comparison between theory and experiment should never be expected to do better than the experimental error. *The various points discussed in this last section are all equally applicable to comparisons between experiment and QM theories as well, and the careful practitioner would do well always to bear them in mind.*

2.6   Force Fields and Docking

Of particular interest in the field of drug design is the prediction of the strength and specificity with which a small to medium sized molecule may bind to a biological macromolecule (Lazaridis 2002; Shoichet *et al.* 2002). Many drugs function by binding to the active sites of particular enzymes so strongly that the normal substrates of these enzymes are unable to displace them and as a result some particular biochemical pathway is stalled.

If we consider a case where the structure of a target enzyme is known, but no structure complexed with the drug (or the natural substrate) exists, one can imagine using computational chemistry to evaluate the energy of interaction between the two for various positionings of the two species. This process is known as 'docking'. Given the size of the total system (which includes a biopolymer) and the very large number of possible arrangements of the drug molecule relative to the enzyme that we may wish to survey, it is clear that speedy methods like molecular mechanics are likely to prove more useful than others. This becomes still more true if the goal is to search a database of, say, 100 000 molecules to see if one can find any that bind still *more* strongly than the current drug, so as to prospect for pharmaceuticals of improved efficacy.

One way to make this process somewhat more efficient is to adopt rigid structures for the various molecules. Thus, one does not attempt to perform geometry optimizations, but simply puts the molecules in some sort of contact and evaluates their interaction energies. To that extent, one needs only to evaluate non-bonded terms in the force field, like those

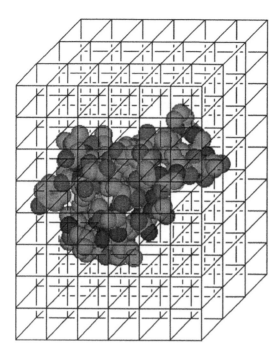

Figure 2.10 Docking grid constructed around a target protein. Each gridpoint can be assigned a force field interaction potential for use in evaluating binding affinities. Note that this grid is very coarse to improve viewing clarity; an actual grid might be considerably finer.

modeled by Eqs. (2.16) and (2.22). Moreover, to further simplify matters, one may consider the rigid enzyme to be surrounded by a three-dimensional grid, as illustrated in Figure 2.10. Given a fixed geometry, one may compute the interaction potential at each grid point for a molecular mechanics atom having unit values of charge and Lennard-Jones parameters. Then, to compute interaction energies, one places a proto-drug molecule at some arbitrary position in space, and assigns each atom to be associated with the grid point nearest it (or one could interpolate if one were willing to pay the computational cost). The potential for each point is then multiplied by the appropriate atomic parameters, and the sum of all atomic interactions defines the docking energy for that particular position and orientation. After a suitable number of random or directed choices of position have been surveyed, the lowest docking energy is recorded, and one moves on to the next molecule in the test set.

Of course, this analysis is rather crude, since it ignores a number of physical phenomena in computing an interaction energy. For instance, we failed to account for the desolvation of the enzyme and the substrate along the surfaces over which they come into contact, and we did not consider the entropy loss associated with binding. As such, the goal of most docking studies tends to be to provide a simple filter that can narrow a vast database down to a merely large database, to which more refined techniques may be applied so as to further winnow down possible leads.

Note that after having made so many approximations in the modeling protocol, there is no particular reason to believe that nonbonded interactions evaluated using particular force field parameters will be better than others that might be developed specifically for the purpose of docking. Thus, other grid-based scoring methods are widely used (see, for example, Meng, Shoichet, and Kuntz 1992), including more recent ones that incorporate some analysis of desolvation penalties (Zou, Sun, and Kuntz 1999; Salichs *et al.* 2002; Li, Chen, and Weng 2003; Kang, Shafer, and Kuntz 2004).

2.7   Case Study: (2*R**,4*S* *)-1-Hydroxy-2,4-dimethylhex-5-ene

Synopsis of Stahl *et al.* (1991), 'Conformational Analysis with Carbon-Carbon Coupling Constants. A Density Functional and Molecular Mechanics Study'.

Many natural products contain one or more sets of carbon backbones decorated with multiple stereogenic centers. A small such fragment that might be found in propiogenic natural products is illustrated in Figure 2.11. From a practical standpoint, the assignment of absolute configuration to each stereogenic center (*R* or *S*), or even of the relative configurations between centers, can be difficult in the absence of single-crystal X-ray data. When many possibilities exist, it is an unpleasant task to synthesize each one.

An alternative means to assign the stereochemistry is to use nuclear magnetic resonance (NMR). Coupling constant data from the NMR experiment can be particularly useful in assigning stereochemistry. However, if the fragments are highly flexible, the interpretation of the NMR data can be complicated when the interconversion of conformers is rapid on the NMR timescale. In that case, rather than observing separate, overlapping spectra for every conformer, only a population-averaged spectrum is obtained.

Deconvolution of such spectra can be accomplished in a computational fashion by (i) determining the energies of all conformers contributing to the equilibrium population, (ii) predicting the spectral constants associated with each conformer, and (iii) averaging over all spectral data weighted by the fractional contribution of each conformer to the equilibrium (the fractional contribution is determined by a Boltzmann average over the energies, see Eq. (10.49)). The authors adopted this approach for (2*R**,4*S**)-1-hydroxy-2,4-dimethylhex-5-ene, where the conformer energies were determined using the MM3 force field and the NMR coupling constants were predicted at the density functional level of theory. As density functional theory is the subject of Chapter 8 and the prediction of NMR data is not discussed until Section 9.4, we will focus here simply on the performance of MM3 for predicting conformer energies and weighting spectral data.

In order to find the relevant conformers, the authors employed a Monte Carlo/minimization strategy that is described in more detail in the next chapter – in practice, (2*R**,4*S**)-1-hydroxy-2,4-dimethylhex-5-ene is sufficiently small that one *could* survey every possible torsional isomer by brute force, but it would be very tedious. Table 2.2 shows, for the nine lowest energy conformers, their predicted energies, their contribution to the 300 K equilibrium population, their individual $^3J_{CC}$ coupling constants between atoms C(2)C(5), C(2)C(8), C(1)C(4), and C(4)C(7), and the mean absolute error in these coupling constants compared to experiment (see Figure 2.11 for atom-numbering convention). In addition, the spectral data predicted from a population-weighted equilibrium average over the nine conformers making up 82% of the equilibrium population are shown.

The population-averaged data are those in best agreement with experiment. Conformer G shows similar agreement (the increased error is within the rounding limit for the table), but is predicted to be sufficiently high in energy that it is unlikely that MM3 could be sufficiently in error for it to be the only conformer at equilibrium. As a separate assessment of this point, the authors carry out *ab initio* calculations at a correlated level of electronic structure theory (MP2/TZ2P//HF/TZ2P; this notation and the relevant theories are discussed in Chapters 6 and 7, but exact details are not important here), and observe what they characterize as very good agreement between the force-field energies and the *ab initio* energies (the data are not provided).

In principle, then, when the relative configurations are not known for a flexible chain in some natural product backbone, the technique outlined above could be used to predict the expected NMR spectra for all possibilities, and presuming one prediction matched to experiment significantly more closely than any other, the assignment would be regarded as reasonably secure. At the least, it would suggest how to prioritize synthetic efforts that would be necessary to provide the ultimate proof.

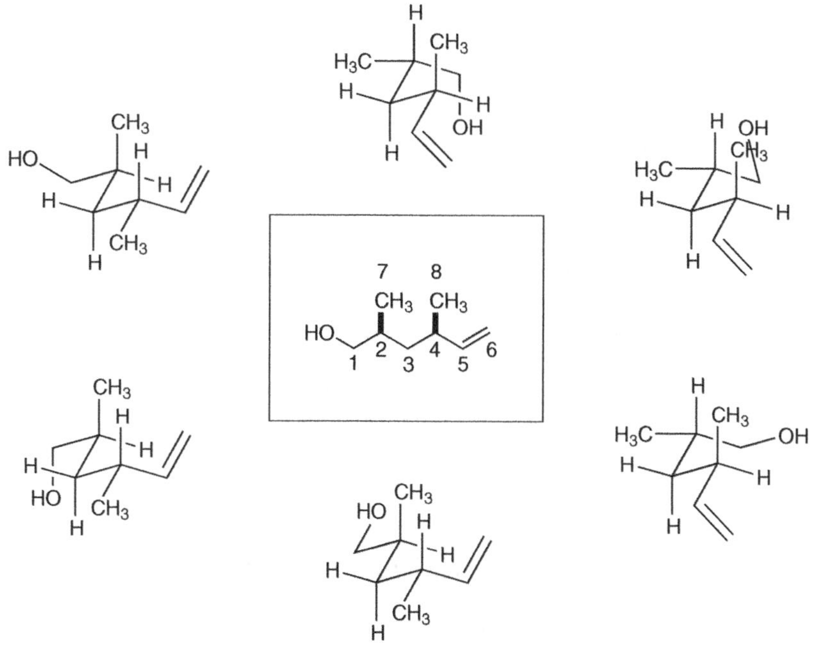

Figure 2.11   Some plausible conformations of (2R*,4S*)-1-hydroxy-2,4-dimethylhex-5-ene. How many different torsional isomers might one need to examine, and how would you go about generating them? [Note that the notation 2R*,4S* implies that the *relative* stereochemical configuration at the 2 and 4 centers is R,S – by convention, when the absolute configuration is not known the first center is always assigned to be R*. However, the absolute conformations that are drawn here are S,R so as to preserve correspondence with the published illustrations of Stahl and coworkers. Since NMR in an achiral solvent does not distinguish between enantiomers, one can work with either absolute configuration in this instance.]

Table 2.2 Relative MM3 energies (kcal mol^{-1}), fractional equilibrium populations F (%), predicted NMR coupling constants (Hz), and mean unsigned error in predicted coupling constants for different conformers and the equilibrium average of (2R^*,4S^*)-1-hydroxy-2,4-dimethyl-hex 5-ene at 300 K.

Conformer	rel E	F	^3J				MUE
			C(2)C(5)	C(2)C(8)	C(1)C(4)	C(4)C(7)	
A	0.0	24	1.1	4.2	3.9	1.3	**0.6**
B	0.1	21	1.1	4.0	5.8	1.2	**1.0**
C	0.2	19	1.0	4.2	4.1	1.2	**0.7**
D	0.9	5	3.8	1.5	1.7	4.5	**2.2**
E	1.1	4	4.1	0.8	1.1	4.4	**2.5**
F	1.3	3	4.1	0.9	0.4	5.3	**2.9**
G	1.4	2	1.2	3.7	3.8	1.5	**0.3**
H	1.4	2	1.4	4.2	5.7	1.4	**0.9**
I	1.5	2	0.1	5.1	0.0	5.3	**2.5**
average		82	*1.4*	*3.7*	*4.1*	*1.8*	*0.3*
experiment			**1.4**	**3.3**	**3.8**	**2.2**	

In that regard, this paper might have been improved by including a prediction (and ideally an experimental measurement) for the NMR coupling data of (2R^*,4R^*)-1-hydroxy-2,4-dimethylhex-5-ene, i.e., the stereoisomer having the R^*,R^* relative configuration between the stereogenic centers instead of the R^*,S^* configuration. If each predicted spectrum matched its corresponding experimental spectrum significantly more closely than it matched the non-corresponding experimental spectrum, the utility of the methodology would be still more convincingly demonstrated. Even in the absence of this demonstration, however, the work of Stahl and his coworkers nicely illustrates how accurate force fields can be for 'typical' C,H,O-compounds, and also how different levels of theory can be combined to address different parts of a computational problem in the most efficient manner. In this case, inexpensive molecular mechanics is used to provide an accurate map of the wells on the conformational potential energy surface and the vastly more expensive DFT method is employed only thereafter to predict the NMR spectral data.

Bibliography and Suggested Additional Reading

Bakken, V. and Helgaker, T. 2002. 'The Efficient Optimization of Molecular Geometries Using Redundant Internal Coordinates', *J. Chem. Phys.*, **117**, 9160.

Bowen, J. P. and Allinger, N. L. 1991. 'Molecular Mechanics: The Art and Science of Parameterization', in *Reviews in Computational Chemistry*, Vol. 2, Lipkowitz, K. B. and Boyd, D. B. Eds., VCH: New York, 81.

Brooijmans, N. and Kuntz, I. D. 2003. 'Molecular Recognition and Docking Algorithms', *Annu. Rev. Biophys. Biomol. Struct.* **32**, 335.

Comba, P. and Hambley, T. W. 2001. *Molecular Modeling of Inorganic Compounds*, 2nd Edn., Wiley-VCH: Weinheim.

Comba, P. and Remenyi, R. 2003. 'Inorganic and Bioinorganic Molecular Mechanics Modeling–the Problem of Force Field Parameterization', *Coord. Chem. Rev.*, **238–239**, 9.

Cramer, C. J. 1994. 'Problems and Questions in the Molecular Modeling of Biomolecules', *Biochem. Ed.* **22**, 140.

Dinur, U. and Hagler, A. T. 1991. 'New Approaches to Empirical Force Fields', in *Reviews in Computational Chemistry*, Vol. 2, Lipkowitz, K. B. and Boyd, D. B., Eds., VCH; New York, 99.

Dykstra, C. E. 1993. 'Electrostatic Interaction Potentials in Molecular Force Fields', *Chem. Rev.* **93**, 2339.

Eksterowicz, J. E. and Houk, K. N. 1993. 'Transition-state Modeling with Empirical Force Fields', *Chem. Rev.* **93**, 2439.

Jensen, F. 1999. *Introduction to Computational Chemistry*, Wiley: Chichester.

Jensen, F. and Norrby, P.-O. 2003. 'Transition States from Empirical Force Fields', *Theor. Chem. Acc.*, **109**, 1.

Kang, X. S., Shafer, R. H., and Kuntz, I. D. 2004. 'Calculation of Ligand-nucleic Acid Binding Free Energies with the Generalized-Born Model in DOCK', *Biopolymers*, **73**, 192.

Landis, C. R., Root, D. M., and Cleveland, T. 1995. 'Molecular Mechanics Force Fields for Modeling Inorganic and Organometallic Compounds' in *Reviews in Computational Chemistry*, Vol. 6, Lipkowitz, K. B. and Boyd, D. B. Eds., VCH: New York, 73.

Lazaridis, T. 2002. 'Binding Affinity and Specificity from Computational Studies', *Curr. Org. Chem.*, **6**, 1319.

Leach, A. R. 2001. *Molecular Modelling*, 2nd Edn., Prentice Hall: London.

Norrby, P.-O. 2001. 'Recipe for an Organometallic Force Field', in *Computational Organometallic Chemistry*, Cundari, T. Ed., Marcel Dekker: New York 7.

Pettersson, I. and Liljefors, T. 1996. 'Molecular Mechanics Calculated Conformational Energies of Organic Molecules: A Comparison of Force Fields', in *Reviews in Computational Chemistry*, Vol. 9, Lipkowitz, K. B. and Boyd, D. B., Eds., VCH: New York, 167.

Schlegel, H. B. 2003. 'Exploring Potential Energy Surfaces for Chemical Reactions: An Overview of Some Practical Methods', *J. Comput. Chem.* **124**, 1514.

References

Allinger, N. L., Zhou, X., and Bergsma, J. 1994. *J. Mol. Struct. (Theochem.)*, **312**, 69.

Barrows, S. E., Storer, J. W., Cramer, C. J., French, A. D., and Truhlar, D. G. 1998. *J. Comput. Chem.*, **19**, 1111.

Bays, J. P. 1992. *J. Chem. Edu.*, **69**, 209.

Beachy, M. D., Chasman, D., Murphy, R. B., Halgren, T. A., and Friesner, R. A. 1997. *J. Am. Chem. Soc.*, **119**, 5908.

Beveridge, D. L. and McConnell, K. J. 2000. *Curr. Opin. Struct. Biol.*, **10**, 182.

Cohen, N. and Benson, S. W. 1993. *Chem. Rev.*, **93**, 2419.

Comba, P. and Remenyi, R. 2002. *J. Comput. Chem.*, **23**, 697.

Dorigo, A. E. and Houk, K. N. 1987. *J. Am. Chem. Soc.*, **109**, 3698.

Fogarasi, G. and Balázs, A. 1985. *J. Mol. Struct. (Thochem)*, **133**, 105.

Greengard, L. and Rokhlin, V. 1987. *J. Comput. Phys.*, **73**, 325.

Gundertofte, K., Liljefors, T., Norrby, P.-O., and Petterson, I. 1996. *J. Comput. Chem.*, **17**, 429.

Gundertofte, K., Palm, J., Petterson, I., and Stamvick, A. 1991. *J. Comput. Chem.*, **12**, 200.

Harvey, S. C., Wang, C., Teletchea, S., and Lavery, R. 2003. *J. Comput. Chem.*, **24**, 1.

Hassinen, T. and Peräkylä, M. 2001. *J. Comput. Chem.*, **22**, 1229.

Hay, B. P. 1993. *Coord. Chem. Rev.*, **126**, 177.

Heramingsen, L., Madsen, D. E., Esbensen, S. L., Olsen, L., and Engelsen, S. B. 2004. *Carboh. Res.*, **339**, 937.

Hill, T. L. 1946. *J. Chem. Phys.*, **14**, 465.

Jacobson, M. P., Friesner, R. A., Xiang, Z. X., and Honig, B. 2002. *J. Mol. Biol.*, **320**, 597.

Jalaie, M. and Lipkowitz, K. B. 2000. *Rev. Comput. Chem.*, **14**, 441.

Jensen, F. 1999. *Introduction to Computational Chemistry*, Wiley: Chichester, Chapter 14 and references therein.

Lazaridis, T. 2002. *Curr. Org. Chem.*, **6**, 1319.

Leach, A. R. 1991. *Rev. Comput. Chem.*, **2**, 1.

Li, L., Chen, R., and Weng, Z. P. 2003. *Proteins*, **53**, 693.

Luo, F., McBane, G. C., Kim, G., Giese, C. F., and Gentry, W. R. 1993. *J. Chem. Phys.*, **98**, 3564.

Meng, E. C., Shoichet, B. K., and Kuntz, I. D. 1992. *J. Comput. Chem.*, **13**, 505.

Menger, F. 1990. *J. Am. Chem. Soc.*, **112**, 8071.

Norrby, P.-O. and Brandt, P. 2001. *Coord. Chem. Rev.*, **212**, 79.

Olsen, P. T. and Jensen, F. 2003. *J. Chem. Phys.*, **118**, 3523.

Petrella, R. J., Andricioaei, I., Brooks, B., and Karplus, M. 2003. *J. Comput. Chem.*, **24**, 222.

Price, D. J. and Brooks, C. L. 2002. *J. Comput. Chem.*, **23**, 1045.

Radom, L., Hehre, W. J., and Pople, J. A. 1971. *J. Am. Chem. Soc.*, **93**, 289.

Reichardt, C. 1990. *Solvents and Solvent Effects in Organic Chemistry*, VCH: New York, 12.

Salichs, A., López, M., Orozco, M., and Luque, F. J. 2002. *J. Comput.-Aid. Mol. Des.*, **16**, 569.

Schlick, T. 1992. *Rev. Comput. Chem.*, **3**, 1.

Shi, S., Yan, L., Yang, Y., Fisher-Shaulsky, J., and Thacher, T. 2003. *J. Comput. Chem.*, **24**, 1059.

Shoichet, B. K., McGovern, S. L., Wei, B. Q., and Irwin, J. J. 2002. *Curr. Opin. Chem. Biol.*, **6**, 439.

Skeel, R. D., Tezcan, I., and Hardy, D. J. 2002. *J. Comput. Chem.*, **23**, 673.

Stahl, M., Schopfer, U., Frenking, G., and Hoffmann, R. W. 1991. *J. Am. Chem. Soc.*, **113**, 4792.

Westheimer, F. H. and Mayer, J. E. 1946. *J. Chem. Phys.*, **14**, 733.

Wolfe, S., Rauk, A., Tel, L. M., and Csizmadia, I. G. 1971. *J. Chem. Soc. B,* **136**.

Woods, R. J. 1996. *Rev. Comput. Chem.*, **9**, 129.

York, D. M., Darden, T. A., and Pedersen, L. G. 1993. *J. Chem. Phys.*, **99**, 8345.

Zou, X. Q., Sun, Y. X., and Kuntz, I. D. 1999. *J. Am. Chem. Soc.*, **121**, 8033.

3

Simulations of Molecular Ensembles

3.1 Relationship Between MM Optima and Real Systems

As noted in the last chapter within the context of comparing theory to experiment, a minimum-energy structure, i.e., a local minimum on a PES, is sometimes afforded more importance than it deserves. Zero-point vibrational effects dictate that, even at 0 K, the molecule probabilistically samples a range of different structures. If the molecule is quite small and is characterized by fairly 'stiff' molecular coordinates, then its 'well' on the PES will be 'narrow' and 'deep' and the range of structures it samples will all be fairly close to the minimum-energy structure; in such an instance it is not unreasonable to adopt the simple approach of thinking about the 'structure' of the molecule as being the minimum energy geometry. However, consider the case of a large molecule characterized by many 'loose' molecular coordinates, say polyethyleneglycol, (PEG, $-(OCH_2CH_2)_n-$), which has 'soft' torsional modes: What is the structure of a PEG molecule having $n = 50$? Such a query is, in some sense, ill defined. Because the probability distribution of possible structures is not compactly localized, as is the case for stiff molecules, the very concept of structure as a time-independent property is called into question. Instead, we have to accept the flexibility of PEG as an intrinsic characteristic of the molecule, and any attempt to understand its other properties must account for its structureless nature. Note that polypeptides, polynucleotides, and polysaccharides all are *also* large molecules characterized by having many loose degrees of freedom. While nature has tended to select for particular examples of these molecules that are less flexible than PEG, nevertheless their utility as biomolecules sometimes derives from their ability to sample a wide range of structures under physiological conditions, and attempts to understand their chemical behavior must address this issue.

Just as zero-point vibration introduces probabilistic weightings to single-molecule structures, so too thermodynamics dictates that, given a large collection of molecules, probabilistic distributions of structures will be found about *different* local minima on the PES at non-zero absolute temperatures. The relative probability of clustering about any given minimum is a function of the temperature and some particular thermodynamic variable characterizing the system (e.g., Helmholtz free energy), that variable depending on what experimental conditions are being held constant (e.g., temperature and volume). Those variables being held constant define the 'ensemble'.

Essentials of Computational Chemistry, 2nd Edition Christopher J. Cramer
© 2004 John Wiley & Sons, Ltd   ISBNs: 0-470-09181-9 (cased); 0-470-09182-7 (pbk)

We will delay a more detailed discussion of ensemble thermodynamics until Chapter 10; indeed, in this chapter we will make use of ensembles designed to render the operative equations as transparent as possible without much discussion of extensions to other ensembles. The point to be re-emphasized here is that the vast majority of *experimental* techniques measure molecular properties as averages – either time averages or ensemble averages or, most typically, both. Thus, we seek computational techniques capable of accurately reproducing these aspects of molecular behavior. In this chapter, we will consider Monte Carlo (MC) and molecular dynamics (MD) techniques for the simulation of real systems. Prior to discussing the details of computational algorithms, however, we need to briefly review some basic concepts from statistical mechanics.

3.2   Phase Space and Trajectories

The state of a classical system can be completely described by specifying the positions and momenta of all particles. Space being three-dimensional, each particle has associated with it six coordinates – a system of N particles is thus characterized by $6N$ coordinates. The $6N$-dimensional space defined by these coordinates is called the 'phase space' of the system. At any instant in time, the system occupies one point in phase space

$$\mathbf{X}' = (x_1, y_1, z_1, p_{x,1}, p_{y,1}, p_{z,1}, x_2, y_2, z_2, p_{x,2}, p_{y,2}, p_{z,2}, \ldots) \tag{3.1}$$

For ease of notation, the position coordinates and momentum coordinates are defined as

$$\mathbf{q} = (x_1, y_1, z_1, x_2, y_2, z_2, \ldots) \tag{3.2}$$

$$\mathbf{p} = (p_{x,1}, p_{y,1}, p_{z,1}, p_{x,2}, p_{y,2}, p_{z,2}, \ldots) \tag{3.3}$$

allowing us to write a (reordered) phase space point as

$$\mathbf{X} = (\mathbf{q}, \mathbf{p}) \tag{3.4}$$

Over time, a dynamical system maps out a 'trajectory' in phase space. The trajectory is the curve formed by the phase points the system passes through. We will return to consider this dynamic behavior in Section 3.2.2.

3.2.1   Properties as Ensemble Averages

Because phase space encompasses every possible state of a system, the average value of a property A at equilibrium (i.e., its expectation value) for a system having a constant temperature, volume, and number of particles can be written as an integral over phase space

$$\langle A \rangle = \int\!\!\int A(\mathbf{q}, \mathbf{p}) P(\mathbf{q}, \mathbf{p}) d\mathbf{q} d\mathbf{p} \tag{3.5}$$

where P is the probability of being at a particular phase point. From statistical mechanics, we know that this probability depends on the energy associated with the phase point according to

$$P(\mathbf{q}, \mathbf{p}) = Q^{-1} e^{-E(\mathbf{q} \cdot \mathbf{p})/k_B T} \tag{3.6}$$

where E is the total energy (the sum of kinetic and potential energies depending on $\mathbf{p}$ and $\mathbf{q}$, respectively) k_B is Boltzmann's constant, T is the temperature, and Q is the system partition function

$$Q = \int \int e^{-E(\mathbf{q} \cdot \mathbf{p})/k_B T} d\mathbf{q} d\mathbf{p} \tag{3.7}$$

which may be thought of as the normalization constant for P.

How might one go about evaluating Eq. (3.5)? In a complex system, the integrands of Eqs. (3.5) and (3.7) are unlikely to allow for analytic solutions, and one must perforce evaluate the integrals numerically. The numerical evaluation of an integral is, in the abstract, straightforward. One determines the value of the integrand at some finite number of points, fits those values to some function that *is* integrable, and integrates the latter function. With an increasing number of points, one should observe this process to converge to a particular value (assuming the original integral is finite) and one ceases to take more points after a certain tolerance threshold has been reached.

However, one must remember just how vast phase space is. Imagine that one has only a very modest goal: One will take only a single phase point from each 'hyper-octant' of phase space. That is, one wants all possible combinations of signs for all of the coordinates. Since each coordinate can take on two values (negative or positive), there are 2^{6N} such points. Thus, in a system having $N = 100$ particles (which is a very small system, after all) one would need to evaluate A and E at 4.15×10^{180} points! Such a process might be rather time consuming ...

The key to making this evaluation more tractable is to recognize that phase space is, for the most part, a wasteland. That is, there are enormous volumes characterized by energies that are far too high to be of any importance, e.g., regions where the positional coordinates of two different particles are such that they are substantially closer than van der Waals contact. From a mathematical standpoint, Eq. (3.6) shows that a high-energy phase point has a near-zero probability, and thus the integrand of Eq. (3.5) will also be near-zero (as long as property A does not go to infinity with increasing energy). As the integral of zero is zero, such a phase point contributes almost nothing to the property expectation value, and simply represents a waste of computational resources. So, what is needed in the evaluation of Eqs. (3.5) and (3.7) is some prescription for picking *important* (i.e., high-probability) points.

The MC method, described in Section 3.4, is a scheme designed to do exactly this in a pseudo-random fashion. Before we examine that method, however, we first consider a somewhat more intuitive way to sample 'useful' regions of phase space.

3.2.2   Properties as Time Averages of Trajectories

If we start a system at some 'reasonable' (i.e., low-energy) phase point, its energy-conserving evolution over time (i.e., its trajectory) seems *likely* to sample relevant regions of phase space.

Certainly, this is the picture most of us have in our heads when it comes to the behavior of a real system. In that case, a reasonable way to compute a property average simply involves computing the value of the property periodically at times t_i and assuming

$$\langle A \rangle = \frac{1}{M} \sum_{i}^{M} A(t_i) \tag{3.8}$$

where M is the number of times the property is sampled. In the limit of sampling continuously and following the trajectory indefinitely, this equation becomes

$$\langle A \rangle = \lim_{t \to \infty} \frac{1}{t} \int_{t_0}^{t_0+t} A(\tau) d\tau \tag{3.9}$$

The 'ergodic hypothesis' assumes Eq. (3.9) to be valid and independent of choice of t_0. It has been proven for a hard-sphere gas that Eqs. (3.5) and (3.9) are indeed equivalent (Ford 1973). No such proof is available for more realistic systems, but a large body of empirical evidence suggests that the ergodic hypothesis is valid in most molecular simulations.

This point being made, we have not yet provided a description of how to 'follow' a phase-space trajectory. This is the subject of molecular dynamics, upon which we now focus.

3.3 Molecular Dynamics

One interesting property of a phase point that has not yet been emphasized is that, since it is defined by the positions and momenta of all particles, it *determines* the location of the next phase point in the absence of outside forces acting upon the system. The word 'next' is used loosely, since the trajectory is a continuous curve of phase points (i.e., between any two points can be found another point) – a more rigorous statement is that the forward trajectory is completely determined by the initial phase point. Moreover, since time-independent Hamiltonians are necessarily invariant to time reversal, a single phase point completely determines a full trajectory. As a result, phase space trajectories cannot cross themselves (since there would then be two *different* points leading away (in both time directions) from a single point of intersection). To illuminate further some of the issues involved in following a trajectory, it is helpful to begin with an example.

3.3.1 Harmonic Oscillator Trajectories

Consider a one-dimensional classical harmonic oscillator (Figure 3.1). Phase space in this case has only two dimensions, position and momentum, and we will define the origin of this phase space to correspond to the ball of mass m being at rest (i.e., zero momentum) with the spring at its equilibrium length. This phase point represents a stationary state of the system. Now consider the dynamical behavior of the system starting from some point other than the origin. To be specific, we consider release of the ball at time t_0 from

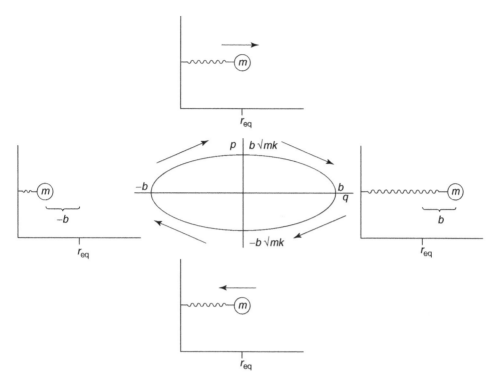

Figure 3.1   Phase-space trajectory (center) for a one-dimensional harmonic oscillator. As described in the text, at time zero the system is represented by the rightmost diagram ($q = b$, $p = 0$). The system evolves clockwise until it returns to the original point, with the period depending on the mass of the ball and the force constant of the spring

a position b length units displaced from equilibrium. The frictionless spring, character-ized by force constant k, begins to contract, so that the position coordinate decreases. The momentum coordinate, which was 0 at t_0, also decreases (momentum is a vector quantity, and we here define negative momentum as movement towards the wall). As the spring passes through coordinate position 0 (the equilibrium length), the *magnitude* of the momentum reaches a maximum, and then decreases as the spring begins resisting further motion of the ball. Ultimately, the momentum drops to zero as the ball reaches position $-b$, and then grows increasingly positive as the ball moves back towards the coordinate origin. Again, after passing through the equilibrium length, the magnitude of the momentum begins to decrease, until the ball returns to the same point in phase space from which it began.

Let us consider the phase space trajectory traced out by this behavior beginning with the position vector. Over any arbitrary time interval, the relationship between two positions is

$$q(t_2) = q(t_1) + \int_{t_1}^{t_2} \frac{p(t)}{m} dt \qquad (3.10)$$

where we have used the relationship between velocity and momentum

$$v = \frac{p}{m} \tag{3.11}$$

Similarly, the relationship between two momentum vectors is

$$p(t_2) = p(t_1) + m \int_{t_1}^{t_2} a(t)dt \tag{3.12}$$

where a is the acceleration. Equations (3.10) and (3.12) are Newton's equations of motion. Now, we have from Newton's Second Law

$$a = \frac{F}{m} \tag{3.13}$$

where F is the force. Moreover, from Eq. (2.13), we have a relationship between force and the position derivative of the potential energy. The simple form of the potential energy expression for a harmonic oscillator [Eq. (2.2)] permits analytic solutions for Eqs. (3.10) and (3.12). Applying the appropriate boundary conditions for the example in Figure 3.1 we have

$$q(t) = b \cos \left(\sqrt{\frac{k}{m}} t \right) \tag{3.14}$$

and

$$p(t) = -b\sqrt{mk} \sin \left(\sqrt{\frac{k}{m}} t \right) \tag{3.15}$$

These equations map out the oval phase space trajectory depicted in the figure.

Certain aspects of this phase space trajectory merit attention. We noted above that a phase space trajectory cannot cross itself. However, it *can* be periodic, which is to say it can trace out the same path again and again; the harmonic oscillator example is periodic. Note that the complete set of *all* harmonic oscillator trajectories, which would completely fill the corresponding two-dimensional phase space, is composed of concentric ovals (concentric circles if we were to choose the momentum metric to be $(mk)^{-1/2}$ times the position metric). Thus, as required, these (periodic) trajectories do not cross one another.

3.3.2   Non-analytical Systems

For systems more complicated than the harmonic oscillator, it is almost never possible to write down analytical expressions for the position and momentum components of the phase space trajectory as a function of time. However, if we *approximate* Eqs. (3.10) and (3.12) as

$$\mathbf{q}(t + \Delta t) = \mathbf{q}(t) + \frac{\mathbf{p}(t)}{m} \Delta t \tag{3.16}$$

and

$$\mathbf{p}(t + \Delta t) = \mathbf{p}(t) + m\mathbf{a}(t)\Delta t \qquad (3.17)$$

(this approximation, Euler's, being exact in the limit of $\Delta t \to 0$) we are offered a prescription for *simulating* a phase space trajectory. [Note that we have switched from the scalar notation of the one-dimensional harmonic oscillator example to a more general vector notation. Note also that although the approximations in Eqs. (3.16) and (3.17) are introduced here from Eqs. (3.10) and (3.12) and the definition of the definite integral, one can also derive Eqs. (3.16) and (3.17) as Taylor expansions of $\mathbf{q}$ and $\mathbf{p}$ truncated at first order; this is discussed in more detail below.]

Thus, given a set of initial positions and momenta, and a means for computing the forces acting on each particle at any instant (and thereby deriving the acceleration), we have a formalism for 'simulating' the true phase-space trajectory. In general, initial positions are determined by what a chemist thinks is 'reasonable' – a common technique is to build the system of interest and then energy minimize it partially (since one is interested in dynamical properties, there is no point in looking for an absolute minimum) using molecular mechanics. As for initial momenta, these are usually assigned randomly to each particle subject to a temperature constraint. The relationship between temperature and momentum is

$$T(t) = \frac{1}{(3N - n)k_{\mathrm{B}}} \sum_{i=1}^{N} \frac{|\mathbf{p}_i(t)|^2}{m_i} \qquad (3.18)$$

where N is the total number of atoms, n is the number of constrained degrees of freedom (vide infra), and the momenta are relative to the reference frame defined by the motion of the center of mass of the system. A force field, as emphasized in the last chapter, is particularly well suited to computing the accelerations at each time step.

While the use of Eqs. (3.16) and (3.17) seems entirely straightforward, the finite time step introduces very real practical concerns. Figure 3.2 illustrates the variation of a single momentum coordinate of some arbitrary phase space trajectory, which is described by a smooth curve. When the acceleration is computed for a point on the true curve, it will be a vector tangent to the curve. If the curve is not a straight line, any mass-weighted step along the tangent (which is the process described by Eq. (3.17)) will necessarily result in a point *off* the true curve. There is no guarantee that computing the acceleration at this new point will lead to a step that ends in the vicinity of the true curve. Indeed, with each additional step, it is quite possible that we will move further and further away from the true trajectory, thereby ending up sampling non-useful regions of phase space. The problem is compounded for position coordinates, since the velocity vector being used is already only an estimate derived from Eq. (3.17), i.e., there is no guarantee that it will even be tangent to the true curve when a point on the true curve is taken. (The atomistic picture, for those finding the mathematical discussion opaque, is that if we move the atoms in a single direction over too long a time, we will begin to ram them into one another so that they are far closer than van der Waals contact. This will lead to huge repulsive forces, so that still larger atomic movements will occur over the next time step, until our system ultimately looks like a nuclear furnace, with

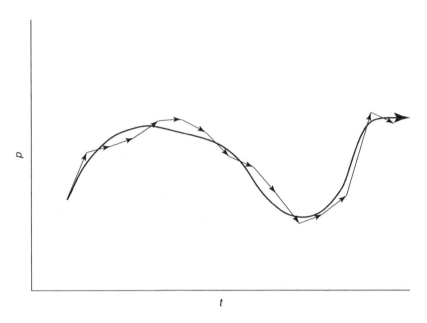

Figure 3.2 An actual phase-space trajectory (bold curve) and an approximate trajectory generated by repeated application of Eq. (3.17) (series of arrows representing individual time steps). Note that each propagation step has an identical Δt, but individual Δp values can be quite different. In the illustration, the approximate trajectory hews relatively closely to the actual one, but this will not be the case if too large a time step is used

atoms moving seemingly randomly. The very high energies of the various steps will preclude their contributing in a meaningful way to any property average.)

Of course, we know that in the limit of an infinitesimally small time step, we will recover Eqs. (3.10) and (3.12). But, since each time step requires a computation of all of the molecular forces (and, presumably, of the property we are interested in), which is computationally intensive, we do not want to take too *small* a time step, or we will not be able to propagate our trajectory for any chemically interesting length of time. What then is the optimal length for a time step that balances numerical stability with chemical utility? The general answer is that it should be at least one and preferably two orders of magnitude smaller than the fastest periodic motion within the system. To illustrate this, reconsider the 1-D harmonic oscillator example of Figure 3.1: if we estimate the first position of the mass after its release, given that the acceleration will be computed to be towards the wall, we will estimate the new position to be displaced in the negative direction. But, if we take too large a time step, i.e., we keep moving the mass towards the wall without ever accounting for the change in the acceleration of the spring with position, we might end up with the mass at a position more negative than $-b$. Indeed, we could end up with the mass behind the wall!

In a typical (classical) molecular system, the fastest motion is bond vibration which, for a heavy-atom–hydrogen bond has a period of about 10^{-14} s. Thus, for a system containing such bonds, an integration time step Δt should not much exceed 0.1 fs. This rather short time step means that modern, large-scale MD simulations (e.g., on biopolymers in a surrounding solvent) are rarely run for more than some 10 ns of simulation time (i.e., 10^7 computations of energies, forces, etc.) That many interesting phenomena occur on the microsecond timescale or longer (e.g., protein folding) represents a severe limitation to the application of MD to these phenomena. Methods to efficiently integrate the equations of motion over longer times are the subject of substantial modern research (see, for instance, Olender and Elber 1996; Grubmüller and Tavan 1998; Feenstra, Hess and Berendsen 1999).

3.3.3 Practical Issues in Propagation

Using Euler's approximation and taking integration steps in the direction of the tangent is a particularly simple integration approach, and as such is not particularly stable. Considerably more sophisticated integration schemes have been developed for propagating trajectories. If we restrict ourselves to consideration of the position coordinate, most of these schemes derive from approximate Taylor expansions in $\mathbf{r}$, i.e., making use of

$$\mathbf{q}(t + \Delta t) = \mathbf{q}(t) + \mathbf{v}(t)\Delta t + \frac{1}{2!}\mathbf{a}(t)(\Delta t)^2 + \frac{1}{3!}\frac{d^3\mathbf{q}(\tau)}{dt^3}\bigg|_{\tau=t}(\Delta t)^3 + \cdots \tag{3.19}$$

where we have used the abbreviations $\mathbf{v}$ and $\mathbf{a}$ for the first (velocity) and second (acceleration) time derivatives of the position vector $\mathbf{q}$.

One such method, first used by Verlet (1967), considers the sum of the Taylor expansions corresponding to forward and reverse time steps Δt. In that sum, all odd-order derivatives disappear since the odd powers of Δt have opposite sign in the two Taylor expansions. Rearranging terms and truncating at second order (which is equivalent to truncating at third-order, since the third-order term has a coefficient of zero) yields

$$\mathbf{q}(t + \Delta t) = 2\mathbf{q}(t) - \mathbf{q}(t - \Delta t) + \mathbf{a}(t)(\Delta t)^2 \tag{3.20}$$

Thus, for any particle, each subsequent position is determined by the current position, the previous position, and the particle's acceleration (determined from the forces on the particle and Eq. (3.13)). For the very first step (for which no position $\mathbf{q}(t - \Delta t)$ is available) one might use Eqs. (3.16) and (3.17).

The Verlet scheme propagates the position vector with no reference to the particle velocities. Thus, it is particularly advantageous when the position coordinates of phase space are of more interest than the momentum coordinates, e.g., when one is interested in some property that is independent of momentum. However, often one wants to control the simulation temperature. This can be accomplished by scaling the particle velocities so that the temperature, as defined by Eq. (3.18), remains constant (or changes in some defined manner), as described in more detail in Section 3.6.3. To propagate the position and velocity vectors in a *coupled* fashion, a modification of Verlet's approach called the leapfrog algorithm has been proposed. In this case, Taylor expansions of the position vector truncated at second order (not third) about $t + \Delta t/2$ are employed, in particular

$$\mathbf{q}\left(t + \frac{1}{2}\Delta t + \frac{1}{2}\Delta t\right) = \mathbf{q}\left(t + \frac{1}{2}\Delta t\right) + \mathbf{v}\left(t + \frac{1}{2}\Delta t\right)\frac{1}{2}\Delta t + \frac{1}{2!}\mathbf{a}\left(t + \frac{1}{2}\Delta t\right)\left(\frac{1}{2}\Delta t\right)^2$$

(3.21)

and

$$\mathbf{q}\left(t + \frac{1}{2}\Delta t - \frac{1}{2}\Delta t\right) = \mathbf{q}\left(t + \frac{1}{2}\Delta t\right) - \mathbf{v}\left(t + \frac{1}{2}\Delta t\right)\frac{1}{2}\Delta t + \frac{1}{2!}\mathbf{a}\left(t + \frac{1}{2}\Delta t\right)\left(\frac{1}{2}\Delta t\right)^2$$

(3.22)

When Eq. (3.22) is subtracted from Eq. (3.21) one obtains

$$\mathbf{q}(t + \Delta t) = \mathbf{q}(t) + \mathbf{v}\left(t + \frac{1}{2}\Delta t\right)\Delta t$$

(3.23)

Similar expansions for $\mathbf{v}$ give

$$\mathbf{v}\left(t + \frac{1}{2}\Delta t\right) = \mathbf{v}\left(t - \frac{1}{2}\Delta t\right) + \mathbf{a}(t)\Delta t$$

(3.24)

Note that in the leapfrog method, position depends on the velocities as computed one-half time step out of phase, thus, scaling of the velocities can be accomplished to control temperature. Note also that no force-field calculations actually take place for the fractional time steps. Forces (and thus accelerations) in Eq. (3.24) are computed at integral time steps, half-time-step-forward velocities are computed therefrom, and these are then used in Eq. (3.23) to update the particle positions. The drawbacks of the leapfrog algorithm include ignoring third-order terms in the Taylor expansions and the half-time-step displacements of the position and velocity vectors – both of these features can contribute to decreased stability in numerical integration of the trajectory.

Considerably more stable numerical integration schemes are known for arbitrary trajectories, e.g., Runge–Kutta (Press *et al.* 1986) and Gear predictor-corrector (Gear 1971) methods. In Runge–Kutta methods, the gradient of a function is evaluated at a number of different intermediate points, determined iteratively from the gradient at the current point, prior to taking a step to a new trajectory point on the path; the 'order' of the method refers to the number of such intermediate evaluations. In Gear predictor-corrector algorithms, higher order terms in the Taylor expansion are used to predict steps along the trajectory, and then the actual particle accelerations computed for those points are compared to those that were predicted by the Taylor expansion. The differences between the actual and predicted values are used to correct the position of the point on the trajectory. While Runge–Kutta and Gear predictor-corrector algorithms enjoy very high stability, they find only limited use in MD simulations because of the high computational cost associated with computing multiple first derivatives, or higher-order derivatives, for every step along the trajectory.

A different method of increasing the time step without decreasing the numerical stability is to remove from the system those degrees of freedom having the highest frequency (assuming, of course, that any property being studied is independent of those degrees of freedom). Thus, if heavy-atom–hydrogen bonds are constrained to remain at a constant length, the next highest frequency motions will be heavy-atom–heavy-atom vibrations; these frequencies are typically a factor of 2–5 smaller in magnitude. While a factor of 2 is of only marginal utility, reducing the number of available degrees of freedom generally offers some savings in time and integration stability. So, when the system of interest is some solute immersed in a large bath of surrounding solvent molecules, it can be advantageous to freeze some or all of the degrees of freedom within the solvent molecules.

A commonly employed algorithm for eliminating these degrees of freedom is called SHAKE (Ryckaert, Ciccotti, and Berendsen 1977). In the context of the Verlet algorithm, the formalism for freezing bond lengths involves defining distance constraints d_{ij} between atoms i and j according to

$$|\mathbf{r}_{ij}|^2 - d_{ij}^2 = 0 \tag{3.25}$$

where $\mathbf{r}_{ij}$ is the instantaneous interatomic distance vector. The position constraints can be applied iteratively in the Verlet algorithm, for example, by first taking an *unconstrained* step according to Eq. (3.20). The constraints are then taken account of according to

$$\mathbf{r}_i(t + \Delta t) = \mathbf{r}_i^0(t + \Delta t) + \Delta \mathbf{r}_i(t) \tag{3.26}$$

where $\mathbf{r}_i^0(t + \Delta t)$ is the position after taking the unconstrained step, and $\Delta \mathbf{r}_i(t)$ is the displacement vector required to satisfy a set of coupled constraint equations. These equations are defined as

$$\Delta \mathbf{r}_i(t) = \frac{2(\Delta t)^2}{m_i} \sum_j \lambda_{ij} \mathbf{r}_{ij}(t) \tag{3.27}$$

where the Lagrange multipliers λ_{ij} are determined iteratively following substitution of Eqs. (3.25) and (3.26) into Eq. (3.20).

Finally, there are a number of entirely mundane (but still very worthwhile!) steps that can be taken to reduce the total computer time required for a MD simulation. As a single example, note that any force on a particle derived from a force-field non-bonded energy term is induced by some *other* particle (i.e., the potential is pairwise). Newton's Third Law tells us that

$$\mathbf{F}_{ij} = -\mathbf{F}_{ji} \tag{3.28}$$

so we can save roughly a factor of two in computing the non-bonded forces by only evaluating terms for $i < j$ and using Eq. (3.28) to establish the rest.

3.3.4 Stochastic Dynamics

When the point of a simulation is not to determine accurate thermodynamic information about an ensemble, but rather to watch the dynamical evolution of some particular system immersed in a larger system (e.g., a solute in a solvent), then significant computational savings can be had by modeling the larger system stochastically. That is, the explicit nature of the larger system is ignored, and its influence is made manifest by a continuum that interacts with the smaller system, typically with that influence including a degree of randomness.

In Langevin dynamics, the equation of motion for each particle is

$$\mathbf{a}(t) = -\zeta \mathbf{p}(t) + \frac{1}{m}[\mathbf{F}_{intra}(t) + \mathbf{F}_{continuum}(t)] \qquad (3.29)$$

where the continuum is characterized by a microscopic friction coefficient, ζ, and a force, $\mathbf{F}$, having one or more components (e.g., electrostatic and random collisional). Intramolecular forces are evaluated in the usual way from a force field. Propagation of position and momentum vectors proceeds in the usual fashion.

In Brownian dynamics, the momentum degrees of freedom are removed by arguing that for a system that does not change shape much over very long timescales (e.g., a molecule, even a fairly large one) the momentum of each particle can be approximated as zero relative to the rotating center of mass reference frame. Setting the l.h.s. of Eq. (3.29) to zero and integrating, we obtain the Brownian equation of motion

$$\mathbf{r}(t) = \mathbf{r}(t_0) + \frac{1}{\zeta}\int_{t_0}^{t}[\mathbf{F}_{intra}(\tau) + \mathbf{F}_{continuum}(\tau)]d\tau \qquad (3.30)$$

where we now propagate only the position vector.

Langevin and Brownian dynamics are very efficient because a potentially very large surrounding medium is represented by a simple continuum. Since the computational time required for an individual time step is thus reduced compared to a full deterministic MD simulation, much longer timescales can be accessed. This makes stochastic MD methods quite attractive for studying system properties with relaxation times longer than those that can be accessed with deterministic MD simulations. Of course, if those properties involve the surrounding medium in some explicit way (e.g., a radial distribution function involving solvent molecules, vide infra), then the stochastic MD approach is not an option.

3.4 Monte Carlo

3.4.1 Manipulation of Phase-space Integrals

If we consider the various MD methods presented above, the Langevin and Brownian dynamics schemes introduce an increasing degree of stochastic behavior. One may imagine carrying this stochastic approach to its logical extreme, in which event there are no equations of motion to integrate, but rather phase points for a system are selected entirely at random. As noted above, properties of the system can then be determined from Eq. (3.5), but the integration converges very slowly because most randomly chosen points will be in chemically meaningless regions of phase space.

One way to reduce the problem slightly is to recognize that for many properties A, the position and momentum dependences of A are separable. In that case, Eq. (3.5) can be written as

$$\langle A \rangle = \int A(\mathbf{q}) \left[\int P(\mathbf{p}, \mathbf{q}) d\mathbf{p} \right] d\mathbf{q} + \int A(\mathbf{p}) \left[\int P(\mathbf{p}, \mathbf{q}) d\mathbf{q} \right] d\mathbf{p} \tag{3.31}$$

Since the Hamiltonian is also separable, the integrals in brackets on the r.h.s. of Eq. (3.31) may be simplified and we write

$$\langle A \rangle = \int A(\mathbf{q}) P(\mathbf{q}) d\mathbf{q} + \int A(\mathbf{p}) P(\mathbf{p}) d\mathbf{p} \tag{3.32}$$

where $P(\mathbf{q})$ and $P(\mathbf{p})$ are probability functions analogous to Eq. (3.6) related only to the potential and kinetic energies, respectively. Thus, we reduce the problem of evaluating a $6N$-dimensional integral to the problem of evaluating two $3N$-dimensional integrals. Of course, if the property is independent of either the position or momentum variables, then there is only one $3N$-dimensional integral to evaluate.

Even with so large a simplification, however, the convergence of Eq. (3.32) for a realistically sized chemical system and a random selection of phase points is too slow to be useful. What is needed is a scheme to select important phase points in a biased fashion.

3.4.2 Metropolis Sampling

The most significant breakthrough in Monte Carlo modeling took place when Metropolis *et al.* (1953) described an approach where 'instead of choosing configurations randomly, then weighting them with $\exp(-E/k_BT)$, we choose configurations with a probability $\exp(-E/k_BT)$ and weight them evenly'.

For convenience, let us consider a property dependent only on position coordinates. Expressing the elegantly simple Metropolis idea mathematically, we have

$$\langle A \rangle = \frac{1}{X} \sum_{i=1}^{X} A(\mathbf{q}_i) \tag{3.33}$$

where X is the total number of points $\mathbf{q}$ sampled according to the Metropolis prescription. Note the remarkable similarity between Eq. (3.33) and Eq. (3.8). Equation (3.33) resembles an ensemble average from an MD trajectory where the order of the points, i.e., the temporal progression, has been lost. Not surprisingly, as time does not enter into the MC scheme, it is not possible to establish a time relationship between points.

The Metropolis prescription dictates that we choose points with a Boltzmann-weighted probability. The typical approach is to begin with some 'reasonable' configuration $\mathbf{q}_1$. The value of property A is computed as the first element of the sum in Eq. (3.33), and then $\mathbf{q}_1$ is randomly perturbed to give a new configuration $\mathbf{q}_2$. In the constant particle number, constant volume, constant temperature ensemble (*NVT* ensemble), the probability p of 'accepting' point $\mathbf{q}_2$ is

$$p = \min\left[1, \frac{\exp(-E_2/k_{\mathrm{B}}T)}{\exp(-E_1/k_{\mathrm{B}}T)}\right] \tag{3.34}$$

Thus, if the energy of point $\mathbf{q}_2$ is not higher than that of point $\mathbf{q}_1$, the point is always accepted. If the energy of the second point *is* higher than the first, p is compared to a random number z between 0 and 1, and the move is accepted if $p \geq z$. Accepting the point means that the value of A is calculated for that point, that value is added to the sum in Eq. (3.33), and the entire process is repeated. If second point is *not* accepted, then the first point 'repeats', i.e., the value of A computed for the first point is added to the sum in Eq. (3.33) a second time and a new, random perturbation is attempted. Such a sequence of phase points, where each new point depends only on the immediately preceding point, is called a 'Markov chain'.

The art of running an MC calculation lies in defining the perturbation step(s). If the steps are very, very small, then the volume of phase space sampled will increase only slowly over time, and the cost will be high in terms of computational resources. If the steps are too large, then the rejection rate will grow so high that again computational resources will be wasted by an inefficient sampling of phase space. Neither of these situations is desirable.

In practice, MC simulations are primarily applied to collections of molecules (e.g., molecular liquids and solutions). The perturbing step involves the choice of a single molecule, which is randomly translated and rotated in a Cartesian reference frame. If the molecule is flexible, its internal geometry is also randomly perturbed, typically in internal coordinates. The ranges on these various perturbations are adjusted such that 20–50% of attempted moves are accepted. Several million individual points are accumulated, as described in more detail in Section 3.6.4.

Note that in the MC methodology, only the energy of the system is computed at any given point. In MD, by contrast, forces are the fundamental variables. Pangali, Rao, and Berne (1978) have described a sampling scheme where forces are used to choose the direction(s) for molecular perturbations. Such a force-biased MC procedure leads to higher acceptance rates and greater statistical precision, but at the cost of increased computational resources.

3.5 Ensemble and Dynamical Property Examples

The range of properties that can be determined from simulation is obviously limited only by the imagination of the modeler. In this section, we will briefly discuss a few typical properties in a general sense. We will focus on structural and time-correlation properties, deferring thermodynamic properties to Chapters 10 and 12.

As a very simple example, consider the dipole moment of water. In the gas phase, this dipole moment is 1.85 D (Demaison, Hütner, and Tiemann 1982). What about water in liquid water? A zeroth order approach to answering this problem would be to create a molecular mechanics force field defining the water molecule (a sizable number exist) that gives the correct dipole moment for the isolated, gas-phase molecule at its equilibrium geometry, which moment is expressed as

$$\mu = \sum_{i=1}^{3} q_i \mathbf{r}_i \tag{3.35}$$

where the sum runs over the one oxygen and two hydrogen atoms, q_i is the partial atomic charge assigned to atom i, and $\mathbf{r}_i$ is the position of atom i (since the water molecule has no net charge, the dipole moment is independent of the choice of origin for $\mathbf{r}$). In a liquid simulation (see Section 3.6.1 for more details on simulating condensed phases), the expectation value of the moment would be taken over *all* water molecules. Since the liquid is isotropic, we are not interested in the average vector, but rather the average magnitude of the vector, i.e.,

$$\langle |\mu| \rangle = \frac{1}{N} \sum_{n=1}^{N} \left| \sum_{i=1}^{3} q_{i,n} \mathbf{r}_{i,n} \right| \tag{3.36}$$

where N is the number of water molecules in the liquid model. Then, to the extent that in liquid water the average geometry of a water molecule changes from its gas-phase equilibrium structure, the expectation value of the magnitude of the dipole moment will reflect this change. Note that Eq. (3.36) gives the ensemble average for a single snapshot of the system; that is, the 'ensemble' that is being averaged over is intrinsic to each phase point by virtue of their being multiple copies of the molecule of interest. By MD or MC methods, we would generate multiple snapshots, either as points along an MD trajectory or by MC perturbations, so that we would finally have

$$\langle |\mu| \rangle = \frac{1}{M \cdot N} \sum_{m=1}^{M} \sum_{n=1}^{N} \left| \sum_{i=1}^{3} q_{i,n,m} \mathbf{r}_{i,n,m} \right| \tag{3.37}$$

where M is the total number of snapshots. [If we were considering the dipole moment of a solute molecule that was present in only one copy (i.e., a dilute solution), then the sum over N would disappear.]

Note that the expectation value compresses an enormous amount of information into a single value. A more complete picture of the moment would be a probability distribution, as depicted in Figure 3.3. In this analysis, the individual water dipole moment magnitudes (all $M \cdot N$ of them) are collected into bins spanning some range of dipole moments. The moments are then plotted either as a histogram of the bins or as a smooth curve reflecting the probability of being in an individual bin (i.e., equivalent to drawing the curve through the midpoint of the top of each histogram bar). The width of the bins is chosen so as to give maximum resolution to the lineshape of the curve without introducing statistical noise from underpopulation of individual bins.

Note that, although up to this point we have described the expectation value of A as though it were a scalar value, it is also possible that A is a function of some experimentally (and computationally) accessible variable, in which case we may legitimately ask about its expectation value at various points along the axis of its independent variable. A good

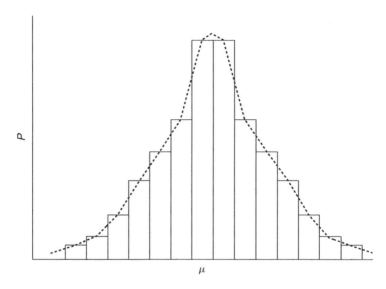

Figure 3.3 Hypothetical distribution of dipole moment magnitudes from a simulation of liquid water. The dashed curve is generated by connecting the tops of histogram bins whose height is dictated by the number of water molecules found to have dipole moments in the range spanned by the bin. Note that although the example is illustrated to be symmetric about a central value (which will thus necessarily be $\langle \mu \rangle$) this need not be the case

example of such a property is a radial distribution function (r.d.f.), which can be determined experimentally from X-ray or neutron diffraction measurements. The r.d.f. for two atoms A and B in a spherical volume element is defined by

$$\frac{1}{V} g_{AB}(r) = \frac{1}{N_A \bullet N_B} \left\langle \sum_{i=1}^{N_A} \sum_{j=1}^{N_B} \delta \left[r - r_{A_i B_j} \right] \right\rangle \tag{3.38}$$

where V is the volume, N is the total number of atoms of a given type within the volume element, δ is the Dirac delta function (the utility of which will become apparent momentarily), and r is radial distance. The double summation within the ensemble average effectively counts for each distance r the number of AB pairs separated by that distance. If we integrate over the full spherical volume, we obtain

$$\frac{1}{V} \int g_{AB}(r) d\mathbf{r} = \frac{1}{N_A \bullet N_B} \left\langle \sum_{i=1}^{N_A} \sum_{j=1}^{N_B} \int \delta \left[r - r_{A_i B_j} \right] d\mathbf{r} \right\rangle \tag{3.39}$$

$$= 1$$

where we have made use of the property of the Dirac delta that its integral is unity. As there are $N_A \bullet N_B$ contributions of unity to the quantity inside the ensemble average, the r.h.s. of Eq. (3.39) is 1, and we see that the $1/V$ term is effectively a normalization constant on g.

We may thus interpret the l.h.s. of Eq. (3.39) as a probability function. That is, we may express the probability of finding two atoms of A and B within some range Δr of distance r from one another as

$$P\{A, B, r, \Delta r\} = \frac{4\pi r^2}{V} g_{AB}(r) \Delta r \qquad (3.40)$$

where, in the limit of small Δr, we have approximated the integral as $g_{AB}(r)$ times the volume of the thin spherical shell $4\pi r^2 \Delta r$.

Note that its contribution to the probability function makes certain limiting behaviors on $g_{AB}(r)$ intuitively obvious. For instance, the function should go to zero very rapidly when r becomes less than the sum of the van der Waals radii of A and B. In addition, at very large r, the function should be independent of r in homogeneous media, like fluids, i.e., there should be an equal probability for any interatomic separation because the two atoms no longer influence one another's positions. In that case, we could move g outside the integral on the l.h.s. of Eq. (3.39), and then the normalization makes it apparent that $g = 1$ under such conditions. Values other than 1 thus indicate some kind of structuring in a medium – values greater than 1 indicate preferred locations for surrounding atoms (e.g., a solvation shell) while values below 1 indicate underpopulated regions. A typical example of a liquid solution r.d.f. is shown in Figure 3.4. Note that with increasing order, e.g., on passing from a liquid to a solid phase, the peaks in g become increasingly narrow and the valleys increasingly wide and near zero, until in the limit of a motionless, perfect crystal, g would be a spectrum of Dirac δ functions positioned at the lattice spacings of the crystal.

It often happens that we consider one of our atoms A or B to be privileged, e.g., A might be a sodium ion and B the oxygen atom of a water and our interests might focus

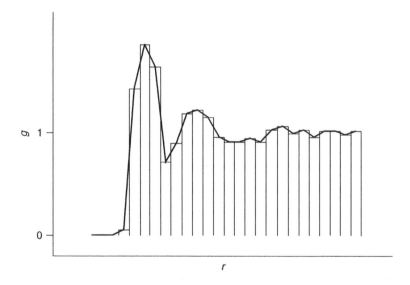

Figure 3.4   A radial distribution function showing preferred ($g > 1$) and disfavored ($g < 1$) interparticle distances. Random fluctuation about $g = 1$ is observed at large r

on describing the solvation structure of water about sodium ions in general. Then, we can define the total number of oxygen atoms n_B within some distance range about *any* sodium ion (atom A) as

$$n_B\{r, \Delta r\} = N_B P\{A, B, r, \Delta r\} \tag{3.41}$$

We may then use Eq. (3.40) to write

$$n_B\{r, \Delta r\} = 4\pi r^2 \rho_B g_{AB}(r) \Delta r \tag{3.42}$$

where ρ_B is the number density of B in the total spherical volume. Thus, if instead of $g_{AB}(r)$ we plot $4\pi r^2 \rho_B g_{AB}(r)$, then the area under the latter curve provides the number of molecules of B for arbitrary choices of r and Δr. Such an integration is typically performed for the distinct peaks in $g(r)$ so as to determine the number of molecules in the first, second, and possibly higher solvation shells or the number of nearest neighbors, next-nearest neighbors, etc., in a solid.

Determining $g(r)$ from a simulation involves a procedure quite similar to that described above for determining the continuous distribution of a scalar property. For each snapshot of an MD or MC simulation, all A−B distances are computed, and each occurrence is added to the appropriate bin of a histogram running from $r = 0$ to the maximum radius for the system (e.g., one half the narrowest box dimension under periodic boundary conditions, vide infra). Normalization now requires taking account not only of the total number of atoms A and B, but also the number of snapshots, i.e.,

$$g_{AB}(r) = \frac{V}{4\pi r^2 \Delta r M N_A N_B} \sum_{m=1}^{M} \sum_{i=1}^{N_A} \sum_{j=1}^{N_B} Q_m \left(r; r_{A_i B_j}\right) \tag{3.43}$$

where Δr is the width of a histogram bin, M is the total number of snapshots, and Q_m is the counting function

$$Q\left(r; r_{A_i B_j}\right) = \begin{cases} 1 & \text{if } r - \Delta r/2 \leq r_{A_i B_j} < r + \Delta r/2 \\ 0 & \text{otherwise} \end{cases} \tag{3.44}$$

for snapshot m.

The final class of dynamical properties we will consider are those defined by time-dependent autocorrelation functions. Such a function is defined by

$$C(t) = \langle a(t_0) a(t_0 + t) \rangle_{t_0} \tag{3.45}$$

where the ensemble average runs over *time* snapshots, and hence can only be determined from MD, *not* MC. Implicit in Eq. (3.45) is the assumption that C does not depend on the value of t_0 (since the ensemble average is over different choices of this quantity), and this will only be true for a system at equilibrium. The autocorrelation function provides a measure of the degree to which the value of property a at one time influences the value at a later time. An autocorrelation function attains its maximum value for a time delay of zero (i.e., no time delay at all), and this quantity, $\langle a^2 \rangle$ (which *can* be determined from MC simulations since no time correlation is involved) may be regarded as a normalization constant.

Now let us consider the behavior of C for long time delays. In a system where property a is not periodic in time, like a typical chemical system subject to effectively random thermal fluctuations, two measurements separated by a sufficiently long delay time should be completely uncorrelated. If two properties x and y are uncorrelated, then $\langle xy \rangle$ is equal to $\langle x \rangle \langle y \rangle$, so at long times C decays to $\langle a \rangle^2$.

While notationally burdensome, the discussion above makes it somewhat more intuitive to consider a reduced autocorrelation function defined by

$$\hat{C}(t) = \frac{\langle [a(t_0) - \langle a \rangle][a(t_0 + t) - \langle a \rangle] \rangle_{t_0}}{\langle [a - \langle a \rangle]^2 \rangle} \qquad (3.46)$$

which is normalized and, because the arguments in brackets fluctuate about their mean (and thus have individual expectation values of zero) decays to zero at long delay times. Example autocorrelation plots are provided in Figure 3.5. The curves can be fit to analytic expressions to determine characteristic decay times. For example, the characteristic decay time for an autocorrelation curve that can be fit to $\exp(-\zeta t)$ is ζ^{-1} time units.

Different properties have different characteristic decay times, and these decay times can be quite helpful in deciding how long to run a particular MD simulation. Since the point of a simulation is usually to obtain a statistically meaningful sample, one does not want to compute an average over a time shorter than several multiples of the characteristic decay time.

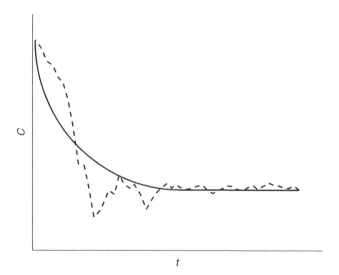

Figure 3.5 Two different autocorrelation functions. The solid curve is for a property that shows no significant statistical noise and appears to be well characterized by a single decay time. The dashed curve is quite noisy and, at least initially, shows a slower decay behavior. In the absence of a very long sample, decay times can depend on the total time sampled as well

As for the properties themselves, there are many chemically useful autocorrelation functions. For instance, particle position or velocity autocorrelation functions can be used to determine diffusion coefficients (Ernst, Hauge, and van Leeuwen 1971), stress autocorrelation functions can be used to determine shear viscosities (Haile 1992), and dipole autocorrelation functions are related to vibrational (infrared) spectra as their reverse Fourier transforms (Berens and Wilson 1981). There are also many useful correlation functions between two *different* variables (Zwanzig 1965). A more detailed discussion, however, is beyond the scope of this text.

3.6   Key Details in Formalism

The details of MC and MD methods laid out thus far can realistically be applied in a rigorous fashion only to systems that are too small to meaningfully represent actual chemical systems. In order to extend the technology in such a way as to make it useful for interpreting (or predicting) chemical phenomena, a few other approximations, or practical simplifications, are often employed. This is particularly true for the modeling of condensed phases, which are macroscopic in character.

3.6.1   Cutoffs and Boundary Conditions

As a spherical system increases in size, its volume grows as the cube of the radius while its surface grows as the square. Thus, in a truly macroscopic system, surface effects may play little role in the chemistry under study (there are, of course, exceptions to this). However, in a typical simulation, computational resources inevitably constrain the size of the system to be so small that surface effects may *dominate* the system properties. Put more succinctly, the modeling of a cluster may not tell one much about the behavior of a macroscopic system. This is particularly true when electrostatic interactions are important, since the energy associated with these interactions has an r^{-1} dependence.

One approach to avoid cluster artifacts is the use of 'periodic boundary conditions' (PBCs). Under PBCs, the system being modeled is assumed to be a unit cell in some ideal crystal (e.g., cubic or orthorhombic, see Theodorouo and Suter 1985). In practice, cut-off distances are usually employed in evaluating non-bonded interactions, so the simulation cell need be surrounded by only one set of nearest neighbors, as illustrated in Figure 3.6. If the trajectory of an individual atom (or a MC move of that atom) takes it outside the boundary of the simulation cell in any one or more cell coordinates, its image simultaneously enters the simulation cell from the point related to the exit location by lattice symmetry.

Thus, PBCs function to preserve mass, particle number, and, it can be shown, total energy in the simulation cell. In an MD simulation, PBCs also conserve linear momentum; since linear momentum is *not* conserved in real contained systems, where container walls disrupt the property, this is equivalent to reducing the number of degrees of freedom by 3. However, this effect on system properties is typically negligible for systems of over 100 atoms. Obviously, PBCs do *not* conserve angular momentum in the simulation cell of an MD simulation, but over time the movement of atoms in and out of each wall of the cell will be such that

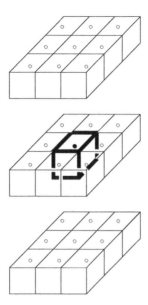

Figure 3.6   Exploded view of a cubic simulation cell surrounded by the 26 periodic images generated by PBCs. If the solid particle translates to a position that is outside the simulation cell, one of its periodic images, represented by open particles, will translate in

fluctuations will take place about a well-defined average. The key aspect of imposing PBCs is that no molecule within the simulation cell sees 'vacuum' within the range of its interaction cutoffs, and thus surface artifacts are avoided. Other artifacts associated with periodicity may be introduced, particularly with respect to correlation times in dynamics simulations (Berne and Harp 1970; Bergdorf, Peter, and Hunenberger 2003), but these can in principle be eliminated by moving to larger and larger simulation cells so that periodicity takes place over longer and longer length scales.

Of course, concerns about periodicity only relate to systems that are *not* periodic. The discussion above pertains primarily to the simulations of liquids, or solutes in liquid solutions, where PBCs are a useful approximation that helps to model solvation phenomena more realistically than would be the case for a small cluster. If the system truly is periodic, e.g., a zeolite crystal, then PBCs are integral to the model. Moreover, imposing PBCs can provide certain advantages in a simulation. For instance, Ewald summation, which accounts for electrostatic interactions to infinite length as discussed in Chapter 2, can only be carried out within the context of PBCs.

An obvious question with respect to PBCs is how large the simulation cell should be. The simple answer is that all cell dimensions must be at least as large as the largest cut-off length employed in the simulation. Otherwise, some interatomic interactions would be at least double counted (once within the cell, and once with an image outside of the cell). In practice, one would like to go well beyond this minimum requirement if the system being modeled is supposedly homogeneous and non-periodic. Thus, for instance, if one is modeling a large, dilute solute in a solvent (e.g., a biomolecule), a good choice for cell size might be the dimensions of the molecule plus at least twice the largest cut-off distance. Thus, no two solute molecules interact with one another nor does any solvent molecule see two copies of the solute. (Note, however, that this does not change the fundamentally periodic nature of the system; it simply increases the number of molecules over which it is made manifest.)

As already noted in Chapter 2, for electrostatic interactions, Ewald sums are generally to be preferred over cut-offs because of the long-range nature of the interactions. For van der Waals type terms, cut-offs do not introduce significant artifacts provided they are reasonably large (typically 8–12 Å).

Because of the cost of computing interatomic distances, the evaluation of non-bonded terms in MD is often handled with the aid of a 'pairlist', which holds in memory all pairs of atoms within a given distance of one another. The pairlist is updated periodically, but less often than every MD step. Note that a particular virtue of MC compared to MD is that the only changes in the potential energy are those associated with a moved particle – all other interactions remain constant. This makes evaluation of the total energy a much simpler process in MC.

3.6.2   Polarization

As noted in Chapter 2, computation of charge–charge (or dipole–dipole) terms is a particularly efficient means to evaluate electrostatic interactions because it is pairwise additive. However, a more realistic picture of an actual physical system is one that takes into account the polarization of the system. Thus, different regions in a simulation (e.g., different functional groups, or different atoms) will be characterized by different local polarizabilities, and the local charge moments, by adjusting in an iterative fashion to their mutual interactions, introduce many-body effects into a simulation.

Simulations including polarizability, either only on solvent molecules or on all atoms, have begun to appear with greater frequency as computational resources have grown larger. In addition, significant efforts have gone into introducing polarizability into force fields in a general way by replacing fixed atomic charges with charges that fluctuate based on local environment (Winn, Ferenczy and Reynolds 1999; Banks *et al.* 1999), thereby preserving the simplicity of a pairwise interaction potential. However, it is not yet clear that the greater 'realism' afforded by a polarizable model greatly improves the accuracy of simulations. There are certain instances where polarizable force fields seem better suited to the modeling problem. For instance, Dang *et al.* (1991) have emphasized that the solvation of ions, because of their concentrated charge, is more realistically accounted for when surrounding solvent molecules are polarizable and Soetens *et al.* (1997) have emphasized its importance in the computation of ion–ion interaction potentials for the case of two guanidinium ions in water.

In general, however, the majority of properties do not yet seem to be more accurately predicted by polarizable models than by unpolarizable ones, provided adequate care is taken in the parameterization process. Of course, if one wishes to examine issues associated with polarization, it must necessarily be included in the model. In the area of solvents, for instance, Bernardo *et al.* (1994) and Zhu and Wong (1994) have carefully studied the properties of polarizable water models. In addition, Gao, Habibollazadeh, and Shao (1995) have developed alcohol force fields reproducing the thermodynamic properties of these species as liquids with a high degree of accuracy, and have computed the polarization contribution to the total energy of the liquids to be 10–20%.

However, the typically high cost of including polarization is not attractive. Jorgensen has argued against the utility of including polarization in most instances, and has shown that bulk liquid properties can be equally well reproduced by fixed-charge force fields given proper care in the parameterization process (see, for instance, Mahoney and Jorgensen 2000). A particularly interesting example is provided by the simple amines ammonia, methylamine, dimethylamine, and trimethylamine. In the gas phase, the basicity of these species increases with increasing methylation in the expected fashion. In water, however, solvation effects compete with intrinsic basicity so that the four amines span a fairly narrow range of basicity, with methylamine being the most basic and trimethylamine and ammonia the least. Many models of solvation (see Chapters 11 and 12 for more details on solvation models) have been applied to this problem, and the failure of essentially all of them to correctly predict the basicity ordering led to the suggestion that in the case of explicit models, the failure derived from the use of non-polarizable force fields. Rizzo and Jorgensen (1999), however, parameterized non-polarizable classical models for the four amines that accurately reproduced their liquid properties and then showed that they further predicted the correct basicity ordering in aqueous simulations, thereby refuting the prior suggestion. [As a point of philosophy, the above example provides a nice illustration that a model's failure to accurately predict a particular quantity does *not* necessarily imply that a more expensive model needs to be developed – sometimes all that is required is a more careful parameterization of the existing model.] At least for the moment, then, it appears that errors associated with other aspects of simulation technology typically continue to be as large or larger than any errors introduced by use of non-polarizable force fields, so the use of such force fields in everyday simulations seems likely to continue for some time.

3.6.3 Control of System Variables

Our discussion of MD above was for the 'typical' MD ensemble, which holds particle number, system volume, and total energy constant – the *NVE* or 'microcanonical' ensemble. Often, however, there are other thermodynamic variables that one would prefer to hold constant, e.g., temperature. As temperature is related to the total kinetic energy of the system (if it is at equilibrium), as detailed in Eq. (3.18), one could in principle scale the velocities of each particle at each step to maintain a constant temperature. In practice, this is undesirable because the adjustment of the velocities, occasionally by fairly significant scaling factors, causes the trajectories to be no longer Newtonian. Properties computed over such trajectories are less likely to be reliable. An alternative method, known as Berendsen coupling (Berendsen *et al.* 1984), slows the scaling process by envisioning a connection between the system and a surrounding bath that is at a constant temperature T_0. Scaling of each particle velocity is accomplished by including a dissipative Langevin force in the equations of motion according to

$$\mathbf{a}_i(t) = \frac{\mathbf{F}_i(t)}{m_i} + \frac{\mathbf{p}_i(t)}{m_i \tau}\left[\frac{T_0}{T(t)} - 1\right] \tag{3.47}$$

where $T(t)$ is the instantaneous temperature, and τ has units of time and is used to control the strength of the coupling. The larger the value of τ the smaller the perturbing force and the more slowly the system is scaled to T_0 (i.e., τ is an effective relaxation time).

Note that, to start an MD simulation, one must necessarily generate an initial snapshot. It is essentially impossible for a chemist to simply 'draw' a large system that actually corresponds to a high-probability region of phase space. Thus, most MD simulations begin with a so-called 'equilibration' period, during which time the system is allowed to relax to a realistic configuration, after which point the 'production' portion of the simulation begins, and property averages are accumulated. A temperature coupling is often used during the equilibration period so that the temperature begins very low (near zero) and eventually ramps up to the desired system temperature for the production phase. This has the effect of damping particle movement early on in the equilibration (when there are presumably very large forces from a poor initial guess at the geometry).

In practice, equilibration protocols can be rather involved. Large portions of the system may be held frozen initially while subregions are relaxed. Ultimately, the entire system is relaxed (i.e., all the degrees of freedom that are being allowed to vary) and, once the equilibration temperature has reached the desired average value, one can begin to collect statistics.

With respect to other thermodynamic variables, many experimental systems are not held at constant volume, but instead at constant pressure. Assuming ideal gas statistical mechanics and pairwise additive forces, pressure P can be computed as

$$P(t) = \frac{1}{V(t)} \left[Nk_\mathrm{B}T(t) + \frac{1}{3} \sum_{i}^{N} \sum_{j>1}^{N} F_{ij}r_{ij} \right] \qquad (3.48)$$

where V is the volume, N is the number of particles, F and r are the forces and distances between particles, respectively. To adjust the pressure in a simulation, what is typically modified is the volume. This is accomplished by scaling the location of the particles, i.e., changing the size of the unit cell in a system with PBCs. The scaling can be accomplished in a fashion exactly analogous with Eq. (3.47) (Andersen 1980).

An alternative coupling scheme for temperature and pressure, the Nosé–Hoover scheme, adds new, independent variables that control these quantities to the simulation (Nosé 1984; Hoover 1985). These variables are then propagated along with the position and momentum variables.

In MC methods, the 'natural' ensemble is the NVT ensemble. Carrying out MC simulations in other ensembles simply requires that the probabilities computed for steps to be accepted or rejected reflect dependence on factors other than the internal energy. Thus, if we wish to maintain constant pressure instead of constant volume, we can treat volume as a variable (again, by scaling the particle coordinates, which is equivalent to expanding or contracting the unit cell in a system described by PBCs). However, in the NPT ensemble, the deterministic thermodynamic variable is no longer the internal energy, but the enthalpy (i.e., $E + PV$) and, moreover, we must account for the effect of a change in system volume (three dimensions) on the total volume of phase space ($3N$ dimensions for position) since probability is related to phase-space volume. Thus, in the *NPT* ensemble, the probability for accepting a new point 2 over an old point 1 becomes

$$p = \min \left\{ 1, \frac{V_2^N \exp[-(E_2 + PV_2)/k_\mathrm{B}T]}{V_1^N \exp[-(E_1 + PV_1)/k_\mathrm{B}T]} \right\} \tag{3.49}$$

(lower case 'p' is used here for probability to avoid confusion with upper case 'P' for pressure).

The choices of how often to scale the system volume, and by what range of factors, obviously influence acceptance ratios and are adjusted in much the same manner as geometric variables to maintain a good level of sampling efficiency. Other ensembles, or sampling schemes other than those using Cartesian coordinates, require analogous modifications to properly account for changes in phase space volume.

Just as with MD methods, MC simulations require an initial equilibration period so that property averages are not biased by very poor initial values. Typically various property values are monitored to assess whether they appear to have achieved a reasonable level of convergence prior to proceeding to production statistics. Yang, Bitetti-Putzer, and Karplus (2004) have offered the rather clever suggestion that the equilibration period can be defined by analyzing the convergence of property values starting from the *end* of the simulation, i.e., the time arrow of the simulation is reversed in the analysis. When an individual property value begins to depart from the value associated with the originally late, and presumably converged, portion of the trajectory, it is assumed that the originally early region of the trajectory should not be included in the overall statistics as it was most probably associated with equilibration. We now focus more closely on this issue.

3.6.4 Simulation Convergence

Convergence is defined as the acquisition of a sufficient number of phase points, through either MC or MD methods, to thoroughly sample phase space in a proper, Boltzmann-weighted fashion, i.e., the sampling is ergodic. While simple to define, convergence is *impossible* to prove, and this is either terribly worrisome or terribly liberating, depending on one's personal outlook.

To be more clear, we should separate the analysis of convergence into what might be termed 'statistical' and 'chemical' components. The former tends to be more tractable than the latter. Statistical convergence can be operatively defined as being *likely* to have been achieved when the average values for all properties of interest appear to remain roughly constant with increased sampling. In the literature, it is fairly standard to provide one or two plots of some particular properties as a function of time so that readers can agree that, to their eyes, the plots appear to have flattened out and settled on a particular value. For instance, in the simulation of macromolecules, the root-mean-square deviation (RMSD) of the simulation structure from an X-ray or NMR structure is often monitored. The RMSD for a particular snapshot is defined as

$$RMSD = \sqrt{\frac{\sum_{i=1}^{N}(r_{i,\text{sim}} - r_{i,\text{expt}})^2}{N}} \qquad (3.50)$$

where N is the number of atoms in the macromolecule, and the positions r are determined in a coordinate system having the center of mass at the origin and aligning the principle moments of inertia along the Cartesian axes (i.e., the simulated and experimental structures are best aligned prior to computing the RMSD). Monitoring the RMSD serves the dual purpose of providing a particular property whose convergence can be assessed and also of offering a quantitative measure of how 'close' the simulated structure is to the experimentally determined one. When no experimental data are available for comparison, the RMSD is typically computed using as a reference either the initial structure or the average simulated structure. A typical RMSD plot is provided in Figure 3.7.

[Note that the information content in Figure 3.7 is often boiled down, when reported in the literature, to a single number, namely ⟨RMSD⟩. However, the magnitude of the fluctuation about the mean, which can be quantified by the standard deviation, is also an important quantity, and should be reported wherever possible. This is true for all expectation values

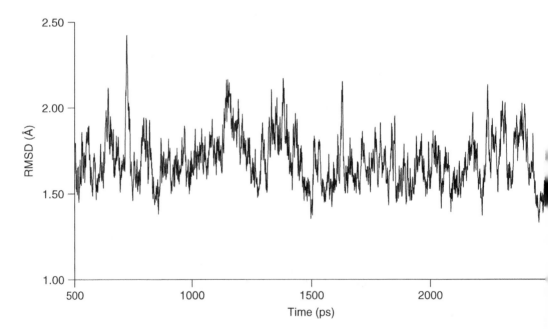

Figure 3.7 RMSD plot after 500 ps of equilibration for a solvated tRNA microhelix relative to its initial structure (Nagan *et al.* 1999)

derived from simulation. The standard deviation can be interpreted as a combination of the statistical noise (deriving from the limitations of the method) and the thermal noise (deriving from the 'correct' physical nature of the system). Considerably more refined methods of error analysis for average values from simulations have been promulgated (Smith and Wells 1984; Straatsma, Berendsen and Stam 1986; Kolafa 1986; Flyvberg and Petersen 1989).]

A more detailed decomposition of macromolecular dynamics that can be used not only for assessing convergence but also for other purposes is principal components analysis (PCA), sometimes also called essential dynamics (Wlodek *et al.* 1997). In PCA the positional covariance matrix $\mathbf{C}$ is calculated for a given trajectory after removal of rotational and translational motion, i.e., after best overlaying all structures. Given M snapshots of an N atom macromolecule, $\mathbf{C}$ is a $3N \times 3N$ matrix with elements

$$C_{ij} = \frac{1}{M} \sum_{k=1}^{M} (q_{i,k} - \langle q_i \rangle) \, (q_{j,k} - \langle q_j \rangle) \tag{3.51}$$

where $q_{i,k}$ is the value for snapshot k of the ith positional coordinate (x, y, or z coordinate for one of the N atoms), and $\langle q_i \rangle$ indicates the average of that coordinate over all snapshots. Diagonalization of $\mathbf{C}$ provides a set of eigenvectors that describe the dynamic motions of the structure; the associated eigenvalues may be interpreted as weights indicating the degree to which each mode contributes to the full dynamics.

Note that the eigenvectors of $\mathbf{C}$ comprise an orthogonal basis set for the macromolecular $3N$-dimensional space, but PCA creates them so as to capture as much structural dynamism as possible with each successive vector. Thus, the first PCA eigenvector may account for, say, 30 percent of the overall dynamical motion, the second a smaller portion, and so on. The key point here is that a surprisingly large fraction of the overall dynamics may be captured by a fairly small number of eigenvectors, each one of which may be thought of as being similar to a macromolecular vibrational mode. Thus, for example, Sherer and Cramer (2002) found that the first three PCA modes for a set of related RNA tetradecamer double helices accounted for 68 percent of the total dynamics, and that these modes were well characterized as corresponding to conceptually simple twisting and bending motions of the helix (Figure 3.8 illustrates the dominant mode). Being able in this manner to project the total macromolecular motion into PCA spaces of small dimensionality can be very helpful in furthering chemical analysis of the dynamics.

Returning to the issue of convergence, as noted above the structure of each snapshot in a simulation can be described in the space of the PCA eigenvectors, there being a coefficient for each vector that is a coordinate value just as an x coordinate in three-dimensional Cartesian space is the coefficient of the $\mathbf{i}$ Cartesian basis vector $(1,0,0)$. If a simulation has converged, the distribution of coefficient values sampled for each PCA eigenvector should be normal, i.e., varying as a Gaussian distribution about some mean value.

Yet another check of convergence in MD simulations, as alluded to in Section 3.5, is to ensure that the sampling length is longer than the autocorrelation decay time for a particular property by several multiples of that time. In practice, this analysis is performed with less regularity than is the simple monitoring of individual property values.

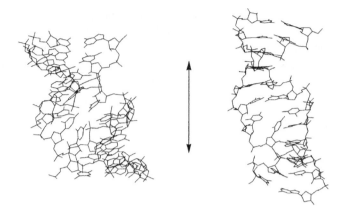

Figure 3.8 Twisted/compressed and untwisted/elongated double helices corresponding to minimal and maximal coefficient values for the corresponding PCA eigenvector.

It must be borne in mind, however, that the typical simulation lengths that can be achieved with modern hardware and software are very, very rarely in excess of 1 μs. It is thus quite possible that the simulation, although it appears to be converged with respect to the analyses noted above, is trapped in a metastable state having a lifetime in excess of 1 μs, and as a result the statistics are not meaningful to the true system at equilibrium. The only way to address this problem is either to continue the simulation for a longer time or to run one or more additional simulations with different starting conditions or both. Entirely separate trajectories are more likely to provide data that are statistically uncorrelated with the original, but they are also more expensive since equilibration periods are required prior to collecting production mode data.

Problems associated with statistical convergence and/or metastability are vexing ones, but more daunting still can be the issue of chemical convergence. This is probably best illustrated with an example. Imagine that one would like to simulate the structure of a protein in water at pH 7 and that the protein contains nine histidine residues. At pH 7, the protein could, in principle, exist in many different protonation states (i.e., speciation) since the pK_a of histidine is quite near 7. Occam's razor and a certain amount of biochemical experience suggest that, in fact, only one or two states are likely to be populated under biological conditions, but how to choose which one(s) for simulation, since most force fields will not allow for protonation/deprotonation to take place? If the wrong state is chosen, it may be possible to acquire very good statistical convergence for the associated region of phase space, but that region is statistically unimportant compared to other regions which were *not* sampled.

3.6.5   The Multiple Minima Problem

A related problem, and one that is commonly encountered, has to do with molecules possessing multiple conformations. Consider N-methylacetamide, which can exist in E and Z forms. The latter stereoisomer is favored over the former by about 3 kcal/mol, but the barrier to interconversion is in excess of 18 kcal/mol. Thus, a simulation of *N*-methylacetamide starting with the statistically less relevant *E* structure is highly unlikely ever to sample the *Z* form, either using MD (since the high barrier implies an isomerization rate that will be considerably slower than the simulation time) or MC (since with small steps, the probability of going so far uphill would be very low, while with large steps it might be possible for the isomers to interconvert, but the rejection rate would be enormous making the simulation intractable). A related example with similar issues has to do with modeling phase transfer by MC methods, e.g., the movement of a solute between two immiscible liquids, or of a molecule from the gas phase to the liquid phase. In each case, the likelihood of moving a molecule in its entirety is low.

A number of computational techniques have been proposed to address these limitations. The simplest approach conceptually, which can be applied to systems where all possible conformations can be readily enumerated, is to carry out simulations for each one and then weight the respective property averages according to the free energies of the conformers (means for estimating these free energies are discussed in Chapter 12). This approach is, of course, cumbersome when the number of conformers grows large. This growth can occur with startling rapidity. For example, 8, 18, 41, 121, and 12 513 distinct minima have been identified for cyclononane, -decane, -undecane, -dodecane, and -heptadecane, respectively (Weinberg and Wolfe 1994). And cycloalkanes are relatively simple molecules compared, say, to a protein, where the holy grail of conformational analysis is prediction of a properly folded structure from only sequence information. Nevertheless, fast heuristic methods continue to be developed to rapidly search low-energy conformational space for small to medium-sized molecules. For example, Smellie *et al.* (2003) have described an algorithm that performed well in generating collections of low-energy conformers for 97 000 drug-like molecules with an average time of less than 0.5 s per stereoisomer.

A different approach to the identification of multiple minima is to periodically heat the system to a very high temperature. Since most force fields do not allow bond-breaking to occur, high temperature simply has the effect of making conformational interconversions more likely. After a certain amount of time, the system is cooled again to the temperature of interest, and statistics are collected. In practice, this technique is often used for isolated molecules in the gas phase in the hope of finding a global minimum energy structure, in which case it is referred to as 'simulated annealing'. In condensed phases, it is difficult to converge the statistical weights of the different accessed conformers. Within the context of MC simulations, other techniques to force the system to jump between minimum-energy wells in a properly energy-weighted fashion have been proposed (see, for instance, Guarnieri and Still 1994; Senderowitz and Still 1998; Brown and Head-Gordon 2003).

An alternative to adjusting the temperature to help the system overcome high barriers is to artificially lower the barrier by adding an external potential energy term that is large and positive in regions where the 'normal' potential energy is large and negative (i.e., in the regions of minima). This summation effectively counterbalances the normal potential energy barrier. For instance, if the barrier is associated with a bond rotation, a so-called 'biasing potential' can be added such that the rotational potential becomes completely flat. The system can now sample freely over the entire range of possible rotations, but computed properties must be corrected for the proper free energy difference(s) in the absence of the biasing potential(s) (Straatsma, T. P.; McCammon, J. A.; Andricioaei and Straub 1996). In the absence of already knowing the shape of the PES, however, it may be rather difficult to construct a useful biasing potential *ab initio*. Laio and Parrinello (2002) have described a protocol whereby the biasing potential is history-dependent, filling in minima as it goes along in a coarse-grained space defined by collective coordinates. Collective coordinates have also been used by Jaqaman and Ortoleva (2002) to explore large-scale conformational changes in macromolecules more efficiently and by Müller, de Meijere, and Grubmüller (2002) to predict relative rates of unimolecular reactions.

Another method to artificially lower barrier heights in certain regions of phase space is to artificially expand that space by a single *extra* coordinate introduced for just that purpose – an idea analogous to the way catalysts lower barrier heights without affecting local minima (Stolovitzky and Berne 2000). In a related fashion, Nakamura (2002) has shown that barriers up to 3000 kcal mol^{-1} can be readily overcome simply by sampling in a logarithmically transformed energy space followed by correction of the resulting probability distribution.

An interesting alternative suggested by Verkhivker, Elber, and Nowak (1992) is to have multiple conformers present *simultaneously* in a 'single' molecule. In the so-called 'locally enhanced sampling' method, the molecule of interest is represented as a sum of different conformers, each contributing fractionally to the total force field energy expression. When combined with 'softened' potentials, Hornak and Simmerling (2003) have shown that this technology can be useful for crossing very high barriers associated with large geometric rearrangements.

Just as with statistical convergence, however, there can be no *guarantee* that any of the techniques above will provide a thermodynamically accurate sampling of phase space, even though on the timescale of the simulation various property values may *appear* to be converged. As with most theoretical modeling, then, it is best to assess the likely utility of the predictions from a simulation by first comparing to experimentally well-known quantities. When these are accurately reproduced, other predictions can be used with greater confidence. As a corollary, the modeling of systems for which few experimental data are available against which to compare is perilous.

3.7 Force Field Performance in Simulations

As discussed in Chapter 2, most force fields are validated based primarily on comparisons to small molecule data and moreover most comparisons involve what might be called static properties, i.e., structural or spectral data for computed fixed conformations. There are a few noteworthy exceptions: the OPLS and TraPPE force fields were, at least for molecular solvents, optimized to reproduce bulk solvent properties derived from simulations, e.g., density, boiling point, and dielectric constant. In most instances, however, one is left with the question of whether force fields optimized for small molecules or molecular fragments will perform with acceptable accuracy in large-scale simulations.

This question has been addressed with increasing frequency recently, and several useful comparisons of the quality of different force fields in particular simulations have appeared. The focus has been primarily on biomolecular simulations. Okur *et al.* (2003) assessed the abilities of the force fields of Cornell *et al.* and Wang, Cieplak, and Kollman (see Table 2.1) to predict correctly folded vs. misfolded protein structures; they found both force fields to suffer from a bias that predicts helical secondary structure to be anomalously too stable and suggested modifications to improve the more recent of the two force fields. Mu, Kosov, and Stock (2003) compared six different force fields in simulations of trialanine, an oligopeptide for which very high quality IR and NMR data are available. They found the most recent OPLS force field to provide the best agreement with experiment for the relative populations of three different conformers, while CHARMM, GROMOS, and force fields coded in the AMBER program systematically overstabilized an α-helical conformer. They also found that the timescales associated with transitions between conformers differed by as much as an order of magnitude between different force fields, although in this instance it is not clear which, if any, of the force fields is providing an accurate representation of reality. Finally, Zamm *et al.* (2003) compared six AA and UA force fields with respect to their predictions for the conformational dynamics of the pentapeptide neurotransmitter Met-enkephalin; they found AA force fields to generally give more reasonable dynamics than UA force fields.

Considering polynucleotides, Arthanari *et al.* (2003) showed that nOe data computed from an unrestrained 12 ns simulation of a double-helical DNA dodecamer using the force field of Cornell *et al.* agreed better with solution NMR experiments than data computed using either the X-ray crystal structure or canonical A or B form structures. Reddy, Leclerc, and Karplus (2003) exhaustively compared four force fields for their ability to model a double-helical DNA decamer. They found the CHARMM22 parameter set to incorrectly favor an A-form helix over the experimentally observed B form. The CHARMM27 parameter set gave acceptable results as did the BMS force field and that of Cornell *et al.* (as modified by Cheatham, Cieplak, and Kollman (1999) to improve performance for sugar puckering and helical repeat).

In conclusion, it appears that the majority of the most modern force fields do well in predicting structural and dynamical properties within wells on their respective PESs. However, their performance for non-equilibrium properties, such as timescales for conformational interconversion, protein folding, etc., have not yet been fully validated. With the increasing speed of both computational hardware and dynamics algorithms, it should be possible to address this question in the near future.

3.8 Case Study: Silica Sodalite

Synopsis of Nicholas *et al.* (1991) 'Molecular Modeling of Zeolite Structure. 2. Structure and Dynamics of Silica Sodalite and Silicate Force Field'.

Zeolites are mesoporous materials that are crystalline in nature. The simplest zeolites are made up of Al and/or Si and O atoms. Also known as molecular sieves, they find use as drying agents because they are very hygroscopic, but from an economic standpoint they are of greatest importance as size-selective catalysts in various reactions involving hydrocarbons and functionalized molecules of low molecular weight (for instance, they can be used to convert methanol to gasoline). The mechanisms by which zeolites operate are difficult to identify positively because of the heterogeneous nature of the reactions in which they are involved (they are typically solids suspended in solution or reacting with gas-phase molecules), and the signal-to-noise problems associated with identifying reactive intermediates in a large background of stable reactants and products. As a first step toward possible modeling of reactions taking place inside the zeolite silica sodalite, Nicholas and co-workers reported the development of an appropriate force field for the system, and MD simulations aimed at its validation.

The basic structural unit of silica sodalite is presented in Figure 3.9. Because there are only two atomic types, the total number of functional forms and parameters required to define a force field is relatively small (18 parameters total). The authors restrict themselves to an overall functional form that sums stretching, bending, torsional, and non-bonded interactions, the latter having separate LJ and electrostatic terms. The details of the force field are described in a particularly lucid manner. The Si−O stretching potential is chosen to be quadratic, as is the O−Si−O bending potential. The flatter Si−O−Si bending potential is modeled with a fourth-order polynomial with parameters chosen to fit a bending potential computed from *ab initio* molecular orbital calculations (such calculations are the subject of Chapter 6). A Urey−Bradley Si−Si non-bonded harmonic stretching potential is added to couple the Si−O bond length to the Si−O−Si bond angle. Standard torsional potentials and LJ expressions are used, although, in the former case, a switching function is applied to allow the torsion energy to go to zero if one of the bond angles in the four-atom link becomes linear (which can happen at fairly low energy). With respect to electrostatic interactions, the authors note an extraordinarily large range of charges previously proposed for Si and O in this and related systems (spanning about 1.5 charge units). They choose a value for Si roughly midway through this range (which, by charge neutrality, determines the O charge as well), and examine the sensitivity of their model to the electrostatics by

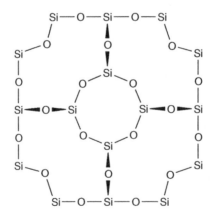

Figure 3.9   The repeating structural unit (with connections not shown) that makes up silica sodalite. What kinds of terms would be required in a force field designed to model such a system?

carrying out MD simulations with dielectric constants of 1, 2, and 5. The simulation cell is composed of 288 atoms (quite small, which makes the simulations computationally simple). PBCs and Ewald sums are used to account for the macroscopic nature of the real zeolite in simulations. Propagation of MD trajectories is accomplished using a leapfrog algorithm and 1.0 fs time steps following 20 ps or more of equilibration at 300 K. Each MD trajectory is 20 ps, which is very short by modern standards, but possibly justified by the limited dynamics available within the crystalline environment.

The quality of the parameter set is evaluated by comparing various details from the simulations to available experimental data. After testing a small range of equilibrium values for the Si–O bond, they settle on 1.61 Å, which gives optimized values for the unit cell Si–O bond length, and O–Si–O and Si–O–Si bond angles of 1.585 Å and 110.1° and 159.9°, respectively. These compare very favorably with experimental values of 1.587 Å and 110.3° and 159.7°, respectively. Furthermore, a Fourier transform of the total dipole correlation function (see Section 3.5) provides a model IR spectrum for comparison to experiment. Again, excellent agreement is obtained, with dominant computed bands appearing at 1106, 776, and 456 cm^{-1}, while experimental bands are observed at 1107, 787, 450 cm^{-1}. Simulations with different dielectric constants showed little difference from one another, suggesting that overall, perhaps because of the high symmetry of the system, sensitivity to partial atomic charge choice was low.

In addition, the authors explore the range of thermal motion of the oxygen atoms with respect to the silicon atoms they connect in the smallest ring of the zeolite cage (the eight-membered ring in the center of Figure 3.9). They determine that motion inward and outward and above and below the plane of the ring takes place with a fair degree of facility, while motion parallel to the Si–Si vector takes place over a much smaller range. This behavior is consistent with the thermal ellipsoids determined experimentally from crystal diffraction.

The authors finish by exploring the transferability of their force field parameters to a different zeolite, namely, silicalite. In this instance, a Fourier transform of the total dipole correlation function provides another model infrared (IR) spectrum for comparison to experiment, and again excellent agreement is obtained. Dominant computed bands appear at 1099, 806, 545, and 464 cm^{-1}, while experimental bands are observed at 1100, 800, 550, and 420 cm^{-1}. Some errors in band intensity are observed in the lower energy region of the spectrum.

As a first step in designing a general modeling strategy for zeolites, this paper is a very good example of how to develop, validate, and report force field parameters and results. The authors are pleasantly forthcoming about some of the assumptions employed in their analysis (for instance, all experimental data derive from crystals incorporating ethylene glycol as a solvent, while the simulations have the zeolite filled only with vacuum) and set an excellent standard for modeling papers of this type.

Bibliography and Suggested Additional Reading

Allen, M. P. and Tildesley, D. J. 1987. *Computer Simulation of Liquids*, Clarendon: Oxford.

Banci, L. 2003. 'Molecular Dynamics Simulations of Metalloproteins', *Curr. Opin. Chem. Biol.*, **7**, 143.

Beveridge, D. L. and McConnell, K. J. 2000. 'Nucleic acids: theory and computer simulation, Y2K' *Curr. Opin. Struct. Biol.*, **10**, 182.

Brooks, C. L., III and Case, D. A. 1993. 'Simulations of Peptide Conformational Dynamics and Thermodynamics' *Chem. Rev.*, **93**, 2487.

Cheatham, T. E., III and Brooks, B. R. 1998. 'Recent Advances in Molecular Dynamics Simulation Towards the Realistic Representation of Biomolecules in Solution' *Theor. Chem. Acc.*, **99**, 279.

Frenkel, D. and Smit, B. 1996. *Understanding Molecular Simulation: From Algorithms to Applications*, Academic Press: San Diego.

Haile, J. 1992. *Molecular Dynamics Simulations*, Wiley: New York.

Jensen, F. 1999. *Introduction to Computational Chemistry*, Wiley: Chichester.

Jorgensen, W. L. 2000. 'Perspective on "Equation of State Calculations by Fast Computing Machines"' *Theor. Chem. Acc.*, **103**, 225.

Lybrand, T. P. 1990. 'Computer Simulation of Biomolecular Systems Using Molecular Dynamics and Free Energy Perturbation Methods' in *Reviews in Computational Chemistry*, Vol. 1, Lipkowitz, K. B. and Boyd, D. B., Eds., VCH: New York, 295.

McQuarrie, D. A. 1973. *Statistical Thermodynamics*, University Science Books: Mill Valley, CA.

Norberg, J. and Nilsson, L. 2003. 'Advances in Biomolecular Simulations: Methodology and Recent Applications', *Quart. Rev. Biophys.*, **36**, 257.

Straatsma, T. P. 1996. 'Free Energy by Molecular Simulation' in *Reviews in Computational Chemistry*, Vol. 9, Lipkowitz, K. B. and Boyd, D. B., Eds., VCH: New York, 81.

References

Andersen, H. C. 1980. *J. Chem. Phys.*, **72**, 2384.

Andricioaei, I. and Straub, J. E. 1996. *Phys. Rev. E*, **53**, R3055.

Arthanari, H., McConnell, K. J., Beger, R., Young, M. A., Beveridge, D. L., and Bolton, P. H. 2003. *Biopolymers*, **68**, 3.

Banks, J. L., Kaminski, G. A., Zhou, R., Mainz, D. T., Berne, B. J., and Friesner, R. A. 1999. *J. Chem. Phys.*, **110**, 741.

Berendsen, H. J. C., Postma, J. P. M., van Gunsteren, W. F., DiNola, A., and Haak, J. R. 1984. *J. Chem. Phys.*, **81**, 3684.

Berens, P. H., and Wilson, K. R. 1981. *J. Chem. Phys.*, **74**, 4872.

Bergdorf, M., Peter, C., and Hunenberger, P. H. 2003. *J. Chem. Phys.*, **119**, 9129.

Bernardo, D. N., Ding, Y., Krogh-Jespersen, K., and Levy, R. M. 1994. *J. Phys. Chem.*, **98**, 4180.

Berne, B. J. and Harp, G. D. 1970. *Adv. Chem. Phys.*, **17**, **63**, 130.

Brown, S. and Head-Gordon, T. 2003. *J. Comput. Chem.*, **24**, 68.

Cheatham, T. E., III, Cieplak, P., and Kollman, P. A. 1999. *J. Biomol. Struct. Dyn.*, **16**, 845.

Dang, L. X., Rice, J. E., Caldwell, J., and Kollman, P. A. 1991. *J. Am. Chem. Soc.*, **113**, 2481.

Demaison, J., Hütner, W., and Tiemann, E. 1982. In: *Molecular Constants, Landolt-Börstein, New Series, Group II*, Vol. 14a, Hellwege, K. -H. and Hellwege, A. M., Eds., Springer-Verlag: Berlin, 584.

Ernst, M. H., Hauge, E. H., and van Leeuwen, J. M. J. 1971. *Phys. Rev. A*, **4**, 2055.

Feenstra, K. A., Hess, B., and Berendsen, H. J. C. 1999. *J. Comput. Chem.*, **20**, 786.

Flyvberg, H. and Petersen, H. G. 1989. *J. Chem. Phys.*, **91**, 461.

Ford, J. 1973. *Adv. Chem. Phys.*, **24**, 155.

Gao, J., Habibollazadeh, D., and Shao, L. 1995. *J. Phys. Chem.*, **99**, 16460.

Gear, C. W. 1971. *Numerical Initial Value Problems in Ordinary Differential Equations*, Prentice-Hall: Englewood Cliffs, N.J.

Grubmüller, H. and Tavan, P. 1998. *J. Comput. Chem.*, **19**, 1534.

Guarnieri, F. and Still, W. C. 1994. *J. Comput. Chem.*, **15**, 1302.

Haile, J. 1992. *Molecular Dynamics Simulations*, Wiley: New York, 291.

Hoover, W. G. 1985. *Phys. Rev. A,* **31**, 1695.

Hornak, V. and Simmerling, C. 2003. *Proteins*, **51**, 577.

Jaqaman, K. and Ortoleva, P. J. 2002. *J. Comput. Chem.*, **23**, 484.

Kolafa, J. 1986. *Mol. Phys.*, **59**, 1035.

Laio, A. and Parrinello, M. 2002. *Proc. Natl. Acad. Sci. USA*, **99**, 12562.

Mahoney, W. and Jorgensen, W. L. 2000. *J. Chem. Phys.*, **112**, 8910.

Metropolis, N., Rosenbluth, A. E., Rosenbluth, M. N., Teller, A. H., and Teller, E. 1953. *J. Chem. Phys.*, **21**, 1087.

Mu, Y., Kosov, D. S., and Stock, G. 2003. *J. Phys. Chem. B*, **107**, 5064.

Müller, E. M., de Meijere, A., and Grubmüller, H. 2002. *J. Chem. Phys.* **116**, 897.

Nagan, M. C., Kerimo, S. S., Musier-Forsyth, K., and Cramer, C. J. 1999. *J. Am. Chem. Soc.*, **121**, 7310.

Nakamura, H. 2002. *J. Comput. Chem.*, **23**, 511.

Nicholas, J. B., Hopfinger, A. J., Trouw, F. R., and Iton, L. E. 1991. *J. Am. Chem. Soc.*, **113**, 4792.

Nosé, S. 1984. *Mol. Phys.*, **52**, 255.

Okur, A, Strockbine, B., Hornak, V., and Simmerling, C. 2003. *J. Comput. Chem.*, **24**, 21.

Olender, R. and Elber, R., 1996. *J. Chem. Phys.*, **105**, 9299.

Pangali, C., Rao, M., and Berne, B. J. 1978. *Chem. Phys. Lett.*, **55**, 413.

Press, W. H., Flannery, B. P., Teukolsky, S. A., and Vetterling, W. T. 1986. *Numerical Recipes*, Cambridge University Press: New York.

Reddy, S. Y., Leclerc, F., and Karplus, M. 2003. *Biophys. J.*, **84**, 1421.

Rizzo, R. C. and Jorgensen, W. L. 1999. *J. Am. Chem. Soc.*, **121**, 4827.

Ryckaert, J. P., Ciccotti, G., and Berendsen, H. J. C. 1977. *J. Comput. Phys.*, **23**, 327.

Senderowitz, H. and Still, W. C. 1998. *J. Comput. Chem.*, **19**, 1736.

Sherer, E. C. and Cramer, C. J. 2002. *J. Phys. Chem. B*, **106**, 5075.

Smellie, A., Stanton, R., Henne, R., and Teig, S. 2003. *J. Comput. Chem.*, **24**, 10.

Smith, E. B. and Wells, B. H. 1984. *Mol. Phys.*, **53**, 701.

Soetens, J.-C., Millot, C., Chipot, C., Jansen, G., Ángyán, J. G., and Maigret, B. 1997. *J. Phys. Chem. B*, **101**, 10910.

Stolovitzky, G. and Berne, B. J. 2000. *Proc. Natl. Acad. Sci. (USA)*, **21**, 11164.

Straatsma, T. P. and McCammon, J. A. 1994. *J. Chem. Phys.*, **101**, 5032.

Straatsma, T. P., Berendsen, H. J. C., and Stam, A. J. 1986. *Mol. Phys.*, **57**, 89.

Theodorouo, D. N. and Suter, U. W. 1985. *J. Chem. Phys.*, **82**, 955.

Verkhivker, G., Elber, R., and Nowak, W. 1992. *J. Chem. Phys.*, **97**, 7838.

Verlet, L. 1967. *Phys. Rev.*, **159**, 98.

Weinberg, N. and Wolfe, S. 1994. *J. Am. Chem. Soc.*, **116**, 9860.

Winn, P. J., Ferenczy, G., and Reynolds, C. A. 1999. *J. Comput. Chem.*, **20**, 704.

Wlodek, S. T., Clard, T. W., Scott, L. R., McCammon, J. A. 1997. *J. Am. Chem. Soc.*, **119**, 9513.

Yang, W., Bitetti-Putzer, R., and Karplus, M. 2004. *J. Chem. Phys.*, **120**, 2618.

Zamm, M. H., Shen, M.-Y., Berry, R. S., and Freed, K. F. 2003. *J. Phys. Chem. B*, **107**, 1685.

Zhu, S.-B. and Wong, C. F. 1994. *J. Phys. Chem.*, **98**, 4695.

Zwanzig, R. 1965. *Ann. Rev. Phys. Chem.*, **16**, 67.

4

Foundations of Molecular Orbital Theory

4.1 Quantum Mechanics and the Wave Function

To this point, the models we have considered for representing microscopic systems have been designed based on classical, which is to say, macroscopic, analogs. We now turn our focus to contrasting models, whose foundations explicitly recognize the fundamental difference between systems of these two size extremes. Early practitioners of chemistry and physics had few, if any, suspicions that the rules governing microscopic and macroscopic systems should be different. Then, in 1900, Max Planck offered a radical proposal that blackbody radiation emitted by microscopic particles was limited to certain discrete values, i.e., it was 'quantized'. Such quantization was essential to reconciling large differences between predictions from classical models and experiment.

As the twentieth century progressed, it became increasingly clear that quantization was not only a characteristic of light, but also of the fundamental particles from which matter is constructed. Bound electrons in atoms, in particular, are clearly limited to discrete energies (levels) as indicated by their ultraviolet and visible line spectra. This phenomenon has no classical correspondence – in a classical system, obeying Newtonian mechanics, energy can vary continuously.

In order to describe microscopic systems, then, a different mechanics was required. One promising candidate was wave mechanics, since standing waves are also a quantized phenomenon. Interestingly, as first proposed by de Broglie, matter can indeed be shown to have wavelike properties. However, it also has particle-like properties, and to properly account for this dichotomy a new mechanics, quantum mechanics, was developed. This chapter provides an overview of the fundamental features of quantum mechanics, and describes in a formal way the fundamental equations that are used in the construction of computational models. In some sense, this chapter is historical. However, in order to appreciate the differences between modern computational models, and the range over which they may be expected to be applicable, it is important to understand the foundation on which all of them are built. Following this exposition, Chapter 5 overviews the approximations inherent

Essentials of Computational Chemistry, 2nd Edition Christopher J. Cramer
© 2004 John Wiley & Sons, Ltd   ISBNs: 0-470-09181-9 (cased); 0-470-09182-7 (pbk)

in so-called semiempirical QM models, Chapter 6 focuses on *ab initio* Hartree–Fock (HF) models, and Chapter 7 describes methods for accounting for electron correlation.

We begin with a brief recapitulation of some of the key features of quantum mechanics. The fundamental postulate of quantum mechanics is that a so-called wave function, Ψ, exists for any (chemical) system, and that appropriate operators (functions) which act upon Ψ return the observable properties of the system. In mathematical notation,

$$\vartheta \Psi = e \Psi \qquad (4.1)$$

where ϑ is an operator and e is a scalar value for some property of the system. When Eq. (4.1) holds, Ψ is called an eigenfunction and e an eigenvalue, by analogy to matrix algebra were Ψ to be an N-element column vector, ϑ to be an $N \times N$ square matrix, and e to remain a scalar constant. Importantly, the product of the wave function Ψ with its complex conjugate (i.e., $|\Psi^*\Psi|$) has units of probability density. For ease of notation, and since we will be working almost exclusively with real, and not complex, wave functions, we will hereafter drop the complex conjugate symbol '*'. Thus, the probability that a chemical system will be found within some region of multi-dimensional space is equal to the integral of $|\Psi|^2$ over that region of space.

These postulates place certain constraints on what constitutes an acceptable wave function. For a bound particle, the normalized integral of $|\Psi|^2$ over all space must be unity (i.e., the probability of finding it somewhere is one) which requires that Ψ be quadratically integrable. In addition, Ψ must be continuous and single-valued.

From this very formal presentation, the nature of Ψ can hardly be called anything but mysterious. Indeed, perhaps the best description of Ψ at this point is that it is an oracle – when queried with questions by an operator, it returns answers. By the end of this chapter, it will be clear the precise way in which Ψ is expressed, and we should have a more intuitive notion of what Ψ represents. However, the view that Ψ is an oracle is by no means a bad one, and will be returned to again at various points.

4.2 The Hamiltonian Operator

4.2.1 General Features

The operator in Eq. (4.1) that returns the system energy, E, as an eigenvalue is called the Hamiltonian operator, H. Thus, we write

$$H\Psi = E\Psi \qquad (4.2)$$

which is the Schrödinger equation. The typical form of the Hamiltonian operator with which we will be concerned takes into account five contributions to the total energy of a system (from now on we will say molecule, which certainly includes an atom as a possibility): the kinetic energies of the electrons and nuclei, the attraction of the electrons to the nuclei, and the interelectronic and internuclear repulsions. In more complicated situations, e.g., in the presence of an external electric field, in the presence of an external magnetic field, in the event of significant spin–orbit coupling in heavy elements, taking account of relativistic effects, etc., other terms are required in the Hamiltonian. We will consider some of these at later points in the text, but we will not find them necessary for general purposes. Casting the Hamiltonian into mathematical notation, we have

$$H = -\sum_i \frac{\hbar^2}{2m_e}\nabla_i^2 - \sum_k \frac{\hbar^2}{2m_k}\nabla_k^2 - \sum_i\sum_k \frac{e^2 Z_k}{r_{ik}} + \sum_{i<j} \frac{e^2}{r_{ij}} + \sum_{k<l} \frac{e^2 Z_k Z_l}{r_{kl}} \qquad (4.3)$$

where i and j run over electrons, k and l run over nuclei, $\hbar$ is Planck's constant divided by 2π, m_e is the mass of the electron, m_k is the mass of nucleus k, ∇^2 is the Laplacian operator, e is the charge on the electron, Z is an atomic number, and r_{ab} is the distance between particles a and b. Note that Ψ is thus a function of $3n$ coordinates where n is the total number of particles (nuclei and electrons), e.g., the x, y, and z Cartesian coordinates specific to each particle. If we work in Cartesian coordinates, the Laplacian has the form

$$\nabla_i^2 = \frac{\partial^2}{\partial x_i^2} + \frac{\partial^2}{\partial y_i^2} + \frac{\partial^2}{\partial z_i^2} \qquad (4.4)$$

Note that the Hamiltonian operator in Eq. (4.3) is composed of kinetic energy and potential energy parts. The potential energy terms (the last three) appear exactly as they do in classical mechanics. The kinetic energy for a QM particle, however, is not expressed as $|\mathbf{p}|^2/2m$, but rather as the eigenvalue of the kinetic energy operator

$$T = -\frac{\hbar^2}{2m}\nabla^2 \qquad (4.5)$$

Note also that, as described in Chapter 1, most of the constants appearing in Eq. (4.3) are equal to 1 when atomic units are chosen.

In general, Eq. (4.2) has *many* acceptable eigenfunctions Ψ for a given molecule, each characterized by a different associated eigenvalue E. That is, there is a complete set (perhaps infinite) of Ψ_i with eigenvalues E_i. For ease of future manipulation, we may assume without loss of generality that these wave functions are orthonormal, i.e., for a one particle system where the wave function depends on only three coordinates,

$$\iiint \Psi_i \Psi_j\, dx\, dy\, dz = \delta_{ij} \qquad (4.6)$$

where δ_{ij} is the Kronecker delta (equal to one if $i = j$ and equal to zero otherwise). Orthonormal actually implies two qualities simultaneously: 'orthogonal' means that the integral in Eq. (4.6) is equal to zero if $i \neq j$ and 'normal' means that when $i = j$ the value of the integral is one. For ease of notation, we will henceforth replace all multiple integrals over Cartesian space with a single integral over a generalized $3n$-dimensional volume element $d\mathbf{r}$, rendering Eq. (4.6) as

$$\int \Psi_i \Psi_j\, d\mathbf{r} = \delta_{ij} \qquad (4.7)$$

Now, consider the result of taking Eq. (4.2) for a specific Ψ_i, multiplying on the left by Ψ_j, and integrating. This process gives

$$\int \Psi_j H \Psi_i d\mathbf{r} = \int \Psi_j E_i \Psi_i d\mathbf{r} \tag{4.8}$$

Since the energy E is a scalar value, we may remove it outside the integral on the r.h.s. and use Eq. (4.7) to write

$$\int \Psi_j H \Psi_i d\mathbf{r} = E_i \delta_{ij} \tag{4.9}$$

This equation will prove useful later on, but it is worth noting at this point that it also offers a prescription for determining the molecular energy. With a wave function in hand, one simply constructs and solves the integral on the left (where i and j are identical and index the wave function of interest). Of course, we have not yet said much about the form of the wave function, so the nature of the integral in Eq. (4.8) is not obvious ... although one suspects it might be unpleasant to solve.

4.2.2   The Variational Principle

The power of quantum theory, as expressed in Eq. (4.1), is that if one has a molecular wave function in hand, one can calculate physical observables by application of the appropriate operator in a manner analogous to that shown for the Hamiltonian in Eq. (4.8). Regrettably, none of these equations offers us a prescription for *obtaining* the orthonormal set of molecular wave functions. Let us assume for the moment, however, that we can pick an arbitrary function, Φ, which is indeed a function of the appropriate electronic and nuclear coordinates to be operated upon by the Hamiltonian. Since we defined the set of orthonormal wave functions Ψ_i to be complete (and perhaps infinite), the function Φ must be some linear combination of the Ψ_i, i.e.,

$$\Phi = \sum_i c_i \Psi_i \tag{4.10}$$

where, of course, since we don't yet know the individual Ψ_i, we certainly don't know the coefficients c_i either! Note that the normality of Φ imposes a constraint on the coefficients, however, deriving from

$$\int \Phi^2 d\mathbf{r} = 1 = \int \sum_i c_i \Psi_i \sum_j c_j \Psi_j d\mathbf{r}$$

$$= \sum_{ij} c_i c_j \int \Psi_i \Psi_j d\mathbf{r}$$

$$= \sum_{ij} c_i c_j \delta_{ij}$$

$$= \sum_i c_i^2 \tag{4.11}$$

Now, let us consider evaluating the energy associated with wave function Φ. Taking the approach of multiplying on the left and integrating as outlined above, we have

$$\int \Phi H \Phi d\mathbf{r} = \int \left(\sum_i c_i \Psi_i \right) H \left(\sum_j c_j \Psi_j \right) d\mathbf{r}$$

$$= \sum_{ij} c_i c_j \int \Psi_i H \Psi_j d\mathbf{r}$$

$$= \sum_{ij} c_i c_j E_j \delta_{ij}$$

$$= \sum_i c_i^2 E_i \qquad (4.12)$$

where we have used Eq. (4.9) to simplify the r.h.s. Thus, the energy associated with the generic wave function Φ is determinable from all of the coefficients c_i (that define how the orthonormal set of Ψ_i combine to form Φ) and their associated energies E_i. Regrettably, we still don't know the values for *any* of these quantities. However, let us take note of the following. In the set of all E_i there must be a lowest energy value (i.e., the set is bounded from below); let us call that energy, corresponding to the 'ground state', E_0. [Notice that this boundedness is a critical feature of quantum mechanics! In a classical system, one could imagine always finding a state lower in energy than another state by simply 'shrinking the orbits' of the electrons to increase nuclear–electronic attraction while keeping the kinetic energy constant.]

We may now combine the results from Eqs. (4.11) and (4.12) to write

$$\int \Phi H \Phi d\mathbf{r} - E_0 \int \Phi^2 d\mathbf{r} = \sum_i c_i^2 (E_i - E_0) \qquad (4.13)$$

Assuming the coefficients to be real numbers, each term c_i^2 must be greater than or equal to zero. By definition of E_0, the quantity $(E_i - E_0)$ must also be greater than or equal to zero. Thus, we have

$$\int \Phi H \Phi d\mathbf{r} - E_0 \int \Phi^2 d\mathbf{r} \geq 0 \qquad (4.14)$$

which we may rearrange to

$$\frac{\int \Phi H \Phi d\mathbf{r}}{\int \Phi^2 d\mathbf{r}} \geq E_0 \qquad (4.15)$$

(note that when Φ is normalized, the denominator on the l.h.s. is 1, but it is helpful to have Eq. (4.15) in this more general form for future use).

Equation (4.15) has extremely powerful implications. If we are looking for the best wave function to define the ground state of a system, we can judge the quality of wave functions that we arbitrarily guess by their associated energies: *the lower the better*. This result is critical because it shows us that we do not have to construct our guess wave function Φ as a linear combination of (unknown) orthonormal wave functions Ψ_i, but we may construct it in any manner we wish. The quality of our guess will be determined by how low a value we calculate for the integral in Eq. (4.15). Moreover, since we would like to find the lowest possible energy within the constraints of how we go about constructing a wave function, we can use all of the tools that calculus makes available for locating extreme values.

4.2.3   The Born–Oppenheimer Approximation

Up to now, we have been discussing many-particle molecular systems entirely in the abstract. In fact, accurate wave functions for such systems are extremely difficult to express because of the correlated motions of particles. That is, the Hamiltonian in Eq. (4.3) contains pairwise attraction and repulsion terms, implying that no particle is moving independently of all of the others (the term 'correlation' is used to describe this interdependency). In order to simplify the problem somewhat, we may invoke the so-called Born–Oppenheimer approximation. This approximation is described with more rigor in Section 15.5, but at this point we present the conceptual aspects without delving deeply into the mathematical details.

Under typical physical conditions, the nuclei of molecular systems are moving much, much more slowly than the electrons (recall that protons and neutrons are about 1800 times more massive than electrons and note the appearance of mass in the denominator of the kinetic energy terms of the Hamiltonian in Eq. (4.3)). For practical purposes, electronic 'relaxation' with respect to nuclear motion is instantaneous. As such, it is convenient to decouple these two motions, and compute electronic energies for *fixed* nuclear positions. That is, the nuclear kinetic energy term is taken to be independent of the electrons, correlation in the attractive electron–nuclear potential energy term is eliminated, and the repulsive nuclear–nuclear potential energy term becomes a simply evaluated constant for a given geometry. Thus, the *electronic* Schrödinger equation is taken to be

$$(H_{el} + V_N)\Psi_{el}(\mathbf{q}_i; \mathbf{q}_k) = E_{el}\Psi_{el}(\mathbf{q}_i; \mathbf{q}_k) \tag{4.16}$$

where the subscript 'el' emphasizes the invocation of the Born–Oppenheimer approximation, H_{el} includes only the first, third, and fourth terms on the r.h.s. of Eq. (4.3), V_N is the nuclear–nuclear repulsion energy, and the electronic coordinates $\mathbf{q}_i$ are independent variables but the nuclear coordinates $\mathbf{q}_k$ are parameters (and thus appear following a semicolon rather than a comma in the variable list for Ψ). The eigenvalue of the electronic Schrödinger equation is called the 'electronic energy'. Note that the term V_N is a constant for a given set of fixed nuclear coordinates. Wave functions are invariant to the appearance of constant terms in the Hamiltonian, so in practice one almost always solves Eq. (4.16) without the inclusion of V_N, in which case the eigenvalue is sometimes called the 'pure electronic energy', and one then adds V_N to this eigenvalue to obtain E_{el}.

In general, the Born–Oppenheimer assumption is an extremely mild one, and it is entirely justified in most cases. It is worth emphasizing that this approximation has very profound consequences from a conceptual standpoint – so profound that they are rarely thought about but simply accepted as dogma. Without the Born–Oppenheimer approximation we would lack the concept of a potential energy surface: The PES is the surface defined by E_{el} over all possible nuclear coordinates. We would further lack the concepts of equilibrium and transition state geometries, since these are defined as critical points on the PES; instead we would be reduced to discussing high-probability regions of the nuclear wave functions. Of course, for some problems in chemistry, we *do* need to consider the quantum mechanical character of the nuclei, but the advantages afforded by the Born–Oppenheimer approximation should be manifest.

4.3 Construction of Trial Wave Functions

Equation (4.16) is simpler than Eq. (4.2) because electron–nuclear correlation has been removed. The remaining correlation, that between the individual electrons, is considerably more troubling. For the moment we will take the simplest possible approach and ignore it; we do this by considering systems with only a single electron. The electronic wave function has thus been reduced to depending only on the fixed nuclear coordinates and the three Cartesian coordinates of the single electron. The eigenfunctions of Eq. (4.16) for a molecular system may now be properly called molecular orbitals (MOs; rather unusual ones in general, since they are for a molecule having only one electron, but MOs nonetheless). To distinguish a one-electron wave function from a many-electron wave function, we will designate the former as ψ_{el} and the latter as Ψ_{el}. We will hereafter drop the subscript 'el' where not required for clarity; unless otherwise specified, all wave functions are electronic wave functions.

The pure electronic energy eigenvalue associated with each molecular orbital is the energy of the electron in that orbital. Experimentally, one might determine this energy by measuring the ionization potential of the electron when it occupies the orbital (fairly easy for the hydrogen atom, considerably more difficult for polynuclear molecules). To measure E_{el}, which includes the nuclear repulsion energy, one would need to determine the 'atomization' energy, that is, the energy required to ionize the electron *and* to remove all of the nuclei to infinite separation. In practice, atomization energies are not measured, but instead we have compilations of such thermodynamic variables as heats of formation. The relationship between these computed and thermodynamic quantities is discussed in Chapter 10.

4.3.1 The LCAO Basis Set Approach

As noted earlier, we may imagine constructing wave functions in any fashion we deem reasonable, and we may judge the quality of our wave functions (in comparison to one another) by evaluation of the energy eigenvalues associated with each. The one with the lowest energy will be the most accurate and presumably the best one to use for computing other properties by the application of other operators. So, how might one go about choosing mathematical functions with which to construct a trial wave function? This is a typical question in mathematics – how can an arbitrary function be represented by a combination of more convenient functions? The convenient functions are called a 'basis set'. Indeed, we have already encountered this formalism – Eq. (2.10) of Chapter 2 illustrates the use of a basis set of cosine functions to approximate torsional energy functions.

In our QM systems, we have temporarily restricted ourselves to systems of one electron. If, in addition, our system were to have only one nucleus, then we would not need to guess wave functions, but instead we could solve Eq. (4.16) *exactly*. The eigenfunctions that are determined in that instance are the familiar hydrogenic atomic orbitals, 1s, 2s, 2p, 3s, 3p, 3d, etc., whose properties and derivation are discussed in detail in standard texts on quantum mechanics. For the moment, we will not investigate the mathematical representation of these hydrogenic atomic orbitals in any detail, but we will simply posit that, as functions, they may be useful in the construction of more complicated *molecular* orbitals. In particular, just as in Eq. (4.10) we constructed a guess wave function as a linear combination of exact wave functions, so here we will construct a guess wave function ϕ as a linear combination of atomic wave functions φ, i.e.,

$$\phi = \sum_{i=1}^{N} a_i \varphi_i \tag{4.17}$$

where the set of N functions φ_i is called the 'basis set' and each has associated with it some coefficient a_i. This construction is known as the linear combination of atomic orbitals (LCAO) approach.

Note that Eq. (4.17) does not specify the locations of the basis functions. Our intuition suggests that they should be centered on the atoms of the molecule, but this is certainly not a requirement. If this comment seems odd, it is worth emphasizing at this point that we should not let our chemical intuition limit our mathematical flexibility. As chemists, we choose to use atomic orbitals (AOs) because we anticipate that they will be efficient functions for the representation of MOs. However, as mathematicians, we should immediately stop thinking about our choices as orbitals, and instead consider them only to be *functions*, so that we avoid being conceptually influenced about how and where to use them.

Recall that the wave function squared has units of probability density. In essence, the electronic wave function is a road map of where the electrons are more or less likely to be found. Thus, we want our basis functions to provide us with the flexibility to allow electrons to 'go' where their presence at higher density lowers the energy. For instance, to describe the bonding of a hydrogen atom to a carbon, it is clearly desirable to use a p function on hydrogen, oriented along the axis of the bond, to permit electron density to be localized in the bonding region more efficiently than is possible with only a spherically symmetric s function. Does this imply that the hydrogen atom is somehow sp-hybridized? Not necessarily – the p function is simply serving the purpose of increasing the flexibility with which the *molecular* orbital may be described. If we took away the hydrogen p function and instead placed an s function *in between* the C and H atoms, we could also build up electron density in the bonding region (see Figure 4.1). Thus, the *chemical* interpretation of the coefficients in Eq. (4.17) should only be undertaken with caution, as further described in Chapter 9.

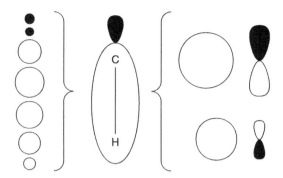

Figure 4.1 Two different basis sets for representing a C–H σ bonding orbital with the size of the basis functions roughly illustrating their weight in the hybrid MO. The set on the right is the more chemically intuitive since all basis functions are centered on the atoms. Note, however, that the use of a p function to polarize the hydrogen density goes beyond a purely minimalist approach. The set on the left is composed entirely of s functions distributed along the bond. Such a basis set may seem odd in concept, but is quite capable of accurately representing the electron density in space. Indeed, the basis set on the left would have certain computational advantages, chief among them the greater simplicity of working with s functions than with p functions

One should also note that the summation in Eq. (4.17) has an upper limit N; we cannot work with an infinite basis in any convenient way (at least not when the basis is AOs). However, the more atomic orbitals we allow into our basis, the closer our basis will come to 'spanning' the true molecular orbital space. Thus, the chemical idea that we would limit ourselves to, say, at most one 1s function on each hydrogen atom is needlessly confining from a mathematical standpoint. Indeed, there may be very many 'true' one-electron MOs that are very high in energy. Accurately describing these MOs may require some unusual basis functions, e.g., very diffuse functions to describe weakly bound electrons, like those found in Rydberg states. We will discuss these issues in much more detail in Section 6.2, but it is worth emphasizing here, at the beginning, that the distinction between orbitals and functions is a critical one in computational molecular orbital theory.

4.3.2 The Secular Equation

All that being said, let us now turn to evaluating the energy of our guess wave function. From Eqs. (4.15) and (4.17) we have

$$
E = \frac{\int \left(\sum_i a_i \varphi_i \right) H \left(\sum_j a_j \varphi_j \right) d\mathbf{r}}{\int \left(\sum_i a_i \varphi_i \right) \left(\sum_j a_j \varphi_j \right) d\mathbf{r}}
$$

$$= \frac{\sum_{ij} a_i a_j \int \varphi_i H \varphi_j d\mathbf{r}}{\sum_{ij} a_i a_j \int \varphi_i \varphi_j d\mathbf{r}}$$

$$= \frac{\sum_{ij} a_i a_j H_{ij}}{\sum_{ij} a_i a_j S_{ij}} \qquad (4.18)$$

where we have introduced the shorthand notation H_{ij} and S_{ij} for the integrals in the numerator and denominator, respectively. These so-called 'matrix elements' are no longer as simple as they were in prior discussion, since the atomic orbital basis set, while likely to be efficient, is no longer orthonormal. These matrix elements have more common names, H_{ij} being called a 'resonance integral', and S_{ij} being called an 'overlap integral'. The latter has a very clear physical meaning, namely the extent to which any two basis functions overlap in a phase-matched fashion in space. The former integral is not so easily made intuitive, but it is worth pointing out that orbitals which give rise to large overlap integrals will similarly give rise to large resonance integrals. One resonance integral which *is* intuitive is H_{ii}, which corresponds to the energy of a single electron occupying basis function i, i.e., it is essentially equivalent to the ionization potential of the AO in the environment of the surrounding molecule.

Now, it is useful to keep in mind our objective. The variational principle instructs us that as we get closer and closer to the 'true' one-electron ground-state wave function, we will obtain lower and lower energies from our guess. Thus, once we have selected a basis set, we would like to choose the coefficients a_i so as to *minimize* the energy for all possible linear combinations of our basis functions. From calculus, we know that a necessary condition for a function (i.e., the energy) to be at its minimum is that its derivatives with respect to all of its free variables (i.e., the coefficients a_i) are zero. Notationally, that is

$$\frac{\partial E}{\partial a_k} = 0 \qquad \forall k \qquad (4.19)$$

(where we make use of the mathematical abbreviation $\forall$ meaning 'for all'). Performing this fairly tedious partial differentiation on Eq. (4.18) for each of the N variables a_k gives rise to N equations which must be satisfied in order for Eq. (4.19) to hold true, namely

$$\sum_{i=1}^{N} a_i (H_{ki} - E S_{ki}) = 0 \qquad \forall k \qquad (4.20)$$

This set of N equations (running over k) involves N unknowns (the individual a_i). From linear algebra, we know that a set of N equations in N unknowns has a non-trivial solution if and only if the determinant formed from the coefficients of the unknowns (in this case the 'coefficients' are the various quantities $H_{ki} - E S_{ki}$) is equal to zero. Notationally again, that is

$$\begin{vmatrix} H_{11} - ES_{11} & H_{12} - ES_{12} & \cdots & H_{1N} - ES_{1N} \\ H_{21} - ES_{21} & H_{22} - ES_{22} & \cdots & H_{2N} - ES_{2N} \\ \vdots & \vdots & \ddots & \vdots \\ H_{N1} - ES_{N1} & H_{N2} - ES_{N2} & \cdots & H_{NN} - ES_{NN} \end{vmatrix} = 0 \qquad (4.21)$$

Equation (4.21) is called a secular equation. In general, there will be N roots E which permit the secular equation to be true. That is, there will be N energies E_j (some of which may be equal to one another, in which case we say the roots are 'degenerate') where each value of E_j will give rise to a different set of coefficients, a_{ij}, which can be found by solving the set of linear Eqs. (4.20) using E_j, and these coefficients will define an optimal wave function ϕ_j within the given basis set, i.e.,

$$\phi_j = \sum_{i=1}^{N} a_{ij} \varphi_i \qquad (4.22)$$

In a one-electron system, the lowest energy molecular orbital would thus define the 'ground state' of the system, and the higher energy orbitals would be 'excited states'. Obviously, as these are different MOs, they have different basis function coefficients. Although we have not formally proven it, it is worth noting that the variational principle holds for the excited states as well: the calculated energy of a guess wave function for an excited state will be bounded from below by the true excited state energy (MacDonald 1933).

So, in a nutshell, to find the optimal one-electron wave functions for a molecular system, we:

1. Select a set of N basis functions.

2. For that set of basis functions, determine all N^2 values of both H_{ij} and S_{ij}.

3. Form the secular determinant, and determine the N roots E_j of the secular equation.

4. For each of the N values of E_j, solve the set of linear Eqs. (4.20) in order to determine the basis set coefficients a_{ij} for that MO.

All of the MOs determined by this process are mutually orthogonal. For degenerate MOs, some minor complications arise, but those are not discussed here.

4.4 Hückel Theory

4.4.1 Fundamental Principles

To further illuminate the LCAO variational process, we will carry out the steps outlined above for a specific example. To keep things simple (and conceptual), we consider a flavor of molecular orbital theory developed in the 1930s by Erich Hückel to explain some of the unique properties of unsaturated and aromatic hydrocarbons (Hückel 1931; for historical insights, see also, Berson 1996; Frenking 2000). In order to accomplish steps 1–4 of the last section, Hückel theory adopts the following conventions:

(a) The basis set is formed entirely from parallel carbon 2p orbitals, one per atom. [Hückel theory was originally designed to treat only planar hydrocarbon π systems, and thus the 2p orbitals used are those that are associated with the π system.]

(b) The overlap matrix is defined by

$$S_{ij} = \delta_{ij} \tag{4.23}$$

Thus, the overlap of any carbon 2p orbital with itself is unity (i.e., the p functions are normalized), and that between any two p orbitals is zero.

(c) Matrix elements H_{ii} are set equal to the negative of the ionization potential of the methyl radical, i.e., the orbital energy of the singly occupied 2p orbital in the prototypical system defining sp^2 carbon hybridization. This choice is consistent with our earlier discussion of the relationship between this matrix element and an ionization potential. This energy value, which is defined so as to be negative, is rarely actually written as a numerical value, but is instead represented by the symbol α.

(d) Matrix elements H_{ij} between neighbors are also derived from experimental information. A 90° rotation about the π bond in ethylene removes all of the bonding interaction between the two carbon 2p orbitals. That is, the (positive) cost of the following process,

$$E = E_\pi \qquad\qquad E = 2E_p$$

is $\Delta E = 2E_p - E_\pi$. The (negative) stabilization energy for the pi bond is distributed equally to the two p orbitals involved (i.e., divided in half) and this quantity, termed β, is used for H_{ij} between neighbors. (Note, based on our definitions so far, then, that $E_p = \alpha$ and $E_\pi = 2\alpha + 2\beta$.)

(e) Matrix elements H_{ij} between carbon 2p orbitals more distant than nearest neighbors are set equal to zero.

4.4.2 Application to the Allyl System

Let us now apply Hückel MO theory to the particular case of the allyl system, C_3H_3, as illustrated in Figure 4.2. Because we have three carbon atoms, our basis set is determined from convention (a) and will consist of 3 carbon 2p orbitals, one centered on each atom. We will arbitrarily number them 1, 2, 3, from left to right for bookkeeping purposes.

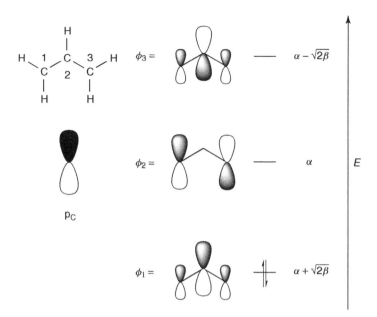

Figure 4.2 Hückel MOs for the allyl system. One p_C orbital per atom defines the basis set. Combinations of these 3 AOs create the 3 MOs shown. The electron occupation illustrated corresponds to the allyl cation. One additional electron in ϕ_2 would correspond to the allyl radical, and a second (spin-paired) electron in ϕ_2 would correspond to the allyl anion

The basis set size of three implies that we will need to solve a 3×3 secular equation. Hückel conventions (b)–(e) tell us the value of each element in the secular equation, so that Eq. (4.21) is rendered as

$$\begin{vmatrix} \alpha - E & \beta & 0 \\ \beta & \alpha - E & \beta \\ 0 & \beta & \alpha - E \end{vmatrix} = 0 \tag{4.24}$$

The use of the Krönecker delta to define the overlap matrix ensures that E appears only in the diagonal elements of the determinant. Since this is a 3×3 determinant, it may be expanded using Cramer's rule as

$$(\alpha - E)^3 + (\beta^2 \cdot 0) + (0 \cdot \beta^2) - [0 \cdot (\alpha - E) \cdot 0] - \beta^2(\alpha - E) - (\alpha - E)\beta^2 = 0 \tag{4.25}$$

which is a fairly simple cubic equation in E that has three solutions, namely

$$E = \alpha + \sqrt{2}\beta, \quad \alpha, \quad \alpha - \sqrt{2}\beta \tag{4.26}$$

Since α and β are negative by definition, the lowest energy solution is $\alpha + \sqrt{2}\beta$. To find the MO associated with this energy, we employ it in the set of linear Eqs. (4.20), together with the various necessary H and S values to give

$$a_1[\alpha - (\alpha + \sqrt{2}\beta) \cdot 1] + a_2[\beta - (\alpha + \sqrt{2}\beta) \cdot 0] + a_3[0 - (\alpha + \sqrt{2}\beta) \cdot 0] = 0$$

$$a_1[\beta - (\alpha + \sqrt{2}\beta) \cdot 0] + a_2[\alpha - (\alpha + \sqrt{2}\beta) \cdot 1] + a_3[\beta - (\alpha + \sqrt{2}\beta) \cdot 0] = 0$$

$$a_1[0 - (\alpha + \sqrt{2}\beta) \cdot 0] + a_2[\beta - (\alpha + \sqrt{2}\beta) \cdot 0] + a_3[\alpha - (\alpha + \sqrt{2}\beta) \cdot 1] = 0 \quad (4.27)$$

Some fairly trivial algebra reduces these equations to

$$a_2 = \sqrt{2}a_1$$

$$a_3 = a_1 \quad (4.28)$$

While there are infinitely many values of a_1, a_2, and a_3 which satisfy Eq. (4.28), the requirement that the wave function be normalized provides a final constraint in the form of Eq. (4.11). The unique values satisfying both equations are

$$a_{11} = \frac{1}{2}, \quad a_{21} = \frac{\sqrt{2}}{2}, \quad a_{31} = \frac{1}{2} \quad (4.29)$$

where we have now emphasized that these coefficients are specific to the *lowest energy* molecular orbital by adding the second subscript '1'. Since we now know both the coefficients and the basis functions, we may construct the lowest energy molecular orbital, i.e.,

$$\varphi_1 = \frac{1}{2}p_1 + \frac{\sqrt{2}}{2}p_2 + \frac{1}{2}p_3 \quad (4.30)$$

which is illustrated in Figure 4.2.

By choosing the higher energy roots of Eq. (4.24), we may solve the sets of linear equations analogous to Eq. (4.27) in order to arrive at the coefficients required to construct ϕ_2 (from $E = \alpha$) and ϕ_3 (from $E = \alpha - \sqrt{2}\beta$). Although the algebra is left for the reader, the results are

$$a_{12} = \frac{\sqrt{2}}{2}, \quad a_{22} = 0, \quad a_{32} = -\frac{\sqrt{2}}{2}$$

$$a_{13} = \frac{1}{2}, \quad a_{23} = -\frac{\sqrt{2}}{2}, \quad a_{33} = \frac{1}{2} \quad (4.31)$$

and these orbitals are also illustrated in Figure 4.2. The three orbitals we have derived are the bonding, non-bonding, and antibonding allyl molecular orbitals with which all organic chemists are familiar.

Importantly, Hückel theory affords us certain insights into the allyl system, one in particular being an analysis of the so-called 'resonance' energy arising from electronic delocalization in the π system. By delocalization we refer to the participation of more than two atoms in a given MO. Consider for example the allyl cation, which has a total of two electrons in the π system. If we adopt a molecular aufbau principle of filling lowest energy MOs first and further make the assumption that each electron has the energy of the one-electron MO that it occupies (ϕ_1 in this case) then the total energy of the allyl cation π system is $2(\alpha + \sqrt{2}\beta)$. Consider the alternative 'fully localized' structure for the allyl system, in which there is a full (doubly-occupied) π bond between two of the carbons, and an empty, non-interacting p orbital on the remaining carbon atom. (This could be achieved by rotating the cationic methylene group $90°$ so that the p orbital becomes orthogonal to the remaining π bond, but that could no longer be described by simple Hückel theory since the system would be non-planar – the non-interaction we are considering here is purely a thought-experiment). The π energy of such a system would simply be that of a double bond, which by our definition of terms above is $2(\alpha + \beta)$. Thus, the Hückel resonance energy, which is equal to $H_\pi - H_{localized}$, is 0.83β (remember β is negative by definition, so resonance is a favorable phenomenon). Recalling the definition of β, the resonance energy in the allyl cation is predicted to be about 40% of the rotation barrier in ethylene.

We may perform the same analysis for the allyl radical and the allyl anion, respectively, by adding the energy of ϕ_2 to the cation with each successive addition of an electron, i.e., $H_\pi(\text{allyl radical}) = 2(\alpha + \sqrt{2}\beta) + \alpha$ and $H_\pi(\text{allyl anion}) = 2(\alpha + \sqrt{2}\beta) + 2\alpha$. In the hypothetical fully π-localized non-interacting system, each new electron would go into the non-interacting p orbital, also contributing each time a factor of α to the energy (by definition of α). Thus, the Hückel resonance energies of the allyl radical and the allyl anion are the same as for the allyl cation, namely, 0.83β.

Unfortunately, while it is clear that the allyl cation, radical, and anion all enjoy some degree of resonance stabilization, neither experiment, in the form of measured rotational barriers, nor higher levels of theory support the notion that in all three cases the magnitude is the same (see, for instance, Gobbi and Frenking 1994; Mo et al. 1996). So, what aspects of Hückel theory render it incapable of accurately distinguishing between these three allyl systems?

4.5   Many-electron Wave Functions

In our Hückel theory example, we derived molecular orbitals and molecular orbital energies using a *one-electron formalism*, and we then assumed that the energy of a many-electron system could be determined simply as the sum of the energies of the occupied one-electron orbitals (we used our chemical intuition to limit ourselves to two electrons per orbital). We further assumed that the orbitals themselves are invariant to the number of electrons in the π system. One might be tempted to say that Hückel theory thus ignores electron–electron repulsion. This is a bit unfair, however. By deriving our Hamiltonian matrix elements from experimental quantities (ionization potentials and rotational barriers) we have implicitly accounted for electron–electron repulsion in some sort of average way, but such an approach, known as an 'effective Hamiltonian' method, is necessarily rather crude. Thus, while Hückel theory continues to find use even today in qualitative studies of conjugated systems, it is rarely sufficiently accurate for quantitative assessments. To improve our models, we need to take a more sophisticated accounting of many-electron effects.

4.5.1 Hartree-product Wave Functions

Let us examine the Schrödinger equation in the context of a one-electron Hamiltonian a little more carefully. When the only terms in the Hamiltonian are the one-electron kinetic energy and nuclear attraction terms, the operator is 'separable' and may be expressed as

$$H = \sum_{i=1}^{N} h_i \tag{4.32}$$

where N is the total number of electrons and h_i is the one-electron Hamiltonian defined by

$$h_i = -\frac{1}{2}\nabla_i^2 - \sum_{k=1}^{M} \frac{Z_k}{r_{ik}} \tag{4.33}$$

where M is the total number of nuclei (note that Eq. (4.33) is written in atomic units).

Eigenfunctions of the one-electron Hamiltonian defined by Eq. (4.33) must satisfy the corresponding one-electron Schrödinger equation

$$h_i \psi_i = \varepsilon_i \psi_i \tag{4.34}$$

Because the Hamiltonian operator defined by Eq. (4.32) is separable, its many-electron eigenfunctions can be constructed as products of one-electron eigenfunctions. That is

$$\Psi_{HP} = \psi_1 \psi_2 \cdots \psi_N \tag{4.35}$$

A wave function of the form of Eq. (4.35) is called a 'Hartree-product' wave function.

The eigenvalue of Ψ is readily found from proving the validity of Eq. (4.35), viz.,

$$H\Psi_{HP} = H\psi_1\psi_2\cdots\psi_N$$

$$= \sum_{i=1}^{N} h_i \psi_1\psi_2\cdots\psi_N$$

$$= (h_1\psi_1)\psi_2\cdots\psi_N + \psi_1(h_2\psi_2)\cdots\psi_N + \ldots + \psi_1\psi_2\cdots(h_N\psi_N)$$

$$= (\varepsilon_1\psi_1)\psi_2\cdots\psi_N + \psi_1(\varepsilon_2\psi_2)\cdots\psi_N + \ldots + \psi_1\psi_2\cdots(\varepsilon_N\psi_N)$$

$$= \sum_{i=1}^{N} \varepsilon_i \psi_1\psi_2\cdots\psi_N$$

$$= \left(\sum_{i=1}^{N} \varepsilon_i\right) \Psi_{HP} \tag{4.36}$$

where repeated application of Eq. (4.34) is used in proving that the energy eigenvalue of the many-electron wave function is simply the sum of the one-electron energy eigenvalues. Note that Eqs. (4.32)–(4.36) provide the mathematical rigor behind the Hückel theory example presented more informally above. Note that if every ψ is normalized then Ψ_{HP} is also normalized, since $|\Psi_{HP}|^2 = |\psi_1|^2 |\psi_2|^2 \cdots |\psi_N|^2$.

4.5.2 The Hartree Hamiltonian

As noted above, however, the Hamiltonian defined by Eqs. (4.32) and (4.33) does *not* include interelectronic repulsion, computation of which is vexing because it depends not on one electron, but instead on all possible (simultaneous) pairwise interactions. We may ask, however, how useful is the Hartree-product wave function in computing energies from the *correct* Hamiltonian? That is, we wish to find orbitals ψ that minimize $\langle \Psi_{HP} | H | \Psi_{HP} \rangle$. By applying variational calculus, one can show that each such orbital ψ_i is an eigenfunction of its own operator h_i defined by

$$h_i = -\frac{1}{2}\nabla_i^2 - \sum_{k=1}^{M} \frac{Z_k}{r_{ik}} + V_i\{j\} \tag{4.37}$$

where the final term represents an interaction potential with all of the other electrons occupying orbitals $\{j\}$ and may be computed as

$$V_i\{j\} = \sum_{j \neq i} \int \frac{\rho_j}{r_{ij}} d\mathbf{r} \tag{4.38}$$

where ρ_j is the charge (probability) density associated with electron j. The repulsive third term on the r.h.s. of Eq. (4.37) is thus exactly analogous to the attractive second term, except that nuclei are treated as point charges, while electrons, being treated as wave functions, have their charge spread out, so an integration over all space is necessary. Recall, however, that $\rho_j = |\psi_j|^2$. Since the point of undertaking the calculation is to determine the individual ψ, how can they be used in the one-electron Hamiltonians before they are known?

To finesse this problem, Hartree (1928) proposed an iterative 'self-consistent field' (SCF) method. In the first step of the SCF process, one *guesses* the wave functions ψ for all of the occupied MOs (AOs in Hartree's case, since he was working exclusively with atoms) and uses these to construct the necessary one-electron operators h. Solution of each differential Eq. (4.34) (in an atom, with its spherical symmetry, this is relatively straightforward, and Hartree was helped by his retired father who enjoyed the mathematical challenge afforded by such calculations) provides a *new* set of ψ, presumably different from the initial guess. So, the one-electron Hamiltonians are formed anew using these presumably more accurate ψ to determine each necessary ρ, and the process is repeated to obtain a still better set of ψ. At some point, the difference between a newly determined set and the immediately preceding set falls below some threshold criterion, and we refer to the final set of ψ as the 'converged' SCF orbitals. (An example of a threshold criterion might be that the total electronic energy change by no more than 10^{-6} a.u., and/or that the energy eigenvalue for each MO change by no more than that amount – such criteria are, of course, entirely arbitrary, and it is typically only by checking computed properties for wave functions computed with varying degrees of imposed 'tightness' that one can determine an optimum balance between convergence and accuracy – the tighter the convergence, the more SCF cycles required, and the greater the cost in computational resources.)

Notice, from Eq. (4.36), that the sum of the individual operators h defined by Eq. (4.37) defines a separable Hamiltonian operator for which Ψ_{HP} is an eigenfunction. This separable Hamiltonian corresponds to a 'non-interacting' system of electrons (in the sense that each individual electron sees simply a constant potential with which it interacts – the nomenclature can be slightly confusing since the potential *does* derive in an average way from the other electrons, but the point is that their interaction is not accounted for instantaneously). The non-interacting Hamiltonian is *not* a good approximation to the true Hamiltonian, however, because each h includes the repulsion of its associated electron with all of the other electrons, i.e., h_i includes the repulsion between electron i and electron j, but so too does h_j. Thus, if we were to sum all of the one-electron eigenvalues for the operators h_i, which according to Eq. (4.36) would give us the eigenvalue for our non-interacting Hamiltonian, we would double-count the electron–electron repulsion. It is a straightforward matter to correct for this double-counting, however, and we may in principle compute $E = \langle \Psi_{HP} | H | \Psi_{HP} \rangle$ not directly but rather as

$$E = \sum_i \varepsilon_i - \frac{1}{2} \sum_{i \neq j} \int\int \frac{|\psi_i|^2 |\psi_j|^2}{r_{ij}} d\mathbf{r}_i d\mathbf{r}_j \tag{4.39}$$

where i and j run over all the electrons, ε_i is the energy of MO i from the solution of the one-electron Schrödinger equation using the one-electron Hamiltonian defined by Eq. (4.37), and we have replaced ρ with the square of the wave function to emphasize how it is determined (again, the double integration over all space derives from the wave function character of the electron – the double integral appearing on the r.h.s. of Eq. (4.39) is called a 'Coulomb integral' and is often abbreviated as J_{ij}). In spite of the significant difference between the non-interacting Hamiltonian and the correct Hamiltonian, operators of the former type have important utility, as we will see in Sections 7.4.2 and 8.3 within the contexts of perturbation theory and density functional theory, respectively.

At this point it is appropriate to think about our Hartree-product wave function in more detail. Let us say we have a system of eight electrons. How shall we go about placing them into MOs? In the Hückel example above, we placed them in the lowest energy MOs first, because we wanted ground electronic states, but we also limited ourselves to two electrons per orbital. Why? The answer to that question requires us to introduce something we have ignored up to this point, namely spin.

4.5.3 Electron Spin and Antisymmetry

All electrons are characterized by a spin quantum number. The electron spin function is an eigenfunction of the operator S_z and has only two eigenvalues, $\pm \hbar/2$; the spin eigenfunctions are orthonormal and are typically denoted as α and β (not to be confused with the α and β of Hückel theory!) The spin quantum number is a natural consequence of the application of relativistic quantum mechanics to the electron (i.e., accounting for Einstein's theory of relativity in the equations of quantum mechanics), as first shown by Dirac. Another consequence of relativistic quantum mechanics is the so-called Pauli exclusion principle, which is usually stated as the assertion that no two electrons can be characterized by the same set of quantum numbers. Thus, in a given MO (which defines all electronic quantum numbers except spin) there are only two possible choices for the remaining quantum number, α or β, and thus only two electrons may be placed in any MO.

Knowing these aspects of quantum mechanics, if we were to construct a ground-state Hartree-product wave function for a system having two electrons of the same spin, say α, we would write

$$^{3}\Psi_{HP} = \psi_{a}(1)\alpha(1)\psi_{b}(2)\alpha(2) \tag{4.40}$$

where the left superscript 3 indicates a triplet electronic state (two electrons spin parallel) and ψ_{a} and ψ_{b} are different from one another (since otherwise electrons 1 and 2 would have all identical quantum numbers) and orthonormal. However, the wave function defined by Eq. (4.40) is fundamentally flawed. The Pauli exclusion principle is an important mnemonic, but it actually derives from a feature of relativistic quantum field theory that has more general consequences, namely that electronic wave functions must *change sign* whenever the coordinates of two electrons are interchanged. Such a wave function is said to be 'antisymmetric'. For notational purposes, we can define the permutation operator P_{ij} as the operator that interchanges the coordinates of electrons i and j. Thus, we would write the Pauli principle for a system of N electrons as

$$P_{ij}\Psi[\mathbf{q}_{1}(1), \ldots, \mathbf{q}_{i}(i), \ldots, \mathbf{q}_{j}(j), \ldots, \mathbf{q}_{N}(N)]$$
$$= \Psi[\mathbf{q}_{1}(1), \ldots, \mathbf{q}_{j}(i), \ldots, \mathbf{q}_{i}(j), \ldots, \mathbf{q}_{N}(N)]$$
$$= -\Psi[\mathbf{q}_{1}(1), \ldots, \mathbf{q}_{i}(i), \ldots, \mathbf{q}_{j}(j), \ldots, \mathbf{q}_{N}(N)] \tag{4.41}$$

where $\mathbf{q}$ now includes not only the three Cartesian coordinates but also the spin function.

If we apply P_{12} to the Hartree-product wave function of Eq. (4.40),

$$P_{12}[\psi_{a}(1)\alpha(1)\psi_{b}(2)\alpha(2)] = \psi_{b}(1)\alpha(1)\psi_{a}(2)\alpha(2)$$
$$\neq -\psi_{a}(1)\alpha(1)\psi_{b}(2)\alpha(2) \tag{4.42}$$

we immediately see that it does *not* satisfy the Pauli principle. However, a slight modification to Ψ_{HP} can be made that causes it to satisfy the constraints of Eq. (4.41), namely

$$^{3}\Psi_{SD} = \frac{1}{\sqrt{2}}[\psi_{a}(1)\alpha(1)\psi_{b}(2)\alpha(2) - \psi_{a}(2)\alpha(2)\psi_{b}(1)\alpha(1)] \tag{4.43}$$

(the reader is urged to verify that $^3\Psi_{SD}$ does indeed satisfy the Pauli principle; for the 'SD' subscript, see next section). Note that if we integrate $|^3\Psi_{SD}|^2$ over all space we have

$$
\begin{aligned}
\int |^3\Psi_{SD}|^2 d\mathbf{r}_1 d\omega_1 d\mathbf{r}_2 d\omega_2 &= \frac{1}{2} \Bigg[\int |\psi_a(1)|^2 |\alpha(1)|^2 |\psi_b(2)|^2 |\alpha(2)|^2 d\mathbf{r}_1 d\omega_1 d\mathbf{r}_2 d\omega_2 \\
&\quad - 2\int \psi_a(1)\psi_b(1) |\alpha(1)|^2 \psi_b(2)\psi_a(2) |\alpha(2)|^2 d\mathbf{r}_1 d\omega_1 d\mathbf{r}_2 d\omega_2 \\
&\quad + \int |\psi_a(2)|^2 |\alpha(2)|^2 |\psi_b(1)|^2 |\alpha(1)|^2 d\mathbf{r}_1 d\omega_1 d\mathbf{r}_2 d\omega_2 \Bigg] \\
&= \frac{1}{2}(1 - 0 + 1) \\
&= 1 \qquad\qquad\qquad\qquad\qquad\qquad\qquad\qquad (4.44)
\end{aligned}
$$

where ω is a spin integration variable, the simplification of the various integrals on the r.h.s. proceeds from the orthonormality of the MOs and spin functions, and we see that the prefactor of $2^{-1/2}$ in Eq. (4.43) is required for normalization.

4.5.4 Slater Determinants

A different mathematical notation can be used for Eq. (4.43)

$$
^3\Psi_{SD} = \frac{1}{\sqrt{2}} \begin{vmatrix} \psi_a(1)\alpha(1) & \psi_b(1)\alpha(1) \\ \psi_a(2)\alpha(2) & \psi_b(2)\alpha(2) \end{vmatrix} \qquad (4.45)
$$

where the difference of MO products has been expressed as a determinant. Note that the permutation operator P applied to a determinant has the effect of interchanging two of the rows. It is a general property of a determinant that it changes sign when any two rows (or columns) are interchanged, and the utility of this feature for use in constructing antisymmetric wave functions was first exploited by Slater (1929). Thus, the 'SD' subscript used in Eqs. (4.43)–(4.45) stands for 'Slater determinant'. On a term-by-term basis, Slater-determinantal wave functions quickly become rather tedious to write down, but determinantal notation allows them to be expressed reasonably compactly as, in general,

$$
\Psi_{SD} = \frac{1}{\sqrt{N!}} \begin{vmatrix} \chi_1(1) & \chi_2(1) & \cdots & \chi_N(1) \\ \chi_1(2) & \chi_2(2) & \cdots & \chi_N(2) \\ \vdots & \vdots & \ddots & \vdots \\ \chi_1(N) & \chi_2(N) & \cdots & \chi_N(N) \end{vmatrix} \qquad (4.46)
$$

where N is the total number of electrons and χ is a spin-orbital, i.e., a product of a spatial orbital and an electron spin eigenfunction. A still more compact notation that finds widespread use is

$$
\Psi_{SD} = |\chi_1 \chi_2 \chi_3 \cdots \chi_N\rangle \qquad (4.47)
$$

where the prefactor $(N!)^{-1/2}$ is implicit. Furthermore, if two spin orbitals differ only in the spin eigenfunction (i.e., together they represent a doubly filled orbital) this is typically represented by writing the spatial wave function with a superscript 2 to indicate double occupation. Thus, if χ_1 and χ_2 represented α and β spins in spatial orbital ψ_1, one would write

$$\Psi_{SD} = |\psi_1^2 \chi_3 \cdots \chi_N\rangle \tag{4.48}$$

Slater determinants have a number of interesting properties. First, note that every electron appears in every spin orbital somewhere in the expansion. This is a manifestation of the indistinguishability of quantum particles (which is *violated* in the Hartree-product wave functions). A more subtle feature is so-called quantum mechanical exchange. Consider the energy of interelectronic repulsion for the wave function of Eq. (4.43). We evaluate this as

$$\int {}^3\Psi_{SD} \frac{1}{r_{12}} {}^3\Psi_{SD} d\mathbf{r}_1 d\omega_1 d\mathbf{r}_2 d\omega_2$$

$$= \frac{1}{2} \left[\int |\psi_a(1)|^2 |\alpha(1)|^2 \frac{1}{r_{12}} |\psi_b(2)|^2 |\alpha(2)|^2 d\mathbf{r}_1 d\omega_1 d\mathbf{r}_2 d\omega_2 \right.$$

$$- 2 \int \psi_a(1)\psi_b(1) |\alpha(1)|^2 \frac{1}{r_{12}} \psi_b(2)\psi_a(2) |\alpha(2)|^2 d\mathbf{r}_1 d\omega_1 d\mathbf{r}_2 d\omega_2$$

$$\left. + \int |\psi_a(2)|^2 |\alpha(2)|^2 \frac{1}{r_{12}} |\psi_b(1)|^2 |\alpha(1)|^2 d\mathbf{r}_1 d\omega_1 d\mathbf{r}_2 d\omega_2 \right]$$

$$= \frac{1}{2} \left[\int |\psi_a(1)|^2 \frac{1}{r_{12}} |\psi_b(2)|^2 d\mathbf{r}_1 d\mathbf{r}_2 \right.$$

$$- 2 \int \psi_a(1)\psi_b(1) \frac{1}{r_{12}} \psi_b(2)\psi_a(2) d\mathbf{r}_1 d\mathbf{r}_2$$

$$\left. + \int |\psi_a(2)|^2 \frac{1}{r_{12}} |\psi_b(1)|^2 d\mathbf{r}_1 d\mathbf{r}_2 \right]$$

$$= \frac{1}{2} \left(J_{ab} - 2 \int \psi_a(1)\psi_b(1) \frac{1}{r_{12}} \psi_a(2)\psi_b(2) d\mathbf{r}_1 d\mathbf{r}_2 + J_{ab} \right)$$

$$= J_{ab} - K_{ab} \tag{4.49}$$

Equation (4.49) indicates that for this wave function the classical Coulomb repulsion between the electron clouds in orbitals a and b is reduced by K_{ab}, where the definition of this integral may be inferred from comparing the third equality to the fourth. This fascinating consequence of the Pauli principle reflects the reduced probability of finding two electrons of the same spin close to one another – a so-called 'Fermi hole' is said to surround each electron.

Note that this property is a correlation effect *unique to electrons of the same spin*. If we consider the contrasting Slater determinantal wave function formed from different spins

$$\Psi_{SD} = \frac{1}{\sqrt{2}} [\psi_a(1)\alpha(1)\psi_b(2)\beta(2) - \psi_a(2)\alpha(2)\psi_b(1)\beta(1)] \tag{4.50}$$

and carry out the same evaluation of interelectronic repulsion we have

$$
\int \Psi_{SD} \frac{1}{r_{12}} \Psi_{SD} d\mathbf{r}_1 d\omega_1 d\mathbf{r}_2 d\omega_2
$$

$$
= \frac{1}{2} \left[\int |\psi_a(1)|^2 |\alpha(1)|^2 \frac{1}{r_{12}} |\psi_b(2)|^2 |\beta(2)|^2 \, d\mathbf{r}_1 d\omega_1 d\mathbf{r}_2 d\omega_2 \right.
$$

$$
- 2 \int \psi_a(1)\psi_b(1)\alpha(1)\beta(1) \frac{1}{r_{12}} \psi_b(2)\psi_a(2)\alpha(2)\beta(2) d\mathbf{r}_1 d\omega_1 d\mathbf{r}_2 d\omega_2
$$

$$
\left. + \int |\psi_a(2)|^2 |\alpha(2)|^2 \frac{1}{r_{12}} |\psi_b(1)|^2 |\beta(1)|^2 \, d\mathbf{r}_1 d\omega_1 d\mathbf{r}_2 d\omega_2 \right]
$$

$$
= \frac{1}{2} \left[\int |\psi_a(1)|^2 \frac{1}{r_{12}} |\psi_b(2)|^2 \, d\mathbf{r}_1 d\mathbf{r}_2 \right.
$$

$$
- 2 \cdot 0
$$

$$
\left. + \int |\psi_a(2)|^2 \frac{1}{r_{12}} |\psi_b(1)|^2 \, d\mathbf{r}_1 d\mathbf{r}_2 \right]
$$

$$
= \frac{1}{2}(J_{ab} + J_{ab})
$$

$$
= J_{ab} \tag{4.51}
$$

Note that the disappearance of the exchange correlation derives from the orthogonality of the α and β spin functions, which causes the second integral in the second equality to be zero when integrated over either spin coordinate.

4.5.5　The Hartree-Fock Self-consistent Field Method

Fock first proposed the extension of Hartree's SCF procedure to Slater determinantal wave functions. Just as with Hartree product orbitals, the HF MOs can be individually determined as eigenfunctions of a set of one-electron operators, but now the interaction of each electron with the static field of all of the other electrons (this being the basis of the SCF approximation) includes exchange effects on the Coulomb repulsion. Some years later, in a paper that was critical to the further development of practical computation, Roothaan described matrix algebraic equations that permitted HF calculations to be carried out using a basis set representation for the MOs (Roothaan 1951; for historical insights, see Zerner 2000). We will forego a formal derivation of all aspects of the HF equations, and simply present them in their typical form for closed-shell systems (i.e., all electrons spin-paired, two per occupied orbital) with wave functions represented as a single Slater determinant. This formalism is called 'restricted Hartree-Fock' (RHF); alternative formalisms are discussed in Chapter 6.

The one-electron Fock operator is defined for each electron i as

$$
f_i = -\frac{1}{2}\nabla_i^2 - \sum_k^{\text{nuclei}} \frac{Z_k}{r_{ik}} + V_i^{HF}\{j\} \tag{4.52}
$$

where the final term, the HF potential, is $2J_i - K_i$, and the J_i and K_i operators are defined so as to compute the J_{ij} and K_{ij} integrals previously defined above. To determine the MOs using the Roothaan approach, we follow a procedure analogous to that previously described for Hückel theory. First, given a set of N basis functions, we solve the secular equation

$$
\begin{vmatrix}
F_{11} - ES_{11} & F_{12} - ES_{12} & \cdots & F_{1N} - ES_{1N} \\
F_{21} - ES_{21} & F_{22} - ES_{22} & \cdots & F_{2N} - ES_{2N} \\
\vdots & \vdots & \ddots & \vdots \\
F_{N1} - ES_{N1} & F_{N2} - ES_{N2} & \cdots & F_{NN} - ES_{NN}
\end{vmatrix} = 0
\tag{4.53}
$$

to find its various roots E_j. In this case, the values for the matrix elements F and S are computed explicitly.

Matrix elements S are the overlap matrix elements we have seen before. For a general matrix element $F_{\mu\nu}$ (we here adopt a convention that basis functions are indexed by lower-case Greek letters, while MOs are indexed by lower-case Roman letters) we compute

$$
F_{\mu\nu} = \left\langle \mu \left| -\frac{1}{2}\nabla^2 \right| v \right\rangle - \sum_k^{\text{nuclei}} Z_k \left\langle \mu \left| \frac{1}{r_k} \right| v \right\rangle
$$
$$
+ \sum_{\lambda\sigma} P_{\lambda\sigma} \left[(\mu\nu|\lambda\sigma) - \frac{1}{2}(\mu\lambda|v\sigma) \right]
\tag{4.54}
$$

The notation $\langle \mu|g|v \rangle$ where g is some operator which takes basis function ϕ_v as its argument, implies a so-called one-electron integral of the form

$$
\langle \mu|g|v \rangle = \int \phi_\mu (g\phi_v) d\mathbf{r}.
\tag{4.55}
$$

Thus, for the first term in Eq. (4.54) g involves the Laplacian operator and for the second term g is the distance operator to a particular nucleus. The notation $(\mu\nu|\lambda\sigma)$ also implies a specific integration, in this case

$$
(\mu\nu|\lambda\sigma) = \int\int \phi_\mu(1)\phi_\nu(1)\frac{1}{r_{12}}\phi_\lambda(2)\phi_\sigma(2)d\mathbf{r}(1)d\mathbf{r}(2)
\tag{4.56}
$$

where ϕ_μ and ϕ_ν represent the probability density of one electron and ϕ_λ and ϕ_σ the other. The exchange integrals $(\mu\lambda|v\sigma)$ are preceded by a factor of 1/2 because they are limited to electrons of the same spin while Coulomb interactions are present for any combination of spins.

The final sum in Eq. (4.54) weights the various so-called 'four-index integrals' by elements of the 'density matrix' $\mathbf{P}$. This matrix in some sense describes the degree to which individual basis functions contribute to the many-electron wave function, and thus how energetically important the Coulomb and exchange integrals should be (i.e., if a basis function fails to contribute in a significant way to any occupied MO, clearly integrals involving that basis function should be of no energetic importance). The elements of **P** are computed as

$$P_{\lambda\sigma} = 2 \sum_{i}^{\text{occupied}} a_{\lambda i} a_{\sigma i} \tag{4.57}$$

where the coefficients $a_{\zeta i}$ specify the (normalized) contribution of basis function ζ to MO i and the factor of two appears because with RHF theory we are considering only singlet wave functions in which all orbitals are doubly occupied.

While the process of solving the HF secular determinant to find orbital energies and coefficients is quite analogous to that already described above for effective Hamiltonian methods, it is characterized by the same paradox present in the Hartree formalism. That is, we need to know the orbital coefficients to form the density matrix that is used in the Fock matrix elements, but the purpose of solving the secular equation is to determine those orbital coefficients. So, just as in the Hartree method, the HF method follows a SCF procedure, where first we guess the orbital coefficients (e.g., from an effective Hamiltonian method) and then we iterate to convergence. The full process is described schematically by the flow chart in Figure 4.3. The energy of the HF wavefunction can be computed in a fashion analogous to Eq. (4.39).

Hartree–Fock theory as constructed using the Roothaan approach is quite beautiful in the abstract. This is not to say, however, that it does not suffer from certain chemical and practical limitations. Its chief chemical limitation is the one-electron nature of the Fock operators. Other than exchange, all electron correlation is ignored. It is, of course, an interesting question to ask just how important such correlation is for various molecular properties, and we will examine that in some detail in following chapters.

Furthermore, from a practical standpoint, HF theory posed some very challenging technical problems to early computational chemists. One problem was choice of a basis set. The LCAO approach using hydrogenic orbitals remains attractive in principle; however, this basis set requires numerical solution of the four-index integrals appearing in the Fock matrix elements, and that is a very tedious process. Moreover, the *number* of four-index integrals is daunting. Since each index runs over the total number of basis functions, there are in principle N^4 total integrals to be evaluated, and this quartic scaling behavior with respect to basis-set size proves to be the bottleneck in HF theory applied to essentially any molecule.

Historically, two philosophies began to emerge at this stage with respect to how best to make further progress. The first philosophy might be summed up as follows: The HF equations are very powerful but still, after all, chemically flawed. Thus, other approximations that may be introduced to simplify their solution, and possibly at the same time improve their accuracy (by some sort of parameterization to reproduce key experimental quantities), are well justified. Many computational chemists continue to be guided by this philosophy today, and it underlies the motivation for so-called 'semiempirical' MO theories, which are discussed in detail in the next chapter.

The second philosophy essentially views HF theory as a stepping stone on the way to exact solution of the Schrödinger equation. HF theory provides a very well defined energy, one which can be converged in the limit of an infinite basis set, and the difference between that

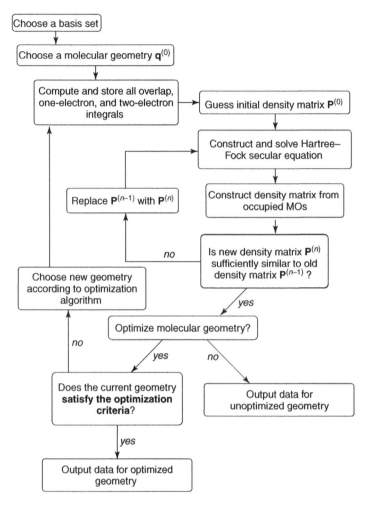

Figure 4.3   Flow chart of the HF SCF procedure. Note that data for an unoptimized geometry is referred to as deriving from a so-called 'single-point calculation'

converged energy and reality is the electron correlation energy (ignoring relativity, spin–orbit coupling, etc.). It was anticipated that developing the technology to achieve the HF limit *with no further approximations* would not only permit the evaluation of the chemical utility of the HF limit, but also probably facilitate moving on from that base camp to the Schrödinger equation summit. Such was the foundation for further research on '*ab initio*' HF theory, which forms the subject of Chapter 6.

Bibliography and Suggested Additional Reading

Frenking, G. 2000. "Perspective on 'Quantentheoretische Beiträge zum Benzolproblem. I. Die Elektronenkonfiguration des Benzols und verwandter Beziehungen" *Theor. Chem. Acc.*, **103**, 187.

Hehre, W. J., Radom, L., Schleyer, P. v. R., and Pople, J. A. 1986. Ab Initio *Molecular Orbital Theory*, Wiley: New York.

Levine, I. N. 2000. *Quantum Chemistry*, 5th Edn., Prentice Hall: New York.

Lowry, T. H. and Richardson, K. S. 1981. *Mechanism and Theory in Organic Chemistry*, 2nd Edn., Harper & Row: New York, 82–112.

Szabo, A. and Ostlund, N. S. 1982. *Modern Quantum Chemistry*, Macmillan: New York.

References

Berson, J. A. 1996. *Angew. Chem., Int. Ed. Engl.*, **35**, 2750.

Frenking. G. 2000. *Theor. Chem. Acc.*, **103**, 187.

Gobbi, A. and Frenking, G. 1994. *J. Am. Chem. Soc.*, **116**, 9275.

Hartree, D. R. 1928. *Proc. Cambridge Phil. Soc.*, **24**, 89, 111, 426.

Hückel, E. 1931. *Z. Phys.*, **70**, 204.

MacDonald, J. K. L. 1933. *Phys. Rev.*, **43**, 830.

Mo, Y. R., Lin, Z. Y., Wu, W., and Zhang, Q. N. 1996. *J. Phys. Chem.*, **100**, 6469.

Roothaan, C. C. J. 1951. *Rev. Mod. Phys.*, **23**, 69.

Slater, J. C., 1930. *Phys. Rev.*, **35**, 210.

Zerner, M. C. 2000. *Theor. Chem. Acc.*, **103**, 217.

5

Semiempirical Implementations of Molecular Orbital Theory

5.1 Semiempirical Philosophy

In the last chapter, the full formalism of Hartree–Fock theory was developed. While this theory is impressive as a physical and mathematical construct, it has several limitations in a practical sense. Particularly during the early days of computational chemistry, when computational power was minimal, carrying out HF calculations without any further approximations, even for small systems with small basis sets, was a challenging task.

In spite of the technical hurdles, however, many chemists recognized the potentially critical role that theory could play in furthering experimental progress on any number of fronts. And the interests of that population of chemists were by no means restricted to molecules composed of only a small handful of atoms. Accepting HF theory as a framework, several research groups turned their attention to implementations of the theory that would make it more tractable, and perhaps more accurate, for molecules of moderate size. These steps 'sideways', if you will, led to a certain bifurcation of effort in the area of molecular orbital theory (although certainly some research groups pursued topics in both directions) that persists to this day. Semiempirical calculations continue to appear in large numbers in the chemical literature; since there will always be researchers interested in molecules that exceed the size of those practically accessible by *ab initio* methods, semiempirical levels of MO theory are certain to continue to be developed and applied. This chapter describes the underlying approximations of semiempirical methods (organizing them roughly in chronological order of appearance) and provides detailed comparisons between methods now in common use for the prediction of various chemical properties. Section 5.7 describes recent developments in the area of improving/extending semiempirical models.

5.1.1 Chemically Virtuous Approximations

Let us consider how one might go about making formal Hartree–Fock theory less computationally intensive without necessarily sacrificing its accuracy. The most demanding step

Essentials of Computational Chemistry, 2nd Edition Christopher J. Cramer
© 2004 John Wiley & Sons, Ltd   ISBNs: 0-470-09181-9 (cased); 0-470-09182-7 (pbk)

of an HF calculation, in terms of computational resources, is the assembly of the two-electron (also called four-index) integrals, i.e., the J and K integrals appearing in the Fock matrix elements defined by Eq. (4.54). Not only is numerical solution of the integrals for an arbitrary basis set arduous, but there are so many of them (formally N^4 where N is the number of basis functions). One way to save time would be to estimate their value accurately in an *a priori* fashion, so that no numerical integration need be undertaken.

For which integrals is it easiest to make such an estimation? To answer that question, it is helpful to keep in mind the intuitive meaning of the integrals. Coulomb integrals measure the repulsion between two electrons in regions of space defined by the basis functions. It seems clear, then, that when the basis functions in the integral for one electron are very far from the basis functions for the other, the value of that integral will approach zero (the same holds true for the one-electron integrals describing nuclear attraction, i.e., if the basis functions for the electron are very far from the nucleus the attraction will go to zero, but these integrals are much less computationally demanding to solve). In a large molecule, then, one might be able to avoid the calculation of a very large number of integrals simply by assuming them to be zero, and one would still have a reasonable expectation of obtaining a Hartree–Fock energy close to that that would be obtained from a full calculation.

Such an approximation is what we might call a numerical approximation. That is, it introduces error to the extent that values employed are not exact, but the calculation can be converged to arbitrary accuracy by tightening the criteria for employing the approximation, e.g., in the case of setting certain two-electron integrals to zero, the threshold could be the average inter-basis-function distance, so that in the limit of choosing a distance of infinity, one recovers exact HF theory. Other approximations in semiempirical theory, however, are guided by a slightly different motivation, and these approximations might be well referred to as 'chemically virtuous approximations'. It is important to keep in mind that HF wave functions for systems having two or more electrons are *not* eigenfunctions of the corresponding non-relativistic Schrödinger equations. Because of the SCF approximation for how each electron interacts with all of the others, some electronic correlation is ignored, and the HF energy is necessarily higher than the exact energy.

How important is the correlation energy? Let us consider a very simple system: the helium atom. The energy of this two-electron system in the HF limit (i.e., converged with respect to basis-set size for the number of digits reported) is $-2.861\,68$ a.u. (Clementi and Roetti 1974). The exact energy for the helium atom, on the other hand, is $-2.903\,72$ a.u. (Pekeris 1959). The difference is $0.042\,04$ a.u., which is about 26 kcal mol^{-1}. Needless to say, as systems increase in size, greater numbers of electrons give rise to considerably larger correlation energies – hundreds or thousands of kcal mol^{-1} for moderately sized organic and inorganic molecules.

At first glance, this is a terrifying observation. At room temperature (298 K), it requires a change of 1.4 kcal mol^{-1} in a free energy of reaction to change an equilibrium constant by an order of magnitude. Similarly, a change of 1.4 kcal mol^{-1} in a rate-determining free energy of activation will change the rate of a chemical reaction by an order of magnitude. Thus, chemists typically would prefer theoretical accuracies to be no worse than 1.4 kcal mol^{-1}, so that room-temperature predictions can be trusted at least to within an order of magnitude (and obviously it would be nice to do much better). How then can we ever hope to use a theory that is intrinsically inaccurate by hundreds or thousands of kilocalories per mole to make chemically useful predictions? Michael J. S. Dewar, who made many contributions in the area of semiempirical MO theory, once offered the following analogy to using HF theory to make chemical predictions: It is like weighing the captain of a ship by first weighing the ship with the captain on board, then weighing the ship without her, and then taking the difference – the errors in the individual measurements are likely to utterly swamp the small difference that is the goal of the measuring.

In practice, as we shall see in Chapter 6, the situation with HF theory is not really as bad as our above analysis might suggest. Errors from neglecting correlation energy cancel to a remarkable extent in favorable instances, so that chemically useful interpretations of HF calculations *can* be valid. Nevertheless, the intrinsic inaccuracy of *ab initio* HF theory suggests that modifications of the theory introduced in order to simplify its formalism may actually *improve on* a rigorous adherence to the full mathematics, provided the new 'approximations' somehow introduce an accounting for correlation energy. Since this improves chemical accuracy, at least in intent, we may call it a chemically virtuous approximation. Most typically, such approximations involve the adoption of a parametric form for some aspect of the calculation where the parameters involved are chosen so as best to reproduce experimental data – hence the term 'semiempirical'.

5.1.2 Analytic Derivatives

If it is computationally demanding to carry out a single electronic structure calculation, how much more daunting to try to optimize a molecular geometry. As already discussed in detail in Section 2.4, chemists are usually interested not in arbitrary structures, but in stationary points on the potential energy surface. In order to find those points efficiently, many of the optimization algorithms described in Section 2.4 make use of derivatives of the energy with respect to nuclear motion – when those derivatives are available analytically, instead of numerically, rates of convergence are typically enhanced.

This is particularly true when the stationary point of interest is a transition-state structure. Unlike the case with molecular mechanics, the HF energy has no obvious bias for minimum-energy structures compared to TS structures – one of the most exciting aspects of MO theory, whether semiempirical or *ab initio*, is that it provides an energy functional from which reasonable TS structures may be identified. However, in the early second half of the twentieth century, it was not at all obvious how to compute analytic derivatives of the HF energy with respect to nuclear motion. Thus, another motivation for introducing semiempirical approximations into HF theory was to facilitate the computation of derivatives so that geometries could be more efficiently optimized. Besides the desire to attack TS geometries, there were also very practical motivations for geometry optimization. In the early days of semiempirical parameterization, experimental structural data were about as widely available as energetic data, and parameterization of semiempirical methods against both kinds of data would be expected to generate a more robust final model.

5.2   Extended Hückel Theory

Prior to considering semiempirical methods designed on the basis of HF theory, it is instructive to revisit one-electron effective Hamiltonian methods like the Hückel model described in Section 4.4. Such models tend to involve the most drastic approximations, but as a result their rationale is tied closely to experimental concepts and they tend to be intuitive. One such model that continues to see extensive use today is the so-called extended Hückel theory (EHT). Recall that the key step in finding the MOs for an effective Hamiltonian is the formation of the secular determinant for the secular equation

$$
\begin{vmatrix}
H_{11} - E S_{11} & H_{12} - E S_{12} & \dots & H_{1N} - E S_{1N} \\
H_{21} - E S_{21} & H_{22} - E S_{22} & \dots & H_{2N} - E S_{2N} \\
\vdots & \vdots & \ddots & \vdots \\
H_{N1} - E S_{N1} & H_{N2} - E S_{N2} & \dots & H_{NN} - E S_{NN}
\end{vmatrix} = 0
\tag{5.1}
$$

The dimension of the secular determinant for a given molecule depends on the choice of basis set. EHT adopts two critical conventions. First, all core electrons are ignored. It is assumed that core electrons are sufficiently invariant to differing chemical environments that changes in their orbitals as a function of environment are of no chemical consequence, energetic or otherwise. *All modern semiempirical methodologies make this approximation.* In EHT calculations, if an atom has occupied d orbitals, typically the highest occupied level of d orbitals is considered to contribute to the set of valence orbitals.

Each remaining valence orbital is represented by a so-called Slater-type orbital (STO). The mathematical form of a normalized STO used in EHT (in atom-centered polar coordinates) is

$$
\varphi(r, \theta, \phi; \zeta, n, l, m) = \frac{(2\zeta)^{n+1/2}}{[(2n)!]^{1/2}} r^{n-1} e^{-\zeta r} Y_l^m (\theta, \phi)
\tag{5.2}
$$

where ζ is an exponent that can be chosen according to a simple set of rules developed by Slater that depend, *inter alia*, on the atomic number (Slater 1930), n is the principal quantum number for the valence orbital, and the spherical harmonic functions $Y_l^m (\theta, \phi)$, depending on the angular momentum quantum numbers l and m, are those familiar from solution of the Schrödinger equation for the hydrogen atom and can be found in any standard quantum mechanics text. Thus, the size of the secular determinant in Eq. (5.2) is dictated by the total number of valence orbitals in the molecule. For instance, the basis set for the MnO_4^- anion would include a total of 25 STO basis functions: one 2s and three 2p functions for each oxygen (for a subtotal of 16) and one 4s, three 4p, and five 3d functions for manganese.

STOs have a number of features that make them attractive. The orbital has the correct exponential decay with increasing r, the angular component is hydrogenic, and the 1s orbital has, as it should, a cusp at the nucleus (i.e., it is not smooth). More importantly, from a practical point of view, overlap integrals between two STOs as a function of interatomic distance are readily computed (Mulliken Rieke and Orloff 1949; Bishop 1966). Thus, in contrast to simple Hückel theory, overlap matrix elements in EHT are not assumed to be equal to the Kronecker delta, but are directly computed in every instance.

The only terms remaining to be defined in Eq. (5.1), then, are the resonance integrals H. For diagonal elements, the same convention is used in EHT as was used for simple Hückel theory. That is, the value for $H_{\mu\mu}$ is taken as the negative of the average ionization potential for an electron in the appropriate valence orbital. Thus, for instance, when μ is a hydrogen 1s function, $H_{\mu\mu} = -13.6$ eV. Of course in many-electron atoms, the valence-shell ionization potential (VSIP) for the ground-state atomic term may not necessarily be the best choice for the atom in a molecule, so this term is best regarded as an adjustable parameter, although one with a clear, physical basis. VSIPs have been tabulated for most of the atoms in the periodic table (Pilcher and Skinner 1962; Hinze and Jaffé 1962; Hoffmann 1963; Cusachs, Reynolds and Barnard 1966). Because atoms in molecular environments may develop fairly large partial charges depending on the nature of the atoms to which they are connected, schemes for adjusting the neutral atomic VSIP as a function of partial atomic charge have been proposed (Rein *et al.* 1966; Zerner and Gouterman 1966). Such an adjustment scheme characterizes so-called Fenske–Hall effective Hamiltonian calculations, which still find considerable use for inorganic and organometallic systems composed of atoms having widely different electronegativities (Hall and Fenske 1972).

The more difficult resonance integrals to approximate are the off-diagonal ones. Wolfsberg and Helmholtz (1952) suggested the following convention

$$H_{\mu\nu} = \frac{1}{2} C_{\mu\nu} (H_{\mu\mu} + H_{\nu\nu}) S_{\mu\nu} \tag{5.3}$$

where C is an empirical constant and S is the overlap integral. Thus, the energy associated with the matrix element is proportional to the average of the VSIPs for the two orbitals μ and ν times the extent to which the two orbitals overlap in space (note that, by symmetry, the overlap between different STOs on the same atom is zero). Originally, the constant C was given a different value for matrix elements corresponding to σ- and π-type bonding interactions. In modern EHT calculations, it is typically taken as 1.75 for *all* matrix elements, although it can still be viewed as an adjustable parameter when such adjustment is warranted.

All of the above conventions together permit the complete construction of the secular determinant. Using standard linear algebra methods, the MO energies and wave functions can be found from solution of the secular equation. Because the matrix elements do not depend on the final MOs in any way (unlike HF theory), the process is not iterative, so it is very fast, even for very large molecules (however, the process *does* become iterative if VSIPs are adjusted as a function of partial atomic charge as described above, since the partial atomic charge depends on the occupied orbitals, as described in Chapter 9).

The very approximate nature of the resonance integrals in EHT makes it insufficiently accurate for the generation of PESs since the locations of stationary points are in general very poorly predicted. Use of EHT is thus best restricted to systems for which experimental geometries are available. For such cases, EHT tends to be used today to generate qualitatively correct MOs, in much the same fashion as it was used by Wolfsberg and Helmholz 50 years ago. Wolfsberg and Helmholz used their model to explain differences in the UV spectroscopies of MnO_4^-, CrO_4^{2-}, and ClO_4^- by showing how the different VSIPs of the central atom and differing bond lengths gave rise to different energy separations between the relevant filled and empty orbitals in spectroscopic transitions. In the 21st century, such a molecular problem has become amenable to more accurate treatments, so the province of EHT is now primarily very large systems, like extended solids, where its speed makes it a practical option for understanding band structure (a 'band' is a set of MOs so densely spread over a range of energy that for practical purposes it may be regarded as a continuum; bands derive from combinations of molecular orbitals in a solid much as MOs derive from combinations of AOs in a molecule).

Thus, for example, EHT has been used by Genin and Hoffmann (1998) to characterize the band structure of a series of organic polymers with the intent of suggesting likely candidates for materials exhibiting organic ferromagnetism. Certain polymers formed from repeating heterocycle units having seven π electrons were identified as having narrow, half-filled valence bands, such bands being proposed as a necessary, albeit not sufficient, condition for ferromagnetism.

Note that one drawback of EHT is a failure to take into account electron spin. There is no mechanism for distinguishing between different multiplets, except that a chemist can, by hand, decide which orbitals are occupied, and thus enforce the Pauli exclusion principle. However, the energy computed for a triplet state is exactly the same as the energy for the corresponding 'open-shell' singlet (i.e., the state that results from spin-flip of one of the unpaired electrons in the triplet) – the electronic energy is the sum of the occupied orbital energies irrespective of spin – such an equality occurs experimentally only when the partially occupied orbitals fail to interact with each other either for symmetry reasons or because they are infinitely separated.

5.3   CNDO Formalism

Returning to the SCF formalism of HF theory, one can proceed in the spirit of an effective Hamiltonian method by developing a recipe for the replacement of matrix elements in the HF secular equation, Eq. (4.53). One of the first efforts along these lines was described by Pople and co-workers in 1965 (Pople, Santry, and Segal 1965; Pople and Segal 1965). The complete neglect of differential overlap (CNDO) method adopted the following conventions:

1.   Just as in EHT, the basis set is formed from valence STOs, one STO per valence orbital. In the original CNDO implementation, only atoms having s and p valence orbitals were addressed.

2.   In the secular determinant, overlap matrix elements are defined by

$$S_{\mu\nu} = \delta_{\mu\nu} \tag{5.4}$$

where δ is the Kronecker delta.

3.   All two-electron integrals are parameterized according to the following scheme. First, define

$$(\mu\nu|\lambda\sigma) = \delta_{\mu\nu}\delta_{\lambda\sigma}(\mu\mu|\lambda\lambda) \tag{5.5}$$

Thus, the only integrals that are non-zero have μ and ν as identical orbitals on the same atom, and λ and σ also as identical orbitals on the same atom, but the second atom might be different than the first (the decision to set to zero any integrals involving overlap of different basis functions gives rise to the model name).

4. For the surviving two-electron integrals,

$$(\mu\mu|\lambda\lambda) = \gamma_{AB} \tag{5.6}$$

where A and B are the atoms on which basis functions μ and λ reside, respectively. The term γ can either be computed explicitly from s-type STOs (note that since γ depends only on the atoms A and B, $(s_A s_A | s_B s_B) = (p_A p_A | s_B s_B) = (p_A p_A | p_B p_B)$, etc.) or it can be treated as a parameter. One popular parametric form involves using the so-called Pariser–Parr approximation for the one-center term (Pariser and Parr 1953).

$$\gamma_{AA} = IP_A - EA_A \tag{5.7}$$

where IP and EA are the atomic ionization potential and electron affinity, respectively. For the two-center term, the Mataga–Nishimoto formalism adopts

$$\gamma_{AB} = \frac{\gamma_{AA} + \gamma_{BB}}{2 + r_{AB}(\gamma_{AA} + \gamma_{BB})} \tag{5.8}$$

where r_{AB} is the interatomic distance (Mataga and Nishimoto 1957). Note the intuitive limits on γ in Eq. (5.8). At large distance, it goes to $1/r_{AB}$, as expected for widely separated charge clouds, while at short distances, it approaches the average of the two one-center parameters.

5. One-electron integrals for diagonal matrix elements are defined by

$$\left\langle \mu \left| -\frac{1}{2}\nabla^2 - \sum_k \frac{Z_k}{r_k} \right| \mu \right\rangle = -IP_\mu - \sum_k (Z_k - \delta_{Z_A Z_k})\gamma_{Ak} \tag{5.9}$$

where μ is centered on atom A. Equation (5.9) looks a bit opaque at first glance, but it is actually quite straightforward. Remember that the full Fock matrix element $F_{\mu\mu}$ is the *sum* of the one-electron integral Eq. (5.9) and a series of two-electron integrals. If the number of valence electrons on each atom is exactly equal to the valence nuclear charge (i.e., every atom has a partial atomic charge of zero) then the repulsive two-electron terms will exactly cancel the attractive nuclear terms appearing at the end of Eq. (5.9) and we will recapture the expected result, namely that the energy associated with the diagonal matrix element is the ionization potential of the orbital. The Kronecker delta affecting the nuclear charge for atom A itself simply avoids correcting for a non-existent two-electron repulsion of an electron in basis function μ with itself. (Removing the attraction to nuclei *other* than A from the r.h.s. of Eq. (5.9) defines a commonly tabulated semiempirical parameter that is typically denoted U_μ.)

6.   The only terms remaining to be defined in the assembly of the HF secular determinant are the one-electron terms for off-diagonal matrix elements. These are defined as

$$\left\langle \mu \left| -\frac{1}{2}\nabla^2 - \sum_k \frac{Z_k}{r_k} \right| \nu \right\rangle = \frac{(\beta_A + \beta_B)S_{\mu\nu}}{2} \tag{5.10}$$

where μ and ν are centered on atoms A and B, respectively, the β values are semiempirical parameters, and $S_{\mu\nu}$ is the overlap matrix element computed using the STO basis set. Note that computation of overlap is carried out for every combination of basis functions, even though in the secular determinant itself S is defined by Eq. (5.4). There are, in effect, two different S matrices, one for each purpose. The β parameters are entirely analogous to the parameter of the same name we saw in Hückel theory – they provide a measure of the strength of through space interactions between atoms. As they are intended for completely general use, it is not necessarily obvious how to assign them a numerical value, unlike the situation that obtains in Hückel theory. Instead, β values for CNDO were originally adjusted to reproduce certain experimental quantities.

While the CNDO method may appear to be moderately complex, it represents a vast simplification of HF theory. Equation (5.5) reduces the number of two-electron integrals having non-zero values from formally N^4 to simply N^2. Furthermore, those N^2 integrals are computed by trivial algebraic formulae, not by explicit integration, and between any pair of atoms all of the integrals have the same value irrespective of the atomic orbitals involved. Similarly, evaluation of one-electron integrals is also entirely avoided, with numerical values for those portions of the relevant matrix elements coming from easily evaluated formulae. Historically, a number of minor modifications to the conventions outlined above were explored, and the different methods had names like CNDO/1, CNDO/2, CNDO/BW, etc.; as these methods are all essentially obsolete, we will not itemize their differences. One CNDO model that does continue to see some use today is the Pariser–Parr–Pople (PPP) model for conjugated π systems (Pariser and Parr 1953; Pople 1953). It is in essence the CNDO equivalent of Hückel theory (only π-type orbitals are included in the secular equation), and improves on the latter theory in the prediction of electronic state energies.

The computational simplifications inherent in the CNDO method are not without chemical cost, as might be expected. Like EHT, CNDO is quite incapable of accurately predicting good molecular structures. Furthermore, the simplification inherent in Eq. (5.6) has some fairly dire consequences; two examples are illustrated in Figure 5.1. Consider the singlet and triplet states of methylene. Clearly, repulsion between the two highest energy electrons in each state should be quite different: they are spin-paired in an sp^2 orbital for the singlet, and spin-parallel, one in the sp^2 orbital and one in a p orbital, for the triplet. However, in each case the interelectronic Coulomb integral, by Eq. (5.6), is simply γ_{CC}. And, just as there is no distinguishing between different types of atomic orbitals, there is also no distinguishing between the orientation of those orbitals. If we consider the rotational coordinate for hydrazine, it is clear that one factor influencing the energetics will be the repulsion of the two lone pairs, one

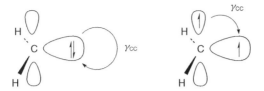

no distinction in two-electron repulsions

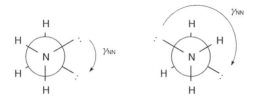

Figure 5.1 The CNDO formalism for estimating repulsive two-electron interactions fails to distinguish in one-center cases between different orbitals (top example for the case of methylene) and in two-center cases either between different orbitals or different orbital orientations (bottom example for the case of hydrazine)

on each nitrogen. However, we see from Eq. (5.8) that this repulsion, γ_{NN}, depends only on the distance separating the two nitrogen atoms, not on the orientation of the lone pair orbitals.

5.4 INDO Formalism

5.4.1 INDO and INDO/S

Of the two deficiencies specifically noted above for CNDO, the methylene problem is atomic in nature – it involves electronic interactions on a single center – while the hydrazine problem is molecular insofar as it involves two centers. Many ultraviolet/visible (UV/Vis) spectroscopic transitions in molecules are reasonably highly localized to a single center, e.g., transitions in mononuclear inorganic complexes. Pople, Beveridge, and Dobosh (1967) suggested modifications to the CNDO formalism to permit a more flexible handling of electron–electron interactions on the same center in order to model such spectroscopic transitions, and referred to this new formalism as 'intermediate neglect of differential overlap' (INDO). The key change is simply to use different values for the unique one-center two-electron integrals. When the atom is limited to a basis set of s and p orbitals, there are five such unique integrals

$$(ss|ss) = G_{ss}$$
$$(ss|pp) = G_{sp}$$
$$(pp|pp) = G_{pp}$$
$$(pp|p'p') = G_{pp'}$$
$$(sp|sp) = L_{sp} \tag{5.11}$$

The G and L values may be regarded as free parameters, but in practice they can be estimated from spectroscopic data. When the atomic valence orbitals include d and f functions, the number of unique integrals increases considerably, and the estimation of appropriate values from spectroscopy becomes considerably more complicated.

One effect of the greater flexibility inherent in the INDO scheme is that valence bond angles are predicted with much greater accuracy than is the case for CNDO. Nevertheless, overall molecular geometries predicted from INDO tend to be rather poor (although preliminary efforts to address this problem have been reported by Da Motta Neto and Zerner 2001). However, if a good molecular geometry is available from some other source (ideally experiment) the INDO method has considerable potential for modeling the UV/Vis spectroscopy of the compound because of its better treatment of one-center electronic interactions.

Ridley and Zerner (1973) first described a careful parameterization of INDO specifically for spectroscopic problems, and designated that model INDO/S. Over the course of many years, Zerner and co-workers extended the model to most of the elements in the periodic table, including the lanthanides (Kotzian, Rösch, and Zerner 1992), although few available modern codes appear to include parameters for elements having f electrons, possibly because of challenges associated with accounting for relativistic effects, especially spin–orbit coupling, which cannot be ignored when such heavy atoms are involved. Table 5.1 lists the energetic separations between various electronic states for three cases studied by INDO/S, ranging from the organic molecule pyridine, to the transition metal complex $Cr(H_2O)_6{}^{3+}$, to the metalloenzyme oxyhemocyanin which has a bimetallic Cu_2O_2 core ligated by enzyme histidine residues. Even just the ligated core of the latter system is daunting in size, but the simplifications intrinsic in semiempirical MO theory render it tractable. All of the geometries used

Table 5.1 Relative state energies (units of 1000 cm^{-1}) as computed by the INDO/S model

System (ground state)	State (transition)	INDO/S prediction	Experiment
Pyridine $(^1A_1)^a$	$^1B_1\ (n \rightarrow \pi^*)$	34.7	35.0
	$^1B_2\ (\pi \rightarrow \pi^*)$	38.6	38.4
	$^1A_2\ (n \rightarrow \pi^*)$	43.9	–
	$^1A_1\ (\pi \rightarrow \pi^*)$	49.7	49.8
	$^1A_1\ (\pi \rightarrow \pi^*)$	56.9	55.0
$Cr(H_2O)_6{}^{3+}\ (^4A_{1g})^b$	$^4T_{2g}\ (t \rightarrow e)$	12.4	12.4
	$^4T_{1g}\ (t \rightarrow e)$	17.5	18.5
	$^2T_{1g}\ (t \rightarrow t)$	13.2	13.1
	$^2E_g\ (t \rightarrow t)$	13.6	13.1
Oxyhemocyanin^c	d→d	15.0	14.3–15.0
	$\pi \rightarrow$ SOMO	17.8	17.5–18.1
	$\pi \rightarrow$ SOMO	18.3	17.5–18.1
	$\pi^* \rightarrow \pi^*$	25.3	23.5–23.6
	$\pi \rightarrow$ SOMO	36.3	29.4–30.4

^{a}Ridley, J. E. and Zerner, M. C. 1973. *Theor. Chim. Acta*, **32**, 111.
^{b}Anderson, W. P., Edwards, W. D., and Zerner, M. C. 1986. *Inorg. Chem.*, **25**, 2728.
^{c}Estiú, G. L. and Zerner, M. C. 1999. *J. Am. Chem. Soc.* **121**, 1893.

were based on experiment, and excited state energies were computed using a CIS formalism (see Chapter 14 for details on CIS).

The INDO/S model is very successful for d → d transitions within transition metal complexes (typical accuracies are within 2000 cm^{-1}), potentially less robust for spectroscopic transitions that are not well localized to a single center (e.g., metal-to-ligand or ligand-to-metal excitations or excitations in extended π systems), and does not do very well in predicting Rydberg excitations since very diffuse orbitals are not a part of the basis set. The INDO/S model also exhibits good accuracy for the prediction of ionization potentials and oscillator strengths for weak electronic transitions (oscillator strengths for strong transitions tend to be overestimated).

It is perhaps appropriate at this point to make a distinction between a semiempirical 'method' and a semiempirical 'model'. The method describes the overall formalism for constructing the relevant secular equation. Within that formalism, however, different choices for free parameters may be made depending on the problem to which the method is being applied. Thus, INDO is a method, but the INDO/S model is a particular parameterization of the method designed for spectroscopic applications. One might say that the model is the set of all of the parameters required by the method. While we make this distinction here in a purely terminological sense, it has profound philosophical origins that should not be taken for granted as they led to the development of the first truly general semiempirical model, as described next.

5.4.2 MINDO/3 and SINDO1

Michael J. S. Dewar was trained as an organic chemist, but very early on he saw the potential for using MO theory, and in particular quantitative formulations of MO theory, to rationalize structure and reactivity in organic chemistry. Prior to Dewar's work, semiempirical models tended to be problem-specific. That is, while there were general methods like CNDO and INDO, most specific parameterizations (i.e., models) were carried out only within the context of narrowly defined chemical problems. Of course, if a particular chemical problem is important, there is certainly nothing wrong with developing a model specific to it, but the generality of the overall methods remained, for the most part, unexplored. Dewar established, as a goal, the development of a parameter set that would be as robust as possible across the widest possible spectrum (at the time, that spectrum was chosen to be primarily organic chemistry, together with a few inorganic compounds comprised of second- and third-row elements).

Dewar also recognized that a truly thorough test of a general model would require the efforts of more than one research group. To that end, he adopted a philosophy that not only should the best parameterization be widely promulgated but so too should computer code implementing it. Dewar's code included geometry optimization routines, which made it particularly attractive to non-developers interested in using the code for chemical purposes.

The first general parameterization to be reported by Dewar and co-workers was a third-generation modified INDO model (MINDO/3; Bingham, Dewar, and Lo, 1975). Some of the specific modifications to the INDO framework included the use of different ζ exponents in s and p type STOs on the same atom, the definition of pair parameters β_{AB} between two atoms A and B that were *not* averages of atomic parameters (actually, four such parameters exist per pair, corresponding to $s_A s_B$, $s_A p_B$, $p_A s_B$, and $p_A p_B$ orbital interactions), adoption of a slightly different form for γ_{AB} than that of Eq. (5.8), and some empirical modifications to the nuclear repulsion energy.

Moreover, rather than following any particular set of rules to generate parameter values (e.g., Slater's rules for orbital exponents), every parameter was treated as a free variable subject to 'chemical common-sense' restraints. That is, parameter values were allowed to vary freely so long as they did not clearly become physically unrealistic; thus, for instance, a parameter set where an atomic U_s value was smaller in magnitude than the corresponding U_p value would not be acceptable, since ionization of a valence s electron on an atom cannot be more favorable than ionization of a valence p electron. To optimize parameter values, Dewar initially took a set of 138 small molecules containing C, H, O, and N, and constructed a penalty function depending on bond distances, valence angles, torsional angles, dipole moments, ionization potentials, and heats of formation (most molecules had experimental data available only for a subset of the penalty function components).

The use of the last experimental observable, the heat of formation, merits a brief digression. Molecular heats of formation are the most widely available thermochemical quantities against which one might imagine carrying out a parameterization (this was even more the case in the early 1970s), but these are enthalpic quantities, not potential energies. The expectation value from a MO calculation is the potential energy to separate all of the electrons and nuclei to infinite separation. How can these two be compared? Dewar's approach was to compute or estimate the MINDO/3 SCF energy for the elements in their standard states (on a per atom basis for those elements whose standard states are not monatomic) and to record this as an atomic parameter. To compute the heat of formation for a molecule, the individual standard-state atomic SCF energies were subtracted from the molecular SCF energy, and the difference was summed with the experimental heats of formation for all the constituent atoms (this is in effect rather similar to the atomic-type-equivalents scheme discussed for force-field calculations in Chapter 2, except that in the semiempirical approach, there is only a single type for each atom). Most computer codes implementing semiempirical models continue to print out the energy in this fashion, as a so-called heat of formation.

Note, however, that zero-point vibrational energy and thermal contributions to the experimental enthalpies have been ignored (or, perhaps more accurately, treated in some sort of average fashion by the parameterization process). At the time, computing those contributions was far from trivial. Now, however, it is quite straightforward to do so (see Chapter 10 for details), leading to some ambiguity in how energies from semiempirical calculations should be reported. As a general rule, it would probably be better to consider the semiempirical SCF energies to have the status of potential energies, and explicitly to account for thermochemical contributions when necessary, but the literature is full of examples where this issue is confused. Of course, if the goal of the calculation is truly to estimate the heat of formation of a particular molecule, then the semiempirical SCF energy should be used uncorrected, since that enthalpic quantity was one of the targets for which parameters were optimized.

The performance of the MINDO/3 model was impressive overall. The mean absolute error in predicted heats of formation was 11 kcal/mol (all molecules), the corresponding error for ionization potentials was 0.7 eV (46 molecules), for heavy-atom bond lengths 0.022 Å (81 molecules), and for dipole moments 0.45 D (31 molecules). While mean errors of this size exceed what would be tolerated today, they were unprecedentedly small in 1975. Dewar's subsequent work on other semiempirical models (see below) rendered MINDO/3 effectively obsolete, but its historical importance remains unchanged.

A modified INDO model that is *not* entirely obsolete is the symmetric orthogonal-ized INDO (SINDO1) model of Jug and co-workers, first described in 1980 (Nanda and Jug 1980). The various conventions employed by SINDO1 represent slightly different modifications to INDO theory than those adopted in the MINDO/3 model, but the more fundamental difference is the inclusion of d functions for atoms of the second row in the periodic table. Inclusion of such functions in the atomic valence basis set proves critical for handling hyper-valent molecules containing these atoms, and thus SINDO1 performs considerably better for phosphorus-containing compounds, for instance, than do other semiempirical models that lack d functions (Jug and Schulz 1988).

5.5 Basic NDDO Formalism

The INDO model extends the CNDO model by adding flexibility to the description of the one-center two-electron integrals. In INDO, however, there continues to be only a single two-center two-electron integral, which takes on the value γ_{AB} irrespective of which orbitals on atoms A and B are considered. As already noted, this can play havoc with the accurate representation of lone pair interactions.

The neglect of diatomic differential overlap (NDDO) method relaxes the constraints on two-center two-electron integrals in a fashion analogous to that for one-center integrals in the INDO method. Thus, all integrals $(\mu\nu|\lambda\sigma)$ are retained provided μ and ν are on the same atomic center and λ and σ are on the same atomic center, but not necessarily the center hosting μ and ν. How many different integrals are permitted? The order of μ and ν does not affect the value of the integral, so we need only worry about combinations, not permutations, in which case there are 10 unique combinations of s, p_x, p_y, and p_z. With 10 unique combinations on each atom, there are 100 possible combinations of combinations for the integrals. If we include d functions, the number of unique integrals increases to 2025.

Although these numbers seem large, this is still a considerable improvement over evaluating *every* possible integral, as would be undertaken in *ab initio* HF theory. Most modern semiempirical models are NDDO models. After examining the differences in their formulation, we will examine their performance characteristics in some detail in Section 5.6.

5.5.1 MNDO

Dewar and Thiel (1977) reported a modified neglect of differential overlap (MNDO) method based on the NDDO formalism for the elements C, H, O, and N. With the conventions specified by NDDO for which integrals to keep, which to discard, and how to model one-electron integrals, it is possible to write the NDDO Fock matrix elements individually for inspection. For the most complex element, a diagonal element, we have

$$
F_{\mu\mu} = U_\mu - \sum_{B \neq A} Z_B(\mu\mu|s_Bs_B) + \sum_{\nu \in A} P_{\nu\nu}\left[(\mu\mu|\nu\nu) - \frac{1}{2}(\mu\nu|\mu\nu)\right]
$$

$$
+ \sum_B \sum_{\lambda \in B} \sum_{\sigma \in B} P_{\lambda\sigma}(\mu\mu|\lambda\sigma) \tag{5.12}
$$

where μ is located on atom A. The first term on the r.h.s. is the atomic orbital ionization potential, the second term the attraction to the other nuclei where each nuclear term is proportional to the repulsion with the valence s electron on that nucleus, the third term reflects the Coulomb and exchange interactions with the other electrons on atom A, and the final term reflects Coulomb repulsion with electrons on other atoms B.

An off-diagonal Fock matrix element for two basis functions μ and ν on the same atom A is written as

$$
F_{\mu\nu} = -\sum_{B \neq A} Z_B(\mu\nu|s_Bs_B) + P_{\mu\nu}\left[\frac{3}{2}(\mu\nu|\mu\nu) - \frac{1}{2}(\mu\mu|\nu\nu)\right] + \sum_B \sum_{\lambda \in B} \sum_{\sigma \in B} P_{\lambda\sigma}(\mu\nu|\lambda\sigma) \tag{5.13}
$$

where each term on the r.h.s. has its analogy in Eq. (5.12). When μ is on atom A and ν on atom B, this matrix element is written instead as

$$
F_{\mu\nu} = \frac{1}{2}(\beta_\mu + \beta_\nu)S_{\mu\nu} - \frac{1}{2}\sum_{\lambda \in A}\sum_{\sigma \in B} P_{\lambda\sigma}(\mu\lambda|\nu\sigma) \tag{5.14}
$$

where the first term on the r.h.s. is the resonance integral that encompasses the one-electron kinetic energy and nuclear attraction terms; it is an average of atomic resonance integrals 'β' times the overlap of the orbitals involved. The second term on the r.h.s. captures favorable exchange interactions. Note that the MNDO model did *not* follow Dewar's MINDO/3 approach of having β parameters specific to pairs of atoms. While the latter approach allowed for some improved accuracy, it made it quite difficult to add new elements, since to be complete all possible pairwise β combinations with already existing elements would require parameterization.

The only point not addressed in Eqs. (5.12) to (5.14) is how to go about evaluating all of the necessary two-electron integrals. Unlike one-center two-electron integrals, it is not easy to analyze spectroscopic data to determine universal values, particularly given the large number of integrals not taken to be zero. The approach taken by Dewar and co-workers was to evaluate these integrals by replacing the continuous charge clouds with classical multipoles. Thus, an ss product was replaced with a point charge, an sp product was replaced with a classical dipole (represented by two point charges slightly displaced from the nucleus along the p orbital axis), and a pp product was replaced with a classical quadrupole (again represented by point charges). The magnitudes of the moments, being one-center in nature, are related to the parameterized integrals in Eq. (5.11). By adopting such a form for the integrals, their evaluation is made quite simple, and so too is evaluation of their analytic derivatives with respect to nuclear motion.

To complete the energy evaluation by the MNDO method, the nuclear repulsion energy is added to the SCF energy. The MNDO nuclear repulsion energy is computed as

$$V_N = \sum_{k<l}^{\text{nuclei}} Z_k Z_l (s_k s_k | s_l s_l) \left(1 + \tau e^{-\alpha_{Z_k} r_{kl}} + e^{-\alpha_{Z_l} r_{kl}}\right) \tag{5.15}$$

where Z is the valence atomic number, α is a parameter having a specific value for each atom type, r is the interatomic distance, and τ is equal to 1 unless the two nuclei k and l are an O/H or N/H pair, in which case it is rXH. Thus, internuclear repulsion is proportional to the repulsion between s electrons on the same centers, and the repulsion is empirically increased slightly at short bond lengths to make up for imbalances in the electronic part of the calculation.

As with MINDO/3, Dewar and Thiel optimized the parameters of the MNDO model against a large test set of molecular properties. Within the assumption of a valence orbital set comprised only of s and p orbitals, MNDO parameters are now available for H, He, Li, Be, B, C, N, O, F, Mg, Al, Si, P, S, Cl, Zn, Ge, Br, Sn, I, Hg, and Pb. The MNDO model is typically not used as often as the NDDO models discussed next, but MNDO calculations still appear in the literature. MNDO forms the foundation for MNDO/d, which is discussed in Section 5.7.2. In addition, a modified MNDO model explicitly adding electron correlation effects (MNDOC) by second-order perturbation theory (see Section 7.4) was described by Thiel in 1981 (Thiel 1981; Schweig and Thiel 1981). By explicitly accounting for electron correlation in the theory, the parameters do not have to absorb the effects of its absence from HF theory in some sort of average way. Thus, in principle, MNDOC should be more robust in application to problems with widely varying degrees of electron correlation. In practice, the model has not yet been compared to other NDDO models to the degree necessary to evaluate whether the formalism lives up to that potential.

5.5.2 AM1

Although a detailed discussion of the performance of MNDO is deferred until Section 5.6, one critical flaw in the method is that is does very poorly in the prediction of hydrogen bonding geometries and energies. Recognizing this to be a major drawback, particularly with respect to modeling systems of biological interest, Dewar and co-workers modified the functional form of their NDDO model; since the primary error was one involving bond lengths, the key modification was to the nuclear repulsion term. In Austin Model 1 (AM1; Dewar, at the time, was a faculty member at the University of Texas, Austin), originally described in 1985 for the elements C, H, O, and N (Dewar *et al.* 1985), the nuclear repulsion energy between any two nuclei A and B is computed as

$$V_N(A, B) = V_{AB}^{\text{MNDO}} + \frac{Z_A Z_B}{r_{AB}} \sum_{i=1}^{4} \lfloor a_{A,i} e^{-b_{A,i}(r_{AB} - c_{A,i})^2} + a_{B,i} e^{-b_{B,i}(r_{AB} - c_{B,i})^2} \rfloor \tag{5.16}$$

where the variables are for the most part those in Eq. (5.15) and in addition every atom has up to 4 each parameters a, b, and c describing Gaussian functions centered at various distances

c that modify the potential of mean force between the two atoms. Simultaneous optimization of the original MNDO parameters with the Gaussian parameters led to markedly improved performance, although the Gaussian form of Eq. (5.16) is sufficiently force-field-like in nature that one may quibble about this method being entirely quantum mechanical in nature.

Since the report for the initial four elements, AM1 parameterizations for B, F, Mg, Al, Si, P, S, Cl, Zn, Ge, Br, Sn, I, and Hg have been reported. Because AM1 calculations are so fast (for a quantum mechanical model), and because the model is reasonably robust over a large range of chemical functionality, AM1 is included in many molecular modeling packages, and results of AM1 calculations continue to be reported in the chemical literature for a wide variety of applications.

5.5.3   PM3

One of the authors on the original AM1 paper and a major code developer in that effort, James J. P. Stewart, subsequently left Dewar's labs to work as an independent researcher. Stewart felt that the development of AM1 had been potentially non-optimal, from a statistical point of view, because (i) the optimization of parameters had been accomplished in a stepwise fashion (thereby potentially accumulating errors), (ii) the search of parameter space had been less exhaustive than might be desired (in part because of limited computational resources at the time), and (iii) human intervention based on the perceived 'reasonableness' of parameters had occurred in many instances. Stewart had a somewhat more mathematical philosophy, and felt that a sophisticated search of parameter space using complex optimization algorithms might be more successful in producing a best possible parameter set within the Dewar-specific NDDO framework.

To that end, Stewart set out to optimize *simultaneously* parameters for H, C, N, O, F, Al, Si, P, S, Cl, Br, and I. He adopted an NDDO functional form identical to that of AM1, except that he limited himself to two Gaussian functions per atom instead of the four in Eq. (5.16). Because his optimization algorithms permitted an efficient search of parameter space, he was able to employ a significantly larger data set in evaluating his penalty function than had been true for previous efforts. He reported his results in 1989; as he considered his parameter set to be the third of its ilk (the first two being MNDO and AM1), he named it Parameterized Model 3 (PM3; Stewart 1989).

There is a possibility that the PM3 parameter set may actually be the global minimum in parameter space for the Dewar-NDDO functional form. However, it must be kept in mind that even if it *is* the global minimum, it is a minimum for a particular penalty function, which is itself influenced by the choice of molecules in the data set, and the human weighting of the errors in the various observables included therein (see Section 2.2.7). Thus, PM3 will not necessarily outperform MNDO or AM1 for any particular problem or set of problems, although it is likely to be optimal for systems closely resembling molecules found in the training set. As noted in the next section, some features of the PM3 parameter set can lead to very unphysical behaviors that were not assessed by the penalty function, and thus were not avoided. Nevertheless, it is a very robust NDDO model, and continues to be used at least as widely as AM1.

In addition to the twelve elements noted above, PM3 parameters for Li, Be, Na, Mg, Ca, Zn, Ga, Ge, As, Se, Cd, In, Sn, Sb, Te, Hg, Tl, Pb, and Bi have been reported. The PM3 methodology is available in essentially all molecular modeling packages that carry out semiempirical calculations.

5.6   General Performance Overview of Basic NDDO Models

Many comparisons of the most widely used semiempirical models have been reported. They range from narrowly focused anecdotal discussions for specific molecules to detailed tests over large sets of molecules for performance in the calculation of various properties. We will discuss here a subset of these comparisons that have the broadest impact – those looking for a more thorough overview are referred to the bibliography at the end of the chapter.

5.6.1   Energetics

5.6.1.1   Heats of formation

The primary energetic observable against which NDDO models were parameterized was heat of formation. Table 5.2 compares the mean unsigned errors for MNDO, AM1, and PM3 for various classes of molecules (the column labeled MNDO/d is discussed in Section 5.7.2). The greater accuracies of AM1 and PM3 compared to MNDO are manifest in every case. PM3 appears to offer a slight advantage over AM1 for estimating the heats of formation of molecules composed of lighter elements (C, H, O, N, F), and a clear advantage for

Table 5.2   Mean unsigned errors (kcal mol^{-1}) in predicted heats of formation from basic NDDO models

Elements (number)	Subset (number)	MNDO	AM1	PM3	MNDO/d
Lighter (181)		7.35	5.80	4.71	
	CH (58)	5.81	4.89	3.79	
	CHN (32)	6.24	4.65	5.02	
	CHNO (48)	7.12	6.79	4.04	
	CHNOF (43)	10.50	6.76	6.45	
	Radicals (14)	9.3	8.0	7.4	
Heavier (488)		29.2	15.3	10.0	4.9
	Al (29)	22.1	10.4	16.4	4.9
	Si (84)	12.0	8.5	6.0	6.3
	P (43)	38.7	14.5	17.1	7.6
	S (99)	48.4	10.3	7.5	5.6
	Cl (85)	39.4	29.1	10.4	3.9
	Br (51)	16.2	15.2	8.1	3.4
	I (42)	25.4	21.7	13.4	4.0
	Hg (37)	13.7	9.0	7.7	2.2
	Normal (421)	11.0	8.0	8.4	4.8
	Hypervalent (67)	143.2	61.3	19.9	5.4
Cations (34)		9.55	7.62	9.46	
Anions (13)		11.36	7.11	8.81	

heavier elements. However, in the latter case, the difference is essentially entirely within the subset of hypervalent molecules included in the test set, e.g., PBr_5, IF_7, etc. Over the 'normal' subset of molecules containing heavy atoms, the performance of AM1 and PM3 is essentially equivalent. Analysis of the errors in predicted heats of formation suggests that they are essentially random, i.e., they reflect the 'noise' introduced into the Schrödinger equation by the NDDO approximations and cannot be corrected for in a systematic fashion without changing the theory. This random noise can be problematic when the goal is to determine the relative energy differences between two or more isomers (conformational or otherwise), since one cannot be as confident that errors will cancel as is the case for more complete quantum mechanical methods.

Errors for charged and open-shell species tend to be somewhat higher than the corresponding errors for closed-shell neutrals. This may be at least in part due to the greater difficulty in measuring accurate experimental data for some of these species, but some problems with the theory are equally likely. For instance, the more loosely held electrons of an anion are constrained to occupy the same STO basis functions as those used for uncharged species, so anions are generally predicted to be anomalously high in energy. Radicals are systematically predicted to be too stable (the mean signed error over the radical test set in Table 5.2 is only very slightly smaller than the mean unsigned error) meaning that bond dissociation energies are usually predicted to be too low. Note that for the prediction of radicals all NDDO methods were originally parameterized with a so-called 'half-electron RHF method', where the formalism of the closed-shell HF equations is used even though the molecule is open-shell (Dewar, Hashmall, and Venier 1968). Thus, while use of so-called 'unrestricted Hartree–Fock (UHF)' technology (see Section 6.3.3) is technically permitted for radicals in semiempirical theory, it tends to lead to unrealistically low energies and is thus less generally useful for thermochemical prediction (Pachkovski and Thiel 1996). Finally, PM3 exhibits a large, non-systematic error in the prediction of proton affinities; AM1 is more successful for the prediction of these quantities.

For the particular goal of computing accurate heats of formation, Repasky, Chandrasekhar, and Jorgensen (2002) have suggested a modification to the original approach taken by Dewar as outlined in Section 5.4.2. Instead of treating a molecule as being composed of atoms as its fundamental building blocks, they propose 61 common bond and group equivalents that may instead be considered as small transferable elements. Each such bond or group is then assigned its own characteristic heat of formation, and a molecular heat of formation is derived from adding the difference between the molecular electronic energy and the sum of the fragment electronic energies to the sum of the bond and group heats of formation. In the form of a general equation we have

$$\Delta H^o_{f,298}(\text{molecule}) = \left[E(\text{molecule}) - \sum_{i=1}^{N} E(\text{fragment}_i) \right] + \sum_{i=1}^{N} \Delta H^o_{f,298}(\text{fragment}_i) \qquad (5.17)$$

where E is the semiempirical electronic energy. In the original Dewar protocol, the fragments were atoms and N was the total number of atoms in the molecule. In the bond/group approach, the fragments are larger and N is smaller than the total number of atoms. Over 583 neutral, closed-shell molecules, Repasky, Chandrasekhar, and Jorgensen found that the mean unsigned errors from MNDO, AM1, and PM3 were reduced from 8.2, 6.6, and 4.2 kcal mol^{-1}, respectively, to 3.0, 2.3, and 2.2 kcal mol^{-1}, respectively, when bond/group equivalents were used in place of atoms as the fundamental fragments.

5.6.1.2 *Other energetic quantities*

Another energetic quantity of some interest is the ionization potential (IP). Recall that in HF theory, the eigenvalue associated with each MO is the energy of an electron in that MO. Thus, a good estimate of the negative of the IP is the energy of the highest occupied MO – this simple approximation is one result from a more general statement known as Koopmans' theorem (Koopmans, 1933; this was Koopmans' only independent paper in theoretical physics/chemistry–immediately thereafter he turned his attention to economics and went on to win the 1975 Nobel Prize in that field). Employing this approximation, all of the semiempirical methods do reasonably well in predicting IPs for organic molecules. On a test set of 207 molecules containing H, C, N, O, F, Al, S, P, Cl, Br, and I, the average error in predicted IP for MNDO, AM1, and PM3 is 0.7, 0.6, and 0.5 eV, respectively. For purely inorganic compounds, PM3 shows essentially unchanged performance, while MNDO and AM1 have errors increased by a few tenths of an electron volt.

With respect to the energetics associated with conformational changes and reactions, a few general comments can be made. MNDO has some well-known shortcomings; steric crowding tends to be too strongly disfavored and small ring compounds are predicted to be too stable. The former problem leads to unrealistically high heats of formation for sterically congested molecules (e.g., neopentane) and similarly too high heats of activation for reactions characterized by crowded TS structures. For the most part, these problems are corrected in AM1 and PM3 through use of Eq. (5.16) to modify the non-bonded interactions. Nevertheless, activation enthalpies are still more likely to be too high than too low for the semiempirical methods because electron correlation energy tends to be more important in TS structures than in minima (see also Table 8.3), and since correlation energy is introduced in only an average way by parameterization of the semiempirical HF equations, it cannot distinguish well between the two kinds of structures.

For intermolecular interactions that are weak in nature, e.g., those arising from London forces (dispersion) or hydrogen bonding, semiempirical methods are in general unreliable. Dispersion is an electron correlation phenomenon, so it is not surprising that HF-based semiempirical models fail to make accurate predictions. As for hydrogen bonding, one of the primary motivations for moving from MNDO to AM1 was to correct for the very weak hydrogen bond interactions predicted by the former. Much of the focus in the parameterization efforts of AM1 and PM3 was on reproducing the enthalpy of interaction of the water dimer, and both methods do better in matching the experimental value of 3.6 kcal mol^{-1} than does MNDO. However, detailed analyses of hydrogen bonding in many different systems have indicated that in most instances the interaction energies are systematically too small by up to 50 percent and that the basic NDDO methods are generally not well suited to the characterization of hydrogen bonded systems (Dannenberg 1997). Bernal-Uruchurtu *et al.* (2000) have suggested that the form of Eq. 5.16 is inadequate for describing hydrogen bonding; by use of an alternative parameterized interaction function, they were able to modify PM3 so that the PES for the water dimer was significantly improved.

Energetic barriers to rotation about bonds having partial double bond character tend to be significantly too low at semiempirical levels. In amides, for instance, the rotation barrier about the C–N bond is underestimated by about 15 kcal/mol. In several computer programs implementing NDDO methods, an *ad hoc* molecular mechanics torsional potential can be added to amide bond linkages to correct for this error. Smaller errors, albeit still large as a fraction of total barrier height, are observed about C–C single bonds in conjugated chains.

With respect to conformational analysis, the NDDO models are not quantitatively very accurate. Hehre has reported calculations for eight different sets of conformer pairs having an average energy difference between pairs of 2.3 kcal mol^{-1}. Predictions from MNDO, AM1, and PM3 gave mean unsigned errors of 1.4, 1.3, and 1.8 kcal mol^{-1}, respectively, although in four of the eight cases AM1 was within 0.5 kcal mol^{-1}. In addition, AM1 and PM3 have been compared for the 11 D-glucopyranose conformers discussed in Chapter 2 in the context of analyzing force field performance; AM1 and PM3 had mean unsigned errors of 1.4 and 0.8 kcal mol^{-1}, respectively, making them less accurate than the better force fields. The PM3 number is misleadingly good in this instance; although the method does reasonably well for the 11 conformers studied, the PM3 PES also includes highly unusual minimum energy structures not predicted by any other method (vide infra).

5.6.2 Geometries

Correct molecular structures are dependent on the proper location of wells in the PES, so they are intimately related to the energetics of conformational analysis. For organic molecules, most gross structural details are modeled with a reasonable degree of accuracy. Dewar, Jie, and Yu (1993) evaluated AM1 and PM3 for 344 bond lengths and 146 valence angles in primarily organic molecules composed of H, C, N, O, F, Cl, Br, and I; the average unsigned errors were 0.027 and 0.022 Å, respectively, for the bond lengths, and 2.3 and 2.8°, respectively, for the angles. In the parameterization of PM3, Stewart (1991) performed a similar analysis for a larger set of molecules, some of them including Al, Si, P, and S. For 460 bond lengths, the mean unsigned errors were 0.054, 0.050, and 0.036 Å for MNDO, AM1, and PM3, respectively. For 196 valence angles, the errors were 4.3, 3.3, and 3.9°, respectively. In the case of MNDO, bond angles at the central O and S atoms in ethers and sulfides, respectively, were found to be up to 9° too large, presumably owing to the overestimation of steric repulsion between the substituting groups.

Comparing all of the sets of comparisons, it is evident that the geometries for the molecules containing second-row elements are considerably more difficult to predict accurately than are those for simpler organics. Furthermore, MNDO and AM1 are less successful when extended to these species than is PM3.

Stewart also carried out an analysis for dihedral angles, and found errors of 21.6, 12.5, and 14.9°, respectively, for MNDO, AM1, and PM3. However, only 16 data points were available and the accurate measurement of dihedral angles is challenging. Nevertheless, there appear to be systematic errors in dihedral angles for small- to medium-sized ring systems, where predicted geometries tend to be too 'flat', again probably because of overestimated steric repulsion between non-bonded ring positions (Ferguson *et al.* 1992). Four-membered rings are typically predicted to be planar instead of puckered.

A few additional geometric pathologies have been discovered over the years for the various semiempirical methods. While many are for species that might be described as exotic, others have considerably more potential to be troublesome.

Heteroatom–heteroatom linkages are often problematic. MNDO and AM1 both predict peroxide O–O bonds to be 0.018 Å too short. In hydrazines, the N–N bond rotamer placing the two nitrogen lone pairs antiperiplanar to one another is usually overstabilized relative to the *gauche* rotamer. Thus, even though experimentally hydrazine has been determined to have a C_2 *gauche* structure, all of the methods predict the global minimum to be the trans C_{2v} structure (PM3 does not find the C_2 structure to be a stationary point at all). In nitroxyl compounds, N–N bonds are predicted to be too short by up to 0.7 Å. AM1 has similar problems with P–P bond lengths. In silyl halides, Si–X bonds are predicted to be too short by tenths of an ångström by PM3.

PM3 shows additional problems that are disturbing. Nitrogen atoms formally possessing a lone pair tend to be significantly biased towards pyramidal geometries. In addition, there is an anomalous, deep well in the non-bonded H–H interaction expressed by Eq. (5.16) at a distance of about 1.4 Å (Csonka 1993), which can lead to such odd situations as hydroxyl groups preferring to interact with one another by H–H contacts instead of typical hydrogen bonding contacts in D-glucopyranose conformers (Barrows *et al.* (1995)). Additional work by Casadesus *et al.* (2004) found that similarly unphysical H–H distances were predicted for cyclodextrin inclusion complexes using AM1, PM3, PDDG/MNDO, or PDDG/PM3 (the last two models are described in Section 5.7.5). Only PM5 (described in Section 5.7.2) was judged to provide acceptable geometries.

As already noted above, the energetics of normal hydrogen bonding is not handled well by any semiempirical method; geometries are similarly problematic. PM3 predicts the expected near-linear single hydrogen bond for most systems, but typically it is too short by as much as 0.2 Å. AM1 predicts heavy-atom–heavy-atom bond distances in hydrogen bonds that are about right, but strongly favors bifurcated hydrogen bonds in those systems where that is possible (e.g., in the water dimer, the water molecule acting as a hydrogen bond donor interacts with the other water molecule through *both* its protons equally). MNDO hydrogen bonds are much, much too long, since the interaction energies at this level are predicted to be far too small.

5.6.3 Charge Distributions

One of the most useful features of a QM model is its ability to provide information about the molecular charge distribution. It is a general rule of thumb that even very low quality QM methods tend to give reasonable charge distributions. For neutral molecules, the dominant moment in the overall charge distribution is the usually dipole moment (unless symmetry renders the dipole moment zero). For a 125-molecule test set including H, C, N, O, F, Al, Si, P, S, Cl, Br, and I functionality, Stewart found mean unsigned errors in dipole moments of 0.45, 0.35, and 0.38 D, respectively, for MNDO, AM1, and PM3 (Stewart 1989). PM3 seems to be somewhat more robust for compounds incorporating phosphorus.

An alternative measure of the charge distribution involves a partitioning into partial atomic charges. While such partitioning is always arbitrary (see Chapter 9) simple methods tend to give reasonably intuitive results when small basis sets are used, as is the case for the NDDO models. While MNDO and AM1 present no particular issues for such analysis, PM3 tends to predict nitrogen atoms to be too weakly electronegative. Thus, in the ammonium cation, PM3 predicts the charge on nitrogen to be $+1.0$ while the charge on each hydrogen is predicted to be 0.0 (Storer *et al.* 1995).

Finally, some attention has been paid to the quality of the complete electrostatic potential about the molecule at the NDDO level. This topic is discussed in Chapter 9, as are additional details associated with the performance of semiempirical models in comparison to other levels of electronic structure theory for a variety of more specialized properties.

5.7   Ongoing Developments in Semiempirical MO Theory

Semiempirical theory is still in widespread use today not because it competes effectively with more sophisticated theories in terms of accuracy, but because it competes effectively in terms of demand for computational resources. Indeed, if one has either an enormously large molecule, or an enormously large number of small molecules to be compared at a consistent level (the next section describes a particular example of this case), semiempirical theory is the only practical option. Of course, with each improvement in technology, the size horizon of the more sophisticated levels expands, but there seems little danger that chemists will not always be able to imagine still larger systems meriting quantum chemical study. Therefore, considerable interest remains in improving semiempirical models in a variety of directions. We close this chapter with a brief overview of some of the most promising of these.

5.7.1   Use of Semiempirical Properties in SAR

This area is a development in the *usage* of NDDO models that emphasizes their utility for large-scale problems. Structure–activity relationships (SARs) are widely used in the pharmaceutical industry to understand how the various features of biologically active molecules contribute to their activity. SARs typically take the form of equations, often linear equations, that quantify activity as a function of variables associated with the molecules. The molecular variables could include, for instance, molecular weight, dipole moment, hydrophobic surface area, octanol–water partition coefficient, vapor pressure, various descriptors associated with molecular geometry, etc. For example, Cramer, Famini, and Lowrey (1993) found a strong correlation ($r = 0.958$) between various computed properties for 44 alkylammonium ions and their ability to act as acetylcholinesterase inhibitors according to the equation

$$\log\left(\frac{1}{K_i}\right) = -2.583 - \frac{0.636}{100}V + \frac{4.961}{0.1}\pi - 2.234q^+ \qquad (5.18)$$

where K_i is the inhibition constant, V is the molecular volume, π derives from the molecular polarizability, q^+ is the largest positive charge on a hydrogen atom, and all of the variables on the r.h.s. of Eq. (5.18) were computed at the MNDO level.

Once a SAR is developed, it can be used to prioritize further research efforts by focusing first on molecules predicted by the SAR to have the most desirable activity. Thus, if a drug company has a database of several hundred thousand molecules that it has synthesized over the years, and it has measured molecular properties for those compounds, once it identifies a SAR for some particular bio-target, it can quickly run its database through the SAR to identify other molecules that should be examined. However, this process is not very useful for identifying *new* molecules that might be better than any presently existing ones. It can be quite expensive to synthesize new molecules randomly, so how can that process be similarly prioritized?

One particularly efficient alternative is to develop SARs not with experimental molecular properties, but with predicted ones. Thus, if the drug company database is augmented with predicted values, and a SAR on predicted values proves useful based on data for compounds already assayed, potential new compounds can be examined in a purely computational fashion to evaluate whether they should be priority targets for synthesis. In 1998, Beck *et al.* (1998) optimized the geometries of a database of 53 000 compounds with AM1 in 14 hours on a 128-processor Origin 2000 computer. Such speed is presently possible only for semiempirical levels of theory. Once the geometries and wave functions are in hand, it is straightforward (and typically much faster) to compute a very wide variety of molecular properties in order to survey possible SARs. Note that for the SAR to be useful, the absolute values of the computed properties do not necessarily need to be accurate – only their variation relative to their activity is important.

5.7.2 d Orbitals in NDDO Models

To extend NDDO methods to elements having occupied valence d orbitals that participate in bonding, it is patently obvious that such orbitals need to be included in the formalism. However, to accurately model even non-metals from the third row and lower, particularly in hypervalent situations, d orbitals are tremendously helpful to the extent they increase the flexibility with which the wave function may be described. As already mentioned above, the d orbitals present in the SINDO1 and INDO/S models make them extremely useful for spectroscopy. However, other approximations inherent in the INDO formalism make these models poor choices for geometry optimization, for instance. As a result, much effort over the last decade has gone into extending the NDDO formalism to include d orbitals.

Thiel and Voityuk (1992, 1996) described the first NDDO model with d orbitals included, called MNDO/d. For H, He, and the first-row atoms, the original MNDO parameters are kept unchanged. For second-row and heavier elements, d orbitals are included as a part of the basis set. Examination of Eqs. (5.12) to (5.14) indicates what is required parametrically to add d orbitals. In particular, one needs U_d and β_d parameters for the one-electron integrals, additional one-center two-electron integrals analogous to those in Eq. (5.11) (there are formally 12 such integrals), and a prescription for handling two-center two-electron integrals including d functions. In MNDO/d, the U, β, and G_{dd} terms are treated as adjustable parameters, the remaining one-center two-electron integrals are analytic functions of G_{dd} and the integrals in Eq. (5.11), and the Dewar convention whereby two-center two-electron integrals are evaluated using classical multipole expansions is retained, except that multipolar representations beyond quadrupole (e.g., a dd cloud would be a hexadecapole) are ignored, since testing indicates they typically contribute negligibly to the total electronic energy. Parameterization of the various new terms proceeds in the same fashion as for prior NDDO models, with a penalty function focused on molecular thermochemical and structural data. The performance of the model for heavy elements is summarized in Table 5.2 (for light elements MNDO/d is identical to MNDO). MNDO/d represents an enormous improvement over AM1 and PM3 in its ability to handle hypervalent molecules, and in most cases the error over the various test sets is reduced by half or more when MNDO/d is used.

It appears, then, that MNDO/d has high utility for thermochemical applications. In addition to the elements specified in Table 5.2, MNDO/d parameters have been determined for Na, Mg, Zn, Zr, and Cd. However, since the model is based on MNDO and indeed identical to MNDO for light elements, it still performs rather poorly with respect to intermolecular interactions, and with respect to hydrogen bonding in particular.

The approach of Thiel and Voityuk has also been adopted by Hehre and co-workers, who have applied it in extending the PM3 Hamiltonian to include d orbitals. This model, which to date has not been fully described in the literature and is only available as part of a commercial software package (SPARTAN), is called PM3(tm), where the 'tm' emphasizes a focus on transition metals. The parameterization philosophy has been different from prior efforts insofar as only geometrical data (primarily from X-ray crystallography) have been included in the penalty function. This choice was motivated at least in part by the general scarcity of thermochemical data for molecules containing transition metals. Thus, the model may be regarded as an efficient way to generate reasonable molecular geometries whose energies may then be evaluated using more complete levels of theory. For example, a study by Goh and Marynick (2001) found that the geometries of metallofullerenes predicted at the PM3(tm) level compared well with those predicted from much more expensive density functional calculations. PM3(tm) includes parameters for Ca, Ti, Cr−Br, Zr, Mo, Ru, Rh, Pd, Cd−I, Hf, Ta, W, Hg, and Gd.

Very recent extensions of the formalism of Thiel and Voityuk to AM1 have been reported by multiple groups. Voityuk and Rösch (2000) first described an AM1/d parameter set for Mo, and, using the same name for the method, Lopez and York (2003) reported a parameter set for P designed specifically to facilitate the study of nucleophilic substitutions of biological phosphates. Winget et al. (2003) described an alternative model, named AM1*, that adds d orbitals to P, S, and Cl. As with MNDO/d, the primary improvement of this model is in its general ability to describe hypervalent molecules more accurately. Subtle differences in the various individual formalisms will not be further delved into here.

A semiempirical model including d orbitals has also been reported by Dewar and co-workers (Dewar, Jie, and Yu 1993; Holder, Dennington, and Jie 1994), although the full details of its functional form still await publication. Semi-*ab initio* model 1 (SAM1, or

SAM1D if d orbitals are included), however, is not quite so straightforward an extension of the NDDO formalism, but represents a rather different approach to constructing the Fock matrix. In SAM1, the valence-orbital basis set is made up not of Slater-type orbitals but instead of Gaussian-type orbitals; in particular the STO-3G basis set is used (see Section 6.2.2). Using this basis set, one- and two-electron integrals not explicitly set to zero in the NDDO formalism are analytically calculated in an *ab initio* fashion (see Section 6.1), but the resulting values are then treated as input to parameterized scaling functions depending on, *inter alia*, interatomic distance. Parameters exist for H, Li, C, N, O, F, Si, P, S, Cl, Fe, Cu, Br, and I. For molecules made up of light elements, SAM1 performs better than AM1 and very slightly better than PM3. The same is true for non-hypervalent molecules made up of heavier elements, while very large improvements are observed for molecules containing hypervalent heavy atoms – across 404 compounds containing Si, P, S, Cl, Br, and/or I as heavy elements, the mean unsigned errors in heats of formation for AM1, PM3, SAM1, and MNDO/d are 16.2, 9.5, 9.3, and 5.1 kcal/mol (Thiel and Voityuk 1996).

Finally, Stewart has also generated a new NDDO parameterization including d orbitals that he has called PM5. As of 2004, a publication describing this method had yet to appear. The model is available in the commercial code MOPAC2002, and a fairly detailed comparison of its performance to earlier semiempirical models may be found at http://www.cachesoftware. com/mopac/Mopac2002manual/node650.html. It would appear that the chief advantage of PM5 lies in its ability to better handle metals, heavy non-metals, and hypervalent species and in its incorporation of dispersion effects via inclusion of Lennard-Jones terms between non-bonded atoms. Further assessments necessarily await the publication of results from specific applications of the model. A noteworthy feature of the comparisons made on the website is that data are listed for MNDO, AM1, and PM3, as well as for PM5, for all of the thousands of molecules examined, many of which include atoms for which parameters have not been published for *any* of the NDDO methods. It would thus appear that MOPAC2002 includes such parameters, whose provenance is uncertain. [Finally, for those wondering about the nomenclature gap between PM3 and PM5, Stewart reserved the name PM4 for a separate, collaborative parameterization effort results for which have not yet been reported. One difference between PM4 and PM5 is that PM4 includes dispersion energies between nonbonded atoms based on computed atomic polarizabilities (Martin, Gedeck, and Clark 2000).]

5.7.3 SRP Models

SRP, a term first coined by Rossi and Truhlar (1995), stands for 'specific reaction (or range) parameters'. An SRP model is one where the standard parameters of a semiempirical model are adjusted so as to foster better performance on a particular problem or class of problems. In a sense, the SRP concept represents completion of a full circle in the philosophy of semiempirical modeling. It tacitly recognizes the generally robust character of some underlying semiempirical model, and proceeds from there to optimize that model for a particular system of interest. In application, then, SRP models are similar to the very first semiempirical models, which also tended to be developed on an *ad hoc*, problem-specific basis. The difference, however, is that the early models typically were developed essentially from scratch, while SRP models may be viewed as perturbations of more general models.

Parameter	AM1	AM1-SRP
C		
U_s	−52.03	−49.85
U_p	−39.61	−40.34
β_s	−15.72	−16.91
β_p	−7.72	−9.19
O		
U_s	−97.83	−99.18
U_p	−78.26	−80.76
β_s	−29.27	−29.00
β_p	−29.27	−29.25

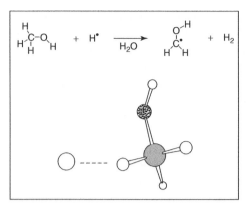

Source	ΔE_{rxn}	$D_e(C–H)$	$D_e(H–H)$
AM1	−28.0	81.4	109.4
AM1-SRP	−4.9	104.4	109.4
Expt.	−5.1	104.4	109.5

Figure 5.2 AM1 and AM1-SRP parameters (eV) optimized to reproduce the C–H bond dissociation energy of methanol, the H–H bond dissociation energy of hydrogen, and the experimental energy for the illustrated hydrogen-atom transfer (kcal mol^{-1}). Note that in all cases but one, the magnitude of the parameter change on going from AM1 to AM1-SRP is less than 10 percent

The concept is best illustrated with an example. Chuang *et al.* (1999) used an AM1-SRP model to study the hydrogen-atom-mediated destruction of organic alcohols in water. As illustrated in Figure 5.2, the AM1 model itself makes a very poor prediction for the dissociation energy of the C–H bond in methanol, and hence for the reaction exothermicity. By minor adjustment of a few of the AM1 parameters, however, the SRP model gives good agreement with experiment. The resulting SRP model in this case was used as a very efficient QM method for generation of a PES from which tunneling contributions to the reaction rate constant could be estimated (see Section 15.3). The very large number of QM calculations required to generate the PES made use of an SRP model preferable to more complete levels of electronic structure theory like those discussed in Chapter 7. Ridder *et al.* (2002) followed a similar protocol in determining SRP AM1 sulfur parameters so as to study the dynamics of the conjugation of glutathione to phenanthrene-9,10-oxide as catalyzed by a rat glutathione S-transferase enzyme (the enzyme was treated using molecular mechanics). Again, the very large number of quantum calculations required during the dynamics made a semiempirical model like AM1 an attractive choice.

A more global SRP reparameterization has been described by Sherer, York, and Cramer (2003). In this case, select PM3 parameters for H, C, N, and O were modified to improve the performance of the resulting SRP model, named PM3$_{BP}$, for the computation of base-pairing energies between hydrogen-bonded nucleic-acid bases. The PM3$_{BP}$ model has a root-mean-square error of about 1.5 kcal mol^{-1} for 31 such base pairing energies compared to either experiment or well benchmarked higher-level theoretical calculations. This compares to RMS errors of about 11, 6, and 6 kcal mol^{-1} for MNDO, AM1, and PM3 over the same test set. Such a reparameterization of a semiempirical Hamiltonian for a focused set of molecules is completely analogous to the targeted parameterization of a force field, and the usual caveats apply with respect to application of a focused model to anything other than molecules falling within the target category.

5.7.4 Linear Scaling

As already touched upon in Section 2.4.2, the development of methods that scale linearly with respect to system size opens the door to the modeling of very large systems with maximal computational efficiency. Because the NDDO approximation is already rather efficient when it comes to forming the Fock matrix (because so many integrals are assumed to be zero, etc.), it serves as an excellent basis on which to build a linear-scaling QM model. Such models have been reported; the details associated with achieving linear scaling are sufficiently technical that interested readers are referred to the original literature (van der Vaart *et al.* 2000; Khandogin, Hu, and York 2000; see also Stewart 1996).

It is worth a pause, however, to consider how such models should best be used. Part of the motivation for developing linear scaling models has been to permit QM calculations to be carried out on biomolecules, e.g., proteins or polynucleic acids. However, one may legitimately ask whether there is any point in such a calculation, beyond demonstrating that it can be done. Because of the relatively poor fashion with which semiempirical models handle non-bonded interactions, there is every reason to expect that such models would be disastrously bad at predicting biomolecular geometries – or at the very least inferior to the far more efficient force fields developed and optimized for this exact purpose.

Instead, the virtue of the semiempirical models when applied to such molecules tends to be that they permit the charge distribution to be predicted more accurately given a particular structure. To the extent that biomolecules often employ charge–charge interactions to enhance reactivity and or specificity in the reaction and recognition of smaller molecules, such predictions can be quite useful. Since the QM calculation intrinsically permits polarization of the overall electronic structure, it is capable of showing greater sensitivity to group–group interactions as they modify the charge distribution than is the case for the typical fixed-atomic-charge, non-polarizable force field.

Of course, one may also be interested in the modeling of a bond-making/bond-breaking reaction that takes place within a very large molecular framework, in which case the availability of appropriate force-field models is extremely limited and one must perforce resort to some QM approach in practice. Recognition of the complementary strengths and weaknesses of QM and MM models has led to extensive efforts to combine them in ways that allow maximum advantage to be taken of the good points of both; such QM/MM hybrid models are the subject of Chapter 13.

5.7.5 Other Changes in Functional Form

Two modifications to the fundamental forms of modern NDDO functions have been reported recently that merit particular attention. First, Weber and Thiel (2000) have reconsidered the

NDDO approximation that all two-electron integrals $(\mu\nu|\lambda\sigma)$ are set to zero unless atomic-orbital basis functions μ and ν are on a single atom and λ and σ are on a single atom. One basis for this approximation is that early studies showed that all other two-electron integrals did indeed tend to have values rather close to zero *provided that the basis set functions are orthogonal to one another*. However, the Slater-type AOs used in NDDO calculations are *not* orthogonal to one another (cf. the first term of Eq. (5.14)). The energetic consequences of the NDDO two-electron integral approximation in the context of a non-orthogonal basis set are thus open to question.

To address this point, Weber and Thiel have developed two levels of orthogonalization corrections to the MNDO model. Orthogonalization Method 1 (OM1) includes corrections to one- and two-center terms, while Orthogonalization Method 2 (OM2) also includes corrections to three-center terms. The details of these corrections are sufficiently technical that we will not examine them closely. It suffices to note that the OM2 model involves simply the addition of two new parameters per atom, these being optimized in the usual semiempirical fashion, and that the various correction terms themselves are expressed entirely using terms already computed in the course of an NDDO calculation. Thus, the OM1 and OM2 models run in about the same time as required for a standard MNDO calculation. Over 81 small molecules composed of C, H, N, and O, the mean unsigned error in predicted heats of formation dropped from 7.8 to 4.4 to 3.4 kcal mol^{-1} on going from MNDO to OM1 to OM2, respectively. In addition, OM2 shows qualitative improvements over MNDO for such phenomena as rotational barriers, ring conformations, relative energies of isomers, hydrogen-bond strengths and geometries, and barriers for pericyclic reactions.

A second modification recently described by Repasky, Chandrasekhar, and Jorgensen (2002) focuses on improving the core-repulsion functions in MNDO and PM3. In particular, they define a pairwise distance directed Gaussian function (PDDG) to compute a contribution to the nuclear repulsion energy between atoms A and B as

$$V_N^{\text{PDDG}}(\text{A,B}) = \left(\frac{1}{n_\text{A} + n_\text{B}}\right) \sum_{i=1}^{2} \sum_{j=1}^{2} (n_\text{A} P_{i,\text{A}} + n_\text{B} P_{j,\text{B}}) \; e^{-[(r_{\text{AB}} - D_{i,\text{A}} - D_{j,\text{B}})^2/10]} \qquad (5.19)$$

where n is the number of atomic valence electrons, parameters P are preexponential factors, and parameters D are distance terms in Å. The total nuclear repulsion energy is then defined to be the sum of the usual MNDO or PM3 terms (Eqs. (5.15) and (5.16), respectively) and the new PDDG component. Repasky, Chandrasekhar, and Jorgensen optimized *all* MNDO and PM3 parameters together with the four new P and D parameters per atom to arrive at final PDDG/MNDO and PDDG/PM3 models that were initially defined for H, C, N, and O only. Over a test set of 622 molecules, they found mean unsigned errors in molecular heats of formation to be 8.4 and 4.4 kcal mol^{-1} for MNDO and PM3, respectively, and these were reduced to 5.2 and 3.2 kcal mol^{-1} for PDDG/MNDO and PDDG/PM3, respectively. Other key energetic improvements were observed for small rings, polyheterocyclic rings, and isomeric equilibria. Subsequent work has extended the two models to the halogens (Tubert-Brohman *et al.* 2004). This work represents a likely limit for what may be accomplished by adjustment of the nuclear repulsion energy.

5.8 Case Study: Asymmetric Alkylation of Benzaldehyde

Synopsis of Goldfuss and Houk (1998) 'Origin of Enantioselectivities in Chiral β-Amino Alcohol Catalyzed Asymmetric Additions of Organozinc Reagents to Benzaldehyde: PM3 Transition State Modeling'.

A major goal of organic synthesis is the preparation of chiral molecules in an optically pure fashion, i.e., as single enantiomers. Any such process must involve discrimination between enantiomerically related transition states based on a chiral environment, and a popular method for establishing such an environment is to employ a so-called chiral auxiliary as part of one or more of the involved reagents. Since the auxiliary must itself be pure in order to be maximally effective, and since optically pure molecules can be expensive, even when derived from natural products, it is especially desirable to design processes where the chiral auxiliary forms part of a catalyst rather than part of a stoichiometric reagent. An example of such a process is the addition of organozinc reagents to aldehydes. In the presence of β-amino alcohols, one equivalent of dialkylzinc reacts with the alcohol to liberate ethane and form an amino-coordinated zinc alkoxide. This alkoxide catalyzes the addition of a second equivalent of dialkylzinc to aldehydes by forming supermolecular complexes like those illustrated in Figure 5.3.

When the β-amino alcohol ligand is chiral and optically pure, there are four potentially low-energy TS structures that may lead to products. Several chiral ligands have been shown

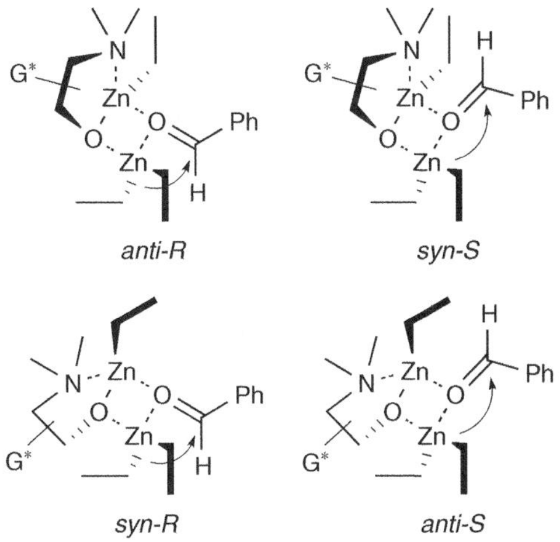

anti-R   syn-S

syn-R   anti-S

Figure 5.3 Four alternative TS structures for catalyzed addition of diethylzinc to benzaldehyde. The descriptors refer to the side of the four-membered ring on which the aldehyde carbon is found relative to the alkoxide carbon – same (*syn*) or opposite (*anti*) – and the absolute configuration of the new stereogenic center formed following ethyl transfer, *R* or *S*. In the absence of chirality in the β-amino alcohol ligand, indicated by the G* group(s), the TS structures at opposite corners would be enantiomeric with one another, and no preference for *R* over *S* product would be observed. At least four other TS structures can be readily imagined while maintaining the atomic connectivities of those shown here. What are they and why might they be intuitively discounted? Are there still other TS structures one might imagine? How does one decide when all relevant TS structures have been considered?

to give high enantioselectivities in the alkyl addition, indicating that either a single one of the four TS structures is significantly lower in energy, or, if not, the two associated with one enantiomer are significantly lower than either of the two for the other enantiomer.

To better determine the specific steric and/or electronic influences giving rise to high observed enantioselectivities, Goldfuss and Houk studied the energies of the four TS structures in Figure 5.3 for different chiral β-amino alcohols at the PM3 level of theory.

One possible concern in such an approach is the quality of the Zn parameters in PM3, since experimental data for zinc compounds are considerably more sparse than for more quotidian organic compounds. Thus, as a first step, Goldfuss and Houk considered the small complex formed from formaldehyde, di*methyl*zinc, and unsubstituted β-aminoethanol. They compared the geometries of the two TS structures predicted at the PM3 level to those previously obtained by another group at the *ab initio* HF/3-21G level (note that since the amino alcohol is not chiral, there are two TS structures, not four); they observed that agreement was reasonable for the gross shapes of the TS structures, although there were fairly substantial differences in various bond lengths – up to 0.2 Å in Zn−O bonds and the forming C−C bond. They also compared the relative energies for the two TS structures at the PM3 level to those previously reported from small, correlated *ab initio* calculations. Agreement was at best fair, with PM3 giving an energy difference between the two structures of 6.8 kcal mol^{-1}, compared to the prior result of 2.9 kcal mol^{-1}.

Comparison between PM3 and the previously reported levels of theory is interesting from a methodological perspective. However, to the extent that there are significant disagreements between the methods, PM3 is as likely to be the most accurate as any, given the rather low levels of *ab initio* theory employed (*ab initio* theory is discussed in detail in the next two chapters). Insofar as the size of the chemical problem makes it impractical to seek converged solutions of the Schrödinger equation, Goldfuss and Houk turned instead to a comparison of PM3 to available experimental data. In particular, they computed product ratios based on the assumption that these would reflect a 273 K Boltzmann distribution of corresponding TS structures (this follows from transition state theory, discussed in Section 15.3, for a reaction under kinetic control). For the TS energies, they employed the relative PM3 electronic energies plus zero-point vibrational energies obtained from frequency calculations (see Section 10.2). Then, for a variety of different chiral β-amino alcohols, they compared predicted enantiomeric excess, defined as

$$\%\text{ee} = |\%R - \%S| \qquad (5.20)$$

to experimental values obtained under a variety of different conditions. This comparison, summarized in Table 5.3, shows remarkably good agreement between PM3 and experiment.

It is worth a brief digression to note that, from a theoretical standpoint, it is rather easy to make predictions in cases where a single product is observed. When experimentalists report a single product, they typically mean that to within the detection limits of their analysis, they observe only a single compound – unless special efforts are undertaken, this might imply no better than 20:1 excess of the observed product over any other possibilities. At 298 K, this implies that the TS structure of lowest energy lies at least 2 kcal mol^{-1} below any competing TS structures. Of course, it might be 20 or 200 kcal mol^{-1} below competing TS structures – when experiment reports only a single product, there is no way to quantify this. Thus, even if theory is badly in error, as long as the correct TS structure is predicted to be lowest by more than 2 kcal mol^{-1}, there will be 'perfect' agreement with

Table 5.3 Comparison of predicted and experimental enantiomeric excesses for diethylzinc addition to benzaldehyde in the presence of various β-amino alcohols

β-Amino alcohol	Configuration	%ee	
		PM3	Experiment
	S	100	99
	R	99	95
	S	100	94
	S	98	81
	R	97	100
	R	82	72
	S	33	49

experiment. If, however, two or more products are reported with a quantitative ratio, the quality of the theoretical results can be much more accurately judged. At 298 K, every error of 1.4 kcal mol^{-1} in predicted relative energies between two TS structures will change the ratios of predicted products by an order of magnitude. Thus, in the case of two competing

TS structures leading to different enantiomers, a %ee of 0 would result from equal TS energies, a %ee of 82 from relative energies of 1.4 kcal mol^{-1}, and a %ee of 98 from relative energies of 2.8 kcal mol^{-1}. Given this analysis, the near quantitative agreement between PM3 and those experimental cases showing %ee values below 90 reflects startlingly good accuracy for a semiempirical level of theory.

Armed with such solid agreement between theory and experiment, Goldfuss and Houk go on to analyze the geometries of the various TS structures to identify exactly which interactions lead to unfavorably high energies and can be used to enhance chiral discrimination. They infer in particular that the optimal situation requires that the alkoxide carbon atom be substituted by two groups of significantly different size, e.g., a hydrogen atom and a bulky alkyl or aryl group. This work thus provides a nice example of how preliminary experimental work can be used to validate an economical theoretical model that can then be used to suggest future directions for further experimental optimization. However, it must not be forgotten that the success of the model must derive in part from favorable cancellation of errors – the theoretical model, after all, fails to account for solvent, thermal contributions to free energies, and various other possibly important experimental conditions. As such, application of the model in a predictive mode should be kept within reasonable limits, e.g., results for new β-amino alcohol structures would be expected to be more secure than results obtained for systems designed to use a substituted 1,2-diaminoethane ligand in place of the β-amino alcohol.

Bibliography and Suggested Additional Reading

Clark, T. 2000. 'Quo Vadis Semiempirical MO-theory?' *J. Mol. Struct. (Theochem)*, **530**, 1.

Dewar, M. J. S. 1975. *The PMO Theory of Organic Chemistry*, Plenum: New York.

Famini, G. R. and Wilson, L. Y. 2002. 'Linear Free Energy Relationships Using Quantum Mechanical Descriptors', in *Reviews in Computational Chemistry*, Vol. 18, Lipkowitz, K. B. and Boyd, D. B., Eds., Wiley-VCH: New York, 211.

Hall, M. B. 2000. 'Perspective on "The Spectra and Electronic Structure of the Tetrahedral Ions MnO_4^-, CrO_4^{2-}, and ClO_4^-"' *Theor. Chem. Acc.*, **103**, 221.

Hehre, W. J. 1995. *Practical Strategies for Electronic Structure Calculations*, Wavefunction: Irvine, CA.

Jensen, F. 1999. *Introduction to Computational Chemistry*, Wiley: Chichester.

Levine, I. N. 2000. *Quantum Chemistry*, 5th Edn., Prentice Hall: New York.

Pople, J. A. and Beveridge, D. A. 1970. *Approximate Molecular Orbital Theory*, McGraw-Hill: New York.

Repasky, M. P., Chandrasekhar, J., and Jorgensen, W. L. 2002. 'PDDG/PM3 and PDDG/MNDO: Improved Semiempirical Methods', *J. Comput. Chem.*, **23**, 1601.

Stewart, J. J. P. 1990. 'Semiempirical Molecular Orbital Methods' in *Reviews in Computational Chemistry*, Vol. 1, Lipkowitz, K. B. and Boyd, D. B., Eds., VCH: New York, 45.

Thiel, W. 1998. 'Thermochemistry from Semiempirical Molecular Orbital Theory' in *Computational Thermochemistry, ACS Symposium Series*, Vol. 677, Irikura, K. K. and Frurip, D. J., Eds., American Chemical Society: Washington, DC, 142.

Thiel, W. 2000. 'Semiempirical Methods' in *Modern Methods and Algorithms of Quantum Chemistry*, Proceedings, 2nd Edn., Grotendorst, J., Ed., NIC Series, Vol. 3, John von Neumann Institute for Computing: Jülich, 261.

Whangbo, M.-H. 2000. "Perspective on 'An extended Hückel theory. I. Hydrocarbons'" *Theor. Chem. Acc.*, **103**, 252.

Zerner, M. 1991. 'Semiempirical Molecular Orbital Methods' in *Reviews in Computational Chemistry*, Vol. 2, Lipkowitz, K. B. and Boyd, D. B., Eds., VCH: New York, 313.

References

Barrows, S. E., Dulles, F. J., Cramer, C. J., French, A. D., and Truhlar, D. G. 1995. *Carbohydr. Res.*, **276**, 219.

Beck, B., Horn, A., Carpenter, J. E., and Clark, T. 1998. *J. Chem. Inf. Comput. Sci.* **38**, 1214.

Bernal-Uruchurtu, M. I., Martins-Costa, M. T. C., Millot, C., and Ruiz-Lopez, M. F. 2000. *J. Comput. Chem.*, **21**, 572.

Bingham, R. C., Dewar, M. J. S., and Lo, D. H. 1975. *J. Am. Chem. Soc.*, **97**, 1285, 1307.

Bishop, D. M. 1966. *J. Chem. Phys.*, **45**, 1880 and references therein.

Casadesus, R., Moreno, M., Gonzalez-Lafont, A., Lluch, J. M., and Repasky, M. P. 2004. *J. Comput. Chem.*, **25**, 99.

Chuang, Y.-Y., Radhakrishnan, M. L., Fast, P. L., Cramer, C. J., and Truhlar, D. G. 1999. *J. Phys. Chem. A*, **103**, 4893.

Clementi, E. and Roetti, C. 1974. *At. Data Nucl. Data Tables.*, **14**, 177.

Cramer, C. J., Famini, G. R., and Lowrey, A. H. 1993. *Acc. Chem. Res.*, **26**, 599.

Csonka, G. I. 1993. *J. Comput. Chem.*, **14**, 895.

Cusachs, L. C., Reynolds, J. W., and Barnard, D. 1966. *J. Chem. Phys.*, **44**, 835.

Da Motta Neto, J. D. and Zerner, M. C. 2001. *Int. J. Quant. Chem.*, **81**, 187.

Dannenberg, J. A. 1997. *J. Mol. Struct. (Theochem)*, **401**, 287.

Dewar, M. J. S., Hashmall, J. A., and Venier, C. G. 1968. *J. Am. Chem. Soc.*, **90**, 1953.

Dewar, M. J. S., Jie, C., and Yu, J. 1993. *Tetrahedron,* **49**, 5003.

Dewar, M. J. S., Zoebisch, E. G., Healy, E. F., and Stewart, J. J. P. 1985. *J. Am. Chem. Soc.*, **107**, 3902.

Ferguson, D. M., Gould, W. A., Glauser, W. A., Schroeder, S., and Kollman, P. A. 1992. *J. Comput. Chem.*, **13**, 525.

Genin, H. and Hoffmann, R. 1998. *Macromolecules,* **31**, 444.

Goh, S. K. and Marynick, D. S. 2001. *J. Comput. Chem.*, **22**, 1881.

Goldfuss, B. and Houk, K. N. 1998. *J. Org. Chem.*, **63**, 8998.

Hall, M. B. and Fenske, R. F. 1972. *Inorg. Chem.*, **11**, 768.

Hinze, J. and Jaffé, H. H. 1962. *J. Am. Chem. Soc.*, **84**, 540.

Hoffmann, R. 1963. *J. Chem. Phys.*, **39**, 1397.

Holder, A., Dennington, R. D., and Jie, C. 1994. *Tetrahedron,* **50**, 627.

Jug, K. and Schulz, J. 1988. *J. Comput. Chem.*, **9**, 40.

Khandogin, L, Hu, A. G., and York, D. M. 2000. *J. Comput. Chem.*, **21**, 1562.

Koopmans, T. 1933. *Physica (Utrecht)*, **1**, 104.

Kotzian, M., Rösch, N., and Zerner, M. C. 1992. *Theor. Chim. Acta*, **81**, 201.

Lopez, X. and York, D. M. 2003. *Theor. Chem. Acc.*, **109**, 149.

Martin, B., Gedeck, P., and Clark, T. 2000. *Int. J. Quant. Chem.*, **77**, 473.

Mataga, N. and Nishimoto, K. 1957. *Z. Phys. Chem.*, **13**, 140.

Mulliken, R. S., Rieke, C. A., and Orloff, H. 1949. *J. Chem. Phys.*, **17**, 1248.

Nanda, D. N. and Jug, K. 1980. *Theor. Chim. Acta*, **57**, 95.

Pachkovski, S. and Thiel, W. 1996. *J. Am. Chem. Soc.*, **118**, 7164.

Pariser, R. and Parr, R. G. 1953. *J. Chem. Phys.*, **21**, 466, 767.

Pekeris, C. L. 1959. *Phys. Rev.*, **115**, 1216.

Pilcher, G. and Skinner, H. A. 1962. *Inorg. Nucl. Chem.*, **24**, 937.

Pople, J. A. 1953. *Trans. Faraday Soc.*, **49**, 1375.

Pople, J. A. and Segal, G. A. 1965. *J. Chem. Phys.*, **43**, S136.

Pople, J. A., Beveridge, D. L., and Dobosh, P. A. 1967. *J. Chem. Phys.*, **47**, 2026.

Pople, J. A., Santry, D. P., and Segal, G. A. 1965. *J. Chem. Phys.*, **43**, S129.

Rein, R., Fukuda, N., Win, H., Clarke, G. A., and Harris, F. E. 1966. *J. Chem. Phys.*, **45**, 4743.

Repasky, M. P., Chandrasekhar, J., and Jorgensen, W. L. 2002. *J. Comput. Chem.*, **23**, 498.

Ridder, L., Rietjens, I. M. C. M., Vervoort, J., and Mulholland, A. J. 2002. *J. Am. Chem. Soc.*, **124**, 9926.

Ridley, J. E. and Zerner, M. C. 1973. *Theor. Chim. Acta,* **32**, 111.

Rossi, I. and Truhlar, D. G. 1995. *Chem. Phys. Lett.*, **233**, 231.

Schweig, A. and Thiel, W. 1981. *J. Am. Chem. Soc.*, **103**, 1425.

Sherer, E. C., York, D. M., and Cramer, C. J. 2003. *J. Comput. Chem.*, **24**, 57.

Slater, J. C. 1930. *Phys. Rev.*, **36**, 57.

Stewart, J. J. P. 1989. *J. Comput. Chem.*, **10**, 209, 221.

Stewart, J. J. P. 1991. *J. Comput. Chem.*, **12**, 320.

Stewart, J. J. P. 1996. *Int. J. Quantum Chem.*, **58**, 133.

Storer, J. W., Giesen, D. J., Cramer, C. J., and Truhlar, D. G. 1995. *J. Comput-Aided Mol. Des.*, **9**, 87.

Thiel, W. 1981. *J. Am. Chem. Soc.*, **103**, 1413, 1420.

Thiel, W. and Voityuk, A. A. 1992, *Theor. Chim. Acta,* **81**, 391.

Thiel, W. and Voityuk, A. A. 1996, *Theor. Chim. Acta,* **93**, 315.

Thiel, W. and Voityuk, A. A. 1996. *J. Phys. Chem.*, **100**, 616.

Tubert-Brohman, I., Guimaraes, C. R. W., Repasky, M. P., Jorgensen, W. L. 2004. *J. Comput. Chem.*, **25**, 138.

van der Vaart, A., Gogonea, V., Dixon, S. L., and Merz, K. M. 2000. *J. Comput. Chem.*, **21**, 1494.

Voityuk, A. A. and Rösch, N. 2000. *J. Phys. Chem. A*, **104**, 4089.

Weber, W. and Thiel, W. 2000. *Theor. Chem. Acc.*, **103**, 495.

Winget, P., Horn, A. H. C., Selçuki, B., Martin, B., Clark, T. 2003. *J. Mol. Model.*, **9**, 408.

Wolfsberg, M. and Helmholz, L. J. 1952. *J. Chem. Phys.*, **20**, 837.

Zerner, M. and Gouterman, M. 1966. *Theor. Chim. Acta*, **4**, 44.

6

Ab Initio Implementations of Hartree–Fock Molecular Orbital Theory

6.1 *Ab Initio* Philosophy

The fundamental assumption of HF theory, that each electron sees all of the others as an average field, allows for tremendous progress to be made in carrying out practical MO calculations. However, neglect of electron correlation can have profound chemical consequences when it comes to determining accurate wave functions and properties derived therefrom. As noted in the preceding chapter, the development of semiempirical theories was motivated in part by the hope that judicious parameterization efforts could compensate for this feature of HF theory. While such compensation has no rigorous foundation, to the extent it permits one to make accurate chemical predictions, it may have great practical utility.

Early developers of so-called '*ab initio*' (Latin for 'from the beginning') HF theory, however, tended to be less focused on making short-term predictions, and more focused on long-term development of a *rigorous* methodology that would be worth the wait (a dynamic tension between the need to make predictions now and the need to make better predictions tomorrow is likely to characterize computational chemistry well into the future). Of course, the ultimate rigor is the Schrödinger equation, but that equation is insoluble in a practical sense for all but the most simple of systems. Thus, HF theory, in spite of its fairly significant fundamental assumption, was adopted as useful in the *ab initio* philosophy because it provides a very well defined stepping stone on the way to more sophisticated theories (i.e., theories that come closer to accurate solution of the Schrödinger equation). To that extent, an enormous amount of effort has been expended on developing mathematical and computational techniques to reach the HF limit, which is to say to solve the HF equations with the equivalent of an infinite basis set, *with no additional approximations*. If the HF limit is achieved, then the energy error associated with the HF approximation for a given system, the so-called electron correlation energy E_{corr}, can be determined as

$$E_{corr} = E - E_{HF} \tag{6.1}$$

Essentials of Computational Chemistry, 2nd Edition Christopher J. Cramer
© 2004 John Wiley & Sons, Ltd   ISBNs: 0-470-09181-9 (cased); 0-470-09182-7 (pbk)

where E is the 'true' energy and E_{HF} is the system energy in the HF limit. Chapter 7 is devoted to the discussion of techniques for estimating E_{corr}.

Along the way it became clear that, perhaps surprisingly, HF energies could be chemically useful. Typically their utility was manifest for situations where the error associated with ignoring the correlation energy could be made unimportant by virtue of comparing two or more systems for which the errors could be made to cancel. The technique of using isodesmic equations, discussed in Section 10.6, represents one example of how such comparisons can successfully be made.

In addition, the availability of HF wave functions made possible the testing of how useful such wave functions might be for the prediction of properties *other* than the energy. Simply because the HF wave function may be arbitrarily far from being an eigenfunction of the Hamiltonian operator does not *a priori* preclude it from being reasonably close to an eigenfunction for some other quantum mechanical operator.

This chapter begins with a discussion of basis sets, the mathematical functions used to construct the HF wave function. Key technical details associated with open-shell vs. closed-shell systems are also addressed. A performance overview and case study are provided in conclusion.

6.2   Basis Sets

The basis set is the set of mathematical functions from which the wave function is constructed. As detailed in Chapter 4, each MO in HF theory is expressed as a linear combination of basis functions, the coefficients for which are determined from the iterative solution of the HF SCF equations (as flow-charted in Figure 4.3). The full HF wave function is expressed as a Slater determinant formed from the individual occupied MOs. In the abstract, the HF limit is achieved by use of an infinite basis set, which necessarily permits an optimal description of the electron probability density. In practice, however, one cannot make use of an infinite basis set. Thus, much work has gone into identifying mathematical functions that allow wave functions to approach the HF limit arbitrarily closely in as efficient a manner as possible.

Efficiency in this case involves three considerations. As noted in Chapter 4, in the absence of additional simplifying approximations like those present in semiempirical theory, the number of two-electron integrals increases as N^4 where N is the number of basis functions. So, keeping the total number of basis functions to a minimum is computationally attractive. In addition, however, it can be useful to choose basis set functional forms that permit the various integrals appearing in the HF equations to be evaluated in a computationally efficient fashion. Thus, a larger basis set can still represent a computational improvement over a smaller basis set if evaluation of the greater number of integrals for the former can be carried out faster than for the latter. Finally, the basis functions must be chosen to have a form that is useful in a chemical sense. That is, the functions should have large amplitude in regions of space where the electron probability density (the wave function) is also large, and small amplitudes where the probability density is small. The simultaneous optimization of these three considerations is at the heart of basis set development.

6.2.1 Functional Forms

Slater-type orbitals were introduced in Section 5.2 (Eq. (5.2)) as the basis functions used in extended Hückel theory. As noted in that discussion, STOs have a number of attractive features primarily associated with the degree to which they closely resemble hydrogenic atomic orbitals. In *ab initio* HF theory, however, they suffer from a fairly significant limitation. There is no analytical solution available for the general four-index integral (Eq. (4.56)) when the basis functions are STOs. The requirement that such integrals be solved by numerical methods severely limits their utility in molecular systems of any significant size. Nevertheless, high quality STO basis sets *have* been developed for atomic and diatomic calculations, where such limitations do not arise (Ema *et al.* 2003).

Boys (1950) proposed an alternative to the use of STOs. All that is required for there to be an analytical solution of the general four-index integral formed from such functions is that the radial decay of the STOs be changed from e^{-r} to e^{-r^2}. That is, the AO-like functions are chosen to have the form of a Gaussian function. The general functional form of a normalized Gaussian-type orbital (GTO) in atom-centered Cartesian coordinates is

$$\phi(x, y, z; \alpha, i, j, k) = \left(\frac{2\alpha}{\pi}\right)^{3/4} \left[\frac{(8\alpha)^{i+j+k}i!j!k!}{(2i)!(2j)!(2k)!}\right]^{1/2} x^i y^j z^k e^{-\alpha(x^2+y^2+z^2)} \tag{6.2}$$

where α is an exponent controlling the width of the GTO, and i, j, and k are non-negative integers that dictate the nature of the orbital in a Cartesian sense.

In particular, when all three of these indices are zero, the GTO has spherical symmetry, and is called an s-type GTO. When exactly one of the indices is one, the function has axial symmetry about a single Cartesian axis and is called a p-type GTO. There are three possible choices for which index is one, corresponding to the p_x, p_y, and p_z orbitals.

When the sum of the indices is equal to two, the orbital is called a d-type GTO. Note that there are six possible combinations of index values (i, j, k) that can sum to two. In Eq. (6.2), this leads to possible Cartesian prefactors of x^2, y^2, z^2, xy, xz, and yz. These six functions are called the Cartesian d functions. In the solution of the Schrödinger equation for the hydrogen atom, only five functions of d-type are required to span all possible values of the z component of the orbital angular momentum for $l = 2$. These five functions are usually referred to as xy, xz, yz, $x^2 - y^2$, and $3z^2 - r^2$. Note that the first three of these canonical d functions are common with the Cartesian d functions, while the latter two can be derived as linear combinations of the Cartesian d functions. A remaining linear combination that can be formed from the Cartesian d functions is $x^2 + y^2 + z^2$, which, insofar as it has spherical symmetry, is actually an s-type GTO. Different Gaussian basis sets adopt different conventions with respect to their d functions: some use all six Cartesian d functions, others prefer to reduce the total basis set size and use the five linear combinations. [Note that if the extra function is kept, the linear combination having s-like symmetry still has the same exponent α governing its decay as the rest of the d set. As d orbitals are more diffuse than s orbitals having the same principal quantum number (which is to say the magnitude of α for the nd GTOs will be smaller than that for the α of the ns GTOs), the extra s orbital does not really contribute at the same principal quantum level, as discussed in more detail below.]

As one increases the indexing, the disparity between the number of Cartesian functions and the number of canonical functions increases. Thus, with f-type GTOs (indices summing to 3) there are 10 Cartesian functions and 7 canonical functions, with g-type 15 and 10, etc. GTOs can be taken arbitrarily high in angular momentum.

6.2.2 Contracted Gaussian Functions

Although they are convenient from a computational standpoint, GTOs have specific features that diminish their utility as basis functions. One issue of key concern is the shape of the radial portion of the orbital. For s type functions, GTOs are smooth and differentiable at the nucleus ($r = 0$), but real hydrogenic AOs have a cusp (Figure 6.1). In addition, all hydrogenic AOs have a radial decay that is exponential in r while the decay of GTOs is exponential in r^2; this results in too rapid a reduction in amplitude with distance for the GTOs.

In order to combine the best feature of GTOs (computational efficiency) with that of STOs (proper radial shape), most of the first basis sets developed with GTOs used them as building blocks to approximate STOs. That is, the basis functions φ used for SCF calculations were not individual GTOs, but instead a linear combination of GTOs fit to reproduce as accurately as possible a STO, i.e.,

$$\varphi(x, y, z; \{\alpha\}, i, j, k) = \sum_{a=1}^{M} c_a \phi(x, y, z; \alpha_a, i, j, k) \qquad (6.3)$$

where M is the number of Gaussians used in the linear combination, and the coefficients c are chosen to optimize the shape of the basis function sum and ensure normalization. When a basis function is defined as a linear combination of Gaussians, it is referred to as a 'contracted' basis function, and the individual Gaussians from which it is formed are called 'primitive' Gaussians. Thus, in a basis set of contracted GTOs, each basis function is defined by the contraction coefficients c and exponents α of each of its primitives. The 'degree of

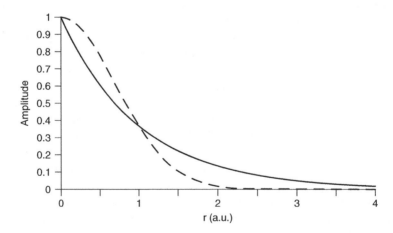

Figure 6.1   Behavior of e^{-x} where $x = r$ (solid line, STO) and $x = r^2$ (dashed line, GTO)

contraction' refers to the total number of primitives used to make all of the contracted functions, as described in more detail below. Contracted GTOs when used as basis functions continue to permit analytical evaluation of all of the four-index integrals.

Hehre, Stewart, and Pople (1969) were the first to systematically determine optimal contraction coefficients and exponents for mimicking STOs with contracted GTOs for a large number of atoms in the periodic table. They constructed a series of different basis sets for different choices of M in Eq. (6.3). In particular, they considered $M = 2$ to 6, and they called these different basis sets STO-MG, for 'Slater-Type Orbital approximated by M Gaussians'. Obviously, the more primitives that are employed, the more accurately a contracted function can be made to match a given STO. However, note that a four-index two-electron integral becomes increasingly complicated to evaluate as each individual basis function is made up of increasingly many primitive functions, according to

$$
\begin{aligned}
(\mu\nu|\lambda\sigma) &= \int\int \varphi_\mu(1)\varphi_\nu(1)\frac{1}{r_{12}}\varphi_\lambda(2)\varphi_\sigma(2)dr_1dr_2 \\
&= \int\int \sum_{a_\mu=1}^{M_\mu} c_{a_\mu}\phi_{a_\mu}(1) \sum_{a_\nu=1}^{M_\nu} c_{a_\nu}\phi_{a_\nu}(1)\frac{1}{r_{12}} \sum_{a_\lambda=1}^{M_\lambda} c_{a_\lambda}\phi_{a_\lambda}(2) \sum_{a_\sigma=1}^{M_\sigma} c_{a_\sigma}\phi_{a_\sigma}(2)dr_1dr_2 \\
&= \sum_{a_\mu=1}^{M_\mu}\sum_{a_\nu=1}^{M_\nu}\sum_{a_\lambda=1}^{M_\lambda}\sum_{a_\sigma=1}^{M_\sigma} c_{a_\mu}c_{a_\nu}c_{a_\lambda}c_{a_\sigma} \int \phi_{a_\mu}(1)\phi_{a_\nu}(1)\frac{1}{r_{12}}\phi_{a_\lambda}(2)\phi_{a_\sigma}(2)dr_1dr_2 \quad (6.4)
\end{aligned}
$$

It was discovered that the optimum combination of speed and accuracy (when comparing to calculations using STOs) was achieved for $M = 3$. Figure 6.2 compares a 1s function using the STO-3G formalism to the corresponding STO and shows also the 3 primitives from which the contracted basis function is constructed. STO-3G basis functions have been defined for most of the atoms in the periodic table.

Gaussian functions have another feature that would be undesirable if they were to be used individually to represent atomic orbitals: they fail to exhibit radial nodal behavior. Thus, no choice of variables permits Eq. (6.2) to mimic a 2s orbital, which is negative near the origin and positive beyond a certain radial distance. Use of a contraction scheme, however, alleviates this problem; contraction coefficients c in Eq. (6.3) can be chosen to have either negative or positive sign, and thus fitting to functions having radial nodal behavior poses no special challenges.

While the acronym STO-3G is designed to be informative about the contraction scheme, it is appropriate to mention an older and more general notation that appears in much of the earlier literature, although it has mostly fallen out of use today. In that notation, the STO-3G H basis set would be denoted (3s)/[1s]. The material in parentheses indicates the number and type of primitive functions employed, and the material in brackets indicates the number and type of contracted functions. If first-row atoms are specified too, the notation for STO-3G would be (6s3p/3s)/[2s1p/1s]. Thus, for instance, lithium would require 3 each (since it is STO-3G) of 1s primitives, 2s primitives, and 2p primitives, so the total primitives are 6s3p, and the contraction schemes creates a single 1s, 2s, and 2p set, so the contracted functions are

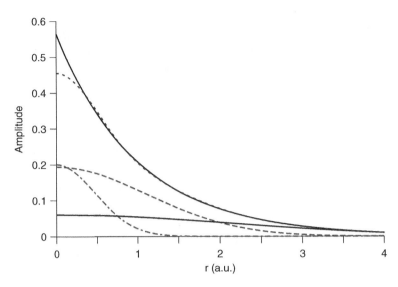

Figure 6.2 The radial behavior of various basis functions in atom-centered coordinates. The bold solid line at top is the STO ($\zeta = 1$) for the hydrogen 1s function; for the one-electron H system, it is also the exact solution of the Schrödinger equation. Nearest it is the contracted STO-3G 1s function (- - - - - -) optimized to match the STO. It is the sum of a set of one each tight (-··-··-··), medium (– – –), and loose (———) Gaussian functions shown below. The respective Gaussian primitive exponents α are 2.227660, 0.405771, and 0.109818, and the associated contraction coefficients c are 0.154329, 0.535328, and 0.444635. Note that from 0.5 to 4.0 a.u., the STO-3G orbital matches the correct orbital closely. However, near the origin there is a notable difference and, were the plot to extend to very large r, it would be apparent that the decay of the STO-3G orbital is more rapid than the correct orbital, in analogy to Figure 6.1

2s1p. These are separated from the hydrogenic details by a slash in each instance. Extensions to higher rows follow by analogy. Variations on this nomenclature scheme exist, but we will not examine them here.

As a final comment on the STO-MG series of basis sets, note that for higher rows than H and He, there is some efficiency to be gained by choosing the exponents used for the primitive Gaussians in the s and p contractions to be the same (then the radial parts of all four-index integrals are identical irrespective of whether they are (ss|ss), (ss|sp), (ss|pp), (sp|sp), etc.). Of course, the shape of s- and p-type functions are different, so the contraction coefficients are *not* identical. When common exponents are chosen in this fashion, the basis functions are sometimes called sp basis functions. Table 6.1 lists the exponents and contraction coefficients for the 2s and 2p functions of oxygen. Note the negative sign of the coefficient for the tightest function in the 2s expansion, thereby providing the proper radial nodal characteristics.

6.2.3 Single-ζ, Multiple-ζ, and Split-Valence

The STO-3G basis set is what is known as a 'single-ζ' basis set, or, more commonly, a 'minimal' basis set. This nomenclature implies that there is one and only one basis function

Table 6.1   STO-3G 2sp basis set for oxygen

α_{2sp}	c_{2s}	c_{2p}
5.0331527	−0.099967	0.155916
1.1695944	0.399513	0.607684
0.3803892	0.700115	0.391957

defined for each type of orbital core through valence. Thus for H and He, there is only a 1s function. For Li to Ne, there are five functions, 1s, 2s, $2p_x$, $2p_y$, and $2p_z$. For Na to Ar, 3s, $3p_x$, $3p_y$, and $3p_z$ are added to the second-row set, making a total of nine functions, etc. This number is the absolute minimum required, and it is certainly nowhere near the infinite basis set limit. Other minimal basis sets include the MINI sets of Huzinaga and co-workers, which are named MINI-1, MINI-2, etc., and vary in the number of primitives used for different kinds of functions.

One way to increase the flexibility of a basis set is to 'decontract' it. That is, we might imagine taking the STO-3G basis set, and instead of constructing each basis function as a sum of three Gaussians, we could construct *two* basis functions for each AO, the first being a contraction of the first two primitive Gaussians, while the second would simply be the normalized third primitive. This prescription would not double the size of our basis set, since we would have all the same individual integrals to evaluate as previously, but the size of our secular equation *would* be increased. A basis set with two functions for each AO is called a 'double-ζ' basis. Of course, we could decontract further, and treat each primitive as a full-fledged basis function, in which case we would have a 'triple-ζ' basis, and we could then decide to add more functions indefinitely creating higher and higher multiple-ζ basis sets. Modern examples of such basis sets are the cc-pCVDZ, cc-pCVTZ, etc. sets of Dunning and co-workers, where the acronym stands for 'correlation-consistent polarized Core and Valence (Double/Triple/etc.) Zeta' (Woon and Dunning 1995); correlation consistency and polarization are described in more detail below.

The advantage of such a scheme is, naturally, that these increasingly large basis sets must come closer and closer to the HF limit. Let us step back for a moment, however, and consider the *chemical* consequences of providing extra basis functions for a given AO. Recall that a final MO from an HF calculation is a linear combination of all of the basis functions. Indeed, if we were to examine the 1s core orbital resulting from an HF calculation on atomic oxygen using the fully uncontracted set of STO-3G Gaussian primitives as a basis (i.e., a triple-ζ basis), we might well find it to be a linear combination of the 1s functions very similar to that *defining* the STO-3G *contracted* oxygen 1s function. And, if we were to look at the MOs resulting from an equivalent calculation on, say, formaldehyde ($H_2C=O$), we would probably find another orbital which we would assign as the oxygen 1s orbital having very similar AO coefficients. Indeed, we would find this same orbital little changed in almost any molecule incorporating oxygen we might choose to examine. The reason for this is that core orbitals are only weakly affected by chemical bonding.

Valence orbitals, on the other hand, can vary widely as a function of chemical bonding. Atoms bonded to significantly more electronegative elements take on partial positive charge from loss of valence electrons, and thus their remaining density is distributed more compactly. The reverse is true when the bonding is to a more electropositive element. From a chemical standpoint, then, there is more to be gained by having flexibility in the valence basis functions than in the core, and recognition of this phenomenon led to the development of so-called 'split-valence' or 'valence-multiple-ζ' basis sets. In such basis sets, core orbitals continue to be represented by a single (contracted) basis function, while valence orbitals are split into arbitrarily many functions.

Amongst the most widely used split-valence basis sets are those of Pople *et al.* These basis sets include 3-21G, 6-21G, 4-31G, 6-31G, and 6-311G. The nomenclature is a guide to the contraction scheme. The first number indicates the number of primitives used in the contracted core functions. The numbers after the hyphen indicate the numbers of primitives used in the valence functions – if there are two such numbers, it is a valence-double-ζ basis, if there are three, valence-triple-ζ. This notation is somewhat more informative than the older style noted in the previous section. Thus, for a calculation on water, for instance, the 6-311G basis would be represented (11s5p/5s)/[4s3p/3s]. The latter notation does not specify how many primitives are devoted to which contracted basis functions, while 6-311G makes this point clear. Like the STO-MG basis sets, the split-valence sets use sp basis functions having common exponents.

An interesting question arises for split-valence and multiple-ζ basis sets: how should one go about choosing exponents and coefficients for the contracted functions? As the basis is no longer minimal, there is no particular virtue in fitting to STOs (which were originally used because they were thought to represent the optimal single-function approximation to an AO). Pople and co-workers, like most other researchers in the field, relied on the variational principle. That is, some test set of atoms and/or molecules was established, and exponents and coefficients were optimized so as to give the minimum energy over the test set. In the end, just as the name of a force field refers to its functional form and a list of all its parameters, so too the name of a basis set refers to its contraction scheme and a list of all of its exponents and coefficients for each atom.

One feature of the Pople basis sets is that they use a so-called 'segmented' contraction. This implies that the primitives used for one basis function are not used for another of the same angular momentum (e.g., no common primitives between the 2s and 3s basis functions for phosphorus). Such a contraction scheme is typical of older basis sets. Other segmented split-valence basis sets include the MIDI and MAXI basis sets of Huzinaga and co-workers, which are named MIDI-1, MIDI-2, etc., MAXI-1, MAXI-2, etc. and vary in the number of primitives used for different kinds of functions.

The Pople basis sets have seen sufficient use in the literature that certain trends have clearly emerged. While a more complete discussion of the utility of HF theory and its basis-set dependence appears at the end of this chapter, we note here that, in general, the 4-31G basis set is inferior to the less expensive 3-21G, so there is little point in ever using it. The 6-21G basis set is obsolete.

An alternative method to carrying out a segmented contraction is to use a so-called 'general' contraction (Raffenetti 1973). In a general contraction, there is a single set of primitives that are used in *all* contracted basis functions, but they appear with different coefficients in each. The general contraction scheme has some technical advantages over the segmented one. One advantage in terms of efficiency is that integrals involving the same primitives, i.e., those occurring in the final line of Eq. (6.4), need in principle be calculated only once, and the value can be stored for later reuse as needed. Examples of split-valence basis sets using general contractions are the cc-pVDZ, cc-pVTZ, etc. sets of Dunning and co-workers, where the acronym stands for 'correlation-consistent polarized Valence (Double/Triple/etc.) Zeta' (Dunning 1989; Woon and Dunning 1993). The 'correlation-consistent' part of the name implies that the exponents and contraction coefficients were variationally optimized not only for HF calculations, but also for calculations including electron correlation, methods for which are described in Chapter 7. The subject of polarization is what we turn to next.

6.2.4 Polarization Functions

The distinction between atomic orbitals and basis functions in molecular calculations has been emphasized several times now. An illustrative example of why the two should not necessarily be thought of as equivalent is offered by ammonia, NH_3. The inversion barrier for interconversion between equivalent pyramidal minima in ammonia has been measured to be 5.8 kcal mol^{-1}. However, a HF calculation with the equivalent of an infinite, atom-centered basis set of s and p functions predicts the planar geometry of ammonia to be a minimum-energy structure!

The problem with the calculation is that s and p functions centered on the atoms do not provide sufficient mathematical flexibility to adequately describe the wave function for the pyramidal geometry. This is true even though the atoms nitrogen and hydrogen can individually be reasonably well described entirely by s and p functions. The *molecular* orbitals, which are eigenfunctions of a Schrödinger equation involving multiple nuclei at various positions in space, require more mathematical flexibility than do the atoms.

Because of the utility of AO-like GTOs, this flexibility is almost always added in the form of basis functions corresponding to one quantum number of higher angular momentum than the valence orbitals. Thus, for a first-row atom, the most useful polarization functions are d GTOs, and for hydrogen, p GTOs. Figure 6.3 illustrates how a d function on oxygen can polarize a p function to improve the description of the O−H bonds in the water molecule. The use of p functions to polarize hydrogen s functions has already been mentioned in Section 4.3.1. [An alternative way to introduce polarization is to allow basis functions not to be centered on atoms. Such floating Gaussian orbitals (FLOGOs) are illustrated on the left-hand side of Figure 4.1. While the use of FLOGOs reduces the need to work with integrals involving high-angular-momentum functions, the process of geometry optimization is rendered considerably more complicated, so they are rarely employed in modern calculations.] Adding d functions to the nitrogen basis set causes HF theory to predict correctly a pyramidal minimum for ammonia, although some error in prediction of the inversion barrier still exists even at the HF limit because of the failure to account for electron correlation.

A variety of other molecular properties prove to be sensitive to the presence of polarization functions. While a more complete discussion occurs in Section 6.4, we note here that d functions on second-row atoms are absolutely required in order to make reasonable

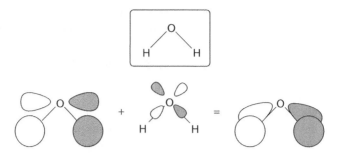

Figure 6.3   The MO formed by interaction between the antisymmetric combination of H 1s orbitals and the oxygen p_x orbital (see also Figure 6.7). Bonding interactions are enhanced by mixing a small amount of O d_{xz} character into the MO

predictions for the geometries of molecules including such atoms in formally hypervalent bonding situations, e.g., phosphates, sulfoxides, siliconates, etc.

A variety of empirical rules exist for choosing the exponent(s) for a set of polarization functions. If only a single set is desired, one possible choice is to make the maximum in the radial density function, $\langle r^2 \rangle$, equal to that for the existing valence set (e.g., the 3d functions that best 'overlap' the 2p functions for a first-row atom – note that the radial density is used instead of the actual overlap integral because the latter, by symmetry, must be zero).

Because of the expense associated with adding polarization functions – the total number of functions begins to grow rather quickly with their inclusion – early calculations typically made use of only a single set. Pople and co-workers introduced a simple nomenclature scheme to indicate the presence of these functions, the '*' (pronounced 'star'). Thus, 6-31G* implies a set of d functions added to polarize the p functions in 6-31G. A second star implies p functions on H and He, e.g., 6-311G** (Krishnan, Frisch, and Pople 1980).

Subsequent work has shown that there is a rough correspondence between the value of adding polarization functions and the value of decontracting the valence basis function(s). In particular, there is a rough equality between each decontraction step and adding one new set of polarization functions, including a new set of higher angular momentum. Put more succinctly, 'balanced' double-ζ basis sets should include d functions on heavy atoms and p functions on H, triple-ζ basis sets should include 1 set of f and 2 sets of d functions on heavy atoms, and 1 set of d and 2 sets of p functions on H, etc. This is the polarization prescription adopted by the cc-pVnZ basis sets of Dunning and co-workers already mentioned above, where n ranges over D (double), T (triple), Q (quadruple), five, and six (Wilson, van Mourik, and Dunning 1996). Thus, for cc-pV6Z, for example, each heavy atom has one i function, two h functions, three g functions, four f functions, five d functions, and six valence s and p functions, in addition to core functions (using the canonical numbers of these functions, we have 140 basis functions for a single second-row atom, so this basis set presently finds use only for the smallest of systems). Note that while it would be an unpleasant exercise to try to draw an i function, it is straightforwardly defined by taking the sum of i, j, and k equal to 6 in Eq. (6.2).

A somewhat more detailed analysis of the correct ratio of number of polarization functions to number of valence functions has been carried out by Jensen (2001) in the context of the polarization consistent basis sets pc-n, where n indicates the largest increment in angular momentum above the valence maximum used in the basis set. The ratio is not fixed, per se, but varies depending on an analysis of energetic convergence over a test set of first-row molecules. The pc-n basis sets are presently defined for H−F for $n = 0$–4.

Recognizing the tendency to use more than one set of polarization functions in modern calculations, the standard nomenclature for the Pople basis sets now typically includes an explicit enumeration of those functions instead of the star nomenclature. Thus, 6-31G(d) is to be preferred over 6-31G* because the former obviously generalizes to allow names like 6-31G(3d2fg,2pd), which implies heavy atoms polarized by three sets of d functions, two sets of f functions, and a set of g functions, and hydrogen atoms by two sets of p functions and one of d (note that since this latter basis set is only valence double-ζ, it is somewhat unbalanced by having so many polarization functions).

A partially polarized basis set, MIDI! (where the '!' is pronounced 'bang'; in some electronic structure programs, the abbreviation MIDIX is employed to avoid complications associated with interpretation of the exclamation point), has been introduced by Cramer and Truhlar and co-workers, who adopted a different philosophy in its development (Easton *et al.* 1996; Li, Cramer, and Truhlar 1998). Rather than optimizing the basis set with respect to molecular energies, they sought to design an economical basis set for geometry optimizations and partial charge calculations on medium-sized molecules, including neutrals, cations, and anions, with special emphasis on functional groups that are important for biomolecules. The MIDI! basis set has d functions on all atoms heavier than H for which it is defined with the exception of carbon (i.e., on heteroatoms). Although much smaller than the 6-31G(d) basis set, for instance, in direct comparisons it yields more accurate geometries and charges as judged by comparison to much higher level calculations. For cases where p polarization functions on H might be expected to be important, the MIDIY basis set includes these as an extension to MIDI! (Lynch and Truhlar 2004).

Finally, an important nomenclature point is that most basis sets are *defined* to use the five spherical d functions, but an important exception is 6-31G* (or 6-31G(d)), which is defined to use the six Cartesian d functions. Some electronic structure programs are not flexible about permitting the user to choose how many d functions are used, so it is important to check that a consistent scheme has been employed when comparing to existing literature data. To avoid ambiguity, it is helpful to modify the basis set name when the number of d functions employed is not the same as that assumed as the default, e.g., MIDI!(6d) to denote use of the six Cartesian d functions instead of the normal spherical five with the MIDI! basis. Another nomenclature issue of importance involves the basis set '3-21G*'. While this notation pervades the literature, it is ambiguous and should be avoided. Pople and co-workers suggested taking from 6-31G* the polarization functions for second-row atoms and beyond and using them directly (i.e., without any reoptimization of exponents) with the smaller basis 3-21G. The motivation for this was to address the serious geometry problems that arise for hypervalent second-row atoms without d functions whilst maintaining a very cheap description of first-row atoms. To distinguish this situation from the normal '*', they named this basis set 3-21G$^{(*)}$, and that is the notation that should always be used to emphasize that no d functions are present on first-row atoms.

6.2.5 Diffuse Functions

The highest energy MOs of anions, highly excited electronic states, and loose supermolecular complexes, tend to be much more spatially diffuse than garden-variety MOs. When a basis set does not have the flexibility necessary to allow a weakly bound electron to localize far from the remaining density, significant errors in energies and other molecular properties can occur. To address this limitation, standard basis sets are often 'augmented' with diffuse basis functions when their use is warranted.

In the Pople family of basis sets, the presence of diffuse functions is indicated by a '+' in the basis set name. Thus, 6-31+G(d) indicates that heavy atoms have been augmented with an additional one s and one set of p functions having small exponents. A second plus indicates the presence of diffuse s functions on H, e.g., 6-311++G(3df,2pd). For the Pople basis sets, the exponents for the diffuse functions were variationally optimized on the anionic one-heavy-atom hydrides, e.g., BH_2^-, and are the same for 3-21G, 6-31G, and 6-311G. In the general case, a rough rule of thumb is that diffuse functions should have an exponent about a factor of four smaller than the smallest valence exponent. Diffuse sp sets have also been defined for use in conjunction with the MIDI! and MIDIY basis sets, generating MIDIX+ and MIDIY+, respectively (Lynch and Truhlar 2004); the former basis set appears particularly efficient for the computation of accurate electron affinities.

In the Dunning family of cc-pVnZ basis sets, diffuse functions on all atoms are indicated by prefixing with 'aug'. Moreover, one set of diffuse functions is added for *each* angular momentum already present. Thus, aug-cc-pVTZ has diffuse f, d, p, and s functions on heavy atoms and diffuse d, p, and s functions on H and He. An identical prescription for diffuse functions has been used by Jensen (2002) in connection with the pc-n basis sets.

Particularly for the calculation of acidities and electron affinities, diffuse functions are absolutely required. For instance, the acidity of HF (not Hartree-Fock in this case, but hydrogen fluoride) increases by 44 kcal/mol when the 6-31+G(d) basis set is used instead of unaugmented 6-31G(d).

6.2.6 The HF Limit

Solution of the HF equations with an infinite basis set is defined as the HF limit. Actually carrying out such a calculation is almost never a practical possibility. However, it is sometimes the case that one may extrapolate to the HF limit with a fair degree of confidence.

Of the basis sets discussed thus far, the cc-pVnZ and cc-pCVnZ examples were designed expressly for this purpose. As they increase in size in a consistent fashion with each increment of n, one can imagine plotting some particular computed property as a function of n^{-1} and extrapolating a curve fit through those points back to the intercept; the intercept corresponds to $n = \infty$, i.e., the infinite basis limit (Figure 6.4).

Note that certain issues do arise in how one should carry out this extrapolation. If the property is sensitive to geometry, should the geometry be optimized at each level, or should a single geometry be chosen, thereby permitting the extrapolating equation to account for basis-set effects only? Are there any fundamental principles dictating what form the extrapolating

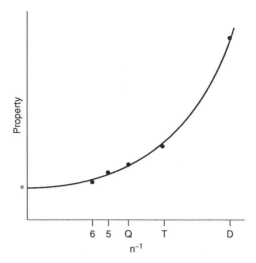

Figure 6.4   Use of an extrapolation procedure to estimate the expectation value for some property at the HF limit. The abscissa is marked off as n^{-1} in cc-pVnZ notation (see page 162). Note the sensitivity of the limiting value, which is to say the ordinate intercept, that might be expected based on the use of different curve-fitting procedures

equation should take, or can any arbitrary curve fitting approach be applied? In general, the answers to these questions are case-dependent, and the chemist cannot be completely removed from the calculation.

Note that the cost of the extrapolation procedure outlined above becomes increasingly large as points for $n = 4$, 5, and 6 are added. For systems having more than five or six atoms, these calculations can be staggeringly demanding in terms of computational resources.

A somewhat more common approach is one that does not try explicitly to extrapolate to the HF limit but uses similar concepts to try to correct for some basis-set incompleteness. The assumption is made that the effects of 'orthogonal' increases in basis set size can be considered to be additive (a substantial amount of work suggests that this assumption is typically not too bad, at least for molecular energies), and thus the individual effects can be summed together to estimate the full-basis-set result. This is best illustrated by example. Consider HF calculations carried out for the chemical warfare agent VX ($C_{11}H_{26}NO_2PS$, Figure 6.5) with the following basis sets: 6-31G, 6-31++G, 6-31G(d,p), 6-311G, and 6-311++G(d,p). With these basis sets, the total number of basis functions for VX are 204, 378, 294, 294, and 542, respectively.

The additivity assumption can be expressed as

$$E[\text{HF}/6\text{-}311{+}{+}G(d,p)] \approx E[\text{HF}/6\text{-}31G]$$

$$+ \{E[\text{HF}/6\text{-}31G(d,p)] - E[\text{HF}/6\text{-}31G]\}$$

$$+ \{E[\text{HF}/6\text{-}311G] - E[\text{HF}/6\text{-}31G]\}$$

$$+ \{E[\text{HF}/6\text{-}31{+}{+}G] - E[\text{HF}/6\text{-}31G]\} \qquad (6.5)$$

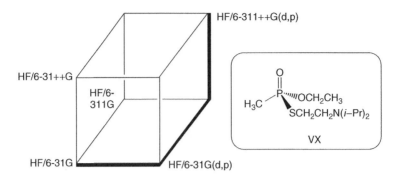

Figure 6.5 The chemical warfare agent VX and a conceptual illustration of the additivity concept embodied in Eq. (6.5). Each boldface line in the additivity cube represents one line on the r.h.s. of the equation

where the notation 'x/y' implies 'level of theory x using basis set y'. Each successive line on the r.h.s. of Eq. (6.5) reflects the incremental contribution from a particular basis set improvement – first polarization functions, then valence decontraction, then diffuse functions. As already noted, the calculation on the l.h.s. requires 542 basis functions. Although there are four *different* calculations on the r.h.s., if we recall that the amount of time for an HF calculation scales formally as the fourth power of the number of basis functions, the amount of time to carry out those four calculations, expressed as a fraction of the amount of time to carry out the full calculation, is $(204^4 + 378^4 + 294^4 + 294^4)/542^4 = 0.43$. That is, evaluation of the r.h.s. of Eq. (6.5) takes less than half the time of evaluation of the l.h.s.

While the above schemes are interesting from a technical standpoint, it must be recalled that chemically there are potentially large errors associated with the HF approximation, so the HF limit is of more interest from a formal standpoint than from a chemical one. Thus, we will defer additional discussion of extrapolation and additivity concepts until Section 7.7, where it is re-examined in the additional context of accounting for electron correlation effects.

6.2.7 Effective Core Potentials

The periodic table is rich and complex, and very heavy elements pose rather distinct challenges to MO theory. First, there is the purely technical hurdle that such elements have large numbers of electrons, and there is thus a concomitant requirement to use a large number of basis functions to describe them. Of course, these extra electrons are mostly core electrons, and thus a minimal representation will probably be adequate. Nevertheless, if one wants to model a small cluster of uranium atoms, for instance, the basis set size quickly becomes intractable. Not surprisingly, more electrons means more energy associated with electron correlation, too.

It was Hellmann (1935) who first proposed a rather radical solution to this problem – replace the electrons with analytical functions that would reasonably accurately, and much more efficiently, represent the combined nuclear–electronic core to the remaining electrons. Such functions are referred to as effective core potentials (ECPs). In a sense, we have already seen ECPs in a very crude form in semiempirical MO theory, where, since only valence electrons are treated, the ECP is a nuclear point charge reduced in magnitude by the number of core electrons.

In *ab initio* theory, ECPs are considerably more complex. They properly represent not only Coulomb repulsion effects, but also adherence to the Pauli principle (i.e., outlying atomic orbitals must be orthogonal to core orbitals having the same angular momentum). This being said, we will not dwell on the technical aspects of their construction. Interested readers are referred to the bibliography at the end of the chapter.

Note that were ECPs to do nothing more than reduce the scope of the electronic structure problem for heavy elements, they would still have great value. However, they have another virtue as well. The core electrons in very heavy elements reach velocities sufficiently near the speed of light that they manifest relativistic effects. A non-relativistic Hamiltonian operator is incapable of accounting for such effects, which can be significant for many chemical properties (see, for example, Kaltsoyannis 2003). A full discussion of modeling relativistic effects, while a fascinating topic, is well beyond the scope of this book, although some details are discussed in Section 7.4.4. We note here simply that, to the extent an ECP represents the behavior of an atomic core, relativistic effects can be folded in, and thereby removed from the problem of finding suitable wave functions for the remaining electrons.

A key issue in the construction of ECPs is just how many electrons to include in the core. So-called 'large-core' ECPs include everything but the outermost (valence) shell, while 'small-core' ECPs scale back to the next lower shell. Because polarization of the sub-valence shell can be chemically important in heavier metals, it is usually worth the extra cost to explicitly include that shell in the calculations. Thus, the most robust ECPs for the elements Sc–Zn, Y–Cd, and La–Hg, employ [Ne], [Ar], and [Kr] cores, respectively. There is less consensus on the small-core vs. large-core question for the non-metals.

Popular pseudopotentials in modern use include those of Hay and Wadt (sometimes also called the Los Alamos National Laboratory (or LANL) ECPs; Hay and Wadt 1985), those of Stevens *et al.* (1992), and the Stuttgart–Dresden pseudopotentials developed by Dolg and co-workers (2002). The Hay–Wadt ECPs are non-relativistic for the first row of transition metals while most others are not; as relativistic effects are usually quite small for this region of the periodic table, the distinction is not particularly important. Lovallo and Klobukowski (2003) have recently provided additional sets of both relativistic and non-relativistic ECPs for these metals. For the p block elements, Check *et al.* (2001) have optimized polarization and diffuse functions to be used in conjunction with the LANL double-ζ basis set.

Another recent set of pseudopotentials for the 4p, 5p, and 6p elements has been developed by Dyall (1998, 2002). These ECPs are designed to be the ECP-equivalent to the correlation-consistent basis sets of Dunning insofar as (i) prescriptions for double-ζ and triple-ζ contractions are provided, (ii) polarization functions of increasingly higher angular momentum are included for the triple-ζ case, and (iii) diffuse functions were optimized for negative ions at a correlated level of electronic structure theory.

6.2.8   Sources

Most electronic structure programs come with a library of built-in basis sets, to include many if not all of those mentioned above. A tremendously useful electronic resource is the Environmental Molecular Sciences Laboratory Gaussian Basis Set Order Form, a website that permits the download of a very large number of different basis sets formatted for a variety of different software packages. Moreover, the site has reference information that typically includes values for test calculations as published by the original authors. Since different software packages may have different conventions for how to deal with certain aspects of the basis set (e.g., five spherical vs. six Cartesian d functions), it is always a good idea to carry out such test calculations to ensure that the basis set is being used in a manner consistent with its definition and, hopefully, with previously reported calculations in the literature.

So, how to choose the 'best' basis set for the problem at hand? Obviously a fair rule of thumb is that bigger is better, keeping in mind issues of balance between valence decontraction and presence of polarization functions. As noted above, diffuse functions are warranted in certain specific situations, but in the absence of those situations, there tends to be no strong reason to include them.

Additionally, access to particular software packages may play some role in motivating the choice of basis set. Some packages are equipped to take advantage of efficiencies possible for such features as combined s and p exponents, or general contractions, while others are not, and there may thus be significant timing issues differentiating basis sets.

Finally, and perhaps most important for the vast majority of chemical problems where saturation of the basis set is not a practical possibility, the choice should consider the degree to which other results from that particular basis set at that particular level of theory are available for comparison. For instance, to the extent that there are an enormous number of HF/6-31G(d) results published, and thus a reasonably firm understanding of the specific successes and failures of the model, this can assist in the interpretation of new results – Pople has referred to the collection of all data from a given theoretical prescription as comprising a 'model chemistry' and emphasized the utility of analyzing theoretical performance (and future model development efforts) within such a framework.

6.3   Key Technical and Practical Points of Hartree–Fock Theory

A deep understanding of the underlying theory is, alas, of only limited value in successfully carrying out a HF calculation with any given software package. This section is not designed to supplant program users' manuals, the utility of reading which cannot be overemphasized, but discusses aspects of practical HF calculations that are often glossed over in formal presentations of the theory.

6.3.1 SCF Convergence

As noted in Chapter 4, there is never any guarantee that the SCF process will actually converge to a stable solution. A fairly common problem is so-called 'SCF oscillation'. This occurs when a particular density matrix, call it $\mathbf{P}^{(a)}$, is used to construct a Fock matrix $\mathbf{F}^{(a)}$ (and thus the secular determinant), diagonalization of which permits the construction of an updated density matrix $\mathbf{P}^{(b)}$; this is a general description of any step in the SCF cycle. In the oscillatory case, however, the diagonalization of the Fock matrix created using $\mathbf{P}^{(b)}$ (i.e., $\mathbf{F}^{(b)}$) gives a density matrix indistinguishable from $\mathbf{P}^{(a)}$. Thus, the SCF simply bounces back and forth from $\mathbf{P}^{(a)}$ to $\mathbf{P}^{(b)}$ and never converges. This behavior can be recognized easily by looking at the SCF energy for each step, which itself bounces back and forth between the two discrete values associated with the two different unconverged wave functions defined by $\mathbf{P}^{(a)}$ and $\mathbf{P}^{(b)}$.

In more pathological cases, the SCF behaves even more badly, with large changes occurring in the density matrix at every step. Again, observation of the energies associated with each step is diagnostic for this problem; they are observed to vary widely and seemingly randomly. Such behavior is not uncommon for the first three or four steps of a typical SCF, but usually beyond this point there is a 'zeroing-in' process that leads to convergence.

In the abstract sense, converging the SCF equations is a problem in applied mathematics, and many algorithms have been developed for this process. While the technical details are not presented here, the process is quite analogous to the process of finding a minimum on a PES as described in Chapter 2. In the SCF problem, instead of a space of molecular coordinates we operate in a space of orbital coefficients (so-called 'Fock space'), and there are certain constraints beyond the purely energetic ones, but many of the search strategies are analogous. Similarly analogous is the degree to which they tend to balance speed and stability. Usually the default optimizer in a given program is the fastest one available, while other methods (e.g., quadratically convergent methods) typically take more steps to converge but are less likely to suffer from oscillation or other problems. Thus, one option for dealing with a system where convergence proves difficult is simply to run through all the different convergence schemes offered by the electronic structure package and hope that one proves sufficiently robust.

In general, however, it is more efficient to solve the problem using chemistry rather than mathematics. If the SCF equations are failing to converge, the problem lies in the initial guess (this is, of course, something of a truism, for if you were to guess the proper eigenfunction, obviously there would be no problem with convergence). Most programs use as their default option a semiempirical method to generate a guess wave function, e.g., EHT or INDO. The resulting wave function (remember that a wave function is simply the list of coefficients describing how the basis functions are put together to form the occupied MOs) is then used to construct a guess for the HF calculation by mapping coefficients from the basis set of the semiempirical method to the basis set for the HF calculation.

When the HF basis set is minimal, this is fairly simple (there is a one-to-one correspondence in basis functions) but when it is larger, some algorithmic choices are made about how to carry out the mapping (e.g., always map to the tightest function or map based on overlap between the semiempirical STO and the large-basis contracted GTO). Thus, it is *usually* easier to converge a small-basis-set HF calculation than a larger one. This suggests a method for bootstrapping one's way to the convergence of a large-basis-set calculation: First, obtain a wave function from a minimal basis set (e.g., STO-3G), then use that as an initial guess for a calculation with a small split-valence basis set (e.g., 3-21G), and repeat this process with increasingly larger basis sets until the target is reached. Because of the exponential scaling, the early calculations typically represent a negligible time investment, especially if they are saving steps in a slowly converging SCF for the full-sized basis set by providing a more accurate initial guess.

The above process has another possible utility that is associated with the molecular geometry. Often when an SCF is difficult to converge, the problem is that the molecular structure is very bad. If that is the case, there can be a very small separation between the highest occupied MO (HOMO) and the lowest unoccupied MO (LUMO). Such small separations wreak havoc on the SCF process, because it is possible that occupation of *either* orbital could lead to HF eigenfunctions of similar energy. In that case, the characters of the two orbitals are very sensitive to all the remaining occupied orbitals, which generate the static potential felt by the highest energy electrons, and their coefficients can undergo large changes that fail to converge (an issue of non-dynamical electron correlation, see Section 7.1). Optimizing the geometry at a low level of theory, where the wave function *can* be coaxed to converge, is typically an efficient way to overcome this problem. Some care must be exercised, however, in systems where the lowest levels of theory may not be reliable for molecular geometries. As a general rule, however, visualization of the structure, and some thoughtful analysis of it by comparison to whatever analogs or prior calculations may be available, is nearly always worth the effort.

Very complete basis sets, or those with many diffuse functions, pose some of the worst problems for SCF convergence because of near-linear dependencies amongst the basis functions. That is, some basis functions may be fairly well described as linear combinations of other basis functions. This is most readily appreciated by considering two very diffuse s orbitals on adjacent atoms; if they have maxima in their radial density at 40 Å but the two atoms are only 1.5 Å apart, the two basis functions are really almost indistinguishable from one another throughout most of space. If a basis set has a *true* linear dependence, then it is necessarily impossible to assure orthogonality of all of the MOs (a division by zero occurs at a particular point of the SCF process), so very near-linear dependence can lead to numerical instabilities. Thus, it is again important to have a good guess. In a case like this, sometimes it is useful not only to carry out bootstrap calculations in terms of basis sets, but in terms of electrons. Thus, if one is interested in an anion, for instance, one can first try to converge a large-basis-set wave function for the neutral (or the cation), to get a good estimate of the more compact MOs, and then import that wave function as a guess for the anionic system, trying thereby to reduce the impact of possible numerical instabilities.

6.3.2 Symmetry

The presence of symmetry in a molecule can be used to great advantage in electronic structure calculations, although some care is required to avoid possible pitfalls that are simultaneously introduced (Appendix B provides a brief overview of nomenclature (e.g., the term "irrep", which is used below) and key principles of group theory as they apply to MO calculations and symmetry). The advantages of symmetry are primarily associated with computational efficiency.

The most obvious advantage is one step removed from the electronic structure problem, namely geometry optimization. The presence of symmetry elements removes some of the $3N - 6$ degrees of molecular freedom (where N is the number of atoms) that would otherwise be present for an asymmetric molecule. This reduction in the dimensionality of the PES can make the search for stationary points more efficient. Consider benzene (C_6H_6), for example. With 12 atoms, the PES formally has 30 dimensions, and a complete representation would be graphically challenging. However, if we restrict ourselves to structures of D_{6h} symmetry, then there are only two degrees of freedom – one's first choice for defining those degrees of freedom might be the C–C and C–H bond lengths; an equally valid choice, which may be more useful as input to a software program that will ensure preservation of symmetry, is the O–C and O–H radial distances, where O is the point at the center of the benzene ring. Finding a minimum on a two-dimensional PES is obviously quite a bit simpler than on a 30-dimensional one (Figure 6.6).

Symmetry is also tremendously useful in several aspects of solving the SCF equations. A key feature is the degree to which it simplifies evaluation of the four-index integrals.

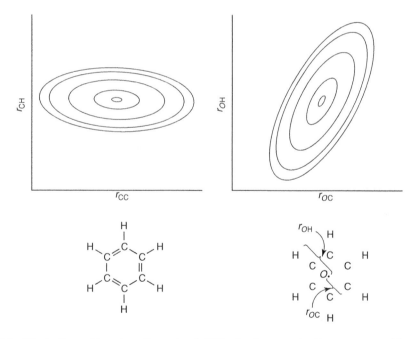

Figure 6.6   Illustration of two two-dimensional PESs for benzene in D_{6h} symmetry. The surfaces differ in choice of coordinates, which may affect optimizer efficiency, ease of input, etc., but will have no effect on the equilibrium structure. Contour lines reflect constant energy intervals of arbitrary magnitude. No attempt is made to illustrate the full 30-dimensional PES, on which it would be considerably more taxing to search for a minimum-energy structure

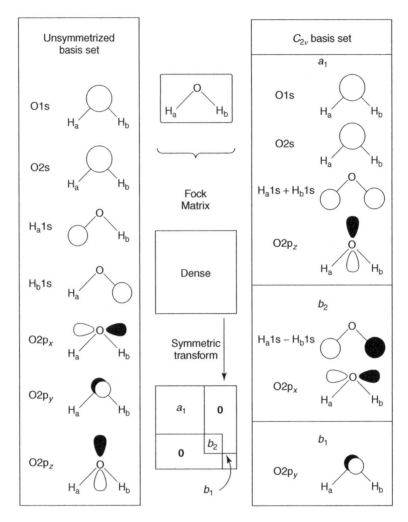

Figure 6.7 Transformation of the minimal water AO basis set to one appropriate for the C_{2v} point group. The effect on the form of the Fock matrix is also illustrated

In particular, if the totally symmetric representation is not included in the product of the irreducible representations of basis functions μ, ν, λ, and σ, then $(\mu\nu|\lambda\sigma) = 0$. The analogous rule holds for the one-electron integrals. In general, of course, the atomic basis functions do not belong to *any* irreducible representation, since they themselves do not transform with all the symmetry elements of the molecule. Thus, as a first step to taking advantage of symmetry, linear combinations of the various basis functions must be formed that *do* belong to irreps of the molecular point group. This process is illustrated in Figure 6.7 for a HF/STO-3G calculation on water, which belongs to the C_{2v} point group. The seven atomic basis functions can be linearly transformed to four, two, and one functions belonging to the

a_1, b_2, and b_1 irreps, respectively (no combination of basis functions belongs to the a_2 irrep with STO-3G; were d functions to be present on oxygen, the d_{xy} function would be a_2).

In the C_{2v} point group, the totally symmetric representation (a_1) is contained only in the product of any irrep with itself. Thus, any matrix element F_{ab} where transformed basis functions a and b belong to *different* irreps is zero, and the Fock matrix expressed in the transformed basis is block diagonal (see Figure 6.7). Recall that the process of solving the secular equation is equivalent to diagonalization of the Fock matrix. Diagonalization of a block diagonal matrix can be accomplished by separate diagonalization of each block. Noting that diagonalization scales as N^3, where N is the dimensionality of the matrix, the total time to diagonalize our symmetrized Fock matrix for water compared to the unsymmetrized alternative is $(4^3 + 2^3 + 1^3)/7^3 = 0.21$, i.e., a saving of almost 80 percent.

Strangely enough, with all of the advantages of symmetry, one often finds in the literature statements by authors that they deliberately did *not* employ symmetry, as though such a protocol has some virtue associated with it. What motivates this choice? For some, it reflects a reluctance to work on a reduced-dimensionality PES because minima on that PES may *not* be minima on the full PES. This is best illustrated with an example. Consider the chloride/methyl chloride system with D_{3h} symmetry imposed upon it (Figure 6.8). The system under these constraints has only two degrees of freedom, the C–H bond length and the C–Cl bond length; the overall structure, however, is that associated with the exchange of one chloride ion for another in a bimolecular nucleophilic substitution (i.e., an S_N2 reaction). Minimizing the energy of the system subject to the D_{3h} constraint will give the best possible energy for this arrangement, and hence the TS structure for the reaction. The reason that it is a TS structure and not a true minimum is that the degree(s) of freedom that would change in order to reach a minimum energy structure, i.e., to generate different C–Cl bond lengths, are not included in the reduced-dimensionality PES. Had no symmetry been imposed, however, the system would eventually have moved in this direction (given a competent optimizer) unless a transition-state search had been specified.

This issue is really of little importance for modern purposes, however. The best way to evaluate the nature of a stationary point, irrespective of whether it was located using symmetry or not, is to carry out a calculation of the full-dimensional Hessian matrix (see Sections 2.4.1 and 9.3.2). Such a calculation is definitive. In the event that a symmetric stationary point is found *not* to have the character desired, it is often of interest in any case (if it is a TS structure) because inspection of the mode(s) having negative force constants permits an efficient start at optimizing to the desired point using lower symmetry. Since higher symmetry calculations tend to be quite efficient in any case, there is little to be lost by imposing symmetry at the start, and deciding along the way whether some or all symmetry constraints must be relaxed. Note, however, that symmetry constraints must arise from *molecular* symmetry, not an erroneous idea of local symmetry. Thus, for instance, the three C–H bonds of a methyl group should not be constrained to have the same length unless they are truly symmetrically related by a molecular C_3 axis.

Another potential pitfall with symmetry constraints involves the nature of the wave function. Consider the nitroxyl radical H_2NO, which has C_s symmetry (Figure 6.9). The unpaired electron can either reside in an MO dominated by an oxygen p orbital that is of a''

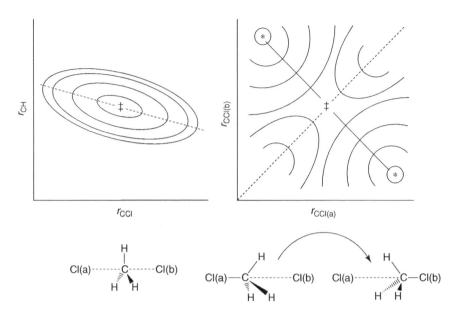

Figure 6.8   Reduced-dimensionality PESs for the chloride/methyl chloride system. On the left is the two-dimensional surface associated with a D_{3h} symmetry constraint. On this surface, the point marked ‡ is a minimum. The simultaneous shortening or lengthening of the C–Cl bonds (simultaneous to preserve D_{3h} symmetry) while allowing the C–H bond lengths to relax is indicated by the dashed line on this surface. The same process is indicated by the dashed line on the surface to the right, whose coordinates are the individual C–Cl bond lengths, and point ‡ again represents the minimum on this line. However, movement *off* the dashed line can lower the energy further. Movement along the *solid* line, which involves lengthening one C–Cl bond whilst shortening the other, corresponds to the reaction path for nucleophilic substitution from one equilibrium structure to another (points marked ∗), and illustrates that the minimum-energy structure under the D_{3h} constraint is actually a TS structure on the full PES

symmetry, or in an MO having π^*_{NO} character that is of a′ symmetry. These two electronic states are fundamentally different. The symmetry of a doublet electronic state is simply the symmetry of the half-filled orbital if all other orbitals are doubly occupied, so we would refer to the two possible electronic states here as $^2A''$ and $^2A'$, respectively. When symmetry is imposed, we will have a block diagonal Fock matrix and the unpaired electron will appear in either the a′ block or the a″ block, depending on the initial guess. Once placed there, most SCF convergence procedures will not provide any means for the electronic state symmetry to change, i.e., if the initial guess is a $^2A'$ wave function, then the calculation will proceed for that state, and if the initial guess is a $^2A''$ wave function, then it will instead be that state that is optimized. The two states both exist, but one is the ground state and the other an excited state, and one must take care to ensure that one is not working with the undesired state.

Typically, one can assess the nature of the state (ground vs. excited) after convergence of the wave function. Continuing with our example, let us say that we have optimized the $^2A'$ state. We can then take that wave function, alter it so that the occupation number of the highest occupied a′ orbital is zero instead of one, and the occupation of the lowest formerly

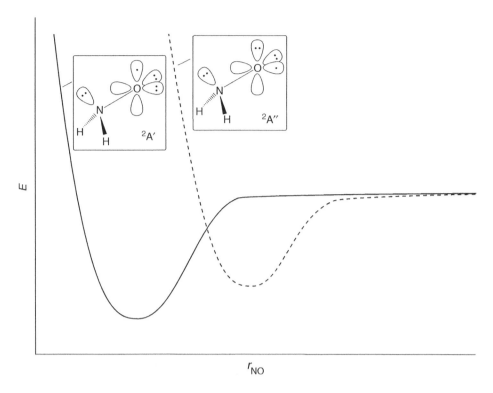

Figure 6.9 Curve crossing of the two lowest energy doublets for C_s H$_2$NO$^\bullet$. The question of which one is the ground state depends on the NO bond length. One can thus be misled in looking for the lowest energy structure if one fails to optimize the geometry for each

unoccupied a$''$ orbital is one instead of zero (i.e., construct a ^{2}A$''$ wave function using the MOs of the ^{2}A$'$ wave function) and carry out an SCF calculation using this construction as the initial guess. If the energy drops relative to the first wave function, then the first was an excited state. Many electronic structure programs offer the option to do this in a systematic fashion, i.e., to consider *every* possible switch of an electron from one orbital to another (such a calculation is really a CIS calculation, see Section 14.2.2). Note that there can be some challenging subtleties in working with systems where many states are close to one another in energy. For instance, it could occur in the nitroxyl example above that the two electronic state PESs cross one another in such a fashion that each of the two states is the ground state *at its respective optimized geometry*. Such a situation can only be determined by a fairly careful analysis of the PESs for both states. Note that failing to impose symmetry on the system does not in any way alleviate this problem. Instead, it obscures it, since no symmetry labels can be applied to the orbitals and thus, in the absence of visualization of the half-filled orbital, there is no simple means to differentiate between the two states.

Note that the problem just discussed above is rarely encountered for closed-shell singlets. That is because any excitation from an orbital of one symmetry type to one of a different symmetry type must be a *double* excitation if the closed-shell character of the wave function is to be preserved. Typically the difference in energy between these two possible configurations is so large that no reasonable means for guessing the initial wave function generates the higher energy possibility. This is one of the advantages of closed-shell states compared to open-shell ones. Certain other aspects of dealing with open-shell systems also merit attention.

6.3.3 Open-shell Systems

The presentation of the HF equations in Chapter 4 assumed a closed-shell singlet for simplicity, but what if there are one or more singly occupied orbitals? Let us proceed with an example to guide the discussion, in this case, the methyl radical, which is planar in its equilibrium structure (Figure 6.10). The most intuitive description of the wave function for this system (ignoring symmetry for ease of discussion) would be

$$^2\Psi = \left| C1s^2 \sigma_{CH_a}^2 \sigma_{CH_b}^2 \sigma_{CH_c}^2 C2p_z^1 \right\rangle \tag{6.6}$$

Thus, there is a doubly occupied carbon 1s core, three C–H bonding orbitals, and the unpaired electron in a carbon 2p orbital. Given this configuration, it might seem natural to envision an extension of HF theory where all of the orbitals continue to be evaluated using essentially the restricted formalism (RHF) for closed-shell systems, but the density matrix elements for the singly occupied orbital(s) are not multiplied by the factor of two appearing in Eq. (4.57). In essence, this describes so-called restricted open-shell HF theory (ROHF). In its completely general form, certain complications arise for systems whose descriptions require more than a single determinant (i.e., unlike Eq. (6.6)), so we will not extend this qualitative description of the nature of the theory to specific equations (such details are available in Veillard (1975)). It suffices to note that most electronic structure packages offer ROHF as an option for open-shell calculations.

Besides being intuitively satisfying, ROHF theory produces wave functions that are eigenfunctions of the operator S^2 (just as the true wave function must be), having eigenvalues $S(S+1)$ where S is the magnitude of the vector sum of the spin magnetic moments for all of the unpaired electrons. However, ROHF theory fails to account for spin polarization in

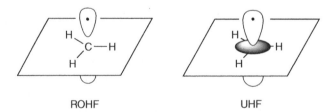

ROHF   UHF

Figure 6.10   In the absence of spin polarization, which corresponds to the ROHF picture, there is zero spin density in the plane containing the atoms of the methyl radical. Accounting for spin polarization, which corresponds to the UHF picture, results in a build-up of negative spin density (represented as a shaded region) in the same plane

the doubly occupied orbitals. To appreciate this point, let us return to the methyl radical. Note that because the unpaired spin in this molecule is in the carbon $2p_z$ orbital, the plane containing the atoms, which is the nodal plane for the $2p_z$ orbital, must have zero spin density. This being the case, electron spin resonance experiments should detect zero hyperfine coupling between the magnetic moments of the hydrogen atoms (or of a ^{13}C nucleus) and the unpaired electron. However, even after correcting for the effects of molecular vibrations, it is clear that there *is* a coupling between the two.

Spin density is found in the molecular plane because of spin polarization, which is an effect arising from exchange correlation. The Fermi hole that surrounds the unpaired electron allows other electrons of the same spin to localize above and below the molecular plane slightly more than can electrons of opposite spin. Thus, if the unpaired electron is α, we would expect there to be a slight excess of β density in the molecular plane; as a result, the ^{1}H hyperfine splitting should be negative (see Section 9.1.3), and this is indeed the situation observed experimentally. An ROHF wave function, because it requires the spatial distribution of both spins in the doubly occupied orbitals to be identical, cannot represent this physically realistic situation.

To permit the α and β spins to occupy different regions of space, it is necessary to treat them individually in the construction of the molecular orbitals. Following this formalism, we would rewrite our methyl radical wave function Eq. (6.6) as

$$^2\Psi = \left| C1s^\alpha C1s'^\beta \sigma^\alpha_{CH_a} \sigma'^\beta_{CH_a} \sigma^\alpha_{CH_b} \sigma'^\beta_{CH_b} \sigma^\alpha_{CH_c} \sigma'^\beta_{CH_c} C2p_z^\alpha \right) \tag{6.7}$$

where the prime notation on each β orbital emphasizes that while it may be spatially similar to the analogous α orbital, it need not be identical. The individual orbitals are found by carrying out separate HF calculations for each spin, with the spin-specific Fock operator now defined as

$$
\begin{aligned}
F^\xi_{\mu\nu} &= \left\langle \mu \left| -\frac{1}{2}\nabla^2 \right| \nu \right\rangle - \sum_k^{\text{nuclei}} Z_k \left\langle \mu \left| \frac{1}{r_k} \right| \nu \right\rangle \\
&+ \sum_{\lambda\sigma} \left[\left(P^\alpha_{\lambda\sigma} + P^\beta_{\lambda\sigma} \right) (\mu\nu|\lambda\sigma) - P^\xi_{\lambda\sigma}(\mu\lambda|\nu\sigma) \right]
\end{aligned}
\tag{6.8}
$$

where ξ is either α or β, and the two spin-density matrices are defined as

$$P^\xi_{\lambda\sigma} = \sum_i^{\xi-\text{occupied}} a^\xi_{\lambda i} a^\xi_{\sigma i} \tag{6.9}$$

where the coefficients a are the usual ones expressing the MOs in the AO basis, but there are separate sets for the α and β orbitals. Notice, then, that in Eq. (6.8), the Coulomb repulsion (the first set of integrals in the double sum) is calculated with both spins, but exchange (the second set of integrals) is calculated only with identical spins. Because the SCF is being carried out separately for each spin, the two density matrices can differ, which is to say the

MOs can be different, and this permits spin polarization. Equations (6.8) and (6.9) define unrestricted Hartree-Fock (UHF) theory.

While UHF wave functions have the desirable feature of including spin polarization, they are *not*, in general, eigenfunctions of S^2. By allowing the spatial parts of the different spin orbitals to differ, the final UHF wave function incorporates some degree of 'contamination' from higher spin states – specifically, states whose high-spin components would derive from flipping the spin of one or more electrons. Thus, doublets are contaminated by quartets, sextets, octets, etc., while triplets are contaminated by pentets, heptets, nonets, etc. The degree of spin contamination can be assessed by inspection of $\langle S^2 \rangle$, which should be 0.0 for a singlet, 0.75 for a doublet, 2.00 for a triplet, 3.75 for a quartet, etc. Values that vary from these proper eigenvalues by more than 5 percent or so should inspire great caution in working with the wave function, since other expectation values will also be skewed by differences between the property for the desired state and those for the contaminating states (see Section 9.1.4 and Appendix C for details on the calculation of $\langle S^2 \rangle$).

Various techniques have been developed to reduce or eliminate the contribution of contaminating states to the UHF wave function or expectation values derived from it. Some of these are described in Appendix C, which contains a more detailed description of spin algebra in general. In general, however, it should be noted that none of these approaches are convenient for geometry optimization, which makes characterization of an open-shell PES quite difficult when spin contamination effects are large. Thus, open-shell systems nearly always require more care than closed-shell singlets, because both the ROHF and the UHF formalisms are subject to intrinsically unphysical behavior. Depending on the nature of the system and the properties being calculated, such behavior may or may not be manifest.

Finally, note that some open-shell systems cannot be described by a single determinant. The classical example is an open-shell singlet, i.e., a system having electrons of α and β spin in different spatial orbitals a and b. The wave function for such a system that is properly antisymmetric and preserves the indistinguishability of particles is

$$^1\Psi = \tfrac{1}{2}[a(1)b(2) + a(2)b(1)][\alpha(1)\beta(2) - \alpha(2)\beta(1)] \tag{6.10}$$

which *cannot* be expressed as a single determinant. Because RHF and UHF are defined to use single-determinantal wave functions, they are formally unable to address this wave function (cf. Appendix C). In its most general form, ROHF *is* defined for multideterminantal systems, but the more typical approach is to use multiconfiguration self-consistent field theory, as described in Section 7.2.

6.3.4   Efficiency of Implementation and Use

We have emphasized up to this point the formal N^4 scaling of HF theory. However, in practice, the situation is never so severe, and indeed linear scaling HF implementations have begun to appear. Of course, one should remember that scaling behavior is different from speed. Thus, for a system of a given size, a HF calculation using algorithms that scale linearly may take significantly longer than conventional algorithms – it is simply true that at *some* point the linear scaling algorithm will become more efficient given a system of large enough size. In any case, because there are several features present in electronic structure programs that allow some control over the efficiency of the calculation, we discuss here the most common ones, recapitulating a few that have already been mentioned above.

First, there is the issue of how to go about computing the four-index integrals, which are responsible for the formal N^4 scaling. One might imagine that the most straightforward approach is to compute every single one and, as it is computed, write it to storage – then, as the Fock matrix is assembled element by element, call back the computed values whenever they are required (most of the integrals are required several times). In practice, however, this approach is only useful when the time required to write to and read from storage is very, very fast. Otherwise, modern processors can actually recompute the integral from scratch faster than modern hardware can recover the previously computed value from, say, disk storage. The process of computing each integral as it is needed rather than trying to store them all is called 'direct SCF'. Only when the storage of all of the integrals can be accomplished in memory itself (i.e., not on an external storage device) is the access time sufficiently fast that the 'traditional' method is to be preferred over direct SCF.

As the size of the system increases, it becomes possible to take advantage of other features of the electronic structure that further improve the efficiency of direct SCF. For instance, it is possible to estimate upper bounds for four-index integrals reasonably efficiently, and if the upper bound is so small that the integral can make no significant contribution, there is no point evaluating it more accurately than assigning it to be zero. Such small integrals are legion in large systems, since if each of the four basis functions is distantly separated from all of the others simple overlap arguments make it clear that the integral cannot be very large.

With very, very large systems, fast-multipole methods analogous to those described in Section 2.4.2 can be used to reduce the scaling of Coulomb integral evaluation to linear (see, for instance, Strain, Scuseria, and Frisch 1996; Challacombe and Schwegler 1997), and linear methods to evaluate the exchange integrals have also been promulgated (Ochensfeld, White, and Head-Gordon 1998). At this point, the bottleneck in HF calculations becomes diagonalization of the Fock matrix (a step having formal N^3 scaling), and early efforts to reduce the scaling of this step have also appeared (Millam and Scuseria 1997).

As already described above, efficiency in converging the SCF for systems with large basis sets can be enhanced by using as an initial guess the converged wave function from a different calculation, one using either a smaller basis set or a less negative charge. This same philosophy can be applied to geometry optimization, which can be quite time-consuming for very large calculations. It is often very helpful to optimize the geometry first at a more efficient level of theory. This is true not just because the geometry optimized with the lower level is probably a good place to start for the higher level, but also because typically one can compute the force constants at the lower level and use them as an initial guess for the higher level Hessian matrix that will be much better than the typical guess generated by the optimizing algorithm. As described in Section 2.4.1, the availability of a good Hessian matrix can make an enormous amount of difference in how quickly a geometry optimization can be induced to converge.

Also as already noted above, taking advantage of molecular symmetry can provide very large savings in time. However, structures optimized under the constraints of symmetry should always be checked by computation of force constants to verify their nature as stationary points on the full PES. Additionally, it is typically worthwhile to verify that open-shell wave functions obtained for symmetric molecules are stable with respect to orbital changes that would generate other electronic states.

Finally, the use of ECP basis sets for heavy elements improves efficiency by reducing the scale of the electronic structure problem. In addition, relativistic effects can be accounted for by construction of the pseudopotential.

One area that takes advantage of many of the above formalisms is the application of HF theory to periodic solids. Periodic HF theory has been most extensively developed within the context of the CRYSTAL code (Dovesi *et al.* 2000) where it is available in RHF, UHF, and ROHF forms. Such calculations can be particularly useful for elucidating band structure in solids, assessing defect effects, etc.

6.4 General Performance Overview of *Ab Initio* HF Theory

6.4.1 Energetics

Because HF theory ignores correlation, and because in its *ab initio* (as opposed to semiempirical) formulation, no attempt is made to correct for this deficiency, HF theory cannot realistically be used to compute heats of formation. Indeed, Feller and Peterson (1998) examined the atomization energies of 66 small molecules at the HF level using the aug-cc-pVnZ basis sets with n = D, T, and Q, and obtained mean unsigned errors of 85, 66, and 62 kcal mol^{-1}, respectively. Thus, even as one approaches the HF limit, the intrinsic error in an absolute molecular energy calculation can be very large. In general, the energy associated with *any* process involving a change in the total number of paired electrons is very poorly predicted at the HF level because of the failure to account for electron correlation (cf. use of isodesmic reactions as described in Section 10.6).

Even if the number of paired electrons remains constant but the nature of the bonds is substantially changed, the HF level can show rather large errors. For instance, the atmospheric reaction converting CO and HO$^{\bullet}$ to H$^{\bullet}$ and CO_2 is known to be exoergic with an energy change of about -23 kcal mol^{-1}. The HF level of theory using the STO-3G, 3-21G, 6-31G(d,p), and near-infinite quality basis sets predicts energy changes of 34.1, 3.1, -5.8, and -7.6 kcal mol^{-1}, respectively, which is quite far from accurate.

Note that isomerization is a process that can change bonding substantially as well. Hehre *et al.* have compared experimental data for 35 isomerization reactions to predictions from the HF/STO-3G, HF/3-21G, and HF/6-31G(d)//HF/3-21G levels (the latter notation, $w/x//y/z$, implies level of theory w with basis set x applied using a geometry optimized at level of theory y using basis set z, i.e., a single point calculation). The isomerizations were quite diverse, including for example acetone to methyl vinyl ether, acetaldehyde to oxetane, formamide to nitrosomethane, and ethanethiol to dimethyl sulfide. The energy differences spanned from 0.2 to 62.6 kcal mol^{-1}, and the dispersion in the data (i.e., the mean absolute error that would be generated by simply guessing the average energy difference over all reactions) was 12.7 kcal mol^{-1}. The mean unsigned errors for the above noted levels were 12.3, 4.8, and 3.2 kcal mol^{-1}, respectively. Thus, the minimal basis set does very badly, HF/3-21G is perhaps qualitatively useful, and the final method has some utility that might well be improved had geometry optimization taken place at the same level as the energy evaluation. Note however that the maximum errors were 51.2, 22.6, and 11.3 kcal mol^{-1}, respectively, which emphasizes that average performances are no guarantee of good behavior in any one system. For comparison, on a subset of 30 of the same isomerizations, MNDO, AM1, and PM3 had mean unsigned errors of 9.1, 7.4, and 5.8 kcal mol^{-1}, and maximal errors of 42, 24, and 23 kcal mol^{-1}, respectively.

The situation continues to improve when the changes in bonding are reduced to those associated with conformational changes. St.-Amant, Cornell, and Kollman (1995) examined 35 different conformational energy differences in a variety of primarily organic molecules, where the average difference in energy between conformers was 1.6 kcal mol^{-1}; at the HF/6-31+G(d,p)//HF/6-31G(d) level, the RMS error in predicted differences was 0.6 kcal mol^{-1}. A similar study on a set of eight organic molecules with an average conformational energy difference of 2.3 kcal mol^{-1} has been reported by Hehre; in that instance, mean unsigned errors of 1.0 and 0.7 kcal mol^{-1} were observed at the HF/3-21G$^{(*)}$ and HF/6-31G(d) levels, respectively (Hehre 1995).

Returning to the 11 glucose conformers already discussed in Chapters 2 and 5 in the context of molecular mechanics and semiempirical models, the performance of several levels of HF theory for predicting the relative conformer energies are listed in Table 6.2. The mean unsigned error associated with assuming all conformers to have the average energy is 1.2 kcal mol^{-1}, so the best HF models do very well by comparison. Note, however, that the small basis sets STO-3G and 3-21G do rather badly. Analysis suggests that these basis sets, which lack polarization functions on heteroatoms, significantly overestimate the energy of hydrogen bonds. Since the glucose conformers are characterized by differing numbers of intramolecular hydrogen bonds, this effect significantly increases the error for these small basis sets. Note also the interesting feature that the polarized double-ζ basis sets 6-31G(d) and cc-pVDZ provide better predictive accuracy than the more complete cc-pVTZ and cc-pVQZ sets. Such a situation is by no means unusual – it is often the case that basis set incompleteness and failure to account for electron correlation introduce errors of opposite sign. If those errors are also of similar *magnitude*, then fortuitously good results can be

Table 6.2 Mean unsigned errors (kcal mol^{-1}) in 11 predicted glucose conformational energies for various basis sets at the HF level in order of basis set size

Basis set	Mean unsigned error
STO-3G	1.1
3-21G	2.0
6-31G(d)	0.2
cc-pVDZ	0.1
cc-pVTZ	0.6
cc-pVQZ	0.8

obtained. A great deal of experience suggests that, very broadly speaking, polarized double-ζ basis sets are the ones most likely to enjoy such a favorable cancellation of errors at the HF level when it occurs. However, it is very risky to *rely* on this phenomenon for any particular calculation *in the absence of prior evidence that it is operative in one or more closely related systems.*

Among the simplest of conformational changes is that associated with rotation about a single bond. Given that this process involves very small changes in bonding, electron correlation effects on the rotation barrier are expected to be small, and indeed, even HF theory with very small basis sets for the most part performs adequately in the prediction of such barriers. For eight rotations about H_mX-YH_n single bonds, where X,Y = {B, C, N, O, Si, S, P}, Hehre *et al.* found mean unsigned errors of 0.6, 0.6, and 0.3 kcal mol^{-1} at the HF/STO-3G, HF/3-21G$^{(*)}$ and HF/6-31G(d) levels, respectively. If H_3B-NH_3 is removed from the set, the error drops to 0.5, 0.2, and 0.2 kcal mol^{-1}, respectively. The dispersion in the data set was 0.6 kcal mol^{-1}.

Although HF theory fares poorly in computing most reaction energies, because of the substantial electron correlation effects associated with making/breaking bonds, it is reasonably robust for predicting protonation/deprotonation energies. Since the proton carries with it no electrons, one may think of these reactions as being considerably less sensitive to differential electron correlation in reactants and products. Provided basis sets of polarized valence-double-ζ quality or better are used, absolute proton affinities of neutral molecules are typically computed to an accuracy of better than 5 percent. Errors increase, however, if the cations are non-classical (e.g., bridging protons are present) since such structures tend to be found as minima only after accounting for electron correlation effects. Deprotonation energies of neutral compounds are computed with similar absolute accuracy (± 8 kcal/mol or so) so long as diffuse functions are included in the basis set to balance the description of the anion. If smaller basis sets are used, very large errors are observed.

Another fairly conservative 'reaction' is the removal or attachment of a single electron from/to a molecule. As already discussed in Chapter 5, Koopmans' theorem equates the energy of the HOMO with the negative of the IP. This approximation ignores the effect of electronic relaxation in the ionized product, i.e., the degree to which the remaining electrons redistribute themselves following the detachment of one from the HOMO. If we were to calculate the IP as the difference in HF energies for the closed-shell neutral and the open-shell product, we would obtain the so-called ΔSCF IP

$$IP_{\Delta SCF} = E_{HF}(A^{+\bullet}) - E_{HF}(A) \tag{6.11}$$

where orbital relaxation *is* included. Including relaxation results in a smaller predicted IP, since relaxation lowers the energy of the cation radical relative to the neutral (the HOMO energy used in Koopmans' theorem derives from orbitals already fully relaxed for the neutral). Note, however, that the neutral species has one more electron than the radical cation, and thus there will be larger electron correlation effects. By ignoring these effects through the use of HF theory, we destabilize the neutral more than the radical cation, and too small an IP is expected in any case. Thus, Koopmans' theorem benefits from a cancellation of errors: the orbital relaxation and the electron correlation effects offset one another (see, for example, Maksic and Vianello 2002). In practice, the cancellation can be remarkably good; Koopmans' theorem IPs are often within 0.3 eV or so of experiment provided basis sets of polarized valence-double-ζ quality or better are used in the HF calculation. However, this favorable cancellation begins to break down if IPs are computed for orbitals other then the HOMO. As more tightly held electrons are ionized, particularly core electrons, the relaxation effects are much larger than the correlation effects, and Koopmans' approximation should not be used.

Koopmans' theorem can be formally applied to electron affinities (EAs) as well, i.e., the EA can be taken to be the negative of the orbital energy of the lowest unoccupied (virtual) orbital. Here, however, relaxation effects and correlation effects both favor the radical anion, so rather than canceling, the errors are additive, and Koopmans' theorem estimates will almost always underestimate the EA. It is thus generally a better idea to compute EAs from a ΔSCF approach whenever possible.

A key point meriting discussion is the use of HF theory to model systems where two or more molecules are in contact, held together by non-bonded interactions. Such interactions in actual physical systems include electrostatic interactions between permanent and induced charge distributions, dispersion, and hydrogen bonding (the latter includes both of the prior two in addition to some possible degree of covalent interaction). It is important to note that HF theory is formally incapable of modeling dispersion, because this phenomenon is entirely a consequence of electron correlation, for which HF theory fails to account. Nevertheless, bimolecular interaction energies are often reasonably well predicted by HF theory, particularly with basis sets like 6-31G(d) and others of similar size. As might be expected based on preceding discussion, this again reflects a cancellation of errors.

Clearly, failure to account for dispersion would be expected to strongly reduce intermolecular interactions, so the remaining errors must be in the direction of overbinding. In this instance, there are two chief contributors to overbinding. The first is that, as noted in Section 6.4.3, HF charge distributions tend to be overpolarized, which gives rise to electrostatic interactions that are somewhat too large. The second effect is more technical in nature, and is referred to as 'basis set superposition error' (BSSE). If we consider a bimolecular interaction, the HF interaction energy can be trivially defined as

$$\Delta E_{\text{bind}} = E_{\text{HF}}^{a \cup b}(\text{A} \bullet \text{B}) - E_{\text{HF}}^{a}(\text{A}) - E_{\text{HF}}^{b}(\text{B}) \tag{6.12}$$

where a and b are the basis functions associated with molecules A and B, respectively. Note that if a and b are not both infinite basis sets, then there are more basis functions employed in the calculation of the complex than in either of the monomers. The greater flexibility of the basis set for the complex can provide an artifactual lowering of the energy when one of the monomers 'borrows' basis functions of the other to improve its own wave function.

One method proposed to correct for this phenomenon is the so-called counterpoise (CP) correction. Although some variations exist, one popular approach defines the CP corrected interaction energy as

$$\Delta E_{\text{bind}}^{\text{CP}} = E_{\text{HF}}^{a\cup b}(A\bullet B)_{A\bullet B} - E_{\text{HF}}^{a\cup b}(A)_{A\bullet B} - E_{\text{HF}}^{a\cup b}(B)_{A\bullet B}$$
$$+ [E_{\text{HF}}^{a}(A)_{A\bullet B} - E_{\text{HF}}^{a}(A)_{A}] + [E_{\text{HF}}^{b}(B)_{A\bullet B} - E_{\text{HF}}^{b}(B)_{B}]$$
(6.13)

where the subscripts appearing after the molecular species describe the geometry employed. Thus, in the first line on the r.h.s., the energy of bringing the two monomers together, each monomer already having the geometry it has in the complex, is computed using a consistent basis set. Thus, in the monomer calculations, basis functions for the missing partner are included in the calculation, even though the nuclei on which those functions are centered are not actually there – such basis functions are sometimes called ghost functions. Since the ghost functions slightly lower the energies of the monomers, the overall binding energy is less than would be the case if they were not to be used. The second line on the r.h.s. of Eq. (6.13) then accounts for the energy required to distort each monomer from its preferred equilibrium structure to the structure found in the complex. Since it is not obvious where to put the ghost functions when the monomer adopts its equilibrium geometry, the geometry-distortion energies are computed using only the nuclei-centered monomer basis sets.

However, it must be noted that the borrowing of basis functions is only partly a mathematical artifact. To the extent that some charge transfer and charge polarization take place as part of forming the bimolecular complex, some of the borrowing simply reflects chemical reality. Thus, CP correction always overestimates the BSSE, and there is no clear way to correct for this overestimation. Indeed, Masamura (2001) has found from analysis of ion-hydrate clusters that interaction energies computed with basis sets of augmented-polarized-double-ζ quality or better were in closer agreement with complete basis-set results before CP correction than after. As a result, there tend to be two schools of thought on how best to deal with BSSE. Some researchers prefer to spend the time that would be required for CP correction instead on the evaluation of Eq. (6.12) with a more saturated basis set. Since, in the limit of an infinite basis, Eqs. (6.12) and (6.13) are equivalent, a demonstration of convergence of Eq. (6.12) with respect to basis-set size is a reasonable indication of accuracy, at least at the HF level.

6.4.2 Geometries

Optimization of the molecular geometry at the HF level appears at first sight to be a daunting task because of the difficulty of obtaining analytic derivatives (see Section 2.4.1). To take the first derivative of Eq. (4.54) with respect to the motion of an atom, we can exhaustively apply the chain rule term by term. Thus, we must determine derivatives of basis functions and operators with respect to a particular coordinate, and this is not so hard, but we also need to know the derivatives of the density matrix elements with respect to atomic motion, and these derivatives are not obvious at all. However, Pulay (1969) discovered an elegant connection between these very complicated derivatives and the much simpler derivatives of the overlap matrix (which depend only on analytically known basis function derivatives). This breakthrough led to rapid developments in computing higher-order derivatives and optimization algorithms, and as a result, HF geometries are now quite efficiently available.

For minimum-energy structures, HF geometries are usually very good when using basis sets of relatively modest size. For basis sets of polarized valence-double-ζ quality, errors in bond lengths between heavy atoms average about 0.03 Å, and between heavy atoms and H about 0.015 Å. Bond angles are predicted to an average accuracy of about 1.5°, and dihedral angles are also generally well predicted, although available experimental data in the gas phase are scarce. Even with the 3-21G$^{(*)}$ basis set, this accuracy is not much degraded.

To the extent that HF theory is in error, it tends to overemphasize occupation of bonding orbitals (see Chapter 7). Thus, errors tend to be in the direction of predicting bonds to be too short, and this effect becomes more pronounced as one proceeds to saturated basis sets; Feller and Peterson (1998) observed predicted geometries at the HF level to *degrade* in quality with increasing basis-set size in the series aug-cc-pVnZ using $n = $ D, T, Q. A good example is the case of the monocyclic singlet diradical 1,3-didehydrobenzene (Figure 6.11). RHF theory erroneously predicts this molecule to be bicyclic with a formal single bond between the radical positions.

There are some additional pathological cases that must be borne in mind in evaluating the quality of predicted HF geometries for minima. As already noted, polarization functions are absolutely required for geometric accuracy in systems characterized by hypervalent bonding; failure to include polarization functions on heteroatoms with single lone pairs can also cause them to be insufficiently pyramidalized. Furthermore, in systems crowding many pairs of non-bonding electrons into small regions of space (e.g., the four oxygen lone pairs in a peroxide) electron correlation effects on geometries, ignored by HF theory, can begin to be large, so some caution is warranted here as well. Finally, dative bonds (i.e., those where both electrons in the bonding pair formally come from only one of the atoms) are often poorly described at the HF level. For instance, at the HF/6-31G(d) level, the B−C and B−N distances in the complexes $H_3B \cdot CO$ and H_3BNH_3 are predicted to be too long by about 0.1 Å.

Geometries of TS structures are not readily available from experiment, but a fairly substantial body of theoretical work permits comparisons to be made with very high-level

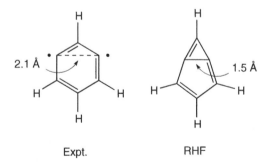

Figure 6.11   Structures of 1,3-didehydrobenzene (*m*-benzyne) from experiment and RHF calculations. Because of its tendency to overemphasize bonding interactions, RHF optimization results in a bicyclic structure. While the RHF error in bond length is very large, it should be noted that the 'bond-stretching' coordinate *is* known to be very flat (for very detailed analyses on the sensitivity of this system to different theoretical levels, see Kraka *et al.* 2001 and Winkler and Sander 2001)

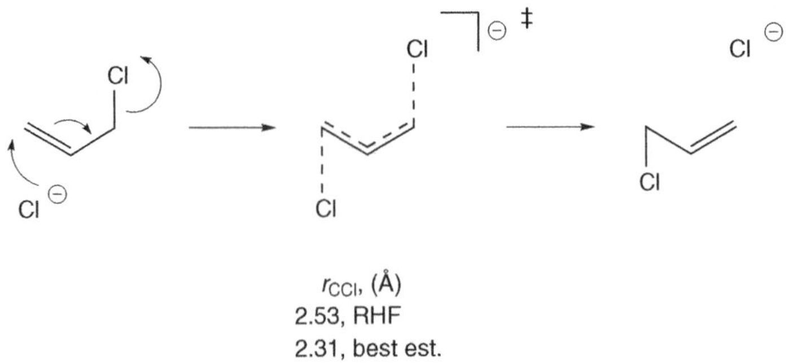

r_{CCl}, (Å)
2.53, RHF
2.31, best est.

Figure 6.12   The anti-S_N2' reaction of chloride with allyl chloride. While RHF theory does well with the reactant/product geometries, it significantly overestimates the C–Cl bond lengths in the C_2 symmetric TS structure based on calculations at more reliable levels of theory

calculations. In the case of TS structures, the failure of HF theory to account for electron correlation can be more problematic, since correlation effects in partial bonds can be large. For example, the difference in C–Cl bond lengths predicted at the HF/6-31G(d) and higher levels of theory is more than 0.2 Å for the anti-S_N2' reaction of chloride anion with allyl chloride (Figure 6.12). Although this single example provides an indication of how large differences can be, Wiest, Montiel, and Houk (1997) have analyzed TS structures for many different organic reactions, particularly electrocyclic reactions, and have inferred that in such instances HF/6-31G(d) TS structures are generally of good quality. Nevertheless, the variation in possible bonding situations in TS structures is such that comparison of HF structures with those obtained at better levels of theory is almost always worthwhile in order to ensure quality.

As for the energies of non-bonded complexes, the failure of HF theory to account for dispersion tends to make such complexes too loose in structure, i.e., intermolecular distances are unrealistically large. Hydrogen bonded structures, on the other hand, are often quite good because errors in overestimating electrostatic interactions cancel the failure to account for dispersion. The HF structures show the expected preference for linear bond angles at hydrogen, when such are possible, and further exhibit reasonable distances between donor and acceptor heavy atoms in most instances.

6.4.3   Charge Distributions

HF dipole moments tend to be fairly insensitive to increases in basis-set size beyond valence-double-ζ. With such basis sets, there is a systematic error in dipole moment estimation – typically the magnitude of the dipole is overestimated by 10–25 percent, i.e., molecules are predicted to be too polar. Individual exceptions to this rule exist, of course. In an absolute sense, Scheiner, Baher, and Andzelm (1997) explored the performance of HF/6-31G(d,p) for 108 molecules and obtained a mean unsigned error of 0.23 D.

Results are erratic with smaller basis sets, in part due to lower quality wave functions and in part due to poorer geometries, which affect the dipole moment. An exception is the economical MIDI! basis set for which, as noted above, heteroatom d exponents were specifically optimized so that high-quality geometries and charge distributions (instead of minimal energies) are obtained from the HF wave function. Electrostatic potentials computed with MIDI! also give good agreement with correlated levels of electronic structure theory.

A more complete discussion of charge distributions is deferred until Chapter 9. The performance of HF theory for other molecular properties is also presented in more detail there.

6.5 Case Study: Polymerization of 4-Substituted Aromatic Enynes

Synopsis of Ochiai, Tomita, and Endo (2001) 'Investigation on Radical Polymerization Behavior of 4-Substituted Aromatic Enynes. Experimental, ESR, and Computational Studies'.

One strategy for making highly functionalized polymers is first to carry out polymerization of a system bearing functionalizable appendages, and then after polymerization to react those appendages to introduce new functionality into the polymer. Such an approach can be advantageous in instances where the monomer that would in principle lead directly to the functionalized polymer fails itself to be useful as a polymerization substrate.

Ochiai and co-workers developed an experimental protocol for the radical polymerization of one such reactive monomer, 4-phenylbut-1-en-3-yne. As illustrated in Figure 6.13, this polymerization creates a polyethylene chain functionalized with phenylethynyl substituents.

A factor that affects the kinetics of the polymerization, and, more critically, the utility of the monomer in copolymerizations with other monomers, e.g., methyl methacrylate, is the stability of the radical formed from addition of the growing polymer chain to the vinyl terminus. In order to gauge the stabilizing effect of the phenylethynyl group, and the sensitivity of the stabilization to substitution at the *para* position of the aromatic ring, Ochiai and co-workers carried out calculations at the UHF/3-21G level to evaluate (i) the spin density in the 1-phenylprop-1-yn-3-yl radical and (ii) the reaction energy for the process

$$RCH_3 + CH_3^\bullet \rightarrow RCH_2^\bullet + CH_4 \tag{6.14}$$

where R was varied over a number of different functional groups. This so-called isodesmic reaction (see Section 10.4.3) essentially computes the $C-H$ bond energy for the substituted system *relative to* the $C-H$ bond energy in methane, thereby reducing absolute errors that would be associated with a small HF calculation for an absolute bond energy.

The spin density calculation, which analyzes the difference between each atom's Mulliken population (see Section 9.1.3.2) of α and β electrons, indicated the unpaired electron to be highly delocalized, with populations on the *ortho* and *para* carbons of the phenyl ring nearly equal to that found on the formal radical position (these large positive spin densities were balanced by large negative spin densities on the intervening carbon atoms, which is a typically observed situation). HF theory tends to overpolarize spin, so the magnitude of the spin polarization is probably not trustworthy, but the large degree of delocalization is probably qualitatively reasonable. The prediction that the ring *para* position carries substantial spin was found to be consistent with copolymerization reactivity

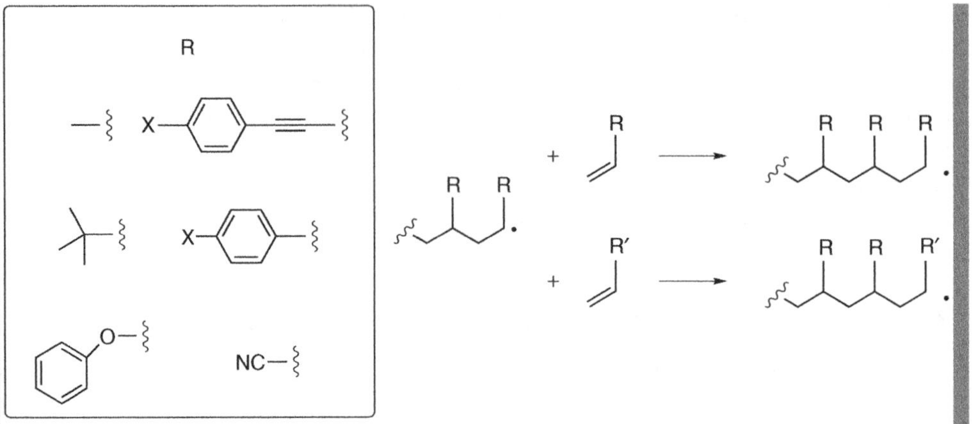

Figure 6.13   Radical polymerization of a growing polymer chain in the presence of two distinct monomers (i.e., copolymerization conditions) can at every step incorporate one monomer or the other. How might one *quantitatively* go about estimating the intrinsic preference for one monomer over the other? What other molecular properties expected to correlate with this discrimination might be subject to computation?

studies that showed substantial sensitivity to the presence of the *para* substituents MeO, Me, Cl, and CF_3.

One measure of the resonance component of radical stability in polymerizations is the so-called Q value of the monomer, which quantifies the resonance stabilization of the radical (Stevens 1990). However, the experimentally determined value of Q can be influenced by other factors unrelated to resonance. To evaluate the extent to which their measured Q values were consistent with resonance stabilization of the monomer radical, the authors compared isodesmic energies from Eq. (6.14) to measured Q values for R = Me, *t*Bu, PhO, CN, Ph, vinyl, and phenylethynyl. The largest stabilization energy was computed for the R = phenylethynyl case, about 101 kJ mol^{-1}, although at the HF/3-21G level the expected linear correlation between $\log Q$ and stabilization energy was only fair ($R^2 = 0.86$; a better correlation for the non-phenylethynyl substituents had been obtained previously at a higher level of theory).

The authors also considered the relative influence of *para* substitution in the phenylethynyl compared to simply phenyl (i.e., compared to the analogous styrenes). They found that over the four substituents noted above, the stabilization energy from Eq. (6.14) varied by 5.2 kJ mol^{-1} for phenylethynyl and 7.0 kJ mol^{-1} for phenyl. Thus, insertion of the acetylene unit between the radical center and the aromatic ring is predicted to decrease the influence of the aryl substituent by only about 25 percent.

This study employs HF theory to answer only very qualitative questions, which is appropriate given the typically rather poor accuracy of the model in the absence of accounting for electron correlation. Future use of HF/3-21G to predict Q values for monomers not yet experimentally characterized might be worthwhile, but quantitative differences between monomers should not be taken particularly seriously except to the extent they may be categorized as large, medium, or small.

Bibliography and Suggested Additional Reading

Almlöf, J. 1994. 'Notes on Hartree-Fock Theory and Related Topics' *Lecture Notes in Quantum Chemistry II*, Roos, B. O., Ed., Springer-Verlag: Berlin, 1.

Almlöf, J. and Gropen, O. 1996. "Relativistic Effects in Chemistry" in *Reviews in Computational Chemistry*, Vol. 8, Lipkowitz, K. B. and Boyd, D. B., Eds., VCH: New York, 211.

Barrows, S. E., Storer, J. W., Cramer, C. J., French, A. D., and Truhlar, D. G. 1998. 'Factors Controlling the Relative Stability of Anomers and Hydroxymethyl Conformers of Glucopyranose' *J. Comput. Chem.*, **19**, 1111.

Carsky, P. and Urban, M. 1980. *Ab Initio Calculations*, Springer-Verlag: Berlin.

Cramer, C. J. 1991. 'The Fluorophosphoranyl Series: Computational Insights into Relative Stabilities and Localization of Spin' *J. Am. Chem. Soc.*, **113**, 2439.

Cundari, T., Benson, M. T., Lutz, M. L., and Sommerer, S. O. 1996. 'Effective Core Potential Approaches to the Chemistry of the Heavier Elements' in *Reviews in Computational Chemistry*, Vol. 8, Lipkowitz, K. B. and Boyd, D. B., Eds., VCH: New York, 145.

Dovesi, R., Orlando, R., Roetti, C., Pisani, C., and Saunders, V. R. 2000. "The Periodic Hartree-Fock Method and Its Implementation in the CRYSTAL Code" *Phys. Stat. Sol. B*, **217**, 63.

Feller, D. and Davidson, E. R. 1990. 'Basis Sets for *Ab Initio* Molecular Orbital Calculations and Intermolecular Interactions' in *Reviews in Computational Chemistry*, Vol. 1, Lipkowitz, K. B. and Boyd, D. B. Eds., VCH: New York, 1.

Frenking, G., Antes, I., Böhme, M., Dapprich, S., Ehlers, A. W., Jonas, V., Neuhaus, A., Otto, M., Stegmann, R., Veldkamp, A., and Vyboishchikov, S. F. 1996. 'Pseudopotential Calculations of Transition Metal Compounds: Scope and Limitations' in *Reviews in Computational Chemistry*, Vol. 8, Lipkowitz, K. B. and Boyd, D. B. Eds., VCH: New York, 63.

Hehre, W. J., Radom, L., Schleyer, P. v. R., and Pople, J. A. 1986. Ab Initio *Molecular Orbital Theory*, Wiley: New York.

Huzinaga, S., Ed. 1984. *Gaussian Basis Sets for Molecular Calculations*, Elsevier: Amsterdam.

Jensen, F. 1999. *Introduction to Computational Chemistry*, Wiley: Chichester.

Levine, I. N. 2000. *Quantum Chemistry*, 5th Edn., Prentice Hall: New York.

Petersson, G. A. 1998. 'Complete Basis-Set Thermochemistry and Kinetics' in *Computational Thermochemistry*, ACS Symposium Series, Volume 677, Irikura, K. K. and Frurip, D. J., Eds., American Chemical Society: Washington, DC, 237.

Schlegel, H. B. 2000. 'Perspective on "*Ab Initio* Calculation of Force Constants and Equilibrium Geometries in Polyatomic Molecules. I. Theory"' *Theor. Chem. Acc.*, **103**, 294.

Szabo, A. and Ostlund, N. S. 1982. *Modern Quantum Chemistry*, Macmillan: New York.

References

Challacombe, M. and Schwegler, E. 1997. *J. Chem. Phys.*, **106**, 5526.

Check, C. E., Faust, T. O., Bailey, J. M., Wright, B. J., Gilbert, T. M., and Sunderlin, L. S. 2001. *J. Phys. Chem. A*, **105**, 8111.

Dolg, M. 2002. *Theor. Comput. Chem.*, **11**, 793.

Dovesi, R., Orlando, R., Roetti, C., Pisani, C., and Saunders, V. R. 2000. *Phys. Stat. Sol. B*, **217**, 63.

Dunning, T. H. 1989. *J. Chem. Phys.*, **90**, 1007.

Dyall, K. G. 1998. *Theor. Chem. Acc.*, **99**, 366.

Dyall, K. G. 2002. *Theor. Chem. Acc.*, **108**, 335. (See also erratum **109**, 284.)

Easton, R. E., Giesen, D. J., Welch, A., Cramer, C. J., and Truhlar, D. G. 1996. *Theor. Chim. Acta*, **93**, 281.

Ema, I., García de la Vega, J. M., Ramírez, G., López, R., Fernández Rico, J., Meissner, H., and Paldus, J. 2003. *J. Comput. Chem.,* **24**, 859.

Feller, D. and Peterson, K. A. 1998. *J. Chem. Phys.*, **108**, 154.

Hay, P. J. and Wadt, W. R. 1985. *J. Chem. Phys.*, **82**, 270.

Hehre, W. J. 1995. *Practical Strategies for Electronic Structure Calculations*, Wavefunction: Irvine, CA, 175.

Hehre, W. J., Stewart, R. F., and Pople, J. A. 1969. *J. Chem. Phys.*, **51**, 2657.

Hellmann, H. 1935. *J. Chem. Phys.*, **3**, 61.

Jensen, F. 2001. *J. Chem. Phys.,* **115**, 9113. (See also erratum **116**, 3502.)

Jensen, F. 2002. *J. Chem. Phys.,* **117**, 9234.

Kaltsoyannis, N. 2003. *Chem. Soc. Rev.*, **32**, 9.

Kraka, E., Anglada, J., Hjerpe, A., Filatov, M., and Cremer, D. 2001. *Chem. Phys. Lett.*, **348**, 115.

Krishnan, R., Frisch, M. J., and Pople, J. A. 1980. *J. Chem. Phys.*, **72**, 4244.

Li, J., Cramer, C. J., and Truhlar, D. G. 1998. *Theor. Chem. Acc.,* **99**, 192.

Lovallo, C. C. and Klobukowski, M. 2003. *J. Comput. Chem.,* **24**, 1009.

Lynch, B. J. and Truhlar, D. G. 2004. *Theor. Chem. Acc.*, **111**, 335.

Maksic, Z. B. and Vianello, R. 2002. *J. Phys. Chem. A*, **106**, 6515.

Masamura, M. 2001. *Theor. Chem. Acc.*, **106**, 301.

Millam, J. M. and Scuseria, G. E. 1997. *J. Chem. Phys.*, **106**, 5569.

Ochensfeld, C., White, C. A., and Head-Gordon, M. 1998. *J. Chem. Phys.*, **109**, 1663.

Ochiai, B., Tomita, I., and Endo, T. 2001. *Macromolecules,* **34**, 1634.

Pulay, P. 1969. *Mol. Phys.*, **17**, 197.

Raffenetti, R. C. 1973. *J. Chem. Phys.*, **58**, 4452.

St.-Amant, A., Cornell, W. D., and Kollman, P. A. 1995. *J. Comput. Chem.*, **16**, 1483.

Scheiner, A. C., Baker, J., and Andzelm, J. W. 1997. *J. Comput. Chem.*, **18**, 775.

Stevens, M. P. 1990. *Polymer Chemistry*, 2nd Edn., Oxford University Press: New York, 225.

Stevens, W. J., Krauss, M., Basch, H., and Jasien, P. G. 1992. *Can. J. Chem.*, **70**, 612.

Strain, M. C., Scuseria, G. E., and Frisch, M. J. 1996. *Science,* **271**, 51.

Veillard, A. 1975. In: *Computational Techniques in Quantum Chemistry and Molecular Physics*, NATO ASI Series C, Vol. 15, Diercksen, G. H. F., Sutcliffe, B. T., and Veillard, A., Eds., Reidel: Dordrecht, 201.

Wiest, O., Montiel, D. C., and Houk, K. N. 1997. *J. Phys. Chem. A*, **101**, 8378.

Wilson, A. K., van Mourik, T., and Dunning, T. H. 1996. *J. Mol. Struct.*, **388**, 339.

Winkler, M. and Sander, W. 2001. *J. Phys. Chem. A*, **105**, 10422.

Woon, D. and Dunning, T. H. 1993. *J. Chem. Phys.*, **98**, 1358.

Woon, D. and Dunning, T. H. 1995. *J. Chem. Phys.*, **103**, 4572.

7

Including Electron Correlation in Molecular Orbital Theory

7.1 Dynamical vs. Non-dynamical Electron Correlation

Hartree–Fock theory makes the fundamental approximation that each electron moves in the static electric field created by all of the other electrons, and then proceeds to optimize orbitals for all of the electrons in a self-consistent fashion subject to a variational constraint. The resulting wave function, when operated upon by the Hamiltonian, delivers as its expectation value the lowest possible energy for a single-determinantal wave function formed from the chosen basis set.

It is important to note that there is a key distinction between the Hamiltonian operator and the Fock operator. The former operator returns the electronic energy for the *many-electron* system; the latter is really not a single operator, but the set of all of the interdependent *one-electron* operators that are used to find the one-electron MOs from which the HF wave function is constructed as a Slater determinant.

So, the question arises of how we might modify the HF wave function to obtain a lower electronic energy when we operate on that modified wave function with the Hamiltonian. By the variational principle, such a construction would be a more accurate wave function. We cannot do better than the HF wave function with a single determinant, so one obvious choice is to construct a wave function as a linear combination of multiple determinants, i.e.,

$$\Psi = c_0 \Psi_{HF} + c_1 \Psi_1 + c_2 \Psi_2 + \cdots \tag{7.1}$$

where the coefficients c reflect the weight of each determinant in the expansion and also ensure normalization. For the moment, we will ignore the nature of the determinants, other than the first one, which is the HF determinant. A general expansion does not *have* to include the HF determinant, but since the HF wave function seems to be a reasonable one for many purposes, it is useful to think of it as a leading term in any more complete wave function.

For the majority of the chemical species we have discussed thus far, the chief error in the HF approximation derives from ignoring the correlated motion of each electron with every other. This kind of electron correlation is called 'dynamical correlation' because it refers

Essentials of Computational Chemistry, 2nd Edition Christopher J. Cramer
© 2004 John Wiley & Sons, Ltd   ISBNs: 0-470-09181-9 (cased); 0-470-09182-7 (pbk)

to the dynamical character of the electron–electron interactions. Empirically, it is observed that for most systems the HF wave function dominates in the linear combination expressed by Eq. (7.1) (i.e., c_0 is much larger than any other coefficient); even though the correlation energy may be large, it tends to be made up from a sum of individually small contributions from other determinants.

However, in some instances, one or more of these other determinants may have coefficients of similar magnitude to that for the HF wave function. It is easiest to illustrate this by consideration of a specific example. Consider the closed-shell singlet wave function for trimethylenemethane (TMM, Figure 7.1). TMM is a so-called non-Kekulé molecule – in D_{3h} symmetry, it has two degenerate frontier orbitals for which only two electrons are available. Following a molecular analog of Hund's rule, the molecule has a triplet ground state (i.e., the lowest energy state has one spin-aligned electron in each degenerate orbital), but here we are concerned with the closed-shell *singlet*.

If we carry out a restricted HF calculation, one or other of the degenerate frontier pair will be chosen to be occupied, the calculation will optimize the shapes of all of the occupied orbitals, and we will end up with a best possible single-Slater-determinantal wave function formed from those MOs. But it should be fairly obvious that an equally good wave function

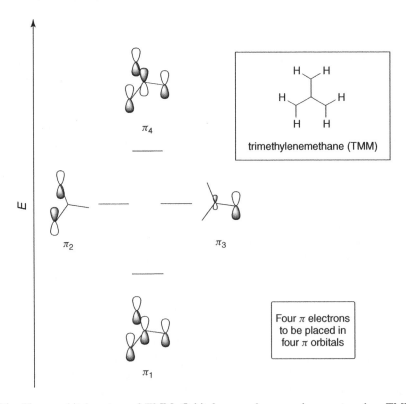

Figure 7.1 The π orbital system of TMM. Orbitals π_2 and π_3 are degenerate when TMM adopts D_{3h} symmetry

might have been formed if the original guess had chosen to populate the *other* of the two degenerate frontier orbitals. Thus, we might expect each of these two different RHF determinants to contribute with roughly equal weight to an expansion of the kind represented by Eq. (7.1). This kind of electron correlation, where different determinants have similar weights because of near (or exact) degeneracy in frontier orbitals, is called 'non-dynamical correlation' to distinguish it from dynamical correlation. This emphasizes that the error here is not so much that the HF approximation ignores the correlated motion of the electrons, but rather that the HF process is constructed in a fashion that is intrinsically single-determinantal, which is insufficiently flexible for some systems.

This chapter begins with a discussion of how to include non-dynamical and dynamical electron correlation into the wave function using a variety of methods. Because the mathematics associated with correlation techniques can be extraordinarily opaque, the discussion is deliberately restricted for the most part to a qualitative level; an exception is Section 7.4.1, where many details of perturbation theory are laid out – those wishing to dispense with those details can skip this subsection without missing too much. Practical issues associated with the employ of particular techniques are discussed subsequently. At the end of the chapter, some of the most modern recipes for accurately and efficiently estimating the exact correlation energy are described, and a particular case study is provided.

7.2 Multiconfiguration Self-Consistent Field Theory

7.2.1 Conceptual Basis

Continuing with our TMM example, let us say that we have carried out an RHF calculation where the frontier orbital that was chosen to be occupied was π_2. The determinant resulting after optimization will be

$$\Psi_{\mathrm{RHF}} = | \cdots \pi_1^2 \pi_2^2 \pi_3^0 \rangle \tag{7.2}$$

and orbital π_3 will be empty (i.e., a virtual orbital). We emphasize this by including it in the Slater determinant with an occupation number of zero, although this notation is not standard. We might generate the alternative determinant by keeping the same MOs but simply switching the occupation numbers, i.e.,

$$\Psi_{\pi_2 \to \pi_3} = | \cdots \pi_1^2 \pi_2^0 \pi_3^2 \rangle \tag{7.3}$$

An alternative, however, would be to require the RHF calculation to populate π_3 in the initial guess, in which case we would determine

$$\Psi'_{\mathrm{RHF}} = | \cdots \pi_1'^2 \pi_2'^0 \pi_3'^2 \rangle \tag{7.4}$$

where the prime on the wave function and orbitals emphasizes that, since different orbitals were occupied during the SCF process, the shapes of *all* orbitals will be different comparing one RHF wave function to the other.

If we were to compare the energies of the wave functions from Eqs. (7.2), (7.3), and (7.4), we would find the energies of the first and third to be considerably lower than that of the second. Since the real system has degenerate frontier orbitals (neglecting Jahn–Teller distortion), it seems reasonable that the energies of wave functions Eq. (7.2) and Eq. (7.4) are similar, but why is the energy of Eq. (7.3) higher? The problem lies in the nature of the SCF process. Only occupied orbitals contribute to the electronic energy – virtual orbitals do not. As such, there is no driving force to optimize the shapes of virtual orbitals; all that is required is that they be orthogonal to the occupied MOs. Thus, the quality of the shape of orbital π_3 depends on whether it is determined as an occupied or a virtual orbital.

From the nature of the system, however, we would really like π_2 and π_3 to be treated *equivalently* during the orbital optimization process. That is, we would like to find the best orbital shapes for these MOs so as to minimize the energy of the *two*-configuration wave function

$$\Psi_{MCSCF} = a_1 |\cdots \pi_1^2 \pi_2^2\rangle + a_2 |\cdots \pi_1^2 \pi_3^2\rangle \tag{7.5}$$

where a_1 and a_2 account for normalization and relative weighting (and we expect them to be equal for D_{3h} TMM). Such a wave function is a so-called 'multiconfiguration self-consistent-field' (MCSCF) one, because the orbitals are optimized for a *combination* of configurations (the particular case where the expansion includes only two configurations is sometimes abbreviated TCSCF).

As a technical point, a 'configuration' or 'configuration state function' (CSF) refers to the molecular spin state and the occupation numbers of the orbitals. For closed-shell singlets, CSFs can always be represented as single determinants, so the terms can be used somewhat loosely. In many open-shell systems, however, proper CSFs can only be represented by a combination of two or more determinants (see Eq. (6.10), for example). MCSCF theory is designed to handle *both* multiple configurations and the possible multi-determinantal character of individual configurations. In that sense, MCSCF is a generalization of ROHF theory, which *can* handle multiple determinants but is *not* capable of handling multiple CSFs.

In general, then, an MCSCF calculation involves a specification of what MOs may be occupied in the CSFs appearing in the expansion of Eq. (7.1). Given that specification, the formalism finds a variational optimum for the shape of each MO (as a linear combination of basis functions) *and* for the weight of each CSF in the MCSCF wave function.

Because a particular 'active' orbital may be occupied by zero, one, or two electrons in any given determinant, these MCSCF orbitals do *not* have unique eigenvalues associated with them, i.e., one cannot discuss the energy of the orbital. Instead, one can describe the 'occupation number' of each such orbital i as

$$(\text{occ. no.})_{i,MCSCF} = \sum_n^{\text{CSFs}} (\text{occ. no.})_{i,n} a_n^2 \tag{7.6}$$

where the sum runs over all CSFs and the occupation number of the orbital in each CSF is multiplied by the percentage contribution of that CSF to the total wave function. Because of the orthogonality of the CSFs, for a normalized MCSCF wave function the sum of the squares of all CSF coefficients is unity and the percent contribution of any CSF to the wave function is simply its expansion coefficient squared.

MCSCF calculations in practice require *much* more technical expertise than do single-configuration HF analogs. One particularly difficult problem is that spurious minima in coefficient space can often be found, instead of the variational minimum. Thus, convergence criteria are met for the self-consistent field, but the wave function is not really optimized. It usually requires a careful inspection of the orbital shapes and, where available, some data on relative energetics between related species or along a reaction coordinate to ascertain if this has happened.

A different issue requiring careful attention is how to go about selecting the orbitals that should be allowed to be partially occupied, and how to specify the 'flexibility' of the CSF expansion. We turn to this issue next.

7.2.2 Active Space Specification

Selection of orbitals to include in an MCSCF requires first and foremost a consideration of the chemistry being examined. For instance, in the TMM example above, a two-configuration wave function is probably not a very good choice in this system. When the orbitals being considered belong to a π system, it is typically a good idea to include *all* of them, because as a rule they are all fairly close to one another in energy. Thus, a more complete active space for TMM would consider all four π orbitals and the possible ways to distribute the four π electrons within them. MCSCF active space choices are often abbreviated as '(m,n)' where m is the number of electrons and n is the number of orbitals, so this would be a (4,4) calculation.

Sometimes reaction coordinates are studied that involve substantial changes in bonding. In such an instance, it is critical that a consistent choice of orbitals be made. For instance, consider the electrocyclization of 1,3-butadiene to cyclobutene (Figure 7.2). The frontier orbitals of butadiene are those associated with the π system, so, as just discussed, a (4,4) approach seems logical. However, the electrocyclization reaction transforms the two π bonds into one different π bond and one new σ bond. Thus, a consistent (4,4) choice in cyclobutene would involve the π and π^* orbitals and the σ and σ^* orbitals of the new single bond.

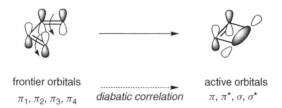

frontier orbitals   ....▶   active orbitals

$\pi_1, \pi_2, \pi_3, \pi_4$ *diabatic correlation* $\pi, \pi^*, \sigma, \sigma^*$

Figure 7.2 The frontier orbitals of s-*cis*-1,3-butadiene are the four π orbitals (π_2 is the specific example shown). If these orbitals are followed in a diabatic sense along the electrocyclization reaction coordinate, they correlate with the indicated orbitals of cyclobutadiene

While these orbitals would be easy to identify in butadiene and cyclobutene, it might be considerably more difficult to choose the corresponding orbitals in a TS structure, where symmetry is lower and mixing of σ and π character might complicate identification.

The next question to consider is how generally to allow the distribution of the electrons in the active space. Returning to TMM, it is clear that we want the CSFs already listed in Eq. (7.5), but in a (4,4) calculation, we might also want to be still more flexible, e.g., considering as the most important four perhaps

$$
\begin{aligned}
\Psi_{\mathrm{MCSCF}} = a_1 | \cdots \pi_1^2 \pi_2^2 \pi_3^0 \pi_4^0 \rangle + a_2 | \cdots \pi_1^2 \pi_2^0 \pi_3^2 \pi_4^0 \rangle \\
+ a_3 | \cdots \pi_1^2 \pi_2^0 \pi_3^0 \pi_4^2 \rangle + a_4 | \cdots \pi_1^0 \pi_2^2 \pi_3^2 \pi_4^0 \rangle
\end{aligned}
\tag{7.7}
$$

where we have again included orbitals with occupation numbers of zero in the notation for clarity. If we ignore symmetry for the moment, we could also take account of possibly important CSFs having partially occupied orbitals, e.g.,

$$
\begin{aligned}
\Psi_{\mathrm{MCSCF}} = a_1 | \cdots \pi_1^2 \pi_2^2 \pi_3^0 \pi_4^0 \rangle + a_2 | \cdots \pi_1^2 \pi_2^0 \pi_3^2 \pi_4^0 \rangle + a_3 | \cdots \pi_1^2 \pi_2^0 \pi_3^0 \pi_4^2 \rangle \\
+ a_4 | \cdots \pi_1^0 \pi_2^2 \pi_3^2 \pi_4^0 \rangle + a_5 \left(| \cdots \pi_1^2 \pi_2^1 \overline{\pi}_3^1 \pi_4^0 \rangle + | \cdots \pi_1^2 \overline{\pi}_2^1 \pi_3^1 \pi_4^0 \rangle \right)
\end{aligned}
\tag{7.8}
$$

where the electron in a singly occupied orbital has α spin unless the orbital has a bar over it, in which case it has β spin. Note again that the open-shell singlet appearing after coefficient a_5 requires two determinants to specify.

If we were to try to decide, based on any more or less rational approach, which CSFs to include in some particular expansion along the lines of Eq. (7.8), this would constitute a general MCSCF calculation. However, an alternative to picking and choosing amongst CSFs is simply to include *all* possible configurations in the expansion. In general, the number N of singlet CSFs that can be formed from the distribution of m electrons in n orbitals is determined as

$$
N = \frac{n!\,(n+1)!}{\left(\dfrac{m}{2}\right)! \left(\dfrac{m}{2}+1\right)! \left(n - \dfrac{m}{2}\right)! \left(n - \dfrac{m}{2}+1\right)!}
\tag{7.9}
$$

In the case of $m = n = 4$, $N = 20$ (it is a mildly diverting exercise to try to generate all 20 by hand). Permitting all possible arrangements of electrons to enter into the MCSCF expansion is typically referred to as having chosen a 'complete active space', and such calculations are said to be of the CASSCF, or just CAS, variety.

Notice that the factorial functions appearing in Eq. (7.9) quickly have daunting consequences. What if we were interested in carrying out a CASSCF calculation on methanol (CH_3OH) including all of the valence electrons in the active space? Such a calculation would be a (14,12) CAS (14 valence electrons, 5 pairs of σ and σ^* orbitals corresponding to the single bonds, and 2 oxygen lone pair orbitals). Neglecting symmetry, the total number of CSFs, from Eq. (7.9), would be 169 884. Recalling that the nature of the MCSCF process is to simultaneously optimize the MO coefficients *and* all of the CSF coefficients, one might imagine that such a calculation would be rather taxing. Indeed, CASSCF calculations on systems having more than 1 000 000 CSFs are extraordinarily demanding of resources and are rarely undertaken.

Various schemes exist to try to reduce the number of CSFs in the expansion in a rational way. Symmetry can reduce the scope of the problem enormously. In the TMM problem, many of the CSFs having partially occupied orbitals correspond to an electronic state symmetry other than that of the totally symmetric irreducible representation, and thus make no contribution to the closed-shell singlet wave function (if symmetry is not used before the fact, the calculation itself will determine the coefficients of non-contributing CSFs to be zero, but no advantage in efficiency will have been gained). Since this application of group theory involves no approximations, it is one of the best ways to speed up a CAS calculation.

An alternative approach is taken with a formalism known as generalized valence bond (GVB). In a typical CASSCF calculation, one first carries out an HF calculation, and then expresses the CSFs in those orbitals for use in the MCSCF process. To improve convergence, one often undertakes a localization of the canonical HF virtual orbitals (which are otherwise rather diffuse, particularly with large basis sets), so that they are more chemically realistic (see Appendix D for more information on orbital localization schemes). Such a transformation of the orbitals is rigorously permitted and has no effect on wave function expectation values. In the case of GVB, in contrast to CASSCF, not only are the virtual orbitals localized but so too are the occupied orbitals. Thus, to the maximum extent possible, the transformed orbitals look like the canonical bonds, lone pairs, and anti-bonds of valence bond theory, i.e., like Lewis structures. In GVB, only excitations from certain occupied to certain unoccupied orbitals are allowed. For instance, in the so-called perfect-pairing (PP) scheme, the pair of electrons in any bonding orbital is allowed to excite only as a pair, and only into the corresponding antibonding orbital (assuming it is empty). The motivation, then, is to try to capture in an efficient and chemically localized way the most important contributions to non-dynamical correlation. Some mathematical difficulties arise in the GVB scheme because the localized orbitals are not necessarily orthogonal, but the method can be quite fast because of the reduced number of configurations, and one hopes that the retained configurations are the most chemically important ones.

Another means to reduce the scale of the problem is to shrink the size of the CAS calculation, but to allow a limited number of excitations from/to orbitals outside of the CAS space. This secondary space is called a 'restricted active space' (RAS), and usually the excitation level is limited to one or two electrons. Thus, while all possible configurations of electrons in the CAS space are permitted, only a limited number of RAS configurations is possible. Remaining occupied and virtual orbitals, if any, are restricted to occupation numbers of exactly two and zero, respectively.

There is one other step sometimes taken to make the CAS/RAS calculation more efficient, and that is to freeze the shapes of the core orbitals to those determined at the HF level. Thus, there may be four different types of orbitals in a particular MCSCF calculation: frozen orbitals, inactive orbitals, RAS orbitals, and CAS orbitals. Figure 7.3 illustrates the situation in detail. Again, symmetry is the theoretician's friend in keeping the size of the system manageable in favorable cases.

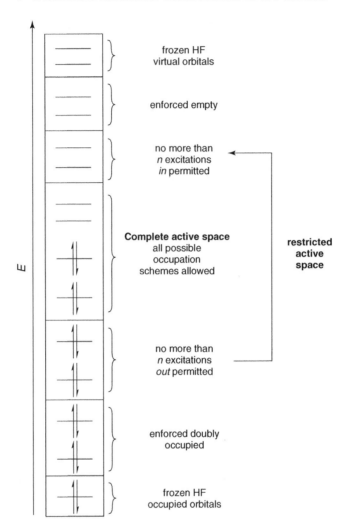

Figure 7.3 Possible assignment of different orbitals in a completely general MCSCF formalism. Frozen orbitals are not permitted to relax from their HF shapes, in addition to having their occupation numbers of zero (virtual) or two (occupied) enforced

While all of the above details are useful for making calculations more efficient, they still are not necessarily very helpful in evaluating just which orbitals should be included in any given space. Typically, a certain amount of trial and error is required in the selection of an active space. After selection of a given active space and convergence of the MCSCF wave function, one should inspect the occupation numbers of the active orbitals. A reasonable rule of thumb is that any orbital having an occupation number greater than 1.98 or less than 0.02 is not important enough to include in the CAS space, and should be removed to avoid instability. In addition, of course, it may be wise to add some orbitals not previously considered to see if *their* occupation numbers justify inclusion in the active space. And, clearly, if one is considering a reaction coordinate or a series of isomers, the active space must be balanced, so any orbital contributing significantly in one calculation should probably be used in all calculations. While methods to include dynamical correlation after an MCSCF calculation *can* help to make up for a less than optimal choice of active space, it is best not to rely on this phenomenon.

7.2.3 Full Configuration Interaction

Having discussed ways to reduce the scope of the MCSCF problem, it is appropriate to consider the other limiting case. What if we carry out a CASSCF calculation for *all* electrons including *all* orbitals in the complete active space? Such a calculation is called 'full configuration interaction' or 'full CI'. Within the choice of basis set, it is the best possible calculation that can be done, because it considers the contribution of every possible CSF. Thus, a full CI with an infinite basis set is an 'exact' solution of the (non-relativistic, Born–Oppenheimer, time-independent) Schrödinger equation.

Note that no reoptimization of HF orbitals is required, since the set of all possible CSFs is 'complete'. However, that is not much help in a computational efficiency sense, since the number of CSFs in a full CI can be staggeringly large. The trouble is not the number of electrons, which is a constant, but the number of basis functions. Returning to our methanol example above, if we were to use the 6-31G(d) basis set, the total number of basis functions would be 38. Using Eq. (7.9) to determine the number of CSFs in our (14,38) full CI we find that we must optimize 2.4×10^{13} expansion coefficients (!), and this is really a rather small basis set for chemical purposes.

Thus, full CI calculations with large basis sets are usually carried out for only the smallest of molecules (it is partly as a result of such calculations that the relative contributions to basis-set quality of polarization functions vs. decontraction of valence functions, as discussed in Chapter 6, were discovered). In larger systems, the practical restriction to smaller basis sets makes full CI calculations less chemically interesting, but such calculations remain useful to the extent that, as an optimal limit, they permit an evaluation of the quality of other methodologies for including electron correlation using the same basis set. We turn now to a consideration of such other methods.

7.3 Configuration Interaction

7.3.1 Single-determinant Reference

If we consider all possible excited configurations that can be generated from the HF determinant, we have a full CI, but such a calculation is typically too demanding to accomplish. However, just as we reduced the scope of CAS calculations by using RAS spaces, what if we were to reduce the CI problem by allowing only a limited number of excitations? How many should we include? To proceed in evaluating this question, it is helpful to rewrite Eq. (7.1) using a more descriptive notation, i.e.,

$$\Psi = a_0 \Psi_{\mathrm{HF}} + \sum_{i}^{\mathrm{occ.}} \sum_{r}^{\mathrm{vir.}} a_i^r \Psi_i^r + \sum_{i<j}^{\mathrm{occ.}} \sum_{r<s}^{\mathrm{vir.}} a_{ij}^{rs} \Psi_{ij}^{rs} + \cdots \qquad (7.10)$$

where i and j are occupied MOs in the HF 'reference' wave function, r and s are virtual MOs in Ψ_{HF}, and the additional CSFs appearing in the summations are generated by exciting an electron from the occupied orbital(s) indicated by subscripts into the virtual orbital(s) indicated by superscripts. Thus, the first summation on the r.h.s. of Eq. (7.10) includes all possible single electronic excitations, the second includes all possible double excitations, etc.

If we assume that we do *not* have any problem with non-dynamical correlation, we may assume that there is little need to reoptimize the MOs even if we do not plan to carry out the expansion in Eq. (7.10) to its full CI limit. In that case, the problem is reduced to determining the expansion coefficients for each excited CSF that *is* included. The energies E of N different CI wave functions (i.e., corresponding to different variationally determined sets of coefficients) can be determined from the N roots of the CI secular equation

$$
\begin{vmatrix}
H_{11} - E & H_{12} & \cdots & H_{1N} \\
H_{21} & H_{22} - E & \cdots & H_{2N} \\
\vdots & \vdots & \ddots & \vdots \\
H_{N1} & N_{N2} & \cdots & H_{NN} - E
\end{vmatrix} = 0
\tag{7.11}
$$

where

$$
H_{mn} = \langle \Psi_m | H | \Psi_n \rangle
\tag{7.12}
$$

H is the Hamiltonian operator and the numbering of the CSFs is arbitrary, but for convenience we will take $\Psi_1 = \Psi_{HF}$ and then all singly excited determinants, all doubly excited, etc. Solving the secular equation is equivalent to diagonalizing $\mathbf{H}$, and permits determination of the CI coefficients associated with each energy. While this is presented without derivation, the formalism is entirely analogous to that used to develop Eq. (4.21).

To solve Eq. (7.11), we need to know how to evaluate matrix elements of the type defined by Eq. (7.12). To simplify matters, we may note that the Hamiltonian operator is composed only of one- and two-electron operators. Thus, if two CSFs differ in their occupied orbitals by 3 or more orbitals, every possible integral over electronic coordinates hiding in the r.h.s. of Eq. (7.12) will include a simple overlap between at least one pair of different, and hence orthogonal, HF orbitals, and the matrix element will necessarily be zero. For the remaining cases of CSFs differing by two, one, and zero orbitals, the so-called Condon–Slater rules, which can be found in most quantum chemistry textbooks, detail how to evaluate Eq. (7.12) in terms of integrals over the one- and two-electron operators in the Hamiltonian and the HF MOs.

A somewhat special case is the matrix element between the HF determinant and a singly excited CSF. The Condon–Slater rules applied to this situation dictate that

$$
H_{1n} = \langle \Psi_{HF} | H | \Psi_i^r \rangle
$$
$$
= \langle \phi_r | F | \phi_i \rangle
\tag{7.13}
$$

where F is the Fock operator and i and r are the occupied and virtual HF orbitals in the single excitation. Since these orbitals are eigenfunctions of the Fock operator, we have

$$
\langle \phi_r | F | \phi_i \rangle = \varepsilon_i \langle \phi_r | \phi_i \rangle
$$
$$
= \varepsilon_i \delta_{ir}
\tag{7.14}
$$

where ε_i is the MO eigenvalue. Thus, all matrix elements between the HF determinants and singly excited determinants are zero, since to be singly excited, r must not be equal to i. This result is known as Brillouin's theorem (Brillouin 1934).

It is *not* the case that arbitrary matrix elements between *other* determinants differing by only one occupied orbital are equal to zero. Nevertheless, the Condon–Slater rules and Brillouin's theorem ensure that the CI matrix in a broad sense is reasonably sparse, as illustrated in Figure 7.4. With that in mind, let us return to the question of which excitations to include in a 'non-full' CI. What if we only keep single excitations? In that case, we see from Figure 7.4 that the CI matrix will be block diagonal. One 'block' will be the HF energy, H_{11}, and the other will be the singles/singles region. Since a block diagonal matrix can be fully diagonalized block by block, and since the HF result is already a block by itself,

	Ψ_{HF}	Ψ_i^a	Ψ_{ij}^{ab}	Ψ_{ijk}^{abc}
Ψ_{HF}	E_{HF}	0	dense	0
Ψ_i^a	0	dense	sparse	very sparse
Ψ_{ij}^{ab}	dense	sparse	sparse	extremely sparse
Ψ_{ijk}^{abc}	0	very sparse	extremely sparse	extremely sparse

Figure 7.4 Structure of the CI matrix as blocked by classes of determinants. The HF block is the (1,1) position, the matrix elements between the HF and singly excited determinants are zero by Brillouin's theorem, and between the HF and triply excited determinants are zero by the Condon–Slater rules. In a system of reasonable size, remaining regions of the matrix become increasingly sparse, but the number of determinants in each block grows to be extremely large. Thus, the (1,1) eigenvalue is most affected by the doubles, then by the singles, then by the triples, etc

it is apparent that the lowest energy root, i.e., the ground-state HF root, is unaffected by inclusion of single excitations. Indeed, one way to think about the HF process is that it is an optimization of orbitals subject to the constraint that single excitations do not contribute to the wave function. Thus, the so-called CI singles (CIS) method finds no use for ground states, although it can be useful for excited states, as described in Section 14.2.2.

So, we might next consider including only double excitations (CID). It is worthwhile to do a very simple example, such as molecular hydrogen in a minimal basis set. In that case, there are only 2 HF orbitals, the σ and the σ^* orbitals associated with the H–H bond, in which case there is only one doubly excited state, corresponding to $|\sigma^{*2}>$. The CID state energies are found from solving

$$\begin{vmatrix} H_{11} - E & H_{12} \\ H_{21} & H_{22} - E \end{vmatrix} = 0 \qquad (7.15)$$

This quadratic equation is simple to solve, and gives root energies

$$E = \frac{1}{2}\left[H_{11} + H_{22} \pm \sqrt{(H_{22} - H_{11})^2 + 4H_{12}} \right] \qquad (7.16)$$

The Condon–Slater rules dictate that H_{12} is an electron-repulsion integral, and it thus has a positive sign (it is actually the exchange integral K_{12}). So, examining Eq. (7.16), we see that to the average of the two pure-state energies (ground and doubly excited) we should either add or subtract a value slightly larger than half the difference between the two state energies. Thus, when we subtract, our energy will be below the HF energy, and the difference will be the correlation energy. In the case of H_2 with the STO-3G basis set at a bond distance of 1.4 a.u., $E_{corr} = -0.02056$ a.u., or about 13 kcal/mol.

In bigger systems, this process can be carried out analogously. However, the size of the CI matrix can quickly become very, very large, in which case diagonalization is computationally taxing. More efficient methods than diagonalization exist for finding only one or a few eigenvalues of large matrices. These methods are typically iterative, and most modern electronic structure programs use them in preference to full matrix diagonalization.

What about triple excitations? While there are no non-zero matrix elements between the ground state and triply excited states, the triples do mix with the doubles, and can through them influence the lowest energy eigenvalue. So, there is some motivation for including them. On the other hand, there are a *lot* of triples, making their inclusion difficult in a practical sense. As a result, triples, and higher-level excitations, are usually not accounted for in truncated CI treatments.

Let us return, however, to singly excited determinants. While, like triples, they fail to interact with the ground state (although in this case because of Brillouin's theorem), they too mix with doubles and thus can have *some* influence on the lowest eigenvalue. In this instance, there are sufficiently few singles compared to doubles that it does not make the problem significantly more difficult to include them, and this level of theory is known as CISD.

The scaling for CISD with respect to system size is, in the large basis limit, on the order of N^6. Such scaling behavior is considerably worse than HF, and thus poses a more stringent limit on the sizes of systems that can be practically addressed. Just as with MCSCF, symmetry can be used to significantly reduce the computational effort by facilitating the evaluation of matrix elements. Similarly, some orbitals can be frozen in the generation of excited states. A popular choice is to leave the core orbitals frozen in CISD.

One of the most appealing features of CISD is that it is variational. Thus, the CISD energy represents an upper bound on the exact energy. However, it has a particularly unattractive feature as well, and that is that it is not 'size consistent'. This property is best explained by example: consider the H_2 molecule case addressed above. We may construct the CID wave function as

$$\Psi_{CID} = (1 - c)^2 \Psi_{HF} + c^2 \Psi_{1\bar{1}}^{2\bar{2}} \qquad (7.17)$$

where the coefficient c is determined from the diagonalization process. Now, consider the CID wave function for two molecules of H_2 separated by, say, 50 Å. For all practical purposes, there is no chemical interaction between them, so we could take the overall wave function simply to be a properly antisymmetrized product of Eq. (7.17) with itself. This expression would include a term, preceded by the coefficient c^4, corresponding to simultaneous double excitation within each molecule. However, that is a quadruply excited configuration. As such, if we carried out a CID calculation on the two molecules as a single system, it would not be permitted. Thus, twice the CID energy of one molecule of H_2 will be lower than the CID energy for two molecules of H_2 at large separations, which is a vexing result.

Various approaches to overcoming the size extensivity problem have been proposed. Owing to its simplicity, one of the more popular methods is that of Langhoff and Davidson (1974), which estimates the energy associated with the missing quadruple excitations as

$$E_Q = (1 - a_0)^2 (E_{CISD} - E_{HF}) \qquad (7.18)$$

where a_0 is the coefficient of the HF determinant in the normalized truncated CISD wave function (which itself is Eq. (7.10) without the ellipsis). This is typically abbreviated as CISD(Q). In modern work, there has been a tendency to avoid single-reference CI calculations in favor of other, size-extensive methods for including electron correlation (vide infra).

A recent variation of CISD that is both variational *and* size-consistent has been proposed by Krylov (2001). In spin-flip CISD (SF-CISD), the reference configuration is always taken to be a high-spin HF configuration, but spin flips are allowed when 'excited' configurations are generated. Thus, for instance, a triplet reference can generate singlet states by spin flip of one electron. The resulting CI matrix is much larger for SF-CISD, but it also has additional sparsity since matrix elements between states of different spin are zero for the standard spin-free Hamiltonian. Diagonalization of the CI matrix provides energies for the various target states. A key virtue of SF-CISD is that the high-spin reference is usually well described as a single determinant, and the CI formalism permits lower-spin states generated by spin flips to be well described irrespective of how much multideterminantal character is present.

The SF-CISD model exhibits timing and scaling behavior equivalent to standard CISD. Significant time savings may be realized in selected instances by estimating the effect of double excitations using perturbation theory (Head-Gordon *et al.* 1994; Section 7.4 presents the basics of perturbation theory); this model is referred to as SF-CIS(D). Preliminary studies on various challenging problems like homolytic bond dissociation energies and singlet–triplet energy separations in biradicals have shown SF-CISD and SF-CIS(D) to be considerably more robust than the corresponding non-spin-flip approaches (Krylov 2001; Krylov and Sherrill 2002; Slipchenko and Krylov 2002).

7.3.2 Multireference

The formalism for multireference configuration interaction (MRCI) is quite similar to that for single-reference CI, except that instead of the HF wave function serving as reference, an MCSCF wave function is used. While it is computationally considerably more difficult to construct the initial MCSCF wave function than a HF wave function, the significant improvement of the virtual orbitals in the former case can make the CI itself more rapidly convergent. Nevertheless, the number of matrix elements requiring evaluation in MRCI calculations is enormous, and they are usually undertaken only for small systems. Typically, MRCI is a useful method to study a large section of a PES, where significant changes in bonding (and thus correlation energy) are taking place so a sophisticated method is needed to accurately predict dynamical and non-dynamical correlation energies.

As with single-reference CI, most MRCI calculations truncate the CI expansion to include only singles and doubles (MRCISD). An analog of Eq. (7.18) has been proposed to make up for the non-size-extensivity this engenders (Bruna, Peyerimhoff, and Buenker, 1980). MRCISD calculations with large basis sets can be better than similarly expensive full CI calculations with smaller basis sets, illustrating that most of the correlation energy can be captured by including only limited excitations, at least in those systems small enough to permit thorough evaluation. Additional efficiencies can be gained by restricting the size of the MCSCF reference to something smaller than a CAS reference and considering only the reduced number of single and double excitations therefrom (Pitarch-Ruiz, Sanchez-Marin, and Maynau 2002).

7.4 Perturbation Theory

7.4.1 General Principles

Often in pseudoeigenvalue equations, the nature of a particular operator makes it difficult to work with. However, it is sometimes worthwhile to create a more tractable operator by removing some particularly unpleasant portion of the original one. Using exact eigenfunctions and eigenvalues of the simplified operator, it is possible to estimate the eigenfunctions and eigenvalues of the more complete operator. Rayleigh–Schrödinger perturbation theory provides a prescription for accomplishing this.

In the general case, we have some operator $\mathbf{A}$ that we can write as

$$\mathbf{A} = \mathbf{A}^{(0)} + \lambda \mathbf{V} \tag{7.19}$$

where $\mathbf{A}^{(0)}$ is an operator for which we can find eigenfunctions, $\mathbf{V}$ is a perturbing operator, and λ is a dimensionless parameter that, as it varies from 0 to 1, maps $\mathbf{A}^{(0)}$ into $\mathbf{A}$. If we expand our ground-state eigenfunctions and eigenvalues as Taylor series in λ, we have

$$\Psi_0 = \Psi_0^{(0)} + \lambda \frac{\partial \Psi_0^{(0)}}{\partial \lambda}\bigg|_{\lambda=0} + \frac{1}{2!}\lambda^2 \frac{\partial^2 \Psi_0^{(0)}}{\partial \lambda^2}\bigg|_{\lambda=0} + \frac{1}{3!}\lambda^3 \frac{\partial^3 \Psi_0^{(0)}}{\partial \lambda^3}\bigg|_{\lambda=0} + \cdots \tag{7.20}$$

and

$$a_0 = a_0^{(0)} + \lambda \frac{\partial a_0^{(0)}}{\partial \lambda}\bigg|_{\lambda=0} + \frac{1}{2!}\lambda^2 \frac{\partial^2 a_0^{(0)}}{\partial \lambda^2}\bigg|_{\lambda=0} + \frac{1}{3!}\lambda^3 \frac{\partial^3 a_0^{(0)}}{\partial \lambda^3}\bigg|_{\lambda=0} + \cdots \tag{7.21}$$

where $a_0^{(0)}$ is the eigenvalue for $\Psi_0^{(0)}$, which is the appropriate normalized ground-state eigenfunction for $\mathbf{A}^{(0)}$. For ease of notation, Eqs. (7.20) and (7.21) are usually written as

$$\Psi_0 = \Psi_0^{(0)} + \lambda \Psi_0^{(1)} + \lambda^2 \Psi_0^{(2)} + \lambda^3 \Psi_0^{(3)} + \cdots \tag{7.22}$$

and

$$a_0 = a_0^{(0)} + \lambda a_0^{(1)} + \lambda^2 a_0^{(2)} + \lambda^3 a_0^{(3)} + \cdots \tag{7.23}$$

where the terms having superscripts (n) are referred to as 'nth-order corrections' to the zeroth order term and are defined by comparison to Eqs. (7.20) and (7.21).

Thus, we may write

$$(\mathbf{A}^{(0)} + \lambda \mathbf{V})|\Psi_0\rangle = a|\Psi_0\rangle \tag{7.24}$$

as

$$(\mathbf{A}^{(0)} + \lambda \mathbf{V})|\Psi_0^{(0)} + \lambda \Psi_0^{(1)} + \lambda^2 \Psi_0^{(2)} + \lambda^3 \Psi_0^{(3)} + \cdots\rangle =$$
$$(a_0^{(0)} + \lambda a_0^{(1)} + \lambda^2 a_0^{(2)} + \lambda^3 a_0^{(3)} + \cdots)|\Psi_0^{(0)} + \lambda \Psi_0^{(1)} + \lambda^2 \Psi_0^{(2)} + \lambda^3 \Psi_0^{(3)} + \cdots\rangle \tag{7.25}$$

Since Eq. (7.25) is valid for any choice of λ between 0 and 1, we can expand the left and right sides and consider only equalities involving like powers of λ. Powers 0 through 3 require

$$\mathbf{A}^{(0)}|\Psi_0^{(0)}\rangle = a_0^{(0)}|\Psi_0^{(0)}\rangle \tag{7.26}$$

$$\mathbf{A}^{(0)}|\Psi_0^{(1)}\rangle + \mathbf{V}|\Psi_0^{(0)}\rangle = a_0^{(0)}|\Psi_0^{(1)}\rangle + a_0^{(1)}|\Psi_0^{(0)}\rangle \tag{7.27}$$

$$\mathbf{A}^{(0)}|\Psi_0^{(2)}\rangle + \mathbf{V}|\Psi_0^{(1)}\rangle = a_0^{(0)}|\Psi_0^{(2)}\rangle + a_0^{(1)}|\Psi_0^{(1)}\rangle + a_0^{(2)}|\Psi_0^{(0)}\rangle \tag{7.28}$$

$$\mathbf{A}^{(0)}|\Psi_0^{(3)}\rangle + \mathbf{V}|\Psi_0^{(2)}\rangle = a_0^{(0)}|\Psi_0^{(3)}\rangle + a_0^{(1)}|\Psi_0^{(2)}\rangle + a_0^{(2)}|\Psi_0^{(1)}\rangle + a_0^{(3)}|\Psi_0^{(0)}\rangle \tag{7.29}$$

where further generalization should be obvious. Our goal, of course, is to determine the various nth-order corrections. Equation (7.26) is the zeroth-order solution from which we are hoping to build, while Eq. (7.27) involves the two unknown first-order corrections to the wave function and eigenvalue.

To proceed, we first impose intermediate normalization of Ψ; that is

$$\langle \Psi_0 | \Psi_0^{(0)} \rangle = 1 \tag{7.30}$$

By use of Eq. (7.22) and normalization of $\Psi_0^{(0)}$, it must then be true that

$$\langle \Psi_0^{(n)} | \Psi_0^{(0)} \rangle = \delta_{n0} \tag{7.31}$$

Now, we multiply on the left by $\Psi_0^{(0)}$ and integrate to solve Eqs. (7.27)–(7.29). In the case of Eq. (7.27), we have

$$\langle \Psi_0^{(0)} | \mathbf{A}^{(0)} | \Psi_0^{(1)} \rangle + \langle \Psi_0^{(0)} | \mathbf{V} | \Psi_0^{(0)} \rangle = a_0^{(0)} \langle \Psi_0^{(0)} | \Psi_0^{(1)} \rangle + a_0^{(1)} \langle \Psi_0^{(0)} | \Psi_0^{(0)} \rangle \tag{7.32}$$

Using

$$\langle \Psi_0^{(0)} | \mathbf{A}^{(0)} | \Psi_0^{(1)} \rangle = \langle \Psi_0^{(1)} | \mathbf{A}^{(0)} | \Psi_0^{(0)} \rangle^* \tag{7.33}$$

and Eqs. (7.26), (7.30), and (7.31), we can simplify Eq. (7.32) to

$$\langle \Psi_0^{(0)} | \mathbf{V} | \Psi_0^{(0)} \rangle = a_0^{(1)} \tag{7.34}$$

which is the well-known result that the first-order correction to the eigenvalue is the expectation value of the perturbation operator over the unperturbed wave function.

As for $\Psi_0^{(1)}$ like *any* function of the electronic coordinates, it can be expressed as a linear combination of the *complete* set of eigenfunctions of $\mathbf{A}^{(0)}$, i.e.,

$$\Psi_0^{(1)} = \sum_{i>0} c_i \Psi_i^{(0)} \tag{7.35}$$

To determine the coefficients c_i in Eq. (7.35), we can multiple Eq. (7.27) on the left by $\Psi_j^{(0)}$ and integrate to obtain

$$\langle \Psi_j^{(0)} | \mathbf{A}^{(0)} | \Psi_0^{(1)} \rangle + \langle \Psi_j^{(0)} | \mathbf{V} | \Psi_0^{(0)} \rangle = a_0^{(0)} \langle \Psi_j^{(0)} | \Psi_0^{(1)} \rangle + a_0^{(1)} \langle \Psi_j^{(0)} | \Psi_0^{(0)} \rangle \tag{7.36}$$

Using Eq. (7.35), we expand this to

$$\left\langle \Psi_j^{(0)} \left| \mathbf{A}^{(0)} \right| \sum_{i>0} c_i \Psi_i^{(0)} \right\rangle + \langle \Psi_j^{(0)} | \mathbf{V} | \Psi_0^{(0)} \rangle =$$
$$a_0^{(0)} \left\langle \Psi_j^{(0)} \left| \sum_{i>0} c_i \Psi_i^{(0)} \right. \right\rangle + a_0^{(1)} \langle \Psi_j^{(0)} | \Psi_0^{(0)} \rangle \tag{7.37}$$

which, from the orthonormality of the eigenfunctions, simplifies to

$$c_j a_j^{(0)} + \langle \Psi_j^{(0)} | \mathbf{V} | \Psi_0^{(0)} \rangle = c_j a_0^{(0)} \tag{7.38}$$

or

$$c_j = \frac{\langle \Psi_j^{(0)} | \mathbf{V} | \Psi_0^{(0)} \rangle}{a_0^{(0)} - a_j^{(0)}} \tag{7.39}$$

With the first-order eigenvalue and wave function corrections in hand, we can carry out analogous operations to determine the second-order corrections, then the third-order, etc. The algebra is tedious, and we simply note the results for the eigenvalue corrections, namely

$$a_0^{(2)} = \sum_{j>0} \frac{|\langle \Psi_j^{(0)} | \mathbf{V} | \Psi_0^{(0)} \rangle|^2}{a_0^{(0)} - a_j^{(0)}} \tag{7.40}$$

and

$$a_0^{(3)} = \sum_{j>0, k>0} \frac{\langle \Psi_0^{(0)} | \mathbf{V} | \Psi_j^{(0)} \rangle [\langle \Psi_j^{(0)} | \mathbf{V} | \Psi_k^{(0)} \rangle - \delta_{jk} \langle \Psi_0^{(0)} | \mathbf{V} | \Psi_0^{(0)} \rangle] \langle \Psi_k^{(0)} | \mathbf{V} | \Psi_0^{(0)} \rangle}{(a_0^{(0)} - a_j^{(0)})(a_0^{(0)} - a_k^{(0)})} \tag{7.41}$$

Let us now examine the application of perturbation theory to the particular case of the Hamiltonian operator and the energy.

7.4.2 Single-reference

We now consider the use of perturbation theory for the case where the complete operator **A** is the Hamiltonian, **H**. Møller and Plesset (1934) proposed choices for $\mathbf{A}^{(0)}$ and **V** with this goal in mind, and the application of their prescription is now typically referred to by the acronym MPn where n is the order at which the perturbation theory is truncated, e.g., MP2, MP3, etc. Some workers in the field prefer the acronym MBPTn, to emphasize the more general nature of many-body perturbation theory (Bartlett 1981).

The MP approach takes $\mathbf{H}^{(0)}$ to be the sum of the one-electron Fock operators, i.e., the non-interacting Hamiltonian (see Section 4.5.2)

$$\mathbf{H}^{(0)} = \sum_{i=1}^{n} f_i \tag{7.42}$$

where n is the number of basis functions and f_i is defined in the usual way according to Eq. (4.52). In addition, $\Psi^{(0)}$ is taken to be the HF wave function, which is a Slater determinant formed from the occupied orbitals. By analogy to Eq. (4.36), it is straightforward to show that the eigenvalue of $\mathbf{H}^{(0)}$ when applied to the HF wave function is the sum of the occupied orbital energies, i.e.,

$$\mathbf{H}^{(0)} \Psi^{(0)} = \sum_{i}^{\text{occ.}} \varepsilon_i \Psi^{(0)} \tag{7.43}$$

where the orbital energies are the usual eigenvalues of the specific one-electron Fock operators. The sum on the r.h.s. thus defines the eigenvalue $a^{(0)}$.

Recall that this is *not* the way the electronic energy is usually calculated in an HF calculation – it is the expectation value for the *correct* Hamiltonian and the HF wave function that determines that energy. The 'error' in Eq. (7.43) is that each orbital energy includes the repulsion of the occupying electron(s) with all of the other electrons. Thus, each electron–electron repulsion is counted twice (once in each orbital corresponding to each pair of electrons). So, the correction term **V** that will return us to the correct Hamiltonian and allow us to use perturbation theory to improve the HF wave function and eigenvalues must be the difference between counting electron repulsion once and counting it twice. Thus,

$$\mathbf{V} = \sum_{i}^{\text{occ.}} \sum_{j>i}^{\text{occ.}} \frac{1}{r_{ij}} - \sum_{i}^{\text{occ.}} \sum_{j}^{\text{occ.}} \left(J_{ij} - \frac{1}{2} K_{ij} \right) \tag{7.44}$$

where the first term on the r.h.s. is the proper way to compute electron repulsion (and is exactly as it appears in the Hamiltonian of Eq. (4.3) and the second term is how it is computed from summing over the Fock operators for the occupied orbitals where J and K are the Coulomb and exchange operators defined in Section 4.5.5. Note that, since we are summing over occupied orbitals, we must be working in the MO basis set, not the AO one.

So, let us now consider the first-order correction $a^{(1)}$ to the zeroth-order eigenvalue defined by Eq. (7.43). In principle, from Eq. (7.34), we operate on the HF wave function $\Psi^{(0)}$ with **V** defined in Eq. (7.44), multiply on the left by $\Psi^{(0)}$, and integrate. By inspection, cognoscenti should not have much trouble seeing that the result will be the negative of the electron–electron repulsion energy. However, if that is not obvious, there is no need to carry through the integrations in any case. That is because we can write

$$a^{(0)} + a^{(1)} = \langle \Psi^{(0)} | \mathbf{H}^{(0)} | \Psi^{(0)} \rangle + \langle \Psi^{(0)} | \mathbf{V} | \Psi^{(0)} \rangle$$

$$= \langle \Psi^{(0)} | \mathbf{H}^{(0)} + \mathbf{V} | \Psi^{(0)} \rangle$$

$$= \langle \Psi^{(0)} | \mathbf{H} | \Psi^{(0)} \rangle$$

$$= E_{\text{HF}} \tag{7.45}$$

i.e., *the Hartree-Fock energy is the energy correct through first-order in Møller-Plesset perturbation theory*. Thus, the second term on the r.h.s. of the first line of Eq. (7.45) must indeed be the negative of the overcounted electron–electron repulsion already noted to be implicit in $a^{(0)}$.

As MP1 does not advance us beyond the HF level in determining the energy, we must consider the second-order correction to obtain an estimate of correlation energy. Thus, we must evaluate Eq. (7.40) using the set of all possible excited-state eigenfunctions and eigenvalues of the operator $\mathbf{H}^{(0)}$ defined in Eq. (7.42). Happily enough, that is a straightforward process, since within a finite basis approximation, the set of all possible excited eigenfunctions is simply all possible ways to distribute the electrons in the HF orbitals, i.e., all possible excited CSFs appearing in Eq. (7.10).

Let us consider the numerator of Eq. (7.40). Noting that $\mathbf{V}$ is $\mathbf{H} - \mathbf{H}^{(0)}$, we may write

$$
\begin{aligned}
\sum_{j>0}\langle\Psi_j^{(0)}|\mathbf{V}|\Psi_0^{(0)}\rangle &= \sum_{j>0}\langle\Psi_j^{(0)}|\mathbf{H} - \mathbf{H}^{(0)}|\Psi_0^{(0)}\rangle \\
&= \sum_{j>0}[\langle\Psi_j^{(0)}|\mathbf{H}|\Psi_0^{(0)}\rangle - \langle\Psi_j^{(0)}|\mathbf{H}^{(0)}|\Psi_0^{(0)}\rangle] \\
&= \sum_{j>0}\left[\langle\Psi_j^{(0)}|\mathbf{H}|\Psi_0^{(0)}\rangle - \sum_i^{\text{occ.}}\varepsilon_i\langle\Psi_j^{(0)}|\Psi_0^{(0)}\rangle\right] \\
&= \sum_{j>0}\langle\Psi_j^{(0)}|\mathbf{H}|\Psi_0^{(0)}\rangle
\end{aligned}
\tag{7.46}
$$

where the simplification of the r.h.s. on proceeding from line 3 to line 4 derives from the orthogonality of the ground- and excited-state Slater determinants. As for the remaining integrals, from the Condon–Slater rules, we know that we need only consider integrals involving doubly and singly excited determinants. However, from Brillouin's theorem, we also know that the integrals involving the singly excited determinants will all be zero. The Condon–Slater rules applied to the remaining integrals involving doubly excited determinants dictate that

$$
\sum_{j>0}\langle\Psi_j^{(0)}|\mathbf{V}|\Psi_0^{(0)}\rangle = \sum_i^{\text{occ.}}\sum_{j>i}^{\text{occ.}}\sum_a^{\text{vir.}}\sum_{b>a}^{\text{vir.}}[(ij|ab) - (ia|jb)]
\tag{7.47}
$$

where the two-electron integrals are those defined by Eq. (4.56).

As for the denominator of Eq. (7.40), from inspection of Eq. (7.43), $a^{(0)}$ for each doubly excited determinant will differ from that for the ground state only by including in the sum the energies of the virtual orbitals into which excitation has occurred and excluding the energies of the two orbitals from which excitation has taken place. Thus, the full expression for the second-order energy correction is

$$
a^{(2)} = \sum_i^{\text{occ.}}\sum_{j>i}^{\text{occ.}}\sum_a^{\text{vir.}}\sum_{b>a}^{\text{vir.}}\frac{[(ij|ab) - (ia|jb)]^2}{\varepsilon_i + \varepsilon_j - \varepsilon_a - \varepsilon_b}
\tag{7.48}
$$

The sum of $a^{(0)}$, $a^{(1)}$, and $a^{(2)}$ defines the MP2 energy.

MP2 calculations can be done reasonably rapidly because Eq. (7.48) can be efficiently evaluated. The scaling behavior of the MP2 method is roughly N^5, where N is the number of basis functions. Analytic gradients and second derivatives are available for this level of theory, so it can conveniently be used to explore PESs. MP2, and indeed all orders of MPn theory, are size-consistent, which is a particularly desirable feature. Finally, Saebø and Pulay have described a scheme whereby the occupied orbitals are localized and excitations out of these orbitals are not permitted if the accepting (virtual) orbitals are too far away (the distance being a user-defined variable; Pulay 1983; Saebø and Pulay 1987). This localized MP2 (LMP2) technique significantly decreases the total number of integrals requiring evaluation in large systems, and can also be implemented in a fashion that leads to linear scaling with system size. These features have the potential to increase computational efficiency substantially.

However, it should be noted that the Møller–Plesset formalism is potentially rather dangerous in design. Perturbation theory works best when the perturbation is small (because the Taylor expansions in Eqs. (7.20) and (7.21) are then expected to be quickly convergent). But, in the case of MP theory, the perturbation is the full electron–electron repulsion energy, which is a rather large contributor to the total energy. So, there is no reason to expect that an MP2 calculation will give a value for the correlation energy that is particularly good. In addition, the MPn methodology is *not* variational. Thus, it is possible that the MP2 estimate for the correlation energy will be too large instead of too small (however, this rarely happens in practice because basis set limitations always introduce error in the direction of underestimating the correlation energy).

Naturally, if one wants to improve convergence, one can proceed to higher orders in perturbation theory (note, however, that even at infinite order, there is no guarantee of convergence when a finite basis set has been used). At third order, it is still true that only matrix elements involving doubly excited determinants need be evaluated, so MP3 is not too much more expensive than MP2. A fair body of empirical evidence, however, suggests that MP3 calculations tend to offer rather little improvement over MP2. Analytic gradients are not available for third and higher orders of perturbation theory.

At the MP4 level, integrals involving triply and quadruply excited determinants appear. The evaluation of the terms involving triples is the most costly, and scales as N^7. If one simply chooses to ignore the triples, the method scales more favorably and this choice is typically abbreviated MP4SDQ. In a small to moderately sized molecule, the cost of accounting for the triples is roughly equal to that for the rest of the calculation, i.e., triples double the time. In closed-shell singlets with large frontier orbital separations, the contributions from the triples tend to be rather small, so ignoring them may be worthwhile in terms of efficiency. However, when the frontier orbital separation drops, the contribution of the triples can become very large, and major errors in interpretation can derive from ignoring their effects. In such a situation, the triples in essence help to correct for the error involved in using a single-reference wave function.

Empirically, MP4 calculations can be quite good, typically accounting for more than 95% of the correlation energy with a good basis set. However, although ideally the MPn results for any given property would show convergent behavior as a function of n, the more typical observation is oscillatory, and it can be difficult to extrapolate accurately from only four points (MP1 = HF, MP2, MP3, MP4). As a rough rule of thumb, to the extent that the results of an MP2 calculation differ from HF, say for the energy difference between two isomers, the difference tends to be overestimated. MP3 usually pushes the result back in the HF direction, by a variable amount. MP4 increases the difference again, but in favorable cases by only a small margin, so that some degree of convergence may be relied upon (He and Cremer 2000a). Additional performance details are discussed in Section 7.6.

7.4.3 Multireference

The generalization of MPn theory to the multireference case involves the obvious choice of using an MCSCF wave function for $\Psi^{(0)}$ instead of a single-determinant RHF or UHF one. However, it is much less obvious what should be chosen for $\mathbf{H}^{(0)}$, as the MCSCF MOs do not diagonalize any particular set of one-electron operators. Several different choices have been made by different authors, and each defines a unique 'flavor' of multireference perturbation theory (see, for instance, Andersson 1995; Davidson 1995; Finley and Freed 1995). One of the more popular choices is the so-called CASPT2N method of Roos and co-workers (Andersson, Malmqvist, and Roos 1992). Often this method is simply called CASPT2 – while this ignores the fact that different methods having other acronym endings besides N have been defined by these same authors (e.g., CASTP2D and CASPT2g1), the other methods are sufficiently inferior to CASPT2N that they are typically used only by specialists and confusion is minimized.

Most multireference methods described to date have been limited to second order in perturbation theory. As analytic gradients are not yet available, geometry optimization requires recourse to more tedious numerical approaches (see, for instance, Page and Olivucci 2003). While some third order results have begun to appear, much like the single-reference case, they do not seem to offer much improvement over second order.

An appealing feature of multireference perturbation theory is that it can correct for some deficiencies associated with an incomplete active space. For instance, the relative energies for various electronic states of TMM (Figure 7.1) were found to vary widely depending on whether a (2,2), (4,4), or (10,10) active space was used; however, the relative energies from corresponding CASPT2 calculations agreed well with one another. Thus, while the motivation for multireference perturbation theory is to address dynamical correlation after a separate treatment of non-dynamical correlation, it seems capable of handling a certain amount of the latter as well.

7.4.4 First-order Perturbation Theory for Some Relativistic Effects

In Møller–Plesset theory, first-order perturbation theory does not improve on the HF energy because the zeroth-order Hamiltonian is not itself the HF Hamiltonian. However, first-order perturbation theory *can* be useful for estimating energetic effects associated with operators that *extend* the HF Hamiltonian. Typical examples of such terms include the mass-velocity and one-electron Darwin corrections that arise in relativistic quantum mechanics. It is fairly difficult to self-consistently optimize wavefunctions for systems where these terms are explicitly included in the Hamiltonian, but an estimate of their energetic contributions may be had from simple first-order perturbation theory, since that energy is computed simply by taking the expectation values of the operators over the much more easily obtained HF wave functions.

The mass-velocity correction is evaluated as

$$E_{\mathrm{mv}} = \left\langle \Psi_{\mathrm{HF}} \left| -\frac{1}{8c^2} \sum_i \nabla_i^4 \right| \Psi_{\mathrm{HF}} \right\rangle \qquad (7.49)$$

where c is the speed of light (137.036 a.u.) and i runs over electrons. The one-electron Darwin correction is evaluated as

$$E_{1D} = \left\langle \Psi_{HF} \left| \frac{\pi}{2c^2} \sum_{ik} Z_k \delta(\mathbf{r}_{ik}) \right| \Psi_{HF} \right\rangle \tag{7.50}$$

where i runs over electrons, k runs over nuclei, and δ is the Dirac delta, which is the integral equivalent of the Kronecker delta in that it integrates to zero everywhere except at the position of its argument, at which point it integrates to one. Thus, the Dirac delta requires only that one know the molecular orbital amplitudes at the nuclear positions, and nowhere else.

The presence of $1/c^2$ in the prefactors for these terms makes them negligible unless the velocities are very, very high (as measured by the del-to-the-fourth-power operator in the mass-velocity term) or one or more orbitals have very large amplitudes at the atomic positions for nuclei whose atomic numbers are also very large (as measured by the one-electron Darwin term). These situations tend to occur only for core orbitals centered on very heavy atoms. Thus, efforts to estimate their energies from first-order perturbation theory are best undertaken with basis sets having core basis functions of good quality. It is the effects of these terms *on* the core orbitals (which could be estimated from the first-order correction to the *wavefunction*, as opposed to the energy) that motivate the creation of relativistic effective core potential basis sets like those described in Section 6.2.7.

7.5 Coupled-cluster Theory

One of the more mathematically elegant techniques for estimating the electron correlation energy is coupled-cluster (CC) theory (Cizek 1966). We will avoid most of the formal details here, and instead focus on intuitive connections to CI and MP*n* theory (readers interested in a more mathematical development may examine Crawford and Schaefer 1996).

The central tenet of CC theory is that the full-CI wave function (i.e., the 'exact' one within the basis set approximation) can be described as

$$\Psi = e^{\mathbf{T}} \Psi_{HF} \tag{7.51}$$

The cluster operator $\mathbf{T}$ is defined as

$$\mathbf{T} = \mathbf{T}_1 + \mathbf{T}_2 + \mathbf{T}_3 + \cdots + \mathbf{T}_n \tag{7.52}$$

where n is the total number of electrons and the various $\mathbf{T}_i$ operators generate all possible determinants having i excitations from the reference. For example,

$$\mathbf{T}_2 = \sum_{i<j}^{\text{occ.}} \sum_{a<b}^{\text{vir.}} t_{ij}^{ab} \Psi_{ij}^{ab} \tag{7.53}$$

where the amplitudes t are determined by the constraint that Eq. (7.51) be satisfied. The expansion of $\mathbf{T}$ ends at n because no more than n excitations are possible.

Of course, operating on the HF wave function with $\mathbf{T}$ is, in essence, full CI (more accurately, in full CI one applies $1 + \mathbf{T}$), so one may legitimately ask what advantage is afforded by the use of the exponential of $\mathbf{T}$ in Eq. (7.51). The answer lies in the consequences associated with truncation of $\mathbf{T}$. For instance, let us say that we only want to consider the double excitation operator, i.e., we make the approximation $\mathbf{T} = \mathbf{T}_2$. In that case, Taylor expansion of the exponential function in Eq. (7.51) gives

$$
\begin{aligned}
\Psi_{\text{CCD}} &= e^{\mathbf{T}} \Psi_{\text{HF}} \\
&= \left(1 + \mathbf{T}_2 + \frac{\mathbf{T}_2^2}{2!} + \frac{\mathbf{T}_2^3}{3!} + \cdots \right) \Psi_{\text{HF}}
\end{aligned}
\tag{7.54}
$$

where CCD implies coupled cluster with only the double-excitation operator. Note that the first two terms in parentheses, $1 + \mathbf{T}_2$, define the CID method described in Section 7.3.1. The remaining terms, however, involve products of excitation operators. Each application of $\mathbf{T}_2$ generates double excitations, so the product of two applications (the square of $\mathbf{T}_2$) generates quadruple excitations. Similarly, the cube of $\mathbf{T}_2$ generates hextuple substitutions, etc. It is exactly the *failure* to include these excitations that makes CI non-size-consistent! So, using the exponential of $\mathbf{T}$ in Eq. (7.51) ensures size consistency. Moreover, through careful analysis of perturbation theory, one can show that CCD is equivalent to including all of the terms involving products of double substitutions out to infinite order, i.e., MP∞D using the notation developed earlier in the context of MP4.

The computational problem, then, is determination of the cluster amplitudes t for all of the operators included in the particular approximation. In the standard implementation, this task follows the usual procedure of left-multiplying the Schrödinger equation by trial wave functions expressed as determinants of the HF orbitals. This generates a set of coupled, non-linear equations in the amplitudes which must be solved, usually by some iterative technique. With the amplitudes in hand, the coupled-cluster energy is computed as

$$
\langle \Psi_{\text{HF}} | \mathbf{H} | e^{\mathbf{T}} \Psi_{\text{HF}} \rangle = E_{\text{CC}}
\tag{7.55}
$$

In practice, the cost of including single excitations (i.e., $\mathbf{T}_1$) in addition to doubles is worth the increase in accuracy, and this defines the CCSD model. The scaling behavior of CCSD is on the order of N^6. Inclusion of connected triple excitations (i.e., those arising with their own unique amplitudes from $\mathbf{T}_3$, not the 'disconnected' triples arising as products of $\mathbf{T}_1$ and $\mathbf{T}_2$) defines CCSDT, but this is very computationally costly (scaling as N^8), making it intractable for all but the smallest of molecules. Various approaches to estimating the effects of the connected triples using perturbation theory have been proposed (each with its own acronym...) Of these, the most robust, and thus most commonly used, is that in the so-called CCSD(T) method, which also includes a singles/triples coupling term (Raghavachari *et al.* 1989). The (T) approach in general slightly overestimates the triples correction, and does so by an amount about equal to the ignored quadruples, i.e., there is a favorable cancellation of errors (Helgaker *et al.* 2001). This makes the CCSD(T) model extremely effective in most instances, and indeed this level has come to be the effective gold standard for single-reference calculations. Analytic gradients (Hald *et al.* 2003) and second derivatives (Kallay and Gauss 2004) are available for CCSD and CCSD(T), which further increases the utility of these methods. Note, however, that truncated coupled-cluster theory is *not* variational.

The CCSD(T) level is reasonably forgiving even in instances where the single-determinant assumption is questionable. Some discussion of this point with examples is provided in the next section. Here, however, we note that one *measure* of the multireference character that is often reported is the so-called T_1 diagnostic of Lee and Taylor (1989), defined as

$$T_1 = \sqrt{\frac{\sum_i^{\text{occ.}} \sum_a^{\text{vir.}} (t_i^a)^2}{n}} \tag{7.56}$$

where n is the number of electrons and the singles amplitudes are defined analogously to those for the doubles appearing in Eq. (7.53). A value above 0.02 has been suggested as warranting some caution in the interpretation of single-reference CCSD results.

The presence of large singles amplitudes can also be problematic for the CCSD(T) method, because the perturbation theory estimate for the triples can become unstable. One possibility to eliminate that instability involves changing the orbitals used to express the reference wave function from the canonical HF orbitals to so-called Brueckner orbitals. The Brueckner orbitals are found as linear combinations of the HF orbitals subject to the constraint that all of the singles amplitudes in the CCSD cluster operator be zero (a process that requires iteration). This approach is sometimes called Brueckner doubles (BD). The energetic effect of connected triples can again be estimated using a perturbative approach, which defines the BD(T) method.

A method that is closely related to coupled cluster theory is quadratic configuration interaction including singles and doubles (QCISD). Originally developed by Pople and co-workers as a way to correct for size-consistency errors in CISD (Pople, Head-Gordon, and Raghavachari 1987), it was later shown to be almost equivalent to CCSD in its construction (He and Cremer 1991). The QCISD(T) method includes the same perturbative correction for contributions from unlinked triples as that used in CCSD(T). Typically, CCSD and QCISD give results closely agreeing with one another and the same holds true for their (T) analogs (although in certain challenging systems the more complete coupled-cluster methods have been found to be more robust). Given their usually close correspondence in quality, any motivation to use one over the other tends to derive from the better features that may be associated with it in any given electronic structure code (e.g., inclusion of analytic gradients, a particularly efficient implementation, etc.) Note, however, that while coupled-cluster methods are in principle well defined for the inclusion of excitations up to any level–and in certain benchmark, small-molecule cases, full inclusion of triples, quadruples, etc., can be undertaken–the development of QCISD and QCISD(T) did not proceed from truncation of a general operator, but rather from augmentation of CISD to correct for size inconsistency.

Thus, there do not exist any 'higher' levels of QCISD, although such levels could be defined to include additional excitations by analogy to CCSD.

Finally, Levchenko and Krylov (2004) have defined spin-flip versions of coupled cluster theories along lines similar to those previously described for SF-CISD. Applications to date have primarily been concerned with the accurate computation of electronically excited states, but the models are equally applicable to computing correlation energies for ground states.

7.6 Practical Issues in Application

The goal of most calculations is to obtain as high a level of accuracy as possible within the constraints of the available computational resources. As including electron correlation in a calculation can be critical to enhancing accuracy, but can also be excruciatingly expensive in large systems, it is important to appreciate the strengths and weaknesses of different correlation techniques with respect to various system characteristics. This section provides some discussion of factors affecting all correlation treatments, and compares and contrasts certain specific issues associated with individual treatments.

7.6.1 Basis Set Convergence

As noted in Chapter 6, basis-set flexibility is key to accurately describing the molecular wave function. When methods for including electron correlation are included, this only becomes more true (see, for instance, He and Cremer 2000b). One can appreciate this in an intuitive fashion from thinking of the correlated wave function as a linear combination of determinants, as expressed in Eq. (7.1). Since the excited determinants necessarily include occupation of orbitals that are virtual in the HF determinant, and since the HF determinant in some sense 'uses up' the best combinations of basis functions for the occupied orbitals (from the requirement that the excited states be orthogonal to the ground state), the excited states are more dependent on basis-set completeness (this generalizes to the MCSCF case as well, although the discussion in this section is primarily focused on single-reference theories).

This differential sensitivity is illustrated in Table 7.1, which compares the convergence of the HF energy for CO with the convergence of the full-CI energy for just the O atom. In this case, the convergence is with respect to adding higher angular momentum basis functions into a set that is saturated with functions of lower angular momentum. Note that even though

Table 7.1 Basis set convergence for HF and full CI energies of CO and O, respectively

Saturated basis functions	$E_{HF}(CO)$ (a.u.)	$E_{CI}(O)$ (a.u.)
s, p	−112.717	−74.935
s, p, d	−112.785	−75.032
s, p, d, f	−112.790	−75.053
s, p, d, f, g		−75.061
Infinite limit	−112.791	−75.069

the O atom has only half as many basis functions as CO (accepting that the practical 'infinite' basis-set limit is still actually finite), the energetic gain derived from adding functions of higher angular momentum in the d to g range is typically 2 to 4 times larger in the former system than the latter. Note also that, although the HF energy of CO is effectively converged by the time a saturated basis including f functions is used, the CI energy is still more than 10 kcal mol^{-1} from being converged. Of course, it is possible for some molecular properties computed at correlated levels to be less sensitive to basis-set-size effects than is the energy. For example, Abrams and Sherrill (2003) found full CI calculations with polarized valence double-ζ basis sets to provide reasonably accurate predictions of spectroscopic constants for six diatomics, e.g., harmonic vibrational frequencies within 1.6 percent of experiment.

The greater dependence on basis-set quality of correlated calculations compared to those of the HF variety has prompted many developers of basis sets to optimize contractions via some scheme that includes evaluating results from the former. For instance, the 'correlation consistent' prefix of the cc-pVnZ basis sets discussed in Chapter 6 highlights this feature.

It has already been mentioned that one way to improve efficiency is to freeze core electrons in correlation treatments. One might think that it represents a more rigorous calculation if one does not freeze them, but unless the basis set includes extra core functions, there is some imbalance in the treatment of core–core vs. core–valence correlation. Put differently, correlating the core electrons requires that basis functions be provided that can be used for this purpose; split-valence basis sets with minimal cores are ill suited in this regard. Instead, basis sets of true multiple-ζ quality should be used. Recent examples of such basis sets include the correlation-consistent polarized core and valence multiple-ζ (cc-pCVnZ) basis sets of Woon and Dunning (1995), where extra core functions are added in increments, and including higher angular momenta, proportional to those employed for the valence space.

The correlation energy is sometimes separated into so-called radial and angular components. Again, an intuitive view of the correlated calculation is that, by considering the contribution of excited determinants having occupation of HF virtual orbitals, one is helping the electrons to get out of one another's way more effectively than they can within the SCF approximation. The radial component derives from decreasing the contraction for functions of a particular angular momentum, i.e., providing tighter and looser functions around each atom. Alternatively, the space around each atom within a given distance range can be made more accessible by adding functions of increasingly higher angular momentum, i.e., polarization functions, and this is the second contributor. In general, the importance of angular correlation increases at the expense of radial correlation as the atomic number increases, but this is mostly an effect associated with the core electrons. As a rule of thumb, the same balance noted for HF theory is true for correlated calculations: each level of decreased contraction in a given shell is worth about as much as adding a set of polarization functions of the next higher angular momentum.

For the very small systems in Table 7.1, it is possible to approach the exact solution of the Schrödinger equation, but, as a rule, convergence of the correlation energy is depressingly slow. Mathematically, this derives from the poor ability of products of one-electron basis functions, which is what Slater determinants are, to describe the cusps in two-electron densities that characterize electronic structure. For the MP2 level of theory, Schwartz (1962) showed that, in the limit of large l, the error in the correlation energy between electrons of opposite spin goes as $(l + 1/2)^{-3}$ where l is the highest angular momentum that is saturated in the basis set. Thus, if we apply this formula to going from an (s,p) saturated basis set to an (s,p,d) basis set, our error drops by only 64%, i.e., we recover a little less than two-thirds of the missing correlation energy. Going from (s,p,d) to (s,p,d,f), the improvement drops to 53%, or, compared to the (s,p) starting point, about five-sixths of the original error. Since the correlation energy can be enormous, and since actually saturating the basis set in these functions of higher angular momentum can be expensive, such convergence behavior is not especially good.

The so-called R12 methods of Klopper and Kutzelnigg (1987) provide an interesting alternative to Slater-determinant-based methods with respect to analysis of convergence behavior. In this methodology, the wave function is not simply a product of one-electron orbitals but includes additionally all interelectronic distances r_{ij} (such wave functions were first pioneered by Hylleraas (1929) in the early days of quantum mechanics in an effort to treat the 2-electron helium atom as accurately as the one-electron hydrogen atom). With the interelectronic cusp explicitly included, the convergence behavior of the MP2 correlation energy improves to $(l + 1/2)^{-7}$. Now the error recovery on going from an (s,p) to an (s,p,d,f) basis set is in principle 98%. In practice, the R12 methods require very large basis sets for technical reasons, and so such calculations continue to be limited to relatively small systems, but they are still quite useful for benchmarking purposes and may see an expanded role with future developments.

Note that the scaling behavior of methods more highly correlated than MP2 is expected in general to be less favorable than MP2. This derives from the greater sensitivity to basis set exhibited by determinants involving excitations beyond double, since still more virtual orbitals must be occupied. Helgaker $et\ al.$ (2001) have shown, however, that extrapolation schemes based on a cubic scaling principle can be highly effective. In particular, they examined the CCSD valence correlation energies for seven small molecules from CCSD calculations using the Dunning cc-pVnZ (n = D, T, Q, 5, 6) basis sets, comparing calculations with individual basis sets to R12-CCSD calculations, and also to calculations assuming that the infinite-basis CCSD valence correlation energy could be computed as

$$E_{\text{corr},\infty} = \frac{x^3 E_{\text{corr},x} - y^3 E_{\text{corr},y}}{x^3 - y^3} \qquad (7.57)$$

where x and y are the highest angular momentum quantum numbers included in sequential Dunning basis sets (e.g., 2 and 3 for cc-pVDZ and cc-pVTZ) and $E_{\text{corr},z}$ is the valence correlation energy computed using the basis with corresponding quantum number $z = x$ or $z = y$ (note that we say valence correlation energy here because the cc-pVnZ basis sets do not have correlating core basis functions like the cc-pCVnZ basis sets). The results of their calculations are in Table 7.2.

Note that compared to the R12 benchmark energies, the CCSD energies with individual basis sets are quite slowly convergent. Even with the staggeringly large cc-pV6Z basis set, the mean unsigned error over the seven molecules remains 4.2 mE_h. The extrapolated values for $E_{\text{corr},\infty}$, on the other hand, are accurate to within about 1 mE_h by the time a cc-pVQZ

Table 7.2 Valence correlation energies $(-E_{corr}, mE_h)$ from standard and R12 CCSD calculations and from extrapolation using Eq. (7.57) for seven closed-shell singlet molecules

	CH_2	H_2O	HF	Ne	CO	N_2	F_2	MUE[a]
Standard calculations[b]								
D	138.0	211.2	206.8	189.0	294.5	309.4	402.7	107.9
T	164.2	267.4	273.9	266.3	358.3	372.0	526.0	39.8
Q	171.4	286.0	297.6	294.7	380.6	393.2	569.7	16.2
5	173.6	292.4	306.3	305.5	388.5	400.7	586.1	7.7
6	174.5	294.9	309.7	309.9	391.7	403.7	592.8	4.2
Extrapolation protocol[c]								
DT	175.2	291.1	302.2	298.9	385.1	398.3	577.9	11.2
TQ	176.7	299.5	314.9	315.4	396.9	408.7	601.6	1.0
Q5	176.0	299.2	315.3	316.8	396.8	408.6	603.4	1.3
56	175.7	298.3	314.4	316.0	396.1	407.9	601.9	0.5
R12 benchmarks								
	175.5	297.9	313.9	315.5	395.7	407.4	601.0	

[a]Mean unsigned error over all molecules compared to R12 energies.
[b]First column gives n for cc-pVnZ basis set.
[c]First column gives x and y equivalents for Eq. (7.57).

calculation has been completed. Indeed, even the simplest DT extrapolation is on average 30 percent more accurate than the single, much more expensive CCSD/cc-pVQZ level.

Having discussed extrapolation in the context of correlation energy, it is appropriate to recognize that if one is going to estimate the infinite basis correlation energy, one wants to add this to the infinite basis HF energy. Parthiban and Martin (2001) have found the analog of Eq. (7.57) involving the fifth power of the angular momentum quantum numbers to be highly accurate, i.e.,

$$E_{HF,\infty} = \frac{x^5 E_{HF,x} - y^5 E_{HF,y}}{x^5 - y^5} \tag{7.58}$$

7.6.2 Sensitivity to Reference Wave Function

For single-reference correlated methods, there are several issues associated with the HF reference that can significantly affect the interpretation of the correlated calculation. First, there is the degree to which the wave function can indeed be reasonably well described by a single configuration, i.e., the extent to which the HF determinant dominates in the expansion of Eq. (7.1). When a non-trivial degree of multireference character exists, perturbation theory is particularly sensitive to this feature, and can give untrustworthy results. To appreciate this, recall the TMM example with which this chapter began (Figure 7.1). Let us take the single-configuration HF wave function represented by Eq. (7.2), and consider the MP2 energy contribution from the double excitation taking both electrons from occupied orbital π_2 to virtual orbital π_3. From Eq. 7.48, we see that the denominator associated with this term is the difference in orbital energies. However, since these orbitals are formally degenerate, the denominator is zero and the perturbation theory expression for the energy associated with this term is infinite! Note that this is not a case of the actual electronic state being degenerate, but is entirely an artifact of using a single-reference wave function.

As a general rule, whenever the frontier orbital separation becomes small, the magnitude of the MPn energy terms will become large because of their inverse dependence on orbital energy separation, and perturbation theory will be very slowly convergent in such instances. Such decreased separations also increase the degree of multireference character in the wave function, so the tight coupling between these two phenomena is rationalized. Figure 7.5 provides an example of this phenomenon for the case of the energy of carbonyl oxide, which has a moderate degree of multireference character, relative to its isomer dioxirane. The comparatively small size of these systems permits the extension of perturbation theory through fifth order with the 6-31G(d) basis set, but even the difference between the MP4 and MP5 results remains a fairly large 1.3 kcal mol^{-1}.

CCSD is similarly sensitive to multireference character, although it is less obvious that this should be so based on the formalism presented above. However, inclusion of triples in the CCSD wave function is usually very effective in correcting for a single-reference treatment of a weakly to moderately multireference problem. Of course, the most common way to include the triples is by perturbation theory, i.e., CCSD(T), and as noted above, this level too can be unstable if singles amplitudes are large. In such an instance, BD(T) calculations, which eliminate the singles amplitudes, can be efficacious.

To illustrate some of these points in greater detail, consider the aryne diradicals p-benzyne and its isoelectronic but charged congener, N-protonated 2,5-pyridyne (Figure 7.6). In these systems, the frontier orbitals of interest are the bonding and antibonding combinations of the 'singly occupied' σ orbitals left after abstraction of two hydrogen atoms from the aromatic ring. Because the orbitals are *para*-related to one another, the interaction between them is weak, and thus the frontier orbital energy separation is small. As such, the closed-shell singlet and triplet states lie relatively near one another; in the case of p-benzyne, negative ion

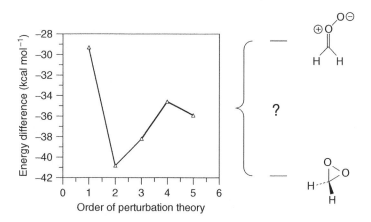

Figure 7.5 Slowly oscillatory behavior of MPn/6-31G(d)//HF/6-31G(d) theory for the energy separation between carbonyl oxide and dioxirane. Accurate extrapolation from this perturbation series is an unlikely prospect

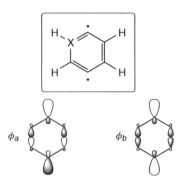

Figure 7.6 Frontier orbitals of a *para* aryne diradical. In the isoelectronic cases of X = C and X = NH$^+$, the energy of orbital ϕ_a is very slightly below that of ϕ_b, leading to a high degree of multiconfigurational character in the singlets. When X = C, the two orbitals belong to the b_{1u} and a_g irreps of the D_{2h} point group, respectively, and thus a single excitation from the former to the latter cannot contribute to the singlet ground state that has overall A$_g$ symmetry; only a double excitation contributes. When X = NH$^+$, however, both orbitals are of a' type symmetry within the C_s point group, so that a single excitation can (and does) make a large contribution to the singlet ground state that has overall A$'$ symmetry. The latter situation can contribute to instability in estimating the energetic effects of triples substitutions in '(T)' methods based on a single-determinantal reference

photoelectron spectroscopy has established the singlet–triplet (S-T) splitting to be -3.8 ± 0.5 kcal mol^{-1} (Wenthold *et al.* 1998). This corresponds to an *energy* splitting of -4.2 ± 0.5 kcal mol^{-1} (i.e., differences in zero-point vibrational energy have been removed). High-level calculations suggest, not surprisingly, that the S-T splitting in the *N*-protonated pyridyne system should be very nearly the same (Debbert and Cramer 2000). Table 7.3 illustrates the results from a variety of different levels of electronic structure theory applied to computing the S-T splitting using the cc-pVDZ basis set.

Notice, first, how spectacularly wrong the HF results are, this being indicative of significant multireference character for the singlets, which should really be described as about 60:40 mixtures of the two determinants corresponding to double occupation of the antibonding and bonding combinations of the σ orbitals. In the *p*-benzyne case, the MP2 calculation correctly predicts the singlet to be the preferred state, but drastically overshoots in doing so. As is typical, MP3 oscillates back to the HF prediction (triplet ground state) but with a smaller margin of error, and then MP4 corrects back again in the proper direction, with a somewhat smaller overestimation than was observed for MP2. Clearly, however, one could not with confidence extrapolate to an infinite-order perturbation theory result from these four points. The situation is much the same for the 2,5-pyridynium ion, except that the MP2 result is very close to experiment. Such fortuitous agreement is obviously entirely coincidental, as the perturbation series is wildly oscillating.

Note also the importance of triple excitations in correcting for the multideterminantal character. The change in the S-T splitting from inclusion of the triples at the MP4 level is as large as or larger than the change in going from MP3 to MP4SDQ. A similar effect is seen with the QCISD and CCSD formalisms – both incorrectly predict triplet ground

Table 7.3 Singlet–triplet splittings (kcal mol^{-1}) for p-benzyne and N-protonated 2,5-pyridyne[a]

Level of Theory	p-Benzyne	N-Protonated 2,5-Pyridyne
HF	87.8	87.4
MP2	−25.3	−4.1
MP2 (cc-pVTZ)	−27.8	11.4
MP3	22.8	42.0
MP4SDQ	14.3	10.1
MP4	−20.9	−21.0
CCSD	16.8	17.3
CCSD (cc-pVTZ)	18.4	18.9
CCSD(T)	−4.5	−29.4
QCISD	16.2	4.7
QCISD(T)	−4.1	−5.9
BD	17.0	17.0
BD(T)	−4.1	−5.1
CAS(8,8)	−2.7	−2.4
CASPT2(8,8)	−5.1	−5.1
Experiment or best estimate	**−4.2**	**−5.0**

[a] Basis set cc-pVDZ unless otherwise indicated; geometries from BPW91/cc-pVDZ density functional calculations (see Chapter 8).

states, but inclusion of triples via the (T) formalism gives for the most part rather good results. A significant exception is the CCSD(T) result for the 2,5-pyridynium ion, where the triples correction drastically overcorrects. This is an example of instability arising from large singles amplitudes in the CCSD expansion. In p-benzyne, the symmetry of the bonding and antibonding combinations of the σ orbitals is different, so a single excitation from one orbital to the other cannot contribute to the closed-shell wave function. In the less symmetric 2,5-pyridynium ion, however, these orbitals are the same symmetry, so such excitations are allowed and are major contributors to the wave function (Figure 7.6). In such instances, BD(T) calculations are to be preferred over CCSD(T), and indeed, the BD(T) level of theory performs very nicely for this problem (in this particular case the QCISD(T) level also seems to be more robust than CCSD(T) with respect to sensitivity to singles, but this is the reverse of the situation that normally obtains).

As for basis-set convergence, triple-ζ calculations at the MP2 and CCSD levels are provided for comparison to the double-ζ results. For this particular property, the results for p-benzyne are not terribly sensitive to improvements in the flexibility of the basis set. In the pyridinium ion case, the CCSD results are also not very sensitive, but a large effect is seen at the MP2 level. This has more to do with the instability of the perturbation expansion than any intrinsic difference between the isoelectronic arynes.

Note that when multiconfigurational character is *explicitly* accounted for, by an MCSCF calculation using a complete active space including the relevant σ orbitals and electrons as well as the six π orbitals and electrons, the results even without accounting for dynamical electron correlation are fairly good. Including dynamical correlation at the CASPT2 level improves them to the point where they are quite good.

Insofar as CASPT2 uses a multiconfigurational reference, one might expect it to be less prone than MP2 to instability. This is entirely true, *so long as the MCSCF reference is adequate*. If the MCSCF has converged to a spurious solution, perturbation theory is often successful in identifying this because a very large contribution from one or more excitations will be observed. Alternatively, if the MCSCF failed to include one or more critical orbitals, again, large contributions will be obtained for corresponding excitations. From a formal standpoint, it is better to include those orbitals in the active space than it is to rely on CASPT2 to correct for both dynamical and non-dynamical behavior, even though in some instances it seems the latter approach can give good results.

An alternative way to account for multiconfigurational character in the singlet is to generate it using SF-CIS(D) from the triplet reference. Slipchenko and Krylov (2002) have done this for *p*-benzyne using the 6-31G(d) basis set and computed a S−T splitting of −2.2 kcal mol^{-1}. The SF-CIS(D) model is thus nearly as accurate as those including estimates for triple excitations even though it is substantially less computationally expensive.

A separate issue that can contribute to instability in correlated calculations is spin contamination. As noted in Chapter 6, spin contamination refers to the inclusion in the wave function of contributions from states of higher spin that mix in when unrestricted methods permit α and β spin orbitals to localize in different regions of space. As a rough rule, the sensitivity of different methods to spin contamination is about what it is to multiconfigurational character: MP*n* methods are to be avoided and inclusion of triples in CCSD or QCISD (or BD) is important. So-called projected MP*n* methods attempt to correct for spin contamination after the fact by projecting out states of higher spin from the correlated wave function (see Appendix B), and these methods tend to be helpful in cases where spin contamination is relatively small, say no more than 10% (Chen and Schlegel 1994). Unfortunately, analytic gradients are not available for spin-projected methods, so they must be applied to geometries the optimization of which may have taken place at a considerably less reliable level.

An issue related to spin contamination is so-called Hartree−Fock instability. Various wave functions can exhibit different kinds of instabilities, often, but not always, associated with trying to describe a multiconfigurational system with a single-determinant approach. Thus, for instance, RHF solutions may be unstable with respect to breaking the identical character of the α and β orbitals – a so-called RHF → UHF instability (the UHF singlet wave function is usually highly contaminated with triplet character after reoptimization). UHF wave functions for symmetric systems can also be unstable, in this case with respect to spatial symmetry breaking of the individual orbitals. The MOs, if allowed to relax, fail to fall into the irreps of the molecular point group, adopting instead lower symmetry shapes even if the molecular framework is held fixed so as to continue to belong to the higher symmetry point group. Such instability tends to be associated with systems having delocalized spin – the allyl radical is a classical example. All of these cases prove very problematic for perturbation theory, but are handled with somewhat greater success by other correlated methods. In certain very highly symmetric systems, the wave function can also be unstable to using complex instead of real MOs, but this situation is rare. Most modern electronic-structure programs allow one to check the stability of the HF wave function with respect to these various phenomena, and such steps are warranted in cases having narrow frontier orbital separations and/or delocalized spin. Resort to MCSCF wave functions can be required in particularly problematic systems.

7.6.3 Price/Performance Summary

For a typical equilibrium structure, the HF level of theory predicts bond lengths that are usually a little too short. It is simple to rationalize this using Eq. (7.1). To the extent that correlated methods include excited configurations in the wave function expansion, and to the extent that the orbitals into which excitations occur typically have some antibonding character, this tends to increase bond lengths in order to lower the energy. As a rule, the MP2 level is an excellent choice for geometry optimizations of *minima* that include correlation energy, and significant improvements can be obtained at fairly reasonable cost. Scheiner *et al.* (1997) examined a large number of bond lengths in 108 molecules containing from two to eight atoms and found that, with the 6-31G(d,p) basis set, the average error in bond length at the MP2 level was 0.015 Å, which may be compared to an error at the HF level of 0.021 Å. An improvement of roughly the same order was obtained by Feller and Peterson (1998) in a separate investigation of 184 small molecules using the aug-cc-pVnZ basis sets. Bond angles are already sufficiently accurate at the HF level that little improvement is observed at the MP2 level.

While analytic derivatives are available for several more highly correlated levels of theory, geometric improvements beyond the MP2 level tend to be so small for equilibrium structures that they are not worth the cost. This is *not* necessarily the case for TS structures, where the accurate description of a partial bond may well require correlation beyond the MP2 level. As a rough rule, if one observes a *large* change in some geometric property on going from the HF to the MP2 level, it is probably worthwhile to investigate the predictions from still higher levels of theory, since clearly the perturbations are large, and there is good reason to believe MP2 does not provide convergence in the property of interest. In some instances, convergence can be hard to achieve using perturbation theory (especially for cases of multiple bonds involving heteroatoms) and coupled-cluster methods are to be preferred, but this typically applies only to calculations seeking the most demanding accuracy (He and Cremer 2000c).

With respect to energetics, MP2 must again be considered a very efficient level of theory for energy differences between minima. In many instances, one finds that the error in such differences is reduced by 25–50% on going from the HF level to the MP2 level. For instance, Hehre reports a sample of 45 isomerizations where errors in isomer energies were reduced from 2.9 to 1.9 kcal mol^{-1} on going from HF/6-31G(d) to MP2/6-31G(d). For the 11 glucose conformers discussed in Chapters 5 and 6, the average error in conformational energy is reduced from 0.6 to 0.4 kcal mol^{-1} on going from HF/cc-pVTZ//MP2/cc-pVDZ to MP2/cc-pVTZ//MP2/cc-pVDZ (Barrows *et al.* 1998). Note, though, that for the same glucose conformers, the error *increases* from 0.1 to 1.0 kcal mol^{-1} on going from HF/cc-pVDZ to MP2/cc-pVDZ, illustrating the degree to which errors in basis set and correlation approximations can sometimes offset one another.

However, the generally robust nature of MP2 in the above examples simply reflects the degree to which most minima are already fairly well described by HF wave functions. When this is not the case, e.g., in TS structures, there are no hard and fast rules that can be cited with respect to the expected quality of any level of theory. Instead, one is thrown back on the twin responsibilities of demonstrating either (a) agreement with known experimental data of one kind or another in the same or related systems or (b) convergence with respect to treatment of electron correlation. A *rough* quality ordering that is often observed is

$$\text{HF} < \text{MP2} \sim \text{MP3} \sim \text{CCD} < \text{CISD}$$

$$< \text{MP4SDQ} \sim \text{QCISD} \sim \text{CCSD} < \text{MP4}$$

$$< \text{QCISD(T)} \sim \text{CCSD(T)} \sim \text{BD(T)} \tag{7.59}$$

Table 7.4 provides a more quantitative feel for the performance summary embodied in Eq. (7.59) using data provided by Bartlett (1995) for the absolute errors in various levels of theory compared to full CI for HB, H_2O, and HF using a polarized double-ζ basis set. In this case, calculations were carried out both at the equilibrium geometries, and also at geometries where the X–H bonds were stretched by 50% and 100%; correlation should become more important in describing these higher energy species (see also Dutta and Sherrill 2003). The ordering of the levels in the table is approximately that listed in Eq. (7.59), although CCD seems to do fortuitously well. As expected, the lower levels of correlation treatment degrade markedly compared to the higher levels when the bonds are stretched. A few levels not generally available in most electronic structure packages are included in the table for completeness, including MP5, MP6, and levels having full inclusion of triples, represented by 'T' instead of '(T)'. The heroically expensive CCSDTQ, which takes full account of all

Table 7.4 Average errors in correlation energies (kcal mol^{-1}) compared to full CI for various methods applied to HB, H_2O, and HF at both equilibrium and bond-stretched geometries

Level of theory	Equilibrium geometry	Equilibrium and stretched geometries
MP2	10.4	17.4
MP3	5.0	14.4
CISD	5.8	13.8
CCD	2.4	8.0
MP4SDQ	2.7	7.1
CCSD	1.9	4.5
QCISD	1.7	4.0
MP4	1.3	3.7
MP5	0.8	3.2
MP6	0.3	0.9
CCSD(T)	0.3	0.6
QCISD(T)	0.3	0.5
CCSDT	0.2	0.5
CCSDTQ	0.01	0.02

Table 7.5 Formal scaling behavior, as a function of basis functions N, of various electronic structure methods

Scaling behavior	Method(s)
N^4	HF
N^5	MP2
N^6	MP3, CISD, MP4SDQ, CCSD, QCISD
N^7	MP4, CCSD(T), QCISD(T)
N^8	MP5, CISDT, CCSDT
N^9	MP6
N^{10}	MP7, CISDTQ, CCSDTQ

triple and quadruple excitations, is also included, and shows the extraordinarily high accuracy one might expect for so complete a treatment, albeit one that can only be applied to the smallest of molecules.

To further judge what level may be appropriate for a given problem, it is critical that cost be taken into account. The scaling behavior of the various levels in Eq. (7.59) varies widely, as indicated in Table 7.5. Given the price/performance ratios implied by comparing Eq. (7.59) with Table 7.5, there is, for instance, usually little point in doing an MP3 or CISD calculation when superior MP4SDQ or CCSD calculations may typically be accomplished at roughly similar cost (note that scaling similarity is *not* the same as overall time similarity, since the times for the benchmark 'one-basis-function' calculations may differ, but for small to moderately sized molecules, the overall times do not tend to be terribly dissimilar). It should also be recalled that in the large molecule limit, all scaling behaviors tend to reduce because prescreening techniques can avoid the calculation of many negligible integrals. In addition, progress continues to be made on linear scaling formalisms, for example, local coupled cluster theory (Li, Ma, and Jiang 2002; Schütz 2002), so that increasingly sophisticated treatments of larger and larger molecules are becoming more and more accessible.

7.7 Parameterized Methods

Having ascended to the heights of theoretical rigor, it is perhaps time for a brief respite and a timely recapitulation of, of all things, the philosophy underlying the development of *semiempirical* methods: wouldn't it be nice to get the right answer for any problem in general? Although methods like full CI and CCSDTQ, when used in conjunction with large and flexible basis sets, are breathtakingly accurate as solutions of the Schrödinger equation, the bottom line is that they simply cannot be applied to more than the smallest fraction of chemically interesting systems because of their computational expense. And, with scaling behaviors on the order of N^{10}, this situation is unlikely to change anytime soon. As a result, particularly within the last decade, practitioners of *ab initio* MO theory have returned to the idea of introducing parameters to improve predictive accuracy, albeit with a considerably lighter touch than that associated with a full-blown semiempirical method. This section describes a variety of different approaches to improving the results from calculations including electron correlation.

7.7.1 Scaling Correlation Energies

The premise behind correlation scaling is particularly simple. Because of basis-set limitations and approximations in the correlation treatment, one is very rarely able to compute the full correlation energy. However, with a given choice of basis set and level of theory, the fraction that *is* calculated is often quite consistent over a fairly large range of structure. Thus, we might define an improved electronic energy as

$$E_{\text{SAC-e.c.m.}} = E_{\text{HF}} + \frac{E_{\text{e.c.m.}} - E_{\text{HF}}}{A} \qquad (7.60)$$

where 'e.c.m.' is a particular electron correlation method, A is an empirical scale factor typically less than one, and thus all of the correlation energy, computed as the difference between E_{ecm} and E_{HF}, is scaled by the constant factor of A^{-1}. SAC emphasizes this 'scaling all correlation' energy assumption. Note that the difference between SAC and the extrapolation schemes of section 7.6.1 is that the latter extrapolate the correlation energy associated with *a given electronic structure model* to an infinite basis set, but SAC attempts to estimate *all* of the correlation energy.

As first proposed by Gordon and Truhlar (1986), typically one would go about selecting A by comparison to known experimental data in a system of interest and/or systems related to it. For example, if the subject of interest is the PES for the reaction of the hydroxyl radical with ethyl chloride, and if the overall energies of reaction are known for the abstraction of the α and β hydrogen atoms (to make water and the corresponding alkyl radicals), then A would be selected for a given electron correlation method (say, MP2) in order to make $E_{\text{SAC-MP2}}$ agree with experiment as closely as possible for those particular data points. This same value of A would then be used for any point on the PES. Of course, the more experimental details that can be included in the choice of A, the better the parameterization (and the better able one is to judge the utility of Eq. (7.60) by examination of the errors in a one-parameter fit).

Note that one particularly attractive feature of Eq. (7.60) is that if the particular electron correlation method has available analytic derivatives, so too must $E_{\text{SAC-e.c.m.}}$, since derivatives for the latter will be simply determined as appropriately scaled sums of the e.c.m. and HF derivatives. Geometry optimization, and indeed the entire calculation, can essentially be carried out for exactly the cost of the e.c.m.

While one might imagine that values of A might best be determined individually within any given system, Siegbahn and co-workers have examined a large number of primarily small inorganic systems and suggested that, for the modified coupled-pair functional (MCPF) treatment of correlation (which is analogous in spirit to coupled cluster) with a polarized double-ζ basis set, a value of 0.80 has broad applicability, and they name this choice PCI-80 (Siegbahn, Blomberg, and Svensson 1994; Blomberg and Siegbahn 1998). A summary of the utility of this level of theory for inorganic systems including comparison to density functional theory (DFT) can be found in Table 8.2.

Gordon and Truhlar (1986) have emphasized that variations on the theme of Eq. (7.60) can be useful in different circumstances. For instance, one might imagine carrying out multireference calculations and assuming two different scale factors, one applying to the non-dynamical correlation energy associated with some increase in active space size, and the other with the dynamical correlation energy associated with a CASPT2 calculation. Alternatively, one could have different scaling factors for the terms associated with different levels of electronic excitation, e.g., scaling the doubles differently than the triples. Choices along these lines should be guided by Occam's parameter-razor: in the absence of significant improvements, fewer is better. Recent developments along these lines are the spin-component-scaled MP2 and MP3 (SCS-MP2 and SCS-MP3) methods of Grimme (2003a; 2003b) where the contributions to the perturbation theory correction from parallel-spin and anti-parallel-spin electron pairs are scaled differently from one another. After this empirical scaling, these methods have been demonstrated to often provide results competitive with models more formally including triple excitations, e.g., QCISD(T).

7.7.2 Extrapolation

The most attractive feature of the SAC methods is their simplicity. A potential contributor to their possible failure, however, is that the factor A, by being based on experiment, hides within it corrections for both basis-set incompleteness and truncation in the correlation operator. It is not obvious over any particular range of chemical space that either one will be constant, in which case it seems particularly unlikely that either *both* will be constant simultaneously, or that their changes will exactly offset one another. There is thus some virtue in attempting to correct for the two approximations separately. As has already been noted in Chapter 6, estimates of the HF limit can be derived by carrying out calculations with increasingly larger basis sets and then assuming some asymptotic behavior as a function of basis-set size (see Figure 6.4 and Section 7.6.1). The same can be done with correlated methods, and many modern basis sets were developed specifically with this goal in mind.

Such a procedure may not seem to be properly classified as a 'parameterized' method, since no individual calculation incorporates a parameter, optimized or otherwise. However, in this instance it is the selection of the functional form for asymptotic behavior that may be considered to be parametric. As noted in Section 7.6.1, for certain levels of theory, like MP2, rigorous convergence behaviors have been derived, but it must be stressed that those behaviors are valid in the limit of a complete basis set, and the ability to fit points obtained with a smaller basis set to the limiting curve is by no means assured (see, for instance, Petersson and Frisch 2000).

In principle, then, in systems where computational costs are not prohibitively expensive, one might try to employ extrapolation so that the energies appearing in Eq. (7.60) represented complete-basis-set (CBS) energies, in which case A corrects only for approximations in the correlation treatment.

7.7.3 Multilevel Methods

In Section 6.2.6, we considered approaches to the HF limit derived under the assumption that various aspects of basis-set incompleteness (radial, angular, etc.) could be accounted for in some additive fashion (see Eq. (6.5)). In essence, multilevel methods carry this approach one step further, and assume a similar behavior for the correlation energy. For instance, the QM energies for glucose conformers that have served as a benchmark for comparison with lower levels of theory in preceding chapters were computed at a composite (C) level as

$$E(C) = E(\text{MP2/cc-pVTZ//MP2/cc-pVDZ})$$
$$+ \{E(\text{CCSD/6-31G(d)//MP2/6-31G(d)})$$
$$- E(\text{MP2/6-31G(d)})\}$$
$$+ \{E(\text{HF/cc-p}^T\text{VQZ//MP2/cc-pVDZ})$$
$$- E(\text{HF/cc-pVTZ//MP2/cc-pVDZ})\} \tag{7.61}$$

Thus, triple-ζ MP2 energies at double-ζ MP2 geometries are augmented with a correction for doubles contributions beyond second order (line 2 on the r.h.s. of Eq. (7.61)) and a correction for basis set size increase beyond triple-ζ (line 3 on the r.h.s. of Eq. (7.61) where the 'T' superscript in the first basis set implies that polarization functions from cc-pVTZ were used in conjunction with valence functions from cc-pVQZ).

While such *ad hoc* multilevel methods have been employed for rather a long time, only in the late 1980s were efforts undertaken to systematize the approach so as to define a model chemistry having broad applicability. The first such effort was the so-called G1 theory of Pople and co-workers, which was followed very rapidly by an improved modification called G2 theory, so that the former may be considered to be obsolete (Curtiss *et al.* 1991). The steps involved in a G2 calculation are detailed in Table 7.6. In this instance, the goal of the calculation is accurate *thermochemistry*, so some of the steps are devoted to computing thermal contributions to the enthalpy, as opposed to the electronic energy, as described in more detail in Chapter 10. Although the philosophy of the remaining steps is essentially the same as that predicating Eq. (7.61), there is considerably more attention to detail in specific aspects of the basis-set problem and the accounting for electron correlation. There is also a completely empirical correction procedure (step 8) to, *inter alia*, account for core-valence correlation and to improve the performance of the model over systems having different numbers of unpaired spins. Note that 'full' following a correlation acronym implies that core electrons were included in the correlation treatment, as opposed to the more typical choice of freezing them to excitation. Over a test set of 148 enthalpies of formation, the average error of G2 theory is 1.6 kcal mol^{-1}.

Over time, many different groups have suggested minor modifications of G2 theory (each spawning a new acronym). Some trade accuracy for computational efficiency in order to permit application to larger systems; the most popular of these has been G2(MP2), which avoids the costly MP4 calculations in G2 at the expense of increasing the error over the test set to 1.8 kcal mol^{-1} (Curtiss, Raghavachari, and Pople 1993). Others emphasize alternative methods for obtaining molecular geometries, or attempt to correct for other deficiencies in G2 applied to specific classes of molecules (G2 does poorly on perfluorinated species, for instance).

Table 7.6 Steps in G2 and G3 theory for molecules[a,b]

Step	G2	G3
(1)	HF/6-31G(d) geometry optimization	HF/6-31G(d) geometry optimization
(2)	ZPVE from HF/6-31G(d) frequencies	ZPVE from HF/6-31G(d) frequencies
(3)	MP2(full)/6-31G(d) geometry optimization (all subsequent calculations use this geometry)	MP2(full)/6-31G(d) geometry optimization (all subsequent calculations use this geometry)
(4)	$E[\text{MP4}/6\text{-}311\text{+G(d,p)}]$ $-E[\text{MP4}/6\text{-}311\text{G(d,p)}]$	$E[\text{MP4}/6\text{-}31\text{+G(d)}] - E[\text{MP4}/6\text{-}31\text{G(d)}]$
(5)	$E[\text{MP4}/6\text{-}311\text{G(2df,p)}]$ $-E[\text{MP4}/6\text{-}311\text{G(d,p)}]$	$E[\text{MP4}/6\text{-}31\text{G(2df,p)}]$ $-E[\text{MP4}/6\text{-}31\text{G(d)}]$
(6)	$E[\text{QCISD(T)}/6\text{-}311\text{G(d)}]$ $-E[\text{MP4}/6\text{-}311\text{G(d)}]$	$E[\text{QCISD(T)}/6\text{-}31\text{G(d)}]$ $-E[\text{MP4}/6\text{-}31\text{G(d)}]$
(7)	$E[\text{MP2}/6\text{-}311\text{+G(3df,2p)}]$ $-E[\text{MP2}/6\text{-}311\text{G(2df,p)}]$ $-E[\text{MP2}/6\text{-}311\text{+G(d,p)}]$ $+E[\text{MP2}/6\text{-}311\text{G(d,p)}]$	$E[\text{MP2(full)}/\text{G3large}^c]$ $-E[\text{MP2}/6\text{-}31\text{G(2df,p)}]$ $-E[\text{MP2}/6\text{-}31\text{+G(d)}]$ $+E[\text{MP2}/6\text{-}31\text{G(d)}]$
(8)	$-0.00481 \times$ (number of valence electron pairs) $-0.00019 \times$ (number of unpaired valence electrons)	$-0.006386 \times$ (number of valence electron pairs) $-0.002977 \times$ (number of unpaired valence electrons)
$E_0 =$	$0.8929 \times (2) + E[\text{MP4}/6\text{-}311\text{G(d,p)}] +$ (4) + (5) + (6) + (7) + (8)	$0.8929 \times (2) + E[\text{MP4}/6\text{-}31\text{G(d)}] +$ (4) + (5) + (6) + (7) + (8)

[a] For atoms, G3 energies are defined to include a spin-orbit correction taken either from experiment or other high-level calculations. In addition, different coefficients are used in step (8).
[b] In the G2 method, the 6-311G basis set and its derivatives are not defined for second-row atoms; instead, a basis set optimized by McLean and Chandler (1980) is used.
[c] Available at http://chemistry.anl.gov/compmat/g3theory.htm. Defined to use canonical 5 d and 7 f functions.

A modification of G2 by Pople and co-workers was deemed sufficiently comprehensive that it is known simply as G3, and its steps are also outlined in Table 7.6. G3 is more accurate than G2, with an error for the 148-molecule heat-of-formation test set of 0.9 kcal mol^{-1}. It is also more efficient, typically being about twice as fast. A particular improvement of G3 over G2 is associated with improved basis sets for the third-row nontransition elements (Curtiss *et al.* 2001). As with G2, a number of minor to major variations of G3 have been proposed to either improve its efficiency or increase its accuracy over a smaller subset of chemical space, e.g., the G3-RAD method of Henry, Sullivan, and Radom (2003) for particular application to radical thermochemistry, the G3(MP2) model of Curtiss *et al.* (1999), which reduces computational cost by computing basis-set-extension corrections at the MP2 level instead of the MP4 level, and the G3B3 model of Baboul *et al.* (1999), which employs B3LYP structures and frequencies.

It should be noted that G2 and G3 potentially fail to be size extensive because of the correction term in step 8. If one is studying a homolytic dissociation into two components, at what point along the reaction coordinate are the formerly paired electrons considered to be unpaired? There will be a discontinuity in the energy at that point. In addition, G3 theory uses a different correction for atoms than for molecules, and this too fails to be size extensive.

Alternative multilevel methods that have some similarities to G2, G3, and their variants, are the CBS methods of Petersson and co-workers (see Bibliography at end of chapter). A key difference between the Gn models and the CBS models is that, rather than assuming basis-set incompleteness effects to be completely accounted for by additive corrections, results for different levels of theory are extrapolated to the complete-basis-set limit in defining a composite energy. Four well-defined CBS models exist, CBS-4, CBS-q, CBS-Q, and CBS-APNO, these being in order of increasing accuracy and, naturally, cost. Over the same 148-molecule test set as used above to evaluate G2 and G3, the average absolute errors of CBS-4, CBS-q, and CBS-Q are 2.7, 2.3, and 1.2 kcal mol^{-1}, respectively. CBS-APNO reduces the error in CBS-Q by a factor of 2 (to only 0.5 kcal mol^{-1} on a somewhat smaller 125-molecule test set), but requires a very expensive QCISD(T)/6-311+G(2df,p) calculation. A particular feature of most of the CBS methods is that they include an (empirical) correction for spin contamination in open-shell species, for which unrestricted treatments potentially sensitive to such contamination are used. In terms of speed, CBS-Q is roughly the speed of G3.

The Weizmann-1 (W1) and Weizmann-2 (W2) models of Martin and de Oliveira (1999) are similar to the CBS models in that extrapolation schemes are used to estimate the infinite basis set limits for SCF and correlation energies. A key difference between the two, however, is that the W1 and W2 models set as a benchmark goal an accuracy of 1 kJ mol^{-1} (0.24 kcal mol^{-1}) on thermochemical quantities. To achieve that kind of accuracy basis sets of size up to cc-pVQZ + 2d1f and cc-pV5Z + 2d1f are used for W1 and W2 theories, respectively, for both SCF *and* CCSD calculations. Other components of the W1 and W2 calculations include accounting for triple excitations, core electron correlation energy, relativistic effects including spin–orbit coupling, and zero-point vibrational energies. W1 and W2 theories predict heats of formation over a 55-molecule subset of the 148-molecule G2/G3 test set mentioned above (this subset is now usually called the G2-1 test set) with mean unsigned errors of 0.6 and 0.5 kcal mol^{-1}, respectively. By comparison, the G2, G3, and CBS-Q results for this subset are 1.2, 1.1, and 1.0 kcal mol^{-1}, respectively. The relative performances of W1 and W2 theories are still more improved for prediction of electron affinities and ionization potentials (Parthiban and Martin 2001). Further development aimed at achieving 'spectroscopic accuracy' (usually defined as energetic accuracy to within 1 cm^{-1}) has resulted in W3 and preliminary W4 theories (Boese *et al.* 2004), but as these protocols include CCSDTQ calculational steps with basis sets of size cc-pVDZ or larger, they are likely to find application only to very small molecules for the foreseeable future.

A somewhat more obviously empirical variation on the multilevel approach is the multi-coefficient method of Truhlar and co-workers. Although many different variations of this approach have now been described, it is simplest to illustrate the concept for the so-called multi-coefficient G3 (MCG3) model (Fast, Sánchez, and Truhlar 1999). In essence, the model assumes a G3-like energy expression, but each term has associated with it a coefficient that is not restricted to be unity, as is the case for G3. Specifically

$$E_{MCG3} = \sum_{i=1}^{9} c_i(i) + E_{SO} + E_{CC} \qquad (7.62)$$

where (i) represents a component of the G3 energy (actually, there are some rather slight variations involved with basis sets and frozen-core approximations that increase efficiency), E_{SO} and E_{CC} are empirically estimated spin-orbit and core-correlation energies, and the coefficients c_i are optimized over the usual G3 thermochemistry test set. One additional important difference in the use of G3 energy components is that the G3 empirical correction, which leads to non-size-extensivity, is *not* included. Thus, MCG3 is size extensive. The performance of MCG3 is very slightly better than G3 itself, but this accuracy is achieved at roughly half the cost in terms of computational resources for molecules having many heavy atoms. Scaling of the G3 components was also reported by Curtis *et al.* (2000) and defines the G3S model. MCG3 and G3S have essentially equivalent accuracy.

The real power in the multi-coefficient models, however, derives from the potential for the coefficients to make up for more severe approximations in the quantities used for (i) in Eq. (7.62). At present, Truhlar and co-workers have codified some 20 different multicoefficient models, some of which they term 'minimal', meaning that relatively few terms enter into analogs of Eq. (7.62), and in particular the optimized coefficients absorb the spin-orbit and core-correlation terms, so they are not separately estimated. Different models can thus be chosen for an individual problem based on error tolerance, resource constraints, need to optimize TS geometries at levels beyond MP2, etc. Moreover, for some of the minimal models, analytic derivatives are available on a term-by-term basis, meaning that analytic derivatives for the composite energy can be computed simply as the sum over terms.

A somewhat more chemically based empirical correction scheme is the bond-additivity correction (BAC) methodology. In the BAC-MP4 approach, for instance, the energy of a molecule is computed as

$$E(\text{BAC-MP4}) = E[\text{MP4/6-31G(d,p)//HF/6-31G(d,p)}]$$

$$+ \sum_{A,B} \Delta E_{A-B} + E_{SC} + E_{MR} \tag{7.63}$$

where E_{SC} and E_{MR} correct for spin contamination (if any) and multireference character (if any) and the summation runs over all atom pairs and each 'bond' correction is a function of bond length (the correction goes to zero at infinite bond length) and a set of parameters, one parameter for each atom and two parameters for each possible pair of atoms. The parameters themselves are determined by fitting to experimental bond dissociation energies, heats of formation (corrected for zero-point vibrational energies and thermal contributions), or other useful thermochemical data. The central assumption of this model, then, is that the error can be decomposed in an additive fashion over the bonds.

In a study of 110 C1 and C2 molecules composed of C, H, O, and F, the average BAC-MP4 unsigned error in predicted heat of formation was 2.1 kcal mol^{-1} (Zachariah *et al.* 1996). As the MP4 calculation uses a relatively modest basis set size, the BAC procedure is quite fast by comparison to some of the multilevel methods described above. On the other hand, as with any method relying on pairwise parameterization, extension to a large number of atoms requires a great deal of parameterization data, and this is a potential limitation of the BAC method when applied to systems containing atoms not already parameterized.

Because they include empirically derived parameters, multilevel models nearly always outperform single-level calculations at an equivalently expensive level of theory. That being said, one should avoid a slavish devotion to any particular multilevel model simply because it has been graced with an acronym defining it. For any given chemical problem, it is quite possible that an individual investigator can construct a specific multilevel model with relatively little effort that will outperform any of the already defined ones. The issue is simply whether sufficient data exist for the particular system of interest in order to make such a focused model possible. When the data do not, then that is the best time to rely on those previously defined models that have been demonstrated to be reasonably robust over relevant swaths of chemical space.

As for the utility of single-level models, it should be recalled that the goal of most multilevel models is *absolute* energy prediction, while many chemical studies are undertaken in order to better understand *relative* energy differences. Cancellation of errors makes the latter studies more tractable at less complete levels of theory, and single-level models can still be useful in both qualitative and quantitative senses. In addition, there is no wave function defined for the typical parametric model; there is only an energy functional that potentially depends on several different wave functions. Should one wish to know the expectation value for some property *other* than the energy, one will either have to devise a separate multilevel expression, or adopt a single-level formalism for which a wave function is indeed defined.

Note that most of the energetic performance data summarized above may also be found in tabular form, compared to density functional models, in Table 8.1

7.8   Case Study: Ethylenedione Radical Anion

Synopsis of Thomas, J. R. *et al.* (1995) 'The Ethylenedione Anion: Elucidation of the Intricate Potential Energy Hypersurface'.

The ground state of ethylenedione, the dimer of carbon monoxide, has been reliably predicted to be a triplet that is bound with respect to dissociation by virtue of its high spin state (two singlet carbon monoxide molecules are lower in energy, but the triplet cannot dissociate into two closed-shell singlets). As such, it has proven an interesting target for synthesis, albeit without success. One possible avenue for its synthesis is to detach electrons from negative ion precursors. This prompted Thomas and co-workers to characterize the radical anion of ethylenedione at a variety of correlated levels of electronic structure theory.

At the UHF level the linear form, which formally has a $^2\Pi_u$ electronic state (see Appendix B for details on group theoretical notation), is predicted to be the minimum energy structure. However, at almost all correlated levels the molecule bends to lift the degeneracy of a pair of a_u and b_u orbitals, leading to a so-called Renner–Teller potential energy surface, as illustrated in Figure 7.7. The lower energy state is ^2A_u and geometric details are provided in the figure for four different correlated levels, all using a large TZP+ basis set.

The details of the molecular structure are difficult to nail down because of the shallow nature of the PES in the vicinity of the linear form. Thus, even with a fairly complete basis set, there are large disagreements between CISD, CCSD, and CCSD(T), although there is a remarkably good (coincidental) agreement between MP2 and CCSD(T). The situation is

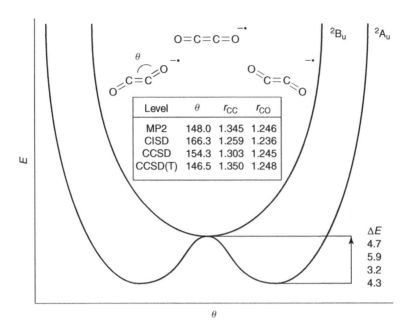

Figure 7.7   Renner–Teller PES for ethylenedione radical anion. Geometrical data for the ^2A_u equilibrium structure are provided for various levels of theory using an augmented polarized triple-ζ basis set (TZP+). Barriers to linearity (ΔE, kcal mol^{-1}) are from CCSD(T) calculations using, from top to bottom, DZP, DZP+, TZP, and TZP+ basis sets. If the initial guess is for the ^2B_u state instead of the ^2A_u state, what will happen?

still more dissatisfying insofar as further increases in basis-set size, in this case adding additional sets of polarization functions, result in bond length changes of up to 0.03 Å and bond angle changes of up to 14° at the MP2, CISD, and CCSD levels. The cost of the CCSD(T) computations is such that use of these larger basis sets is not practical, and thus it is not clear what the effect will be at this formally most complete level of theory.

To further clarify the situation, the authors examined two other quantities dependent on the shape of the PES in the vicinity of the linear form. First, they computed the barrier to double-inversion through the linear form. The data are listed in Figure 7.7., and show some basis set dependence. Note that the CCSD(T)/TZP+ result is approximated to within 0.1 kcal mol^{-1} by summing the CCSD(T)/TZP barrier with the difference between the CCSD(T)/DZP+ and CCSD(T)/DZP barriers. That is, the effect of diffuse functions evaluated with a double-ζ basis set can be treated as additive to the non-augmented triple-ζ results, along the lines described in Section 7.7.3.

The authors made a more exacting comparison for vibrational frequencies, where experimental data were available for the matrix isolated radical anion. Focusing on one fundamental and one combination band, the CCSD(T)/TZP+ predictions of 1527 and 1955 cm^{-1} compared reasonably well to the experimental values of 1518 and 2042. Again, the flat nature of the PES in the vicinity of the linear form makes things difficult for theory, since this introduces potentially large anharmonicity that is not accounted for in the usual harmonic approximation employed to compute vibrational frequencies (see Section 9.3.2).

Isotope shifts in the frequencies, however, showed very close agreement between theory and experiment, all data agreeing to within 5% for seven different isotopomers.

The authors did examine whether significant non-dynamical correlation effects complicated the system, but MCSCF calculations with large active spaces failed to identify any configurations other than the dominant one that entered with coefficients in excess of 0.09, suggesting that the use of single-reference methods was well justified. Part of the challenge for this particular system simply derives from its negative charge, which imposes a greater demand on basis-set saturation. In any case, this example illustrates how deceptively difficult it can be to converge solution of the Schrödinger equation even for seemingly simple chemical systems – a mere four heavy atoms.

Bibliography and Suggested Additional Reading

Bartlett, R. J. 2000. 'Perspective on "On the Correlation Problem in Atomic and Molecular Systems. Calculations of Wavefunction Components in Ursell-type Expansion Using Quantum-field Theoretical Methods"' *Theor. Chem. Acc.*, **103**, 273.

Cioslowski, J., Ed. 2001. *Quantum-Mechanical Prediction of Thermochemical Data*, Kluwer: Dordrecht.

Cramer, C. J. 1998. 'Bergman, Aza-Bergman, and Protonated Aza-Bergman Cyclizations and Intermediate 2,5-Arynes: Chemistry and Challenges to Computation' *J. Am. Chem. Soc.*, **120**, 6261.

Cramer, C. J. and Smith, B. A. 1996. 'Trimethylenemethane. Comparison of Multiconfiguration Self-consistent Field and Density Functional Methods for a Non-Kekulé Hydrocarbon' *J. Phys. Chem.* **100**, 9664.

Curtiss, L. A., Raghavachari, K., Redfern, P. C., Rassolov, V., and Pople, J. A. 1998. 'Gaussian-3 (G3) Theory for Molecules Containing First and Second-row Atoms', *J. Chem. Phys.* **109**, 7764.

Feller, D. and Davidson, E. R. 1990. 'Basis Sets for Ab Initio Molecular Orbital Calculations and Intermolecular Interactions' in *Reviews in Computational Chemistry*, Vol. 1, Lipkowitz, K. B. Boyd, D. B., Eds., VCH: New York, 1.

Hehre, W. J. 1995. *Practical Strategies for Electronic Structure Calculations*, Wavefunction: Irvine, CA.

Hehre, W. J., Radom, L., Schleyer, P. v. R., and Pople, J. A. 1986. Ab Initio *Molecular Orbital Theory*, Wiley: New York.

Jensen, F. 1999. *Introduction to Computational Chemistry*, Wiley: Chichester.

Levine, I. N. 2000. *Quantum Chemistry*, 5th Edn., Prentice Hall: New York.

Lynch, B. J. and Truhlar, D. G. 2003. 'Robust and Affordable Multicoefficient Methods for Thermochemistry and Thermochemical Kinetics: The MCCM/3 Suite and SAC/3', *J. Phys. Chem. A*, **107**, 3898.

Martin, J. M. L. 1998. 'Calibration of Atomization Energies of Small Polyatomics' in *Computational Thermochemistry,* ACS Symposium Series, Vol. 677, Irikura, K. K. and Frurip, D. J. Eds., American Chemical Society, Washington, DC, 212.

Petersson, G. A. 1998. 'Complete Basis-set Thermochemistry and Kinetics' in *Computational Thermochemistry,* ACS Symposium Series, Vol. 677, Irikura, K. K. and Frurip, D. J., Eds., American Chemical Society, Washington, DC, 237.

Petersson, G. A., Malick, D. K., Wilson, W. G., Ochterski, J. W., Montgomery, J. A., Jr., and Frisch, M. J. 1998. 'Calibration and Comparison of the Gaussian-2, Complete Basis Set, and Density Functional Methods for Computational Thermochemistry' *J. Chem. Phys.*, **109**, 10 570.

Slipchenko, L. V. and Krylov, A. I. 2003. 'Electronic Structure of the Trimethylenemethane Diradical in its Ground and Electronically Excited States: Bonding, Equilibrium Geometries, and Vibrational Frequencies', *J. Chem. Phys.*, **118**, 6874.

Szabo, A. and Ostlund, N. S. 1982. *Modern Quantum Chemistry*, Macmillan: New York.

Werner, H.-J. 2000. 'Perspective on "Theory of Self-consistent Electron Pairs. An Iterative Method for Correlated Many-electron Wavefunctions"' *Theor. Chem. Acc.*, **103**, 322.

Zachariah, M. R. and Melius, C. F. 1998. 'Bond-additivity Correction of *Ab Initio* Computations for Accurate Prediction of Thermochemistry' in *Computational Thermochemistry*, ACS Symposium Series, Vol. 677, Irikura, K. K. and Frurip, D. J., Eds., American Chemical Society, Washington, DC, 162.

References

Abrams, M. L. and Sherrill, C. D. 2003. *J. Chem. Phys.*, **118**, 1604.

Andersson, K. 1995. *Theor. Chim. Acta*, **91**, 31.

Andersson, K., Malmqvist, P. -Å., and Roos, B. O. 1992. *J. Chem. Phys.*, **96**, 1218.

Baboul, A. G., Curtiss, L. A., Redfern, P. C., and Raghavachari, K. 1999. *J. Chem. Phys.*, **110**, 7650.

Barrows, S. E., Storer, J. W., Cramer, C. J., French, A. D., and Truhlar, D. G. 1998. *J. Comput. Chem.*, **19**, 1111.

Bartlett, R. J. 1981. *Ann. Rev. Phys. Chem.*, **32**, 359.

Bartlett, R. J. 1995. In: *Modern Electronic Structure Theory*, Yarkony, D. R., Ed., World Scientific: New York, Part 2, Chapter 6.

Blomberg, M. R. A. and Siegbahn, P. E. M. 1998. In: *Computational Chemistry*, ACS Symposium Series, Vol. 677, Irikura, K. K. and Frurip, D. J., Eds., American Chemical Society: Washington, DC, 197.

Boese, A. D., Oren, M., Atasoylu, O., Martin, J. M. L., Kállay, M., and Gauss, J. 2004. *J. Chem. Phys.*, **120**, 4129.

Brillouin, L. 1934. *Actualities Sci. Ind.*, **71**, 159.

Bruna, P. J., Peyerimhoff, S. D., and Buenker, R. J. 1980. *Chem. Phys. Lett.*, **72**, 278.

Chen, W. and Schlegel, H. B. 1994. *J. Chem. Phys.*, **101**, 5957.

Cizek, J. 1966. *J. Chem. Phys.*, **45**, 4256.

Crawford, T. D. and Schaefer, H. F., III, 1996. In: *Reviews in Computational Chemistry*, Vol. 14, Lipkowitz, K. B. and Boyd, D. B., Eds., Wiley-VCH: New York, 33 and references therein.

Curtiss, L. A., Raghavachari, K., and Pople, J. A. 1993. *J. Chem. Phys.*, **98**, 1293.

Curtiss, L. A., Raghavachari, K., Redfern, P. C., and Pople, J. A. 2000. *J. Chem. Phys.*, **112**, 7374.

Curtiss, L. A., Raghavachari, K., Trucks, G. W., and Pople, J. A. 1991. *J. Chem. Phys.*, **94**, 7221.

Curtiss, L. A., Redfern, P. C., Raghavachari, K., Rassolov, V., and Pople, J. A. 1999. *J. Chem. Phys.*, **110**, 4703.

Curtiss, L. A., Redfern, P. C., Rassolov, V., Kedziora, G., and Pople, J. A. 2001. *J. Chem. Phys.*, **114**, 9287.

Davidson, E. R. 1995. *Chem. Phys. Lett.*, **241**, 432.

Debbert, S. L. and Cramer, C. J. 2000. *Int. J. Mass Spectrom.*, **201**, 1.

Dutta, A. and Sherrill, C. D. 2003. *J. Chem. Phys.*, **118**, 1610.

Fast, P. L., Sánchez, P. L., and Truhlar, D. G. 1999. *Chem. Phys. Lett.*, **306**, 407.

Feller, D. and Peterson, K. A. 1998. *J. Chem. Phys.*, **108**, 154.

Finley, J. P. and Freed, K. F. 1995. *J. Chem. Phys.*, **102**, 1306.

Gordon, M. S. and Truhlar, D. G. 1986. *J. Am. Chem. Soc.*, **108**, 5412.

Grimme, S. 2003a. *J. Chem. Phys.*, **118**, 9095.

Grimme, S. 2003b. *J. Comput. Chem.*, **24**, 1529.

Hald, K., Halkier, A., Jørgensen, P., Coriani, S., Hättig, C., and Helgaker, T. 2003. *J. Chem. Phys.*, **118**, 2985.

He, Z. and Cremer, D. 1991. *Int. J. Quantum Chem., Quantum Chem. Symp.*, **25**, 43.

He, Y. and Cremer, D. 2000a. *Mol. Phys.*, **98**, 1415.

He, Y. and Cremer, D. 2000b. *Theor. Chem. Acc.*, **105**, 110.

He, Y. and Cremer, D. 2000c. *J. Phys. Chem. A*, **104**, 7679.

Head-Gordon, M., Rico, R. J., Oumi, M., and Lee, T. J. 1994. *Chem. Phys. Lett.*, **219**, 21.

Helgaker, T., Klopper, W., Halkier, A., Bak, K. L., Jørgensen, P., and Olsen, J. 2001. in *Quantum Mechanical Prediction of Thermodynamic Data*, Cioslowski, J., Ed., Kluwer: Dordrecht, 1.

Henry, D. J., Sullivan, M. B., and Radom, L. 2003. *J. Chem. Phys.*, **118**, 4849.

Hylleraas, E. A. 1929. *Z. Phys.*, **54**, 347.

Kallay, M. and Gauss, J. 2004. *J. Chem. Phys.*, **120**, 6841.

Klopper, W. and Kutzelnigg, W. 1987. *Chem. Phys. Lett.*, **134**, 17.

Krylov, A. I. 2001. *Chem. Phys. Lett.*, **350**, 522.

Krylov, A. I. and Sherrill, C. D. 2002. *J. Chem. Phys.*, **116**, 3194.

Langhoff, S. R. and Davidson, E. R. 1974. *Int. J. Quantum Chem.*, **8**, 61.

Lee, T. J. and Taylor, P. R. 1989. *Int. J. Quantum Chem.*, **S23**, 199.

Levchenko, S. V. and Krylov, A. I. 2004. *J. Chem. Phys.*, **120**, 175.

Li, S. H., Ma, J., and Jiang, Y. S. 2002. *J. Comput. Chem.*, **23**, 237.

Martin, J. M. L. and de Oliveira, G. 1999. *J. Chem. Phys.*, **111**, 1843.

McLean, A. D. and Chandler, G. S. 1980. *J. Chem. Phys.*, **72**, 5639.

Møller, C. and Plesset, M. S. 1934. *Phys. Rev.*, **46**, 618.

Page, C. S. and Olivucci, M. 2003. *J. Comput. Chem.*, **24**, 298.

Parthiban, S. and Martin, J. M. L. 2001. *J. Chem. Phys.*, **114**, 6014.

Petersson, G. A. and Frisch, M. J. 2000. *J. Phys. Chem. A*, **104**, 2183.

Pitarch-Ruiz, J., Sanchez-Marin, J., and Maynau, D. 2002. *J. Comput. Chem.*, **23**, 1157.

Pople, J. A., Head-Gordon, M., and Raghavachari, K. 1987. *J. Chem. Phys.*, **87**, 5968.

Pulay, P. 1983. *Chem. Phys. Lett.*, **100**, 151.

Raghavachari, K., Trucks, G. W., Pople, J. A., and Head-Gordon, M. 1989. *Chem. Phys. Lett.*, **157**, 479.

Saebø, S. and Pulay, P. 1987. *J. Chem. Phys.*, **86**, 914.

Scheiner, A. C., Baker, J., and Andzelm, J. W. 1997. *J. Comput. Chem.*, **18**, 775.

Schütz, M. 2002. *Phys. Chem. Chem. Phys.*, **4**, 3941.

Schwartz, C. 1962. *Phys. Rev.*, **126**, 1015.

Siegbahn, P. E. M., Blomberg, M. R., and Svensson, M. 1994. *Chem. Phys. Lett.*, **223**, 35.

Slipchenko, L. V. and Krylov, A. I. 2002. *J. Chem. Phys.*, **117**, 4694.

Thomas, J. R., DeLeeuw, B. J., O'Leary, P., Schaefer, H. F., III, Duke, B. J., and O'Leary, B. 1995. *J. Chem. Phys.*, **102**, 6525.

Wenthold, P. G., Squires, R. R., and Lineberger, W. C. 1998. *J. Am. Chem. Soc.*, **120**, 5279.

Woon, D. E. and Dunning, T. H., Jr., 1995 *J. Chem. Phys.*, **103**, 4572.

Zachariah, M. R., Westmoreland, P. R., Burgess, D. R., Jr., Tsang, W., and Melius, C. F. 1996. *J. Phys. Chem.*, **100**, 8737.

8

Density Functional Theory

8.1 Theoretical Motivation

8.1.1 Philosophy

What a strange and complicated beast a wave function is. This function, depending on one spin and three spatial coordinates for every electron (assuming fixed nuclear positions), is not, in and of itself, particularly intuitive for systems of more than one electron. Indeed, one might approach the HF approximation as not so much a mathematical tool but more a philosophical one. By allowing the wave function to be expressed as a Slater determinant of one-electron orbitals, one preserves for the chemist some semblance of clarity by permitting each electron still to be thought of as being loosely independent. It was noted in Section 7.6.1 that wave functions can be dramatically improved in quality by removing this rough independence and including in the wave functional form a dependence on interelectronic distances (R12 methods). The key disadvantage? The wave function itself is essentially uninterpretable – it is an inscrutable oracle that returns valuably accurate answers when questioned by quantum mechanical operators, but it offers little by way of sparking intuition.

One may be forgiven for stepping back from the towering edifice of molecular orbital theory and asking, shouldn't things be simpler? For instance, rather than having to work with a wave function, which has rather odd units of probability density to the one-half power, why can't we work with some physical *observable* in determining the energy (and possibly other properties) of a molecule? That such a thing should be possible would probably not have surprised physicists before the discovery of quantum mechanics, insofar as such simple formalisms are widely available in classical systems.

However, we *may* take advantage of our knowledge of quantum mechanics in asking about what particular physical observable might be useful. Having gone through the exercise of constructing the Hamiltonian operator and showing the utility of its eigenfunctions, it would be sufficient to our task simply to find a physical observable that permitted the *a priori* construction of the Hamiltonian operator. What then is needed? The Hamiltonian depends only on the positions and atomic numbers of the nuclei and the total number of electrons. The dependence on total number of electrons immediately suggests that a useful physical observable would be the electron density ρ, since, integrated over all space, it gives the total

Essentials of Computational Chemistry, 2nd Edition Christopher J. Cramer
© 2004 John Wiley & Sons, Ltd   ISBNs: 0-470-09181-9 (cased); 0-470-09182-7 (pbk)

number of electrons N, i.e.,

$$N = \int \rho(\mathbf{r})d\mathbf{r} \tag{8.1}$$

Moreover, because the nuclei are effectively point charges, it should be obvious that their positions correspond to local maxima in the electron density (and these maxima are also cusps), so the only issue left to completely specify the Hamiltonian is the assignment of nuclear atomic numbers. It can be shown that this information too is available from the density, since for each nucleus A located at an electron density maximum $\mathbf{r}_A$

$$\left.\frac{\partial \overline{\rho}(r_A)}{\partial r_A}\right|_{r_A=0} = -2Z_A\rho(\mathbf{r}_A) \tag{8.2}$$

where Z is the atomic number of A, r_A is the radial distance from A, and $\overline{\rho}$ is the spherically averaged density.

Of course, the arguments above do not provide any simpler *formalism* for finding the energy. They simply indicate that given a known density, one could form the Hamiltonian operator, solve the Schrödinger equation, and determine the wave functions and energy eigenvalues. Nevertheless, they suggest that some simplifications might be possible.

8.1.2   Early Approximations

Energy is separable into kinetic and potential components. If one decides *a priori* to try to evaluate the molecular energy using only the electron density as a variable, the simplest approach is to consider the system to be classical, in which case the potential energy components are straightforwardly determined. The attraction between the density and the nuclei is

$$V_{ne}[\rho(\mathbf{r})] = \sum_{k}^{nuclei} \int \frac{Z_k}{|\mathbf{r} - \mathbf{r}_k|}\rho(\mathbf{r})d\mathbf{r} \tag{8.3}$$

and the self-repulsion of a classical charge distribution is

$$V_{ee}[\rho(\mathbf{r})] = \frac{1}{2}\iint \frac{\rho(\mathbf{r}_1)\rho(\mathbf{r}_2)}{|\mathbf{r}_1 - \mathbf{r}_2|}d\mathbf{r}_1 d\mathbf{r}_2 \tag{8.4}$$

where $\mathbf{r}_1$ and $\mathbf{r}_2$ are dummy integration variables running over all space.

The kinetic energy of a continuous charge distribution is less obvious. To proceed, we first introduce the fictitious substance 'jellium'. Jellium is a system composed of an infinite number of electrons moving in an infinite volume of a space that is characterized by a uniformly distributed positive charge (i.e., the positive charge is not particulate in nature, as it is when represented by nuclei). This electronic distribution, also called the uniform electron gas, has a constant non-zero density. Thomas and Fermi, in 1927, used fermion statistical mechanics to derive the kinetic energy for this system as (Thomas 1927; Fermi 1927)

$$T_{ueg}[\rho(\mathbf{r})] = \frac{3}{10}(3\pi^2)^{2/3} \int \rho^{5/3}(\mathbf{r})d\mathbf{r} \qquad (8.5)$$

Note that the various T and V terms defined in Eqs. (8.3)–(8.5) are functions of the density, while the density itself is a function of three-dimensional spatial coordinates. A function whose argument is also a function is called a 'functional', and thus the T and V terms are 'density functionals'. The Thomas–Fermi equations, together with an assumed variational principle, represented the first effort to define a density functional theory (DFT); the energy is computed with no reference to a wave function. However, while these equations are of significant historical interest, the underlying assumptions are sufficiently inaccurate that they find no use in modern chemistry (in Thomas–Fermi DFT, all molecules are unstable relative to dissociation into their constituent atoms...)

One large approximation is the use of Eq. (8.4) for the interelectronic repulsion, since it ignores the energetic effects associated with correlation and exchange. It is useful to introduce the concept of a 'hole function', which is defined so that it corrects for the energetic errors introduced by assuming classical behavior. In particular, we write

$$\left\langle \Psi \left| \sum_{i<j}^{electrons} \frac{1}{r_{ij}} \right| \Psi \right\rangle = \frac{1}{2} \int\int \frac{\rho(\mathbf{r}_1)\rho(\mathbf{r}_2)}{|\mathbf{r}_1 - \mathbf{r}_2|}d\mathbf{r}_1 d\mathbf{r}_2 + \frac{1}{2} \int\int \frac{\rho(\mathbf{r}_1)h(\mathbf{r}_1;\mathbf{r}_2)}{|\mathbf{r}_1 - \mathbf{r}_2|}d\mathbf{r}_1 d\mathbf{r}_2 \qquad (8.6)$$

The l.h.s. of Eq. (8.6) is the exact QM interelectronic repulsion. The second term on the r.h.s. corrects for the errors in the first term (the classical expression) using the hole function h associated with ρ (the notation $h(\mathbf{r}_1;\mathbf{r}_2)$ emphasizes that the hole is centered on the position of electron 1, and is evaluated from there as a function of the remaining spatial coordinates defining $\mathbf{r}_2$; note, then, that not only does the value of h vary as a function of $\mathbf{r}_2$ for a given value of $\mathbf{r}_1$, but the precise *form* of h itself can vary as a function of $\mathbf{r}_1$).

The simplest way to gain a better appreciation for the hole function is to consider the case of a one-electron system. Obviously, the l.h.s. of Eq. (8.6) must be zero in that case. However, just as obviously, the first term on the r.h.s. of Eq. (8.6) is *not* zero, since ρ must be greater than or equal to zero throughout space. In the one-electron case, it should be clear that h is simply the negative of the density, but in the many-electron case, the exact form of the hole function can rarely be established. Besides the self-interaction error, hole functions in many-electron systems account for exchange and correlation energy as well.

By construction, HF theory avoids any self-interaction error and *exactly* evaluates the exchange energy (it is only the correlation energy that it approximates); however, it is time-consuming to evaluate the four-index integrals from which these various energies are calculated. While Slater (1951) was examining how to speed up HF calculations he was aware that one consequence of the Pauli principle is that the Fermi exchange hole is larger than the correlation hole, i.e., exchange corrections to the classical interelectronic repulsion are significantly larger than correlation corrections (typically between one and two orders of magnitude). So, Slater proposed to ignore the latter, and adopted a simple approximation for the former. In particular, he suggested that the exchange hole about any position could be approximated as a sphere of constant potential with a radius depending on the magnitude of the density at that position. Within this approximation, the exchange energy E_x is determined as

$$E_x[\rho(\mathbf{r})] = -\frac{9\alpha}{8}\left(\frac{3}{\pi}\right)^{1/3}\int \rho^{4/3}(\mathbf{r})d\mathbf{r} \qquad (8.7)$$

Within Slater's derivation, the value for the constant α is 1, and Eq. (8.7) defines so-called 'Slater exchange'.

Starting from the uniform electron gas, Bloch and Dirac had derived a similar expression several years previously, except that in that case $\alpha = \frac{2}{3}$ (Bloch, F. 1929 and Dirac, P. A. M. 1930). The combination of this expression with Eqs. (8.3)–(8.5) defines the Thomas–Fermi–Dirac model, although it too remains sufficiently inaccurate that it fails to see any modern use.

Given the differing values of α in Eq. (8.7) as a function of different derivations, many early workers saw fit to treat it as an empirical value, and computations employing Eq. (8.7) along these lines are termed Xα calculations (or sometimes Hartree–Fock–Slater calculations in the older literature). Empirical analysis in a variety of different systems suggests that $\alpha = \frac{3}{4}$ provides more accurate results than either $\alpha = 1$ or $\alpha = \frac{2}{3}$. This particular DFT methodology has largely fallen out of favor in the face of more modern functionals, but still sees occasional use, particularly within the inorganic community.

8.2   Rigorous Foundation

The work described in the previous section was provocative in its simplicity compared to wave-function-based approaches. As a result, early DFT models found widespread use in the solid-state physics community (where the enormous system size required to mimic the properties of a solid puts a premium on simplicity). However, fairly large errors in molecular calculations, and the failure of the theories to be rigorously founded (no variational principle had been established), led to their having little impact on chemistry. This state of affairs was set to change when Hohenberg and Kohn (1964) proved two theorems critical to establishing DFT as a legitimate quantum chemical methodology. Each of the two theorems will be presented here in somewhat abbreviated form.

8.2.1   The Hohenberg–Kohn Existence Theorem

In the language of DFT, electrons interact with one another and with an 'external potential'. Thus, in the uniform electron gas, the external potential is the uniformly distributed positive charge, and in a molecule, the external potential is the attraction to the nuclei given by the usual expression. As noted previously, to establish a dependence of the energy on the density, and in the Hohenberg–Kohn theorem it is the ground-state density that is employed, it is sufficient to show that this density determines the Hamiltonian operator. Also as noted previously, integration of the density gives the number of electrons, so all that remains to define the operator is determination of the external potential (i.e., the charges and positions of the nuclei). The proof that the ground-state density determines the external potential proceeds via *reductio ad absurdum*, that is, we show that an assumption to the contrary generates an impossible result.

Thus, let us assume that two *different* external potentials can each be consistent with the same *nondegenerate* ground-state density ρ_0. We will call these two potentials v_a and v_b and the different Hamiltonian operators in which they appear H_a and H_b. With each Hamiltonian will be associated a ground-state wave function Ψ_0 and its associated eigenvalue E_0. The variational theorem of molecular orbital theory dictates that the expectation value of the Hamiltonian a over the wave function b must be higher than the ground-state energy of a, i.e.,

$$E_{0,a} < \langle \Psi_{0,b} | H_a | \Psi_{0,b} \rangle \tag{8.8}$$

We may rewrite this expression as

$$
\begin{aligned}
E_{0,a} &< \langle \Psi_{0,b} | H_a - H_b + H_b | \Psi_{0,b} \rangle \\
&< \langle \Psi_{0,b} | H_a - H_b | \Psi_{0,b} \rangle + \langle \Psi_{0,b} | H_b | \Psi_{0,b} \rangle \\
&< \langle \Psi_{0,b} | v_a - v_b | \Psi_{0,b} \rangle + E_{0,b}
\end{aligned}
\tag{8.9}
$$

Since the potentials v are one-electron operators, the integral in the last line of Eq. (8.9) can be written in terms of the ground-state density

$$E_{0,a} < \int [v_a(\mathbf{r}) - v_b(\mathbf{r})] \rho_0(\mathbf{r}) d\mathbf{r} + E_{0,b} \tag{8.10}$$

As we have made no distinction between a and b, we can interchange the indices in Eq. (8.10) to arrive at the equally valid

$$E_{0,b} < \int [v_b(\mathbf{r}) - v_a(\mathbf{r})] \rho_0(\mathbf{r}) d\mathbf{r} + E_{0,a} \tag{8.11}$$

Now, if we add inequalities (8.10) and (8.11), we have

$$
\begin{aligned}
E_{0,a} + E_{0,b} &< \int [v_b(\mathbf{r}) - v_a(\mathbf{r})] \rho_0(\mathbf{r}) d\mathbf{r} + \int [v_a(\mathbf{r}) - v_b(\mathbf{r})] \rho_0(\mathbf{r}) d\mathbf{r} + E_{0,b} + E_{0,a} \\
&< \int [v_b(\mathbf{r}) - v_a(\mathbf{r}) + v_a(\mathbf{r}) - v_b(\mathbf{r})] \rho_0(\mathbf{r}) d\mathbf{r} + E_{0,b} + E_{0,a} \\
&< E_{0,b} + E_{0,a}
\end{aligned}
\tag{8.12}
$$

where the assumption that the ground-state densities associated with wave functions a and b were the same permits us to eliminate the integrals as they must sum to zero. However, we are left with an impossible result (that the sum of the two energies is less than itself), which must indicate that our initial assumption was incorrect. So, the non-degenerate ground-state density must determine the external potential, and thus the Hamiltonian, and thus the wave function. Note moreover that the Hamiltonian determines not just the ground-state wave function, but all excited-state wave functions as well, so there is a tremendous amount of information coded in the density.

The non-degenerate ground-state character of the density was used to ensure the validity of the variational inequalities. A question that naturally arises is: of what utility, if any, are the densities of excited states? Using group theory, Gunnarsson and Lundqvist (1976a,b) proved that the Hohenberg–Kohn existence theorem can be extended to the lowest energy (non-degenerate) state within each irreducible representation of the molecular point group. Thus, for instance, the densities of the lowest energy A_g and B_{1u} states of p-benzyne each uniquely determine their respective wave functions, energies, etc. (these states are singlet and triplet, respectively, see Figure 7.5), but nothing can be said about the density of the triplet A' state of N-protonated 2,5-didehydropyridine, since there is a lower energy singlet state belonging to the same A' irrep (to which the Hohenberg–Kohn existence theorem *does* apply, see again Figure 7.5). The development of DFT formalisms to handle arbitrary excited states remains a subject of active research, as discussed in more detail in Chapter 14.

8.2.2   The Hohenberg–Kohn Variational Theorem

The first theorem of Hohenberg and Kohn is an existence theorem. As such, it is provocative with potential, but altogether unhelpful in providing any indication of how to *predict* the density of a system. Just as with MO theory, we need a means to optimize our fundamental quantity. Hohenberg and Kohn showed in a second theorem that, also just as with MO theory, the density obeys a variational principle.

To proceed, first, assume we have some well-behaved candidate density that integrates to the proper number of electrons, N. In that case, the first theorem indicates that this density determines a candidate wave function and Hamiltonian. That being the case, we can evaluate the energy expectation value

$$\langle \Psi_{cand} | H_{cand} | \Psi_{cand} \rangle = E_{cand} \geq E_0 \tag{8.13}$$

which, by the variational principle of MO theory, must be greater than or equal to the true ground-state energy.

So, in principle, we can keep choosing different densities and those that provide lower energies, as calculated by Eq. (8.13), are closer to correct. Such a procedure is, of course, rather unsatisfying on at least two levels. First, we have no prescription for how to go about choosing improved candidate densities rationally, and second, insofar as the motivation for DFT was to avoid solving the Schrödinger equation, computing the energy as the expectation value of the Hamiltonian is no advance – we know how to do *that* already.

The difficulty lies in the nature of the functional itself. Up to this point, we have indicated that there are mappings from the density onto the Hamiltonian and the wave function, and hence the energy, but we have not suggested any mechanical means by which the density can be used as an argument in some general, characteristic variational equation, e.g., with terms along the lines of Eqs. (8.5) and (8.7), to determine the energy directly without recourse to the wave function. Such an approach first appeared in 1965.

8.3 Kohn–Sham Self-consistent Field Methodology

The discussion above has emphasized that the density determines the external potential, which determines the Hamiltonian, which determines the wave function. And, of course, with the Hamiltonian and wave function in hand, the energy can be computed. However, if one attempts to proceed in this direction, there is no simplification over MO theory, since the final step is still solution of the Schrödinger equation, and this is prohibitively difficult in most instances. The difficulty derives from the electron–electron interaction term in the correct Hamiltonian. In a key breakthrough, Kohn and Sham (1965) realized that things would be considerably simpler if only the Hamiltonian operator were one for a *non-interacting* system of electrons (Kohn and Sham 1965). Such a Hamiltonian can be expressed as a sum of one-electron operators, has eigenfunctions that are Slater determinants of the individual one-electron eigenfunctions, and has eigenvalues that are simply the sum of the one-electron eigenvalues (see Eq. (7.43) and surrounding discussion).

The crucial bit of cleverness, then, is to take as a starting point a *fictitious* system of *non-interacting* electrons that have for their overall ground-state density the *same* density as some *real* system of interest where the electrons *do* interact (note that since the density determines the position and atomic numbers of the nuclei (see Eq. (8.2)), these quantities are necessarily identical in the non-interacting and in the real systems). Next, we divide the energy functional into specific components to facilitate further analysis, in particular

$$E[\rho(\mathbf{r})] = T_{ni}[\rho(\mathbf{r})] + V_{ne}[\rho(\mathbf{r})] + V_{ee}[\rho(\mathbf{r})] + \Delta T[\rho(\mathbf{r})] + \Delta V_{ee}[\rho(\mathbf{r})] \qquad (8.14)$$

where the terms on the r.h.s. refer, respectively, to the kinetic energy of the non-interacting electrons, the nuclear–electron interaction (Eq. (8.3)), the classical electron–electron repulsion (Eq. (8.4)), the *correction* to the kinetic energy deriving from the interacting nature of the electrons, and *all* non-classical corrections to the electron–electron repulsion energy.

Note that, for a non-interacting system of electrons, the kinetic energy is just the sum of the individual electronic kinetic energies. Within an orbital expression for the density, Eq. (8.14) may then be rewritten as

$$
\begin{aligned}
E[\rho(\mathbf{r})] = \sum_{i}^{N} & \left(\left\langle \chi_i \left| -\frac{1}{2}\nabla_i^2 \right| \chi_i \right\rangle - \left\langle \chi_i \left| \sum_{k}^{\text{nuclei}} \frac{Z_k}{|\mathbf{r}_i - \mathbf{r}_k|} \right| \chi_i \right\rangle \right) \\
& + \sum_{i}^{N} \left\langle \chi_i \left| \frac{1}{2} \int \frac{\rho(\mathbf{r}')}{|\mathbf{r}_i - \mathbf{r}'|} d\mathbf{r}' \right| \chi_i \right\rangle + E_{xc}[\rho(\mathbf{r})]
\end{aligned}
\qquad (8.15)
$$

where N is the number of electrons and we have used that the density for a Slater-determinantal wave function (which is an exact eigenfunction for the non-interacting system) is simply

$$\rho = \sum_{i=1}^{N} \langle \chi_i | \chi_i \rangle \qquad (8.16)$$

Note that the 'difficult' terms ΔT and ΔV_{ee} have been lumped together in a term E_{xc}, typically referred to as the exchange-correlation energy. This is something of a misnomer, insofar as it is less than comprehensive – the term includes not only the effects of quantum mechanical exchange and correlation, but also the correction for the classical self-interaction energy (discussed in Section 8.1.2) and for the difference in kinetic energy between the fictitious non-interacting system and the real one.

If we undertake in the usual fashion to find the orbitals χ that minimize E in Eq. (8.15), we find that they satisfy the pseudoeigenvalue equations

$$h_i^{KS} \chi_i = \varepsilon_i \chi_i \tag{8.17}$$

where the Kohn–Sham (KS) one-electron operator is defined as

$$h_i^{KS} = -\frac{1}{2}\nabla_i^2 - \sum_k^{\text{nuclei}} \frac{Z_k}{|\mathbf{r}_i - \mathbf{r}_k|} + \int \frac{\rho(\mathbf{r}')}{|\mathbf{r}_i - \mathbf{r}'|}d\mathbf{r}' + V_{xc} \tag{8.18}$$

and

$$V_{xc} = \frac{\delta E_{xc}}{\delta \rho} \tag{8.19}$$

V_{xc} is a so-called functional derivative. A functional derivative is analogous in spirit to more typical derivatives, and V_{xc} is perhaps best described as the one-electron operator for which the expectation value of the KS Slater determinant is E_{xc}.

Note that because the E of Eq. (8.14) that we are minimizing is exact, the orbitals χ must provide the *exact* density (i.e., the minimum must correspond to reality). Further note that it is these orbitals that form the Slater-determinantal eigenfunction for the separable non-interacting Hamiltonian defined as the sum of the Kohn–Sham operators in Eq. (8.18), i.e.,

$$\sum_{i=1}^{N} h_i^{KS} |\chi_1 \chi_2 \cdots \chi_N\rangle = \sum_{i=1}^{N} \varepsilon_i |\chi_1 \chi_2 \cdots \chi_N\rangle \tag{8.20}$$

so there is internal consistency in the Kohn–Sham approach of positing a non-interacting system with a density identical to that for the real system. It is therefore justified to use the first term on the r.h.s. of Eq. (8.15) to compute the kinetic energy of the non-interacting electrons, which turns out to be a large fraction of the kinetic energy of the actual system.

As for determination of the KS orbitals, we may take a productive approach along the lines of that developed within the context of MO theory in Chapter 4. Namely, we express them within a basis set of functions $\{\phi\}$, and we determine the individual orbital coefficients by solution of a secular equation entirely analogous to that employed for HF theory, except that the elements $F_{\mu\nu}$ are replaced by elements $K_{\mu\nu}$ defined by

$$K_{\mu\nu} = \left\langle \phi_\mu \left| -\frac{1}{2}\nabla^2 - \sum_k^{\text{nuclei}} \frac{Z_k}{|\mathbf{r} - \mathbf{r}_k|} + \int \frac{\rho(\mathbf{r}')}{|\mathbf{r} - \mathbf{r}'|}d\mathbf{r}' + V_{xc} \right| \phi_\nu \right\rangle \tag{8.21}$$

Indeed, the similarities with HF theory extend well beyond the mathematical technology offered by a common variational principle. For instance, the kinetic energy and nuclear attraction components of matrix elements of $\mathbf{K}$ are identical to those for $\mathbf{F}$. Furthermore, if the density appearing in the classical interelectronic repulsion operator is expressed in the same basis functions used for the Kohn–Sham orbitals, then the result is that the same four-index electron-repulsion integrals appear in $\mathbf{K}$ as are found in $\mathbf{F}$ (historically, this made it fairly simple to modify existing codes for carrying out HF calculations to also perform DFT computations). Finally, insofar as the density is required for computation of the secular matrix elements, but the density is determined using the orbitals derived from *solution* of the secular equation (according to Eq. (8.16)), the Kohn–Sham process must be carried out as an iterative SCF procedure.

Of course, there *is* a key difference between HF theory and DFT – as we have derived it so far, DFT contains no approximations: it is *exact*. All we need to know is E_{xc} as a function of $\rho \ldots$ Alas, while Hohenberg and Kohn proved that a functional of the density must *exist*, their proofs provide no guidance whatsoever as to its form. As a result, considerable research effort has gone into finding functions of the density that may be expected to reasonably approximate E_{xc}, and a discussion of these is the subject of the next section. We close here by emphasizing that the key contrast between HF and DFT (in the limit of an infinite basis set) is that HF is a deliberately *approximate* theory, whose development was in part motivated by an ability to solve the relevant equations *exactly*, while DFT is an *exact* theory, but the relevant equations must be solved *approximately* because a key operator has unknown form.

It should also be pointed out that although exact DFT is variational, this is not true once approximations for E_{xc} are adopted. Thus, for instance, the BPW91 functional described in Section 8.4.2 predicts an energy for the H atom of -0.5042 E_h, but the exact result is -0.5. Note that the H atom is a one-electron system for which the Schrödinger solution can be solved exactly – there is no correlation energy. However, because the BPW91 E_{xc} for this system slightly exceeds the classical self-interaction energy (third term on the r.h.s. of Eq. (8.15)), which is 100 percent in error for this one-electron system, the energy is predicted to be slightly below the exact result. Both exact *and* approximate DFT are size-consistent.

The Kohn–Sham methodology has many similarities, and a few important differences, to the HF approach. We will, however, delay briefly a full discussion of how exactly to carry out a KS calculation, as it is instructive first to consider how to go about determining E_{xc}.

8.4 Exchange-correlation Functionals

As already emphasized above, in principle E_{xc} not only accounts for the difference between the classical and quantum mechanical electron–electron repulsion, but it also includes the difference in kinetic energy between the fictitious non-interacting system and the real system. In practice, however, most modern functionals do not attempt to compute this portion explicitly. Instead, they either ignore the term, or they attempt to construct a hole function that is analogous to that of Eq. (8.6) except that it also incorporates the kinetic energy difference between the interacting and non-interacting systems. Furthermore, in many functionals empirical parameters appear, which necessarily introduce some kinetic energy correction if they are based on experiment.

In discussing the nature of various functionals, it is convenient to adopt some of the notation commonly used in the field. For instance, the functional dependence of E_{xc} on the electron density is expressed as an interaction between the electron density and an 'energy density' ε_{xc} that is *dependent* on the electron density, viz.

$$E_{xc}[\rho(\mathbf{r})] = \int \rho(\mathbf{r})\varepsilon_{xc}[\rho(\mathbf{r})]d\mathbf{r} \qquad (8.22)$$

The energy density ε_{xc} is always treated as a sum of individual exchange and correlation contributions. Note that there is some potential for nomenclature confusion here because two different kinds of densities are involved: the electron density is a per unit volume density, while the energy density is a per particle density. In any case, within this formalism, it is clear from inspection of Eq. (8.7) that the Slater exchange energy density, for example, is

$$\varepsilon_{x}[\rho(\mathbf{r})] = -\frac{9\alpha}{8}\left(\frac{3}{\pi}\right)^{1/3}\rho^{1/3}(\mathbf{r}) \qquad (8.23)$$

Another convention expresses the electron density in terms of an effective radius such that exactly one electron would be contained within the sphere defined by that radius were it to have the same density throughout as its center, i.e.,

$$r_{S}(\mathbf{r}) = \left(\frac{3}{4\pi\rho(\mathbf{r})}\right)^{1/3} \qquad (8.24)$$

Lastly, we have ignored the issue of spin up to this point. Spin can be dealt with easily enough in DFT – one simply needs to use individual functionals of the α and β densities – but there is again a notational convention that sees widespread use. The spin densities at any position are typically expressed in terms of ζ, the normalized spin polarization

$$\zeta(\mathbf{r}) = \frac{\rho^{\alpha}(\mathbf{r}) - \rho^{\beta}(\mathbf{r})}{\rho(\mathbf{r})} \qquad (8.25)$$

so that the α spin density is simply one-half the product of the *total* ρ and $(\zeta + 1)$, and the β spin density is the difference between that value and the total ρ.

8.4.1 Local Density Approximation

The term local density approximation (LDA) was originally used to indicate any density functional theory where the value of ε_{xc} at some position $\mathbf{r}$ could be computed exclusively from the value of ρ at that position, i.e., the 'local' value of ρ. In principle, then, the only requirement on ρ is that it be single-valued at every position, and it can otherwise be wildly ill-behaved (recall that there *are* cusps in the density at the nucleus, so some ill-behavior in ρ has already been noted). In practice, the only functionals conforming to this definition that have seen much application are those that derive from analysis of the uniform electron gas (where the density has the same value at *every* position), and as a result LDA is often used more narrowly to imply that it is these exchange and correlation functionals that are being employed.

The distinction is probably best indicated by example. Following from Eq. (8.7) and the discussion in Section 8.1.2, the exchange energy for the uniform electron gas can be computed exactly, and is given by Eq. (8.23) with the constant α equal to $\frac{2}{3}$. However, the Slater approach takes a value for α of 1, and the Xα model most typically uses $\frac{3}{4}$. All of these models have the same 'local' dependence on the density, but only the first is typically referred to as LDA, while the other two are referred to by name as Slater (S) and Xα .

The LDA, Slater, and Xα methods can all be extended to the spin-polarized regime using

$$\varepsilon_x[\rho(\mathbf{r}), \zeta] = \varepsilon_x^0[\rho(\mathbf{r})] + \{\varepsilon_x^1[\rho(\mathbf{r})] - \varepsilon_x^0[\rho(\mathbf{r})]\} \left[\frac{(1+\zeta)^{4/3} + (1-\zeta)^{4/3} - 2}{2(2^{1/3} - 1)} \right] \quad (8.26)$$

where the superscript-zero exchange energy density is given by Eq. (8.23) with the appropriate value of α (referring here to the empirical constant, not the electron spin), and the superscript-one energy is the analogous expression derived from consideration of a uniform electron gas composed only of electrons of like spin. Noting that $\zeta = 0$ everywhere for an unpolarized system, it is immediately apparent that the second term in Eq. (8.26) is zero for that special case. Systems including spin polarization (e.g., open-shell systems) must use the spin-polarized formalism, and its greater generality is sometimes distinguished by the sobriquet 'local spin density approximation' (LSDA).

As for the correlation energy density, even for the 'simple' uniform electron gas no analytical derivation of this functional has proven possible (although some analytical details about the zero- and infinite-density limits can be established). However, using quantum Monte Carlo techniques, Ceperley and Alder (1980) computed the total energy for uniform electron gases of several different densities to very high numerical accuracy. By subtracting the analytical exchange energy for each case, they were able to determine the correlation energy in these systems. Vosko, Wilk, and Nusair (1980) later designed local functionals of the density fitting to these results (and the analytical low and high density limits). In particular, they proposed one spin-polarized functional completely analogous to Eq. (8.26) in terms of its dependence on ζ, but with the unpolarized and fully polarized energy densities expressed (now in terms of r_S instead of ρ, see Eq. (8.24)) as

$$\varepsilon_c^i(r_S) = \frac{A}{2} \left\{ \ln \frac{r_S}{r_S + b\sqrt{r_S} + c} + \frac{2b}{\sqrt{4c - b^2}} \tan^{-1} \left(\frac{\sqrt{4c - b^2}}{2\sqrt{r_S} + b} \right) \right.$$

$$\left. - \frac{bx_0}{x_0^2 + bx_0 + c} \left\{ \ln \left[\frac{(\sqrt{r_S} - x_0)^2}{r_S + b\sqrt{r_S} + c} \right] + \frac{2(b + 2x_0)}{\sqrt{4c - b^2}} \tan^{-1} \left(\frac{\sqrt{4c - b^2}}{2\sqrt{r_S} + b} \right) \right\} \right\}$$

$$(8.27)$$

where different sets of empirical constants A, x_0, b, and c are used for $i = 0$ and $i = 1$. Vosko, Wilk, and Nusair actually proposed several different fitting schemes, varying the functional forms of both Eq. (8.26) and (8.27). The two forms that have come to be most widely used tend to be referred to as VWN and VWN5, and in most cases give reasonably similar results. LSDA calculations that employ a combination of Slater exchange and the VWN correlation energy expression are sometimes referred to as using the SVWN method.

It is fairly obvious that Eq. (8.27) represents an utter violation of the promise with respect to intuitive equations found in the preface to this book; not only can every term not be assigned an intuitive meaning, it is rather difficult to assign *any* term such a meaning. However, the virtue of this momentary failure of authorial fidelity is that it allows the highlighting of several important details associated with DFT in general. First, it is apparent just how complex the correlation energy functional in a completely general system may be expected to be, and how difficult a task a first principles analysis may be. Secondly, it indicates the extent to which most modern DFT approaches can legitimately be described as semiempirical methods, in that they include empirically optimized constants and functional forms (albeit there are considerably fewer of these constants and they tend to be more globally used than in, say, semiempirical MO theory – in this respect they are rather like the parameterized electron correlation methods discussed in Section 7.7). Lastly, it should be fairly apparent that solution of the integral in Eq. (8.22) employing the VWN correlation functional is highly unlikely to be accomplished analytically.

In regard to this latter point, the evaluation of the integrals involving the exchange and correlation energy densities in DFT poses something of a mathematical challenge. Most modern electronic structure codes carry out this integration numerically on a grid (much along the lines already discussed in Section 3.4 in the context of Monte Carlo methods, except that the grid points are not sampled randomly, but exhaustively). Through the use of efficient quadratures, grid sizes can be kept manageable from the standpoint of computational resources. Usually, some default grid density is employed unless a user specifies otherwise – it must be noted that in certain situations the numerical noise associated with the default grid density can lead to problems, particularly when first and second derivatives are computed in order to optimize geometries (see Section 2.4), compute vibrational spectra (see Section 9.3.2), etc. For most calculations, however, the numerical noise falls very comfortably below the level of chemical interpretation, and no special care need be taken. So-called 'grid-free' integration schemes have been proposed, where in essence the exchange-correlation energy density is expressed in a basis set and advantage is taken of linear algebraic techniques to replace the numerical problem with a 'smooth' error associated with basis-set truncation. However, developmental work has not necessarily indicated this approach to be any more robust because very large basis sets are required to maintain reasonable accuracy (Zheng and Almlöf 1993).

Returning from this mathematical digression, let us make clear the steps involved in a LSDA calculation. These are summarized in Figure 8.1, and they are for the most part quite similar to those associated with a HF calculation. There are some important differences, however. For instance, step 1 is to choose a basis set. In DFT, there are sometimes several *different* basis sets involved in a calculation. First, there is the basis set from which the KS

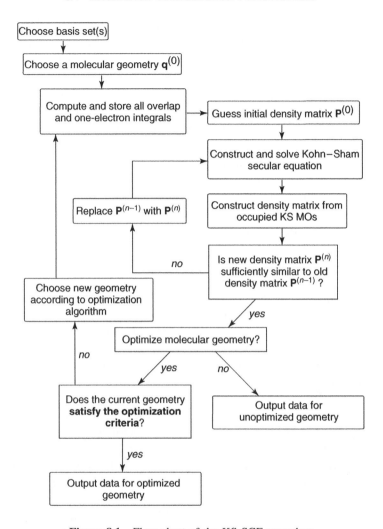

Figure 8.1 Flow chart of the KS SCF procedure

orbitals are formed. In addition, the density itself may be expanded in an 'auxiliary' basis set. Such a concept may at first seem strange, since we know the density can be represented as the product of AO basis functions and density matrix elements, as it is in the Coulomb integrals of HF theory. However, in HF theory this is a natural choice because one needs to evaluate both Coulomb and exchange integrals, and in the latter the interchange of electronic coordinates requires that orbitals be used, not densities (which are the product of orbitals). In Eq. (8.21), however, there are no exchange integrals, so it is computationally convenient to represent $\rho(\mathbf{r})$ with an auxiliary basis set, i.e.,

$$\rho(\mathbf{r}) = \sum_{i=1}^{M} c_i \Omega_i(\mathbf{r}) \tag{8.28}$$

where the M basis functions Ω have units of probability density (not the square root of probability density, as orbitals do) and the coefficients c_i are determined by a least-square fitting to the density that *is* determined from the KS orbitals using Eq. (8.16). Note that the number of Coulomb integrals requiring evaluation in order to compute all the KS matrix elements in Eq. (8.21) is then formally N^2M, where N is the number of KS AO basis functions, instead of N^4, as is true for HF theory. As a result, the formal bottleneck in solving the KS SCF equations is matrix diagonalization, which scales as N^3, and one frequently sees in the literature reference to this reduced scaling behavior associated with DFT: compared to HF theory, DFT includes electron correlation and does so in a fashion that scales more favorably with respect to system size. However, many electronic structure programs do *not* employ auxiliary basis sets to represent the density, choosing instead to compute it in the HF-like way as a product of KS-orbital basis functions (the motivation for this choice was primarily historical: existing HF codes could be easily modified to carry out DFT calculations with this choice), in which case formal N^4 scaling is not reduced.

Continuing with our analysis of steps in the KS SCF procedure, after choice of molecular geometry, the overlap integrals and the kinetic-energy and nuclear-attraction integrals are computed. The latter two kinds of integrals are called 'one-electron' integrals in HF theory to distinguish them from the 'two-electron' Coulomb and exchange integrals. In KS theory, such an appellation is less clearly informative: *all* integrals can in some sense be regarded as one-electron integrals since every one reflects the interactions of each one electron with external potentials, but we will not dwell on the semantics. In any case, to evaluate the remaining integrals, we must guess an initial density, and this density can be constructed as a matrix entirely equivalent to the density matrix used in HF theory (Eq. (4.57)). With our guess density in hand, we can construct V_{xc} (and determine fitting coefficients for our auxiliary basis set if we are using the approach of Eq. (8.28)) and evaluate the remaining integrals in each KS matrix element. After this point, the KS and HF SCF schemes are essentially identical. New orbitals are determined from solution of the secular equation, the density is determined from those orbitals, and it is compared to the density from the preceding iteration. Once convergence of the SCF is achieved, the energy is computed by plugging the final density into Eq. (8.14) – this is in contrast to HF theory, where the energy is evaluated as the expectation value of the Hamiltonian operator acting on the HF Slater determinant. At this point either the calculation is finished, or, if geometry optimization is the goal, a determination of whether the structure corresponds to a stationary point is made.

Having reviewed the mechanics of the KS calculation, we now return to a discussion of how best to represent the exchange-correlation functional. We should be entirely clear on the nature of the LSDA approximation applied to a molecule. Invoking the uniform electron gas as the source of the energy expressions is *not* equivalent to assuming that the electron density of the *molecule* is a constant throughout space. Instead, it is an assumption that the exchange-correlation energy density at every position in space for the molecule is the same as it would be for the uniform electron gas having the *same* density as is found at that position.

8.4.2 Density Gradient and Kinetic Energy Density Corrections

In a molecular system, the electron density is typically rather far from spatially uniform, so there is good reason to believe that the LDA approach will have limitations. One obvious way to improve the correlation functional is to make it depend not only on the local value of the density, but on the extent to which the density is locally changing, i.e., the gradient of the density. Such an approach was initially referred to as 'non-local' DFT because the Taylor-expansion-like formalism implies reliance on values of the density at more than a single position. Mathematically speaking, however, the first derivative of a function at a single position is a local property, so the more common term in modern nomenclature for functionals that depend on both the density and the gradient of the density is 'gradient-corrected'. Including a gradient correction defines the 'generalized gradient approximation' (GGA).

Most gradient corrected functionals are constructed with the correction being a term added to the LDA functional, i.e.,

$$\varepsilon_{x/c}^{GGA}[\rho(\mathbf{r})] = \varepsilon_{x/c}^{LSD}[\rho(\mathbf{r})] + \Delta\varepsilon_{x/c}\left[\frac{|\nabla\rho(\mathbf{r})|}{\rho^{4/3}(\mathbf{r})}\right] \tag{8.29}$$

Note that the dependence of the correction term is on the dimensionless reduced gradient, not the absolute gradient.

The first widely popular GGA exchange functional was developed by Becke. Usually abbreviated simply 'B', this functional adopts a mathematical form that has correct asymptotic behavior at long range for the energy density, and it further incorporates a single empirical parameter the value of which was optimized by fitting to the exactly known exchange energies of the six noble gas atoms He through Rn (Table 8.7 at the end of this chapter provides references and additional details for B as well as for a reasonably complete menagerie of other functionals developed to date, including all of those discussed below). Other exchange functionals similar to the Becke example in one way or another have appeared, including CAM, FT97, O, PW, mPW, and X, where X is a particular combination of B and PW found to give improved performance over either.

Alternative GGA exchange functionals have been developed based on rational function expansions of the reduced gradient. These functionals, which contain *no* empirically optimized parameters, include B86, LG, P, PBE, and mPBE.

With respect to correlation functionals, corrections to the correlation energy density following Eq. (8.29) include B88, P86, and PW91 (which uses a different expression than Eq. (8.27) for the LDA correlation energy density and contains no empirical parameters). Another popular GGA correlation functional, LYP, does not correct the LDA expression but instead computes the correlation energy *in toto*. It contains four empirical parameters fit to the helium atom. Of all of the correlation functionals discussed, it is the only one that provides an exact cancellation of the self-interaction error in one-electron systems.

Typically in the literature, a complete specification of the exchange and correlation functionals is accomplished by concatenating the two acronyms in that order. Thus, for instance, a BLYP calculation combines Becke's GGA exchange with the GGA correlation functional of Lee, Yang, and Parr.

[As an aside to the exasperated reader, this mishmash of acronyms for referring to density functionals has resulted in part from failures on the parts of early developers to clearly specify their own preferences for names for their methods. Thus, a tendency to refer to methods by authors' initials arose and has persisted. However, with multiple methods developed by the same authors, occasionally described over more than one article with different co-authors, and with some articles introducing both exchange and correlation functionals, the situation can be extremely confusing. Worse still, different codes may adopt different keyword abbreviations for the same method, so that comparable calculations may appear in the literature under different acronyms. Thus, DFT suffers to some extent from the same problem as molecular mechanics: reproducibility may depend on using the same code to ensure the same functional specification. It is to be hoped that in the future careful definitions of nomenclature will always be made, possibly even including program-specific keywords to generate particular functionals.]

Given the Taylor-function-expansion justification for the importance of the gradient of the density in Eq. (8.29), it is obvious that a logical next step in functional improvement might be to take account of the second derivative of the density, i.e., the Laplacian. Becke and Roussel were the first to proposed an exchange functional (BR) having such dependence while work of Proynov, Salahub, and co-workers examined the same idea for the correlation functional (Lap). Such functionals are termed meta-GGA (MGGA) functionals as they go beyond simply the gradient correction. However, numerically stable calculations of the Laplacian of the density pose something of a technical challenge, and the somewhat improved performance of MGGA functionals over GGA analogs is balanced by this slight drawback.

An alternative MGGA formalism that is more numerically stable is to include in the exchange-correlation potential a dependence on the kinetic-energy density τ, defined as

$$\tau(\mathbf{r}) = \sum_i^{\text{occupied}} \frac{1}{2} |\nabla \psi_i(\mathbf{r})|^2 \qquad (8.30)$$

where the ψ are the self-consistently determined Kohn–Sham orbitals. The BR functional includes dependence on τ in addition to its already noted dependence on the Laplacian of the density. The same is true of the $\tau 1$ correlation functional of Proynov, Chermette, and Salahub. Other developers, however, have tended to discard the Laplacian in their MGGA functionals, retaining only a dependence on τ. Various such MGGA functionals for exchange, correlation, or both have been developed including B95, B98, ISM, KCIS, PKZB, τHCTH, TPSS, and VSXC. The cost of an MGGA calculation is entirely comparable to that for a GGA calculation, and the former is typically more accurate than the latter for a pure density functional. Prior to a more detailed analysis of performance, however, we must consider at least one additional wrinkle in functional design, namely, the inclusion of HF exchange.

8.4.3 Adiabatic Connection Methods

Imagine that one could control the extent of electron–electron interactions in a many-electron system. That is, imagine a switch that would smoothly convert the non-interacting KS reference system to the real, interacting system. Using the Hellmann–Feynman theorem, one can show that the exchange-correlation energy can then be computed as

$$E_{xc} = \int_0^1 \langle \Psi(\lambda)|V_{xc}(\lambda)|\Psi(\lambda)\rangle d\lambda \tag{8.31}$$

where λ describes the extent of interelectronic interaction, ranging from 0 (none) to 1 (exact). To evaluate this integral, it is helpful to adopt a geometric picture, as illustrated in Figure 8.2. We seek the area under the curve defined by the expectation value of V_{xc}. While we know very little about V and Ψ as functions of λ in general, we can evaluate the left endpoint of the curve. In the non-interacting limit, the only component of V is exchange (deriving from antisymmetry of the wave function). Moreover, as discussed in Section 8.3, the Slater determinant of KS orbitals is the *exact* wave function for the non-interacting Hamiltonian operator. Thus, the expectation value is the exact exchange for the non-interacting system, which can be computed just as it is in HF calculations except that the KS orbitals are used. The total area under the expectation value curve thus contains the rectangle having the curve's left endpoint as its upper left corner, which has area E_x^{HF}. The remaining area is

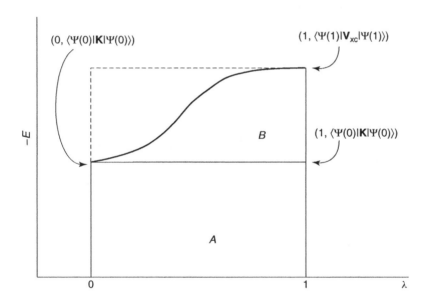

Figure 8.2 Geometrical schematic for the evaluation of the integral in Eq. (8.31). The area under the curve is the sum of the areas of regions A and B. As region A is a rectangle, its area is trivially computed to be one times the expectation value of the HF exchange operator acting on the Slater determinantal wave function for the non-interacting system, $\Psi(0)$. The area of region B is less easily determined. One simplification is to assume (i) that $\langle\Psi(1)|V_{xc}(1)|\Psi(1)\rangle$ is equal to the corresponding computed value from an approximate DFT calculation and (ii) that area B will be some characteristic fraction of the area of the rectangle having dotted lines for 2 sides (e.g., if the curve is well approximated by a line, clearly the characteristic fraction would be 0.5). Note that if the curve rises very steeply from its left endpoint, i.e., the value of z in Eq. (8.32) is very close to 1, then the adiabatic connection method is of limited value

some fraction z of the area corresponding to the rectangle sitting immediately on top of the first; the second rectangle has area $\langle \Psi(1)|V_{xc}(1)|\Psi(1)\rangle - E_x^{HF}$. Unfortunately, not only do we not know z, we also do not know the expectation value of the fully interacting exchange-correlation potential applied to the fully interacting wave function. However, we may regard z as an empirical constant to be optimized. In that case, we may as well approximate the unknown right endpoint, and a convenient approximation is the E_{xc} computed directly by some choice of DFT functional. We see then that the total area under the expectation value curve can be written as

$$E_{xc} = E_x^{HF} + z(E_{xc}^{DFT} - E_x^{HF}) \tag{8.32}$$

In practice, Eq. (8.32) is usually written using another variable, a, defined as $1 - z$, providing

$$E_{xc} = (1 - a)E_{xc}^{DFT} + aE_x^{HF} \tag{8.33}$$

This analysis forms the basis of the so-called 'adiabatic connection method' (ACM), because it connects between the non-interacting and fully interacting states.

If we adopt the assumption that the expectation value curve in Figure 8.2 is a line, simple geometry allows us to determine that $z = 0.5$. Use of this value defines a so-called 'half-and-half' (H&H) method. Using LDA exchange-correlation, Becke showed that the H&H approach had an error of 6.5 kcal mol^{-1} over a subset of the G3 enthalpy of formation test set mentioned in Chapter 7. This compared quite favorably with the GGA method BPW91, which had an error of 5.7 kcal mol^{-1} over the same set.

One may in some sense regard the ACM approach as being similar in spirit to the KS SCF scheme. In the latter case, one does not know the exact kinetic energy as a function of the density, so one employs a scheme where a large portion of it is computed exactly (as the expectation value of the kinetic energy operator over the KS determinant) and worries later about the small remainder. So too, the ACM approach computes a large fraction of the total exchange-correlation energy, and then worries later about the difference between the total and the exact (HF) exchange.

Of course, if one is forced to estimate a constant like a, one might just as well choose a value that maximizes the utility of the method. And one may legitimately ask whether inclusion of additional empirical parameters results in sufficient improvement to make such inclusion worthwhile. Becke was the first to do this, developing the 3-parameter functional expression

$$E_{xc}^{B3PW91} = (1 - a)E_x^{LSDA} + aE_x^{HF} + b\Delta E_x^B + E_c^{LSDA} + c\Delta E_c^{PW91} \tag{8.34}$$

where a, b, and c were optimized to 0.20, 0.72, and 0.81, respectively. The name of the functional, B3PW91, implies its use of a three-parameter scheme, as well as the GGA exchange and correlation functionals B and PW91, respectively (recall that the expression for the LSDA correlation energy in PW91 is different from eq. 8.27).

Subsequently, Stevens *et al.* modified this functional to use LYP instead of PW91. Because LYP is designed to compute the full correlation energy, and not a correction to LSDA, the

B3LYP model is defined by

$$E_{xc}^{\text{B3LYP}} = (1-a)E_x^{\text{LSDA}} + aE_x^{\text{HF}} + b\Delta E_x^{\text{B}} + (1-c)E_c^{\text{LSDA}} + cE_c^{\text{LYP}} \qquad (8.35)$$

where a, b, and c have the same values as in B3PW91. Of all modern functionals, B3LYP has proven the most popular to date. Its overall performance, as described in more detail in Section 8.6, is remarkably good, particularly insofar as the three parameters were not optimized! Such serendipity is rare in computational chemistry. The O3LYP functional is similar in character to B3LYP, with $a = 0.1161$, $b = 0.9262$ (multiplying O exchange instead of B exchange), and $c = 0.8133$. The two differ, however, in that B3LYP uses the VWN3 LSDA correlation functional while O3LYP uses the VWN5 version. The X3LYP functional also uses the form of Eq. (8.35), with $a = 0.218$, $b = 0.709$ (multiplying a combination of 76.5% B exchange and 23.5% PW exchange instead of pure B exchange), and $c = 0.129$.

Because they incorporate HF and DFT exchange, ACM methods are also called 'hybrid' methods. Some interest has developed in so-called 'parameter-free' hybrid methods, but this terminology must be regarded with some skepticism. Analysis of simple one- and two-electron systems like H_2 and H_2^+ make it clear that the correct amount of HF exchange to include in any hybrid model *using a GGA functional* cannot be a constant over all species (or even all geometries of a single species; see, for instance, Gritsenko, Schipper, and Baerends 1996). In any case, besides the B3 methods a number of one-parameter models, restricting themselves to adjusting the percentage of HF exchange included in the functional, have been proposed. These include B1PW91 and B1LYP ($a = 0.25$, $b = (1-a)$, $c = 1$ in Eqs. (8.34) and (8.35), respectively), B1B95, mPW1PW91, and PBE1PBE (sometimes called PBE0 because the parameter dictating the percentage contribution of HF exchange, 0.25, was not empirically optimized, but instead chosen based on perturbation theory arguments, thus there are 'zero' parameters). Overall, the performance of these functionals, as well as other ACM functionals listed in Table 8.7, tends to be fairly comparable to the B3 methods.

From a careful comparison of DFT densities to those generated from highly correlated wave functions, He *et al.* (2000) concluded that inclusion of HF exchange in a hybrid functional makes up for an underestimation by pure functionals of the importance of ionic terms in describing polar bonds. However, one may also adopt a less formal viewpoint of the hybrid methods. Experience indicates that GGA functionals have certain systematic errors. For instance, they tend to underestimate barrier heights to chemical reactions. Hartree–Fock theory, on the other hand, tends to overestimate barrier heights. To the extent that errors in total HF energies track with errors in HF exchange energy, one may regard the addition of HF exchange to 'pure' DFT results as something of a back-titration to accuracy. With this in mind, Lynch *et al.* have reoptimized the percent HF exchange in the mPW1PW91 model against a database of energies of activation and reaction for hydrogen-atom transfer reactions, referring to this model as MPW1K ('K' for 'kinetics'). MPW1K increases the percentage HF contribution in the functional from the 'default' value of 25% to 42.8%; this increase in HF exchange leads to significantly improved performance over the chosen test set, although Boese, Martin, and Handy (2003) have noted that so high a fraction of HF exchange degrades the functionals performance for geometries and atomization energies. Zhao, Lynch, and Truhlar performed an equivalent optimization of percent HF exchange for B1B95 (42%, thereby generating BB1K) and observed that this kinetics model slightly outperformed MPW1K while at the same time reducing the error in atomization energies compared to MPW1K by 40%.

Other variations on this optimization scheme have also appeared, including mPW1N, which uses a value of 40.6% in conjunction with the 6-31+G(d) basis set to maximize accuracy over a set of halide/alkyl halide nucleophilic substitution reactions, and MPW1S, which employs a value of 6% in conjunction with the 6-31+G(d,p) basis set to improve the relative accuracy of computed conformational energies for sugars and sugar analogs. Within the context of B3LYP, Salomon, Reiher, and Hess have shown that changing a in Eq. (8.33) from 0.20 to 0.15 (which they dub B3LYP*) significantly improves energy separations predicted for high and low spin states of molecules containing first-row transition metals; heavier transition metals do not appear, however, to be as energetically sensitive to the fraction of HF exchange (Poli and Harvey 2003).

In a similar spirit, Poater et al. have reoptimized the B3LYP parameters in order to minimize differences between computed electron densities from this modified DFT level and calculated at the QCISD level for a series of 16 small molecules (Poater, Duran, and Solà 2001). They observed, as already emphasized above, that different molecules require different amounts of exact HF exchange for optimal agreement between the two methods.

Finally, ACM definitions involving MGGA functionals have also begun to appear and these are listed in Table 8.7. Improvements associated with inclusion of HF exchange in the ACM functional appear to be diminished in magnitude for MGGA functionals compared to GGA functionals, but are still noticeable. Detailed comparisons between models are provided in Section 8.6.

8.4.4   Semiempirical DFT

Adamson, Gill, and Pople have proposed a parameterized pure GGA functional (i.e., no HF exchange is included) designed specifically to give good results with small basis sets; their Empirical Density Functional 1 (EDF1) is thus essentially a semiempirical model, where all limitations in theory and numerical accuracy are folded into the parameters. Of course, one may legitimately claim that *any* DFT model, even if used with an infinite basis set, is semiempirical if it includes any optimized parameters. It is easy to get bogged down in this argument (particularly with individuals who treat 'semiempirical' as a pejorative term), but one may turn the issue around and ask, can one develop DFT models with drastically improved efficiency that, while semiempirical, may be particularly applicable to problems still outside the range of more rigorous functionals?

One such model that has promise is density functional tight-binding (DFTB) theory. In DFTB, we begin by expressing the energy associated with a reference density $\rho_0(\mathbf{r})$ as

$$
E[\rho_0(\mathbf{r})] = \sum_i^{\text{occupied}} \langle \psi_i(\mathbf{r}) \left| h_i^{\text{KS}}[\rho_0(\mathbf{r})] \right| \psi_i(\mathbf{r}) \rangle - \frac{1}{2} \int \int \frac{\rho_0(\mathbf{r}_1)\rho_0(\mathbf{r}_2)}{|\mathbf{r}_1 - \mathbf{r}_2|} d\mathbf{r}_1 d\mathbf{r}_2
$$

$$
+ E_{\text{xc}}[\rho_0(\mathbf{r})] - \int V_{\text{xc}}[\rho_0(\mathbf{r})]\rho_0(\mathbf{r})d\mathbf{r} + E_{\text{N}}
$$

$$(8.36)$$

A quick comparison of this equation with Eqs. (8.15) and (8.18) should make clear that this expression is indeed valid; its form is reminiscent of Eq. (4.39) insofar as the energy is expressed as a sum of orbital energies (the first term on the r.h.s.) corrected for double-counting of electron–electron interactions (the next three terms on the r.h.s.). In addition the nuclear repulsion term, which is constant for a given geometry, is explicitly written here for reasons that will become clear shortly.

Consider now the possibility that we are interested in minimizing the energy of Eq. (8.36), but *only* subject to the shape of the KS orbitals, not to changing the density. That is, having picked a density, we optimize the orbitals by variational minimization of basis set coefficients, but we do not then recompute a new density from those orbitals (note that with a fixed density and geometry, *all* terms on the r.h.s. of Eq. (8.36) are constants except for the KS orbitals in the initial sum). Such a process is analogous to extended Hückel theory in the sense that it is *non-self-consistent*, i.e., we solve a secular equation analogous to Eq. (4.21) a single time in order to derive our KS orbital basis set coefficients and then we are finished. This protocol defines the Harris functional approach, where the fixed density is usually chosen to be the sum of unperturbed atomic densities computed by whatever manner is deemed most appropriate. Energies from Harris functional calculations are obviously unlikely to be particularly good, but the orbitals may themselves be useful either for qualitative analysis in systems where charge transfer between atoms is small (e.g., in pure solid metals) or as very good starting-guess orbitals for a follow-on calculation at some self-consistent level of theory (e.g., HF or KS DFT); see Cullen (2004) for other applications of the Harris functional.

To speed this process up further for very large systems, one can make some further assumptions entirely analogous to those found in semiempirical MO theory. In particular, one may assume

$$\langle \mu | h^{KS} | \nu \rangle = \begin{cases} \varepsilon_\mu, & \mu = \nu \\ \langle \mu | T + v_{\text{eff}}[\rho_{0,A}(\mathbf{r}) + \rho_{0,B}(\mathbf{r})] | \nu \rangle, & \mu \in A, \nu \in B \end{cases} \tag{8.37}$$

where ε is the KS energy of the atomic orbital basis function (usually Slater-type orbitals) in the neutral atom (or one could use an experimental ionization potential to replace this number, in a semiempirical spirit), and the off-diagonal matrix elements depend only on the two atoms involved, A and B, and T is the kinetic-energy operator and v_{eff} is the effective potential deriving only from the electron densities and nuclei of atoms A and B. The off-diagonal matrix elements then depend only on interatomic separation and can be computed once and then either fit to analytic functions or interpolated from tabulations over various distances.

This same restriction of consideration to no more than pairwise interactions can be adopted for the remaining terms on the r.h.s. of Eq. (8.36), which are usually grouped together and referred to collectively as the repulsive energy E_{rep}, so that this energy component is computed as

$$E_{\text{rep}}[\rho_0(\mathbf{r})] = \sum_A^{\text{atoms}} E_{\text{rep}}[\rho_{0,A}(\mathbf{r})] + \sum_{A<B}^{\text{atoms}} E_{\text{rep}}^{(2)}[\rho_{0,A}(\mathbf{r}), \rho_{0,B}(\mathbf{r})] \tag{8.38}$$

where, again, the pairwise terms can be computed once for every pair of atoms and then either fit to analytic functions or tabulated for future reference.

With these further simplifications, enormously large systems may be handled fairly easily to include geometry optimization. This non-self-consistent protocol defines DFTB.

The critical assumption of DFTB, however, is that the charge density of a composite system is well represented by the sum of the charge densities of its unperturbed constituent atoms. Clearly such a situation is inconsistent with the polarization that occurs in bonds between elements having significantly different electronegativities. To address such polarized systems, we consider a generalization of Eq. (8.36) that is valid to second order in the density fluctuation $\delta\rho(\mathbf{r})$ about the fixed density $\rho_0(\mathbf{r})$, namely

$$E[\rho_0(\mathbf{r}) + \delta\rho(\mathbf{r})] = E[\rho_0(\mathbf{r})]$$

$$+ \frac{1}{2} \int \int \left[\left(\frac{1}{|\mathbf{r}_1 - \mathbf{r}_2|} + \frac{\delta^2 E_{\mathrm{xc}}}{\delta\rho(\mathbf{r}_1)\delta\rho(\mathbf{r}_2)} \bigg|_{\rho_0} \right) \delta\rho(\mathbf{r}_1)\delta\rho(\mathbf{r}_2) \right] d\mathbf{r}_1 d\mathbf{r}_2$$

$$(8.39)$$

In the spirit of DFTB, one may consider $\delta\rho(\mathbf{r})$ to be decomposable into atomic contributions according to

$$\delta\rho(\mathbf{r}) = \sum_{A}^{\mathrm{atoms}} \Delta q_A \qquad (8.40)$$

where q is used for the atomic contribution to emphasize the analogy to partial atomic charge. The second-order term in Eq. (8.39) may then be written as

$$\frac{1}{2} \int \int \left[\left(\frac{1}{|\mathbf{r}_1 - \mathbf{r}_2|} + \frac{\delta^2 E_{\mathrm{xc}}}{\delta\rho(\mathbf{r}_1)\delta\rho(\mathbf{r}_2)} \bigg|_{\rho_0} \right) \delta\rho(\mathbf{r}_1)\delta\rho(\mathbf{r}_2) \right] d\mathbf{r}_1 d\mathbf{r}_2 = \frac{1}{2} \sum_{A,B}^{\mathrm{atoms}} \Delta q_A \Delta q_B \gamma_{AB}$$

$$(8.41)$$

If the atomic charge distributions are assumed to be spherically symmetric, the effective inverse distance γ is computed as

$$\gamma_{AB} = \begin{cases} 2\eta_A, & A = B \\ (aa|bb), & A \neq B \end{cases} \qquad (8.42)$$

where η is an atomic hardness (formally the second derivative of the atomic energy with respect to a change from neutrality in charge; η is well approximated as $(\mathrm{IP} - \mathrm{EA})/2$, where IP and EA are the atomic ionization potential and electron affinity, respectively) and a and b are Slater-type s orbitals on atoms A and B, respectively, in which case the electron-repulsion integral has an analytic solution. Once again this function, for which the off-diagonal terms depend on the distance between atoms A and B, may be approximated with a simpler analytic form or tabulated for all relevant atomic pairs.

The presence of the partial atomic charges in Eq. (8.41), however, poses the question of how they are to be computed. A popular choice is to compute them from Mulliken population analysis (see Section 9.1.3.2), in which case the partial atomic charges depend on the KS orbitals. Such dependence re-introduces a self-consistency requirement into the minimization of the energy from Eq. (8.39), and this approach is called self-consistent charge density-functional tight-binding (SCC-DFTB) theory. Thus, for a realistic representation of charge redistribution, one must sacrifice the higher efficiency of DFTB for an SCF approach. Nevertheless, SCC-DFTB is about as fast as a semiempirical NDDO model and many promising applications have begun to appear. For example, Elstner *et al.* (2003) found SCC-DFTB to compare favorably to B3LYP and MP2 calculations with the 6-311+G(d,p) basis set for structural and energetic properties associated with biological model systems coordinating zinc. One feature requiring further attention, however, is that in very large molecules like biopolymers there are likely to be non-bonded interactions, e.g., dispersion, between different sections of the molecule. Dispersion is not well treated by SCC-DFTB; in a QM MD study of the protein crambin, Liu *et al.* found that inclusion of an ad hoc scaled r^{-6} potential between non-bonded atoms (i.e., the attractive portion of a Lennard-Jones potential, cf. Eq. (2.14)) was required to maintain a structure in acceptable agreement with experiment.

8.5 Advantages and Disadvantages of DFT Compared to MO Theory

Since 1990 there has been an enormous amount of comparison between DFT and alternative methods based on the molecular wave function. The bottom line from all of this work is that, as a rule, DFT is the most cost-effective method to achieve a given level of accuracy, sometimes by a very wide margin. There are, however, significant exceptions to this rule, deriving either from inadequacies in modern functionals or intrinsic limitations in the KS approach for determining the density. This section describes some of these cases.

8.5.1 Densities vs. Wave Functions

The most fundamental difference between DFT and MO theory must never be forgotten: DFT optimizes an electron density while MO theory optimizes a wave function. So, to determine a particular molecular property using DFT, we need to know how that property depends on the density, while to determine the same property using a wave function, we need to know the correct quantum mechanical operator. As there are more well-characterized operators then there are generic property functionals of the density, wave functions clearly have broader utility. As a simple example, consider the total energy of interelectronic repulsion. Even if we had the exact density for some system, we do not know the exact exchange-correlation energy functional, and thus we cannot compute the exact interelectronic repulsion. However, with the exact wave function it is a simple matter of evaluating the expectation value for the interelectronic repulsion operator to determine this energy,

$$
E_{ee} = \left\langle \Psi \left| \sum_{i<j} \frac{1}{r_{ij}} \right| \Psi \right\rangle
\tag{8.43}
$$

where i and j run over all electrons.

Another key example is in the area of dynamics, where transition probabilities depend on matrix elements between *different* wave functions. Because densities do not have phases as wave functions do, multistate resonance effects, interference effects, etc., are not readily evaluated within a DFT formalism.

Because of the mechanical details of the KS formalism, it is easy to become confused about whether there is a KS 'wave function'. Early work in the field tended to resist any attempts to interpret the KS orbitals, viewing them as pure mathematical constructs useful only in construction of the density. In practice, however, the shapes of KS orbitals tend to be remarkably similar to canonical HF MOs, and they can be quite useful in qualitative analysis of chemical properties. If we think of the procedure by which they are generated, there are indeed a number of reasons to *prefer* KS orbitals to HF orbitals. For instance, all KS orbitals, occupied and virtual, are subject to the *same* external potential. HF orbitals, on the other hand, experience varying potentials, and, in particular, HF virtual orbitals experience the potential that would be felt by an *extra* electron being added to the molecule. As a result, HF virtual orbitals tend to be too high in energy and anomalously diffuse compared to KS virtual orbitals. (In *exact* DFT, it can also be shown that the eigenvalue of the highest KS MO is the exact first ionization potential, i.e., there is a direct analogy to Koopmans' theorem for this orbital – in practice, however, *approximate* functionals are quite bad at predicting IPs in this fashion without applying some sort of correction scheme, e.g., an empirical linear scaling of the eigenvalues).

In point of fact, there *is* a DFT wave function; it is just not clear how useful it should be considered to be. Recall that the Slater determinant formed from the KS orbitals is the exact wave function for the fictional *non-interacting* system having the same density as the real system. This KS Slater determinant has certain interesting properties by comparison to its HF analogs. In open-shell systems, KS determinants usually show extremely low levels of spin contamination, even for cases where HF determinants are pathologically bad (Baker, Scheiner, and Andzelm 1993). For instance, the spin contamination in planar triplet phenylnitrenium cation (PhNH$^+$) is very high at the UHF/cc-pVDZ level (see Section 6.3.3) as judged by an expectation value for S^2 of 2.50. At the BLYP/cc-pVDZ level, on the other hand the expectation value for S^2 over the KS determinant is 2.01, very close to the proper eigenvalue of 2.0. The high spin contamination at the UHF level leads to the planar structure being erroneously determined to be a minimum, while at the BLYP level it is correctly identified as a TS structure for rotation about the C–N bond (Cramer, Dulles, and Falvey 1994).

While it is by no means guaranteed that the expectation value for S^2 over the KS determinant has any bearing at all on its expectation value over the *exact* wave function corresponding to the KS density (see Gräfenstein and Cremer 2001), it is an empirical fact that DFT is generally much more robust in dealing with open-shell systems where HF methods show high spin contamination (recall that high HF spin contamination makes post-HF methods of questionable utility, so DFT can be a happy last resort). Note, incidentally, that expectation values of S^2 are sensitive to the amount of HF exchange in the functional. A 'pure' functional nearly always shows very small spin contamination, and each added percent of HF exchange tends to titrate in a corresponding percentage of the spin contamination exhibited by the HF wave function. This behavior can mitigate the utility of hybrid functionals in some open-shell systems.

Finally, one clear utility of a wave function is that excited states can be generated as linear combinations of determinants derived from exciting one or more electrons from occupied to virtual orbitals (see Section 14.1). Although the Hohenberg–Kohn theorem makes it clear that the density alone carries sufficient information to determine the excited-state wave functions, it is only very recently that progress has been made on applying DFT to excited states (the exception being in symmetric molecules, where the lowest energy state in each spatial irreducible representation is amenable to a simple SCF treatment as already noted in Section 8.2.1). Additional discussion on this subject is deferred to Section 14.2.1.

8.5.2 Computational Efficiency

The formal scaling behavior of DFT has already been noted to be in principle no worse than N^3, where N is the number of basis functions used to represent the KS orbitals. This is better than HF by a factor of N, and very substantially better than other methods that, like DFT, also include electron correlation (see Table 7.4). Of course, scaling refers to how time increases with size, but says nothing about the *absolute* amount of time for a given molecule. As a rule, for programs that use approximately the same routines and algorithms to carry out HF and DFT calculations, the cost of a DFT calculation on a moderately sized molecule, say 15 heavy atoms, is double that of the HF calculation with the same basis set.

However, it is possible to do very much better than that in programs optimized for DFT. One area where DFT enjoys a clear advantage over HF is in its ability to use basis functions that are *not* necessarily contracted Gaussians. Recall that the motivation for using contracted GTOs is that arbitrary four-center two-electron integrals can be solved analytically. In most electronic structure programs where DFT was added as a new feature to an existing HF code, the representation of the density in the classical electron-repulsion operator is carried out using the KS orbital basis functions. Thus, the net effect is to create a four-index integral, and these codes inevitably continue to use contracted GTOs as basis functions. However, if the density is represented using an auxiliary basis set, or even represented numerically, other options are readily available for the KS orbital basis set, including Slater-type functions. STOs enjoy the advantage that fewer of them are required (since, *inter alia*, they have correct cusp behavior at the nuclei) and certain advantages associated with symmetry can more readily be taken, so they speed up calculations considerably. The widely used Amsterdam Density Functional code (ADF) makes use of STO basis functions covering atomic numbers 1 to 118 (Snijders, Baerends, and Vernooijs 1982; van Lenthe and Baerends 2003; Chong *et al.* 2004).

Another interesting possibility is the use of plane waves as basis sets in periodic infinite systems (e.g., metals, crystalline solids, or liquids represented using periodic boundary conditions). While it takes an enormous number of plane waves to properly represent the decidedly aperiodic densities that are possible within the unit cells of interesting chemical systems, the necessary integrals are particularly simple to solve, and thus this approach sees considerable use in dynamics and solid-state physics (Dovesi *et al.* 2000).

Even in cases where contracted GTOs are chosen as basis sets, DFT offers the advantage that convergence with respect to basis-set size tends to be more rapid than for MO techniques (particularly correlated MO theories). Thus, polarized valence double-ζ basis sets are quite adequate for a wide variety of calculations, and very good convergence in many properties can be seen at the level of employing polarized triple-ζ basis sets. Extensive studies of basis set effects on functional performance and parameterization have been carried out by Jensen (2002a, 2002b, 2003) and Boese, Martin, and Handy (2003). They found, *inter alia*, that for most functionals Pople-type basis sets provide much better accuracy than cc-pVnZ basis sets of similar size, that adding diffuse functions offers substantial improvement over using the non-augmented analog basis (a point also made by Lynch, Zhao, and Truhlar (2003), particularly for the computation of barrier heights or conformational energies in molecules containing multiple lone pairs of electrons), that satisfactory convergence is generally arrived at for most properties of interest by the time triple-ζ basis sets are used, and finally that the optimal values for parameters that are included in various functionals are sensitive to choice of basis set size. Thus, the optimal percent HF exchange for HCTH/407 was about 28% with double-ζ basis sets, but about 18% with triple-ζ basis sets. Jensen (2002b, 2003) found that, with reoptimization of the polarization exponents for DFT, the pc-n basis sets were always able to provide the best accuracy for a given basis set size.

Besides issues associated with basis sets, considerable progress has been made in developing linear-scaling algorithms for DFT. In this regard, DFT is somewhat simpler than MO theoretical techniques because all potentials are local (this refers to 'pure' DFT – incorporation of HF exchange introduces the non-local exchange operator). Thus, one promising technique is the 'divide-and-conquer' formalism of Yang and co-workers, where a large system is divided up into a number of smaller regions, within each of which a KS SCF is carried out representing the other regions in a simplified fashion (Yang and Lee 1995). The total cost of matrix diagonalization is thereby reduced from N^3 scaling to $M(N/M)^3$ scaling where M is the number of sub-regions. Since the number of basis functions in each sub-region (N/M) tends to be close to some fixed value irrespective of N, the overall scaling goes as order M, i.e., linear. Of course, all the algorithms developed to facilitate linear scaling in computing Coulomb interactions in HF and MD calculations (e.g., fast multipole methods) can be used in DFT calculations as well.

As a final point with regard to efficiency, note that SCF convergence in DFT is sometimes more problematic than in HF. Because of the similarities between the KS and HF orbitals, this problem can often be very effectively alleviated by using the HF orbitals as an initial guess for the KS orbitals. Because the HF orbitals can usually be generated quite quickly, the extra step can ultimately be time-saving if it sufficiently improves the KS SCF convergence.

8.5.3 Limitations of the KS Formalism

It is important to emphasize that nearly all applications of DFT to molecular systems are undertaken within the context of the Kohn–Sham SCF approach. The motivation for this choice is that it permits the kinetic energy to be computed as the expectation value of the kinetic-energy operator over the KS single determinant, avoiding the tricky issue of determining the kinetic energy as a functional of the density. However, as has already been discussed in the context of MO theory, some chemical systems are not well described by a single Slater determinant. The application of DFT to such systems is both technically and conceptually problematic.

To illustrate this point, let us return to the cases of *p*-benzyne and *N*-protonated 2,5-pyridyne already discussed at length in Section 7.6.2. When restricted DFT is applied to the closed-shell singlet states of these molecules, the predicted splittings between the singlet and triplet states at the BPW91/cc-pVDZ level are 3.1 and 3.7 kcal mol^{-1}, respectively. Comparing to the last line of Table 7.2, we see that these predictions are in error by about 8 kcal mol^{-1} and are qualitatively incorrect about which state is the ground state. A careful analysis indicates that there is no problem with the triplet state, but that the singlet state is predicted to be insufficiently stable as a consequence of enforcing a single-determinantal description as part of the KS formalism (this also results in rather poor predicted geometries for the singlets).

In cases like this, showing high degrees of non-dynamical correlation, there are two primary approaches to correcting for inadequacies in the KS treatment. In the first approach, the remedy is fairly simple: an unrestricted KS formalism is applied and the wave function for the singlet is allowed to break spin symmetry. That is, even though the singlet is closed-shell, the α and β orbitals are permitted to be spatially different. When this unrestricted formalism is applied to *p*-benzyne and *N*-protonated 2,5-pyridyne, the S−T splittings are predicted to be −3.6 and −3.9 kcal mol^{-1}, respectively, in dramatically improved agreement with experiment/best estimates (singlet geometries are also improved).

Similar results have been obtained in transition-metal compounds containing two metal atoms that are antiferromagnetically coupled. An adequate description of the singlet state sometimes requires a broken-symmetry SCF, and inspection of the KS orbitals afterwards typically indicates the highest energy α electron(s) to be well localized on one metal atom while the corresponding highest energy β electron(s) can be found on the other metal atom. The transition from a stable restricted DFT solution to a broken-symmetry one takes place as the distance between the metal atoms increases and covalent-like bonding gives way to more distant antiferromagnetic interactions (Lovell *et al.* 1996; Cramer, Smith, and Tolman 1996; Adamo *et al.* 1999).

What is to be made of these broken-symmetry singlet KS wave functions? One interpretation is to invoke the variational principle and assert that, insofar as they lower the energy, they must provide better densities and one is fully justified in using them. While pragmatic, this view is somewhat unsatisfying in a number of respects. One troubling issue is that the expectation value of the total spin operator for the KS determinant is often significantly in excess of the expected exact value. Thus, in the case of the singlet arynes discussed above, $\langle S^2 \rangle$ values of zero are expected, but computed values are on the order of 0.2 for broken-symmetry solutions. If these were HF wave functions, we would take such a value as being indicative of a fair degree of spin contamination. However, it is by no means obvious that $\langle S^2 \rangle$ for the non-interacting KS wave function is in any way indicative of what $\langle S^2 \rangle$ may be for the *interacting* wave function corresponding to the final KS density. It may not be spin contaminated at all. On the other hand, DFT energies where broken-symmetry KS wave functions have $\langle S^2 \rangle$

values of 1.0 (i.e., they are equal mixtures of singlet and triplet) can be usefully interpreted as being the average of the singlet and triplet state energies (see below and Section 14.4), so it does not seem that one can ignore spin contamination as an issue entirely.

Indeed, if we have a situation where a higher-spin state is lower in energy than a corresponding lower-spin state generated by a single spin flip (e.g., triplet and singlet), it is almost always the case that a wave function corresponding to the higher-energy lower-spin state will be unstable to symmetry breaking that can mix in character of the lower-energy higher-spin state. A good example is phenylnitrene (PhN) whose triplet ground state and first three singlet excited states are depicted in Figure 8.3. To properly model the S2 state, which is closed-shell in character, we must use a restricted DFT formalism, and when this is done the resulting state energy splitting agrees well with experiment (Johnson and Cramer 2001). Note that the S3 state cannot be handled by standard DFT formalisms because, since it has the same spatial symmetry as the S2 state, restricted DFT variationally collapses to the latter. As for the broken-symmetry approach, when that is applied to a system having an equal number of α and β electrons, a KS wave function is obtained having an $\langle S^2 \rangle$ value of 1.0 which, as mentioned above, is best interpreted as a 50:50 mixture of T0 and S1.

An important point with respect to phenylnitrene is that the S1 state is not multideterminantal because of the mixing of different electronic configurations having similar weights. Instead, it is *intrinsically* two-determinantal because it is an open-shell singlet. In order to better accommodate systems like this, some efforts have been undertaken to develop multideterminantal DFT formalisms. Gräfenstein, Kraka, and Cremer (1998) have proposed a restricted open-shell singlet (ROSS) methodology specifically for such states, while Filatov and Shaik (1998) have advanced a more general MCSCF-like formalism named restricted open-shell Kohn–Sham (ROKS) theory. Khait and Hoffman (2004) have also described a general multireference spin-adapted DFT. The details of these methods are sufficiently

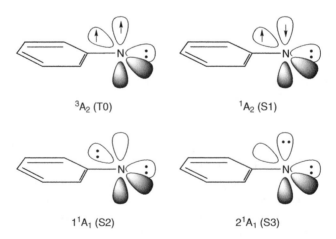

Figure 8.3 Configuration cartoons for the ground-state triplet and first three singlet states of phenylnitrene. Note that the cartoon for S1 glosses over its two-determinantal character

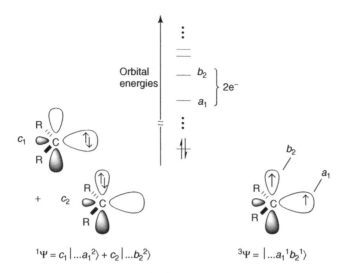

Figure 8.4 Illustration of the two-configurational character of singlet carbenes compared to their triplet congeners. Single-determinantal methodologies are entirely appropriate for the triplet, but begin to fail for the singlets as the weights c_1 and c_2 grow closer to one another. Empirically, KS-DFT methods are less sensitive to this instability than HF, but they ultimately fail as well, at least within a restricted formalism

technical and the range of their applications still sufficiently narrow that we will not discuss them at any more length.

As a practical point, however, returning to non-dynamical correlation as it gives rise to wave functions best described in MO theory as a combination of configurations, DFT seems to be less sensitive to this issue than HF theory. Thus, for example, as has already been mentioned, HF theory usually does very poorly with singlet–triplet splittings in carbenes and isoelectronic analogs because the singlet states are best described as mixtures of two (or more) configurations (Figure 8.4). Speaking very roughly, the weights of the two dominant configurations are in the range of 85:15 for some of the more typical cases. Interestingly, DFT usually does very well indeed for these same systems, and it is not until the weights of the individual configurations become much nearer one another (e.g., the 60:40 balance for p-benzyne discussed above) that DFT too begins to badly underestimate the stability of the singlet. In one case, the ring opening of methylenecyclopropane to closed-shell singlet trimethylenemethane (see Section 7.1) was examined and the BP86/cc-pVDZ level of theory was found to be accurate up to about a 75:25 mixing of the dominant configurations for TMM, at which point it began to diverge from a multireference description of the reaction coordinate (Cramer and Smith 1996).

Note that the generally lower sensitivity of DFT to multireference character is again dependent on the amount of HF exchange included in the functional. Pure DFT functionals seem to be most robust, and inclusion of HF exchange introduces a proportional degree of instability for systems with multiconfigurational character. Thus, the generally higher accuracy of the ACM methods can fail to extend to systems where they are methodologically less stable owing to their Hartree–Fock component.

8.5.4   Systematic Improvability

In molecular orbital theory, there is a clear and well defined path to the exact solution of the Schrödinger equation. All we need do is express our wave function as a linear combination of all possible configurations (full CI) and choose a basis set that is infinite in size, and we have arrived. While such a goal is essentially never practicable, at least the path to it can be followed unambiguously until computational resources fail.

With density functional theory, the situation is much less clear when it comes to evaluating how to do a 'better' calculation. One thing that seems fairly clear is that, as a general rule, results from MGGA functionals tend to improve on those from GGA functionals, which in turn drastically improve on those from LSDA. Somewhat less clear is the status of hybrid functionals. The best ones are competitive in quality with the best MGGA functionals (and B3LYP seems to continue to be the 'magic' functional in project after project) subject to the caveat that in certain situations the presence of HF exchange may cause problems that are associated with the single-determinant KS formalism to become manifest more quickly. As for basis-set effects, just as with MO theory one *can* examine convergence with respect to basis-set size, but there is no guarantee that this may not lead to *increased* errors since errors associated with basis-set incompleteness may offset errors associated with approximate functionals.

All that being said, experience dictates that, across a surprisingly wide variety of systems, DFT tends to be remarkably robust. Thus, unless a problem falls into one of a few classes of well characterized problems for DFT, there is good reason to be optimistic about any particular calculation.

Finally, it seems clear that routes to further improve DFT must be associated with better defining hole functions in arbitrary systems. In particular, the current generation of functionals has reached a point where finding efficient algorithms for correction of the classical self-interaction error are likely to have the largest qualitative (and quantitative) impact.

8.5.5   Worst-case Scenarios

Certain failures of modern DFT should be anticipated, and others are readily explained after some thought about the forms of current functionals. One clear problem with modern functionals is that they make the energy a function entirely of the local density and possibly the density gradient. As such, they are incapable of properly describing London dispersion forces, which, as noted in Section 2.2.4, derive entirely from electron correlation at 'long range'. Adding HF exchange to the DFT functional cannot entirely alleviate this problem, since the HF level of theory, while non-local, does not account in any way for opposite-spin electron correlation.

So, even though noble-gas dimers like He_2, Ne_2, etc., exhibit potential energy minima at van der Waals contact, DFT predicts the potential energy curve for these diatomics to be purely repulsive, at least as long as saturated basis sets are used. If an incomplete basis set is used, it is possible for BSSE to introduce a spurious minimum in the association curve at about the right position, but this is purely fortuitous – the physics of dispersion is simply not included in the functional(s). This is an area of active developmental research (see, for instance, Lein, Dobson, and Gross 1999) and indeed Adamo and Barone (2002) have reported that *m*PBE does reasonably well for noble-gas-dimer geometries and energies with saturated basis sets, and Xu and Goddard (2004a) have shown that XLYP and X3LYP give reasonable results for the dimers of He and Ne.

Other problems with non-bonded complexes have also been documented with DFT. In hydrogen bonded systems, heavy-atom–heavy-atom distances tend to be rather variable as a function of functional. Hobza and Sponer (1999) have examined a large number of nucleic acid base pairs and found, by comparison to X-ray crystal structures and high levels of correlated MO theory, that heavy-atom–heavy-atom distances predicted by GGA DFT functionals are typically too short by about 0.1 Å. Nevertheless, interaction energies are often reasonably well predicted at these levels. Critical to accurate prediction, however, is that a basis set including diffuse functions be employed, as large errors can otherwise be observed, particularly for intramolecular hydrogen bonds (an interesting comparison is provided by Ma, Schaefer, and Allinger 1998 and Lii, Ma, and Allinger 1999; see also Lynch, Zhao, and Truhlar 2003). Staroverov *et al.* (2003) have provided an analysis of 16 different functionals for the energetics and geometrics of 11 hydrogen-bonded systems with the 6-311++G(3df,3pd) basis set; their results suggest that B3LYP and TPSS are both fairly robust, with mean unsigned errors for dissociation of about 0.5 kcal mol^{-1} and for H-bond lengths of about 0.02 Å. Xu and Goddard (2004a, 2004b) observe similarly good performance for X3LYP applied to the water dimer.

Somewhat more problematic are intermolecular complexes bound together by charge transfer interactions. Modern DFT functionals have a tendency to predict such interactions to be stronger than they should be. Thus, Ruiz, Salahub, and Vela (1995) showed that some pure DFT functionals overestimated the binding of ethylene and molecular fluorine by as much as 20 kcal mol^{-1}. Including HF exchange in the functional alleviates the problem to some extent, but only by cancellation of errors, since HF theory incorrectly predicts the interaction between ethylene and molecular fluorine to be purely repulsive. Dative bonds have also been found to be problematic for many functionals. Gilbert (2004) found that heterolytic B−N bond dissociation energies in amine-boranes were underestimated by standard functionals, and that inclusion of a substantial fraction of HF exchange (e.g., as in MPW1K) was required to improve agreement with experiment for this quantity.

A problem with DFT that is not restricted to intermolecular complexes is what might be called 'overdelocalization'. In part because of problems in correcting for the classical self-interaction energy, many functionals overstabilize systems having more highly delocalized densities over more localized alternatives. Such an imbalance can lead to erroneous predictions of higher symmetry structures being preferred over lower symmetry ones, as has been observed, for instance, for phosphoranyl radical structures (Lim *et al.* 1996), transition-state structures for cationic [4+3] cycloadditions (Cramer and Barrows 1998), and in the comparison of cumulenes to poly-ynes (Woodcock, Schaefer, and Schreiner 2002). It can also lead to very poor predictions along coordinates for bond dissociation (Bally and Sastry 1997; Zhang and Wang 1998; Gräfenstein, Kraka, and Cremer 2004), nucleophilic substitution (Adamo and Barone 1998; Gritsenko *et al.* 2000), competing cycloaddition pathways (Jones *et al.* 2002), and rotation about single bonds in conjugated systems, like the benzylic bond in styrene (Choi, Kertesz, and Karpfen 1997).

Note that, because electron correlation often stabilizes delocalized electronic structures over localized ones, HF theory tends to be inaccurate for such systems in the opposite direction from DFT, and thus, again, hybrid ACM functionals tend to show improved performance by an offsetting of errors.

A number of different methods have been proposed to introduce a self-interaction correction into the Kohn–Sham formalism (Perdew and Zunger 1981; Kümmel and Perdew 2003; Gräfenstein, Kraka, and Cremer 2004). This correction is particularly useful in situations with odd numbers of electrons distributed over more than one atom, e.g., in transition-state structures (Patchkovskii and Ziegler 2002). Unfortunately, the correction introduces an additional level of self-consistency into the KS SCF process because it depends on the KS orbitals, and it tends to be difficult and time-consuming to converge the relevant equations. However, future developments in non-local correlation functionals may be able to correct for self-interaction error in a more efficient manner.

8.6 General Performance Overview of DFT

While the cases noted in the immediately preceding section illustrate certain pathological failures of current DFT functionals, the general picture for DFT is really quite bright. For the 'average' problem, DFT is the method of choice to achieve a particular level of accuracy at lowest cost. With the appearance of each new functional, there has tended to be at least one paper benchmarking the performance of that functional on a variety of standard test sets (for energies, structures, etc.) and there is now a rather large body of data that is somewhat scattered and disjoint with respect to individual functional performance. The comparisons made below are designed to provide as broad a coverage as possible without becoming unwieldy, and as such are not necessarily exhaustive.

8.6.1 Energetics

Exact DFT is an *ab initio* theory (even if most modern implementations may be regarded as having a semiempirical flavor) and like other such theories its quality with respect to energetic predictions is usually judged based on its performance for atomization energies. Table 8.1 collects average unsigned and maximum absolute errors in atomization energies as computed for various functionals, and for some other computational methodologies, over several different test sets of increasing complexity. The G2/97 and G3/99 test sets (columns D and E) include substituted hydrocarbons, radicals, inorganic hydrides, unsaturated ring hydrocarbons, and polyhalogenated organics and inorganics. While effort is made to describe all levels of theory accurately, it should be noted that in many cases geometries are optimized (and zero-point vibrational energies computed) using a basis set smaller than that used to compute the atomization energies. In addition, some results for the larger test sets include

Table 8.1 Mean absolute errors and maximum errors, where available, in atomization energies for different methods over several different test sets (kcal mol^{-1})

Level of theory	Test sets[a]				
	A	B	C	D	E
MO theoretical methods					
MNDO				9.3	
				(116.7)	
AM1				7.8	
				(58.2)	
PM3				7.0	
				(32.2)	
MNDO/d				7.3	
				(33.9)	
HF/6-31G(d)	85.9	80.5	150.6[b]		
		(184.3)			
HF/6-31G(d,p)			119.2		
HF/6-311G(2df,p)	82.0		146.2[b]		
HF/6-311+G(3df,2p)		74.5	144.4[b]	148.3[c]	211.5[c]
		(170.0)		(344.1)	(582.2)
MP2/6-31G(d)	22.4	16.0	38.1[b]		
		(40.3)			
MP2/6-31G(d,p)	23.7		22.0		
MP2/6-311+G(3df,2p)		7.3	9.7[b]		
		(25.4)			
MP4/6-31G(2df,p)			13.5[b]		
QCISD/6-31G(d)	28.8		51.7[b]		
MC-QCISD			1.7[b]		
CCSD(T)/6-311G(2df,p)	11.5				
G2(MP2)				1.8	
				(8.8)	
CBS-4				2.7	
				(12.9)	
G2	1.2	1.2		1.4	
		(4.9)		(10.6)	
CBS-q				2.3	
				(11.4)	
CBS-Q		1.0	1.4[b]	1.2	
		(3.3)		(8.1)	
G3(MP2)				1.2	1.2
					(9.3)
G3		1.1	0.9[b]	0.9	1.1
		(4.0)		(4.9)	(7.1)
G3S(MP2)				1.2	1.3
G3S			0.9[b]	1.0	1.1
MCG3			1.0[b]		

(continued overleaf)

Table 8.1 (*continued*)

Level of theory	Test sets[a]				
	A	B	C	D	E
W1		0.6			
		(2.2)			
W2		0.5			
		(1.9)			
LSDA functionals					
SVWN/6-31G(d)	35.7	36.4		83.8	
		(84.0)			
SVWN/6-31G(d,p)			52.2		
SVWN/TZ2P			50.1		
SVWN/6-311+G(2df,p)	43.5			83.7[c]	121.9[c]
				(207.7)	(347.5)
GGA and MGGA functionals					
BB95/numeric		8.6			
		(28.6)			
BISM/6-311+G(3df,2p)		5.9			
		(21.8)			
BVWN/6-31G(d)	4.4				
BKCIS/6-311+G(3df,2p)		8.2		14.6	
		(25.9)		(39.6)	
BLYP/6-31G(d)	5.6	5.3			
		(18.8)			
BLYP/6-31G(d,p)			7.0		
BLYP/6-31+G(d)		4.4			
		(16.3)			
BLYP/6-311G(2df,p)	9.6				
BLYP/6-311+G(3df,2p)		5.0		7.3	9.3
		(15.8)		(28.4)	
BLYP/6-311++G(3df,3pd)	9.6			7.3[c]	9.5[c]
				(28.1)	(41.0)
BLYP/pVQZ			7.0[d]		
			(27.8)		
BP86/6-31G(d)		7.2			
		(24.0)			
BP86/6-311+G(3df,2p)		10.3		20.1[c]	26.3[c]
		(25.4)		(48.7)	(72.7)
BPW91/6-31G(d,p)			7.4		
BPW91/TZ2P			7.3		
BPW91/6-311+G(3df,2p)				7.8	
				(32.2)	
BPW91/6-311++G(3df,3pd)	6.0			8.0[c]	9.0[c]
				(32.4)	(28.0)
EDF1/6-31+G(d)		3.2			
		(15.3)			

Table 8.1   (*continued*)

Level of theory	Test sets[a]				
	A	B	C	D	E
HCTH/6-311++G(3df,3pd)				5.6[c]	7.2[c]
				(28.0)	(28.0)
OLYP/pVQZ			5.5[d]	4.8[c]	5.9[c]
			(23.6)	(27.0)	(27.0)
PBE/6-311+G(3df,2p)		8.2		16.9[c]	22.2[c]
		(29.1)		(50.5)	(79.7)
*m*PBE/6-311++G(3df,3pd)		4.6		6.3	
		(18.9)		(27.5)	
PBEKCIS/6-311+G(3df,2p)		11.9		24.9	
		(35.2)		(66.5)	
PKZB/6-311+G(3df,2p)		3.6		4.8[c]	7.0[c]
		(11.0)		(39.8)	(39.8)
PKZBKCIS/6-311+G(3df,2p)		4.1		9.7	
		(14.4)		(38.8)	
PWPW91/6-311++G(3df,3pd)	8.6			17.7[c]	23.6[c]
				(52.7)	(81.1)
*m*PWPW91/6-311++G(3df,3pd)	6.7			15.0	
TPSS/6-311++G(3df,3pd)				6.0[c]	5.8[c]
				(22.9)	(22.9)
VSXC/6-311+G(3df,2p)		2.5		2.8[c]	3.5[c]
		(10.0)		(11.5)	(12.0)
XLYP/6-311+G(3df,2p)				7.6[e]	
Hybrid functionals					
BH&HLYP/6-311++G(3df,3pd)	11.7			21.7[e]	
	(23.4)				
B0KCIS/6-311+G(3df,2p)		3.0		5.3	
		(10.5)		(28.6)	
B1B95/numeric		2.0			
		(7.5)			
B1LYP/6-311++G(3df,3pd)	3.1				
	(12.2)				
B1PW91/6-311++G(3df,3pd)	5.4				
	(14.3)				
MPW1K/MG3S			11.0[b]		
PBE1PBE/6-311+G(3df,2p)		3.1	4.4[b]	4.9[c]	6.7[c]
		(10.7)		(21.3)	(35.6)
PW1PW91/6-311+G(3df,2p)				5.3[e]	
*m*PW1PW91/6-311++G(3df,3pd)	3.5		4.2[b]	3.9[e]	
	(6.7)				
TPSSh/6-311++G(3df,3pd)				4.2[c]	3.9[c]
				(22.0)	(16.2)

(continued overleaf)

Table 8.1   (*continued*)

Level of theory	Test sets[a]				
	A	B	C	D	E
B97/numeric		1.8 (5.5)			
B97-1/pVQZ			3.2[d] (9.8)		
B3LYP/6-31G(d)		5.2 (31.5)	6.8		
B3LYP/6-31+G(d)		5.9 (35.9)			
B3LYP/6-311+G(3df,2p)		2.4 (9.9)		3.1 (20.2)	4.8 (21.6)
B3LYP/aug-cc-pVTZ		2.6 (18.2)			
B3LYP/6-311++G(3df,3pd)	3.3		4.2[b]	3.1[c] (20.1)	4.9[c] (20.8)
B3LYP/pVQZ			3.4[d] (22.2)		
B3LYP*/TZVPP		3.0 (11.1)			
B3P86/6-31G(d)		5.9 (22.6)			
B3P86/6-311+G(3df,2p)		7.8 (22.7)		18.2[c] (48.1)	26.1[c] (79.2)
B3PW91/6-31G(d,p)			6.8		
B3PW91/TZ2P			6.5		
B3PW91/6-311+G(3df,2p)				3.5 (21.8)	
B3PW91/6-311++G(3df,3pd)	4.8			3.4[c] (21.6)	3.9[c] (21.6)
O3LYP/pVQZ			3.9[d] (11.8)	4.2[e]	
*m*PW3PW91/6-311++G(3df,3pd)	2.7 (7.4)				
X3LYP/6-311+G(3df,2p)				2.8[e]	

[a] A: G2 subset (32 molecules containing only first-row atoms, see Johnson, Gill, and Pople 1993); B: G2 set (55 molecules including first- and second-row atoms, see Curtiss *et al.* 1991); C: (108 molecules including first- and second-row atoms, see Scheiner, Baker, and Andzelm 1997); D: G2/97 set (148 molecules including first- and second-row atoms, see Curtiss *et al.* 1998 and also http://chemistry.anl.gov/compmat/g3theory.htm); E: G3/99 set (223 molecules including first- and second-row atoms, larger organics, and problematic inorganic molecules, see Curtiss *et al.* 2000).
[b] Zero-point-exclusive atomization energies for 109 molecules having high overlap with test set C, MG3S basis set for DFT values, see Lynch and Truhlar 2003a.
[c] Geometries and ZPVE obtained at the B3LYP/6-31G(2df,p) level with scaling of vibrational frequencies by 0.9854; electronic energies use 6-311++G(3df,3pd) basis set; see Staroverov *et al.* (2003).
[d] 105 molecules having high overlap with test set C, see Hoe, Cohen, and Handy (2001).
[e] MP2 geometries and scaled HF thermal contributions; electronic energies use 6-311+G(3df,2p) basis set; see Xu and Goddard (2004a, 2004b).

spin-orbit corrections and updated experimental data for the heats of formation of the silicon and beryllium atoms and carbonyl difluoride, while others do not. Readers may refer to the original literature listed in the bibliography for full details.

Some key points may be inferred from Table 8.1.

1. For a given average level of accuracy, hybrid and meta-GGA DFT methods are obviously the most efficient, showing mean unsigned errors almost equal in quality to the much more expensive multilevel correlated methods. However, the maximum absolute errors are larger with the former methods than the latter, indicating a slightly lower generality even in the most current generation of functionals.

2. Hybrid and meta-GGA DFT functionals usually offer some improvement over corresponding pure DFT functionals.

3. Increasing basis-set size does not always improve the accuracy of the DFT models, although it must, of course, ultimately lead to a converged prediction.

4. Of the currently available DFT models, it is clear that GGA models offer a major improvement over the older LSDA model. It further appears that the P86 functional should be avoided. With respect to a more specific ranking of functionals, Boese, Martin and Handy (2003) have carried out a careful evaluation of a large number of functionals for various energetic quantities (atomization enthalpies, IPs, EAs, geometries, etc.) using polarized triple-ζ basis sets. They conclude that for pure GGA functionals an error ranking (i.e., lower is better) is HCTH < OLYP < BPW91 < BLYP < mPWPW91 < BP86 < PWPW91 < PBE. They further conclude that for hybrid and meta-GGA functionals, which are overall better than pure functionals, an analogous error ranking is hybrid τ-HCTH < B97-1 < τ-HCTH < VSXC ~ B97-2 ~ B3LYP ~ B98 < PBE1PBE << MPW1K < PKZB (note that these comparisons do *not* take into account predicted reaction activation enthalpies). Prior work by Proynov, Chermette, and Salahub (2000) suggests that Bmτ1 and BLap3 have qualities intermediate between B3LYP and PBE1PBE in the latter rank ordering.

While fewer data are available, the utility of DFT in computing the bond strengths between transition metals and hydrides, methyl groups, and methylene groups has also been demonstrated (Table 8.2). Because of the non-dynamical correlation problem associated with the partially filled metal d orbitals, such binding energies are usually very poorly predicted by MO theory methods, until quite high levels are used to account for electron correlation.

In the area of reaction energetics, Baker, Muir, and Andzelm have compared six levels of theory for the enthalpies of forward activation and reaction for 12 organic reactions: the unimolecular rearrangements vinyl alcohol $\rightarrow$ acetaldehyde, cyclobutene $\rightarrow$ *s-trans* butadiene, *s-cis* butadiene $\rightarrow$ *s-trans* butadiene, and cyclopropyl radical $\rightarrow$ allyl radical; the unimolecular decompositions tetrazine $\rightarrow$ 2HCN + N_2 and trifluoromethanol $\rightarrow$ carbonyl difluoride + HF; the bimolecular condensation reactions butadiene + ethylene $\rightarrow$ cyclohexene (the Diels–Alder reaction), methyl radical + ethylene $\rightarrow$ propyl radical, and methyl radical + formaldehyde $\rightarrow$ ethoxyl radical; and the bimolecular exchange reactions FO + H_2 $\rightarrow$ FOH + H,   HO + H_2 $\rightarrow$ H_2O + H,   and   H + acetylene $\rightarrow$ H_2 + HC_2. Their results are summarized in Table 8.3 (Reaction Set 1). One feature noted by these authors is

Table 8.2 Mean absolute errors in metal–ligand binding energies for different methods (kcal mol^{-1})[a]

Level of theory[b]	Ligand		
	H	CH$_3$	CH$_2$
MO theoretical methods			
MCPF	6	9	20
QCISD(T)		6	
PCI-80	2	2	4
Density functional methods			
SVWN	12		
BP86	8		
BH&HLYP			16
BH&HLYP/ECP		5	9
B3LYP	5	6	4
B3LYP/ECP		9	7

[a] Complexes MX$^+$ where M = Sc, Ti, V, Cr, Mn, Fe, Co, Ni, Cu, and X = ligand.
[b] Basis sets are for the most part double-ζ polarized in quality. Use of metal effective core potentials is indicated by ECP.

that the pure DFT functional BLYP badly underestimated the activation enthalpies for the H-atom transfer reactions. This behavior has since been noted by many other authors, and is part of the motivation for the MPW1K model discussed in Section 8.4.3.

In a different analysis of hydrogen-atom transfer reaction barriers, Lynch and Truhlar (2003b) considered the forward and reverse barriers for 22 such reactions, and then demonstrated that the performance of various models on only six data (the forward and reverse barrier heights for HO + CH$_4$ → H$_2$O + CH$_3$, H + OH → O + H$_2$, and H + H$_2$S → H$_2$ + HS) was highly predictive of their accuracy for all 44 barrier heights. In this and subsequent work by Zhao *et al.* (2004), they assayed the accuracy of a substantial number of different models for this test set and found that modern functionals, and particularly those including a dependence on the kinetic energy density, gave the best results (Table 8.3, Reaction Set 2). However, all of the geometries used in this analysis were obtained at the QCISD/MG3 level (partly in recognition of the multideterminantal character of many H-atom transition-state structures). So, the reported errors might be expected to change somewhat were geometries to be located at the DFT levels of theory. In this regard, it is noteworthy that Kang and Musgrave (2001) examined 29 different barrier heights for H-atom transfer using geometries optimized at the same level as used to compute energies, and reported mean unsigned errors of 0.9, 3.3, 3.2, and 2.9 kcal mol^{-1}, respectively, for KMLYP, B3LYP, BH&HLYP, and G2.

With respect to chemical reactions not involving H-atom transfer, Guner *et al.* (2003, 2004) examined a set of 11 pericyclic organic reactions having experimental data available for nine enthalpies of activation and six enthalpies of forward reaction. They found the B3LYP and MPW1K functionals to be about as accurate for the activation enthalpies as the more expensive CBS-QB3 and CASPT2 levels, with other functionals doing somewhat less well (Table 8.3, Reaction Set 3). Interestingly, they found the performance of the B3LYP

Table 8.3 Mean and maximum absolute errors (kcal mol^{-1}) in enthalpies of activation and forward reaction for different methods

Level of theory	Activation		Reaction	
	Mean	Maximum	Mean	Maximum
Reaction set 1[a,b]				
MNDO	23.4	51.8	10.9	57.7
AM1	9.3	34.2	7.5	22.1
HF/6-31G(d)[c]	13.6	30.6	10.5	24.8
MP2	9.9	28.8	6.3	26.0
BLYP	5.9	21.9	5.9	13.0
BLYP/6-311G(2df,2pd)			5.8	16.0
B3LYP			5.0	9.4
B3LYP/6-311G(2df,2pd)			3.7	8.5
B3PW91	3.7	12.9	6.8	17.7
OLYP			4.6	12.7
OLYP/6-311G(2df,2pd)			3.6	10.6
O3LYP			4.6	11.6
O3LYP/6-311G(2df,2pd)			2.8	7.1
Reaction set 2[d]				
HF	12.4		149.5	
MP2	5.5		24.4	
QCISD	3.9		38.6	
QCISD(T)	3.1		32.3	
CBS-Q	0.8		1.3	
MC-QCISD	0.9		1.1	
MCG3	0.8		0.8	
BB95	8.3		8.2	
BLYP	8.3		6.8	
BP86	9.4		13.0	
G96LYP	6.9		11.1	
HCTH	5.4		5.3	
*m*PWPW91	8.6		7.2	
OLYP	6.0		3.4	
PBE	9.5		12.1	
PWPW91	9.8		12.1	
VSXC	5.1		3.4	
B1B95	3.7		3.4	
B97-1	4.2		5.8	
B97-2	3.1		3.9	
B98	4.1		6.3	
KMLYP	2.9		1.0	
*m*PW1PW91	3.9		8.0	
MPW1K	1.4		14.9	
PBE1PBE	4.6		7.1	
B3LYP	5.0		7.2	

(*continued overleaf*)

Table 8.3   (*continued*)

Level of theory	Activation		Reaction	
	Mean	Maximum	Mean	Maximum
B3PW91	4.4		5.8	
O3LYP	4.7		5.8	
Reaction set 3[b,e]				
HF	18.7	26.7	3.8	6.5
CASSCF	16.0	34.6	14.7	20.6
MP2	4.6	7.6	6.0	9.6
CASPT2//CASSCF	2.4	5.7	1.6	4.5
CBS-QB3	1.9	4.3	1.6	2.5
BPW91	3.7	6.9	3.4	7.4
KMLYP	3.2	10.3	12.7	19.8
OLYP	3.4	9.0	6.2	12.9
OLYP/6-311+G(2d,p)	4.4	13.0	9.8	20.5
MPW1K/6-31+G(d,p)	2.2	6.9	6.2	10.0
B3LYP	1.7	6.0	4.1	8.6
B3LYP/6-31+G(d,p)	2.4	8.1	7.0	13.6
B3LYP/6-311+G(2d,p)	2.9	10.1	8.2	15.9
O3LYP//OLYP	3.0	9.0	3.9	8.3

[a] See Baker, Muir, and Andzelm (1995).
[b] 6-31G(d) basis set unless otherwise indicated.
[c] Using five spherical d functions instead of the usual six Cartesian functions implied by this basis set name.
[d] See Lynch and Truhlar (2003a) and Zhao *et al.* (2004); 6-31+G(d,p) basis set; the Reaction column refers to the atomization enthalpies for six molecules chosen to be representative of a larger set in a fashion analogous to the H-atom transfer reactions, namely, SiO, S_2, silane, propyne, glyoxal, and cyclobutane.
[e] See Guner *et al.* (2003, 2004).

functional to become systematically worse with increasing basis-set size, so some cancellation of errors appears to be responsible for the excellent performance of this functional when used with the 6-31G(d) basis set. This analysis is supported as well by the relatively poor performance of B3LYP for the forward reaction enthalpies. Indeed, although the best MO methods continue to perform well for this latter test, the best DFT functional was found to be BPW91, which was the worst for the activation enthalpies.

St-Amant *et al.* (1995) have analyzed the utility of HF, MP2, LSDA, and BP86, using basis sets of DZP to TZP quality, for the prediction of 35 conformational energy differences in small to medium-sized organic molecules (e.g., axial–equatorial disposition of substituents on cyclohexanes, E vs. Z amide rotamers, *s-cis* vs. *s-trans* rotamers of carboxylic acids and conjugated systems, etc.) The mean unsigned errors over the data for these four methods are 0.5, 0.4, 0.6, and 0.3 kcal mol^{-1}, respectively; the average conformational energy difference is 1.6 kcal mol^{-1}.

A substantial body of data exists evaluating the utility of DFT (and other methods) for computing ionization potentials and electron affinities following a ΔSCF approach. These data are summarized over four different test sets in Table 8.4. The conclusions one may

Table 8.4 Mean absolute errors and maximum errors, where available, in IPs and EAs for different methods over several different test sets (eV)

Level of theory	IP test sets		EA test sets	
	G2[a]	G2/97[b]	G2	G2/97
MO theoretical methods				
HF		1.03[c]		1.10[c]
		(2.60)		(2.21)
G2(MP2)	0.1		0.1	
G2	0.06	0.06	0.06	0.06
	(0.19)	(0.19)	(0.14)	(0.17)
CBS-QB3	0.05		0.05	
	(0.12)		(0.12)	
G3	0.04	0.04	0.05	0.04
	(0.18)	(0.18)	(0.18)	(0.18)
W1	0.01	0.02	0.02	0.02
	(0.06)	(0.13)	(0.05)	(0.08)
W2	0.01		0.01	
	(0.05)		(0.04)	
LSDA functionals				
SVWN/6-311+G(2df,p)	0.7	0.6	0.7	0.7
	(1.2)	(1.7)	(1.2)	(1.3)
SVWN/aug-cc-pVTZ	0.7		0.8	
SVWN5/6-311+G(2df,p)	0.2	0.23[c]	0.3	0.24[c]
	(0.6)	(1.18)	(0.7)	(0.88)
GGA and MGGA functionals				
BB95/numeric	0.2			
	(0.5)			
BLYP/6-311+G(3df,2p)	0.19	0.29[c]	0.11	0.12[c]
	(0.6)	(1.06)	(0.4)	(0.70)
BLYP/aug-cc-pVTZ	0.2		0.1	
BP86/6-311+G(3df,2p)	0.18	0.22[c]	0.21	0.19[c]
		(1.21)		(0.89)
BP86/aug-cc-pVTZ	0.2		0.2	
BPW91/6-311+G(3df,2p)	0.16	0.24[c]	0.09	0.12[c]
		(1.14)		(0.78)
HCTH/6-311++G(3df,3pd)		0.23[c]		
		(1.34)		
OLYP/6-311++G(3df,3pd)		0.29[c]		0.15[c]
		(1.11)		(0.60)
PBE/6-311+G(3df,2p)	0.16	0.24[c]	0.12	0.12[c]
	(0.5)	(1.11)	(0.3)	(0.78)
PKZB/6-311++G(3df,3pd)		0.31[c]		0.15[c]
		(1.31)		(0.57)
PWPW91/6-311++G(3df,3pd)	0.16	0.22[c]	0.14	
		(1.19)		

(*continued overleaf*)

Table 8.4 (*continued*)

Level of theory	IP test sets		EA test sets	
	G2[a]	G2/97[b]	G2	G2/97
mPWPW91/6-311+G(3df,2p)	0.16		0.12	
TPSS/6-311++G(3df,3pd)		0.24[c]		0.14[c]
		(1.22)		(0.82)
VSXC/6-311+G(3df,2p)	0.1	0.23[c]		0.13[c]
	(0.4)	(1.20)		(0.78)
XLYP/6-311+G(3df,2p)	0.19		0.12	
Hybrid functionals				
BH&HLYP/6-311+G(3df,2p)	0.21		0.25	
B1B95/numeric	0.1			
	(0.4)			
B97/numeric	0.1		0.1	
	(0.6)		(0.4)	
B3LYP/cc-pVDZ	0.2			
B3LYP/aug-cc-pVDZ			0.2	
B3LYP/6-31+G(d)		0.2		0.2
B3LYP/6-311+G(3df,2p)	0.17	0.18[c]	0.10	0.12[c]
	(0.8)	(1.57)	(0.5)	(1.10)
B3LYP/aug-cc-pVTZ	0.2		0.1	
B3LYP*/TZVPP	0.4		0.1	
	(0.7)		(0.4)	
B3P86/6-311+G(3df,2p)	0.64	0.55[c]	0.59	0.59[c]
		(2.13)		(1.63)
B3PW91/6-311+G(3df,2p)	0.16	0.19[c]	0.10	0.14[c]
		(1.58)		(1.08)
O3LYP/6-311+G(3df,2p)	0.14		0.11	
PBE1PBE/6-311+G(3df,2p)	0.16	0.20[c]	0.13	0.16[c]
	(0.7)	(1.61)	(0.3)	(1.09)
PW1PW91/6-311+G(3df,2p)	0.16		0.11	
mPW1PW91/6-311+G(3df,2p)	0.16		0.11	
TPSSh/6-311++G(3df,3pd)		0.23[c]		0.16[c]
		(1.41)		(0.95)
X3LYP/6-311+G(3df,2p)	0.15		0.09	

[a]38 IPs and 25 EAs, see Curtiss *et al.* (1991).
[b]83 IPs and 58 EAs, see Curtiss *et al.* (1998) and also http://chemistry.anl.gov/compmat/g3theory.htm.
[c]IP set includes three additional data for toluene, aniline, and phenol; geometries and ZPVE obtained at the B3LYP/6-31G(2df,p) level with scaling of vibrational frequencies by 0.9854; electronic energies use 6-311++ G(3df,3pd) basis set; see Staroverov *et al.* (2003).

draw from the table are for the most part similar to those noted above based on atomization energies, except that there is now much less, if any, preference for hybrid functionals over pure functionals, so long as P86 is avoided. In analyses of other systems, including metal atoms, Rienstra-Kiracofe *et al.* (2002) and Bauschlicher and Gutsev (2002) have separately noted that B3LYP with large basis sets seems to be particularly robust.

Finally, atomic and molecular proton affinities (PAs) have also been evaluated for various functionals for ammonia, water, acetylene, silane, phosphine, silylene, hydrochloric acid, and molecular hydrogen. For G2 and G3 theories, the mean unsigned error in PAs is 1.1 and 1.3 kcal mol^{-1}, respectively. At the SVWN, BLYP, BP86, BPW91, B3LYP, B3P86, and B3PW91 levels (using the 6-311+G(3df,2p) basis set), the corresponding errors are 5.8, 1.8, 1.5, 1.5, 1.3, 1.1, and 1.2 kcal mol^{-1}, respectively (quantitatively similar results have also been obtained with more modern functionals). The much cheaper hybrid DFT methods are thus entirely competitive with G2 and G3, although the data set is perhaps too small to come to any firm conclusions on this topic (cf. Pokon *et al.* 2001).

8.6.2 Geometries

Analytic first derivatives are available for almost all density functionals, and as a result geometry optimization can be carried out with facility. The performance of the various functionals is usually quite good when it comes to predicting minimum energy structures. As summarized in Table 8.5, bond lengths at the LDA level for molecules composed of first- and second-row atoms are typically as good as those predicted from MP2 optimizations, with both these levels being somewhat improved over HF theory. The use of GGA functionals does not usually result in much improvement over the LDA level. However, the GGA functionals tend to systematically overestimate bond lengths. As noted in Section 6.4.2, the HF level tends to systematically *under*estimate bond lengths. Thus, it should come as no surprise that the hybrid ACM functionals, which mix the two, give noticeable improvement in predicted bond lengths (of course, the improvements are on the order of 0.005 Å, and it should be noted that most molecular properties are very little affected by such small variations in bond lengths). Very small improvements in geometrical accuracy are usually noted with increasing basis-set size beyond those listed in Table 8.5. Accuracies in bond angles for all flavors of DFT average about 1°, the same as is found for HF and MP2. Similarly, the limited amount of data available for dihedral angles suggests that HF, MP2, and DFT all perform equivalently in this area.

Table 8.5 also indicates that the most popular functionals fail to be as accurate for molecules containing third-row main-group elements as they are for molecules made up of elements from the first two rows. The LYP correlation functional seems to perform particularly badly, while the PW91 functional is more robust.

It is, however, for the transition metals themselves that DFT has proven to be a tremendous improvement over HF and post-HF methods, particularly for cases where the metal atom is coordinatively unsaturated. The narrow separation between filled and empty d-block orbitals typically leads to enormous non-dynamical correlation problems with an HF treatment, and DFT is much less prone to analogous problems. Even in cases of a saturated coordination sphere, DFT methods typically significantly outperform HF or MP2. Jonas and Thiel (1995) used the BP86 functional to compute geometries for the neutral hexacarbonyl complexes of Cr, Mo, and W, the pentacarbonyl complexes of Fe, Ru, and Os, and the tetracarbonyl

Table 8.5 Mean absolute errors in bond lengths for different methods over several different test sets (Å)

Level of theory	Test sets[a]		
	A	B	C
MO theoretical methods			
HF	0.022	0.021	
MP2	0.014[b]	0.014	0.022
QCISD	0.013[b]		
CCSD(T)	0.005[b]		
LSDA functionals			
SVWN	0.017	0.016	
		0.013[d]	
GGA and MGGA functionals			
BLYP	0.014	0.021	0.048
		0.019[c]	
		0.022[d]	
BP86		0.018[d]	
BPW91	0.014	0.017	0.020
		0.017[d]	
HCTH		0.013[c]	
		0.014[d]	
OLYP		0.018[d]	
PBE	0.012	0.016[d]	
PKZB		0.027[d]	
PWPW91	0.012	0.014[d]	
mPWPW91	0.012		
TPSS		0.014[d]	
VSXC		0.013[d]	
Hybrid functionals			
BH&HLYP	0.015		
B1LYP	0.005		
B1PW91	0.010		
B97-1		0.008[c]	
mPW1PW91	0.010		
PBE1PBE	0.012	0.010[d]	
TPSSh		0.010[d]	
B3LYP	0.004	0.008[c]	0.030
		0.010[d]	

Table 8.5   (*continued*)

Level of theory	Test sets[a]		
	A	B	C
B3P86		0.008^d	
B3PW91	0.008	0.011	0.020
		0.009^d	
*m*PW3PW91	0.008		

[a] A: G2 subset (32 molecules containing only first-row atoms, see Johnson, Gill, and Pople 1993), 6-311G(d,p) basis set unless otherwise specified; B: (108 molecules including first- and second-row atoms, see Scheiner, Baker, and Andzelm 1997), 6-31G(d,p) basis set; C: (40 molecules containing third-row atoms Ga-Kr, see Redfern, Blaudeau, and Curtiss 1997).
[b] 6-31G(d,p) basis set.
[c] A 40-molecule subset with a polarized triple-ζ basis set, see Hamprecht *et al.* (1998).
[d] A 96-molecule set with the 6-311++G(3df,3pd) basis set, see Staroverov *et al.* (2003).

complexes of Ni, Pd, and Pt. Over the 10 unique metal–carbon bond lengths for which experimental data are available, they observed no error in excess of 0.01 Å except for W, where the error was 0.017 Å. At the HF and MP2 levels using equivalent basis sets, the corresponding *average* absolute errors are 0.086 and 0.028 Å, and the maximum deviations are 0.239 and 0.123 Å (Frenking *et al.* 1996).

To the extent DFT shows systematic weaknesses in geometries, it is in those areas where it similarly does poorly for energetics. Thus, van der Waals complexes tend to have interfragment distances that are too large because the dispersion-induced attraction is not properly modeled (although it may accidentally be mimicked by BSSE). Hydrogen bonds are somewhat too short as a rule, and indeed, most charge transfer complexes have their polarities overestimated so that they are too tightly bound. Finally, the tendency noted above in Section 8.5.6 for DFT to overdelocalize structures can show up in geometrical predictions. Thus, for instance, in 1,3-butadiene DFT tends to predict the formal single bond to be a bit too short and the formal double bonds to be somewhat too long (and this extends to other conjugated π systems). As already noted above in Section 8.5.6, this can also lead to a tendency to favor higher symmetry structures over ones of lower symmetry since the former tend to have more highly delocalized frontier orbitals (see also Section 9.1.6). Finally, loose transition state structures can result from this phenomenon; for instance, the C–Cl bond lengths in the TS structure illustrated in Figure 6.12 are 2.45 and 2.39 Å at the BLYP/6-31G(d) and B3LYP/6-31G(d) levels of theory, respectively. Of course, this is a significant improvement over HF theory, and insofar as TS structures tend to be fairly floppy, the remaining geometrical errors may have only small energetic consequences.

Wiest, Montiel, and Houk (1997) have studied carefully a large number of TS structures for organic electrocyclic reactions and, based on comparison to experiment (particularly including kinetic isotope effect studies) and very high levels of electronic structure theory, concluded that the B3LYP functional is particularly robust for predicting geometries in this area. This is consistent with the good behavior of this functional when applied to minimum-energy structures composed only of first-row atoms as already noted above. Cramer and Barrows (1998) have emphasized, however, that overdelocalization problems can arise in ionic examples of such electrocyclic reactions, and caution may be warranted in these instances.

8.6.3 Charge Distributions

Over the 108 molecules in Test Set B of Table 8.5, Scheiner, Baker, and Andzelm computed the mean unsigned errors in predicted dipole moments to be 0.23, 0.20, 0.23, 0.19, and 0.16 D at the HF, MP2, SVWN, BPW91, and B3PW91 levels of theory, respectively, using the 6-31G(d,p) basis set. These results were improved somewhat for the DFT levels of theory when more complete basis sets were employed.

Cohen and Tantirungrotechai (1999) compared HF, MP2, BLYP, and B3LYP to one another with respect to predicting the dipole moments of some very small molecules using a very large basis set, and their results are summarized in Table 8.6. In general the performances of MP2, the pure BLYP functional, and the hybrid B3LYP functional are about equal, although both DFT functionals do very slightly better than MP2 for several cases. HF theory shows its typical roughly 10–15 percent overestimation of dipole moments, and its historically well-known reversal of moment for carbon monoxide.

In addition to the moments of the charge distribution, molecular polarizabilities have also seen a fair degree of study comparing DFT to conventional MO methods. While data on molecular polarizabilities are less widely available, the consensus appears to be that for this property DFT methods, pure or hybrid, fail to do as well as the MP2 level of theory, with conventional functionals typically showing errors only slightly smaller than those predicted by HF (usually about 1 a.u.), while the MP2 level has errors only 25 percent as large. In certain instances, ACM functionals have been more competitive with MP2, but still not quite as good.

Table 8.6 Dipole moments (D) for eight small molecules at four levels of theory using the very large POL basis set[a]

Molecule	HF	MP2	BLYP	B3LYP	Experiment
NH_3	1.62	1.52	1.48	1.52	1.47
H_2O	1.98	1.85	1.80	1.86	1.85
HF	1.92	1.80	1.75	1.80	1.83
PH_3	0.71	0.62	0.59	0.62	0.57
H_2S	1.11	1.03	0.97	1.01	0.97
HCl	1.21	1.14	1.08	1.12	1.11
CO	−0.25	0.31	0.19	0.10	0.11
SO_2	1.99	1.54	1.57	1.67	1.63

[a] From Cohen and Tantirungrotechai 1999.

Table 8.7 Density Functionals[a]

Abbreviation	Comments	Reference(s)
B	Becke's 1988 GGA exchange functional containing one empirical parameter and showing correct asymptotic behavior.	Becke, A. D. 1988. *Phys. Rev. A*, **38**, 3098.
B0KCIS	One-parameter hybrid functional of B and KCIS incorporating 25% HF exchange (B1KCS optimizes the percent HF exchange to 23.9%).	Toulouse, J., Savin, A., and Adamo, C. 2002. *J. Chem. Phys.*, **117**, 10465.
B1B95	One-parameter hybrid functional of B and B95 incorporating 28% HF exchange.	Becke, A. D. 1996. *J. Chem. Phys.*, **104**, 1040.
B1LYP	One-parameter hybrid functional of B and LYP incorporating 25% HF exchange.	Adamo, C. and Barone, V. 1997. *Chem. Phys. Lett.*, **274**, 242.
B1PW91	One-parameter hybrid functional of B and PW91 incorporating 25% HF exchange.	Adamo, C. and Barone, V. 1997. *Chem. Phys. Lett.*, **274**, 242.
B3LYP	ACM functional discussed in more detail in Section 8.4.3.	Stephens, P. J., Devlin, F. J., Chabalowski, C. F., and Frisch, M. J. 1994. *J. Phys. Chem.*, **98**, 623.
B3LYP*	ACM functional discussed in more detail in Section 8.4.3.	Salomon, O., Reiher, M., and Hess, B. A. 2002. *J. Chem. Phys.*, **117**, 4729.
B3PW91	ACM functional discussed in more detail in Section 8.4.3.	Becke, A. D. 1993b. *J. Chem. Phys.*, **98**, 5648.
B86	Becke's 1986 GGA exchange functional.	Becke, A. D. 1986. *J. Chem. Phys.*, **84**, 4524.
B88	Becke's 1988 GGA correlation functional.	Becke, A. D. 1988. *J. Chem. Phys.*, **88**, 1053.
B95	Becke's 1995 (sic) MGGA correlation functional.	Becke, A. D. 1996. *J. Chem. Phys.*, **104**, 1040.
B97	Becke's 1997 GGA exchange-correlation functional containing 10 optimized parameters including incorporating 19.43% HF exchange.	Becke, A. D. 1997 *J. Chem. Phys.*, **107**, 8554.
B97-1	Hamprecht, Cohen, Tozer, and Handy hybrid GGA exchange-correlation functional based on a reoptimization of empirical parameters in B97 and incorporating 21% HF exchange.	Hamprecht, F. A., Cohen, A. J., Tozer, D. J., and Handy, N. C. 1998. *J. Chem. Phys.*, **109**, 6264.
B97-2	Wilson, Bradley, and Tozer's hybrid GGA exchange-correlation functional based on further reoptimization of empirical parameters in B97 and incorporating 21% HF exchange.	Wilson, P. J., Bradley, T. J., and Tozer, D. J. 2001. *J. Chem. Phys.*, **115**, 9233.
B98	Schmider and Becke's 1998 revisions to the B97 hybrid GGA exchange-correlation functional to create a hybrid MGGA incorporating 21.98% HF exchange.	Schmider, H. L. and Becke, A. D. 1998. *J. Chem. Phys.*, **108**, 9624.
BB1K	Optimization of B1B95 primarily for kinetics of H-atom abstractions by using 40% HF exchange instead of default 28%.	Zhao, Y., Lynch, B. J., and Truhlar, D. G. 2004. *J. Phys. Chem. A*, **108**, 2715.

(continued overleaf)

Table 8.7   (*continued*)

Abbreviation	Comments	Reference(s)
Bm	A modification of B88 to optimize its performance with the $\tau1$ correlation functional.	Proynov, E., Chermette, H., and Salahub, D. R. 2000. *J. Chem. Phys.*, **113**, 10013.
BR	Becke and Roussel's 1989 MGGA exchange functional that includes a dependence on the Laplacian of the density in addition to its gradient.	Becke, A. D. and Roussel, M. R. 1989. *Phys. Rev. A*, **39**, 3761.
CAM	Cambridge GGA exchange functional (denoted as either CAM(A) or CAM(B))	Laming, G. J., Termath, V., and Handy, N. C. 1993. *J. Chem. Phys.*, **99**, 8765.
CS	Colle and Salvetti's correlation functional (depending on more than only the density) parameterized to be exact for the He atom.	Colle, R. and Salvetti, O. 1975. *Theor. Chim. Acta*, **37**, 329.
EDF1	Empirical density functional 1 designed as a pure GGA exchange-correlation functional to be used with small basis sets.	Adamson, R. D., Gill, P. M. W., and Pople, J. A. 1998. *Chem. Phys. Lett.*, **284**, 6.
FT97	Filatov and Thiel's GGA exchange functional.	Filatov, M. and Thiel, W. 1997. *Mol. Phys.*, **91**, 847.
G96	Gill's 1996 GGA exchange functional.	Gill, P. M. W. 1996. *Mol. Phys.*, **89**, 433.
H&H	One-parameter hybrid exchange functional combining 50% LSDA with 50% HF exchange.	Becke, A. D. 1993b. *J. Chem. Phys.*, **98**, 1372.
HCTH	Hamprecht, Cohen, Tozer, and Handy GGA exchange-correlation functional based on a reoptimization/extension of empirical parameters in B97 and a removal of HF exchange. Now a family of functionals with optimizations over different numbers of test-set molecules, typically denoted HCTH/n where n is the number of molecules in the test set, e.g., HCTH/93, HCTH/120, HCTH/147, and HCTH/407.	Hamprecht, F. A., Cohen, A. J., Tozer, D. J., and Handy, N. C. 1998. *J. Chem. Phys.*, **109**, 6264. (Most recent refinement, Boese, A. D., Martin, J. M. L., and Handy, N. C. 2003. *J. Chem. Phys.*, **119**, 3005.)
ISM	Imamura, Scuseria, and Martin's MGGA correlation functional based on CS.	Imamura, Y., Scuseria, G. E., and Martin, R. M. 2002. *J. Chem. Phys.*, **116**, 6458.
KCIS	Kriger, Chen, Iafrate, and Savin's MGGA correlation functional including a self-interaction correction.	Krieger, J. B., Chen, J., Iafrate, G. J., and Savin, A. 1999. In *Electron Correlations and Materials Properties*, Gonis, A. and Kioussis, N., Eds., Plenum: New York, 463.
KMLYP	Kang and Musgrave two-parameter hybrid exchange-correlation functional using a mixture of Slater and HF (55.7%) exchange and a mixture of LSDA and LYP (44.8%) correlation functionals.	Kang, J. K. and Musgrave, C. B. 2001. *J. Chem. Phys.*, **115**, 11040.

Table 8.7   (*continued*)

Abbreviation	Comments	Reference(s)
Lap	MGGA correlation functionals that include a dependence on the Laplacian of the density in addition to its gradient, typically denoted either Lap1 or Lap3 depending on version.	Proynov, E. I., Sirois, S., and Salahub, D. R. 1997. *Int. J. Quantum Chem.*, **64**, 427.
LG	Lacks and Gordon's GGA correlation functional.	Lacks, D. J. and Gordon, R. G. 1993. *Phys. Rev. A*, **47**, 4681.
LT2A	Local square kinetic energy density exchange functional depending *only* on the kinetic energy density (i.e., not at all on the electron density).	Maximoff, S. N., Enzerhof, M., Scuseria, G. E. 2002. *J. Chem. Phys.*, **117**, 3074.
LYP	Lee, Yang, and Parr's GGA correlation functional based on the CS functional but depending only on the density.	Lee, C., Yang, W., and Parr, R. G. 1988. *Phys. Rev. B*, **37**, 785.
*m*PBE	Adamo and Barone's modification of PBE exchange with PBE correlation.	Adamo, C. and Barone, V. 2002. *J. Chem. Phys.*, **116**, 5933.
*m*PW	Adamo and Barone's modification of PW.	Adamo, C. and Barone, V. 1998. *J. Chem. Phys.*, **108**, 664.
MPW1K	Optimization of *m*PW1PW91 for kinetics of H-atom abstractions by using 42.8% HF exchange instead of default 25% and the 6-31+G(d,p) basis set.	Lynch, B. J., Fast, P. L., Harris, M., and Truhlar, D. G. 2000. *J. Phys. Chem. A*, **104**, 4811.
*m*PW1N	Optimization of *m*PW1PW91 for halide/haloalkane nucleophilic substitution reactions by using 40.6% HF exchange instead of default 25% and the 6-31+G(d) basis set.	Kormos, B. L. and Cramer, C. J. 2002. *J. Phys. Org. Chem.*, **15**, 712.
MPW1S	Optimization of *m*PW1PW91 for sugar conformational analysis by using 6% HF exchange instead of default 25% and the 6-31+G(d,p) basis set.	Lynch, B. J., Zhao, Y., and Truhlar, D. G. 2003. *J. Phys. Chem. A*, **107**, 1384.
O	Handy and Cohen OPTX GGA exchange functional including two optimized parameters	Handy, N. C. and Cohen, A. J. 2001. *Mol. Phys.*, **99**, 403.
O3LYP	ACM functional discussed in more detail in Section 8.4.3.	Hoe, W.-M., Cohen, A. J., and Handy, N. C. 2001. *Chem. Phys. Lett.*, **341**, 319.
P	Perdew's 1986 GGA exchange functional.	Perdew, J. P. 1986. *Phys. Rev. B*, **33**, 8822
P86	Perdew's 1986 GGA correlation functional.	Perdew, J. P. 1986. *Phys. Rev. B*, **33**, 8822
PBE	Perdew, Burke, and Enzerhof GGA exchange-correlation functional.	Perdew, J. P., Burke, K., and Enzerhof, M. 1996. *Phys. Rev. Lett.*, **77**, 3865 and erratum 1997. *ibid.*, **78**, 1396.

(*continued overleaf*)

Table 8.7   (*continued*)

Abbreviation	Comments	Reference(s)
PBE1PBE	One-parameter hybrid PBE functional incorporating 25% HF exchange (sometimes alternatively called PBE0, PBE0PBE, or PBE1).	Adamo, C., Cossi, M., and Barone, V. 1999. *J. Mol. Struct. (Theochem)*, **493**, 145.
PKZB	Perdew, Kurth, Zupan, and Blaha's MGGA exchange-correlation functional developed primarily for solids.	Perdew, J. P., Kurth, S., Zupan, A., and Blaha, P. 1999. *Phys. Rev. Lett.*, **82**, 2544.
PW	Perdew and Wang's GGA exchange functional.	Perdew, J. P. and Wang, Y. 1986. *Phys. Rev. B*, **33**, 8800.
PW91	Perdew and Wang's (sic) 1991 GGA correlation functional.	Perdew, J. P. 1991. In: *Electronic Structure of Solids '91*, Ziesche, P. and Eschrig, H., Eds., Akademie Verlag: Berlin, 11.
$\tau 1$	A MGGA correlation functional.	Proynov, E., Chermette, H., and Salahub, D. R. 2000. *J. Chem. Phys.*, **113**, 10013.
τHCTH	MGGA extension of the HCTH functional. Comes in both hybrid and pure DFT variations.	Boese, A. D. and Handy, N. C. 2002. *J. Chem. Phys.*, **116**, 9559.
TPSS	Tao, Perdew, Staroverov, and Scuseria's MGGA exchange-correlation functional.	Tao, J., Perdew, J. P., Staroverov, V. N., and Scuseria, G. E. 2003. *Phys. Rev. Lett.*, **91**, 146401.
TPSSh	One-parameter ACM of TPSS incorporating 10% HF exchange.	Staroverov, V. N., Scuseria, G. E., Tao, J., Perdew, J. P. 2003. *J. Chem. Phys.*, **119**, 12129.
VSXC	van Voorhis and Scuseria's MGGA exchange-correlation functional.	van Voorhis, T. and Scuseria, G. E. 1998. *J. Chem. Phys.*, **109**, 400.
VWN	Local correlation functional of Vosko, Wilk, and Nusair fit to the uniform electron gas. Note that VWN proposed several different forms for this functional, usually identified by a trailing number, e.g., VWN3 or VWN5. Different gradient-corrected and hybrid functionals built onto the VWN local correlation functional may use different versions. For example, B3LYP is defined to use VWN3, while O3LYP is defined to use VWN5.	Vosko, S. H., Wilk, L., and Nussair, M. 1980. *Can. J. Phys.*, **58**, 1200.
X	GGA exchange functional defined as a combination of one part LSDA, 0.722 parts B, and 0.347 parts PW91.	Xu, X. and Goddard, W. A., III. 2004. *Proc. Natl. Acad. Sci (USA)*, **101**, 2673.
X3LYP	ACM functional discussed in more detail in Section 8.4.3.	Xu, X. and Goddard, W. A., III. 2004. *Proc. Natl. Acad. Sci (USA)*, **101**, 2673.

[a]Exchange, correlation, and specific, specially defined combinations or hybrid functionals are contained herein. Routine combinations of exchange and correlation functionals (e.g., BLYP, OP86, or PWPW91) are not included.

8.7   Case Study: Transition-Metal Catalyzed Carbonylation of Methanol

Synopsis of Kinnunen and Laasonen (2001), 'Reaction Mechanism of the Reductive Elimination in the Catalytic Carbonylation of Methanol. A Density Functional Study'.

Acetic acid is made industrially by the condensation of methanol and carbon monoxide catalyzed by either a diiododicarbonylrhodium species or the corresponding iridium complex. The proposed catalytic cycle for this process is illustrated in Figure 8.5. Experimentally establishing a complete catalytic mechanism can be quite challenging, since reactive intermediates in the cycle may be present at such low concentrations that they are very difficult to detect. Theory can therefore play a useful role in establishing the energetic profiles for proposed catalytic steps, with the ultimate goal being the design of improved catalysts based on a fundamental understanding of the mechanism.

To that end, Kinnunen and Laasonen model the reductive elimination pathways from the anionic acetyltriiododicarbonyl rhodium and iridium anions, and from the acetyldiiodotricarbonyl iridium neutral using the B3LYP functional in combination with an unpolarized

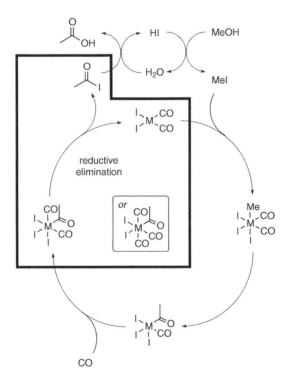

Figure 8.5   Catalytic cycle for the metal-catalyzed carbonylation of methanol, with the reductive elimination step highlighted. In the case of iridium, the diiodotricarbonyl species has also been suggested as a possible precursor to reductive elimination. What are the issues of stereochemistry associated with the intermediates? What special basis-set requirements will be involved in modeling this system?

double-ζ basis set on C, H, and O, and a valence basis set of similar size for I and the metals combined with relativistic effective core potentials. Happily, from a simplicity standpoint, all species are predicted to be ground-state singlets by large margins, so a restricted DFT formalism can be employed. In this instance, some experimental data are available for species involved in the reductive elimination step, so the adequacy of the theoretical level can be evaluated.

The authors begin by characterizing the relative energies of all possible stereoisomers in the octahedral complexes. For acetyltriiododicarbonyl metal complexes, there are *mer,trans*, *mer,cis*, and *fac,cis* possibilities (*mer* implies two iodides to be *trans* to one another while *fac* implies all I–M–I bond angles to be about 90°; *trans* and *cis* refer to whether the central iodine atom in the *mer* arrangement is opposite the acetyl group or adjacent to it) as well as acetyl rotamers to consider. For both rhodium and iridium, a single *mer,trans* geometry is predicted to be lower than all other possibilities by at least 2.7 kcal mol^{-1}. Experimental IR and NMR data for the Rh system are in accord with this prediction, while IR data for the Ir system suggest the presence of *fac,cis*, which is the next lowest energy species predicted from the computations. Kinnunen and Laasonen suggest that weak IR bands for the *mer,trans* isomer may make it difficult to detect experimentally, and infer that it is possible that both may be present experimentally.

Of course, while the intermediate energies are of interest, so long as interconversion between stereoisomers takes place at lower energy than reductive elimination, the latter process may potentially go through *any* stereoisomer on the way to the lowest energy TS structure for the reaction (the Curtin–Hammett principle). For the Rh system the lowest energy TS structure, which follows from a *mer,cis* reactant, has an associated 298 K free energy of activation of 20.1 kcal mol^{-1}, which compares well with an experimental value of about 18. In the case of Ir, a *fac,cis* TS structure is computed to be slightly lower than the *mer,cis* structure, and the overall free energy of activation is about 8 kcal mol^{-1} higher than was the case for Rh. In both cases, iodide dissociation is predicted to proceed with a lower barrier than reductive elimination, so stereoisomer scrambling via elimination/addition should be possible prior to reductive elimination.

Kinnunen and Laasonen carry out a similarly thorough analysis for the diiodotricarbonyliridium case. Consideration of all possibilities is complicated (and will depend experimentally on the iodide ion concentration and carbon monoxide pressure) but in essence 'all' stationary points corresponding to stereoisomeric minima and transition state structures for dissociation/association and reductive elimination steps are found and characterized energetically ('all' in quotes here because in such complicated systems it is essentially impossible to be entirely certain that every stationary point has been found). This exhaustive mapping of the PES provides insight into the catalytic process in a fashion typically not available experimentally, and takes good advantage of DFT's ability to handle transition metal systems in an efficient manner.

Bibliography and Suggested Additional Reading

Adamo, C. and Barone, V. 1998. 'Exchange Functionals with Improved Long-range Behavior and Adiabatic Connection Methods Without Adjustable Parameters: The *m*PW and *m*PW1PW Models', *J. Chem. Phys.*, **108**, 664.

Ban, F., Rankin, K. N., Gauld, J. W., and Boyd, R. J. 2002. 'Recent Applications of Density Functional Theory Calculations to Biomolecules', *Theor. Chem. Acc.*, **108**, 1.

Cramer, C. J. 1998. 'Bergman, Aza-Bergman, and Protonated Aza-Bergman Cyclizations and Intermediate 2,5-Arynes: Chemistry and Challenges to Computation' *J. Am. Chem. Soc.*, **120**, 6261.

Frauenheim, T., Seifert, G., Elstner, M., Hajnal, Z., Jungnickel, G., Porezag, D., Suhai, S., and Scholz, R. 2000. 'A Self-consistent Charge Density-functional Based Tight-binding Method for Predictive Materials Simulations in Physics, Chemistry and Biology', *Phys. Stat. Sol. B*, **217**, 41.

Geerlings, P., de Proft, F. and Langenaeker, W. 2003. 'Conceptual Density Functional Theory', *Chem. Rev.*, **103**, 1793.

Ghosh, A. and Taylor, P. R. 2003. 'High-level ab initio Calculations on the Energetics of Low-lying Spin States of Biologically Relevant Transition Metal Complexes: A First Progress Report', *Curr. Opin. Chem. Biol.*, **7**, 113.

Jensen, F. 1999. *Introduction to Computational Chemistry*, Wiley: Chichester.

Koch, W. and Holthausen, M. C. 2001. *A Chemist's Guide to Density Functional Theory*, 2nd Edn., Wiley, Chichester.

Laird, B. B., Ross, R. B., and Ziegler, T., Eds., 1996. *Chemical Applications of Density-functional Theory*, ACS Symposium Series, Volume 629, American Chemical Society: Washington, DC.

Levine, I. N. 2000. *Quantum Chemistry*, 5th Edn., Prentice Hall: New York.

Perdew, J. P., Kurth, S., Zupan, A., and Blaha, P. 1999. 'Accurate Density Functional with Correct Formal Properties: A Step Beyond the Generalized Gradient Approximation', *Phys. Rev. Lett.*, **82**, 2544.

Raghavachari, K. 2000. 'Perspective on "Density Functional Thermochemistry. III. The Role of Exact Exchange"' *Theor. Chem. Acc.*, **103**, 361.

St-Amant, A. 1996. 'Density Functional Methods in Biomolecular Modeling' in *Reviews in Computational Chemistry*, Vol. 1, Lipkowitz, K. B. and Boyd, D. B., Eds., VCH: New York, 217.

Stowasser, R. and Hoffmann, R. 1999. 'What Do the Kohn-Sham Orbitals and Eigenvalues Mean?' *J. Am. Chem. Soc.*, **121**, 3414.

Worthington, S. E. and Cramer, C. J. 1997. 'Density Functional Calculations of the Influence of Substitution on Singlet-triplet Gaps in Carbenes and Vinylidenes' *J. Phys. Org. Chem.*, **10**, 755.

Ziegler, T. 1991. 'Approximate Density Functional Theory as a Practical Tool in Molecular Energetics and Dynamics' *Chem. Rev.*, **91**, 651.

References

Adamo, C. and Barone, V. 1998. *J. Chem. Phys.*, **108**, 664.

Adamo, C. and Barone, V. 2002. *J. Chem. Phys.*, **116**, 5933.

Adamo, C., Barone, V., Bencini, A., Totti, F., and Ciofini, I. 1999. *Inorg. Chem.*, **38**, 1996.

Baker, J., Muir, M., and Andzelm, J. 1995. *J. Chem. Phys.*, **102**, 2063.

Baker, J., Scheiner, A., and Andzelm, J. 1993. *Chem. Phys. Lett.*, **216**, 380.

Bally, T. and Sastry, G. N. 1997. *J. Phys. Chem. A*, **101**, 7923.

Bauschlicher, C. W., Jr, and Gutsev, G. L. 2002. *Theor. Chem. Acc.*, **108**, 27.

Boese, A. D., Martin, J. M. L., and Handy, N. C. 2003. *J. Chem. Phys.*, **119**, 3005.

Bloch, F. 1929. *Z. Physik*, **57**, 545.

Burke, K., Enzerhof, M., and Perdew, J. P. 1997. *Chem. Phys. Lett.*, **265**, 115.

Burke, K., Perdew, J. P., and Wang, Y. 1998. In: *Electronic Density Functional Theory. Recent Progress and New Directions*, Dobson, J. F., Vignale, G., and Das, M. P., Eds., Plenum Press: New York, 81.

Ceperley, D. M. and Alder, B. J. 1980. *Phys. Rev. Lett.*, **45**, 566.

Choi, C. H., Kertesz, M., and Karpfen, A. 1997. *J. Chem. Phys.*, **107**, 6712.

Chong, D. P., van Lenthe, E., van Gisbergen, S., and Baerends, E. J. 2004. *J. Comput. Chem.*, **25**, 1030.

Cohen, A. J. and Tantirungrotechai, Y. 1999. *Chem. Phys. Lett.*, **299**, 465.

Cramer, C. J. and Barrows, S. E. 1998. *J. Org. Chem.*, **63**, 5523.

Cramer, C. J. and Smith, B. A. 1996. *J. Phys. Chem.*, **100**, 9664.

Cramer, C. J., Dulles, F. J., and Falvey, D. E. 1994. *J. Am. Chem. Soc.*, **116**, 9787.

Cramer, C. J., Smith, B. A., and Tolman, W. B. 1996. *J. Am. Chem. Soc.*, **118**, 11283.

Cullen, J. 2004. *J. Comput. Chem.*, **25**, 637.

Curtiss, L. A., Raghavachari, K., Redfern, P. C., and Pople, J. A. 2000. *J. Chem. Phys.*, **112**, 7374.

Curtiss, L. A., Raghavachari, K., Trucks, G. W., and Pople, J. A. 1991. *J. Chem. Phys.*, **94**, 7221.

Curtiss, L. A., Redfern, P. C., Raghavachari, K., and Pople, J. A. 1998. *J. Chem. Phys.*, **109**, 42.

Dirac, P. A. M. 1930. *Proc. Camb. Phil. Soc.*, **26**, 376.

Dovesi, R., Orlando, R., Roetti, C., Pisani, C., and Saunders, V. R. 2000. *Phys. Stat. Sol. B*, **217**, 63.

Elstner, M., Qui, C., Munih, P., Kaxiras, E., Frauenheim, T., and Karplus, M. 2003. *J. Comput. Chem.*, **24**, 565.

Fermi, E. 1927. *Rend. Accad. Lincei.*, **6**, 602.

Filatov, M. and Shaik, S. 1998. *Chem. Phys. Lett.*, **288**, 689.

Frenking, G., Antes, I., Böhme, M., Dapprich, S., Ehlers, A. W., Jonas, V., Neuhaus, A., Otto, M., Stegmann, R., Veldkamp, A., and Vyboishchikov, S. F. 1996. In: *Reviews in Computational Chemistry*, Vol. 8, Lipkowitz, K. B. and Boyd, D. B., Eds., VCH: New York, 63.

Gilbert, T. M. 2004. *J. Phys. Chem. A*, **108**, 2550.

Gräfenstein, J. and Cremer, D. 2001. *Mol. Phys.*, **99**, 981.

Gräfenstein, J., Kraka, E., and Cremer, D. 1998. *Chem. Phys. Lett.*, **288** 593.

Gräfenstein, J., Kraka, E., and Cremer, D. 2004. *J. Chem. Phys.*, **120**, 524.

Gritsenko, O. V., Ensing, B., Schipper, P. R. T., and Baerends, E. J. 2000. *J. Phys. Chem. A*, **104**, 8558.

Gritsenko, O. V., Schipper, P. R. T., and Baerends. E. J. 1996. *Int. J. Quantum Chem., Quantum Chem. Symp.*, **30**, 1375.

Guner, V., Khuong, K. S., Houk, K. N., Chuma, A., and Pulay, P. 2004. *J. Phys. Chem. A*, **108**, 2959.

Guner, V., Khuong, K. S., Leach, A. G., Lee, P. S., Bartberger, M. D., and Houk, K. N. 2003. *J. Phys. Chem. A*, **107**, 11445.

Gunnarsson, O. and Lundqvist, B. I. 1976a. *Phys. Rev. B*, **13**, 4274.

Gunnarsson, O. and Lundqvist, B. I. 1976b. *Phys. Rev. B*, **15**, 6006 (erratum).

Hamprecht, F. A., Cohen, A. J., Tozer, D. J., and Handy, N. C. 1998 *J. Chem. Phys.*, **109**, 6264.

He, Y., Gräfenstein, J., Kraka, E., and Cremer, D. 2000. *Mol. Phys.*, **98**, 1639.

Hobza, P. and Sponer, J. 1999. *Chem. Rev.*, **99**, 3247.

Hoe, W.-M., Cohen, A. J., and Handy, N. C. 2001. *Chem. Phys. Lett.*, **341**, 319.

Hohenberg, P. and Kohn, W. 1964. *Phys. Rev.*, **136**, B864.

Jensen, F. 2002a. *J. Chem. Phys.*, **116**, 7372.

Jensen, F. 2002b. *J. Chem. Phys.*, **117**, 9234.

Jensen, F. 2003. *J. Chem. Phys.*, **118**, 2459.

Johnson, W. T. G. and Cramer, C. J. 2001. *Int. J. Quantum Chem.*, **85**, 492.

Johnson, B. G., Gill, P. M. W., and Pople, J. A. 1993. *J. Chem. Phys.*, **98**, 5612.

Jonas, V. and Thiel, W. 1995. *J. Chem. Phys.*, **102**, 8474.

Jones, G. A., Paddon-Row, M. N., Sherburn, M. S., and Turner, C. I. 2002. *Org. Lett.*, **4**, 3789.

Kang, J. K. and Musgrave, C. B. 2001. *J. Chem. Phys.*, **115**, 11040.

Khait, Y. G. and Hoffmann, M. R. 2004. *J. Chem. Phys.*, **120**, 5005.

Kinnunen, T. and Laasonen, K. 2001. *J. Organomet. Chem.*, **628**, 222.

Kohn, W. and Sham, L. J. 1965. *Phys. Rev.*, **140**, A1133.

Kormos, B. L. and Cramer, C. J. 2002. *J. Phys. Org. Chem.*, **15**, 712.

Kümmel, S. and Perdew, J. P. 2003. *Mol. Phys.*, **101**, 1363.

Lein, M., Dobson, J. F., and Gross, E. K. U. 1999. *J. Comput. Chem.*, **20**, 12.

Lii, J.-H., Ma, B., and Allinger, N. L. 1999. *J. Comput. Chem.*, **20**, 1593.

Lim, M. H., Worthington, S. E., Dulles, F. J., and Cramer, C. J. 1996. In: *Density-Functional Methods in Chemistry*, ACS Symposium Series, Vol. 629, Laird, B. B., Ross, R. B., and Ziegler, T., Eds., American Chemical Society: Washington, DC, 402.

Liu, H., Elstner, M., Kaxiras, E., Frauenheim, T., Hermans, J., and Yang, W. 2001. *Proteins*, **44**, 484.

Lovell, T., McGrady, J. E., Stranger, R., and Macgregor, S. 1996. *Inorg. Chem.*, **35**, 3079.

Lynch, B. J. and Truhlar, D. G. 2003a. *J. Phys. Chem. A*, **107**, 3898.

Lynch, B. J. and Truhlar, D. G. 2003b. *J. Phys. Chem. A*, **107**, 8996.

Lynch, B. J., Zhao, Y., and Truhlar, D. G. 2003. *J. Phys. Chem. A*, **107**, 1384.

Ma, B., Schaefer, H. F., III, and Allinger, N. L. 1998. *J. Am. Chem. Soc.*, **120**, 3411.

Patchkovskii, S. and Ziegler, T. 2002. *J. Chem. Phys.*, **116**, 7806.

Perdew, J. P. and Zunger, A. 1981. *Phys. Rev. B*, **23**, 5048.

Perdew, J. P., Chevary, J. A., Vosko, S. H., Jackson, K. A., Pederson, M. R., Singh, D. J., and Fiolhais, C. 1992. *Phys. Rev. B*, **46**, 6671.

Poater, J., Duran, M., and Solà, M. 2001. *J. Comput. Chem.*, **22**, 1666.

Pokon, E. K., Liptak, M. D., Feldgus, S., and Shields, G. C. 2001. *J. Phys. Chem. A*, **105**, 10483.

Poli, R. and Harvey, J. N. 2003. *Chem. Soc. Rev.*, **32**, 1.

Proynov, E., Chermette, H., and Salahub, D. R. 2000. *J. Chem. Phys.*, **113**, 10013.

Redfern, P. C., Blaudeau, J. -P., and Curtiss, L. A. 1997. *J. Phys. Chem. A*, **101**, 8701.

Rienstra-Kiracofe, J. C., Tschumper, G. S., Schaefer, H. F., III, Nandi, S., and Ellison, G. B. 2002. *Chem. Rev.*, **102**, 231.

Ruiz, E., Salahub, D., and Vela, A. 1995. *J. Am. Chem. Soc.*, **117**, 1141.

Scheiner, A. C., Baker, J., and Andzelm, J. W. 1997. *J. Comput. Chem.*, **18**, 775.

Snijders, G. J., Baerends, E. J., and Vernooijs, P. 1982. *At. Data Nucl. Data Tables.*, **26**, 483.

St-Amant, A., Cornell, W. D., Halgren, T. A., and Kollman, P. A. 1995. *J. Comput. Chem.*, **16**, 1483.

Staroverov, V. N., Scuseria, G. E., Tao, J. and Perdew, J. P. 2003. *J. Chem. Phys.*, **119**, 12129.

Thomas, L. H. 1927. *Proc. Camb. Phil. Soc.*, **23**, 542.

van Lenthe, E. and Baerends, E. J. 2003. *J. Comput. Chem.*, **24**, 1142.

Vosko, S. J., Wilk, L., and Nussair, M. 1980. *Can. J. Phys.*, **1980**, 1200.

Wiest, O., Montiel, D. C., and Houk, K. N. 1997. *J. Phys. Chem. A*, **101**, 8378.

Woodcock, H. L., Schaefer, H. F., III, and Schreiner, P. R. 2002. *J. Phys. Chem. A*, **106**, 11923.

Xu, X. and Goddard, W. A., III. 2004a. *Proc. Natl. Acad. Sci (USA)*, **101**, 2673.

Xu, X. and Goddard, W. A., III. 2004b. *J. Phys. Chem. A*, **108**, 2305.

Yang, W. and Lee, T.-S. 1995. *J. Chem. Phys.*, **103**, 5674.

Zhang, Y. and Wang, W. 1998. *J. Chem. Phys.*, **109**, 2604.

Zhao, Y., Pu, J., Lynch, B. J., and Truhlar, D. G. 2004. *Phys. Chem. Chem. Phys.*, **6**, 673.

Zheng, Y. C. and Almlöf, J. *Chem. Phys. Lett.*, **214**, 397.

9

Charge Distribution and Spectroscopic Properties

9.1 Properties Related to Charge Distribution

We know from the fundamental theorems of DFT that the charge distribution (i.e., the density) determines the external potential, and that this determines the Hamiltonian operator, and that this in turn ultimately determines the wave function. So, in a formal sense, one might say that the charge distribution determines *all* molecular properties. For purposes of presentation, however, we will distinguish between properties that are rather direct measures of the charge distribution, and others, such as rotational and vibrational frequencies, that may be regarded as being decoupled from the charge distribution except to the extent that the molecular potential energy surface is intimately connected with it.

This chapter concerns itself with the prediction of a variety of measurable spectroscopic properties, including molecular multipole moments and polarizabilities, electron spin resonance (ESR) hyperfine coupling constants, nuclear magnetic resonance (NMR) chemical shifts and spin–spin coupling constants, and molecular rotational, vibrational, and photoelectron spectra. Because the modeling of electronic spectroscopy – the prediction of energy separations and transition probabilities between distinct electronic states – is somewhat more complex than these others in implementation, its discussion is deferred until Chapter 14. In addition, some 'unmeasurable' properties are examined, the most important of which, because it forms the basis for much chemical reasoning, is the concept of partial atomic charge.

9.1.1 Electric Multipole Moments

In Cartesian coordinates, the expectation values of multipole moment operators are computed as

$$\langle \mathbf{x}^k \mathbf{y}^l \mathbf{z}^m \rangle = \sum_i^{\text{atoms}} Z_i x_i^k y_i^l z_i^m - \int \Psi(\mathbf{r}) \left(\sum_j^{\text{electrons}} x_j^k y_j^l z_j^m \right) \Psi(\mathbf{r}) d\mathbf{r} \qquad (9.1)$$

Essentials of Computational Chemistry, 2nd Edition Christopher J. Cramer
© 2004 John Wiley & Sons, Ltd   ISBNs: 0-470-09181-9 (cased); 0-470-09182-7 (pbk)

where the sum of k, l, and m determines the type of moment ($0 = $ monopole, $1 = $ dipole, $2 = $ quadrupole, etc.), Z_i is the nuclear charge on atom i, and the integration variable $\mathbf{r}$ contains the x, y, and z coordinates of all of the electrons j. When Ψ is expressed as a single Slater determinant, we may write

$$\langle \mathbf{x}^k \mathbf{y}^l \mathbf{z}^m \rangle = \overset{\text{atoms}}{\underset{i}{\sum}} Z_i x_i^k y_i^l z_i^m - \overset{\text{electrons}}{\underset{j}{\sum}} \int \psi_j(\mathbf{r}_j)(x_j^k y_j^l z_j^m)\psi_j(\mathbf{r}_j)d\mathbf{r}_j \tag{9.2}$$

where ψ_j and $\mathbf{r}_j$ are the molecular orbital occupied by electron j and its Cartesian coordinate system, respectively. Equation (9.2) is also valid for DFT, where the various ψ are occupied KS orbitals. For ease of notation, we will restrict our discussion to situations where Eq. (9.2) holds, but all of the qualitative details we will consider are equally valid within the more general formalism of Eq. (9.1).

The simplest moment to evaluate is the monopole moment, which has only the component $k = l = m = 0$, so that the operator becomes $\mathbf{1}$ and, independent of coordinate system, we have

$$\langle \mathbf{1} \rangle = \overset{\text{atoms}}{\underset{i}{\sum}} Z_i - \overset{\text{electrons}}{\underset{j}{\sum}} \int \psi_j(\mathbf{r}_j)\psi_j(\mathbf{r}_j)d\mathbf{r}_j$$

$$= \overset{\text{atoms}}{\underset{i}{\sum}} Z_i - N \tag{9.3}$$

where N is the total number of electrons (the simplification of the second term on the r.h.s. follows from the normalization of the MOs). The monopole moment is thus the difference between the sum of the nuclear charges and the number of electrons, i.e., it is the molecular charge.

For the dipole moment, there are three possible components: x, y, or z depending on which of k, l, or m is one (with the others set equal to zero). These are written μ_x, μ_y, and μ_z. Experimentally, however, one rarely measures the separate components of the dipole moment, but rather the total magnitude, μ, which can be determined as

$$\langle \mu \rangle = \sqrt{\langle \mu_x \rangle^2 + \langle \mu_y \rangle^2 + \langle \mu_z \rangle^2} \tag{9.4}$$

The dipole moment measures the degree to which positive and negative charge are differentially distributed relative to one another, i.e., overall molecular polarity. Thus, for instance, if the electronic wave function has a large amplitude at some positive x value while the nuclear charge is concentrated at some negative x value, inspection of Eq. (9.2) indicates that the dipole moment in the x direction will be negative. If they are both concentrated at the same position and the total electronic charge is equal to the total nuclear charge, the first and second terms on the r.h.s. of Eq. (9.2) cancel, and the dipole moment is zero. Figure 9.1 illustrates the concept for the case of the water molecule. The nuclear charges are shown here lying flat in the xy plane, and entirely at the nuclear positions. The electrons of the

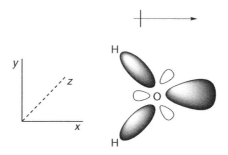

Figure 9.1 Contributions to the dipole moment of the water molecule

hydrogen atoms, however, are pulled to the positive x direction relative to the H nuclei by bonding interactions with the more electronegative O atom (the polarization is exaggerated here by depicting the σ orbitals with surfaces that fail to encompass the nuclear positions). In addition, the oxygen atom contributes two electrons into its in-plane lone pair, the orbital for which is localized at large, positive values of x, while only contributing a single electron each to the σ orbitals, resulting in another net polarization of negative charge in the positive x direction. The sum of these and other effects is such that water has a dipole moment of 1.8 D in the direction indicated (parallel with the x axis by symmetry). Note that the out-of-plane p orbital may be thought of as 'cancelling' two protons in the oxygen nucleus when the dipole moment is computed, since it is circularly symmetric about the nucleus when projected into the xy plane. (Since it is also symmetric above and below the xy plane, as are all other orbitals, there is no z component to the dipole moment.)

Note that, if the total number of electrons is equal to the total nuclear charge, then the dipole moment is independent of the choice of origin. This is again fairly obvious from inspection of Eq. (9.2), since any change in origin will affect the total contribution of the first term on the r.h.s. by the same amount as the second term, and the terms have opposite sign; thus, so long as the sum of the atomic numbers is equal to the number of the electrons the net effect of moving the origin is zero. However, if it is *not* the case that the two are equal, i.e., the system carries a positive or negative charge, then the dipole moment is *not* independent of origin, since moving it will cause a larger change in the magnitude of the first term on the r.h.s. compared to the second for cations, and of the second term compared to the first for anions. One can show in general that *only the first non-zero electric moment of a molecule is independent of origin.* For a charged molecule, this is the monopole, for a neutral molecule, the dipole, for a neutral molecule with zero dipole moment (perhaps by symmetry, e.g., CO_2), it is the quadrupole, etc. Any other, *higher* moment that is reported from a calculation *must* specify the origin that was chosen in order to be meaningful. The most common choices are the center of charge and the center of mass.

Electrical moments are useful because at long distances from a molecule the total electronic distribution can be increasingly well represented as a truncated multipole expansion, and thus molecular interactions can be approximated as multipole–multipole interactions (charge–charge, charge–dipole, dipole–dipole, etc.), which are computationally particularly simple to evaluate. At short distances, however, the multipole expansion may be very slowly convergent, and the multipole approximation has less utility.

9.1.2 Molecular Electrostatic Potential

A (truncated) multipole expansion is a computationally convenient single-center formalism that allows one to quantitatively compute the degree to which a positive or negative test charge is attracted to or repelled by the molecule that is being represented by the multipole expansion. This quantity, the molecular electrostatic potential (MEP), can be computed *exactly* for any position $\mathbf{r}$ as

$$V_{\text{MEP}}(\mathbf{r}) = \sum_{k}^{\text{nuclei}} \frac{Z_k}{|\mathbf{r} - \mathbf{r}_k|} - \int \Psi(\mathbf{r}') \frac{1}{|\mathbf{r} - \mathbf{r}'|} \Psi(\mathbf{r}') d\mathbf{r}' \tag{9.5}$$

Note that this assumes no polarization of the molecule in response to the test charge. The MEP is an observable, although in practice it is rather difficult to design appropriate experiments to measure it. Computationally, it is usually evaluated within the formalism of either HF or DFT theories, in which case one may write

$$V_{\text{MEP}}(\mathbf{r}) = \sum_{k}^{\text{nuclei}} \frac{Z_k}{|\mathbf{r} - \mathbf{r}_k|} - \sum_{r,s} P_{rs} \int \varphi_r(\mathbf{r}') \frac{1}{|\mathbf{r} - \mathbf{r}'|} \varphi_s(\mathbf{r}') d\mathbf{r}' \tag{9.6}$$

where r and s run over the indices of the AO basis set, $\mathbf{P}$ is the one-electron density matrix defined by Eq. (4.57) or its appropriate analog for UHF and DFT, and the orbitals φ are those comprising the basis set.

The MEP is particularly useful when visualized on surfaces or in regions of space, since it provides information about local polarity. Typically, after having chosen some sort of region to be visualized, a color-coding convention is chosen to depict the MEP. For instance, the most negative potential is assigned to be red, the most positive potential is assigned to be blue, and the color spectrum is mapped to all other values by linear interpolation. If this is done on the molecular van der Waals surface, one can immediately discern regions of local negative and positive potential, which may be informative for purposes of predicting chemical reactivity. Figure 9.2 provides an example of this procedure for a particular example.

Mapping of the electrostatic potential to grid points in three dimensions has also proven useful for comparative molecular field analysis (CoMFA). This procedure is used to identify common features in the electrostatic potentials of several molecules when the goal is to correlate such commonality with another chemical property, e.g., pharmaceutical activity. In the latter instance, a group of molecules having high activity are oriented about a common origin (finding the correct orientation for each molecule is a key challenge in CoMFA) and the electrostatic potential of each is evaluated at each grid point. A subsequent averaging of the values at every point may identify key regions of positive or negative potential associated

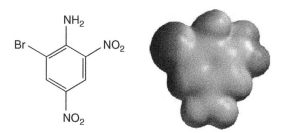

Figure 9.2 MEP of the radical anion produced by one-electron reduction of the dinitroaromatic shown at left. The spectrum is mapped so that red corresponds to maximum negative charge density and deep blue to minimum (here shown in grayscale). This depiction indicates that the buildup of negative charge density is larger on the nitro group *ortho* to the amino group than on that *para* to NH_2. Such polarization is consistent with the observed reactivity of the molecule under reducing conditions (Barrows *et al.* 1996)

with activity and that information may be used for the design of future drug candidates. Full scale CoMFA typically considers more fields than only the electrostatic potential (e.g., steric or non-polar fields), but there is a strong tendency for electrostatic effects to influence pharmaceutical activity and specificity.

For very large molecules (like biopolymers) the ESP can be very useful for analyzing function, but direct calculation from quantum mechanics is restricted to fairly low levels of theory (Khandogin, Hu, and York 2000). An alternative method that shows promise is to assemble the full molecular density from high-level calculations on small constituent fragments (Exner and Mezey 2002).

9.1.3 Partial Atomic Charges

A very old concept in chemistry is to associate molecular polarity with charge build-up or depletion on the individual atoms. In rationalizing hydrogen bonding in water, for instance, we speak of the oxygen being 'negative' and the hydrogen atoms 'positive'. Part of the driving force for this conceit is that it allows one to conveniently ignore the wave character of the electrons and deal only with the pleasantly more particulate atoms, these atoms reflecting electronic distribution by the degree to which they carry positive or negative charge. Absolutely critical to the efficiency of most force-field methodologies is that they compute electrical interactions as atom-centered charge–charge interactions, so the development of methods for assigning accurate partial charges to atoms in molecules has seen considerable research.

The concept of a partial atomic charge, however, is ill defined. One often sees it written that the atomic partial charge is not a quantum mechanical observable. This is, however, a bit misleading. One *can* define unambiguous procedures making use of well-defined quantum mechanical operators for computing the partial atomic charge, and such procedures are in principle subject to experimental realization (e.g., the atoms-in-molecules charges defined in Section 9.1.3.3). However, there is no universally agreed upon 'best' procedure for computing partial atomic charge. This failure to agree is, in some sense, inevitable, because partial atomic charges are used in different ways within the context of different quantitative and qualitative models in chemistry, so there is no reason to expect a single procedure for determining such charges to be optimal for all purposes. Thus, many methodologies have been promulgated for computing partial charges, and we will examine the most prevalent ones here. For taxonomical purposes, it is helpful to categorize different partial charge methodologies into four classes, and the remainder of this section will be subdivided along these lines.

9.1.3.1 Class I charges

A Class I charge is one which is not determined from quantum mechanics, but through some arbitrary or intuitive approach. For instance, in a neutral diatomic molecule having a known dipole moment, one atomic charge must be $+q$ and the other, to preserve charge balance, must be $-q$. One obvious choice for q will be that value that, in conjunction with the experimental bond length r_e, causes the classical dipole moment qr_e to be equal to the experimental one.

Class I charges have been very popular historically because of the enormous speed with which they may in principle be computed. If one is interested in obtaining atomic partial charges for, say, 100 000 drug-like molecules (perhaps because one wants to carry out CoMFA analyses as described in Section 9.1.2), one might well look to Class I charges for succor. One widely used Class I approach is the partial equalization of orbital electronegativity (PEOE) method as codified by Gasteiger and Marsilli (1980). This model assumes that the electronegativity χ of an atom type k, where types tend to be dependent upon atomic number and also hybridization, is a quadratic function of the atomic partial charge

$$\chi_k = a_k + b_k(Z_k - Q_k) + c_k(Z_k - Q_k)^2 \qquad (9.7)$$

where Z is the atomic number, Q is the number of electrons on the atom (thus the partial charge $q = Z - Q$), and a, b, and c are parameters to be optimized.

The partial equalization of orbital electronegativity then proceeds as a convergent, iterative process. At step 0, all atoms are assigned charges based on their atomic type (usually zero, but possibly not if the atom is part of an intrinsically charged functional group). Then, for each subsequent step n, electronic charge is transferred from atoms of lower electronegativity k to atoms of higher electronegativity k' within every bonded pair according to

$$\Delta Q^{(n)}_{k \to k'} = \frac{\chi^{(n-1)}_{k'} - \chi^{(n-1)}_{k}}{a_k + b_k + c_k} f^n_{kk'} \qquad (9.8)$$

where iteration numbers appearing in parentheses as superscripts are simply used for indexing, the denominator is the electronegativity of the less electronegative atom's cation (cf. Eq. (9.7)), and f is a damping factor raised to the nth power. New electronic populations are then computed according to

$$Q_k^{(n)} = Q_k^{(n-1)} - \sum_{\substack{k' \text{ bonded} \\ \text{to } k \\ \chi_{k'} > \chi_k}} \Delta Q_{k \to k'}^{(n)} + \sum_{\substack{k' \text{ bonded} \\ \text{to } k \\ \chi_k > \chi_{k'}}} \Delta Q_{k' \to k}^{(n)} \tag{9.9}$$

In the original PEOE method, the damping function f was simply taken to be the constant 0.5, which led to practical convergence in the atomic partial charges within about five iterations. No *et al.* (1990a, 1990b) subsequently proposed a modification in which different damping factors were used for different bonds (MPEOE) and observed that this, together with some other minor changes, gave improved charge distributions when compared to known multipole moments. More recently, Cho *et al.* (2001) proposed computing the damping factor as

$$f_{kk'} = \min\left[0, \left(1 - \frac{r_{kk'}}{r_k^{\text{vdw}} + r_{k'}^{\text{vdw}}}\right)\right] \tag{9.10}$$

where $r_{kk'}$ is the distance between the two atoms and r_{vdw} is a parametric van der Waals radius. The use of Eq. (9.10) delivers geometry-dependent atomic charge (GDAC) values, which were found to improve additionally on computed electrical moments.

Another Class I charge model that is also sensitive to geometry is the QEq charge equilibration model of Rappé and Goddard (1991). From representing the energy u of an isolated atom k as a Taylor expansion in its charge truncated at second order, one can derive

$$u_k = \tilde{u}_k + \chi_k q_k + \frac{1}{2} J_{kk} q_k^2 \tag{9.11}$$

where $\tilde{u}$ is the energy of the neutral isolated atom, χ is the electronegativity (experimentally the average of the atomic IP and EA), and J is the idempotential, which is formally equal to IP − EA. With this formula in hand, we may write the electrostatic energy of a *collection* of N atoms as

$$U = \sum_{k=1}^{N} (\tilde{u}_k + \chi_k q_k) + \frac{1}{2} \sum_{k=1}^{N} \sum_{k'=1}^{N} J_{kk'} q_k q_{k'} \tag{9.12}$$

where $\mathbf{J}$ is a matrix of Coulomb integrals for which we have already defined the diagonal elements as the idempotentials. The off-diagonal elements are computed as $(aa|bb)$ where a and b are STOs on the centers k and k', respectively (thereby introducing geometry dependence). QEq charges q are then determined from minimization of U subject to the constraint that the total molecular charge remain constant. Note the close conceptual similarities between QEq and SCC-DFTB described in Section 8.4.4.

Eq. (9.12) does not require any specification of bonding – all atoms electrically interact with all other atoms. Sefcik *et al.* (2002) have combined QEq electrostatics with Morse potentials for non-electrostatic non-bonded interactions between all atom pairs to create a 'connectivity-free' force field for zeolites that provides highly realistic structural and dynamic data for these species. Such connectivity-free force fields are in principle equally well suited to modeling systems where bonds are being made and broken as they are to modeling stable structures, although in practice parameter optimization for such a global force field tends to be hampered by a scarcity of data for high-energy regions of phase space.

9.1.3.2 Class II charges

Class II charge models involve a direct partitioning of the molecular wave function into atomic contributions following some arbitrary, orbital-based scheme. The first such scheme was proposed by Mulliken (1955), and this method of population analysis now bears his name. Conceptually, it is very simple, with the electrons being divided up amongst the atoms according to the degree to which different atomic AO basis functions contribute to the overall wave function. Starting from the expression used for the total number of electrons in Eq. (9.3), and expanding the wave function in its AO basis set, we have

$$
N = \sum_j^{\text{electrons}} \int \psi_j(\mathbf{r}_j)\psi_j(\mathbf{r}_j)d\mathbf{r}_j
$$

$$
= \sum_j^{\text{electrons}} \sum_{r,s} \int c_{jr}\varphi_r(\mathbf{r}_j)c_{js}\varphi_s(\mathbf{r}_j)d\mathbf{r}_j
$$

$$
= \sum_j^{\text{electrons}} \left(\sum_r c_{jr}^2 + \sum_{r \neq s} c_{jr}c_{js}S_{rs} \right) \tag{9.13}
$$

where r and s index AO basis function φ, c_{jr} is the coefficient of basis function r in MO j, and S is the usual overlap matrix element defined in Eq. (4.18).

From the last line of Eq. (9.13), we see that we may divide the total number of electrons up into two sums, one including only squares of single AO basis functions, the other including products of two different AO basis functions. Clearly, electrons associated with only a single basis function (i.e., terms in the first sum in parentheses on the r.h.s. of the last line of Eq. (9.13)) should be thought of as belonging entirely to the atom on which that basis function resides. As for the second term, which represents the electrons 'shared' between basis functions, Mulliken suggested that one might as well divide these up evenly between the two atoms on which basis functions r and s reside. If we follow this prescription and furthermore divide the basis functions up over atoms k so as to compute the atomic population N_k, Eq. (9.13) becomes

$$
N_k = \sum_j^{\text{electrons}} \left(\sum_{r \in k} c_{jr}^2 + \sum_{r,s \in k, r \neq s} c_{jr}c_{js}S_{rs} + \sum_{r \in k, s \notin k} c_{jr}c_{js}S_{rs} \right) \tag{9.14}
$$

Note that the orthonormality of basis functions of different angular momentum both residing on the same atom k causes many terms in the second sum of Eq. (9.14) to be zero. The

Mulliken partial atomic charge is then defined as

$$q_k = Z_k - N_k \tag{9.15}$$

where Z is the nuclear charge and N_k is computed according to Eq. (9.14).

With minimal or small split-valence basis sets, Mulliken charges tend to be reasonably intuitive, certainly in sign if not necessarily in magnitude. Analysis of *changes* in charge as a function of substitution or geometric change tends to be the best use to which Mulliken charges may be put, and this can often provide chemically meaningful insight, as illustrated in Figure 9.3.

The use of a non-orthogonal basis set in the Mulliken analysis, however, can lead to some undesirable results. For instance, if one divides up the total number of electrons over AO basis functions (in a fashion exactly analogous to that used for atoms), it is possible for individual basis functions to have occupation numbers greater than 1 (which would be greater than 2 in a restricted theory) or less than 0, and such a situation obviously can have no physical meaning. In addition, the rule that all shared electrons should be divided up equally between the atoms on which the sharing basis functions reside would seem to ignore the possibly very different electronegativities of these atoms. Finally, Mulliken partial charges prove to be very sensitive to basis-set size, so that comparisons of partial charges

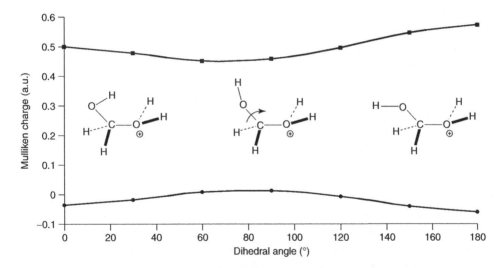

Figure 9.3 AM1 Mulliken charges of the hydroxyl (circles) and oxonium (squares) oxygen atoms in protonated dihydroxymethane as a function of HOCO$^+$ dihedral angle. Standard precepts of conformational analysis suggest that hyperconjugation of hydroxyl oxygen lone-pair density (acting as a donor) into the C–O$^+$ σ^* orbital (acting as an acceptor) may occur, and the effect is expected to be maximal at a dihedral angle of 90°, and minimal at 0° and 180°. The computed Mulliken charges on the oxygen atoms support hyperconjugation being operative, with about one-tenth of a positive charge being transferred from the oxonium oxygen to the hydroxyl oxygen at a dihedral angle of 90° compared to 180° (an interpretation also consistent with geometric and energetic analysis, see Cramer 1992)

from different levels of theory are in no way possible. Moreover, with very complete basis sets, Mulliken charges have a tendency to become unphysically large.

To alleviate a number of these problems, Löwdin proposed that population analysis not be carried out until the AO basis functions φ were transformed into an orthonormal set of basis functions χ using a symmetric orthogonalization scheme (Löwdin 1970; Cusachs and Politzer 1968)

$$\chi_r = \sum_s S_{rs}^{-1/2} \varphi_s \tag{9.16}$$

where r and s run over the total number of basis functions, and $S^{-1/2}$ is the inverse of the square root of the overlap matrix. When the MOs ϕ are expressed in the new orthonormal basis set, the result is

$$\phi_j = \sum_r a_{jr} \chi_r$$
$$= \sum_r \sum_s S_{rs}^{1/2} c_{jr} \chi_r \tag{9.17}$$

where the new coefficients a can be easily determined from the old coefficients c. Note that Mulliken analysis applied with the adoption of the orthogonalized basis set has no problem with shared electrons, because the overlap matrix in the new basis set is the unit matrix, so all terms in the last two sums on the r.h.s. of Eq. (9.14) are zero.

Löwdin population analysis enjoys much better stability than Mulliken analysis in terms of the predicted atomic partial charges as a function of basis set. For instance, the Mulliken charge on the central carbon atom of the allenyl anion ($C_3H_3^-$) changes from -0.17 at the HF/3-21G level to 2.47 when diffuse functions are added to the basis set. By contrast, the Löwdin charge changes only from -0.09 to -0.21. Nevertheless, even Löwdin charges can eventually become unstable with very large basis sets, although Thompson et al. (2002) have proposed a renormalized Löwdin population analysis (RLPA) that reduces the sensitivity of the procedure to the presence of diffuse functions.

The shortcoming in the Löwdin procedure derives from the symmetric nature of the orthogonalization. In a very large basis set, only a few AOs are really very important, but in the Löwdin process all AOs are distorted in a similar fashion to achieve orthonormality. A considerably more complicated procedure for achieving orthogonality is used in the Natural Population Analysis (NPA) scheme of Reed, Weinstock, and Weinhold (1985). Ignoring the exact details, orthogonalization takes place in a four-step process in such a way that the electron density around each atom is initially rendered as compact as possible, and further diagonalization is carried out so as to preserve the shape of the strongly occupied atomic orbitals to as large an extent as possible. Following orthogonalization, again, a Mulliken-like analysis in the new basis gives the atomic populations with no contributions from off-diagonal terms.

The most appealing feature of the NPA scheme is that each atomic partial charge effectively converges to a stable value with increasing basis-set size. In comparison to other schemes, however, including some of those yet to be discussed, NPA charges tend to be amongst the largest in magnitude, which can be mildly disquieting. However, as with any population analysis method, a focus on absolute partial atomic charges is usually much less profitable than an analysis of trends in charge(s) as a function of some variable (see, for instance, Gross, Seybold, and Hadad 2002).

Note that *all* of the Class II charge models discussed here suffer from the disadvantage of their population analyses being orbital-based. To illustrate this point, consider a calculation on the water molecule using an *infinite* basis, but one with every basis function defined so as to be centered on the oxygen atom. Insofar as the basis set is infinite, we should be able to obtain an arbitrarily good representation of the density, but in this case, Mulliken and Löwdin analyses, for instance, are equivalent, and both predict that the oxygen atom charge is −2 and the hydrogen atom charges are +1, since all electrons necessarily reside on oxygen, that being the only atom with basis functions. Nevertheless, the great speed with which Class II charges can be computed (Mulliken charges are the fastest, followed by Löwdin, and then by NPA) suggests that they will remain useful tools for qualitative analysis for some time to come.

9.1.3.3 Class III charges

Rather than being determined from an (arbitrary) analysis of the wave function itself, Class III charges are computed based on analysis of some *physical observable* that is *calculated* from the wave function. As already noted in Section 9.1.3.1, there is an obvious relationship that may be proposed between atomic partial charges and dipole moments in diatomics. Cioslowski (1989) has generalized this idea for polyatomic molecules, defining the generalized atomic polar tensor (GAPT) charge as

$$q_k = \frac{1}{3} \left(\frac{\partial \mu_x}{\partial x_k} + \frac{\partial \mu_y}{\partial y_k} + \frac{\partial \mu_z}{\partial z_k} \right) \qquad (9.18)$$

where the quantities evaluated on the r.h.s. are the changes in the molecular dipole moment as a function of moving atom k in each of the three Cartesian directions (the GAPT charge is independent of coordinate system).

While GAPT charges converge quickly with respect to basis-set size, it is important to note that a level of theory that fails to give good dipole moments (e.g., HF) is then unlikely to give useful charges. In addition, GAPT charges are relatively expensive to compute – equivalent to the cost of a vibrational frequency calculation, as described in Section 9.3.2.2 – and as such they have seen only moderate use in the literature.

An alternative physical observable that has been used to define partial atomic charges is the electron density. In X-ray crystallography, the electron density is directly measured, and by comparison to, say, spherically symmetric neutral atoms, atomic partial charges may be defined experimentally, following some decisions about what to do with respect to partitioning space between the atoms (Coppens 1992). Bader and co-workers have adopted a particular partitioning scheme for use with electronic structure calculations that defines the atoms-in-molecules (AIM) method (Bader 1990). In particular, an atomic volume is

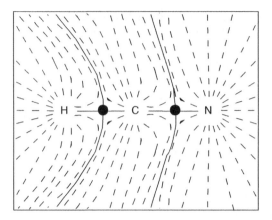

Figure 9.4   Electron density gradient paths in a plane containing the atoms of the HCN molecule. The solid lines are the intersections of the zero-flux surfaces with the plane. The large black dots are the bond critical points

defined as that region of space including the nucleus that lies within all zero-flux surfaces surrounding the nucleus.

To make this more clear, let us consider the electron density more closely. As already noted in Section 8.1.1, local maxima in the density occur at the positions of the nuclei. Now, imagine following some path outwards from the nucleus, where the direction we take is opposite to the gradient of the density. Two possibilities exist and are illustrated in Figure 9.4 for the case of HCN. Either we will proceed outward from the molecule indefinitely, with the density dropping off exponentially but in principle never reaching zero, or, on rare occasions, we will come to a point where the gradient does reach zero, because it passes from a negative value (falling back towards the nucleus we started from) to a positive value (falling towards some other nucleus). These latter points are called 'bond critical points' and in AIM theory it is their existence that defines whether a bond between two atoms exists or not. In any case, we may define the zero-flux surface mathematically as the union of all points for which

$$\nabla \rho \cdot \mathbf{n} = 0 \qquad (9.19)$$

where ρ is the density and $\mathbf{n}$ is the unit vector normal to the surface. Note that to satisfy this condition, either the gradient of the density must run along the surface itself (in which case it is orthogonal to the unit vector) or it must be zero (i.e., the density has a critical point). Bond critical points are minima in the density for the direction to and from the two nuclei defining the bond, but maxima for the density within the zero-flux surface itself. Two other kinds of critical points can exist, so-called 'ring critical points', which are found in the interiors of rings and are minima in two dimensions but maxima in one, and 'cage critical points', which can be found in the middle of polyhedral structures, and are local minima in all directions.

The analysis of bond, ring, and cage critical points, and of the behavior of the electron density in their vicinity, is a subject of considerable interest for the analysis of chemical structure and reactivity, but beyond the scope of this text (interested readers are directed to Bader 1991). For our purposes, we will restrict ourselves to consideration of partial atomic charge which, within the AIM theory, is defined as nuclear charge less the total number of electrons residing within the atomic basin. That is

$$q_k = Z_k - \int_{\Omega_k} \rho(\mathbf{r}) d\mathbf{r} \qquad (9.20)$$

where the integral is marked to indicate that it is over the spatial volume Ω_k encompassed by the zero-flux surface of k.

Partial atomic charges from the AIM method are derived from a formalism that is really quite elegant, but in practice they are of little chemical utility. At times, AIM charges can even seem rather bizarre – for instance, saturated hydrocarbons are predicted to have weakly positive carbon atoms and weakly negative hydrogen atoms, in disagreement with essentially every other method for assigning partial atomic charges. This odd behavior does not derive from any particular flaw within the methodology, but more from it being inconsistent with the purpose to which partial charges are usually meant to be put. The problem is that the charge within an atomic basin may be *very* non-uniformly distributed, as illustrated in Figure 9.5 for the methyldiazonium cation. In such an instance, the electron density may be 'piled up' rather far from the nucleus, but Eq. (9.20) does not distinguish that situation from the charge being spherically symmetric about the nucleus. In Figure 9.5, the polarization of the molecule is such that the electronic charge associated with the basin of the terminal nitrogen localizes predominantly in the region labeled a, while there is proportionately much less in the region labeled b. Thus, there is a rather large local dipole moment associated with this basin. This dipole is not well represented when only partial atomic charges are considered and the partial atomic charge itself is obtained by simply summing the total density in the basin with the nuclear charge. Moreover, the position of the zero-flux surface between two

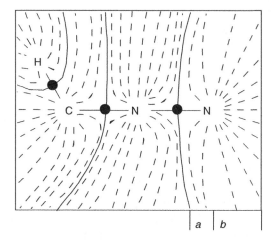

Figure 9.5 AIM partitioning of space in a plane containing four atoms of $CH_3N_2^+$ (the other two hydrogen atoms are symmetrically above and below the plane)

identical atoms (e.g., the two nitrogen atoms) is very sensitive to their substitution, since they otherwise have equal electronegativities. Small movements of the surface (e.g., when methyl is changed to ethyl) can cause large changes in the total density assigned to each basin, even though the density *itself* does not change much. This leads to unphysically large changes in partial atomic charge that are simply artifacts if higher electrical moments in the basins are not also taken into consideration.

The solution to this problem is to compute not simply the AIM charges, but also atomic multipole moments (defined over the atomic basins in a fashion analogous to their definition over all space for molecules; Laidig 1993). However, while this provides an accurate picture of the electron density distribution, it is inconsistent with the simplification that is the goal of using only partial atomic charges in the first place. As a rule, then, AIM partial atomic charges should not be used for analysis without some indication that the lower atomic multipole moments are quantitatively unimportant. As the necessary determinations of atomic volumes and the integrations over them are time-consuming, AIM analysis is not an entirely routine undertaking (Biegler-Konig and Schonbohm 2002).

Voronoi cells have also been used for the partitioning of space into atomic volumes. A Voronoi cell is essentially defined by the volume enclosed by intersecting planes each of which bisects the vector defined by any two neighboring atoms. Guerra *et al.* (2004) have found analysis of the deformation density in Voronoi cells (i.e., the degree to which the density differs from that expected for the unbonded atom) to provide chemically intuitive partial atomic charges that in particular tend not to overstate ionic character, as is sometimes the case for Bader or NPA charges. These Voronoi deformation density (VDD) charges also prove to be largely insensitive to choice of basis set.

The final observable from which charges are typically derived is the MEP. In the field of partial charges, the more common notation is to replace MEP with ESP, for 'electrostatic potential' (and *not* 'extrasensory perception'), and we will follow that convention from this point on. The ESP is perhaps the most obvious property to reproduce if one wants partial atomic charges that will be useful in modeling molecule–molecule interactions at short to long range, as is the case, say, in molecular mechanics simulations (Williams 1988). All ESP charge-fitting schemes involve determining atomic partial charges q_k that, when used as a monopole expansion according to

$$V_{\text{ESP}}(\mathbf{r}) = \sum_k^{\text{nuclei}} \frac{q_k}{|\mathbf{r} - \mathbf{r}_k|} \tag{9.21}$$

minimize the difference between V_{ESP} and the correct V_{MEP} calculated according to Eq. (9.5). Typical algorithms select a large number of points spaced evenly on a cubic grid surrounding the van der Waals surface of the molecule. To ensure rotational invariance, a reasonable density of points is required. The two algorithms in widest use are CHELPG (Breneman and Wiberg 1990), which is a modification designed to improve the stability of the charges from electrostatic potentials (CHELP) algorithm of Chirlian and Francl (1987), and the scheme of Besler, Merz, and Kollman (1990), which is sometimes slightly less robust than CHELPG, but usually gives very similar results.

When ESP charges are to be used in molecular simulations for flexible molecules, it is something of a problem to decide what to do about the conformational dependence of the partial charges. Thus, for instance, no more than two of the three methyl hydrogen atoms in methanol are ever symmetrically related for any reasonable conformation, so there will always be at least two different hydrogen partial atomic charges. Since the methyl group is freely rotating in a simulation, however, it is unreasonable to distinguish between the hydrogens. A modification of the ESP method advanced by Cornell *et al.* (1993) to address this issue is the restrained ESP (RESP) approach, where such dilemmas are erased by fiat (Bayly *et al.* 1993). Of course, if resources permit, a more accurate simulation *can* take account of the conformational dependence of partial atomic charges, but such a protocol is very rarely adopted since it is not only intrinsically expensive, but complicates force-field energy derivatives enormously. One compromise solution is to use fixed partial charges that are adjusted to reflect some weighted average over accessible conformations (see, for example, Basma *et al.* 2001).

A more serious problem with ESP methods is that the resulting partial charges have been shown to be ill conditioned. That is, the statistical reliability for some or many of the charges may be very low. This problem is particularly true for atoms in molecules that are not near the molecular surface. Wide variations in partial atomic charge for such atoms have minimal impact on the ESP at any point, particularly if the charges for atoms nearer to the surface are allowed to adjust slightly, so the final value from the minimization process is not particularly meaningful (and indeed, numerical instabilities in the fitting procedure may arise in unfavorable cases). For example, for two different conformations of glycerylphosphorylcholine, the variations computed for the partial atomic charges of the carbons in the ethanolamine bridge were about 0.5 charge units, and that for the ammonium nitrogen was 0.3 charge units. However, use of either charge set was observed to have negligible impact on the ESP about the functionality in question for either conformation (Francl and Chirlian 2000). While this problem is not necessarily anything to worry about when computing intermolecular interactions within a force-field calculation (if the ESP is insensitive to that partial charge, then so is the interaction energy), to the extent *intra*molecular interactions are also computed using the ESP charges, problems may develop.

9.1.3.4   *Class IV charges*

A hallmark of Class II and Class III charges is that they are derived from analysis of *computed* wave functions and physical observables, respectively. Thus, to the extent that an employed level of theory is in error for the particular quantity computed, the partial charges will faithfully reflect that error. A Class IV charge, on the other hand, is one that is derived by a semiempirical mapping of a precursor charge (either from a Class II or Class III model), in order to reproduce an *experimentally determined* observable.

Charge Model 1 (CM1) was the first method developed to compute Class IV charges (Storer *et al.* 1995). In this model, the input charges are Mulliken charges determined at a semiempirical level – different CM1 mappings are available for both the AM1 and

PM3 Hamiltonians – and the experimental observable for which the mappings were optimized was the molecular dipole moment as computed from the atomic partial charges according to

$$\mu = \left[\left(\sum_k q_k x_k \right)^2 + \left(\sum_k q_k y_k \right)^2 + \left(\sum_k q_k z_k \right)^2 \right]^{1/2} \tag{9.22}$$

The form of the mapping itself is relatively simple, with the CM1 charge defined as

$$q_k^{CM1} = q_k^{(0)} + B_k \Delta q_k - \sum_{k' \neq k} B_{kk'} \Delta q_{k'}, \tag{9.23}$$

where $q_k^{(0)}$ is the input Mulliken charge, $B_{kk'}$ is the bond order between atoms k and k', and B_k is defined as the sum of the bond orders of atom k to all other atoms. The quantity Δq_k is where the mapping comes in, and it is computed as

$$\Delta q_k = c_k q_k^{(0)} + d_k \tag{9.24}$$

where parameters c and d are optimized for each atom k so as to minimize errors in the predicted dipole moments. The form of Eq. (9.23) is such that (i) charge redistribution is local (since charge is passed between atoms based on the bond order between them) and (ii) total charge is preserved. The bond order is determined, as originally suggested by Mulliken in conjunction with population analysis, as

$$B_{kk'} = \sum_{\mu \in k} \sum_{\nu \in k'} P_{\mu\nu}^2 \tag{9.25}$$

The CM1 models for AM1 and PM3 yield root-mean-square errors of 0.30 and 0.26 D, respectively, in the dipole moments of 195 neutral molecules consisting of 103 molecules containing H, C, N, and O covering variations of multiple common organic functional groups, 68 fluorides, chlorides, bromides, and iodides, 15 compounds containing H, C, Si or S, and 9 compounds containing C–S–O or C–N–O linkages. Duffy and Jorgensen have demonstrated the utility of using CM1-AM1 partial atomic charges for arbitrary solutes in force-field simulations (Duffy and Jorgensen 2000).

In subsequent work, Li et al. (1998, 1999) and Winget et al. (2002) defined the next generation Charge Models 2 (CM2) and 3 (CM3), respectively. CM2 and CM3 differ from one another for the most part only with respect to the size and diversity of the training sets and levels of theory for which they were parameterized (CM3 is more diverse in both instances). Both models have a common functional form that is different in several ways from that for CM1. Most critically, in order to minimize sensitivity to basis set size at *ab initio* and density functional levels of theory, Löwdin starting charges and Mayer bond orders are used in place of their Mulliken analogs in the CM2 and CM3 charge mapping schemes.

The Mayer (1983) bond order is defined as

$$B_{kk'} = \sum_{\mu \in k} \sum_{v \in k'} (\mathbf{PS})_{\mu v} (\mathbf{PS})_{v\mu} \tag{9.26}$$

where $\mathbf{P}$ and $\mathbf{S}$ are the usual density and overlap matrices, respectively. This definition of bond order proves to be quite robust across a wide variety of bonding situations (for an example of its use to settle a controversy over alternative modes of bonding in a silylpalladium complex, see Sherer *et al.* 2002).

CM2 and CM3 charges are then defined as

$$q_k^{\text{CM2/CM3}} = q_k^{(0)} + \sum_{k \neq k'} B_{kk'} (C_{Z_k Z_{k'}} + D_{Z_k Z_{k'}} B_{kk'}) \tag{9.27}$$

where C and D are model parameters specific to *pairs* of atoms (as opposed to individual atoms, as in CM1), and charge normalization is assured simply by taking

$$C_{Z_k Z_{k'}} = -C_{Z_{k'} Z_k} \quad \text{and} \quad D_{Z_k Z_{k'}} = -D_{Z_{k'} Z_k} \tag{9.28}$$

CM2 and CM3 mappings have to date been defined for many different levels of theory, including AM1, PM3, SCC-DFTB, HF, and DFT.

Table 9.1 provides several molecular dipole moments as computed by a variety of different charge models and electronic structure methods, and compares them to experiment. The expectation value of the dipole moment operator evaluated for MP2/6-31G(d) wave functions has an RMS error compared to experiment of 0.21 D. The same expectation value at the HF level shows the expected increase in error from the tendency of the HF level to overestimate dipole moments. Dipole moments computed using Eq. (9.22) and ESP charges have about the same accuracy as the operator expectation value (indeed, it is possible to constrain the ESP fit so that the expectation value of the dipole moment is exactly reproduced; of course, this is not necessarily desirable if one knows the expectation value to suffer from a systematic error because of the level of theory). Eq. (9.22) used with either Mulliken or NPA charges shows rather high errors. Indeed, the error associated with the NPA charges is larger than the dispersion in the data (the dispersion is the RMS error for the simple model that assumes every dipole moment to be 2.11 D, which is the mean of the experimental data). At the PM3 level, not only are the Mulliken charges rather bad, but the expectation value of the dipole moment operator is not particularly good either. However, the CM1 mapping corrects for the errors in the PM3 electronic structures sufficiently well that the RMS error for the CM1P model is lower than that for the MP2 expectation value. The CM1 model with the AM1 Hamiltonian, the CM2 model for the BPW91/MIDI! level of theory, and the CM3 model for a tight-binding DFT level also do well. Note also that the CM*x* models, the last four columns of the table, represent the four fastest methodologies listed.

Jakalian, Jack, and Bayly (2002) have described a scheme similar in spirit to the CM*n* models insofar as AM1 charges are corrected in a bond-dependent fashion. In their AM1 bond charge corrections (AM1-BCC) model, however, each bond is assigned by the chemist

Table 9.1 Dipole moments (D) computed from different methods and RMS error compared to experiment

Molecule	Expt.	MP2[a] $\langle\mu\rangle$	HF/6-31G(d) Mullik	ESP	NPA	$\langle\mu\rangle$	PM3 Mullik	CM1P	AM1 CM1A	BPW91[b] CM2	SCC-DFTB[c] CM3
Water	1.85	2.16	2.39	2.25	2.63	1.74	0.97	1.92	2.02	1.85	1.61
Methanol	1.70	1.77	2.68	1.85	3.00	1.49	0.93	1.59	1.63	1.65	1.57
Methylformate	1.77	1.87	1.85	2.03	2.24	1.52	1.49	1.70	1.56	1.90	1.71
Formaldehyde	2.33	2.17	3.15	2.67	3.97	2.16	1.83	2.41	2.42	2.49	2.28
Acetone	2.88	2.64	3.80	3.15	4.65	2.77	2.33	2.92	2.95	3.11	3.08
Cyclopropanone	2.67	2.58	3.65	3.19	2.43	2.21	1.67	2.28	2.37	2.62	2.50
Acetic acid	1.70	1.46	2.00	1.83	2.37	1.84	2.16	1.96	1.98	1.83	2.06
Dimethyl ether	1.30	1.50	2.72	1.63	3.04	1.49	1.11	1.44	1.53	1.63	1.60
Tetrahydrofuran	1.63	1.84	2.96	1.92	3.36	1.71	1.32	1.67	1.76	1.73	1.89
Furan	0.66	0.60	1.96	0.74	1.75	0.22	0.07	0.54	0.66	0.45	0.51
Ammonia	1.47	1.92	1.77	1.96	1.98	1.52	0.00	1.32	1.75	1.71	1.34
Methylamine	1.31	1.50	1.60	1.48	1.81	1.39	0.08	1.21	1.40	1.40	1.13
Hydrogen cyanide	2.99	2.96	3.65	3.16	3.08	2.70	1.54	3.08	2.99	2.76	3.09

Acetonitrile	3.93	3.75	4.04	4.75	4.03	3.79	3.21	2.29	3.89	3.79	3.90	4.02
Formamide	3.73	3.73	4.10	4.27	4.10	5.16	3.11	2.70	3.43	3.10	3.71	3.74
Acetamide	3.76	3.63	4.03	4.26	4.05	5.21	3.28	2.86	3.63	3.26	3.76	3.83
Cyanamide	4.32	4.36	4.56	3.91	4.58	3.92	3.44	2.43	4.44	3.75	4.40	4.49
Fluoromethane	1.86	1.78	1.99	3.41	2.00	3.68	1.44	1.14	1.38	1.48	2.07	[d]
Methylsilane	0.74	0.72	0.68	0.03	0.66	0.78	0.43	0.18	0.94	0.73	0.77	[d]
Hydrogen sulfide	0.97	1.49	1.41	0.96	1.52	1.14	1.78	0.14	0.61	1.12	1.03	1.20
Methanethiol	1.52	1.78	1.79	−0.86	1.75	0.71	1.95	0.55	1.23	1.28	1.33	1.40
Thioformaldehyde	1.65	1.70	2.23	1.64	2.27	0.88	2.07	0.40	1.69	1.70	1.59	1.76
Chloromethane	1.89	2.08	2.25	1.99	2.30	2.00	1.38	0.81	1.76	1.74	1.73	[d]
RMS error	**1.01**[e]	**0.21**	**0.31**	**0.93**	**0.33**	**1.05**	**0.43**	**1.00**	**0.20**	**0.27**	**0.15**	**0.18**

[a] MP2/6-31G(d)//HF/6-31G(d). [b] BPW91/MIDI!//HF/MIDI!. [c] Kalinowski et al. (2004). [d] The underlying SCC-DFTB model does not yet contain parameters for halogens or silicon. [e] Dispersion of the experimental data.

to be one of a large number of possible types, and a fixed correction determined from having fit these parameters on a 2700-molecule test set of HF/6-31G(d) ESP charges is applied. While this protocol is robust for most molecules, it cannot be readily applied to structures not characterized by standard bonding, like transition states or structures along a reaction pathway.

9.1.4   Total Spin

Well-behaved wave functions are eigenfunctions of the total spin operator S^2, having eigenvalues of $s(s+1)$, where the quantum number s is 0 for a singlet, 1/2 for a doublet, 1 for a triplet, etc. One sometimes sees it written that s is equal to the sum of the s_z values for all of the electrons, where s_z is the expectation value of the corresponding operator S_z (spin angular momentum along the z coordinate) and takes on values in a.u. of $+1/2$ for an α electron and $-1/2$ for a β electron. This is incorrect, however. In fact, s is equal to the magnitude of the *vector* sum of the individual electronic angular momenta, and thus s can take on values according to

$$s = \frac{|n^\alpha - n^\beta|}{2}, \frac{|n^\alpha - n^\beta|}{2} + 1, \ldots, \frac{n^\alpha + n^\beta}{2} \qquad (9.29)$$

where n^ξ is the number of *unpaired* electrons of spin ξ. Thus, for instance, a system having an α and a β electron that are not paired with one another in the same MO can be either a singlet or a triplet (the so-called $S_z = 0$ triplet), reflecting the ability of s to take on values of either 0 or 1.

As described in more detail in Appendix C, the $S_z = 0$ triplet cannot be expressed as a single determinant over spin orbitals, so it cannot be represented in HF or KS theory. Of course, this is not usually a concern, since it is trivial to construct one of the other two degenerate representations of the triplet state (having an excess of either two α or two β electrons), and these *can* be approximated as single-determinantal wave functions, so we work with them instead. The point, however, is that one cannot arbitrarily sum together the s_z eigenvalues for all the unpaired electrons in a single determinant and assign to that determinant a unique spin state with s equal to the sum. Unless all of the unpaired electrons have the same spin (in which case inspection of Eq. (9.29) indicates that s can only take on one value), a single determinant is usually a mixture of states, and any properties determined as expectation values over that determinant reflect this mixing.

In UHF theory, the expectation value of the total spin operator over the single-determinantal UHF wave function is computed as

$$\langle S^2 \rangle = \left(\frac{|N^\alpha - N^\beta|}{2} \right) \left(\frac{|N^\alpha - N^\beta|}{2} + 1 \right) + \min\{N^\alpha, N^\beta\} - \sum_{i=1}^{\alpha_{occ.}} \sum_{j=1}^{\beta_{occ.}} \langle \phi_i^\alpha | \phi_j^\beta \rangle \qquad (9.30)$$

where N^ξ is the *total* number of electrons of spin ξ and the ϕ^ξ are the UHF MOs for spin ξ. Note that if all of the 'doubly' occupied orbitals are identical in shape for the α and

β electrons, then the final term on the r.h.s. will be equal to the total number of doubly occupied orbitals, since the overlap integrals can then be computed as the Krönecker δ. If all of the electrons having the minority spin occupy such orbitals, then the third term on the r.h.s. will exactly cancel the second, and the wave function will be a high-spin eigenfunction of S^2 – indeed, this describes exactly the nature of an ROHF high-spin wave function. If, on the other hand, the orbitals of the electrons of minority spin have high amplitude in regions of space occupied to a lesser extent by electrons of the opposite spin, then the sum of the overlap integrals will be less than the second term on the r.h.s., and S^2 will be greater than the presumably desired eigenvalue corresponding to the first term on the r.h.s. The degree to which the expectation value exceeds this eigenvalue reflects the spin contamination. Since the expectation value is *larger* than the expected value, the deviation derives from higher spin states contaminating the UHF wave function. So-called 'spin-projection' techniques can be used to remove these contaminating states, as discussed in more detail in Appendix C.

In DFT, there is no formal way to evaluate spin contamination for the (unknown) interacting wave function. As has already been discussed in Sections 8.5.1 and 8.5.3, however, the expectation value of S^2 computed from Eq. (9.30) over the KS determinant can nevertheless sometimes provide qualitative information about the likely utility of the DFT results with respect to their interpretation as corresponding to a pure spin state compared to a mixture of different spin states.

9.1.5 Polarizability and Hyperpolarizability

In Section 9.1.1, we discussed the molecular dipole moment as a measure of the inhomogeneity of the charge distribution. The dipole moment for an isolated molecule in a vacuum, which corresponds to that which would be computed in a typical electronic structure calculation, is often referred to as the 'permanent' electric dipole, μ_0. However, if an electric field **E** is applied to the molecule, since the charge distribution interacts with the electric field through a new term in the Hamiltonian, the dipole moment will change. The magnitude of that change per unit of electric field strength defines the electric polarizability α, i.e.,

$$\alpha = \frac{\partial \mu}{\partial \mathbf{E}} \tag{9.31}$$

Note that since both μ and **E** are vector quantities, α is a second-rank tensor. The elements of α can be computed through differentiation of Eqs. (9.1) and (9.2). The difference between the permanent electric dipole moment and that measured in the presence of an electric field is referred to as the 'induced' dipole moment.

Experimentally, the dipole moment is usually determined by measuring the change in energy for a molecule when an electric field is applied – the so-called Stark effect. At low electric-field strength, the energy change is linear in field strength, and the slope of the line is the permanent electric dipole moment. At larger field strengths, the energy change becomes quadratic because the dipole moment begins to increase proportional to the polarizability, and this permits measurement of that quantity. For still larger field strengths, a cubic contribution to the energy change can be measured (although technically it becomes increasingly challenging to fit the data reliably) and this change can be used to define the first hyperpolarizability, β (now a third-rank tensor).

It is possible to generalize this discussion in a useful way. Spectral measurements invariably assess how a molecular system changes in energy in response to some sort of external perturbation. The example presently under discussion involves application of an external electric field. If we write the energy as a Taylor expansion in some generalized vector perturbation $\mathbf{X}$, we have

$$E(\mathbf{X}) = E(0) + \left.\frac{\partial E}{\partial \mathbf{X}}\right|_{\mathbf{X}=0} \cdot \mathbf{X} + \frac{1}{2!} \left.\frac{\partial^2 E}{\partial \mathbf{X}^2}\right|_{\mathbf{X}=0} \cdot \mathbf{X}^2 + \frac{1}{3!} \left.\frac{\partial^3 E}{\partial \mathbf{X}^3}\right|_{\mathbf{X}=0} \cdot \mathbf{X}^3 + \cdots \tag{9.32}$$

Thus, Eq. (9.32) makes more clear the measurement of the Stark effect, for instance. At low electric field strengths, the only expansion term having significant magnitude involves the first derivative, and it defines the permanent dipole moment. At higher field strengths, the second derivative term begins to be noticeable, and it contributes to the energy quadratically and defines the polarizability. Finally, we see naturally how additional terms in the Taylor expansion can be used to define the first hyperpolarizability, the second hyperpolarizability γ, etc. (Note that conventions differ somewhat on whether the $1/n!$ term preceding the corresponding nth derivative term is included in the value of the physical constant or not, so that care should be exercised in comparing values reported from different sources to ensure consistency in this regard.)

Analogous quantities to the electric moments can be defined when the external perturbation takes the form of a magnetic field. In this instance the first derivative defines the permanent magnetic moment (always zero for non-degenerate electronic states), the second derivative the magnetizability or magnetic susceptibility, etc.

Equation (9.32) is also useful to the extent it suggests the general way in which various spectral properties may be computed. The energy of a system represented by a wave function is computed as the expectation value of the Hamiltonian operator. So, differentiation of the energy with respect to a perturbation is equivalent to differentiation of the expectation value of the Hamiltonian. In the case of first derivatives, if the energy of the system is minimized with respect to the coefficients defining the wave function, the Hellmann–Feynman theorem of quantum mechanics allows us to write

$$\frac{\partial}{\partial \mathbf{X}} \langle \Psi | \mathbf{H} | \Psi \rangle = \left\langle \Psi \left| \frac{\partial \mathbf{H}}{\partial \mathbf{X}} \right| \Psi \right\rangle \tag{9.33}$$

Note that $\mathbf{H}$ here is the *complete* Hamiltonian, that is, it presumably includes new terms dependent on the nature of $\mathbf{X}$. It is occasionally the case that the integral on the r.h.s. of Eq. (9.33) can be readily evaluated. Indeed, it is choice of $\mathbf{X} = \mathbf{E}$ that leads to the definition of the dipole moment operator presented in Eq. (9.1).

However, even when it is *not* convenient to solve the integral on the r.h.s. of Eq. (9.33) analytically, or when Eq. (9.33) does not hold because the wave function is not variationally optimized, it is certainly always possible to carry out the differentiation numerically. That is, one can compute the energy in the absence of the perturbation, then modify the Hamiltonian to include the perturbation (e.g., introduce an electric-field term), then compute the property as

$$\frac{\partial}{\partial X}\langle\Psi|\mathbf{H}|\Psi\rangle = \lim_{X\to 0}\frac{\langle\Psi|\mathbf{H}|\Psi\rangle - \langle\Psi|\mathbf{H}^{(0)}|\Psi\rangle}{X} \tag{9.34}$$

where $\mathbf{H}$ is again the complete Hamiltonian and $\mathbf{H}^{(0)}$ is the perturbation-free Hamiltonian. This procedure is called the 'finite-field' approach. In practice, one must take some care to ensure that computed values are numerically converged (a balance must be struck between using a small enough value of the perturbation that the limit holds but a large enough value that the numerator does not suffer from numerical noise).

Note that Eq. (9.34) can be generalized for higher derivatives, but numerical stability now becomes harder to achieve. Moreover, the procedure can be rather tedious, since in practice one must carry out a separate computation for each component associated with properties that are typically tensors. It is computationally much more convenient when analytic expressions can be found that permit direct calculation of these higher-order derivatives in a fashion that generalizes the procedure by which Eq. (9.33) is derived (not shown here).

As for the utility of different levels of theory for computing the polarizability and hyperpolarizability, the lack of high-quality gas-phase experimental data available for all but the smallest of molecules makes comparison between theory and experiment rather limited. As a rough rule of thumb, *ab initio* HF theory seems to do better for these properties than for dipole moments – at least there does not appear to be any particular systematic error. Semiempirical levels of theory are less reliable. DFT and correlated levels of MO theory do well, but it is not obvious for the latter that the improvement over HF necessarily justifies the cost, at least for routine purposes.

9.1.6 ESR Hyperfine Coupling Constants

When a molecule carries a net electronic spin, that spin interacts with the (non-zero) spins of the individual nuclei. The energy difference between the two possibilities of the electronic and nuclear spins being either aligned or opposed in the z direction can be measured by electron spin resonance (ESR) spectroscopy and defines the isotropic hyperfine splitting (h.f.s.) or hyperfine coupling constant. If we were to pursue computation of this quantity using the approach outlined in the last section, we would modify the Hamiltonian to introduce a spin magnetic dipole at a particular nuclear position. The integral that results when Eq. (9.33) is used to evaluate the necessary perturbation is known as a Fermi contact integral. Isotropic h.f.s. values are determined as

$$a_X = (4\pi/3)\langle S_z\rangle^{-1}gg_X\beta\beta_X\rho(X) \tag{9.35}$$

where $\langle S_z\rangle$ is the expectation value of the operator S_z (1/2 for a doublet, 1 for a triplet, etc.), g is the electronic g factor (typically taken to be 2.0, the approximate value for a free electron), β is the Bohr magneton, g_X and β_X are the corresponding values for nucleus X, and $\rho(X)$ is the Fermi contact integral which, when the wave function can be expressed as a Slater determinant, can be computed as

$$\rho(X) = \sum_{\mu\nu} P_{\mu\nu}^{\alpha-\beta} \varphi_\mu(\mathbf{r}_X)\varphi_\nu(\mathbf{r}_X) \tag{9.36}$$

where $\mathbf{P}^{\alpha-\beta}$ is the one-electron spin-density-difference matrix (computed as the difference between the two separate density matrices for the α and β electrons), and evaluation of the overlap between basis functions φ_μ and φ_ν is only at the nuclear position, $\mathbf{r}_X$.

We have previously defined the one-electron spin-density matrix in the context of standard HF methodology (Eq. (6.9)), which includes semiempirical methods and both the UHF and ROHF implementations of Hartree–Fock for open-shell systems. In addition, it is well defined at the MP2, CISD, and DFT levels of theory, which permits straightforward computation of h.f.s. values at many levels of theory. Note that if the one-electron density matrix is *not* readily calculable, the finite-field methodology outlined in the last section allows evaluation of the Fermi contact integral by an appropriate perturbation of the quantum mechanical Hamiltonian.

For Eq. (9.35) to be useful the density matrix employed must be accurate. In particular, localization of excess spin must be well predicted. ROHF methods leave something to be desired in this regard. Since all doubly occupied orbitals at the ROHF level are spatially identical, they make no contribution to $\mathbf{P}^{\alpha-\beta}$; only singly occupied orbitals contribute. As discussed in Section 6.3.3, this can lead to the incorrect prediction of a zero h.f.s. for all atoms in the nodal plane(s) of the singly occupied orbital(s), since their interaction with the unpaired spin(s) arises from spin polarization. In metal complexes as well, the importance of spin polarization compared to the simple analysis of orbital amplitude for singly occupied molecular orbitals (SOMOs) has been emphasized (Braden and Tyler 1998).

UHF, on the other hand, does optimize the α and β orbitals so that they need not be spatially identical, and thus is able to account for both spin polarization and some small amount of configurational mixing. As a result, however, UHF wave functions are generally *not* eigenfunctions of the operator S^2, but are contaminated by higher spin states.

The challenge with unrestricted methods is the simultaneous minimization of spin 'contamination' and accurate prediction of spin 'polarization'. The projected UHF (PUHF, see Appendix C) spin density matrix can be employed in Eq. (9.36), usually with somewhat improved results.

A complicating factor is that each spin density matrix element is multiplied by the corresponding basis function overlap at the nuclear positions. The orbitals having maximal amplitude at the nuclear positions are the core s orbitals, which are usually described with less flexibility than valence orbitals in typical electronic structure calculations. Moreover, actual atomic s orbitals are characterized by a cusp at the nucleus, a feature accurately modeled by STOs, but only approximated by the more commonly used GTOs. As a result, there are basis sets in the literature that systematically improve the description of the core orbitals in order to improve prediction of h.f.s., e.g. IGLO-III (Eriksson *et al.* 1994) and EPR-III (Barone 1995).

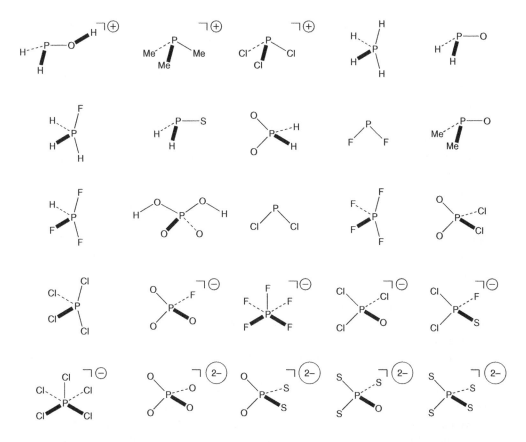

Figure 9.6 P-containing radicals for which experimental ESR data are available

A considerable body of data exists for the evaluation of different methods with respect to computing h.f.s. values. One particularly interesting data set is composed of 25 different radicals all of which contain one phosphorus atom (Figure 9.6). The experimental data that have been taken for this set, primarily as condensed-phase measurements, include 20 isotropic h.f.s. values for ^{31}P, 8 for ^{19}F, 7 for ^{35}Cl, and 5 for ^{1}H, spanning a range of about 1650 G. Cramer and co-workers (Cramer and Lim 1994; Lim *et al.* 1996) and Nguyen, Creve, and Vanquickenborne (1997) have examined the accuracy of a number of different levels of theory over these data, primarily using the 6-311G(d,p) basis set.

When geometries optimized at the MP2/6-31G(d,p) were employed, the mean unsigned errors at the ROHF, UHF, PUHF, and MP2 levels of theory were 35.8, 45.6, 24.8, and 21.1 G, respectively. ROHF theory is more accurate than UHF theory in this case, presumably owing to moderate spin contamination in the latter. Projecting out the spin contamination at the PUHF level reduces the error by almost one half, while going to second-order perturbation theory (which introduces electron correlation and also probably reduces the spin contamination compared to UHF) provides an improvement of about the same order.

Of these four levels, the computation of the MP2 spin-density matrix is considerably more time-consuming than the other three. It is thus of interest to examine the accuracy of DFT methods, which by construction include electron correlation directly into their easily computed spin-density matrices. For the same geometries, the mean unsigned errors for the BVWN, BLYP, B3P86, and B3LYP levels of theory were 32.6, 32.6, 29.7, and 28.9 G. Somewhat surprisingly, these errors increased in every case when geometries were optimized at the corresponding DFT level, to 60.5, 54.4, 30.9, and 34.3 G. For this particular data set, several of the radicals seem prone to the DFT overdelocalization problem noted in Section 8.5.6. Guerra (2000) has shown similarly poor performance of the B3LYP functional in the context of vinylacyl radicals, where the functional strongly overestimates the stability of π delocalized radicals relative to σ alternatives, in contravention of experimental data.

In cases where overdelocalization is *not* a problem, however, DFT methods have proven to be quite robust for computing h.f.s. constants. For instance, Adamo, Cossi, and Barone (1999) have reported results for h.f.s. constants in the methyl radical using PW, B3LYP, and PBE1PBE that are competitive with correlated MO methods (Chipman 1983; Cramer 1991; Barone *et al.* 1993). Moreover, if a given system suffers from heavy spin contamination at the UHF level of theory, DFT may be the only reasonable recourse.

In general, then, DFT methods provide the best combination of accuracy and efficiency so long as overdelocalization effects do not poison their performance. The MP2 level of theory also provides a reasonably efficient way of carrying out h.f.s. calculations at a correlated level of theory. More highly correlated levels of MO theory are generally more accurate, but can be prohibitively expensive in large systems.

As a final note, although we have focused here on the computation of isotropic h.f.s. values, it is also straightforward to compute anisotropic hyperfine couplings, although these cannot be observed experimentally unless the system can be prevented from random tumbling (e.g., by freezing in a matrix or single crystal). Similarly, it is possible to calculate the electronic *g* value. These subjects are beyond the scope of the text, however, and interested readers are referred to relevant titles in the bibliography.

9.2   Ionization Potentials and Electron Affinities

As the general utility of semiempirical, HF, and DFT methods for the computation of IPs and EAs has already been discussed in some detail in Sections 5.6.1, 6.4.1, and 8.6.1, this section is restricted to a very brief recapitulation of the most important points relative to these properties.

Koopmans' theorem suggests that the ionization energies for any orbital (usually 'IP' refers specifically to the ionization potential associated with the HOMO) will be equal to the negative of the eigenvalue of that orbital in HF theory. This provides a particularly simple method for estimating IPs, and because of canceling errors in basis-set incompleteness and failure to adequately account for electron correlation, the approach works reasonably well for the occupied orbitals in the highest energy range in *ab initio* HF wave functions (with semiempirical methods, performance is spottier). However, as one ionizes from orbitals that are lower in energy, relaxation effects become large, and Koopmans' approximation breaks down.

A more rigorous alternative, at least in formulation, is to explicitly calculate the energy of the radical cation resulting from ionization, either at the neutral geometry (in which case the energy difference with the neutral is the vertical IP) or at its own optimized geometry (in which case the energy difference with the neutral is the adiabatic IP). However, in order to obtain reasonable accuracy with this so-called ΔSCF approach, it is critical to employ a level of theory capable of accurately capturing the differential correlation energies in these systems having different numbers of electrons. It must be noted, incidentally, that Koopmans' approximation should really be regarded only as an approach to the *vertical* IP – good agreement with adiabatic IPs when it occurs is purely fortuitous insofar as neither electronic nor geometric relaxation effects are accounted for in any way.

In DFT, Koopmans' theorem does not apply, but the eigenvalue of the highest KS orbital has been proven to be the IP *if* the functional is exact. Unfortunately, with the prevailing approximate functionals in use today, that eigenvalue is usually a rather poor predictor of the IP, although use of linear correction schemes can make this approximation fruitful. ΔSCF approaches in DFT can be successful, but it is important that the radical cation not be subject to any of the instabilities that can occasionally plague the DFT description of open-shell species.

Koopmans' theorem also implies that the eigenvalue associated with the HF LUMO may be equated with the EA. However, in the case of EAs errors associated with basis set incompleteness and differential correlation energies do not cancel, but instead they reinforce one another, and as a result EAs computed by this approach are usually entirely untrustworthy.

Although ΔSCF methods are more likely to be successful, it is critical that diffuse functions be included in the basis set so that the description of the radical anion is adequate with respect to the loosely held extra electron. In general, correlated methods are to be preferred, and DFT represents a reasonably efficient choice that seems to be robust so long as the radical anion is not subject to overdelocalization problems. Semiempirical methods do rather badly for EAs, at least in part because of their use of minimal basis sets.

9.3 Spectroscopy of Nuclear Motion

Within the context of the Born–Oppenheimer approximation, the potential energy surface may be regarded as a 'property' of an empirical molecular formula. With a defined PES, it is possible to formulate and solve Schrödinger equations for *nuclear* motion (as opposed to electronic motion)

$$\left[-\sum_{i}^{N} \frac{1}{2m_i} \nabla_i^2 + V(\mathbf{q}) \right] \Xi(\mathbf{q}) = E \Xi(\mathbf{q}) \tag{9.37}$$

where N is the number of atoms, m is the atomic mass, V is the potential energy from the PES as a function of the $3N$ nuclear coordinates $\mathbf{q}$, and Ξ is the nuclear wave function that is expressed in those coordinates. Solution of Eq. (9.37) provides entry into the realms of rotational and vibrational spectroscopy. The following subsections describe the relevant theory and detail the applicability of different methodologies for such computations.

9.3.1 Rotational

The simplest approach to modeling rotational spectroscopy is the so-called 'rigid-rotor' approximation. In this approximation, the geometry of the molecule is assumed to be constant at the equilibrium geometry q_{eq}. In that case, $V(q_{eq})$ in Eq. (9.37) becomes simply a multiplicative constant, so that we may write the rigid-rotor rotational Schrödinger equation as

$$-\sum_i^N \frac{1}{2m_i} \nabla_i^2 \Xi(q) = E \Xi(q) \tag{9.38}$$

where E_0, the eigenvalue for Eq. (9.38) corresponding to the lowest-energy rotational state, is taken to be the electronic energy for the equilibrium geometry.

Equation (9.38), if restricted to two particles, is identical in form to the radial component of the electronic Schrödinger equation for the hydrogen atom expressed in polar coordinates about the system's center of mass. In the case of the hydrogen atom, solution of the equation is facilitated by the simplicity of the two-particle system. In rotational spectroscopy of polyatomic molecules, the kinetic energy operator is considerably more complex in its construction. For purposes of discussion, we will confine ourselves to two examples that are relatively simple, presented without derivation, and then offer some generalizations therefrom. More advanced treatises on rotational spectroscopy are available to readers hungering for more.

The simplest possible case is a non-homonuclear diatomic (non-homonuclear because a dipole moment is required for a rotational spectrum to be observed). In that case, solution of Eq. (9.38) is entirely analogous to solution of the corresponding hydrogen atom problem, and indicates the eigenfunctions Ξ to be the usual spherical harmonics $Y_J^m(\theta, \phi)$, with eigenvalues given by

$$E_J = \frac{J(J+1)\hbar^2}{2I} \tag{9.39}$$

where the moment of inertia I about a given axis is defined as

$$I = \sum_k^{\text{nuclei}} m_k r_k^2 \tag{9.40}$$

In the special case of a heteronuclear diatomic, rotation occurs exclusively about a single axis passing through the center of mass and perpendicular to the bond, and I is simply μr_{eq}^2, where the reduced mass μ is computed as

$$\mu = \frac{m_1 m_2}{m_1 + m_2} \tag{9.41}$$

Because the wave functions are the spherical harmonics, each rotational level is $(2J + 1)$-fold degenerate (over the quantum number m); note that the lowest rotational level has a rotational energy of zero, consistent with the earlier statement that the *total* energy associated with this level is just the electronic energy of the equilibrium structure. Selection rules dictate that transitions occur only between adjacent levels, i.e., $\Delta J = \pm 1$ (see Section 14.5), in which case the energy change observed for transition from level J to level $J + 1$ is

$$\Delta E = \frac{\{(J + 1)[(J + 1) + 1] - J(J + 1)\}\hbar^2}{2I}$$

$$= \frac{2(J + 1)\hbar^2}{2I} \tag{9.42}$$

When probed spectroscopically, the absorption frequency ν can be determined as

$$\nu_J = \frac{\Delta E}{h}$$

$$= \frac{2(J + 1)\hbar^2}{2hI}$$

$$= 2(J + 1)B \tag{9.43}$$

where B, the molecular rotational constant, is

$$B = \frac{h}{8\pi^2 I} \tag{9.44}$$

Non-linear molecules are more complicated than linear ones because they are characterized by three separate moments of inertia. In highly symmetric cases, however, relatively simple solutions of Eq. (9.38) continue to exist. For instance, in molecules possessing an axis of rotation that is three-fold or higher in symmetry, the two moments of inertia for rotation about the two axes perpendicular to the high-symmetry axis will be equal. For example, in fluoromethane, which is C_{3v}, there is one moment of inertia, I_A, about the symmetry axis A, and there are two equal moments of inertia, I_B and I_C, about the axes perpendicular to axis A. In this particular case, the magnitude of the latter two moments is larger than that of the former moment because the heavy atoms have displacements of 0 from axis A but not from the other two, and such a molecule is called a prolate top. In the case of a prolate top, the rotational eigenvalues are given by

$$E_J^K = \frac{J(J + 1)\hbar^2}{2I_B} + K^2 \left(\frac{1}{I_A} - \frac{1}{I_B} \right) \frac{\hbar^2}{2} \tag{9.45}$$

where K is the quantum number, running over $-J, -J + 1, \ldots, J-1, J$, expressing the component of the angular momentum along the highest symmetry axis. The selection rules for a rotational transition in this case are $\Delta J = \pm 1$ and $\Delta K = 0$, and thus Eqs. (9.43) and (9.44) continue to be valid for absorption frequencies using $I = I_B$.

Less symmetric molecules require a considerably more complicated treatment, but in the end their spectral transitions are functions of their three moments of inertia (see Section 10.3.5). From a computational standpoint, then, prediction of rotational spectral lines depends *only* on the moments of inertia, and hence only on the molecular geometry. Thus, any method which provides good geometries will permit an accurate prediction of rotational spectra within the regime where the rigid-rotor approximation is valid.

Since even very low levels of theory can give fairly accurate geometries, rotational spectra are quite simple to address computationally, at least over low rotational quantum numbers. For higher-energy rotational levels, molecular centrifugal distortion becomes an issue, and more sophisticated solutions of Eq. (9.37) are required.

9.3.2   Vibrational

When thinking about chemical thermodynamics and kinetics, it is a convenient formalism to picture a molecule as being a ball rolling on a potential energy surface. In this simple model, the exact position of the ball determines the molecular geometry and the potential energy, and its speed as it rolls in a frictionless way determines its kinetic energy. Of course, quantum mechanical particles are different than classical ones in many ways; one of the more important differences is that they are subject to the uncertainty principle. One consequence of the uncertainty principle is that polyatomic molecules, even at absolute zero, must vibrate – within the simple ball and surface picture, the ball must always be moving, with a sum of potential and kinetic energy that exceeds the energy of the nearest minimum by some non-zero amount. This energy is contained in molecular vibrations.

Transitions in molecular vibrational energy levels typically occur within the IR range of the frequency spectrum. Because vibrational motions tend to be highly localized within molecules, and the energy spacings associated with individual linkages tend to be reasonably similar irrespective of remote molecular functionality, IR spectroscopy has a long history of use in structure determination. Vibrational frequencies also have other important uses, for example in kinetics (Section 14.3) and computational geometry optimization (Section 2.4.1), so their accurate prediction has been a long-standing computational goal. We now examine different approaches towards that goal, and the utility of different levels of theory in application.

9.3.2.1   One-dimensional Schrödinger equation

It is again useful to begin with the simplest possible case, the diatomic molecule. Equation (9.37), when restricted to the vibrational motion alone, is clearly a function of only a single variable, the interatomic distance r. Solutions of differential equations of only a single variable are typically reasonably straightforward. Our only challenge here is that we do not know exactly what the potential energy function V looks like as a function of r. Given a level of theory, however, we can compute V point by point to an arbitrary level of fineness (i.e., simply compute the electronic energy of the system for various fixed values of r). Those points may then be fit to any convenient analytic function – polynomial,

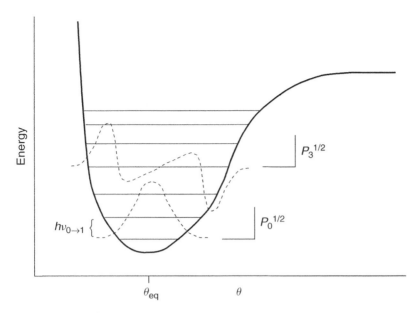

Figure 9.7 Vibrational energy levels determined from solution of the one-dimensional Schrödinger equation for some arbitrary variable θ (some higher levels not shown). In addition to the energy levels (horizontal lines across the potential curve), the vibrational wave functions are shown for levels 0 and 3. Conventionally, the wave functions are plotted in units of (probability)$^{1/2}$ with the same abscissa as the potential curve and an individual ordinate having its zero at the same height as the location of the vibrational level on the energy ordinate – those coordinate systems are explicitly represented here. Note that the absorption frequency typically measured by infrared spectroscopy is associated with the $0 \rightarrow 1$ transition, as indicated on the plot. For the harmonic oscillator potential, *all* energy levels are separated by the same amount, but this is not necessarily the case for a more general potential

Morse, etc. – and the one-dimensional Schrödinger equation solved using standard numerical recipes to yield eigenfunctions and eigenvalues. A typical representation of the results from such a calculation is provided in Figure 9.7. Assuming a high level of theory is used in the generation of V, very high accuracy can be achieved for the energies of all of the vibrational levels, and hence for the energies of transition between them.

In a more complicated polyatomic molecule, Eq. (9.37) is a function of $3N$ variables (now including translation and rotation in the overall motion), for instance, the x, y, and z coordinates of each atom in some laboratory frame. If N is more than a very small number of atoms, however, it becomes quite tedious to generate V pointwise as a function of all of these variables. Moreover, neither the fitting of V to an analytic form nor the solution of the resulting multi-dimensional differential equation is at all trivial, so the approach described above for diatomics is rarely used to compute vibrational data for larger molecules. The exception is in certain instances of high symmetry, where group theory may sometimes be used to reduce the dimensionality of the problem by separating vibrational coordinates according to their irreducible representations. Such efforts are not, however, routine.

One way to simplify the problem is to recognize that most chemical systems of interest are at sufficiently low temperature that only their lowest vibrational levels are significantly populated. Thus, from a spectroscopic standpoint, only the transition from the zeroth vibrational level to the first is observed under normal conditions, and so it is these transitions that we are most interested in predicting accurately. Another way of thinking about this situation is that we are primarily concerned only with regions of the PES relatively near to the minimum, since these are the regions sampled by molecules in their lowest and first excited vibrational states. Once we restrict ourselves to regions of the PES near minima, we may take advantage of Taylor expansions to simplify our construction of V.

9.3.2.2 Harmonic oscillator approximation

Let us consider again our simple diatomic case. Using Eq. (2.2) for the potential energy from a Taylor expansion truncated at second order, Eq. (9.37) transformed to internal coordinates becomes

$$\left[-\frac{1}{2\mu} \frac{\partial^2}{\partial r^2} + \frac{1}{2} k (r - r_{eq})^2 \right] \Xi(r) = E \Xi(r) \tag{9.46}$$

where μ is the reduced mass from Eq. (9.41), r is the bond length, and k is the bond force constant, i.e., the second derivative of the energy with respect to r at r_{eq} (see Eq. (2.1)). Eq. (9.46) is the quantum mechanical harmonic oscillator equation, which is typically considered at some length in elementary quantum mechanics courses. Its eigenfunctions are products of Hermite polynomials and Gaussian functions, and its eigenvalues are

$$E = \left(n + \frac{1}{2} \right) h\omega \tag{9.47}$$

where n is the vibrational quantum number and

$$\omega = \frac{1}{2\pi} \sqrt{\frac{k}{\mu}} \tag{9.48}$$

The selection rules for the QM harmonic oscillator permit transitions only for $\Delta n = \pm 1$ (see Section 14.5). As Eq. (9.47) indicates that the energy separation between any two adjacent levels is always $h\omega$, the predicted frequency for the $n = 0$ to $n = 1$ absorption (or indeed any allowed absorption) is simply $\nu = \omega$. So, in order to predict the stretching frequency within the harmonic oscillator equation, all that is needed is the second derivative of the energy with respect to bond stretching computed at the equilibrium geometry, i.e., k. The importance of k has led to considerable effort to derive analytical expressions for second derivatives, and they are now available for HF, MP2, DFT, QCISD, CCSD, MCSCF and select other levels of theory, although they can be quite expensive at some of the more highly correlated levels of theory.

Prior to proceeding, it is important to address the errors introduced by the harmonic approximation. These errors are intrinsic to the truncation of the Taylor expansion, and will remain even for an exact level of electronic structure theory. The most critical difference is that real bonds dissociate as they are stretched to longer and longer values. Thus, as indicated in Figure 9.7, the separation between vibrational levels narrows with increasing vibrational quantum number, and the total number of levels is finite. By contrast, the harmonic oscillator has an infinite number of levels, all equally spaced.

While the differences between the harmonic oscillator approximation and the true system are largest for higher vibrational levels, even at very short distances beyond the equilibrium bond length the true potential energy of the bond stretch curve is lower than that predicted by the parabolic potential of the harmonic approximation. Since the more shallow correct potential generates a lower vibrational frequency than that associated with the parabola, this means that an 'exact' harmonic frequency will always be greater than the true frequency. Over the few data available for diatomics that are sufficiently complete so that the PES can be constructed and harmonic frequencies inferred, the difference averages about 3%. Any level of theory that exceeds this accuracy using the harmonic approximation is presumably simply benefiting from a fortuitous cancellation of errors.

What about the polyatomic case? In that case, we must carry out a multi-dimensional Taylor expansion analogous to Eq. (2.26). This leads to the multi-dimensional analog of Eq. (9.46)

$$\left[-\sum_i^{3N} \frac{1}{2m_i} \frac{\partial^2}{\partial x_i^2} + \frac{1}{2}(\mathbf{x} - \mathbf{x}_{eq})^\dagger \mathbf{H}(\mathbf{x} - \mathbf{x}_{eq}) \right] \Xi(\mathbf{x}) = E\,\Xi(\mathbf{x}) \tag{9.49}$$

where $\mathbf{x}$ is the vector of atomic coordinates, $\mathbf{x}_{eq}$ defines the equilibrium structure, and $\mathbf{H}$ is the Hessian matrix defined by Eq. (2.37).

While Eq. (9.49) has a well-defined potential energy function, it is quite difficult to solve in the indicated coordinates. However, by a clever transformation into a unique set of mass-dependent spatial coordinates $\mathbf{q}$, it is possible to separate the $3N$-dimensional Eq. (9.49) into $3N$ one-dimensional Schrödinger equations. These equations are identical to Eq. (9.46) in form, but have force constants and reduced masses that are defined by the action of the transformation process on the original coordinates. Each component of $\mathbf{q}$ corresponding to a molecular vibration is referred to as a 'normal mode' for the system, and with each component there is an associated set of harmonic oscillator wave functions and eigenvalues that can be written entirely in terms of square roots of the force constants found in the Hessian matrix and the atomic masses.

Note that because Eq. (9.49) is over the full $3N$ coordinates, the transformed coordinate system $\mathbf{q}$ includes three translational and three rotational (two for linear molecules) 'modes'. The eigenvalues associated with these modes are typically very close to zero, and indeed, the degree to which they *are* close to zero can be regarded as a diagnostic of how well optimized the structure is in terms of being at the local minimum geometry.

A few last technical points merit some discussion prior to an assessment of the relative utilities of different theoretical levels for prediction of IR spectra. First, note that the first derivatives in the Taylor expansion disappear only when the potential is expanded about a critical point on the PES (since then the gradients are all zero). Thus, the form of Eq. (9.49) is not valid if the level of theory used in the computation of the Hessian matrix differs from that used for geometry optimization, since the two different levels of theory will almost inevitably have different minimum energy structures. Put more succinctly, there is little value in a frequency calculation for a particular geometry under the harmonic oscillator approximation unless the geometry in question was optimized at the same level of theory. (Note that mathematically one could certainly include the gradient term in the potential in Eq. (9.49), but the resulting differential equation is not worth working with.)

Another interesting point in this regard is that the form of Eq. (9.49) *is* valid for other stationary points that are *not* minima on the PES. However, in this instance there will be one or more normal mode force constants that will be negative, corresponding to motion along modes that lead to energy lowering. Insofar as the frequencies are computed from the square roots of the force constants, this leads to an imaginary frequency (one often sees these called negative frequencies in the literature, but this is simply sloppy). Frequency calculations thus are diagnostic as to the nature of stationary points. All positive frequencies implies a (local) minimum, one imaginary frequency implies a transition state structure, and two or more imaginary frequencies refers to stationary points characterized by additional negative force constants. Such structures are sometimes useful in searching for TS structures by following the various energy-lowering modes, but they have no chemical significance.

The utility of Eq. (9.49) depends on the ease with which the Hessian matrix may be constructed. Methods that allow for the analytic calculation of second derivatives are obviously the most efficient, but if analytic first derivatives are available, it may still be worth the time required to determine the second derivatives from finite differences in the first derivatives (where such a calculation requires that the first derivatives be evaluated at a number of perturbed geometries at least equal to the number of independent degrees of freedom for the molecule). If analytic first derivatives are not available, it is rarely practical to attempt to construct the Hessian matrix.

A technical point in this regard with respect to DFT is that when one refers to 'analytic' derivatives, what is actually meant is analytic derivatives to the quadrature schemes that are used to *approximate* the solution of the complicated integrals defining the exchange-correlation energy; analytic solutions to these integrals are not in general available, and hence neither are their derivatives. In practice, failure to converge the quadrature schemes has a considerably larger effect on second derivatives than it does on energies, and it is not uncommon to see potentially rather large changes in computed vibrational frequencies when switching from default to more dense quadrature-point densities (sometimes also called 'grid' densities) in standard electronic-structure packages. This effect can be particularly troubling with low frequencies, since the error can cause the frequencies to switch from real to imaginary and vice versa, and some care should be exercised where such issues are important.

With respect to absolute accuracy, Table 9.2 provides the mean unsigned errors in harmonic vibrational frequencies for a number of levels of theory over the 32 molecules in the reduced G2 test set. HF theory shows the poorest performance (AM1 and PM3 are in general somewhat worse than HF with a moderate basis set, however data are not available for this particular test set). MP2 shows significant improvement over HF, but substantial

Table 9.2 Mean absolute errors in harmonic vibrational frequencies over a 32-molecule G2 subset $(cm^{-1})^a$

Level of theory	Error
MO theoretical methods	
HF/6-311G(3df,2p)	144
MP2/6-31G(d,p)	99
CCSD(T)/6-311G(3df,2p)	31
LSDA functionals	
SVWN/6-31G(d,p)	75
GGA functionals	
BLYP/6-311G(d,p)	59
BPW91/6-311G(d,p)	69
PWPW91/6-311G(d,p)	66
mPWPW91/6-311G(d,p)	66
Hybrid functionals	
BH&HLYP/6-311G(d,p)	100
B1LYP/6-311G(d,p)	33
B1PW91/6-311G(d,p)	48
mPW1PW91/6-311G(d,p)	39
B3LYP/6-311G(d,p)	31
B3PW91/6-311G(d,p)	45
mPW3PW91/6-311G(d,p)	37

[a]Test set includes 32 molecules containing only first-row atoms, see Johnson, Gill, and Pople (1993). Data from Adamo and Barone (1998).

error remains. CCSD(T) and some of the hybrid levels of density functional theory show the highest accuracies. In general, the BLYP combination seems to be more accurate than BPW91, whether pure or hybrid in formulation, but PWPW91 is nearly as accurate as BLYP, again whether pure or hybrid in formulation.

Of some interest in the error analysis is the degree to which the error is systematic. Although HF errors are large, they are very systematic. HF overemphasizes bonding, so all force constants are too large, and thus so are all frequencies. However, application of a constant scaling factor to the HF frequencies improves their accuracy enormously (Pople *et al.* 1993). Scott and Radom studied this issue in detail for eight different levels of theory using a database of 122 molecules and 1066 fundamentals (i.e., measured, anharmonic vibrational frequencies) and a summary of their results, together with a few other recommended scaling factors, is provided in Table 9.3 (Scott and Radom 1996; see also, Wong 1996). Note that even though the scale factor required for the HF/6-31G(d) level of theory is substantial, reducing every frequency by more than 10%, the final accuracy is quite high – better than the considerably more expensive MP2. Note also that the pure DFT functional BLYP requires essentially no scaling, i.e., its errors are random about the experimental values,

Table 9.3 Scale factors and post-scaling errors in vibrational frequencies from different levels of theory[a]

Level of theory	Scale factor	RMS error (cm^{-1})	Outliers $(\%)^b$
AM1	0.9532	126	15
PM3	0.9761	159	17
HF/3-21G	0.9085	87	9
HF/6-31G(d)	0.8953	50	2
HF/6-31G(d,p)	0.8992	53	3
HF/6-311G(d)^c	0.9361	32	
HF/6-311G(d,p)	0.9051	54	3
HF/LANL2DZ^c	0.9393	49	
MP2/6-31G(d)	0.9434	63	4
MP2/6-31G(d,p)^d	0.9646		
MP2/pVTZ^e	0.9649	70	
QCISD/6-31G(d)	0.9537	37	2
BLYP/6-31G(d)	0.9945	45	2
BLYP/6-311G(d)^c	1.0160	38	
BLYP/LANL2DZ^c	1.0371	47	
BP86/6-31G(d)	0.9914	41	2
B3LYP/6-31G(d)	0.9614	34	1
	0.9664^f	46	
	0.9800^g		
B3LYP/6-311G(d)^c	0.9739	38	
B3LYP/pVTZ^d	0.9726	42	
B3LYP/6-311+G(3df,2p)^g	0.9890		
B3LYP/6-311++G(3df,3pd)^f	0.9542	31	
B3LYP/LANL2DZ^c	0.9978	45	
B3PW91/6-31G(d)	0.9573	34	2
B3PW91/pVTZ^d	0.9674	43	
VSXC/6-31G(d)	0.9659	48	
VSXC/6-311++G(3df,3pd)^f	0.9652	37	

[a] Data from Scott and Radom (1996) unless otherwise indicated. [b] Number of frequencies still in error by more than 20% of the experimental value after scaling. [c] From analysis of 511 frequencies in 50 *inorganic* molecules (Bytheway and Wong 1998). [d] Pople *et al.* (1993). [e] From analysis of 900 frequencies for 111 molecules comprised of first- and second-row atoms and hydrogen (Halls, Velkovski, and Schlegel 2001). [f] From analysis of 110 frequencies for 31 small molecules having only first-row atoms and hydrogen (Jaramillo and Scuseria 1999). [g] Bauschlicher and Partridge (1995).

while the hybrid functionals require scale factors consistent with their inclusion of some HF character. Thus, including HF character results in proportionately too high predictions in vibrational frequencies, although the scaling procedure is very effective here as well. Finally, the errors in the semiempirical levels are quite high, and scaling is only modestly helpful. For those looking for the highest accuracy, the U.S. National Institute of Standards and Technology (NIST) maintains a web facility that permits users to select a focused set of molecules from NIST's computational chemistry database (presumably based upon the user's interest in a structurally related unknown) and then to compute least-squares best scaling factors for specific levels of theory based only on those molecules (srdata.nist.gov/cccbdb/). One example of such an approach, albeit not using the NIST website, was provided by Yu, Srinivas, and Schwartz (2003) who optimized scale factors just for the C$-$O stretch of metal bound carbonyls.

Results from molecular mechanics can also be of reasonable accuracy, so long as the molecules addressed contain only functionality well represented in the force field training set. While extensive compilations of data are not available, Halgren has compared MM3 and MMFF94 over a test set of 157 frequencies from organic molecules and found RMS errors of 57 and 60 cm^{-1}, respectively.

An interesting alternative to scaling the frequencies is instead to scale the force constants in the Hessian, which permits some sensitivity to different kinds of vibrations, e.g., stretches, bends, and torsions (see, for example, Grunenberg and Herges 1997; Baker, Jarzecki, and Pulay 1998; Arenas *et al.* 2000). Of course, as with any parameterization procedure, as the number of parameters increases so too does the requirement for additional data to ensure statistical reliability, and this approach has not yet seen wide application.

One final caveat with respect to comparing experimental IR spectra with theoretically predicted frequencies is that the latter do not account for such experimental complications as Fermi resonances (where two nearby fundamentals are shifted to higher and lower frequencies, respectively), overtones, etc. Such details require case-by-case evaluation.

In comparing complete theoretical spectra to complete experimental spectra in molecules of moderate to large size, there can be a large number of lines. To ensure proper correspondence of the normal modes, it is helpful to compare not only the absorption frequencies themselves but also the intensities of the absorptions. For a typical experimental spectrum, such intensities are usually reported simply as strong, medium, or weak, although in careful experiments absorption cross-sections can be measured accurately. From a computational standpoint, the prediction of IR intensities can be accomplished using the mixed second derivatives of the energy with respect to geometric motion and an external electric field (thereby permitting estimation of the changes in the dipole moment as a function of the vibrations, which is what IR intensities are proportional to). These mixed second derivatives are available analytically for all levels of theory for which analytic second derivatives with respect to the geometry are available, so it is a straightforward matter to compute IR intensities. The actual computed values tend to be no better than qualitative in the absence of using a very complete basis set and accounting for electron correlation, but insofar as most experimental intensities are essentially qualitative, this is not typically much of a drawback. Being able to line up strong absorptions in computed and experimental spectra is often quite helpful for assessing the validity of the comparison.

An alternative experiment that measures the same vibrational fundamentals subject to different selection rules is Raman spectroscopy. Raman intensities, however, are more difficult to compute than IR intensities, as a mixed *third* derivative is required to approximate the change in the molecular polarizability with respect to the vibration that is measured by the experiment. The sensitivity of Raman intensities to basis set and correlation is even larger than it is for IR intensities. However, Halls, Velkovski, and Schlegel (2001) have reported good results from use of the large polarized valence-triple-ζ basis set of Sadlej (1992) and have determined frequency scaling factors for its use in conjunction with several levels of theory (see Table 9.3).

9.3.2.3   *Vibrationally averaged expectation values*

With fairly few exceptions, all discussion of computed molecular properties up to this point has proceeded under the assumption that the value computed for the stationary equilibrium structure is relevant in comparison to experiment. However, the experimental population is in constant vibrational motion, even at 0 K, so the experimental measurement actually samples structures having a distribution dictated by the molecular vibrational wave function. Thus, for some property A, the measured value is the expectation value given by

$$\langle A \rangle = \int \Xi(\mathbf{q}) A(\mathbf{q}) \Xi(\mathbf{q}) d\mathbf{q} \tag{9.50}$$

where Ξ is the wave function for nuclear motion and $\mathbf{q}$ is the coordinate system.

Some general analysis of Eq. (9.50) is warranted. Note that the zeroth vibrational level for a normal mode within the harmonic approximation is characterized by a Gaussian wave function. The total molecular vibrational wave function is a product of the wave functions of all of the individual modes, so if every vibration is in its ground state, Ξ is an even function if the origin is taken to be $\mathbf{q}_{eq}$ ('even' meaning that the function has the same value for equal displacements in the positive and negative directions along any axis). It is now helpful to consider not the expectation value of A, but the expectation value for the *deviation* of A from the value at the equilibrium position, i.e.,

$$\langle A - A(\mathbf{q}_{eq}) \rangle = \int \Xi(\mathbf{q}) \left[A(\mathbf{q}) - A(\mathbf{q}_{eq}) \right] \Xi(\mathbf{q}) d\mathbf{q} \tag{9.51}$$

If the deviation of A from $A(\mathbf{q}_{eq})$ is coupled with only a single component of $\mathbf{q}$, and if its dependence on displacement from the equilibrium structure is linear, then ΔA is an odd function of $\mathbf{q}$ ('odd' meaning now that the function takes on positive and negative values of equal magnitude when displaced an equal distance along any axis). From elementary calculus, we know that the product of two even functions and an odd function is an odd function, and that the integral over all space of an odd function is zero, so under the conditions outlined above the expectation value defined by Eq. (9.51) is zero and the value of $\langle A \rangle$ must be $A(\mathbf{q}_{eq})$. An example of such a situation would be a harmonic oscillator having a dipole moment. The change in dipole moment is linear in the displacement from the equilibrium bond length, so the expectation value of the dipole moment over the first vibrational wave function (indeed, over *any* of the vibrational wave functions in this case, since the system is harmonic) is exactly equal to the dipole moment at the equilibrium bond length.

Of course, in a real system with many atoms, the coupling of the property to the individual degrees of freedom is more complicated, and there is no guarantee that ΔA will be an odd function. Nevertheless, the assumption that Eq. (9.51) is equal to zero is often sufficiently accurate for everyday computational predictions.

To illustrate a case where this is not true, consider the methyl radical $CH_3{}^\bullet$. The equilibrium structure for this system is planar, and the unpaired electron occupies the out-of-plane p_z orbital on carbon. Because this orbital has a node at the carbon atom, in the absence of polarization the ^{13}C isotropic hyperfine splitting should be zero (only spin polarization makes it non-zero). However, one of the normal modes of the methyl radical is the so-called 'umbrella' mode that simultaneously bends all the hydrogen atoms to one side of the plane or the other. This motion rehybridizes the singly occupied molecular orbital (SOMO) so that it includes some s character, and thus the ^{13}C h.f.s. value should become increasingly positive. Moreover, this is true irrespective of the side to which the umbrella motion takes place. That is, if we now take ΔA to be the change in ^{13}C h.f.s., it is an *even* function about the equilibrium structure (Figure 9.8). As such, we expect from Eq. (9.51) that the expectation value of the h.f.s. splitting over the umbrella mode vibrational wave function should be significantly different from the value at the equilibrium position (the other vibrational modes of $CH_3{}^\bullet$ do not cause the molecule to deviate from planarity, so they have minimal impact on the expectation value).

Table 9.4 compares to experiment the isotropic h.f.s. values computed for ^{13}C and ^1H in the methyl radical at the UMP2/6-311G(d,p) level both (i) at the UHF/6-31G(d) equilibrium geometry and (ii) as the expectation value over the umbrella mode vibrational wave function computed at this level. Also included are data for the monofluoromethyl radical $CH_2F^\bullet$, which is even more affected by vibrational averaging because it has a very shallow double-well potential along the umbrella mode (i.e., the equilibrium structure is pyramidal, but the barrier to inversion is less than 1 kcal mol^{-1}), so that its vibrational wave function has large amplitude around a planar structure with *smaller* ^{13}C h.f.s. than for the equilibrium structure.

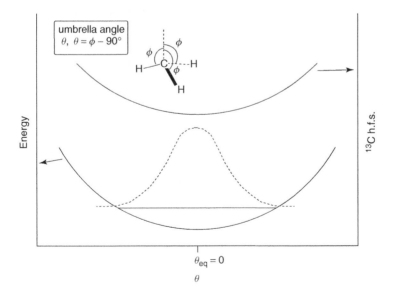

Figure 9.8 Potential energy and zeroth vibrational level with associated wave function (reference left ordinate) and ^{13}C h.f.s. (reference right ordinate) as a function of umbrella angle for $CH_3{}^\bullet$

Table 9.4 Isotropic hyperfine splittings (G) in the methyl and fluoromethyl radicals

Radical	Nucleus	$A(\mathbf{q}_{eq})$	$\langle A \rangle$, Eq. (9.41)	$\langle A \rangle$, expt.
$CH_3^{\bullet}$	^{13}C	22.6	32.8	38.3
	^{1}H	−27.6	−25.6	−25.0
$CH_2F^{\bullet}$	^{19}F	73.7	71.7	64.3
	^{13}C	72.6	49.6	54.8
	^{1}H	−15.4	−22.7	−21.1
RMS error		11.8	4.8	

Considering all five h.f.s. values, the agreement with experiment improves in every case when vibrational averaging is taken into account, and the RMS error drops from 11.8 G to 4.8 G.

9.4 NMR Spectral Properties

Nuclear magnetic resonance (NMR) is probably the most widely applied spectroscopic technique in modern chemical research. Its high sensitivity and the mild conditions required for its application render it peerless for structure determination and kinetics measurements in many instances. As an experimental technique, its use is extraordinarily widespread.

Until quite recently, however, theoretical prediction of NMR spectral properties significantly lagged experimental work. The ultimate factor slowing theoretical work has been simply that it is more difficult to model the interactions of a wave function with a magnetic field than it is to model interactions with an electric field. Nevertheless, great progress has been made over the last decade, particularly with respect to DFT, and calculation of chemical shifts is becoming much more routine than had previously been true.

This section begins with a very brief summary of some of the technical issues associated with NMR spectral calculations. Subsequent subsections address the various utilities of modern methods for predicting chemical shifts and nuclear coupling constants.

9.4.1 Technical Issues

NMR measurements assess the energy difference between a system in the presence and absence of an external magnetic field. For a chemical shift measurement on a given nucleus, there are two magnetic fields of interest: the external field of the instrument and the internal field of the nucleus. The chemical shift is proportional to the second derivative of the energy with respect to these two fields, and it can be computed using second-derivative analogs of Eqs. (9.33) or (9.34). However, the integrals in question are more complex because, unlike the electric field, which perturbs the potential energy term of the Hamiltonian, the magnetic field perturbs the *kinetic* energy term (it is the motion of the electrons that generates electronic magnetic moments). The nature of the perturbed kinetic energy operator is such that an origin must be specified defining a coordinate system for the calculation. This origin is called the 'gauge origin'.

The magnetic field is independent of the choice of the gauge origin. So too are the computed magnetic properties *if* the wave function used is exact. Regrettably, we are not often afforded the opportunity to work with exact wave functions. For HF wave functions, one can also achieve independence of the gauge by using an infinite basis set, but that is hardly a practical option either.

To reduce artifacts associated with the gauge origin, two different approaches have seen extensive use in the literature. The older method employs gauge-including atomic orbitals (GIAOs) as a basis set (London 1937). By a clever incorporation of the gauge origin into the basis functions themselves, all matrix elements involving the basis functions can be arranged to be independent of it. An alternative is the 'individual gauge for localized orbitals' (IGLO) method, where different gauge origins are used for each localized MO in order to minimize error introduced by having the gauge origin far from any particular MO (Schindler and Kutzelnigg 1982). Of the two methods, modern implementations of GIAO are probably somewhat more robust, but it is possible to obtain good results with either.

Much of the benchmark work in the area of NMR calculations has been carried out with very large basis sets, and recommendations have tended to call for at least triple-ζ quality with diffuse and polarization functions aplenty. Of course, such basis sets are simply not practical for larger molecules, even when used solely in the context of a single-point calculation following geometry optimization with some more economical basis (note that the single-point calculation, being a second-derivative property, has timing requirements rather similar to the more routinely carried out calculation of vibrational frequencies). Some early work has begun to appear aimed at identifying scale factors, or linear regressions, that may be applied to computational results from less well-converged calculations, this work being very similar in spirit to the scaling of IR frequencies discussed in Section 9.3.2.2.

A separate basis set issue is associated with calculations for molecules including heavy atoms. If the core electrons of the heavy atom are represented by an ECP, then it is not in general possible to predict the chemical shift for that nucleus, since the remaining basis functions will have incorrect behavior at the nuclear position (note that it is mostly the 'tails' of the valence orbitals at the nucleus that influence the chemical shift, not the core orbitals themselves, since they are filled shells). However, ECPs may be an efficient choice if the only chemical shifts of interest are computed for *other* nuclei.

A different issue associated with NMR chemical shifts for heavy atoms is the influence of relativistic effects. In terms of computing *absolute* chemical shifts, relativistic effects can be very large in heavy elements. For *relative* chemical shifts, since relativistic effects are primarily associated with core orbitals, and core orbitals do not change much from one chemical environment to the next, the effect is typically markedly reduced. Nevertheless, accurate calculations involving atoms beyond the first row of transition metals are still a particular challenge.

9.4.2 Chemical Shifts and Spin–spin Coupling Constants

Experimental chemical shifts are reported in parts per million (ppm) so as to make them independent of the external magnetic field strength. Moreover, they are usually not reported as absolute values, but instead as values relative to some standard compound, e.g., tetramethyl-silane, which is often used for ^{1}H and ^{13}C. Comparison between computed and experimental numbers requires some care to ensure the same convention is being used for reporting data. To compute a relative chemical shift, obviously one must carry out a separate calculation for the reference compound.

For molecules composed of only first-row atoms, heavy-atom chemical shifts can be computed with a fair degree of accuracy, as indicated in Table 9.5. Happily, even HF theory gives acceptable accuracy in most instances, although some improvements are available in favorable instances from DFT (provided it is neither LDA nor B3LYP) and MP2. The latter level of theory is quite accurate, but at relatively high cost in terms of demand for computational resources. Various groups have demonstrated that errors from levels having lower accuracy are sufficiently systematic that errors may be significantly reduced by application of a simple linear regression equation (Sebag, Forsyth, and Plante 2001; Giesen and Zumbulyadis 2002). Thus, for instance, scaling ^{13}C shieldings computed at the B3LYP/MIDI! level by -1.16 and adding 225.1 ppm provides an RMS error of only 3.6 ppm over a diverse test set of experimental values measured in solution.

Note that the mean unsigned errors listed in Table 9.5 for absolute chemical shifts are larger than the errors for relative chemical shifts, as expected. The errors in the relative shifts must be considered to be rather good given the range of experimental values spanned. Note also that the high anisotropy of multiple bonds makes the chemical shifts of the atoms involved quite sensitive to the level of theory, particularly for nitrogen and oxygen atoms.

Table 9.5 contains a relative paucity of data for ^{1}H. This nucleus is somewhat more difficult to work with because it spans a fairly modest spectral range, perhaps 15 ppm in typical chemical environments. Rablen, Pearlman, and Finkbiner (1999), however, have carried out calculations of ^{1}H chemical shifts for 80 organic molecules, and demonstrated reasonable results from various DFT functionals with large basis sets; they also identified scaling factors that improved agreement with experiment (for a similar study focusing on aromatic proton chemical shifts, see Wang *et al.* 2001). Wang, Hinton, and Pulay (2002) similarly reported good success from both HF and DFT calculations for the prediction of ^{1}H chemical shifts in eight cyclic amides for which experimental data in both DMSO and D_2O were available. Finally, Patchkovskii and Thiel (1999) have reported a reparameterization for H, C, N, and O within the MNDO model with the goal of better predicting chemical shifts; they applied their modification B with three-center terms (MB3) MNDO to 384 common organic molecules and obtained errors consistent with those listed in Table 9.5, which must be regarded as fairly good given the tremendous efficiency of the model (of course, making predictions for other nuclei would require further reparameterization).

Computed and experimental data for the chemical shifts of heavy elements have been less extensively compared. Table 9.6 lists some results for ^{77}Se that are illustrative of the wide range of chemical shifts typically possible for such nuclei (here more than 2000 ppm) as well as the degree to which the chemical phase may affect the comparisons. The calculations are gas phase, although in Chapters 11 and 12 we will discuss techniques for including condensed-phase effects in computational predictions.

Table 9.5   Absolute chemical shifts (ppm) from various levels of theory.[a]

Molecule	Nucl	MB3[b]	HF	MP2	LDA	BLYP	BP86[c]	B3LYP	PBE1PBE[d]	B97-2[e]	Expt.
CH_4	^{13}C	189.4	195.7	201.5	193.7	187.5	191.2	189.6	194.0	190.7	195.1
	^1H	29.9					31.4				30.6
C_2H_2	^{13}C	100.4	113.9	123.3	100.0	105.7	110.4	106.3	114.0	113.9	117.2
	^1H	27.3					30.4				29.3
C_2H_4	^{13}C	63.2	59.9	71.2	42.3	47.1	48.7		58.4	57.2	64.5
C_2H_6	^{13}C	175.5	184.0	188.0	176.7	169.7	173.6		179.7		180.9
H_2CCCH_2	^{13}C	94.4	114.0	120.9	103.2	103.0		104.5	112.5		115.2
	^{13}C	−2.9	−44.3	−26.0	−53.0	−51.7		−51.7	−36.6		−28.9
C_6H_6	^{13}C	58.4	55.0	64.0	41.7	43.7	50.0	45.2	55.3		57.2
N_2	^{15}N	−87.7	−128.7	−44.9	−104.8	−97.1	−72.9	−105.4	−76.8	−64.0	−61.6
NH_3	^{15}N	264.7	262.6	276.2	266.1	259.2	262.0	260.3	263.1	261.3	264.5
	^1H	30.3			31.2[c]		31.6				32.4
HCN	^{13}C	82.7	68.1	87.3	63.0	68.7	91.5	67.2	76.6	78.3	82.1
	^{15}N	−55.4	−56.0	1.0	−60.2	−49.2	8.4	−53.1	−34.9	−22.2	−20.4
CH_3NH_2	^{13}C	160.1	163.8	164.9	151.1	145.3		150.1	157.1		158.3
	^{15}N	253.5	250.0	261.2	244.7	233.1		238.4	244.0		249.5
CH_3CN	^{13}C	175.8	190.9	193.6	182.3	177.1		180.4	187.7		187.7
	^{13}C	79.4	60.6	76.1	54.7	57.8		57.4	68.2		73.8
	^{15}N	−57.3	−46.6	−13.2	−44.7	−36.5		−40.7	−24.4		−8.1
H_2O	^{17}O	281.2	326.9	344.8	332.3	326.4	331.5	325.7	328.9	329.8	344.0
	^1H	29.4					31.2				30.1
CO	^{13}C	−40.0	−29.2	11.1	−23.9	−17.3	−9.3	−21.7	−7.8	−2.4	1.0
	^{17}O	−74.0	−95.0	−47.4	−93.7	−82.9	−68.4	−87.8	−70.0	−45.4	−42.3
CO_2	^{13}C	76.9	47.8	63.5	47.2	47.9	56.1	46.9	56.8	58.7	58.5
	^{17}O	92.5	214.8	241.0	203.3	206.5		206.9	220.0	225.2	243.4
$H_2C=O$	^{13}C	−4.9	−9.2	6.7	−41.0	−27.7	−15.7	−25.4	−11.1	−12.4	−8.4
	^{17}O		−461.2	−341.9	−509.2	−459.7	−418.8	−469.8	−422.2	−348.3	−312.1
	^1H						20.7				18.3
CH_3OH	^{13}C	125.7	143.7	142.2	126.1	122.0		127.4	136.5		136.6
	^{17}O	283.5	274.7	350.6	334.5	313.9		321.6	334.7		
$(CH_3)_2C=O$	^{13}C	169.7	163.5	164.5	148.8	146.9		150.4	157.0		158.0
	^{13}C	−14.1	−23.2	−5.8	−44.4	−37.4		−35.7	−11.1		−13.1
	^{17}O	−172.6	−340.5	−279.8	−375.5	−351.5		−358.1	−330.2		
F_2	^{19}F				−310.2[c]			−282.7		−202.7	−232.8
HF	^{19}F				415.1[c]			412.5		414.4	410.0
	^1H				29.4[c]			30.0			28.7
CH_3F	^{13}C		124.5	121.8	103.2	101.2	111.4	106.6	116.5		116.8
	^{19}F						462.3				471.6
	^1H						27.2				26.6
CF_4	^{13}C		79.2	64.4	39.2	38.3		46.5	59.2		64.5
mue_{abs}[f]		21.7	19.4	7.9	27.2	23.1	17.1	23.5	10.3	8.8	
mue_{rel}[g]		17.2	16.9	3.8	24.9	14.5	13.7	16.2	7.4	4.7	

[a]Unless otherwise indicated, a quadruple-ζ basis set with double polarization functions is used (Cheeseman *et al.* 1996; the quoted experimental data are mostly taken from this reference as well). [b]MNDO modification B including three-center terms (Patchkovskii and Thiel 1999). [c]Using a basis set of STOs triple-ζ in the valence region and double-ζ in the core (Schreckenbach and Ziegler 1995). [d]Using the 6-311++G(2d,p) basis set (Adamo, Cossi, and Barone 1999). [e]Using the IGLO IV basis set and the multiplicative Kohn–Sham (MKS; Wilson and Tozer 2001) method to compute the chemical shifts (Wilson, Bradley, and Tozer 2001). [f]Mean unsigned error in heavy-atom absolute shieldings. [g]Mean unsigned error in heavy-atom shieldings relative to CH_4 (^{13}C), NH_3 (^{15}N), H_2O (^{17}O), and HF (^{19}F).

Table 9.6　Chemical shifts for ^{77}Se (ppm) relative to $(CH_3)_2Se$

Molecule	Phase	Calculated[a]	Experiment
$(CH_3)_2Se$		0	0
H_2Se	gas	−427	−345
	neat liquid		−226
CSe_2	gas	225	331
	liquid solution		299
SeF_6	gas	678	610
	neat liquid		631
$Se_4{}^+$	gas	1836	
	liquid solution		1940

[a]BP86 using a basis set of STOs triple-ζ in the valence region and double-ζ in the core (Schreckenbach *et al.* 1996).

Table 9.7　Spin−spin coupling constants (Hz) from LDA calculations and experiment

Molecule	Coupling	LDA[a]	B3LYP[b]	Experiment
CH_4	$^2J_{HH}$	−6.4	−10.9	−12.6
	$^1J_{CH}$	122.4	123.5	125.3
NH_3	$^2J_{HH}$	−8.3	−8.5	−10.4
	$^1J_{NH}$	46.5	41.9	43.4
H_2O	$^2J_{HH}$	−10.8	−6.5	−7.2
	$^1J_{OH}$	−80.5	−71.7	−78.2
HF	$^1J_{FH}$	388.9(494.1)[c]	422.8	530.3
C_2H_2	$^3J_{HH}$	2.5	10.2	9.6
	$^2J_{CH}$	47.4	51.5	49.3
	$^1J_{CH}$	239.0	254.4	248.7
	$^1J_{CC}$	204.9	201.7	171.5
C_2H_4	$^3J_{HH}$ cis	6.5	10.3	11.7
	$^3J_{HH}$ trans	12.1	15.2	19.1
	$^2J_{CH}$	1.8	−3.4	−2.4
	$^1J_{CH}$	145.3	163.1	156.4
	$^1J_{CC}$	68.6	58.4	67.6
C_2H_6	$^3J_{HH}$ gauche	6.6[d]	4.0	8.0[d]
	$^3J_{HH}$ anti	6.6[d]	13.8	8.0[d]
	$^2J_{CH}$	−1.8	−4.1	−4.5
	$^1J_{CH}$	123.9	127.5	124.9
	$^1J_{CC}$	30.2	23.7	34.6
CO	$^1J_{CO}$	27.3	18.8	16.4
CO_2	$^1J_{CO}$	21.7	22.7	16.1
CH_3F	$^2J_{HH}$	−2.8	−7.7	−9.6
	$^2J_{FH}$	33.2	50.8	46.4
	$^1J_{CH}$	142.3	144.9	149.1
	$^1J_{CF}$	−262.2	−227.1	−161.9
$V(CO)_6^-$	$^1J_{VC}$	101.0		116.2
$Fe(CO)_5$	$^1J_{FeC}$	20.9[d]		23.4[d]
$Co(CO)_4^-$	$^1J_{CoC}$	252.4		286.0

[a]Using a basis set of STOs triple-ζ in the valence region and double-ζ in the core unless otherwise indicated (Schreckenbach *et al.* 1996). [b]Using a (11s7p2d/6s2p)[7s6p2d/4s2p] basis set (Sychrovsky, Gräfenstein, and Cremer 2000). [c]Using a basis set with the core expanded to triple-ζ. [d](Pseudo)rotationally averaged.

The calculation of spin–spin coupling is less straightforward than the calculation of chemical shift, in part because of the additional complications associated with two local magnetic moments, as opposed to one moment and one external, uniform field. Moreover, the most commonly reported couplings in the experimental literature are proton–proton couplings in organic and biological molecules, and these are amongst the more difficult to predict because they tend to be small in magnitude, so absolute errors are magnified when considered in a relative sense. Some representative calculations are provided in Table 9.7.

Computed coupling constants show moderate to large sensitivity to basis set, and accurate predictions require very flexible bases (see, for example, the hydrogen fluoride (HF) data in Table 9.7). In addition, DFT is much more robust than HF theory for predicting coupling constants, and the latter level of theory simply should not be used for this purpose.

9.5 Case Study: Matrix Isolation of Perfluorinated *p*-Benzyne

Synopsis of Wenk *et al.* (2001) 'Matrix Isolation of Perfluorinated *p*-Benzyne'.

The class of antitumor-antibiotics known as enediynes undergo *in vivo* Bergman cyclization of the enediyne functionality to generate *p*-benzyne reactive intermediates that damage genetic material. Because the damage results in double-stranded DNA cleavage, they are extraordinarily cytotoxic, and this has sparked interest in better understanding *p*-benzynes in general (this species has already been discussed at some length in Chapters 7 and 8). One issue associated with the parent *p*-benzyne is that it is thermochemically unstable relative to its enediyne precursor, making its isolation more challenging. In this case, Wenk and co-workers sought to identify a precursor not suffering from this problem, and determined from DFT and CASSCF calculations that perfluorinated *p*-benzyne was roughly 8 kcal mol^{-1} more stable than the enediyne that would be produced from retro-Bergman ring opening, and moreover that the barrier to that ring opening was nearly 38 kcal mol^{-1}, this being nearly double the barrier in the unfluorinated case. Girded with this thermochemical armor, they set out to synthesize the diradical by UV photolysis of 1,4-diiodo-2,3,5,6-tetrafluorobenzene (Figure 9.9).

When this precursor is photolyzed at 3 K in a neon matrix, IR spectroscopy indicates rapid formation of a new species A. Prolonged photolysis creates a second product B whose IR bands are distinct from the first. And, if the matrix containing the second product is irradiated with UV light of somewhat longer wavelength, IR analysis indicates that a third product C is generated. All of the IR bands observed for A, B, and C are listed in Table 9.8. These bands are compared to frequencies computed at the B3LYP/6-311G(d,p) level for 4-iodo-2,3,5,6-tetrafluorophenyl radical (ITFP) and perfluorohex-3-en-1,5-diyne (PFHED) and to frequencies computed at the CASSCF(8,8)/cc-pVDZ level of theory and scaled by 0.91 for perfluorinated *p*-benzyne (PFPB). The authors do not explain their recourse to two different levels of theory, but presumably they were not comfortable with the DFT model, even used unrestricted, for the multiconfigurational *p*-benzyne.

In any event, the generally excellent agreement between the experimental and computed spectra permits the secure assignment of the bands for A to ITFP, the bands for B to PFPB, and the bands for C to PFHED (note that scaling of the DFT bands by the scale factor for

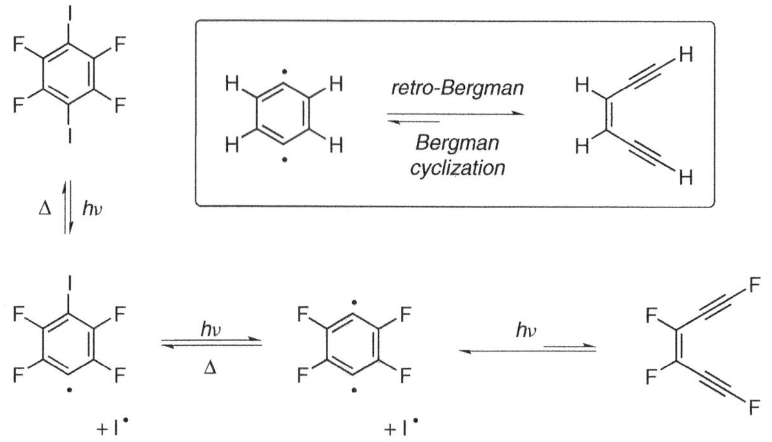

Figure 9.9 Bergman cycloaromatization reactions for hex-3-en-1,5-diyne and its perfluorinated congener, as well as a photochemical reaction scheme for generating the perfluorinated diradical from an iodinated precursor. What spectral features would be expected to be most diagnostic of the different intermediates? What levels of theory would be appropriate for predicting these spectral signatures? (Note that equilibrium arrows of unequal length indicate which species predominates at equilibrium.)

Table 9.8 Experimental and computed IR spectra (cm^{-1}) for A, B, and C, and ITFP, PFPB, and PFHED, respectively

A	ITFP	B	PFPB	C	PFHED
693	693		677		576
834	834		690		591
942/956	954	925	911		680
1138	1145		1148	912	918
1188	1195	1117	1151	1072	1067
1259	1298		1154		1151
1352	1400		1360		1363
1428	1441	1407	1421	1398	1414
1472	1487	1502/1516	1499	1678	1707
1574	1585		1560	2337	2419
			1610		2426

B3LYP/6-31G(d) in Table 9.3 would result in slightly improved agreement for A and C) as would be suggested by the synthetic scheme in Figure 9.9. Intensity data were also used, although those are not shown here; importantly, the 'missing' bands in the experimental IR spectra are all predicted to be of very low intensity in the computed spectra. Interestingly, both CASSCF and unrestricted B3LYP predict the singlet and triplet states of the diradical to be essentially degenerate, leaving the question open as to which (if either) is the lower in energy.

> The use of computed spectra to bolster structural assignments has seen heavy use in matrix isolation experiments. This is a slightly atypical example, insofar as the species involved actually require some careful attention to non-dynamical correlation, but represents an excellent example of how theory can aid experiment in the identification of short-lived reactive species.

Bibliography and Suggested Additional Reading

Bachrach, S. 1994. 'Population Analysis and Electron Densities from Quantum Mechanics', in *Reviews in Computational Chemistry*, Vol. 5, Lipkowitz, K. B. and Boyd, D. B. Eds., VCH: New York, 171.

Cramer, C. J. 1991. 'Dependence of Isotropic Hyperfine Coupling in the Fluoromethyl Radical Series on Inversion Angle' *J. Org. Chem.*, **56**, 5229.

Hehre, W. J., Radom, L., Schleyer, P. v. R., and Pople, J. A. 1986. Ab Initio *Molecular Orbital Theory*, Wiley: New York.

Helgaker, T., Jaszunski, M., and Ruud, K. 1999. '*Ab Initio* Methods for the Calculation of NMR Shielding and Indirect Spin–spin Coupling Constants' *Chem. Rev.*, **99**, 293.

Jensen, F. 1999. *Introduction to Computational Chemistry*, Wiley: Chichester.

Koch, W. and Holthausen, M. C. 2000. *A Chemist's Guide to Density Functional Theory*, Wiley-VCH: Weinheim.

Kubinyi, H. 2003. 'Comparative Molecular Field Analysis (CoMFA)', in *Handbook of Chemoinformatics. From Data to Knowledge*, Vol. 4, Gasteiger, J., Ed., Wiley-VCH: Weinheim, 1555.

Leach, A. R. 2001. *Molecular Modelling*, 2nd Edn., Prentice Hall: London.

Levine, I. N. 1975. *Molecular Spectroscopy*, Wiley, New York.

Malkin, V. G., Malkina, O. L., Eriksson, L. A., and Salahub, D. R. 1995. 'The Calculation of NMR and ESR Spectroscopy Parameters Using Density Functional Theory', in *Modern Density Functional Theory; A Tool for Chemistry*, Politzer, P. and Seminario, J., Eds., Elsevier: Amsterdam, 273.

Thompson, J. D., Cramer, C. J., and Truhlar, D. G. 2003. 'Parameterization of Charge Model 3 for AM1, PM3, BLYP, and B3LYP', *J. Comput. Chem.*, **24**, 1291.

Wiberg, K. B. and Rablen, P. R. 1993. 'Comparison of Atomic Charges Derived via Different Procedures' *J. Comput. Chem.*, **14**, 1504.

Wilson, E. B., Jr., Decius, J. C., and Cross, P. C. 1955. *Molecular Vibrations*, Dover: New York.

References

Adamo, C. and Barone, V. 1998. *J. Chem. Phys.*, **108**, 664.

Adamo, C., Cossi, M., and Barone, V. 1999. *J. Mol. Struct. (Theochem)*, **493**, 145.

Arenas, J. F., Centeno, S. P., Marcos, J. I., Otero, J. C., Soto, J. 2000. *J. Chem. Phys.*, **113**, 8472.

Bader, R. W. F. 1990. *Atoms in Molecules – A Quantum Theory*. Oxford University Press: Oxford.

Bader, R. W. F. 1991. *Chem. Rev.*, **91**, 893.

Baker, J., Jarzecki, A. A., and Pulay, P. 1998. *J. Phys. Chem. A*, **102**, 1412.

Barone, V. 1995. In: *Recent Advances in Density Functional Methods, Part 1*, Chong, D. P., Ed., World Scientific: Singapore, 278.

Barone, V., Grand, A., Minichino, C., and Subra, R. 1993. *J. Chem. Phys.*, **99**, 6787.

Barrows, S. E., Cramer, C. J., Truhlar, D. G., Weber, E. J., and Elovitz, M. S. 1996. *Environ. Sci. Technol.*, **30**, 3028.

Basma, M., Sundara, S., Çalgan, D., Vernali, T., and Woods, R. J. 2001. *J. Comput. Chem.*, **22**, 1125.

Bauschlicher, C. W. and Partridge, H. 1995. *J. Chem. Phys.*, **103**, 1788.

Bayly, C. I., Cieplak, P., Cornell, W. D., and Kollman, P. A. 1993. *J. Phys. Chem.*, **97**, 10269.

Besler, B. H., Merz, K. M., and Kollman, P. A. 1990. *J. Comput. Chem.*, **11**, 431.

Biegler-Konig, F. and Schonbohm, J. 2002. *J. Comput. Chem.*, **23**, 1489.

Braden, D. A. and Tyler, D. R. 1998. *Organometallics*, **17**, 4060.

Breneman, C. M. and Wiberg, K. B. 1990. *J. Comput. Chem.*, **11**, 361.

Bytheway, I. and Wong, M. W. 1998. *Chem. Phys. Lett.*, **282**, 219.

Cheeseman, J. R., Trucks, G. W., Keith, T. A., and Frisch, M. J. 1996. *J. Chem. Phys.*, **104**, 5497.

Chipman, D. M. 1983. *J. Chem. Phys.*, **78**, 3112.

Chirlian, L. E. and Francl, M. M. 1987. *J. Comput. Chem.*, **8**, 894.

Cho, K.-H., Kang, Y. K., No, K. T., and Scheraga, H. A. 2001. *J. Phys. Chem. B*, **105**, 3624.

Cioslowski, J. 1989. *J. Am. Chem. Soc.*, **111**, 8333.

Cornell, W. C., Cieplak, P., Bayly, C. I., and Kollman, P. A. 1993. *J. Am. Chem. Soc.*, **115**, 9620.

Cramer, C. J. 1991. *J. Org. Chem.*, **56**, 5229.

Cramer, C. J. and Lim, M. H. 1994. *J. Phys. Chem.*, **98**, 5024.

Cusachs, L. C. and Politzer, P. 1968. *Chem. Phys. Lett.*, **1**, 529.

Duffy, E. M. and Jorgensen, W. L. 2000. *J. Am. Chem. Soc.*, **122**, 2878.

Eriksson, L. A., Malkina, O. L., Malkin, V. G., and Salahub, D. R. 1994. *J. Chem. Phys.*, **100**, 5066.

Exner, T. E. and Mezey, P. G. 2002. *J. Phys. Chem. A*, **106**, 11791.

Francl, M. M. and Chirlian, L. E. 2000. In: *Reviews in Computational Chemistry*, Vol. 14, Lipkowitz, K. B. and Boyd, D. B., Eds., Wiley: New York, 1.

Gasteiger, J. and Marsilli, M. 1980. *Tetrahedron*, **36**, 3219.

Giesen, D. J. and Zumbulyadis, N. 2002. *Phys. Chem. Chem. Phys.*, **4**, 5498.

Gross, K. C., Seybold, P. G., and Hadad, C. M. 2002. *Int. J. Quant. Chem.*, **90**, 445.

Grunenberg, J. and Herges, R. 1997. *J. Comput. Chem.*, **18**, 2050.

Guerra, M. 2000. *Theor. Chem. Acc.*, **104**, 455.

Guerra, C. F., Handegraaf, J. W., Baerends, E. J., and Bickelhaupt, F. M. 2004. *J. Comput. Chem.*, **25**, 189.

Halls, M. D., Velkovski, J., and Schlegel, H. B. 2001. *Theor. Chem. Acc.*, **105**, 413.

Jakalian, A., Jack, D. B., and Bayly, C. I. 2002. *J. Comput. Chem.*, **23**, 1623.

Jaramillo, J. and Scuseria, G. E. 1999. *Chem. Phys. Lett.*, **312**, 269.

Johnson, B. G., Gill, P. M. W., and Pople, J. A. 1993. *J. Chem. Phys.*, **98**, 5612.

Kalinowski, J. A., Lesyng, B., Thompson, J. D., Cramer, C. J., and Truhlar, D. G. 2004. *J. Phys. Chem. A*, **108**, 2545.

Khandogin, J., Hu, A. G., and York, D. M. 2000. *J. Comput. Chem.*, **21**, 1562.

Laidig, K. E. 1993. *J. Phys. Chem.*, **97**, 12760.

Li, J., Xing, J., Cramer, C. J., and Truhlar, D. G. 1999. *J. Chem. Phys.*, **111**, 885.

Li, J., Zhu, T., Cramer, C. J., and Truhlar, D. G. 1998. *J. Phys. Chem. A*, **102**, 1820.

Lim, M. H., Worthington, S. E., Dulles, F. J., and Cramer, C. J. 1996. In: *Chemical Applications of Density-functional Theory*, ACS Symposium Series, Vol. 629, Laird, B. B., Ross, R. B., and Ziegler, T., Eds., American Chemical Society: Washington, DC, 402.

London, F. 1937. *J. Phys. Radium*, **8**, 397.

Löwdin, P.-O. 1970. *Adv. Quantum Chem.*, **5**, 185.

Mayer, I. 1983. *Chem. Phys. Lett.*, **97**, 270.

Mulliken, R. S. 1955. *J. Chem. Phys.*, **23**, 1833, 1841, 2338, 2343.

Nguyen, M. T., Creve, S., and Vanquickenborne, L. G. 1997. *J. Phys. Chem. A*, **101**, 3174.

No, K. T., Grant, J. A., and Scheraga, H. A. 1990a. *J. Phys. Chem.*, **94**, 4732.

No, K. T., Grant, J. A., Jhon, M. S., and Scheraga, H. A. 1990b. *J. Phys. Chem.*, **94**, 4740.

Patchkovskii, S. and Thiel, W. 1999. *J. Comput. Chem.*, **20**, 1220.

Pople, J. A., Scott, A. P., Wong, M. W., and Radom, L. 1993. *Isr. J. Chem.*, **33**, 345.

Rablen, P. R., Pearlman, S. A., and Finkbiner, J. 1999. *J. Phys. Chem. A*, **103**, 7357.

Rappé, A. K. and Goddard, W. A. 1991. *J. Phys. Chem.*, **95**, 3358.

Reed, A. E., Weinstock, R. B., and Weinhold, F. 1985. *J. Chem. Phys.*, **83**, 735.

Sadlej, A. J. 1992. *Theor. Chem. Acc.*, **79**, 123.

Schindler, M. and Kutzelnigg, W. 1982. *J. Chem. Phys.*, **76**, 1919.

Schreckenbach G. and Ziegler, T. 1995. *J. Phys. Chem.*, **99**, 606.

Schreckenbach, G., Dickson, R. M., Ruiz-Morales, Y., and Ziegler, T. 1996. In: *Chemical Applications of Density-functional Theory*, Laird, B. B., Ross, R. B., and Ziegler, T., Eds., ACS Symposium Series, Vol. 629, American Chemical Society: Washington, DC, 328.

Scott, A. P. and Radom, L. 1996. *J. Phys. Chem.*, **100**, 16502.

Sebag, A. B., Forsyth, D. A., and Plante, M. A. 2001 *J. Org. Chem.*, **66**, 7967.

Sefcik, J., Demiralp, E., Cagin, T., and Goddard, W. A., III, 2002. *J. Comput. Chem.*, **23**, 1507.

Sherer, E. C., Kinsinger, C. R., Kormos, B. L., Thompson, J. D., and Cramer, C. J. 2002. *Angew. Chem., Int. Ed. Engl.*, **41**, 1953.

Storer, J. W., Giesen, D. J., Cramer, C. J., and Truhlar, D. G. 1995. *J. Comput.-Aid. Mol. Des.*, **9**, 87.

Sychrovsky, V., Gräfenstein, J., and Cremer, D. 2000. *J. Chem. Phys.*, **113**, 3530.

Thompson, J. D., Xidos, J. D., Sonbuchner, T. M., Cramer, C. J., and Truhlar, D. G. 2002. *Phys. Chem. Comm.*, **5**, 117.

Wang, B., Fleischer, U., Hinton, J. F., and Pulay, P. 2001. *J. Comput. Chem.*, **22**, 1887.

Wang, B., Hinton, J. F., and Pulay, P. 2002. *J. Comput. Chem.*, **23**, 492.

Wenk, H. H., Balster, A., Sander, W., Hrovat, D. A., and Borden, W. T. 2001. *Angew. Chem., Int. Ed. Engl.*, **40**, 2295.

Williams, D. E. 1988. *J. Comput. Chem.*, **9**, 745.

Wilson, P. J., Bradley, T. J., and Tozer, D. J. 2001. *J. Chem. Phys.*, **115**, 9233.

Wilson, P. J. and Tozer, D. J. 2001. *Chem. Phys. Lett.*, **337**, 341.

Winget, P., Thompson, J. D., Xidos, J. D., Cramer, C. J., Truhlar, D. G. 2002. *J. Phys. Chem. A*, **106**, 10707.

Wong, M. W. 1996. *Chem. Phys. Lett.*, **256**, 391.

Yu, L., Srinivas, G. N., and Schwartz, M. 2003. *J. Mol. Struct. Theochem*, **625**, 215.

10

Thermodynamic Properties

10.1 Microscopic–macroscopic Connection

In the very recent past, it has become possible under certain circumstances to observe single molecules in the laboratory. Nevertheless, the vast majority of chemical research concerns itself not with individual molecules, but instead with macroscopic quantities of matter that are made up of unimaginably large numbers of molecules. The behavior of such ensembles of molecules is governed by the empirically determined laws of thermodynamics, and most chemical reactions and many chemical properties are defined in terms of some of the fundamental variables of thermodynamics, such as enthalpy, entropy, free energy, and others.

Until now, we have for the most part concerned ourselves only with the potential and kinetic energies of individual electrons and nuclei in *single* molecules, and one of our rare connections to thermodynamics has been in some sense a misleading one, namely that we have often converted atomic units into other units more typically associated with macroscopic quantities, e.g., kcal mol^{-1} or kJ mol^{-1}. This sometimes leads newcomers to the field to think of the atomic unit of energy, the hartree, as being enormously large, since 1 E_h is equal to 627.51 kcal mol^{-1}. In reality, however, the hartree is a tremendously *tiny* unit, since 'kcal mol^{-1}', as its name makes clear, refers to the energy associated with *one mole* (i.e., 6.0221×10^{23}) of molecules, not with the single molecule that is the typical subject of an electronic-structure calculation. Moreover, when we make comparisons between two different calculations, say to determine the relative energies of two isomers, and we carry out such simple unit conversions, we tacitly (and often incorrectly) assume that the potential energy difference determined from the calculation is all that matters when comparing to a *measured* energy difference, that is almost always in the form of some thermodynamic quantity, most typically enthalpy or free energy.

In this chapter, the most common procedures for augmenting electronic-structure calculations in order to convert single-molecule potential energies to ensemble thermodynamic variables will be detailed, and key potential ambiguities and pitfalls described. Within the context of certain assumptions, this connection can be established in a rigorous way.

Note that the situation is less clear-cut for molecular mechanics calculations. As already discussed in Chapter 2, the 'strain energy' from a typical MM calculation must be thought

Essentials of Computational Chemistry, 2nd Edition Christopher J. Cramer
© 2004 John Wiley & Sons, Ltd   ISBNs: 0-470-09181-9 (cased); 0-470-09182-7 (pbk)

of as a potential energy with a rather particular zero of energy, namely, one uniquely defined by all of the atom types found in the molecule. To make a connection with thermodynamics, the most common approach is to associate with each atom type a heat of formation, and then assume that the strain energies may be thought of as enthalpies (see Figure 2.8). In this instance, the connection between the microscopic and macroscopic regimes is hidden in the fitting procedure by which the heats of formation of the atomic types are assigned, and thermodynamic aspects must be considered to be wholly empirical. As such, we will not consider this point further.

10.2   Zero-point Vibrational Energy

The first step in moving from the microscopic regime to the macroscopic is to recognize that the Born–Oppenheimer potential energy surface is fundamentally a classical construct (although the energies of various points whose coordinates are defined by the fixed nuclear positions are determined from quantum mechanical calculations of the electronic energy). As already discussed in Chapter 9, when the motion of the nuclei on this surface is also accounted for in a quantum mechanical way, energy is 'tied up' in molecular vibrations. This is true even at a temperature arbitrarily close to absolute zero, since the lowest vibrational energy level for any bound vibration is *not* zero.

Within the harmonic oscillator approximation, the energy of the lowest vibrational level can be determined from Eq. (9.47) as $h\omega/2$ where h is Planck's constant (6.6261×10^{-34} J s) and ω is the vibrational frequency. The sum of all of these energies over all molecular vibrations defines the zero-point vibrational energy (ZPVE). We may then define the internal energy at 0 K for a molecule as

$$U_0 = E_{\text{elec}} + \sum_i^{\text{modes}} \tfrac{1}{2} h\omega_i, \tag{10.1}$$

where E_{elec} is the energy for the *stationary point* on the Born–Oppenheimer PES. Note that U_0 is also often written as E_0 in thermochemical literature.

One may legitimately ask what errors may be implicit in Eq. (10.1), introduced by invocation of the harmonic oscillator approximation. To answer that question it is instructive to consider which modes will be likely to be *least* harmonic in character. In general, we expect these modes to be the softest ones, e.g., weakly hindered torsions. However, such weak modes will also tend to have very small vibrational frequencies associated with them, and thus they will not contribute much to the ZPVE in any case (since ZPVE is linear in the frequencies). As a rule, then, the harmonic approximation does fairly well in computing ZPVE. Note, of course, that the frequencies themselves must be computed at a level of electronic-structure theory that ensures their acceptable accuracy. Thus, if one uses HF theory with some basis set that is known in general to require a scaling factor of 0.9 to bring computed frequencies into line with experiment, that same scaling factor should be used to compute the ZPVE (or, equivalently, the ZPVE should be computed using the scaled frequencies).

One can generalize the concept of ZPVE to non-stationary points on the PES (although some convention must be adopted for dealing with the non-zero first derivatives of the energy at such points). The new surface that is generated by summing the Born–Oppenheimer surface with this generalized ZPVE is called the zero-point-including energy surface. This surface thus includes the quantum mechanical character of the nuclear motion at 0 K, and can be very useful in reaction dynamics simulations, as described in more detail in Chapter 14.

A key feature of the ZPVE is that it is isotope dependent, since the vibrational frequencies themselves are isotope dependent (see Eq. (9.48) and recall that the reduced mass μ for any mode is a function of the atomic masses for the nuclei involved in the motion). Thus, if one is considering a large ensemble of molecules, it must be kept in mind that the computed ZPVE refers to an ensemble of isotopically *pure* molecules, *not* to an ensemble composed from isotopes at natural abundance. Most electronic structure programs default to using the atomic isotopes of highest natural abundance, and permit use of other isotopes in some keyword-driven way. Older versions of some semiempirical programs originally used atomic masses derived from natural-abundance averaging over isotopes, but this is a fundamentally flawed approach, since the influence of the atomic masses on the frequencies is not linear. Typically, in part because many elements have a single dominant isotope, one does not need to worry about the rather small inaccuracies introduced by assuming isotopically pure samples (Jensen 2003). In those rare instances where true natural abundance results are desired, it is a straightforward if somewhat tedious task to construct multiple ensembles differing in isotopic composition and weight them appropriately in an overall mixture.

10.3 Ensemble Properties and Basic Statistical Mechanics

Statistical mechanics is, obviously, a course unto itself in the standard chemistry/physics curriculum, and no attempt will be made here to introduce concepts in a formal and rigorous fashion. Instead, some prior exposure to the field is assumed, or at least to its thermodynamical consequences, and the fundamental equations describing the relationships between key thermodynamic variables are presented without derivation. From a computational-chemistry standpoint, many simplifying assumptions make most of the details fairly easy to follow, so readers who have had minimal experience in this area should not be adversely affected.

In order to deal with collections of molecules in statistical mechanics, one typically requires that certain macroscopic conditions be held constant by external influence. The enumeration of these conditions defines an 'ensemble'. We will confine ourselves in this chapter to the so-called 'canonical ensemble', where the constants are the total number of particles N (molecules, and, for our purposes, identical molecules), the volume V, and the temperature T. This ensemble is also sometimes referred to as the (N, V, T) ensemble.

Just as there is a fundamental function that characterizes the microscopic system in quantum mechanics, i.e., the wave function, so too in statistical mechanics there is a fundamental function having equivalent status, and this is called the partition function. For the canonical ensemble, it is written as

$$Q(N, V, T) = \sum_i e^{-E_i(N,V)/k_{\mathrm{B}}T} \tag{10.2}$$

where i runs over all possible energy states of the system having energy E_i and k_B is Boltzmann's constant (1.3806×10^{-23} J K^{-1}).

Within the canonical ensemble, and using established thermodynamic definitions, all of the following are true

$$U = k_B T^2 \left(\frac{\partial \ln Q}{\partial T} \right)_{N,V} \tag{10.3}$$

$$H = U + PV \tag{10.4}$$

$$S = k_B \ln Q + k_B T \left(\frac{\partial \ln Q}{\partial T} \right)_{N,V} \tag{10.5}$$

$$G = H - TS \tag{10.6}$$

where the notation associated with the partial derivatives in Eqs. (10.3) and (10.5) implies N and V held constant during differentiation with respect to T, H is enthalpy, P is pressure, S is entropy, and G is (Gibbs) free energy.

Of course, the elegance of Eqs. (10.3)–(10.6) is somewhat muted by the daunting prospect of finding an explicit representation of Q that permits the necessary partial differentiations in Eqs. (10.3) and (10.5) to be carried out. For a true ensemble, Q must be some fantastically complex many-body function involving a staggeringly enormous number of energy levels. So, in order to make progress, we indulge in a number of simplifying assumptions.

10.3.1 Ideal Gas Assumption

We begin by assuming that our ensemble is an ideal gas. The first consequence of this assumption, since ideal gas molecules do not interact with one another, is that we may rewrite the partition function as

$$Q(N, V, T) = \frac{1}{N!} \sum_i e^{-[\varepsilon_1(V)+\varepsilon_2(V)+\cdots+\varepsilon_N(V)]_i/k_B T}$$

$$= \frac{1}{N!} \left[\sum_{j(1)} e^{-\varepsilon_{j(1)}(V)/k_B T} \right] \left[\sum_{j(2)} e^{-\varepsilon_{j(2)}(V)/k_B T} \right] \cdots \left[\sum_{j(N)} e^{-\varepsilon_{j(N)}(V)/k_B T} \right]$$

$$= \frac{1}{N!} \left[\sum_k^{\text{levels}} g_k e^{-\varepsilon_k(V)/k_B T} \right]^N$$

$$= \frac{[q(V, T)]^N}{N!} \tag{10.7}$$

where the factor of $1/N!$ derives from the quantum mechanical indistinguishability of the particles, ε is the total energy of an individual molecule, and the change on going from the first to the second line derives from expressing the exponential of all possible sums of energies as a product of all possible sums of exponentials of individual energies. On going from the second to the third line, the sum has been changed so that it goes over discrete energy levels, rather than individual states, and g_k is the degeneracy of level k. The quantity in brackets on the third line defines the *molecular* partition function q.

A second consequence of the ideal gas assumption is that PV in Eq. (10.4) may be replaced by Nk_BT. In the special case where we are working with one mole of molecules, in which case $N = N_A$ (Avogadro's number), we may replace PV with RT, where R is the universal gas constant (8.3145 J mol^{-1} K^{-1}).

10.3.2 Separability of Energy Components

We have thus reduced the problem from finding the ensemble partition function Q to finding the molecular partition function q. In order to make further progress, we assume that the molecular energy ε can be expressed as a separable sum of electronic, translational, rotational, and vibrational terms, i.e.,

$$q(V, T) = \sum_k^{\text{levels}} g_k e^{-\varepsilon_k(V)/k_BT}$$

$$= \sum_k^{\text{levels}} g_k e^{-[\varepsilon_{\text{elec}}+\varepsilon_{\text{trans}}(V)+\varepsilon_{\text{rot}}+\varepsilon_{\text{vib}}]_k/k_BT}$$

$$= \left[\sum_i^{\text{elec}} g_i e^{-\varepsilon_i/k_BT}\right]\left[\sum_j^{\text{trans}} g_j e^{-\varepsilon_j(V)/k_BT}\right]\left[\sum_k^{\text{rot}} g_k e^{-\varepsilon_k/k_BT}\right]\left[\sum_l^{\text{vib}} g_l e^{-\varepsilon_l/k_BT}\right]$$

$$= q_{\text{elec}}(T)q_{\text{trans}}(V, T)q_{\text{rot}}(T)q_{\text{vib}}(T) \tag{10.8}$$

where again advantage is taken of the ability to express an exponential of sums as a product of sums of exponentials, and the separate lines make clear that the degeneracy of a total molecular energy level is simply the product of the degeneracies of each of its contributing components.

Note in Eqs. (10.3) and (10.5), Q always appears as the argument of the natural logarithm function. Using Eqs. (10.7) and (10.8), our assumptions to this point allow us to write

$$\ln[Q(N, V, T)] = \ln\left\{\frac{[q_{\text{elec}}(T)q_{\text{trans}}(V, T)q_{\text{rot}}(T)q_{\text{vib}}(T)]^N}{N!}\right\}$$

$$= N\{\ln[q_{\text{elec}}(T)] + \ln[q_{\text{trans}}(V, T)] + \ln[q_{\text{elec}}(T)]$$

$$+ \ln[q_{\text{vib}}(T)]\} - \ln(N!)$$

$$\approx N\{\ln[q_{\text{elec}}(T)] + \ln[q_{\text{trans}}(V, T)]$$

$$+ \ln[q_{\text{elec}}(T)] + \ln[q_{\text{vib}}(T)]\} - N\ln N + N \tag{10.9}$$

where going from the second to the third equality makes use of Stirling's approximation for $\ln(N!)$ when N is large. This separation of terms by the logarithm function makes evident that we may speak of individual components – electronic, translational, rotational, and vibrational – of thermodynamic functions within our approximate treatment. All that remains is to express the various components of the molecular partition function in some useful, preferably analytic, form.

10.3.3   Molecular Electronic Partition Function

The electronic partition function is usually the simplest to compute. For a typical, closed-shell singlet molecule, the degeneracy of the ground state is unity, and the various excited states are so high in energy that, at least at temperatures below thousands of degrees, they make no significant contribution to the partition function, so that we might effectively take

$$q_{elec} = e^{-E_{elec}/k_B T} \tag{10.10}$$

If we evaluate the electronic component of U using Eq. (10.3), we determine that it is, independent of temperature, simply E_{elec}.

In practice, it proves more convenient to work within a convention where we define the ground state for each energy component to have an energy of zero. Thus, we view U_{elec} as the internal energy that must be *added* to U_0, which already includes E_{elec} (see Eq. (10.1)), as the result of additional available electronic levels. One obvious simplification deriving from this convention is that the electronic partition function for the case just described is simply $q_{elec} = 1$. Inspection of Eq. (10.5) then reveals that the electronic component of the entropy will be zero (ln of 1 is zero, and the constant 1 obviously has no temperature dependence, so both terms involving q_{elec} are individually zero).

Another commonly encountered situation involves a ground state of higher spin multiplicity than singlet, but with excited states still sufficiently high in energy that they play no role in the electronic partition function. In that case, the definition of E_{elec} as the zero of energy continues to make the exponential part of the partition function equal to 1, but the degeneracy is now $2S + 1$, where S is the spin multiplicity ($\frac{1}{2}$ for doublet, 1 for triplet, etc.) Thus, the partition function is also $2S + 1$. This still has no temperature dependence, so it makes no contribution to the internal energy, but it is no longer unity, so it makes a contribution to the entropy. In general, then, for the electronic components of U and S we simply use

$$U_{elec} = 0 \tag{10.11}$$

and

$$S_{elec} = Nk_B \ln(2S + 1) \tag{10.12}$$

where it should be emphasized to avoid confusion that the S on the l.h.s. of Eq. (10.12) refers to entropy, while that on the r.h.s. refers to spin. The usual choice that is made is to compute molar quantities of the thermodynamic functions, so that N is N_A, in which case Eq. (10.12) becomes

$$S_{elec} = R \ln(2S + 1) \tag{10.13}$$

We will assume molar quantities from this point forward.

In particular instances, the above approximation is insufficiently accurate because one or more excited electronic states lie close in energy to the ground state. A typical example occurs for heavy halogen atoms, where spin–orbit coupling creates $^2P_{1/2}$ and $^2P_{3/2}$ states having only a narrow energy separation. In such cases, explicit formation of the partition function cannot be avoided, but even then one typically need only include a small number of terms in the total sum, and evaluation is not unduly taxing.

10.3.4 Molecular Translational Partition Function

To evaluate q_{trans}, we assume that the molecule acts as a particle in a three-dimensional cubic box of dimension a^3 where a is the length of one side of the cube. The energy levels for this elementary quantum mechanical system are given by

$$\varepsilon_{trans}(n_x, n_y, n_z) = \frac{h^2}{8Ma^2}(n_x^2 + n_y^2 + n_z^2) \tag{10.14}$$

where M is the molecular mass, and each energy level has associated with it the three unique quantum numbers n_x, n_y, and n_z. Because the energy levels for the particle in a box are very, very closely spaced (at least for a box of macroscopic dimensions), the partition function sum may be replaced by an indefinite integral, and this integral can be evaluated analytically as

$$q_{trans}(V, T) = \left(\frac{2\pi Mk_BT}{h^2}\right)^{3/2} V \tag{10.15}$$

where the volume of the box is now written V as opposed to a^3. Note that it is only the translational partition function that depends on volume, and it does so because particle-in-a-box wave functions cannot be normalized without choice of a specific, finite, non-zero volume (it can be shown that it is only the volume that matters, and not the shape of the box). It is this term, then, that dictates the necessity of choosing a 'standard state' volume to ensure comparison of thermodynamic values in a consistent fashion. We will have more to say about standard states below, but here we simply note that, because we have chosen to model our substance as an ideal gas, we may replace V by RT/P and specify a standard-state pressure instead. This is typically the language that is used in electronic structure calculations, and the usual choice for standard state is 1 atm pressure (corresponding to a standard-state molar volume of 24.5 L at 298 K).

Evaluating Eqs. (10.3) and (10.5) for a molar quantity of particles using Eq. (10.15) for the translational partition function gives

$$U_{trans} = \tfrac{3}{2}RT \tag{10.16}$$

and

$$S_{trans}^o = R\left\{\ln\left[\left(\frac{2\pi Mk_BT}{h^2}\right)^{3/2}V^o\right] + \frac{3}{2}\right\} \tag{10.17}$$

(The superscript 'o' indicates that a 'standard state' is being referred to (see below).) Conventionally, however, the last two terms in the final line of Eq. (10.9), i.e., those deriving from Stirling's approximation, are typically assigned to the translational partition function as well. As they have no temperature dependence, this has no impact on U_{trans}; however, the entropy of translation becomes

$$S^o_{trans} = R \left\{ \ln \left[\left(\frac{2\pi M k_B T}{h^2} \right)^{3/2} \frac{V^o}{N_A} \right] + \frac{5}{2} \right\} \tag{10.18}$$

Note that, because we are working under the assumption of an ideal gas, V^o/N_A in the term in brackets can be replaced by $k_B T/P^o$.

A noteworthy aspect of Eqs. (10.16) and (10.18) is that they are altogether free of any requirement to carry out an electronic structure calculation. Equation (10.16) is well known for an ideal gas and is entirely independent of the molecule in question, and Eq. (10.18) can be computed trivially as soon as the molecular weight is specified. Note, however, that the units chosen for the various quantities must be such that the argument of the logarithm in Eq. (10.18) (i.e., the partition function), is unitless.

The translational partition function is a function of both temperature and volume. However, none of the other partition functions have a volume dependence. It is thus convenient to eliminate the volume dependence of S_{trans} by agreeing to report values that use exclusively some volume that has been agreed upon by convention. The choices of the numerical value of V and its associated units define a 'standard state' (or, more accurately, they contribute to an overall definition that may be considerably more detailed, as described further below). The most typical standard state used in theoretical calculations of entropies of translation is the volume occupied by one mole of ideal gas at 298 K and 1 atm pressure, namely, $V^o = 24.5$ L.

10.3.5   Molecular Rotational Partition Function

In Section 9.3.1, the approach that is taken to solving the rigid-rotor nuclear Schrödinger equation in order to compute rotational wave functions and energy levels was outlined, and the particular cases of diatomic molecules and polyatomic prolate tops were explicitly presented. The solution for the diatomic case is general for any linear molecule, so long as the molecular moment of inertia I is computed in the appropriate fashion for more than 2 atoms [Eq. (9.40)]. When the energy levels from Eq. (9.39) are used in the rotational partition function with their appropriate degeneracies, as usual taking the lowest energy rotational eigenvalue as the zero of energy, the sum can again be well approximated as an indefinite integral at 'normal' temperatures, and solving that integral we find for a linear molecule that

$$q^{linear}_{rot}(T) = \frac{8\pi^2 I k_B T}{\sigma h^2} \tag{10.19}$$

where σ is 1 for asymmetric linear molecules and 2 for symmetric linear molecules (i.e., belonging to the $C_{\infty v}$ and $D_{\infty h}$ point groups, respectively). To be more precise, the validity of Eq. (10.19) requires that the temperature be such that $q_{rot} \gg 1$, but this is almost always the case unless one is dealing with *very* low temperatures (say, 10 K and below) or very light molecules (like diatomics) or both.

Evaluation of the rotational components of the internal energy and entropy using the partition function of Eq. (10.19) gives

$$U_{rot}^{linear} = RT \tag{10.20}$$

and

$$S_{rot}^{linear} = R \left[\ln \left(\frac{8\pi^2 I k_B T}{\sigma h^2} \right) + 1 \right] \tag{10.21}$$

As was also previously noted in Section 9.3.1, the completely general rigid-rotor Schrödinger equation for a molecule characterized by three unique axes and associated moments of inertia does not lend itself to easy solution. However, by pursuing a generalization of the *classical* mechanical rigid-rotor problem, one can derive a quantum mechanical approximation that is typically quite good. Within that approximation, the rotational partition function becomes

$$q_{rot}(T) = \frac{\sqrt{\pi I_A I_B I_C}}{\sigma} \left(\frac{8\pi^2 k_B T}{h^2} \right)^{3/2} \tag{10.22}$$

where I_A, I_B, and I_C are the principal moments of inertia, and σ is again a symmetry number. In this case, σ is the number of pure rotations that carry the molecule into itself. Table 10.1 lists the specific values of σ for all chemically relevant symmetry point groups. Note that

Table 10.1 Rotational symmetry numbers for molecular point groups

Point Group[a]	σ
C_1	1
C_i	1
C_s	1
$C_{\infty v}$	1
$D_{\infty h}$	2
S_n, $n = 2, 4, 6, \ldots$	$n/2$
C_n, $n = 2, 3, 4, \ldots$	n
C_{nh}, $n = 2, 3, 4, \ldots$	n
C_{nv}, $n = 2, 3, 4, \ldots$	n
D_n, $n = 2, 3, 4, \ldots$	$2n$
D_{nh}, $n = 2, 3, 4, \ldots$	$2n$
D_{nd}, $n = 2, 3, 4, \ldots$	$2n$
T	12
T_d	12
O_h	24
I_h	60

[a] See Appendix B for explanations of point groups.

the presence of symmetry in a molecule may also require a structural degeneracy correction to the molecular partition function for certain conformers, as described in more detail in Appendix B.

Evaluation of the rotational components of the internal energy and entropy using the partition function of Eq. (10.22) for more typically encountered non-linear molecules gives

$$U_{rot} = \tfrac{3}{2}RT \tag{10.23}$$

and

$$S_{rot} = R\left\{\ln\left[\frac{\sqrt{\pi I_A I_B I_C}}{\sigma}\left(\frac{8\pi^2 k_B T}{h^2}\right)^{3/2}\right] + \frac{3}{2}\right\} \tag{10.24}$$

Again, it must be noted that evaluating the rotational components of U and S requires relatively little in the way of molecular information. All that is required is the principal moments of inertia, which derive only from the molecular structure. Thus, any methodology capable of predicting accurate geometries should be useful in the construction of rotational partition functions and the thermodynamic variables computed therefrom. Also again, the units chosen for quantities appearing in the partition function must be consistent so as to render q dimensionless.

10.3.6   Molecular Vibrational Partition Function

In a polyatomic molecule with many vibrations, we simplify the vibrational partition function much as the original molecular partition function was simplified: we assume that the total vibrational energy can be expressed as a sum of individual energies associated with each mode, in which case, for a non-linear molecule, we have

$$q_{vib}(T) = \sum_i e^{-[\varepsilon_1 + \varepsilon_2 + \cdots + \varepsilon_{3N-6}]_i / k_B T}$$

$$= \left[\sum_{j(1)} e^{-\varepsilon_{j(1)}/k_B T}\right]\left[\sum_{j(2)} e^{-\varepsilon_{j(2)}/k_B T}\right] \cdots \left[\sum_{j(3N-6)} e^{-\varepsilon_{j(3N-6)}/k_B T}\right] \tag{10.25}$$

where the energies ε_k are the vibrational energy levels associated with each mode k, and there are $3N - 6$ such modes in a non-linear molecule ($3N - 5$ in a linear molecule) where N is the number of atoms.

To evaluate the sums associated with each mode, we assume that the modes can be approximated as quantum mechanical harmonic oscillators (QMHOs), in which case the energy levels are given by Eqs. (9.47) and (9.48). In this case, we are offered a choice with respect to convention. We may either take the zero of energy as the bottom of the potential energy well on the PES, in which case the zeroth vibrational level has energy $\frac{1}{2}h\omega$, or we may take the zero of energy as the energy of the equilibrium structure plus the ZPVE, in which case the energy of the zeroth vibrational energy level is zero for every mode. Both conventions are used routinely, and one must simply be careful to ensure consistency – the entropy is independent of the choice of convention, and the internal energy varies by the ZPVE as a function of which convention is chosen.

Here, we will adopt the convention of including the ZPVE in the zero of energy [Eq. (10.1)], so that each zeroth vibrational level has an energy of zero. In that case, any individual mode's partition function can be written as

$$q_{\text{vib}}^{\text{QMHO}}(T) = \sum_{k=0}^{\infty} e^{-k\hbar\omega/k_{\text{B}}T} \tag{10.26}$$

The sum in Eq. (10.26) is well known as a convergent geometric series, so that we may write

$$q_{\text{vib}}^{\text{QMHO}}(T) = \frac{1}{1 - e^{-\hbar\omega/k_{\text{B}}T}} \tag{10.27}$$

This is a serendipitous result, insofar as the energy level spacing for most molecular vibrations is sufficiently large that significant errors would be introduced by replacing the sum by the corresponding indefinite integral as we did successfully for translation and rotation (such a replacement actually would amount to assuming a *classical* harmonic oscillator, for which $q_{\text{vib}} = k_{\text{B}}T/\hbar\omega$; by expanding the exponential in Eq. (10.27) as its corresponding power series, one can see that the classical and quantum partition functions agree only when $k_{\text{B}}T \gg \hbar\omega$).

Using Eq. (10.27) for each mode, the full vibrational partition function of Eq. (10.25) can be expressed as

$$q_{\text{vib}}(T) = \prod_{i=1}^{3N-6} \left(\frac{1}{1 - e^{-\hbar\omega_i/k_{\text{B}}T}} \right) \tag{10.28}$$

where $\prod$ implies a product series (the multiplicative analogy of a sum), and the upper limit would be $3N - 5$ for a linear molecule. Evaluation of the vibrational components of the internal energy and entropy using the partition function of Eq. (10.28) provides

$$U_{\text{vib}} = R \sum_{i=1}^{3N-6} \frac{\hbar\omega_i}{k_{\text{B}}(e^{\hbar\omega_i/k_{\text{B}}T} - 1)} \tag{10.29}$$

and

$$S_{\text{vib}} = R \sum_{i=1}^{3N-6} \left[\frac{\hbar\omega_i}{k_{\text{B}}T(e^{\hbar\omega_i/k_{\text{B}}T} - 1)} - \ln(1 - e^{-\hbar\omega_i/k_{\text{B}}T}) \right] \tag{10.30}$$

Note that Eqs. (10.29) and (10.30) take the vibrational frequencies as independent variables, and as such cannot be calculated *ab initio* without first optimizing a structure at some level of theory and then computing the second derivatives in order to obtain the frequencies within the harmonic oscillator approximation. (Of course, one could avoid the harmonic oscillator approximation (see, for example, Barone 2004), but the necessary calculations and the less tractable vibrational partition functions restrict this choice to only the most ambitious of calculations.)

In practice, then, it is fairly straightforward to convert the potential energy determined from an electronic structure calculation into a wealth of thermodynamic data – all that is required is an optimized structure with its associated vibrational frequencies. Given the many levels of electronic structure theory for which analytic second derivatives are available, it is usually worth the effort required to compute the frequencies and then the thermodynamic variables, especially since experimental data are typically measured in this form. For one such quantity, the absolute entropy $S°$, which is computed as the sum of Eqs. (10.13), (10.18), (10.24) (for non-linear molecules), and (10.30), theory and experiment are directly comparable. Hout, Levi, and Hehre (1982) computed absolute entropies at 300 K for a large number of small molecules at the MP2/6-31G(d) level and obtained agreement with experiment within 0.1 e.u. for many cases. Absolute heat capacities at constant volume can also be computed using the thermodynamic definition

$$C_V = \left(\frac{\partial U}{\partial T} \right)_V \tag{10.31}$$

and the various equations for components of U above.

Absolute internal energies, enthalpies, and free energies, on the other hand, are somewhat less straightforward. From a theoretical standpoint, using the electronic energy as something to which thermodynamic components are added is equivalent to setting the absolute zero of energy as corresponding to all nuclei and electrons infinitely separated one from another and at rest. In the laboratory, this is a very inconvenient zero, since the relevant elementary particles are not easily handled. The alternative conventions in common use for reporting H and G as determined from experiment, and the steps which must be taken so that theory and experiment may be consistently compared, are addressed next.

10.4 Standard-state Heats and Free Energies of Formation and Reaction

The experimental convention for assigning a zero to an enthalpy or free-energy scale is that this is the value that corresponds to the heat or free energy of formation associated with every element in its most stable, pure form under *standard* conditions (273 K, 1 atm). Thus, for instance, the elemental standard states for the first few elements are hydrogen gas (diatomic), helium gas (monatomic), solid lithium, solid beryllium, solid boron, solid carbon as its graphite allostere, nitrogen gas (diatomic), oxygen gas (diatomic), fluorine gas (diatomic), and neon gas (monatomic). Following this convention, the meaning of an experimental heat of formation for a molecule is that it is the (molar) enthalpy change associated with removing each of the atoms in the molecule from its elemental standard state and assembling them into the molecule.

Put in this manner, it is easy to imagine this as a two-step procedure. There is first an enthalpy cost to pull each atom out of its elemental standard state – always a non-negative quantity, since the elemental standard states are chosen to be the most stable forms. This is followed by the enthalpy change for combining them into the molecular structure, which is the negative of the enthalpy of atomization. As an example, the 0 K heat of formation of 2-butanone (methyl ethyl ketone, a widely used industrial solvent) is tabulated as -51.9 kcal mol^{-1}. The molecule is composed of eight atoms of hydrogen, four of carbon, and one of oxygen. The enthalpy cost to split 4 moles of hydrogen gas to create 8 moles of hydrogen atoms at 0 K is 413.5 kcal, the cost to strip 4 moles of carbon atoms from an infinite graphite block at 0 K is 1066.8 kcal, and the cost to split one-half mole of oxygen gas to create 1 mole of oxygen atoms at 0 K is 59.0 kcal. The 0 K atomization enthalpy of 1 mole of 2-butanone is 1591.2 kcal. Thus, the tabulated 0 K heat of formation cited above is determined as $(413.5 + 1066.8 + 59.0 - 1591.2)$.

An important technical point that must be mentioned here is that some attention must be paid to the states of the atoms to ensure that the difference between the molecular atomization enthalpy and the enthalpies of formation of the atoms is carried out consistently. When one atomizes a species like 2-butanone *experimentally*, each resulting atom will typically contain a number of spin-unpaired electrons equal to its number of formal bonds in the molecule (because each bond is being ruptured into two unpaired electrons, one on each atom formerly involved in the bond). Thus, for instance, each carbon atom will have four unpaired electrons, corresponding to the quintet S (^{5}S) term of the atom. However, this is *not* the ground state of the carbon atom (the ground state is ^{3}P), so the value of 1066.8 kcal noted above for the enthalpy change associated with stripping 4 moles of carbon atoms from a graphite block may, under some experimental conditions, be *measured* as 680.6 kcal, the cost to generate the 4 moles of C atoms in their ^{3}P ground state, plus 4×96.5 kcal mol^{-1}, where the latter energy is the molar enthalpy cost to excite an atom of C from ^{3}P to ^{5}S.

When one speaks of a *computational* atomization energy for a molecule, it should be carefully specified whether the energies of the product atoms are being computed in their ground states or in excited states that may be more convenient to work with for one reason or another. This specification is also critical to determining a computed heat or free energy of formation, as described next.

10.4.1   Direct Computation

Direct computation of a molecular heat or free energy of formation is something of a misnomer, since it would imply computing the difference in H or G for some molecule compared to the reference elemental standard states. Such a calculation might readily be imagined for a molecule like HF, because the standard states of H and F are gaseous diatomics. However, carrying out a high-level quantum mechanical calculation on an infinite block of graphite is another matter altogether. As a result, almost all so-called direct computations of heats of formation are carried out as illustrated in Figure 10.1. All quantities in the large inset region are *computed* relative to the theoretical zero of energy (all nuclei and electrons infinitely separated and at rest). To determine a *standard-state* molecular thermodynamic quantity, the computed energy *difference* between the molecule and its constituent atoms is added to the *experimental* thermodynamic value determined for the identical atoms. For instance, if we wanted to predict the 298 K heat of formation above

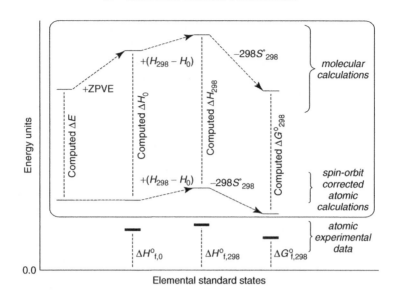

Figure 10.1   Graphical illustration of the procedure for predicting enthalpies and free energies of formation from computation

we would slide the entire inset region down until the computed H_{298} and experimental $\Delta H^\circ_{f,298}$ bars for the atoms overlapped; then, the position of the *molecular* H_{298} would be the predicted value for the molecular $\Delta H^\circ_{f,298}$. (Note that, for ease of illustration, all molecular energies are shown here as higher than the corresponding atomic energies, but this is certainly not a requirement, and negative molecular heats and free energies of formation are common.) Thus, a theoretical level is chosen for the computation of the electronic energies of the molecule and its constituent atoms. Either the same or potentially a different level of theory is chosen for the calculation of the thermal contributions to the enthalpy and entropy. The difference between the computed quantities for the molecule and its constituent atoms is then added to the *experimental* quantity associated with the atoms to determine the final theoretical value.

The three quantities that are most routinely employed in practice are the 0 K enthalpy of formation, the 298 K enthalpy of formation, and the 298 K free energy of formation. Enthalpies of formation are also commonly known as heats of formation. The values of these quantities for most of the elements in the first two rows of the periodic table are provided in Table 10.2. Also listed in the table are the spin–orbit corrections for atoms having P ground states. Experimental data refer to the lower energy spin–orbit state, but calculations that do not include a (relativistic) spin–orbit operator in the Hamiltonian fail to account for the energy lowering associated with this coupling. To make the experimental and computational levels for the atoms consistent with one another (see Figure 10.1) the spin–orbit correction must be added to theoretical atomic energies.

A technical point associated with Figure 10.1 bears mention. If one overlaps the computed and experimental atomic values for any one quantity, there is no guarantee (and indeed it is

Table 10.2 Experimental $\Delta H_{f,0}^{\circ}$, $\Delta H_{f,298}^{\circ}$, $\Delta G_{f,298}^{\circ}$ values and spin–orbit corrections (kcal mol^{-1}) for the atoms

Atom	$\Delta H_{f,0}^{\circ}$ [a]	$\Delta H_{f,298}^{\circ}$ [a]	$\Delta G_{f,298}^{\circ}$ [a]	Spin–orbit[b]
H (^{2}S)	51.63 ± 0.001	52.103	43.93	
Li (^{2}S)	37.69 ± 0.20	38.074	28.19	
Be (^{1}S)	76.48 ± 1.20	77.438	67.73	
	75.8 ± 0.8^c	76.75 ± 0.8^c	66.04 ± 1.2^c	
B (^{2}P)	136.2 ± 0.2	133.84	122.91 ± 0.2	-0.03
C (^{3}P)	169.98 ± 0.10	171.29	160.03	-0.09
N (^{4}S)	112.53 ± 0.02	112.97	102.05 ± 0.02	
O (^{3}P)	58.99 ± 0.02	59.553	48.08	-0.23
F (^{2}P)	18.47 ± 0.07	18.97	7.66	-0.38
Na (^{2}S)	25.69 ± 0.17	25.645	14.70	
Mg (^{1}S)	34.87 ± 0.20	35.158	24.57	
Al (^{2}P)	78.23 ± 1.00	78.800	67.08	-0.21
Si (^{3}P)	106.6 ± 1.9	107.55	95.59	-0.43
	108.1 ± 0.5^c	109.0 ± 0.5^c	97.1 ± 0.5^c	
P (^{4}S)	75.42 ± 0.20	75.619	64.00	
S (^{3}P)	65.66 ± 0.06	66.200	54.25	-0.56
Cl (^{2}P)	28.59 ± 0.001	28.991	17.23	-0.84

[a] All data, unless otherwise noted, are from the JANAF tables, see Chase, M. W., Jr. 1998. *J. Phys. Chem. Ref. Data, Monograph 9*, 1, for most recent versions.
[b] Amount by which lower energy spin–orbit state lies below unsplit term, see Moore, C. Natl. Bur. Stand. (US) Circ 467, 1952.
[c] Estimates considered to improve on experimental values, see Ochterski, J. W.; Petersson, G. A.; Wiberg, K. B. 1995. *J. Am. Chem. Soc.*, **117**, 11299.

unlikely) that the atomic levels will overlap for the other two quantities. As a consequence, one cannot compute, say, $\Delta H_{f,298}^{\circ}$ for the molecule by overlapping the atomic levels for H_0 and then taking the level of H_{298} as $\Delta H_{f,298}^{\circ}$. From inspection of Figure 10.1, this would be equivalent to computing the 298 K heat of formation as

$$\Delta H_{f,298}^{\circ}(M) = E(M) + ZPVE(M) + [H_{298}(M) - H_0(M)] - \sum_z^{\text{atoms}} E(X_z) + \sum_z^{\text{atoms}} \Delta H_{f,0}^{\circ}(X_z)$$

(10.32)

where molecule M is composed of constituent atoms X. *This is not valid.* The correct procedure is instead to overlap the theoretical and experimental atomic H_{298} values and then read off the molecular 298 K heat of formation, as detailed in the caption to Figure 10.1. This procedure is expressed mathematically as

$$\Delta H_{f,298}^{\circ}(M) = E(M) + ZPVE(M) + [H_{298}(M) - H_0(M)]$$
$$- \sum_z^{\text{atoms}} \{E(X_z) + [H_{298}(X_z) - H_0(X_z)]\} + \sum_z^{\text{atoms}} \Delta H_{f,298}^{\circ}(X_z) \quad (10.33)$$

The reason the former procedure fails is that the *theoretical* reference state is taken to be a constant temperature (0 K, by virtue of particles being taken to be at rest), but the *elemental* standard states are not, and thus the combinations of the last two terms on the r.h.s.s of Eqs. (10.32) and (10.33) are not equal. It is probably simplest to see this by considering the example of molecular hydrogen. Its translational enthalpy is given by Eq. (10.16) as $(\frac{3}{2})RT$. So, at 0 K it has no translational enthalpy and at 298 K it has roughly 0.9 kcal mol^{-1} of such enthalpy. Analogous changes are associated with rotation and vibration. However, molecular hydrogen is the elemental standard state, so it is defined experimentally to have a zero heat of formation at *whatever* temperature. Thus, when we compute the 298 K thermal contributions to the enthalpy of two H atoms, we determine from theory an absolute translational contribution of $3RT$ (again from Eq. (10.16) now applied to two separate particles), but experimentally we would only obtain $\frac{3}{2}RT$ for this term, since the reference elemental standard state also has increased *absolute* enthalpy at 298 K.

Having discussed in detail how to go about computing heats and free energies of formation, we should now consider how useful typical electronic-structure methods are for that purpose. The somewhat disappointing answer is that most single levels of theory are disastrously bad, with the problem lying primarily in the computation of ΔE between the molecule and its constituent atoms (the leftmost vertical line in Figure 10.1). As there is vastly more correlation energy in a molecule, with its collection of bonded pairs of electrons, than there is in a collection of atoms, and as practically affordable correlated electronic-structure methods capture at best perhaps 70–90% of the correlation energy, the differential error can be very large. Only with very, very small molecules is it possible to apply a *single* sufficiently high level of theory to accurately compute heats and free energies of formation *ab initio*. However, a number of different approaches employing varying degrees of semiempiricism have been promulgated to improve on this situation.

10.4.2 Parametric Improvement

In Section 7.7, parametric methods for improving the quality of correlated electronic-structure calculations were discussed in detail. Similarly, in Section 8.4.3, the mild parameterization of density functional methods to give maximal accuracy was described. Given that background, and the substantial data presented in those earlier chapters, this section will only recapitulate in a rough categorical fashion the various approaches whose development was motivated by a desire to compute more accurate thermochemical quantities.

Most attention has been focused on the computation of E_{elec}, because even fairly modest levels of theory can compute molecular geometries and vibrational frequencies sufficiently accurately to give good ZPVEs and thermal contributions, particularly if the frequencies are scaled by an appropriate factor (see Section 9.3). The simplest approach to improved E_{elec} estimation is to scale it as a raw value as well, and this is the formalism implicit in the PCI-80 and SAC methods described in Section 7.7.1.

At a higher level of complexity, correlation energies are computed assuming that effects associated with basis-set incompleteness and, say, truncated levels of perturbation theory, can be corrected for in a piecewise fashion. Such extrapolation schemes are described in Section 7.7.2, and specific recipes for extrapolation are at the heart of the G2 and G3 methods and the various CBS methods. Of course, scaling and extrapolation need not be mutually exclusive, and the combination of the two is the hallmark of the multicoefficient methods described in Section 7.7.3.

The G2 and G3 methods go beyond extrapolation to include small and entirely general empirical corrections associated with the total numbers of paired and unpaired electrons. When sufficient experimental data are available to permit more constrained parameterizations, such empirical corrections can be associated with more specific properties, e.g., with individual bonds. Such bond-specific corrections are employed by the BAC method described in Section 7.7.3. Note that this approach is different from those above insofar as the fundamentally modified quantity is not E_{elec}, but rather ΔH. That is, the goal of the method is to predict improved heats of formation, not to compute more accurate electronic energies, *per se*. Irikura (2002) has expanded upon this idea by proposing correction schemes that depend not only on types of bonds, but also on their lengths and their electron densities at their midpoints. Such detailed correction schemes can offer very high accuracy, but require extensive sets of high quality experimental data for their formulation.

Finally, hybrid DFT methods have a somewhat murky status with respect to their parameters, with some being founded on theoretical arguments while others are unabashedly empirical in their design to give improved agreement with experiment. From a practical standpoint, the hybrid DFT methods tend to offer the lowest overhead with respect to bookkeeping: all computed quantities in Figure 10.1 can usually be determined conveniently from a single level of theory. As noted in Chapter 8, however, the best DFT results are still somewhat less reliably accurate than the best multilevel models, although it must be borne in mind that the latter tend to be considerably more expensive than the former. As might be expected given their success in the context of MO theoretical methods, the use of bond additivity correction schemes to improve DFT performance has begun to be explored (see, for instance, Cloud and Schwartz 2003 and Winget and Clark 2004), as has the use of multicoefficient models (Zhao, Lynch, and Truhlar 2004).

One point meriting additional discussion concerns dispersion. Most of the databases used to validate the predictive ability of different theoretical models for heats of formation have been restricted to fairly small molecules. As such, there are few examples of molecules having different portions that interact with one another through London dispersion forces (as might be expected for a coiled long-chain alkane in the gas-phase, for instance). While the highly correlated MO-theory based models should perform acceptably for such cases (if cost is not prohibitive), DFT models would be expected to do less well, since they do poorly in general in predicting weak non-bonded interaction energies. This is also true for NDDO models, but these are already sufficiently inaccurate on average that failure to account for dispersion may not necessarily lead to substantially increased error. In any case, the magnitude of the error for DFT and NDDO models would be expected to increase with molecular size, so this is a source of some concern. Resolution of this issue will require greater attention to large molecules for which accurate data are available (see, for example, Winget and Clark 2004).

10.4.3   Isodesmic Equations

An alternative method for computing heats (or free energies) of formation involves consideration of a balanced chemical equation, e.g.,

$$m\text{A} + n\text{B} \longrightarrow r\text{C} + s\text{D} \tag{10.34}$$

where A, B, C, and D are molecules and m, n, r, and s indicate the number of moles of each in the balanced equation. The heat of reaction for a chemical transformation is defined as the difference between the heats of formation of the products and those of the reactants when these are defined relative to consistent standard states. For the reaction of Eq. (10.34), we would have

$$\Delta H^{\circ}_{\text{rxn},298} = [r\Delta H^{\circ}_{\text{f},298}(\text{C}) + s\Delta H^{\circ}_{\text{f},298}(\text{D})] - [m\Delta H^{\circ}_{\text{f},298}(\text{A}) + n\Delta H^{\circ}_{\text{f},298}(\text{B})] \tag{10.35}$$

where we have arbitrarily selected 298 K as the temperature of interest. Note that the standard-state symbol on the heat of reaction (as opposed to the heats of formation) does *not* imply the use of elemental standard states to assign a zero of enthalpy. Because the reaction is balanced, the standards used to define the zeroes for the heats of formation must cancel out on the two sides of the equation. So it is equally valid to write

$$\Delta H^{\circ}_{\text{rxn},298} = [r H_{298}(\text{C}) + s H_{298}(\text{D})] - [m H_{298}(\text{A}) + n H_{298}(\text{B})] \tag{10.36}$$

where H_{298} is the quantity typically addressed theoretically, i.e., the enthalpy relative to all nuclei and electrons infinitely separated and at rest.

Insofar as the r.h.s. of Eq. (10.35) must then be equal to the r.h.s. of Eq. (10.36), if the experimental heats of formation for all but one of the species in Eq. (10.34) are known (say B), we may rearrange our equality to determine this quantity as

$$\Delta H^{\circ}_{\text{f},298}(\text{B}) = -\frac{1}{n}\{[r H_{298}(\text{C}) + s H_{298}(\text{D})] - [m H_{298}(\text{A}) + n H_{298}(\text{B})]$$
$$- [r\Delta H^{\circ}_{\text{f},298}(\text{C}) + s\Delta H^{\circ}_{\text{f},298}(\text{D})] + m\Delta H^{\circ}_{\text{f},298}(\text{A})\} \tag{10.37}$$

This technique at first seems rather cumbersome, since we must perforce compute H_{298} for four different species in this example, but it has one great advantage over the apparently simpler *a priori* calculation of a single heat of formation, and that is that the difficulty in computing heats of atomization can be avoided. As noted above, computed heats of atomization tend to be highly inaccurate unless heroic levels of theory are employed, because the correlation energies for the electrons in the atoms and in the molecule are so enormously different. However, assuming experimental data are available, we may select our balanced chemical Eq. (10.34) in such a way that the various bonds on the left- and right-hand sides are essentially identical. That being the case, we would expect bond-by-bond errors in correlation energy to largely cancel in the computed heat of reaction (the top line on the r.h.s. of Eq. (10.37)) so that any error might be no larger than the errors associated with the experimentally measured heats. Such a reaction is called 'isodesmic'.

As a specific example, we might seek the heat of formation of 6-methylquinoline, the size of which is such that application of a methodology like G3 would be computationally rather intensive. However, consider the isodesmic reaction

$$(10.38)$$

where 6-methylquinoline is now molecule 'B' of Eq. (10.34). So long as heats of formation for the common molecules naphthalene, quinoline, and 2-methylnaphthalene are known, we may then compute enthalpies for all four species and predict the heat of formation of 6-methylquinoline using Eq. (10.37). Note that, by construction, all of the bonds on the l.h.s. of Eq. (10.38) are essentially identical to those on the r.h.s. (they are only non-identical once one starts to define them not only according to the two atoms which are bonded, but based on their distant neighbor atoms as well). As such, we might expect a much more affordable level of theory, say a DFT calculation, to be useful in the evaluation of Eq. (10.37).

Note that the construction of an isodesmic equation is something of an art, depending on chemical intuition and available experimental data. In the above situation, if an experimental heat of formation for quinoline were not available, we might decide to resort to a reaction like

$$(10.39)$$

While this reaction is still balanced, it is less ideal than Eq. (10.38). For instance, on the r.h.s. of Eq. (10.39), there are two aromatic C–H bonds where the carbon atom is bonded to a nitrogen, but on the l.h.s. there is only one. As a result, we might be forced to go to higher levels of theory to ameliorate any error this might introduce. In the extreme, one can imagine balanced reactions like

$$17 H_2 + \quad \longrightarrow \quad 10 CH_4 + NH_3 \qquad (10.40)$$

The necessary experimental heats of formation are known to exquisite accuracy (or defined as zero, in the case of H_2), and the calculations will be trivial for such small molecules, but accurately accounting for the enormous differences in the natures of the bonds on the two sides of Eq. (10.40) will require levels of electronic-structure theory nearly as high as those that would be necessary for a direct or parametric computation on 6-methylquinoline alone. The one virtue of Eq. (10.40), which is an example of a 'bond separation reaction', is that the total amount of unpaired electron spin on the two sides of the reaction is the same (in this case, zero); such a reaction is called 'isogyric'. Note that atomization processes are generally *not* isogyric, and this is an important factor in the large change in correlation energy associated with atomization.

Note that the above discussion can be rephrased in a very transparent way: a good isodesmic equation should predict a near-zero heat of reaction. The larger the predicted change in enthalpy, the greater the chance that lower levels of theory will fail to accurately account for energetic differences between dissimilar bonds. Note that just because a reaction *does* predict an overall enthalpy change near zero does not necessarily imply that the bonds on both sides are similar – large changes in one type may be offset by similarly large changes in another type – thus a near-zero heat of reaction is a necessary but not sufficient condition for an ideal isodesmic equation (for a mathematically more sophisticated approach to employing various isodesmic reactions, see Fishtik, Datta, and Liebman 2003).

These points are illustrated in more detail for the case of singlet *p*-benzyne, which has already been the subject of some discussion in preceding chapters. Consider the following three isodesmic reactions that might be used to determine its heat of formation:

$$\text{(ethylene)} + \text{(}p\text{-benzyne)} \longrightarrow \text{H}\!\equiv\!\text{H} + \text{(benzene)} \qquad (10.41)$$

$$2\,\text{CH}_4 + \text{(}p\text{-benzyne)} \longrightarrow 2\,\text{CH}_3^{\bullet} + \text{(benzene)} \qquad (10.42)$$

$$\text{(benzene)} + \text{(}p\text{-benzyne)} \longrightarrow 2\,\text{(phenyl radical)} \qquad (10.43)$$

The issue of isogyricity is a bit tricky in this instance, since *p*-benzyne is a ground-state singlet, but the coupling between the highest energy pair of electrons is very small. Table 10.3 indicates the heats of reaction computed for each of Eqs. (10.41)–(10.43) and the heats of formation determined for *p*-benzyne (using Eq. (10.37) and the experimentally available data for the methyl radical, methane, acetylene, ethylene, the phenyl radical, and benzene) at the CASPT2 and CCSD(T) levels; in each case, the equivalent of a basis set roughly triple-ζ in quality was used. Note that Eq. (10.43) is predicted to be the most nearly thermoneutral (which seems intuitively reasonable) and using it both levels of theory make predictions within the experimental error for $\Delta H^{\circ}_{f,298}(p\text{-benzyne})$. Equation (10.41) is predicted to be highly exothermic, because the r.h.s. has an extra π bond in acetylene compared to ethylene,

Table 10.3 Predicted heats of reaction and *p*-benzyne heats of formation (kcal mol^{-1}) using isodesmic Eqs. (10.41)–(10.43)

| Theory | Quantity | Isodesmic equation | | | Experiment |
		(10.41)	(10.42)	(10.43)	
CASPT2	$\Delta H^{\mathrm{o}}_{\mathrm{rxn,298}}$	−68.1	−10.8	4.4	
	$\Delta H^{\mathrm{o}}_{\mathrm{f,298}}$	129.6	136.1	138.2	138.0 ± 1.0
CCSD(T)	$\Delta H^{\mathrm{o}}_{\mathrm{rxn,298}}$	−76.0	−15.1	5.4	
	$\Delta H^{\mathrm{o}}_{\mathrm{f,298}}$	137.5	140.5	137.2	138.0 ± 1.0

and this degrades the performance of CASPT2. The CCSD(T) level, on the other hand, captures enough correlation energy (or enjoys some fortuitous cancellation of errors) that this equation gives an accurate heat of formation as well. Finally, Eq. (10.42), which involves exchanging aromatic C–H bonds for sp^3 C–H bonds causes both levels of theory to fall outside the experimental error bars by about 1 kcal mol^{-1}.

10.5 Technical Caveats

10.5.1 Semiempirical Heats of Formation

Recall that semiempirical methods were parameterized in such a way that the computed *electronic energies* were equated with heats of formation, *not* computed *enthalpies*. Thus, when a semiempirical electronic structure program reports a 298 K heat of formation for AM1, for instance, the reported value derives from adding the atomization energy ΔE to the experimental 298 K heats of formation of the atoms. Inspection of Figure 10.1 indicates that this will differ from the rigid-rotor-harmonic-oscillator computed result by ZPVE and the differential thermal contributions to the enthalpy of the molecule compared to the atoms.

As a result, the 'correct' way to compute a heat of formation with a semiempirical Hamiltonian is somewhat ambiguous. Since experimental data were used to optimize the parameters, the ZPVE and differential thermal contributions have been absorbed into the semiempirical parameters, so one is not necessarily improving things by adding these quantities *post facto*. On the other hand, to the extent ZPVE and thermal contributions *are* included in the parameters, it is in a very average way, and by no means consistent with rigorous statistical mechanics. In the end, individual investigators must decide for themselves, on the basis of what they are studying, whether to compute thermodynamic variables at the semiempirical level or simply to accept the electronic energies as having the status of enthalpies.

10.5.2 Low-frequency Motions

In the limit of a particular vibration going to zero, we see from Eq. (10.1) that it ceases to contribute to the zero-point vibrational energy. However, it is less obvious what happens to the molar internal energy, since in Eq. (10.29), both the numerator and the denominator of the term associated with the vanishing frequency go to zero. However, if we examine this behavior using a power series expansion for the exponential, we see that

$$
\lim_{\omega \to 0} \left[R \frac{\hbar\omega}{k_B(e^{\hbar\omega/k_BT} - 1)} \right] = \lim_{\omega \to 0} \left\{ \frac{R\hbar\omega}{k_B \left[1 + \dfrac{\hbar\omega}{k_BT} + \dfrac{1}{2!}\left(\dfrac{\hbar\omega}{k_BT}\right)^2 + \cdots - 1 \right]} \right\}
$$

$$
= \lim_{\omega \to 0} \left[\frac{R\hbar\omega}{k_B \left(\dfrac{\hbar\omega}{k_BT}\right)} \right]
$$

$$
= RT \tag{10.44}
$$

Thus, each vanishing frequency contributes a factor of RT to the molar internal energy (and thus the enthalpy).

Equation (10.44) can also be used to indicate that the first term in brackets on the r.h.s. of Eq. (10.30) goes to 1 as the frequency vanishes. However, if we examine the *second* term in brackets on the r.h.s. of Eq. (10.30), we discover

$$
\lim_{\omega \to 0} [-R \ln(1 - e^{-\hbar\omega_i/k_BT})] = \lim_{\omega \to 0} \left\{ -R \ln \left[1 - 1 + \frac{\hbar\omega}{k_BT} - \frac{1}{2!}\left(\frac{\hbar\omega}{k_BT}\right)^2 + \cdots \right] \right\}
$$

$$
= \lim_{\omega \to 0} \left[-R \ln \left(\frac{\hbar\omega}{k_BT} \right) \right]
$$

$$
= \infty \tag{10.45}
$$

which is certainly not a very pleasant result, since free energies will become infinitely negative with infinitely positive entropies. A careful analysis of Eq. (10.45) also indicates that small errors in very small non-zero frequencies can lead to very large errors in entropies. Unfortunately, it is precisely for low-frequency motions that we typically expect the harmonic oscillator approximation to be most poor. Thus, when a molecule is characterized by one or more very low-frequency vibrations, it is usually best not to discuss the molecular free energy, but instead restrict oneself to enthalpy or internal energy.

Note that there is nothing 'wrong' with Eq. (10.45). The entropy of a quantum mechanical harmonic oscillator really does go to infinity as the frequency goes to zero. What is wrong is that one usually should not apply the harmonic oscillator approximation to describe those modes exhibiting the smallest frequencies. More typically than not, such modes are torsions about single bonds characterized by very small or vanishing barriers. Such situations are known as hindered and free rotors, respectively.

More accurately, 'free rotor' is used to imply any torsion having a barrier substantially below k_BT. In such a situation, the contribution of the free rotor to the molar internal energy is

$$U_{\text{free rotor}} = \tfrac{1}{2}RT \tag{10.46}$$

i.e., only half that computed for a harmonic oscillator using Eq. (10.44). The contribution of a free rotor to the entropy is given by

$$S_{\text{free rotor}} = R\left\{\ln\left[\frac{(8\pi^3 I_{\text{int}}k_B T)^{1/2}}{\sigma_{\text{int}}h}\right] + \frac{1}{2}\right\} \tag{10.47}$$

where σ_{int} and I_{int} are the reduced symmetry numbers and moments of inertia associated with the free rotor. The definitions for these quantities may be found in the definitive work of Pitzer and Gwinn (1942) on this subject.

Pitzer and Gwinn have also provided tables to determine the thermodynamic contributions of hindered rotors (those having torsional barriers on the order of $k_B T$) when such rotors are well described by the torsional potential

$$E_{\text{hindered rotor}} = \tfrac{1}{2}V(1 - \cos\sigma_{\text{int}}\theta) \tag{10.48}$$

where V is the torsional barrier height and θ is the torsion angle. For the most careful work, this is the appropriate treatment to employ.

Note that if the torsional barrier is considerably greater than $k_B T$, then the harmonic oscillator approximation is as valid as for any vibration, and no special precautions need be taken.

10.5.3 Equilibrium Populations over Multiple Minima

It is not uncommon for a single molecule to have multiple populations. At non-zero temperatures, the population of different conformations will be dictated by Boltzmann statistics. If we make the approximation that we may neglect the continuous character of conformational space and simply work with discrete potential energy minima, we can replace a statistical mechanical probability integral with a discrete sum, and the equilibrium fraction F of any given conformer A at temperature T may be computed as

$$F(\text{A}) = \frac{e^{-G_{\text{A}}^{\circ}/RT}}{\sum_i e^{-G_i^{\circ}/RT}} \tag{10.49}$$

where i runs over all possible conformers, each characterized by its own free energy G°. In measurements on systems at equilibrium, it is rarely possible to determine the free energies of individual components of the equilibrium. Rather, one refers to the free energy of the whole equilibrium population, which may be written

$$G_{\{\text{A}\}}^{\circ} = -RT\ln\sum_{i\in\{\text{A}\}} e^{-G_i^{\circ}/RT} \tag{10.50}$$

where {A} emphasizes computation over the population of all conformers of A (where this set can include structures differing only by atom labeling, as detailed further in Appendix B). Free-energy changes, then, between two species each of which exist as populations over multiple conformers, must be computed as the difference between their averages. Note that the formalism of Eq. (10.50) may also be applied to determine averaged transition state free energies provided multiple transition state structures exist all of which lead to the same product; the difference between an averaged reactant free energy and an averaged transition state free energy defines a free energy of activation.

In fortunate instances, one conformer in a population has a free energy that is much lower than that of any of the other possibilities. Inspection of Eq. (10.50) makes clear that in that instance, only the low-energy term contributes significantly to the sum, in which case that free energy may be taken as the population free energy.

10.5.4 Standard-state Conversions

Two issues associated with thermodynamic standard states bear some further attention. The first is associated with the enthalpy of ions. Ion heats of formation may be defined based on the heats of ionization of neutral molecules (or electron attachments thereto). For example, one might consider a reaction like

$$A \longrightarrow A^{+\bullet} + e^- \tag{10.51}$$

and define the heat of formation of the radical cation $A^{+\bullet}$ as the sum of the heat of formation of A and the enthalpy change for Eq. (10.51). In that case, however, one needs to assign a heat of formation to the free electron. The thermal electron convention takes the free electron as the 'electron standard state', i.e., its enthalpy of formation is always zero. The so-called ion convention, on the other hand, takes the electron at rest to be the standard state (this is the usual theoretical convention as well, recall), and predominates in the mass spectrometric literature. The conversion between the two is straightforward, namely

$$\Delta H_{f,T}^{o}(X^q) = \Delta H_{f,T}^{o'}(X^q) + \tfrac{5}{2}qRT \tag{10.52}$$

where superscript 'o' represents the thermal electron standard state, superscript 'o'' represents the ion convention standard state, and q is the signed charge on X.

A separate issue arises in the discussion of standard-state free energies. Recall that the entropy of translation requires a concentration specification to be included as part of the standard-state conditions. Different tabulations of data often adopt different concentration conventions, and it is very important that care be taken to ensure consistent comparisons. To convert from one convention to another, we write

$$\Delta G^{o'} = \Delta G^{o} + RT \ln\left(\frac{Q^{o'}}{Q^{o}}\right) \tag{10.53}$$

where Q is the reaction quotient (i.e., the ratio of concentrations that appear in the equilibrium constant) evaluated with all species at their standard-state concentrations, expressed so that the logarithm is dimensionless. As an example, consider the gas-phase condensation reaction

$$A + B \longrightarrow C \qquad (10.54)$$

where we will define the 'o' standard state to imply all species at 1 atm pressure and the 'o'' standard state to imply all species in the gas phase at 1 mol L^{-1}. If A, B, and C are ideal gases, their concentration at 1 atm may be derived from the ideal gas law as $\frac{1}{24.5}$ mol L^{-1} at 298 K. Since the reaction quotient Q is [C]/[A][B], Eq. (10.53) becomes

$$\Delta G^{o'} = \Delta G^o + RT \ln \left(\frac{\frac{1}{1 \cdot 1}}{\frac{24.5 \cdot 24.5}{24.5}} \right)$$

$$= \Delta G^o - RT \ln(24.5) \qquad (10.55)$$

Additional standard-state issues can arise in condensed phases, and these will be dealt with in subsequent chapters.

10.5.5 Standard-state Free Energies, Equilibrium Constants, and Concentrations

While our focus has been primarily on thermodynamic quantities, like free energy, it should be borne in mind that the ultimate motivation for computing free energy differences is usually to permit calculation of chemical concentrations in actual systems. To accomplish this for a generic equilibrium is straightforward. For example, consider the following reaction (chosen in a completely arbitrary fashion)

$$A + B + C \rightleftharpoons 2D + E \qquad (10.56)$$

From the relationship between the equilibrium constant and the free energies of the reactants and the products we may write

$$\frac{[D]^2[E]}{[A][B][C]} = e^{-\Delta G^o / RT} \qquad (10.57)$$

where the standard-state symbol on the free energy change dictates the units used for the concentrations of the species. Thus, if we were carrying out all free energy calculations for gas-phase species at 1 atm pressure, we would express the reactant and product concentrations in those units. Stoichiometry then permits Eq. (10.57) to be rewritten as

$$\frac{(2x)^2 x}{(p_{0,A} - x)(p_{0,B} - x)(p_{0,C} - x)} = e^{-\Delta G^o / RT} \qquad (10.58)$$

where x is the concentration of E (and half the concentration of D) in units of partial pressure at equilibrium, and the initial partial pressures of reactants A, B, and C appear as constants in the denominator on the l.h.s. of Eq. (10.58). Note that the sign of the free energy change for Eq. (10.56) is only predictive of the side to which the equilibrium shifts when all species are initially present at their unit standard-state concentrations. All other situations require explicit evaluation of equations like Eq. (10.58) in order to determine the final concentrations predicted at equilibrium.

One variation on this theme that should be borne in mind when analyzing actual chemical situations is that certain species in the real system may be 'buffered'. That is, their concentrations may be held constant by external means. A good example of this occurs in condensed phases, where solvent molecules may play explicit roles in chemical equilibria but the concentration of the free solvent is so much larger than that for any other species that it may be considered to be effectively constant. Modeling solvation phenomena in general is covered in detail in the next two chapters, but it is instructive to consider here a particular case as it relates to computing equilibria. Consider such a reaction as

$$3(A \cdot S) \rightleftharpoons B \cdot 2S + S \tag{10.59}$$

That is, three monosolvates of A are in equilibrium with a disolvate of trimeric B (i.e., $B = A_3$) and a liberated solvent molecule. A rather typical protocol for evaluating the ratio of monomer to trimer in solution would be the following: (i) compute the gas-phase free energies of A·S, B·2S, and S at the appropriate temperature and a partial pressure of 1 atm (the default in most electronic structure programs), (ii) add to these gas-phase free energies the appropriate solvation free energies (usually computed assuming no change in standard-state concentration, as described in Chapters 11 and 12), and (iii) convert the free-energy change on going from reactants to products to standard-state units of 1 M concentration following the protocol of Eq. (10.55) because this is the more conventional standard state in solution. Having accomplished this, we would then be able to write

$$\frac{(x/3)[S]_0}{(x_{0,A \cdot S} - x)^3} = e^{-\Delta G^{o'}/RT} \tag{10.60}$$

where x is the moles of A monosolvate converted at equilibrium to $x/3$ moles of trimeric B disolvate and $[S]_0$ is the concentration of the solvent (determined from its density and molecular weight). To cement this example with actual values, imagine the solvent to be water ($[S]_0 = 55.56$ M) and, for a particular choice of A and B, the free energy change in solution (i.e., for the 'o' standard state) to be -3.0 kcal mol^{-1}. If we take the starting concentration of A monosolvate ($x_{0,A \cdot S}$) to be 0.2 M, we determine from solving the cubic Eq. (10.60) that at equilibrium x is 0.037 M, which is to say that there is about one molecule of B·2S for every 16 molecules of A·S. The failure of the reasonably large negative free-energy change to lead to substantial trimerization seems paradoxical only if one forgets that that negative number *refers specifically to all species being at their standard-state concentrations* (1 M)–actual systems may be quite far from that reference point.

10.6   Case Study: Heat of Formation of H_2NOH

Synopsis of Saraf *et al.* (2003) 'Theoretical Thermochemistry: *ab initio* Heat of Formation for Hydroxylamine'.

Hydroxylamine (H_2NOH) is a highly reactive molecule. As such, handling bulk quantities poses significant safety concerns, and indeed serious accidents occurred in 1999 and 2000 with this molecule in industrial settings. A direct measurement of the 298 K gas-phase enthalpy of formation of hydroxylamine is not available. Data from the solid phase have been interpreted to suggest a value of -12.0 ± 2.4 kcal mol^{-1}, but so large an uncertainty suggests that theory might prove useful in providing an improved estimate of this quantity, and this in turn might aid in the design of reaction conditions for reactors containing hydroxylamine. With that goal in mind, Saraf *et al.* surveyed a very large number of different levels of theory, including composite levels, to assess their likely utility for this task.

We consider here three different reaction protocols for predicting the enthalpy of formation of H_2NOH:

$$H_2NOH \rightleftharpoons 3H + N + O \tag{10.61}$$

$$H_2 + H_2NOH \rightleftharpoons NH_3 + H_2O \tag{10.62}$$

$$H_2O + H_2NOH \rightleftharpoons NH_3 + H_2O_2 \tag{10.63}$$

The latter two equations were used by Saraf *et al.* since the 298 K gas-phase enthalpies of formation of hydrogen, water, and ammonia are all known to very high accuracy. Thus, the procedures outlined in Section 10.4.3 may be used to compute the unknown hydroxylamine enthalpy of formation. As isodesmic equations go, Eq. (10.62) is not particularly good. The H—H and N—O bonds appearing on the l.h.s. are replaced by new N—H and O—H bonds on the r.h.s., and there is not much reason to expect these bonds to have similar errors in computed correlation energies. Eq. (10.63) is an improvement to the extent that the only major difference in bonding from the l.h.s. to the r.h.s. is the change of an N—O bond to an O—O bond. As both bonds are heteroatom to heteroatom for first-row atoms, we may expect a much more favorable cancellation of errors. Saraf *et al.* did not discuss the atomization energy, Eq. (10.61), which is in some sense the worst possible isodesmic reaction (perhaps one should call it the nihildesmic reaction) However, in the limit of perfect accuracy there is no need for the systematic cancellation of errors that isodesmic reactions are designed to provide, so we will consider Eq. (10.61) here for comparison.

As can be seen in Table 10.4, AM1 semiempirical theory is poorly suited for this application. With a polarized valence-double-ζ basis set, HF theory provides surprisingly good agreement with much higher levels of theory, but this is a case of fortuitous cancellation of errors, since use of a polarized valence-quadruple-ζ basis set decreases that agreement. The B3LYP model with a good basis set provides predictions that are not overall particularly much of an improvement over HF theory. The MP2 level with a large basis set does better for the more balanced isodesmic equation (10.63), but fares poorly with Eq. (10.62). Some improvement can be had with CCSD(T), but the cost of such a

Table 10.4   Predicted 298 K enthalpies of formation (kcal mol^{-1}) for hydroxylamine

Level of theory	From		
	Eq. (10.61)	Eq. (10.62)	Eq. (10.63)
AM1		−32.34	−31.31
HF/cc-pVDZ		−12.14	−12.02
HF/cc-pVQZ		−8.83	−13.06
B3LYP/aug-cc-pVTZ		−12.18	−9.69
MP2/cc-pVQZ		−8.61	−12.09
CCSD(T)/cc-pVQZ// CCSD(T)/cc-pVDZ		−11.56	−10.61
BAC-MP4		−12.98	−11.09
G2	−10.60	−11.78	−11.53
G2MP2	−11.46	−11.69	−11.67
G3B3	−10.02	−11.51	−11.53
G3	−9.36	−11.15	−11.28
CBS-Q	−10.03	−12.18	−11.16
Statistical^a	−10.29 ± 0.70	−11.66 ± 0.34	−11.43 ± 0.19

^{a}Average ± standard deviation from G2, G2MP2, G3B3, G3, and CBS-Q.

calculation far exceeds every other entry in the table. Since the G2, G2MP2, G3B3, G3, and CBS-Q models (all discussed in Chapter 7) are cheaper than CCSD(T)/cc-pVQZ and moreover designed specifically for the purpose of computing enthalpies of formation, there is ample reason to focus more closely on their performance.

Rather than attempting to rationalize why any one of these composite levels might be more or less good than another, let us examine their joint performance. The final row of Table 10.4 provides the means and standard deviations of the predicted $\Delta H^{\circ}_{f,298}$ (H$_2$NOH) values from these levels for Eqs. (10.61) to (10.63). The largest standard deviation is associated with Eq. (10.61), the next largest with Eq. (10.62), and the smallest, only 0.19 kcal mol^{-1}, with Eq. (10.63). This trend is entirely consistent with the above discussion of the relative quality of the three isodesmic equations, and provides some quantitative feel for how difficult the accurate computation of an atomization energy really is. Given this analysis, it appears reasonable to take the average value from the last five methods and Eq. (10.63) as a best estimate: −11.4 kcal mol^{-1}. Further support for this choice comes from considering a different reaction, namely

$$H_2 + H_2O_2 \rightleftharpoons 2H_2O \tag{10.64}$$

Note that this is the analog to Eq. (10.62) with hydrogen peroxide replacing hydroxylamine. In this case, all enthalpies of formation are known experimentally to high accuracy, so the performance of the various theoretical models may be directly assessed. Applying the same averaging procedure, one finds that the models predict an enthalpy of formation for H$_2$O$_2$ that is too negative by 0.3 kcal mol^{-1}. Note that if one assumes that this correction may be applied to the results from Eq. (10.62) for hydroxylamine, one predicts −11.4 kcal mol^{-1}, in perfect agreement with the uncorrected results from Eq. (10.63).

Note that the average atomization energy prediction differs from −11.4 kcal mol^{-1} by only 1.1 kcal mol^{-1}, which is about the range of accuracy typically quoted for the models involved. However, some of the *individual* models are off by still more (e.g., G3), which illustrates the utility of exploring multiple methods to ensure that one is not victimized by an otherwise unusual failure in accuracy.

Finally, although not discussed by Saraf *et al.*, it is noteworthy that H_2NOH has two local minima, one with the O–H bond anti to the N lone pair and one with the O–H bond eclipsing it. The latter is computed to be 4.38 kcal mol^{-1} lower in free energy than the former at 298 K, a result that is entirely in concert with intuition. All of the results discussed here are indeed for the correct local (global) minimum, but one should always be aware that as systems become more complex more effort may need to be expended in order to ensure that one is indeed working with the global minimum in one's computations.

Bibliography and Suggested Additional Reading

Cioslowski, J., Ed. 2001. *Quantum-Mechanical Prediction of Thermochemical Data*, Understanding Chemical Reactivity, Vol. 22, Kluwer: Dordrecht.

Cramer, C. J., Nash, J. J., and Squires, R. R. 1997. 'A Reinvestigation of Singlet Benzyne Thermochemistry Predicted by CASPT2, Coupled-cluster and Density Functional Calculations', *Chem. Phys. Lett.*, **277**, 311.

Curtiss, L. A., Raghavachari, K., Redfern, P. C., Rassolov, V., and Pople, J. A. 1998. 'Gaussian-3 (G3) Theory for Molecules Containing First and Second-row Atoms', *J. Chem. Phys.*, **109**, 7764.

Curtiss, L. A., Redfern, P. C., and Frurip, D. J. 2000. 'Theoretical Methods for Computing Enthalpies of Formation of Gaseous Compounds', in *Reviews in Computational Chemistry*, Vol. 15, Boyd, D. B. and Lipkowitz, K. B., Eds., Wiley-VCH: New York, 147.

Hehre, W. J., Radom, L., Schleyer, P. v. R., and Pople, J. A. 1986. Ab Initio *Molecular Orbital Theory*, Wiley: New York.

Irikura, K. K. and Frurip, D. J., Eds., 1998. *Computational Thermochemistry*, ACS Symposium Series, Volume 677, American Chemical Society: Washington, DC.

Jensen, F. 1999. *Introduction to Computational Chemistry*, Wiley: Chichester.

Martin, J. M. L. 1998. 'Calibration of Atomization Energies of Small Polyatomics' in *Computational Thermochemistry*, ACS Symposium Series, Vol. 677, Irikura, K. K. and Frurip, D. J., Eds., American Chemical Society: Washington, DC, 212.

McQuarrie, D. A. 1973. *Statistical Thermodynamics*, University Science Books: Mill Valley, CA.

Petersson, G. A. 1998. 'Complete Basis-set Thermochemistry and Kinetics' in *Computational Thermochemistry*, ACS Symposium Series, Vol. 677, Irikura, K. K., and Frurip, D. J., Eds., American Chemical Society: Washington, DC, 237.

Petersson, G. A., Malick, D. K., Wilson, W. G., Ochterski, J. W., Montgomery, J. A., Jr., and Frisch, M. J. 1998. 'Calibration and Comparison of the Gaussian-2, Complete Basis Set, and Density Functional Methods for Computational Thermochemistry' *J. Chem. Phys.*, **109**, 10570.

Zachariah, M. R. and Melius, C. F. 'Bond-additivity Correction of *Ab Initio* Computations for Accurate Prediction of Thermochemistry' in *Computational Thermochemistry*, ACS Symposium Series, Vol. 677, Irikura, K. K., and Frurip, D. J., Eds., American Chemical Society: Washington, DC, 162.

References

Barone, V. 2004. *J. Chem. Phys.*, **120**, 3059.

Chase, M. W., Jr., Davies, C. A., Downey, J. R., Jr., Frurip, D. J., McDonald, R. A., and Syverud, A. N. 1985. *J. Phys. Chem., Data Suppl.*, **14**, 1.

Cloud, C. F., III, and Schwartz, M. 2003. *J. Comput. Chem.*, **24**, 640.

Cossi, M. and Crescenzi, O. 2003. *J. Chem. Phys.*, **118**, 8863.

Fishtik, I., Datta, R., and Liebman, J. F. 2003. *J. Phys. Chem. A*, **107**, 695.

Hout, R. F., Jr., Levi, B. A., and Hehre, W. J. 1982. *J. Comput. Chem.*, **3**, 234.

Irikura, K. K. 2002. *J. Phys. Chem. A*, **106**, 9910.

Jensen, F. 2003. *Mol. Phys.*, **101**, 2315.

Moore, C. 1952. Natl. Bur. Stand. (US), Circ. 467.

Ochterski, J. W., Petersson, G. A., and Wiberg, K. B. 1995. *J. Am. Chem. Soc.*, **117**, 11299.

Pitzer, K. S. and Gwinn, W. D. 1942. *J. Chem. Phys.*, **10**, 428.

Saraf, S. R., Rogers, W. J., Mannan, M. S., Hall, M. B., and Thomson, L. M. 2003. *J. Phys. Chem. A*, **107**, 1077.

Winget, P. and Clark, T. 2004. *J. Comput. Chem.*, **25**, 725.

Zhao, Y., Lynch, B. J., and Truhlar, D. G. 2004. *J. Phys. Chem. A*, **108**, 4786.

11

Implicit Models for Condensed Phases

11.1 Condensed-phase Effects on Structure and Reactivity

The gas phase is delightful in its simplicity. At low to moderate pressures, molecules may be treated as isolated, non-interacting species, and this facilitates theoretical modeling enormously, insofar as the system of interest is entirely defined by the molecule itself. Were theory to restrict itself to the gas phase, however, it would be inapplicable to vast tracts of chemistry, to include essentially *all* of biochemistry.

Of course, one can carry out accurate gas-phase calculations and then make broad generalizations about how one might expect a surrounding condensed phase to affect the results. Indeed, this *modus operandi* was much employed well into the 1980s and still sees modest use today. Provided one can be reasonably confident that condensed-phase effects are small for the particular properties being studied (either in an absolute sense or through cancellation by judicious comparisons), such an approach can still be useful, particularly in a qualitative sense. However, significant developmental efforts over the last two decades combined with growth in the computational power required to implement them have resulted in the widespread availability of condensed-phase models designed to more accurately describe the physical nature of condensed-phase systems. This chapter considers one such class of these models, namely, implicit solvation models, which are also often called continuum solvation models.

At first thought, of course, it might seem that the modeling of a condensed-phase system should be trivial. Take for example a liquid solution (which we will take as our 'default' condensed phase in this and the next chapter, although others will be discussed). If our solution is dilute, then the 'obvious' way to construct a model is to surround our solute with a number of solvent molecules. But a critical question is, how many? If we want to consider glucose in water, for instance, it seems clear that we would want at least the entire surface of the glucose molecule to be covered. This might take, say, 14 water molecules, which we could place approximately at the corners and faces of an imaginary cube about our solute. However, it would be something of a stretch of faith to imagine this as true aqueous solvation – none of the water molecules is interacting with a second solvation

Essentials of Computational Chemistry, 2nd Edition Christopher J. Cramer
© 2004 John Wiley & Sons, Ltd   ISBNs: 0-470-09181-9 (cased); 0-470-09182-7 (pbk)

shell, making many otherwise plausible hydrogen bonding arrangements inaccessible. So, we might add another solvation shell. Since the surface area of a sphere increases as the square of the radius, it is clear that we will need many more than 14 waters this time. Will two solvation shells be enough? Almost certainly not – to eliminate spurious 'edge' effects associated with the water molecules on the outside, we really should add hundreds or thousands of waters. But if we do so, our system size becomes enormous – a quantum mechanical calculation on a single geometry will be staggeringly expensive. Worse still, that single geometry is of very limited value. As noted in Chapter 3 and discussed further in Chapter 12, with so many molecules and so many possible minima we need to compute statistical mechanical averages in order to determine equilibrium properties. Thus, we need to carry out possibly millions of these staggeringly expensive quantum calculations, and such a situation is simply not practical within the confines of present resources.

The assumption underlying continuum solvation models, which are the subject of this chapter, is that one may remove the huge number of individual solvent molecules from the model, as long as one modifies the space those molecules used to occupy so that, modeled as a continuous medium, it has properties consistent with those of the solvent itself. To determine how to define such a medium, one must consider the solvation process itself.

11.1.1   Free Energy of Transfer and Its Physical Components

The most important fundamental quantity describing the interaction of a solute with a surrounding solvent is the free energy of solvation ΔG_S^o. This quantity is sometimes also called the free energy of transfer, and refers to the change in free energy for a molecule A leaving the gas phase and entering a condensed phase. This free energy may be determined from the equilibrium constant describing the partitioning of A between the gas and condensed phases according to

$$\Delta G_S^o(A) = \lim_{[A]_{sol} \to 0} \left\{ -RT \ln \frac{[A]_{sol}}{[A]_{gas}} \bigg|_{eq} \right\} \tag{11.1}$$

where the limit is applied to ensure 'ideal solution' behavior. As with all free-energy quantities, attention must be paid to the standard-state concentrations. Most theoretical work makes use of standard-state concentrations of 1 M in both the gas phase and the condensed phase. In that case, there is no intrinsic change in the entropy of translation of the solute associated with a change in standard-state volume. Common experimental conventions, however, include expressing the gas-phase concentration as a partial pressure, with 1 atm defining the standard state, and/or expressing the solution concentration as mole fraction, with unit mole fraction defining the standard state (conversion between different concentration standard states for free energies of solvation can be accomplished using Eq. (10.53) in the same fashion already described for reaction equilibrium constants).

Experimental free energies of solvation span a wide range of values, from positive tens to negative hundreds of kilocalories per mole (for those values where the solution/gas equilibrium constants fall outside the range of about 10^{-6} to 10^6, experimental techniques other than simply measuring the concentrations in the two distinct phases are typically required). Different physical effects contribute to the overall solvation process; of these, the most important components are electrostatic interactions, cavitation, changes in dispersion, and changes in bulk solvent structure.

Equilibrium electrostatic interactions between a solute and a solvent are always non-positive – they are zero if the solute is characterized by no electrical moments (e.g., a noble gas atom) and negative otherwise, i.e., attractive. It is easiest to visualize the electrostatic interactions as developing in a stepwise fashion. Consider a solute A characterized by electrical moments; for simplicity, consider only the dipole moment. When A passes from the gas phase into a solvent, the solvent molecules, if they have permanent moments of their own, reorient so that, averaged over thermal fluctuations, their own dipole moments oppose that of the solute. In an isotropic liquid with solvent molecules undergoing random thermal motion, the average electric field at any point will be zero; however, the net orientation induced by the solute changes this, and the field induced by introduction of the solute is sometimes called the 'reaction field'.

Of course, the presence of an electric field means that a term accounting for the interactions of charged particles with this field should be included in the solute Hamiltonian. When it *is* included, the effect is to increase the solute polarity in a fashion proportional to the solute polarizability and the strength of the external field. Thus, the dipole moment of A increases. The solvent, seeing this increase, itself polarizes and moreover increases its own orientation to oppose A's dipole, and so on.

However, neither the orientation/polarization of the solvent nor the electronic polarization of A is without cost. In the first instance, since solvent molecules are oriented to oppose the dipole moment of A, they each interact in an *un*favorable sense with the reaction field they create. Moreover, to the extent they have lost some configurational freedom, there is an associated free-energy cost. As for the cost of electronic polarization, this may be viewed as the gas-phase cost (as computed with the gas-phase Hamiltonian) associated with distortion of the wave function away from the gas-phase minimum. As a result of these opposing energetics, the polarization of the solute/solvent system stops at that point where any energy gain from additional polarization is exactly balanced by the energy cost to achieve that polarization. Under some fairly mild assumptions from so-called linear response theory, one can show that this occurs when the energy cost up to a certain point becomes equal to one half of the total interaction energy between the solute and the solvent.

It cannot be overemphasized that solvation changes the solute electronic structure. As noted above, dipole moments in solution are larger than the corresponding dipole moments in the gas phase. Indeed, *any* property that depends on the electronic structure will tend to have a different expectation value in solution than in the gas phase. How large will the difference be? That depends on the strength of the solute–solvent interactions. Table 11.1 lists dipole moments computed in the gas phase, chloroform, and water for six nucleic acid bases at the HF/6-31G(d) level using the SM5.42R continuum solvation model that is described in more detail below. Note that the increases in dipole moments on going from the gas phase to water range from about 25 to 33 percent for these molecules; a smaller but still substantial increase is predicted in chloroform solution.

Table 11.1 Nucleic acid base dipole moments (D) at
the SM5.42R/HF/6-31G(d) level

Molecule	Dipole moment		
	Gas	Chloroform	Water
Adenine	2.4	2.9	3.1
Cytosine	6.5	8.0	8.5
Guanine	5.3	6.7	7.1
Hypoxanthine	6.4	7.8	8.2
Thymine	4.4	5.6	6.0
Uracil	4.5	5.6	6.0

Another physical effect associated with solvation is cavitation. It is again helpful to visualize the solvation process as a stepwise procedure. Here, we imagine the first step as being creation of a cavity of vacuum within the solvent into which the solute will be inserted as a second step. The energy cost of the cavity creation is the cavitation energy. Note that energy is always required to create the cavity – if it were favorable to create 'bubbles' of vacuum in the liquid, the solvent would not remain in the liquid phase.

As for what holds liquids together in the first place, the majority of the interaction energy in uncharged fluids derives from dispersion forces between solvent molecules that are in contact with one another. This is true even for liquids composed of very polar molecules: dispersion accounts for 70–90 percent of the total cohesion energy in liquid HCl or 2-butanone. Recalling the discussion in Section 2.2.4, dispersion refers to the always favorable interaction between the simultaneous induced dipoles in adjacent molecules that are a consequence of the correlated motion of electrons. When a solute is inserted into a pre-existing cavity into which it exactly fits, it will experience favorable dispersion interactions with the surrounding solvent. Note that while *formally* dispersion is an electrostatic interaction, it is usually discussed separately from the solute–solvent polarization described above in deference to its short-range character and its non-classical origin. Note also that it is dispersion alone that can account for a favorable transfer free energy of a solute into a solvent when neither of the two is characterized by any permanent electrical moments.

Finally, under certain conditions, the introduction of a solute molecule may significantly alter the equilibrium structure/dynamics of a solvent in the near vicinity of the solute, and this phenomenon will have associated with it a free-energy change. The most widely documented example is the hydrophobic effect, where loss of orientational freedom for water molecules in the first solvation shell about hydrocarbon fragments of solutes carries with it a free-energy cost that is responsible for the increasingly positive aqueous free energies of solvation of alkanes as they increase in chain length.

Having enumerated the various processes involved in the transfer of a solute molecule from the gas phase to solution, it must be emphasized that it is not possible to separately measure their contributions to the fundamental observable, ΔG_S^o. One can, of course, attempt to design systems where one expects only a single contribution to dominate, in the hopes of learning more about the nature of that contribution from experimental measurements, but inferences drawn therefrom become less certain as they are applied to systems less like those originally measured. This point is made insofar as we will discuss below, for example, theoretical predictions for the electrostatic component of the free energy of solvation. However, insofar as this quantity is *not* an experimental observable, absolute judgments of quality comparing one level of theory to another are necessarily model-dependent.

11.1.2 Solvation as It Affects Potential Energy Surfaces

In order to visualize the effects of solvation on structure and reactivity, it is helpful to consider the potential energy surface created by adding the free energy of solvation point by point to the gas-phase PES, as illustrated in Figure 11.1. (To be rigorous, one really should use a gas-phase *free*-energy surface so as not to be haphazardly mixing potential and free energies, but for qualitative purposes, we may ignore this technical point.) Processes in solution may be regarded as occurring on the lower surface, and all of the phase-space dimensions associated with solvent molecules have been averaged over in computing its energies.

Figure 11.1 illustrates several critical concepts associated with solvation. First, note that the reaction process depicted on the gas-phase surface joins two minima of roughly equal energy, while that on the lower surface is quite exergonic in the left-to-right direction. This derives from the minimum-energy structure at the larger x coordinate having a more negative free energy of solvation. Differential solvation of two (or more) minima implies a different equilibrium constant in solution than in the gas phase. Many examples of this

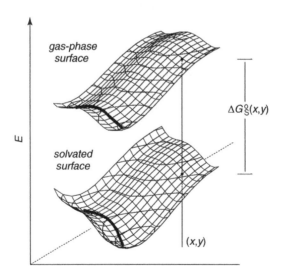

Figure 11.1 A two-dimensional gas-phase PES and the corresponding PES derived from adding the free energy of solvation to every point. This process is illustrated for point (x,y). Thick lines on the two surfaces indicate some chemical reaction proceeding from one minimum-energy structure to another. Note that there is no requirement for the x and y coordinates of equivalent stationary points on the two surfaces to be the same

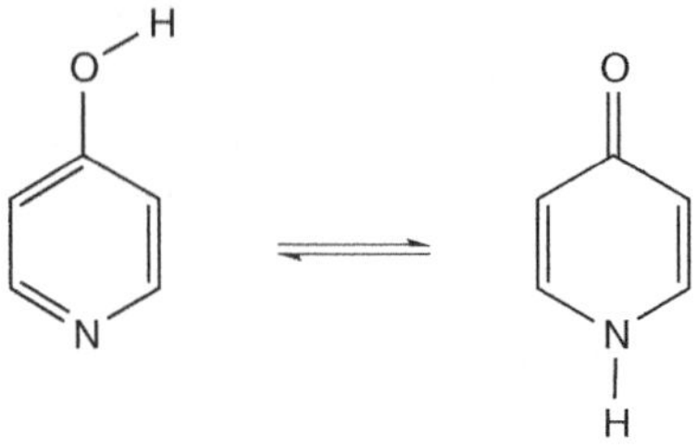

Figure 11.2 4-Pyridone is considerably more polar than its hydroxypyridine tautomer, and its extended π system also renders it highly polarizable. As a result, polar solvents shift the equilibrium between the two strongly to the right in comparison to the gas phase

phenomenon are known; one of the most striking is for the tautomeric equilibrium between 4-hydroxypyridine and 4-pyridone, where aqueous solvation changes the 298 K equilibrium constant by some six orders of magnitude (Figure 11.2; see Beak 1977).

Figure 11.1 also indicates that the free energy of activation for the left-to-right reaction is lower on the solvated surface than on the gas-phase surface, so that the rate will be increased in solution compared to the gas phase. Thus, differential solvation of minima and connected TS structures can affect relative rates. Again, many examples are known. One of the most carefully studied is the effect of solvation on the S_N2 reaction of chloride ion with chloromethane: While the gas-phase activation free energy is around 3 kcal mol^{-1}, the diffuse negative charge associated with the S_N2 transition state structure compared to a chloride ion makes the TS structure much less well solvated than the reactants, and aqueous solvation decreases the 298 K rate by more than 15 orders of magnitude (Chandresekhar, Smith, and Jorgensen 1985).

Given our picture of the free energy surface in solution deriving from addition of solvation free energy to the gas-phase PES, and noting that equilibria and kinetics can be well estimated based only on knowledge of the relative energies of appropriate stationary points, we may represent a protocol for computing these relative energies from the thermodynamic cycle in Figure 11.3. In order to compute the lower horizontal leg of the cycle, corresponding to

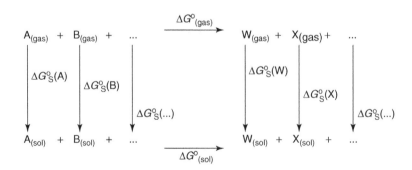

Figure 11.3 Cycle for computation of a free-energy change in solution

the process in solution, one need only take appropriate sums and differences of the upper horizontal leg and the vertical legs, viz.

$$\Delta G^\circ_{(sol)} = \Delta G^\circ_{(gas)} + \left[\Delta G^\circ_S (W) + \Delta G^\circ_S (X) + \cdots\right] - \left[\Delta G^\circ_S (A) + \Delta G^\circ_S (B) + \cdots\right] \quad (11.2)$$

As the upper leg is a gas-phase quantity, it can be computed taking advantage of all of the technology discussed in earlier chapters. The two vertical legs, on the other hand, consist exclusively of free energies of solvation. Thus, the development of models to efficiently compute molecular solvation free energies has been a high priority.

A compromise representation of our discussion thus far is to consider the effects of solvation on a one-dimensional slice through the energy surface–what we normally call the reaction coordinate–as illustrated in Figure 11.4. This representation is more informative than the free-energy cycle in showing how the *structures* of the stationary points differ in the gas phase and solution, in addition to their relative energies. A change in structure is indicated by a movement of the stationary point along the coordinate axis. Particularly for TS structures, which may be characterized by one or more soft normal modes and/or a soft reaction coordinate, changes in structure induced by solvation may be important.

Note, however, that this one-dimensional representation can be somewhat misleading if it is taken to be a computational protocol. The trouble is that the one-dimensional slice of

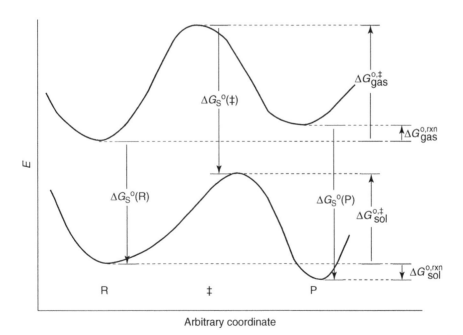

Figure 11.4 Gas-phase (upper) and solution (lower) reaction coordinates, and the thermodynamic cycles that connect them via free energies of solvation of the various stationary points (vertical lines). Note the significant left to right movement of all stationary points, and particularly the TS structure, on going from the gas phase to solution

the energy surface that generates the gas-phase reaction coordinate may be quite different from the one-dimensional slice that generates the reaction coordinate on the solvated surface. Put differently, solvation can move a stationary point not only along the gas-phase reaction coordinate but in other directions as well. Thus, if one constructs a solution coordinate by computing solvation free energies point-by-point for the gas-phase coordinate, one may miss important effects associated with movement *off* the gas-phase reaction coordinate. This is illustrated by the example of the Claisen rearrangement of allyl vinyl ether in Figure 11.5. Here the reaction coordinate may be thought of as the difference in distance between the initially bonded O3 and C4 atoms and the ultimately bonded C1 and C6 atoms. In water, aldehydes are better solvated than ethers, and this differential solvation is felt by the TS structure, so that it shifts to the right along the reaction coordinate (new TS location not shown). However, it also moves along an orthogonal coordinate best described by the distance between the C1C2O3 and C4C5C6 fragments. Polar solvents interact more strongly with a TS structure having greater interfragment separation because of the zwitterionic character associated with this structure. This move off the reaction coordinate can significantly lower the activation free energy.

Another situation that can complicate the interpretation of a reaction in solution by comparison to the gas phase involves a reaction that fails to have corresponding stationary points in the gas phase. If there is no stationary point in the gas phase, there is no real sense in talking about the free energy of solvation of the structure that exists in solution. A good example

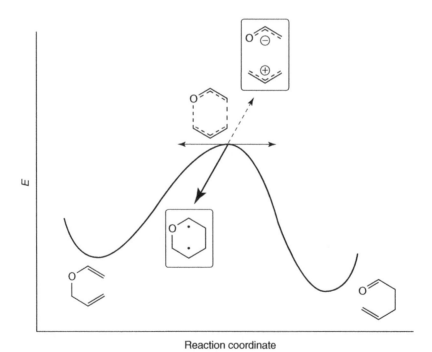

Reaction coordinate

Figure 11.5   Gas-phase reaction coordinate for the Claisen rearrangement of allyl vinyl ether

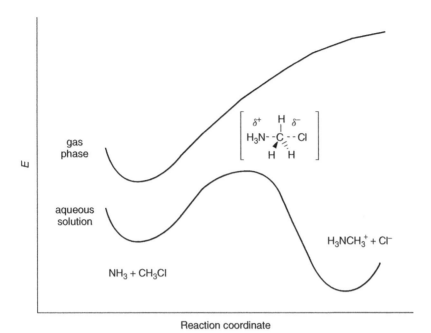

Figure 11.6   Menschutkin reaction of ammonia and chloromethane. In the gas phase nucleophilic displacement fails to take place, while in water solvation of the anions allows the reaction to proceed

of this situation is the organic Menschutkin reaction, where nucleophilic displacement of a halide by an amine generates a halide anion and an alkylammonium cation, as shown in Figure 11.6. In the gas phase, such separation of opposite charges is so unfavorable that no stationary point exists for the separated ions. In highly polar solvents, on the other hand, the solvation free energy of the ions is sufficiently high that not only are the products stationary points, but the reaction is exergonic. This is a situation where invoking a free-energy cycle is not particularly useful, although the direct computation of the lower leg (i.e., the solvated process) by one or another solvation model is a perfectly valid option.

Besides affecting equilibria and kinetics on single energy surfaces, differential solvation effects on distinct electronic states can cause significant changes in UV-Vis absorption spectra. Such so-called solvatochromic effects are discussed in more detail in Chapter 14.

The various effects of solvation discussed above may in principle be modeled in different ways. For the remainder of this chapter, we will focus on the utility of continuum solvation models in this regard. Having identified the importance and utility of the free energy of solvation, we will pay special attention to prediction of this quantity as a measure of quantitative accuracy.

11.2   Electrostatic Interactions with a Continuum

When a solute is immersed in a solvent, its charge distribution interacts with that of the solvent. In a continuum model, rather than representing the charge distribution of the solvent explicitly, we replace it by a continuous electric field that represents a statistical average over all solvent degrees of freedom at thermal equilibrium. This field is usually called the 'reaction field' in the regions of space occupied by the solute, since it derives from reaction of the solvent to the presence of the solute. The electric field at a given point in space is the gradient of the electrostatic potential ϕ at that point, and the work required to create the charge distribution may be determined from the interaction of the solute charge density ρ with the electrostatic potential according to

$$G = -\frac{1}{2} \int \rho\,(\mathbf{r})\phi\,(\mathbf{r})\,d\mathbf{r} \tag{11.3}$$

The charge density ρ of the solute may be expressed either as some continuous function of $\mathbf{r}$ or as discrete point charges, depending on the theoretical model used to represent the solute. The polarization energy, G_P, discussed above, is simply the difference in the work of charging the system in the gas phase and in solution. Thus, in order to compute the polarization free energy, all that is needed is the electrostatic potential in solution and in the gas phase (the latter may be regarded as a dielectric medium characterized by a dielectric constant of 1).

11.2.1 The Poisson Equation

At the heart of all continuum solvent models is a reliance on the Poisson equation of classical electrostatics to express the electrostatic potential as a function of the charge density and the dielectric constant. The Poisson equation, valid for situations where a surrounding dielectric medium responds in a linear fashion to the embedding of charge, is written

$$\nabla^2\phi\,(\mathbf{r}) = -\frac{4\pi\rho\,(\mathbf{r})}{\varepsilon} \tag{11.4}$$

where ε is the dielectric constant of the medium. Insofar as continuum solvation involves representing the solute explicitly and the solvent implicitly, the charge distribution of the solute is thought of as being inside a cavity that displaces an otherwise homogeneous dielectric medium. Thus, there are really two regions, one inside and one outside the cavity, in which case the Poisson equation is properly written as

$$\nabla\varepsilon\,(\mathbf{r}) \cdot \nabla\phi\,(\mathbf{r}) = -4\pi\rho\,(\mathbf{r}) \tag{11.5}$$

The Poisson equation is valid under conditions of zero ionic strength. If dissolved, mobile electrolytes are present in the solvent, the Poisson–Boltzmann (PB) equation applies instead

$$\nabla\varepsilon\,(\mathbf{r}) \cdot \nabla\phi\,(\mathbf{r}) - \varepsilon\,(\mathbf{r})\,\lambda\,(\mathbf{r})\,\kappa^2 \frac{k_B T}{q} \sinh\left[\frac{q\phi\,(\mathbf{r})}{k_B T}\right] = -4\pi\rho\,(\mathbf{r}) \tag{11.6}$$

where q is the magnitude of the charge of the electrolyte ions, λ is a simple switching function which is zero in regions inaccessible to the electrolyte and one otherwise, and κ^2

is the Debye–Hückel parameter given by

$$\kappa^2 = \frac{8\pi q^2 I}{\varepsilon k_B T} \tag{11.7}$$

where I is the ionic strength of the electrolyte solution. The inverse of κ is also called the Debye length.

Thus, in order to determine the electrostatic potential in solvents containing either non-electrolytes or electrolytes, we need only solve Eqs. (11.5) or (11.6), respectively, using the known charge density of the solute and some cavity defining how the dielectric constant varies about the solute. As differential equations go, Eq. (11.5) is straightforward, but Eq. (11.6) is fairly unpleasant. As a result, it is often simplified at low ionic strength by using a truncated power expansion for the hyperbolic sine, giving the so-called linearized PB equation

$$\nabla\varepsilon\,(\mathbf{r}) \cdot \nabla\phi\,(\mathbf{r}) - \varepsilon\,(\mathbf{r})\,\lambda\,(\mathbf{r})\,\kappa^2\phi\,(\mathbf{r}) = -4\pi\rho\,(\mathbf{r}) \tag{11.8}$$

Note that it is fairly common in the literature for continuum solvation calculations to be reported as having been carried out using Poisson–Boltzmann electrostatics even when no electrolyte concentration is being considered, i.e., the Poisson equation is considered a special case of the PB equation and not named separately.

For certain ideal cavity shapes, the relevant PB equations have particularly simple analytic solutions. While such ideal cavities are not typically to be expected for arbitrary solute molecules, consideration of some examples is instructive in illustrating how more sophisticated modeling may be undertaken by generalization therefrom.

11.2.1.1   Ideal cavities

Consider a conducting sphere bearing charge q, which may be taken as an approximation to a monatomic ion. The charge on such an object spreads out uniformly on the surface, and the charge density at any point on the surface may thus be expressed as

$$\rho\,(\mathbf{s}) = \frac{q}{4\pi a^2} \tag{11.9}$$

where $\mathbf{s}$ is a surface point and a is the radius of the sphere. So, in order to evaluate Eq. (11.3), we will need to integrate only on the surface of the sphere (since the charge density is zero everywhere else). To determine the electrostatic potential at the surface we must approach from the outside (the dielectric constant of a conductor is infinite and the electrostatic potential everywhere inside is zero, so there is a formal discontinuity in the potential at the surface). From the outside, the electrostatic potential is well known to be equivalent to that for a point charge q at the origin, giving the central field result

$$\phi\,(\mathbf{r}) = -\frac{q}{\varepsilon\,|\mathbf{r}|} \tag{11.10}$$

where ε is the *exterior* dielectric constant. Taking $\mathbf{r}$ on the surface of the sphere, i.e., $|\mathbf{r}| = a$, Eq. (11.3) becomes

$$G = -\frac{1}{2} \int \left(\frac{q}{4\pi a^2}\right)\left(-\frac{q}{\varepsilon a}\right) d\mathbf{s} = \frac{q^2}{2\varepsilon a} \tag{11.11}$$

As the square of the charge, the dielectric constant, and the ionic radius must all be positive, work must be expended to charge the sphere, but the work is less for higher exterior dielectric constants, as expected. Recalling that the polarization energy is the difference in the required work in the gas phase and solution, we may write

$$G_P = -\frac{1}{2}\left(1 - \frac{1}{\varepsilon}\right)\frac{q^2}{a} \tag{11.12}$$

which is the so-called Born equation for the polarization energy of a monatomic ion in atomic units.

If instead of carrying a charge, our sphere appears to be characterized by a perfectly dipolar distribution having dipole moment μ, an analogous analysis provides

$$G_P = -\frac{1}{2}\left[\frac{2(\varepsilon - 1)}{(2\varepsilon + 1)}\right]\frac{\mu^2}{a^3} \tag{11.13}$$

which is the so-called Kirkwood–Onsager equation in atomic units.

An important difference between the Born and Kirkwood–Onsager formulae is that the former depends on the charge, which is a property of the system restricted to integral values, while the latter depends on the dipole moment, which can potentially vary in different environments. In the context of quantum mechanical calculations, let us define the Kirkwood–Onsager polarization energy *operator* by invoking μ as the dipole moment operator in Eq. (11.13). In that case, the Schrödinger equation in solution becomes

$$\left\{H - \frac{1}{2}\left[\frac{2(\varepsilon - 1)}{(2\varepsilon + 1)}\right]\frac{\langle\Psi|\mu|\Psi\rangle}{a^3}\mu\right\}\Psi = E\Psi \tag{11.14}$$

where H is the usual gas-phase Hamiltonian. Written in this fashion, the components of the second term on the l.h.s. that precede the final dipole moment operator may be regarded as the reaction field.

Equation (11.14) is an example of a non-linear Schrödinger equation. It can be solved in the usual HF fashion by construction of a Slater determinant formed from MOs ψ that are optimized using a modified Fock operator according to

$$\left\{F_i - \left[\frac{2(\varepsilon - 1)}{(2\varepsilon + 1)}\right]\frac{1}{a^3}\langle\Psi|\mu|\Psi\rangle^2\right\}\psi_i = e_i\psi_i \tag{11.15}$$

where F_i is the usual gas-phase Fock operator for MO i (Ángyán 1992). A critical feature of Eq. (11.15) is that it involves an additional level of iteration compared to the standard HF approach. Not only must the final wave function render the density matrix and Fock operator stationary, but it must also lead to a stationary dipole moment. Solution of the HF equations (or the equivalent Kohn–Sham DFT equations) in such a fashion, where accounting for solvation leads to a non-linear Schrödinger equation, is referred to as a self-consistent reaction field (SCRF) calculation.

Inspection of Eq. (11.14) should make it clear that the manner in which the dipole-dependence enters into the equations will lead to an *increase* in dipole moments in increasingly polar solvents. As noted in Section 11.1, the increase in the dipole moment in such an SCRF formalism provides an energy lowering that is counterbalanced by an increase in the energy computed from the 'usual' Hamiltonian H (the first operator on the l.h.s.) so that a stationary solution is reached when additional distortion costs associated with H exactly balance additional energy lowering associated with further increasing the dipole moment.

In describing the results from SCRF calculations, it is useful to keep careful track of the various components of the energy. The electrostatic component of the solvation free energy is the difference between the energy in the gas phase and the energy in solution. This may be written

$$\Delta G_{ENP} = \left[\langle \Psi^{(\mathrm{sol})} | H | \Psi^{(\mathrm{sol})} \rangle + \langle \Psi^{(\mathrm{sol})} | G_P | \Psi^{(\mathrm{sol})} \rangle \right] - \langle \Psi^{(\mathrm{gas})} | H | \Psi^{(\mathrm{gas})} \rangle$$

$$= \Delta E_{EN} + G_P \tag{11.16}$$

where the difference between the first and third expectation values on the r.h.s. in the first line of Eq. 11.16 defines the distortion energy ΔE_{EN}, which must be positive since $\Psi^{(\mathrm{gas})}$ minimizes H. The 'EN' subscript on this term emphasizes it is associated with the electronic and nuclear components of the total energy; in the absence of any geometry reoptimization, the N subscript is superfluous. As written, Eq. (11.16) mixes potential and free energies, but we will ignore this issue for now.

The Kirkwood–Onsager equations can be generalized to include multipole moments higher than the dipole, leading to the expression

$$G_P = -\frac{1}{2} \sum_{l=0}^{L} \sum_{m=-l}^{l} \sum_{l'=0}^{L} \sum_{m'=-l'}^{l'} M_l^m f_{ll'}^{mm'} M_{l'}^{m'} \tag{11.17}$$

where each component m of every molecular multipole M of order l interacts with the reaction field, which is itself expressed as a multipole expansion equal and opposite to the molecular multipoles, through the reaction field factors f that carry the dependence on dielectric constant and cavity radius. In principle, the multipole expansion may be carried out to infinite order, but in practice, some judicious choice of l is made in Eq. (11.17) to keep things tractable. A fairly typical choice is $l = 6$ (note that $l = 0$ and $l = 1$ define the Born and Born–Kirkwood–Onsager (BKO) approaches, respectively).

The simplicity of the BKO approach to computing polarization free energies led to its widespread use for the qualitative analysis of solvation effects on various properties for many years (including in the absence of any explicit theoretical calculations). For quantitative purposes, however, it suffers from a number of undesirable features. One such feature is the slow nature of the convergence of Eq. (11.17) with respect to l. Table 11.2 lists ΔG_{EP}

Table 11.2 ΔG_{EP} values (kcal mol^{-1}) for *trans* 1,2-dichloroethane as a function of the truncation point in the multipole moment expansion[a]

l	ΔG_{EP}
1	0.00
2	−0.93
5	−1.14
8	−1.70
10	−1.79
20	−1.82

[a]From Christiansen and Mikkelsen 1999

values for *trans* 1,2-dichloroethane computed at the MCSCF level for various choices of l. Note that, since the *trans* conformation has no dipole moment by symmetry, a simple BKO calculation must predict a polarization free energy of zero, which represents a very large error. Including the quadrupole moment captures 50 percent of the total, but further convergence initially proceeds slowly (note that there is no requirement for convergence to proceed in a monotonic fashion), and it is not until $l = 8$ that the result is converged to within about 5 percent. Since 1,2-dichloroethane overall has a rather simple electronic distribution, it is disturbing to consider how much larger l may have to be to accurately treat more complex molecules.

Worse still, however, is that even well-converged values are unlikely to be meaningful in the absence of the solutes in question being well described as spherical. When they are not, and very few molecules are, the value that should be chosen for the radius a is highly ambiguous. Since the dipole term has an inverse cubic dependence on this parameter, small variations can have large effects on solvation free energies, and the literature is replete with examples where obviously non-physical values have been chosen, rendering interpretation of subsequent results highly suspect.

This situation can be somewhat ameliorated by choosing a regular ellipsoid instead of a sphere for the solute cavity. In that case, Eq. (11.17) can still be solved in a simple fashion, with the reaction field factors depending on the ellipsoidal semiaxes (Rinaldi, Rivail, and Rguini 1992). However, while this is clearly an improvement on a spherical cavity, the small number of solutes that may be well described as ellipsoidal does not make this a particularly satisfactory solution.

So, while derivations of SCRF theory using ideal cavities are very useful for conceptual purposes, they are insufficiently accurate for all but the most crude analyses. Modern applications of continuum models almost invariably use arbitrary cavity shapes, typically constructed from overlapping atomic spheres, and we turn to examples of these models next.

11.2.1.2 Arbitrary cavities

The concept of molecular shape with which most chemists are comfortable is almost certainly that represented by space-filling models constructed from the overlap of atomic spheres having appropriate van der Waals radii. For such arbitrary, lumpy cavities, analytic solutions of the PB equation are no longer possible, and the reaction field must be determined numerically. The approach taken by most classical PB software – classical implying the charge distribution is not allowed to change – is formally to

1. Divide space according to a three-dimensional grid.

2. Define the molecular cavity and assign gridpoints the appropriate dielectric constant – in classical calculations, the interior is often assigned a dielectric constant between two and four to mimic solute polarizability.

3. 'Discretize' the solute charge distribution onto interior grid points using some algorithm – e.g., divide every atomic partial charge equally over the nearest grid point and its 14 nearest neighbors.

4. Determine the electrostatic potential at each grid point by numerical solution of the PB equation; this process is typically iterative.

5. Once the potential is available, evaluate Eq. (11.3) as a pointwise sum over points carrying non-zero charge.

There are many technical challenges associated with this process that are worth keeping in mind. Like any numerical method, it is most successful when the density of discrete points is very high. However, as we are working with three-dimensional space, an order of magnitude decrease in the spacing between adjacent points increases the total number of points by three orders of magnitude, making the solution of the PB equation much more computationally demanding. So-called 'focusing' methods have been developed to try to move from coarse grids to finer grids in an efficient manner. With most grid densities in everyday use, the density remains sufficiently coarse that different orientations of the solute in space can give rise to non-trivially different values for G_P. Reported values are sometimes averaged over several random orientations.

A related issue is that the potential can be very sensitive to grid points very near the cavity surface, where the dielectric constant is changing instantaneously. By construction, the cavity is actually defined only to within the grid-point spacing. The van der Waals radii defining the cavity surface that determines whether a given gridpoint is inside or outside the solute may either be chosen arbitrarily or optimized for a particular computational model (see, for example, Banavali and Roux 2002).

The primary area where classical PB equations find application is to biomolecules, whose size for the most part precludes application of quantum chemical methods. The dynamics of such macromolecules in solution is often of particular interest, and considerable work has gone into including PB solvation effects in the dynamics equations (see, for instance, Lu and Luo 2003). Typically, force-field atomic partial charges are used for the primary solute charge distribution.

It is noteworthy that with biomolecules it is often the electrostatic potential itself that is of primary interest, not its use to compute solvation free energies. Since the PB potential is presumably a more accurate picture of the potential in solution than one that would be derived from a vacuum calculation, as described in Chapter 9, the method is often employed for this purpose. When potentials are visualized on the molecular van der Waals surface, many enzymes, for instance, show large regions of uniformly positive or negative potential, suggesting preferred binding sites for ligands having opposite charge, or channels for directing in substrates having opposite charge, etc.

Rather than solving the PB equation on a three-dimensional grid, the differential equation can be recast into a boundary element problem by representing the potential using a charge density spread over the molecular surface (see, for instance, Zauhar and Varnek, 1996). To make the calculation more convenient, the surface is usually tesselated into spherical triangles, and the charge density on each element is collapsed into a point charge in the center of the triangle. The charge–potential integral of Eq. (11.3) is thus replaced by a sum over charge–charge interactions. As a solution of what amounts to a surface integral instead of a volume integral, this procedure is somewhat less sensitive to numerical noise, but still requires some care to ensure sufficiently small surface tesserae are employed. A problem of some concern can arise when the centroids of spherical triangles associated with two different atoms are very near one another in space. In that case, the short-range charge–charge interaction can be so large as to introduce significant instabilities. As a result, some procedures delete regions of the surface near where atomic spheres overlap and accept a reduced accuracy in being able to represent the potential as a consequence.

Coming back to quantum mechanical continuum models, in the most general sense we now seek to solve the non-linear Schrödinger equation

$$\left(H - \frac{1}{2} V \right) \Psi = E \Psi \tag{11.18}$$

where V is a general reaction field inside the cavity that depends upon Ψ. As shown for the special case of the Kirkwood–Onsager model above, when Ψ is expressed as a Slater determinant, the orbitals minimizing Eq. (11.18) can be determined from

$$(F_i - V) \psi_i = e_i \psi_i \tag{11.19}$$

where F is the Fock operator. Entirely analogous expressions exist for DFT.

In formalism, this is really no different than the classical situation just described, except that the electronic-charge distribution is continuous, as opposed to discretized, and the non-linear character of the equations introduces an iterative component to the SCRF procedure that goes hand in hand with permitting relaxation of the charge distribution. That being the case, the methods used to represent the reaction field are essentially the same as those used in the classical situation. For example, SCRF schemes solving for the reaction field on a three-dimensional grid have been described by both Chen et al. (1994) and Tannor et al. (1994).

Perhaps the most widely used scheme for SCRF implementations of the Poisson equation is the surface area boundary element approach. This was first formalized by Miertus, Scrocco, and Tomasi (1981), and these authors referred to their construction as the polarized continuum model (PCM). While that name continues to find ample use in the literature, MST (the initials of the authors' last names) finds roughly equal usage, and some authors use PCM to refer generically to any continuum SCRF scheme.

A number of variations on the PCM formalism have appeared since its first publication. Some are purely technical in nature, designed to improve the computational performance of the method, e.g., an integral equation formalism for solving the relevant SCRF equations which facilitates computation of gradients and molecular response properties (IEF-PCM; Cossi *et al.* 2002), an extension to permit application to infinite periodic systems in one and two dimensions (Cossi 2004), and an extension to liquid/liquid and liquid/vapor interfaces (Frediani *et al.* 2004). Others reflect differences in how the molecular cavity is defined. For the most part, Tomasi and co-workers have maintained a strategy where the cavity is constructed from overlapping atomic spheres having radii 20% larger than their tabulated van der Waals radii, with a special distinction being made between 'polar' and 'non-polar' hydrogen atoms. As an alternative, Foresman *et al.* (1996) suggested defining the cavity as that region of space surrounded by an arbitrary isodensity surface, i.e., a surface characterized by a constant value of the electron density. That surface can either be located from the gas-phase density, and held fixed (IPCM) or determined self-consistently, adding yet another iterative level to the SCRF process (SCIPCM). Part of the motivation for these latter two modifications was to decrease the number of cavity parameters from one per atom to one total. However, insofar as the modeling of a molecular solvent by a continuum is by nature a fictional construct, it is not obvious that such a decrease in parameters can be regarded as a virtue. A further discussion of cavity definitions is deferred to Section 11.4.1, and it suffices to note here that the IPCM and SCIPCM methods tend to be considerably less stable in implementation than the original PCM process, and can be subject to erratic behavior in charged systems, so their use cannot be recommended (Cossi *et al.* 1996).

A third possibility that has received extensive study in the SCRF regime is one that has seen less use at the classical level, at least within the context of general cavities, and that is representation of the reaction field by a multipole expansion. Rinaldi and Rivail (1973) presented this methodology in what is arguably the first paper to have clearly defined the SCRF procedure. While the original work focused on ideal cavities, this group later extended the method to cavities of arbitrary shape. In formalism, Eq. (11.17) is used for any choice of cavity shape, but the reaction field factors f must be evaluated numerically when the cavity is not a sphere or ellipsoid (Dillet *et al.* 1993). Analytic derivatives for this approach have been derived and implemented (Rinaldi *et al.* 2004).

Most of the models described above have also been implemented at correlated levels of theory, including perturbation theory, CI, and coupled-cluster theory (of course, the DFT SCRF process is correlated by construction of the functional). Unsurprisingly, if a molecule is subject to large correlation effects, so too is the electrostatic component of its solvation free energy.

Note that, insofar as all of the above models simply represent alternative mathematical approaches to solving the Poisson equation, in the limit of converging them with respect to grid density, tesserae density, multipole expansion, etc., they should all give identical answers for identical molecules in identical cavities. Thus, any choice between them should largely be predicated on computational convenience and efficiency of implementation.

11.2.2 Generalized Born

In order to solve the Poisson equation for an arbitrary cavity, recourse to numerical methods is required. An alternative approach that has seen considerable development involves computing the polarization free energy using an approximation to the Poisson equation that can be solved analytically, and this is the Generalized Born (GB) approach. As its name implies, the GB method extends the Born Eq. (11.12) to polyatomic molecules. The fundamental equation of the GB method expresses the polarization energy as

$$G_P = -\frac{1}{2}\left(1 - \frac{1}{\varepsilon}\right) \sum_{k,k'}^{\text{atoms}} q_k q_{k'} \gamma_{kk'} \tag{11.20}$$

where k and k' run over atoms, each of which is characterized by a partial charge q, and γ has units of inverse length, i.e., it is an effective Coulomb integral. In order for the GB equation to be an accurate approximation to the PB equation, a suitable functional form for γ must be chosen. A very good functional form is given by

$$\gamma_{kk'} = \left(r_{kk'}^2 + \alpha_k \alpha_{k'} e^{-r_{kk'}^2/d_{kk'}\alpha_k\alpha_{k'}}\right)^{-1/2} \tag{11.21}$$

where $r_{kk'}$ is the interatomic distance, α_k is the effective Born radius of atom k, and d is a parameter that may in principle vary from one atom pair to the next, but which is typically taken to have a universal value of 4 (Still *et al.* 1990).

While the full form of γ is not necessarily intuitively obvious, it is noteworthy that it has appropriate limiting behavior. Thus, for large interatomic distance, γ becomes simply r^{-1}, which is the expected result from Coulomb's law. For diagonal terms in the summation (i.e., $k = k'$ so $r_{kk'} = 0$), γ is simply α^{-1}, so the Born equation is recovered. However, there is a distinction between the effective Born radius of an atom in a molecule and its Born radius as a monatomic ion. Clearly, the 'self' solvation free energy of an atom in a molecule should be less than for the isolated atom, since the rest of the molecule displaces ('descreens') the dielectric medium in certain regions of space. As depicted in Figure 11.7, the manner in which α is typically determined is to solve the Poisson equation and Eq. (11.3) for each atom in the molecule using the full molecular cavity but with all partial charges *other* than that for the particular atom in question set to zero. With that value of G_P in hand for partial charge q, one solves Eq. (11.12) for α; once each effective Born radius has been determined in this fashion, Eq. (11.21) may be used to determine γ for any pair of k and k'.

So, the steps in a GB calculation to determine the polarization free energy given a particular molecular geometry are essentially:

1. Assign atomic radii to all atoms for purposes of defining the cavity.

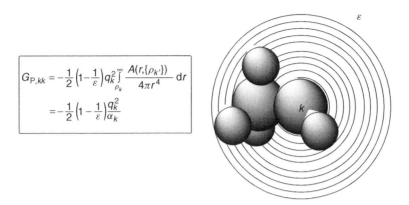

$$G_{P,kk} = -\frac{1}{2}\left(1-\frac{1}{\varepsilon}\right)q_k^2 \int_{\rho_k}^{\infty} \frac{A(r,\{\rho_{k'}\})}{4\pi r^4}\, dr$$

$$= -\frac{1}{2}\left(1-\frac{1}{\varepsilon}\right)\frac{q_k^2}{\alpha_k}$$

Figure 11.7 To determine α for atom k in the GB approach, the interaction of the charge q on atom k with the surrounding continuum is determined by the radial integration shown, where ρ_k is the cavity radius (sometimes called the Coulomb radius) of atom k and A is the area of expanding spherical shells used in the integration which depend on all other atomic radii $\rho_{k'}$. (Note that, once the outermost shell encompasses the entire molecule, the remaining integral may be solved analytically using the Born formula, since the system is simply a sphere having charge q_k.) The effective Born radius α is then determined by requiring the equality of the first and second lines on the r.h.s. of the indicated equation

2. Compute effective Born radii α for all atoms using the procedure outlined above.

3. Using those effective Born radii, compute all values of $\gamma_{kk'}$.

4. Compute or arbitrarily assign the atomic partial charges.

5. Evaluate Eq. (11.20).

Insofar as both GB and PB depend parametrically on the atomic radii used to define the cavity, direct comparisons between the two methodologies must be made using identical choices. Comparisons under such conditions have been made, and the agreement between the two models has been found to be excellent for small to medium-sized molecules (see, for instance, Edinger *et al.* 1997; Onufriev, Bashford, and Case 2000), so for practical purposes they may be taken as essentially equivalent in terms of predictive utility, and choice of model will usually be dictated by matters of computational convenience. For larger molecules, like biopolymers, agreement is still generally good but some technical care is required to ensure that the GB protocol does not assign empty volume inside the biomolecule to be characterized erroneously by the dielectric constant of the solvent (Feig *et al.* 2004). Some additional issues associated with comparison of the two approaches are discussed further in Section 11.4.

A modification of GB that includes the effects of dissolved electrolytes in the formalism, i.e., an extension analogous to the Poisson–Boltzmann extension of the Poisson equation, has been proposed by Srinivasan *et al.* (1999). In this model, the dielectric constant is a function of the interatomic distance and the Debye–Huckel parameter (Eq. (11.7)).

It may seem that using the Poisson equation to determine the effective Born radii, as described above, defeats the purpose of developing GB as an alternative to PB. Actually, the solution of the Poisson equation for a single charge is vastly simpler than for a complete charge distribution, but the procedure is still computationally intensive, and subject to possible numerical noise. An analytical approximation to this procedure, known as pairwise descreening (PD), has been described by Hawkins, Cramer, and Truhlar (1995), and has been shown to increase computational efficiency with very little cost in accuracy. Gallicchio and Levy (2004) have described a modified PD algorithm that has improved sensitivity to conformational changes in biological macromolecules and emphasized its potential utility in docking calculations. A procedure that is similar in spirit to the PD approach but also takes advantage of the molecular connectivity that must be defined in a force-field calculation (and is thus limited to such applications) has also been described (Qiu *et al.* 1997).

Note that the GB approach describes the charge distribution of the solute using atom-centered atomic partial charges. In that sense it may be called a distributed monopole representation. A key issue, obviously, is how those partial charges are computed. In force-field GB implementations, all models to date simply use the partial charges already defined for the atom types for use in solving the charge–charge interaction term in the molecular mechanics energy, and parameters in the GB model, like the Coulomb radius, are optimized with respect to this choice (see, for example, Cheng *et al.* 2000, Onufriev, Case, and Bashford 2002, and Zhang *et al.* 2003). For quantum mechanical calculations, the charges may in principle be determined from any one of the many methods described in Section 9.3. However, it must be kept in mind that at the QM level, the calculation is of the SCRF variety. That is, the atomic partial charges will be free to change as the wave function polarizes in response to the surrounding dielectric medium.

In order to implement the reaction field conveniently into the SCF equations, it is helpful if the partial charges have a relatively simply dependence on elements of the density matrix. Thus, for instance, early versions of the QM SCRF GB solvation models of Cramer and Truhlar (so-called SM*x* models, where *x* is essentially a version number) used Mulliken charges. As noted in Section 9.3, however, Mulliken charges provide a rather poor approximation of the molecular charge distribution. Later generations of these models, to include the most modern versions SM5.42 and SM5.43, use the CM2 and CM3 Class IV charge models, respectively, to assign the atomic partial charges (hence the '.42' and '.43' suffixes in the model names). As the charge models are designed to predict 'good' partial atomic charges irrespective of the underlying wave function, there is a leveling of the electrostatics across methods and SM*x* models for different levels of theory tend to have very similar parameters. The parameters themselves are primarily the Coulomb radii ρ_k, as defined in Fig 11.7. A GB SCRF implementation has also been reported for an SCC-DFTB model (Xie and Liu 2002).

11.2.3 Conductor-like Screening Model

When the Poisson equation is solved using a boundary element approach, the charges on the tesselated molecular surface are determined so that they provide an equivalent representation to the electrostatic potential that is distributed throughout space when the Poisson equation is solved using a volume element approach. However, if the surrounding space were to be characterized by an *infinite* dielectric constant, i.e., if the medium were conducting, then no potential exists in the medium, and instead image charge develops on the conductor surface in contact with the solute. Such a situation considerably simplifies the necessary electrostatic equations for the calculation of polarization free energy (and also associated energy derivatives) and this approximation is made in the so-called conductor-like screening model (COSMO), first described in detail at the semiempirical SCRF level by Klamt and Schüürmann (1993).

Of course, the response of a conductor to a solute charge distribution is 'complete', while that of a dielectric medium is not. So, in COSMO models, the more simply evaluated conductor-polarization free energy is scaled by a factor of $2(\varepsilon - 1)/(2\varepsilon + 1)$ after its computation (i.e., by the Onsager factor; in the case of the SM5C model, however, the scaling factor is $(\varepsilon - 1)/\varepsilon$ – see Section 11.3.3).

Since its original description at the semiempirical level, COSMO has also been generalized to the *ab initio* and density functional levels of theory as well (Klamt *et al.* 1998). In addition, conductor-like modifications of the PCM formalism have also been described, and to distinguish between the conductor-like version and the original (dielectric) version, the acronyms C-PCM and D-PCM have been adopted for the two, respectively (Barone and Cossi 1998).

From a chemical perspective, dielectric- and conductor-like continuum models give sufficiently similar electrostatic results that the differences in their underlying assumptions appear to have no impact. Conductor-like models seem to be slightly more computationally robust in some instances, which may make them a better choice if instability is manifest in an SCRF calculation. Some concerns were raised initially that the *post facto* correction for dielectric behavior might render the models appropriate only for media having reasonably high dielectric constants, but a systematic study by Dolney *et al.* (2000) indicated non-polar solvents to be equally amenable to treatment by a COSMO model.

Moving beyond computation of the electrostatic component of the solvation free energy, Klamt (1995) has also described using the results of COSMO calculations to model 'real solvents' (COSMO-RS). In this model, a molecule in solution is considered to be entirely defined by the screening charge density on its cavity surface, which is called its σ profile. That surface is then shattered into a discrete number of fragments (each carrying its own characteristic charge density σ) and a chemical potential is defined in a statistical mechanical formalism by considering the optimal matching of all fragments with partners having charge densities of opposite sign for the collection of all fragments in the liquid. In spite of the loss of structural information associated with breaking the molecular surface into completely independent fragments, this model has proven to be particularly effective for describing the thermodynamic properties of mixtures of molecules that are not too dissimilar, for example, vapor–liquid equilibria in binary solvent systems (Spuhl and Arlt 2004). Extending this idea to charged solutes, however, has proven more challenging.

It is important to re-emphasize that the electrostatic component of the solvation free energy is not a physical observable. Thus, it is impossible to assert on any basis other than intuition that one continuum modeling algorithm is more or less accurate than another in the computation of this quantity (Curutchet *et al.* 2003a). One may take as a standard for comparison numerically converged solutions of the Poisson equation, but the Poisson equation is itself a model, and not necessarily the optimal one. In order to make comparisons against experiment, it is necessary to supplement the polarization (and distortion) energies with terms corresponding to cavitation, dispersion, structural rearrangement, etc. Models that purport to compute the *full* free energy of solvation may then be compared one to another using experimental free energies of transfer as a common yardstick.

11.3   Continuum Models for Non-electrostatic Interactions

Just as the electrostatic component of the free energy of solvation cannot be measured, neither can the non-electrostatic components. That being said, various experimental systems may be biased so as to make one component or another likely to heavily dominate the solvation free energy. For example, the solvation free energies of charged species would be expected to be dominated by the electrostatic component, and solvation free energies for ions can be helpful in the assignment of parametric Born radii to atoms. To assess the free-energy changes associated with cavitation, dispersion, and other physical effects, different neutral model systems have been studied, and we examine some of these next.

11.3.1   Specific Component Models

Noble gas atoms have no permanent electrical moments, and the lighter ones are amongst the least polarizable of chemical systems. Thus, their transfer into a solvent may be regarded as a process reasonably cleanly associated with cavitation, i.e., the introduction of the noble gas atom is like introducing a vacuum of equivalent size into the solvent. By examining solvation data for the noble gases and certain other systems, Pierotti (1976) developed a formula for the cavitation free energy, associated with a spherical cavity volume, that depends on the radius of the sphere to the first, second, and third powers. Simulation data have been used to supplement noble-gas experimental data and refine constants appearing in the Pierotti formula (Höfinger and Zerbetto 2003). By viewing a non-spherical solute as a collection of atomic spheres where overlapping volumes are only accounted for once, Pierotti's formula has been generalized to molecular cavities (Claverie 1978; Colominas *et al.* 1999).

Dispersion is a considerably more difficult modeling task. As first noted in Section 2.2.4, dispersion is a purely quantum mechanical effect associated with the interactions between instantaneous local moments favorably arranged owing to correlation in electronic motions. In order to compute dispersion at the QM level, it is necessary to include electron correlation between interacting fragments, which immediately sets a potentially rather high price on direct computation. More difficult still, however, is that the continuum model by construction does not include the solvent molecules in the first place.

As a result, some approaches to computing dispersion energy have involved using either experimental or theoretical data for gas-phase clusters to estimate the strength of dispersion interactions between different possible solute and solvent functional groups. However, when the cluster interaction involves molecules with permanent electrical moments, it can be quite difficult to separate out the dispersion interaction from the overall interaction. In any case, the typical approach deriving from this work is to develop a set of atomic (or group) polarizabilities that will be used together with bulk solvent polarizabilities to estimate dispersion interactions; usually, these are combined with other methods for estimating exchange-repulsion, i.e., repulsive van der Waals effects, to come up with a complete short-range term (see, for example, Floris, Tomasi, and Pascual-Ahuir 1991).

In practice, models that directly calculate cavitation and dispersion/repulsion tend to predict that both effects are quite large in magnitude, but with opposite sign so that there is a large degree of cancellation. This suggests the unfortunate possibility that errors in the individual models may be larger than the net result.

Other energetic components associated with the solvation process include non-electrostatic aspects of hydrogen bonding and solvent-structural rearrangements like the hydrophobic effect. Despite many years of study, the fundamental physics associated with both of these processes remains fairly controversial, and physically based models have not been applied with any regularity in the context of continuum solvation models.

11.3.2 Atomic Surface Tensions

Given the somewhat *ad hoc* nature of most specific schemes for evaluating the non-electrostatic components of the solvation free energy, a reliance on a simpler, if somewhat more empirical, scheme has become widely accepted within available continuum models. In essence, the more empirical approach assumes that the free energy associated with the non-electrostatic solvation of any atom will be characteristic for that atom (or group) and proportional to its solvent-exposed surface area. Thus, the molecular G_{CDS} may be computed simply as

$$G_{CDS} = \sum_k A_k \sigma_k \qquad (11.22)$$

where k runs over atoms or groups, A is the exposed surface area, and σ is the characteristic 'surface tension' associated with the atom or group. Note that here the use of the term surface tension refers to the unit dimensionality of energy per area, and the atomic terms should not be confused with the surface tension of the solvent, which is a macroscopic property.

Part of the motivation behind so straightforward an approach derives from its ready application to certain simple systems, such as the solvation of alkanes in water. Figure 11.8 illustrates the remarkably good linear relationship between alkane solvation free energies and their exposed surface area. Insofar as the alkane data reflect cavitation, dispersion, and the hydrophobic effect, this seems to provide some support for the notion that these various terms, or at least their sum, can indeed be assumed to contribute in a manner proportional to solvent-accessible surface area (SASA).

It should be noted that SASA itself can be defined in many ways (see, for instance, Pascual-Ahuir, Silla, and Tuñon 1994). In the simplest approach, one imagines solvent molecules to be spheres having some characteristic radius. The SASA is then generated by the center of

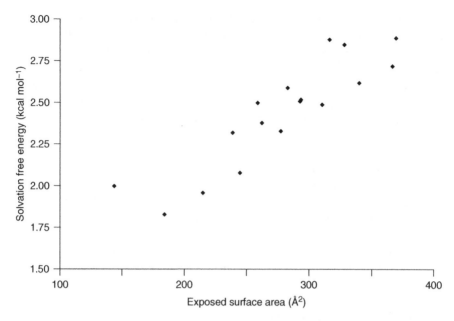

Figure 11.8 Approximately linear relationship between solvation free energy and solvent-accessible surface area for linear and branched alkanes. A best fit line passing through zero has a slope of 8.6 cal mol^{-1} Å^{-2}, which may be taken as the σ value for alkane surface area in Eq. (11.22) (Giesen, Cramer, and Truhlar 1994)

a solvent sphere rolling over the van der Waals surface of the solute, which is equivalent to constructing the SASA as the surface of solute atomic spheres having radii equal to the atomic van der Waals radius plus the solvent radius. Such a surface has sharp concave intersections between atoms, which is sometimes considered undesirable, in which case one can take instead of the surface mapped out by the sphere center the surface actually accessible to the surface of the solvent sphere. Other variations have also been presented, but in practice the utilities of the various surfaces are comparable once optimized surface tensions have been developed specifically for them.

In the majority of continuum solvation models incorporating a surface-tension approach to estimating the non-electrostatic solvation components, the index k in Eq. (11.22) runs over a list of atom types, and the user assigns the appropriate type to each atom of the solute. This is particularly straightforward for MM models, like the Generalized Born/Surface Area (GB/SA) model (Still *et al.* 1990; see also Best, Merz, and Reynolds 1997), since atom types are already intrinsic to the force field approach. This same formalism has been combined with the CHARMM and Cornell *et al.* force fields (see Table 2.1) to define GB models for proteins and nucleic acids (Dominy and Brooks 1999; Jayaram, Sprous, and Beveridge 1998). Considering this approach applied within the QM arena, the MST-ST models of Orozco and Luque have been the most extensively developed (see, for instance, Curutchet, Orozco, and Luque 2001).

The surface tensions themselves in the GB/SA and MST-ST models were developed by taking collections of experimental data for the free energy of solvation in a specific solvent, removing the electrostatic component as calculated by the GB or MST model, and fitting the surface tensions to best reproduce the residual free energy given the known SASA of the solute atoms. Such a multilinear regression procedure requires a reasonably sized collection of data to be statistically robust, and limitations in data have thus restricted these models to water, carbon tetrachloride, chloroform, and octanol as solvents.

In order to be more generally applicable, the SMx models of Cramer and Truhlar address the issue of data scarcity by making the atomic surface tensions a function of quantifiable *solvent* properties, i.e.,

$$\sigma_k = \sum_j \Gamma_j \eta_k^{\Gamma_j} \tag{11.23}$$

where j runs over the property list, Γ is the value of a particular property in convenient units, and the quantities $\eta_k^{\Gamma_j}$ become the parameters needing to be fit by multilinear regression. Although this introduces multiple parameters per atom type k, it permits regression over a single data set containing solvation free energies into *any* solvent, so long as its required solvent properties are known. In the SM5 versions of the models, the macroscopic solvent properties include surface tension, index of refraction, hydrogen bonding acidity and basicity, and percent composition of aromatic carbon atoms and electronegative halogen atoms, and the parameterization set involves more than 2500 data in 91 different solvents (Li *et al.* 1999).

A separate flexibility built into the SMx models compared to most other QM continuum models augmented with surface tensions is that no assignment of atom type need be made. Instead, the SMx surface tensions are functions of local geometry, so that, for instance, a carbon-bound hydrogen atom is distinguished from an oxygen-bound hydrogen atom and assigned a different surface tension to reflect its different character. The surface tension functions are smooth and differentiable, which facilitates their use in modeling situations where an atom may change from one type to another along a reaction coordinate, for instance.

Surface-tension augmented continuum models permit the computation of full free energies of solvation and may thus be used to construct solvated potential energy surfaces in the spirit of Figure 11.1. Insofar as the solvation free energy itself and any equilibrium or kinetic quantities computed for the solvated PES are physical observables, the accuracy of the solvation models may be assessed by comparison to experimental data. We consider several such comparisons in the next section in addition to addressing certain important technical details. Prior to doing so, however, it must be mentioned that the use of atomic surface tensions has been carried to the extreme of assuming that they can account for the *entire* solvation free energy, i.e., the electrostatics are completely implicit and the parameters in Eq. (11.23) are fit to the full solvation free energy (recent examples include Hawkins *et al.* 1998, Wang *et al.* 2001, and Hou *et al.* 2002). Such models are typically designed for use with biopolymers, where there is a need for extreme efficiency and the range of atom types is rather limited. An approach that is similar in its conceptual simplicity, albeit not entirely devoid of electrostatics, is the solvation free energy density (SFED) approach of No *et al.* (1999) where the full free energy of solvation is computed from the accessible volume (as opposed to surface area) of a finite shell surrounding each atom. This model, too, is primarily designed for use with biomolecular simulations, although its performance for more general small neutral solutes is perfectly acceptable.

11.4 Strengths and Weaknesses of Continuum Solvation Models

11.4.1 General Performance for Solvation Free Energies

For neutral solutes experimental free energies of solvation between the range of +5 and −15 kcal mol^{-1} are typically amenable to measurement with an accuracy of ±0.1 kcal mol^{-1}. Carefully parameterized surface-tension-augmented continuum models typically exhibit average errors over large data sets on the order of 0.5 kcal mol^{-1}. Ionic solutes pose more difficulties experimentally, since measurement of a gas/solution equilibrium is no longer a viable methodology. However, for singly charged species, solvation free energies ranging from −40 to −110 kcal mol^{-1} can be obtained with accuracies of ±2–5 kcal mol^{-1}, depending on the experimental technique. Well parameterized continuum models achieve mean absolute errors at the high end of the experimental error range, which is perhaps the best that can be expected. Reliable data for more highly charged species are extremely scarce, so no legitimate comparison can be made.

It is worth noting that the solvation free energy of the proton is a somewhat special case. Determining the solvation free energy of the proton is equivalent to determining the absolute potential of the normal hydrogen electrode (NHE), which is a tricky issue in electrochemistry (Trasatti 1986). In 1986, the International Union of Pure and Applied Chemistry (IUPAC) recommended an absolute value of 4.44 V for the NHE which corresponds to a 1 M gas phase to 1 M solution standard-state aqueous proton solvation free energy of −261.7 kcal mol^{-1}. In the 1990s, however, Tissandier *et al.* (1998) used ion-cluster measurements to establish a value of −264.0 kcal mol^{-1} for the same standard-state process, which corresponds to an NHE potential of 4.36 V (Lewis *et al.* 2004). Subsequent experimental and theoretical work has been supportive of the greater accuracy of the newer value and its use can be recommended. Note that most methods for determining ionic solvation free energies experimentally rely on having a benchmark value for the proton solvation free energy, so a change in the benchmark changes *all* ionic solvation free energies. Thus, care should be employed in comparing tabulations of such values in the literature to ensure common standard-state conventions *and* proton solvation free energies.

One of the reasons that it is hard to predict accurate solvation free energies for charged species is that such predictions tend to be very sensitive to the size of the solute cavity, leading to many proposals in the literature for how to go about choosing the 'best' electrostatic cavity. However, insofar as the electrostatic component of the solvation free energy is *not* an observable, there is not much weight to these arguments. The essentially equivalent performances of surface-tension augmented models like MST-ST and SM*x* for full free energies of solvation, even though they use very different cavity radii in some cases and therefore determine very different electrostatic free energies of solvation (Curutchet *et al.* 2003a), speak to the ability of the parameterization process to mask any lack of physicality in the cavity definitions.

Table 11.3   Absolute free energies of solvation (kcal mol^{-1}) and chloroform/water partition coefficients (log$_{10}$ units) for nucleic acid bases at the SM5.4/AM1 level^a

Solute	Frozen		Relaxed		logK_{CHCl_3/H_2O}		
	H$_2$O	CHCl$_3$	H$_2$O	CHCl$_3$	Frozen	Relaxed	Experiment
9-Methyladenine	−12.8	−11.9	−15.8	−13.6	−0.7	−1.6	−0.8
9-Methylguanine	−13.1	−11.8	−22.3	−16.7	−0.9	−4.1	−3.5
9-Methylhypoxanthine	−15.0	−13.1	−19.5	−14.6	−1.4	−3.5	−2.5
1-Methylcytosine	−12.2	−11.1	−22.6	−16.8	−0.8	−4.3	−3.0
1-Methylthymine	−6.9	−7.7	−9.6	−9.2	0.6	−0.3	−0.5
1-Methyluracil	−7.6	−7.3	−10.5	−8.9	−0.3	−1.2	−1.2
Mean unsigned error					1.3	0.6	

^{a}Computational results from Giesen *et al.* (1997); experimental results from Cullis and Wolfenden (1981)

In the future, analysis of this problem at the SCRF level will necessarily have to focus on molecular properties other than the solvation free energy to assess the greater accuracy of one cavity compared to another. Thus, *differences in the gas-phase and solvated wave functions*, and their corresponding effects on such properties as NMR, IR, and UV spectral transitions, may prove useful in identifying optimal methods for handling the electrostatics.

Such differences may in principle be quite large, as already illustrated in Table 11.1. Even the solvation free energies themselves may be strongly influenced by the relaxation of the wave function in solution. In Table 11.3 are listed the SM5.4/AM1 solvation free energies of six methylated nucleic acid bases, both in chloroform and in water, computed using either the charge distribution from the gas-phase wave function or from the relaxed wave function. As discussed further in Section 11.4.2, the difference between the two may be expressed as a partition coefficient, and the two sets of partition coefficients (frozen and relaxed) are compared to experimental values. Agreement is significantly better using the relaxed solvation free energies rather than the so-called 'no solute polarization' solvation free energies.

Note that one implication of the importance of solute polarization is that intrinsically non-SCRF methods, like continuum solvation models associated with force fields or other fixed-charge-density representations of the solute, must somehow include the energetic effect of polarization by other means. For instance, often atomic partial charges are chosen from calculations at the HF/6-31G(d) level. This level overestimates charge separation (as judged by a consistent roughly 10% overestimation of dipole moments), but this may be regarded as a virtue, not a failing, when used with non-SCRF continuum models, because the solute polarization is 'built-in' through the gas-phase wave function errors. Alternatively, fixed-charge models can use cavity radii slightly smaller than those for SCRF models to offset the use of unrelaxed charges.

11.4.1.1   pK_a values

Returning to ionic solvation free energies, such quantities play important roles in the computation of two common properties of interest, namely pK_a values and relative redox

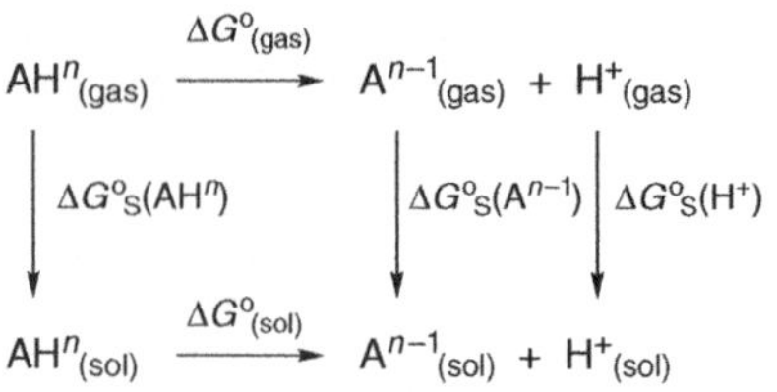

$$\Delta G^{\circ}_{(gas)}$$

$$AH^{n}_{(gas)} \xrightarrow{} A^{n-1}_{(gas)} + H^{+}_{(gas)}$$

$$\Bigg\downarrow \Delta G^{\circ}_{S}(AH^{n}) \qquad \Bigg\downarrow \Delta G^{\circ}_{S}(A^{n-1}) \Bigg| \Delta G^{\circ}_{S}(H^{+})$$

$$\Delta G^{\circ}_{(sol)}$$

$$AH^{n}_{(sol)} \xrightarrow{} A^{n-1}_{(sol)} + H^{+}_{(sol)}$$

Figure 11.9 Free energy cycle for computation of pK_a values (where n is an integer). This cycle is sometimes referred to as a Born–Haber cycle

potentials. Computation of the former is accomplished by employing the free-energy cycle depicted in Figure 11.9. Thus, gas-phase free energies of AH^n and A^{n-1} may be computed at arbitrarily high levels of theory to establish as accurately as possible the deprotonation free energy of AH^n. Note that if n and/or $n - 1$ are negative numbers then the basis set will need to include diffuse functions in order to obtain even modest quantitative accuracy. As for the proton, its electronic energy is obviously zero, and its gas-phase free energy derives entirely from a PV enthalpy term and a translational free energy that may be computed from Eqs. (10.16) and (10.17). At 298 K in the usual 1 atm standard state the free energy of the proton is -0.00999 a.u.

To compute the deprotonation free energy in solution, we take the gas-phase free energy change, add the free energies of solvation of A^{n-1} and H^+ (see above for the latter), and subtract the free energy of solvation of AH^n. However, note that most continuum solvation models compute the free energy of solvation assuming the same standard-state concentration in the gas phase as in solution. As most pK_a values adopt a standard-state concentration of 1 M, we need then to compute the free energy change associated with adjusting the concentrations of all of the gas-phase species from 1 mol per 24.5 L (the concentration of an ideal gas at 1 atm pressure and 298 K) to 1 mol per 1 L. As described in Section 10.5.4, this change is $RT \ln(24.5)$ for every species. As there are two products and only one reactant in the deprotonation reaction, the net effect is to make deprotonation less favorable by 1.9 kcal mol^{-1} in the 1 M standard state compared to the 1 atm standard state at 298 K.

Having computed the free-energy change in solution by this protocol, we may then compute K_a as

$$K_a = e^{-\Delta G^{\circ}_{(sol)}/RT} \tag{11.24}$$

and pK_a as

$$pK_a = -\log\left(e^{-\Delta G^{\circ}_{(sol)}/RT}\right) \tag{11.25}$$

As errors in ionic solvation free energies are often on the order of 5 kcal mol^{-1}, and as errors in the gas-phase deprotonation free energies may be of similar magnitude even with reasonably good levels of theory, errors in predicted absolute pK_a values of 5 or more pK units are not terribly unusual, which is not particularly satisfying insofar as experimental measurements can be accurate to 0.01 pK units.

One approach for reducing the errors associated with the prediction of pK_a values is to employ an isodesmic reaction. To illustrate with a specific example, it may be very hard to correctly predict the free energy change for the aqueous reaction

$$(11.26)$$

However, if the theoretical target is instead the free energy change for the isodesmic equation

$$(11.27)$$

one may well expect this to be computed far more accurately, since errors in levels of theory should largely cancel from left to right. Provided experimental data are available for the unsubstituted case

$$(11.28)$$

then the free energy change for Eq. (11.26) may be estimated from the difference between the computed value for Eq. (11.27) and the experimental value for Eq. (11.28). Chen and MacKerell (2000) have provided a more detailed demonstration of the utility of this approach for a series of substituted pyridines using a variety of different levels of theory for the gas phase and computed solvation free energies.

An alternative approach for improving predicted pK_a values has been suggested by Klicic et al. (2002), who developed functional-group-specific linear regression corrections for pK_a values computed from a particular DFT SCRF PB formalism. Correction of the raw computed pK_as increases the model's accuracy to about 0.5 pK units for those acidic functional groups well represented in their parameterization set.

11.4.1.2 Redox potentials

Oxidation and reduction potentials in solution are also computed via reference to particular thermodynamic cycles as illustrated in Figure 11.10. In this case, however, the

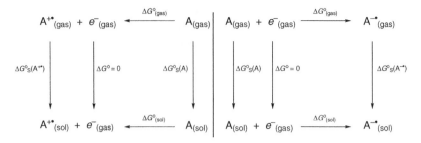

Figure 11.10 Thermodynamic cycles for one-electron oxidation (left) and reduction (right) potentials in solution

thermodynamic cycles perforce involve open-shell species and free electrons. Note that the oxidation and reduction cycles in the gas phase correspond conceptually to ionization potentials and electron affinities, respectively, except that IPs and EAs are enthalpies, not free energies, so thermal and entropic terms must be included therein. For the free electron, like the free proton, the electronic energy is zero, but the sum of the PV and translational terms leads to a total gas-phase free energy at 298 K and 1 atm of -0.00001 a.u. (it is a coincidence that for this standard state the free energy associated with the translational entropy almost exactly cancels the enthalpy).

Another key feature of redox thermodynamic cycles is that the free energy change in solution is still defined to involve a *gas-phase* electron, that is, the solvation free energy of the electron is happily not an issue. And, once again, redox potentials in solution typically assume 1 M standard states for all species (but not always; in this chapter's case study, for instance, all redox potentials were measured and computed for chloride ion concentrations buffered to 0.001 M). So, free energy changes associated with concentration adjustments must also be properly taken into account.

Once the free energy change in solution has been computed, the absolute redox potential E° may be computed as

$$E^\circ = -\frac{\Delta G^\circ}{nF} \tag{11.29}$$

where n is the number of electrons transferred and F is the Faraday constant equal to 23.061 kcal mol^{-1} V^{-1}. Note that while Figure 11.10 presents thermodynamic cycles for one-electron processes, analogous cycles involving multiple electrons are readily constructed and may sometimes be more amenable to experimental determination.

In practice, experimental redox potentials are reported relative to a standard electrode. If the standard is the NHE, one subtracts 4.36 V from the absolute reduction potential (the 'cost' of the free electron) or adds 4.36 V to the absolute oxidation potential (the 'return' from the removed electron) in order to determine the relative potential. Adjustment to other standard electrodes is straightforward, since their potentials relative to the NHE are well known.

With respect to accuracy, it is again important to employ basis sets including diffuse functions when anions are present as either reactants or products. With large well balanced basis sets, B3LYP for gas-phase energetics, and a PB SCRF solvation model, Baik and

Friesner (2002) have reported average errors of about 150 mV for various organometallic species in different organic solutions. As already discussed for pK_as, still better accuracy in redox potentials can often be achieved through the use of isodesmic equations or functional-group-specific correction schemes (see, for example, Winget *et al.* 2000).

11.4.1.3 *Supermolecular solutes*

Some final technical points merit attention. In SCRF models that use the full electronic distribution as part of the representation of the density, a problem arises in that the wave function has non-zero amplitude in the space *outside* the cavity. Thus, the construction of the cavity truncates the charge distribution, so that, for instance, neutral molecules have a small net positive charge inside the cavity. To return to integral charge values, the charge inside the cavity must somehow be renormalized. There are many different approaches to rectifying this problem; early methods tended to introduce considerable instability into the solvation computation, although more modern approaches seem reasonably robust (see, for example, Curutchet *et al.* 2004). Methods that do *not* suffer from the charge-penetration problem include all those that represent the density as either a single- or multi-center multipole or monopole expansion (this then includes GB methods). In addition, approaches have been developed that specifically handle, as a separate physical component, the polarization energy associated with penetration of charge into the solvent, and these models too seem to be well balanced (Chipman 2002).

The charge-penetration problem is in some sense related to a specific drawback of current continuum models, namely, that they have no mechanism to account for possible charge transfer between the solute and the surrounding solvent. It is not yet clear to what extent such solute/solvent charge transfer is important.

Of course, the simplest way to account for charge transfer would be to 'materialize' one or more solvent molecules around the solute and to treat the resulting cluster as a supermolecule embedded in the continuum. Pliego and Riveros (2002) and Fu *et al.* (2004) have recently suggested that such an approach provides a more robust protocol for the computation of accurate pK_a values, for instance. However, while this model has conceptual merits, it can introduce significant computational overhead. First, the supermolecule is obviously bigger than the solute, and depending on the level of theory employed the difference in computational time for a single SCRF calculation may be large. Second, clusters tend to generate fairly complex PESs, with many minima, and any attempt to compute free energy must sample over all of the minima in a statistically correct fashion. Since part of the motivation for using a continuum model is to avoid the sampling issues associated with explicit models, the representation of specific solvent molecules is usually not undertaken in the absence of compelling need.

One case, however, where materialization of a specific solvent molecule out of the continuum is indeed critical is when that solvent molecule loses its 'solvent' character. For instance, a water molecule tightly bound as both a hydrogen-bond donor and acceptor in a chain involving two solute functional groups clearly should be regarded as a unique fragment in what is fundamentally a two-piece supermolecule. Unfortunately, it is not always clear when such situations will arise, and modeling must simply be carried out with great care to ensure that the possibility is not overlooked.

11.4.2   Partitioning

The free energy of solvation may be regarded as the quantity describing the partitioning of a solute between the gas phase and a particular solvent. However, one is often interested in the free energy associated with the partitioning of a solute between two *different* condensed phases. Continuum solvent models may still conveniently be used to predict such partitioning behavior, based on the procedure outlined in Figure 11.11. The validity of the thermodynamic cycle is secure as drawn; however, under certain experimental conditions, the cycle is not representative of the experiment and the performance of the model may thus be degraded. For instance, the two solvents may be immiscible in a bulk sense, so the experiment may be carried out simply by dissolving a solute into a container holding both solvents, shaking the system until equilibrium is reached, and then measuring the solute concentration in the two phases. However, in spite of the phases being immiscible in bulk, the small percentages of each solvent that dissolve into the other may significantly affect the bulk's ability to solvate the solute in question. The continuum model, on the other hand, necessarily assumes a homogeneous 'pure' medium.

One major motivation for studying partitioning behavior has been a desire to understand the fashion in which drug molecules pass through largely non-polar (lipid) biomembranes that separate largely aqueous biocompartments. Historically, the octanol/water partition coefficient has been useful in this regard, as have some others. Such partition coefficients are usually not expressed as free energies of transfer, but as the logarithm of the associated equilibrium constant P. From a modeling perspective, using the formalism embodied in Figure 11.11, one computes

$$\log P_{A/B} = -\frac{\Delta G^{\circ}_{S_A} - \Delta G^{\circ}_{S_B}}{2.303RT} \tag{11.30}$$

where A and B are the solvents of interest and the two terms in the numerator are the free energies of transfer from the gas phase into solvent A and solvent B, respectively.

The similar accuracies of different well-parameterized continuum models implies that they will also perform similarly for the computation of partition coefficients, and that has proven to be the case in most studies to date (see, for example, Bordner, Cavasotto, and Abagyan 2002 and Curutchet *et al.* 2003b). In Table 11.4 the previously presented SM*x* results for the chloroform/water partitioning of the methylated canonical nucleic acid bases are compared to results from the MST-ST/HF/6-31G* method, and also to purely electrostatic results obtained using a multipole expansion SCRF method. As the latter does not include any accounting for non-electrostatic effects, its performance is significantly degraded compared to the other two.

11.4.3   Non-isotropic Media

All of the continua discussed thus far have been isotropic in nature. An interesting question arises as to the ability of the continuum approximation to model *non*-isotropic media. In

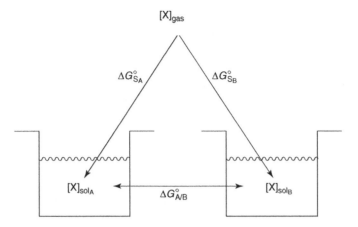

Figure 11.11   Thermodynamic relationship between partitioning free energies and free energies of solvation. Knowledge of any two free energies permits prediction of the third since any cycle around the free-energy triangle must sum to zero

Table 11.4   Chloroform/water partition coefficients ($\log_{10}$ units) for nucleic acid bases at different SCRF levels

Solute	$\log K_{CHCl_3/H_2O}$			
	SM5.4[a]	MST-ST[b]	SCRF[c]	Experiment[a]
9-Methyladenine	−1.6	−0.3	−0.6	−0.8
9-Methylguanine	−4.1	−4.8	−1.3	−3.5
9-Methylhypoxanthine	−3.5	−1.4		−2.5
1-Methylcytosine	−4.3	−3.4	−1.1	−3.0
1-Methylthymine	−0.3	−0.4	−0.8	−0.5
1-Methyluracil	−1.2	−1.0		−1.2
Mean unsigned error	0.6	0.6	1.2	

[a] See Table 11.3.
[b] Orozco, Colominas, and Luque 1996.
[c] Young and Hillier 1993; Young, Hillier, and Gould 1994.

certain instances, this has proven relatively straightforward. One example is the extension of the PCM model to include handling liquid crystals as solvents. In the case of a liquid crystal, the ordering of the solvent gives rise to a dielectric tensor as opposed to a single uniform dielectric constant. In order to extend the continuum model, an absolute reference frame is chosen and the x, y, and z components of the PCM equations are solved separately using the appropriate dielectric constant values; in addition, non-isotropic effects on cavitation energies have been considered (Mennucci, Cossi, and Tomasi 1996).

Another interesting case is a supercritical fluid. Near their critical points, supercritical fluids can exhibit very large changes in density (and density-related properties) in response to very small changes in conditions. By including possible density changes and their effects into the

SCRF equations governed according to the experimentally measured solvent compressibility, Luo and Tucker (1995) have been able to model these effects efficiently.

While both of the above examples are gas/condensed-phase solvation phenomena, there are also many interesting cases of partitioning phenomena where one phase is non-isotropic. Perhaps the most common is the case where one phase is a pure solid and the other a liquid solvent, in which case the partitioning phenomenon corresponds to solubility. Within the context of the free energy cycle of Figure 11.11, if we consider A to be the solid phase for solute X, the free-energy change from the solid to the gas corresponds to sublimation. To compute the free energy of sublimation rigorously, one must know the crystal packing energy. However, even when the unit cell geometry of the crystal is known (which it often is *not*), it is by no means trivial to compute the free energy of interaction of one monomer with the full crystal. A very rough estimate can be had by assuming that organic non-electrolytes have crystal packing energies similar to the solvation energies that the solute would have in a solvent 'similar' to itself. Thus, for instance, a highly non-polar hydro-carbon would be assumed to have a crystal packing energy equal to its solvation free energy in *n*-hexadecane. In essence, this treats solid/liquid partitioning as just a typical liquid/liquid partitioning problem (see, for instance, Reinwald and Zimmermann 1998 and Thompson, Cramer, and Truhlar 2003). While this approach can work well for non-polar solutes, it is less secure when more complicated functionality is present. In such instances, modern work typically includes some combination of solvation free-energy estimates combined with statistical analysis over data sets of molecules having similar functionality for which solu-bilities have been measured in order to make predictions (generating a so-called quantitative structure–property relationship (QSPR); see, for example, Lipinski *et al.* 1997).

Liquid/liquid partition constants within pharmaceutical chemistry have been of primary interest because of their correlation with liquid/membrane partitioning behavior. A suffi-ciently fluid membrane may, in some sense, be regarded as a solvent. With such an outlook, the partitioning phenomenon may again be regarded as a liquid/liquid example, amenable to treatment with standard continuum techniques. Of course, accurate continuum solvation models typically rely on the availability of solvation free energies or bulk solvent properties in order to develop useful parameterizations, and such data may be sparse or non-existent for membranes. Some success, however, has been demonstrated for predicting such data either by intuitive or statistical analysis (see, for example, Chambers *et al.* 1999).

Indeed, the utility of the continuum approach for modeling non-homogeneous phases has even been extended to the modeling of soil. The partitioning behavior of organic compounds between aqueous phases and soil is an important factor affecting the persistence of organic contaminants in the environment. Thus, environmental chemists define P_{OC} as the partition constant of a solute between water and soil, where the mass of the soil is normalized by organic carbon content (such normalization has the effect of making the partition coefficient remarkably constant over wide ranges of soil types). Using Eq. (11.30), then, one can deter-mine a free energy of transfer into the organic carbon component of soil. An SM*x* model trained on a data set of a few hundred molecules in order to determine the necessary bulk 'solvent' properties to define a soil phase has been shown to be capable of predicting P_{OC} values to within about one log unit (Winget, Cramer, and Truhlar, 2000). Thus, continuum models may be regarded as very useful initial approximations in describing transfer free energies into phases of arbitrary complexity, provided sufficient data are available to permit their robust parameterization.

11.4.4 Potentials of Mean Force and Solvent Structure

Consider the association of two molecules in a solution. When association is favorable, one may imagine generating a curve similar to that shown in Figure 11.12, where the association is reduced to a one-dimensional 'reaction coordinate'. The ordinate of the graph, in this case, is not potential energy (as it would be in a similarly shaped graph for, say, a bond dissociation) but free energy. This implies that each point on the curve reflects a proper statistical average over all of the solvent configurations that may solvate the 'reacting' system. Such a curve is referred to as a potential of mean force (PMF). The generation of accurate PMFs is one of the most significant challenges facing continuum models, in some cases for technical reasons and also because of a lack of experimental data for paradigmatic systems.

With respect to the technical challenges involved, continuum models that use a cavity-based approach to solve the Poisson equation are not well suited to computing PMFs. The problem is that it is quite difficult to solve the necessary equations when there are *two* cavities. Moreover, when the two cavities first touch one another and begin to penetrate, the narrow 'neck' of the joined cavities can lead to numerical instabilities.

Generalized Born models do not suffer from the multiple cavity problem, which is a particular advantage of that methodology. However, initial studies have suggested that

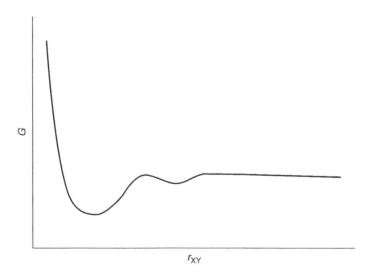

Figure 11.12   Potential of mean force between solutes X and Y as a function of the distance between their centers of mass. This particular PMF indicates that both tight and solvent-separated pairs exist as distinct minima

while GB models are very robust for the prediction of solvation free energies, they are less successful in the generation of potentials of mean force (Rankin, Sulea, and Purisima 2003). Lack of high-quality data makes it difficult to evaluate this possibility at present, although ongoing comparisons between different theoretical models are helping to further illuminate the issue (see, for example, Jayaram, Liu, and Beveridge 1998 and Gohlke and Case 2004).

Note that one feature of Figure 11.12 is a solvent-separated minimum for the X–Y pair. Insofar as solvent-separated minima involve intervening solvent molecules that typically differ significantly in their behavior from normal bulk solvent as a consequence of being isolated between the two solutes, such situations are unlikely to be handled accurately by continuum models in general.

It is sometimes the case that the structure of the first shell (or shells) of solvent is a property of primary interest for a given modeling study. It is perhaps stating the obvious to note that in such an instance, continuum models cannot be used, since by construction they ignore the molecular nature of the solvent and assume a homogeneous surrounding medium.

Of course, if one is interested only in the free-energy well associated with full complexation, many technical aspects of the calculation are simplified. The tremendous speed of continuum solvent models has made them attractive tools in evaluating solvation effects on docking, especially insofar as they permit more extensive sampling of varying target-receptor geometries to be carried out in an efficient manner (see, for example, Gouda *et al.* 2003; Taylor, Jewsbury, and Essex 2003; and Zoete, Michielin, and Karplus 2003).

11.4.5 Molecular Dynamics with Implicit Solvent

A large fraction of the expense of a typical MD simulation involving a solute in solution (discussed in much more detail in the next chapter) is associated with the hundreds or thousands of solvent molecules that are explicitly represented in the full simulation cell. However, when the fine details of the solvation process are not of primary interest, it can be about an order of magnitude more efficient to propagate a trajectory for the solute within the context of continuum solvation. The methodology that has been most extensively explored for this process to date has tended to involve GB solvation models developed for biomolecular force fields (although PB models have also seen substantial use). To maximize speed, Born radii are computed either from a PD algorithm or are set to constant values determined from initial PB calculations (Onufriev, Case, and Bashford 2002). For truly enormous systems, additional algorithms allowing certain portions of the solute to be held frozen while others are dynamical have been described (Banavali, Im, and Roux 2002; Guvench *et al.* 2002).

A particular advantage of MD with implicit solvation is that solvent friction is not an issue with respect to the solute being able to explore phase space. That is, no solvent molecules need to be pushed out of the way in order for otherwise energetically accessible large-scale motions to take place. As long as the energy landscape for the solute is as accurately predicted with the continuum solvent as with an explicit solvent, this feature leads to much more rapid achievement of converged sampling (Okur *et al.* 2003) especially when LES is used (Cheng, Hornak, and Simmerling 2004). This behavior has been successfully exploited for modeling pathways associated with protein folding (see, for instance, Jang, Shin, and Pak 2002 and Chowdhury *et al.* 2003), the timescale for which would have made simulations with explicit solvent prohibitively expensive. Of course, if one is interested in the *kinetics* of the process in question, then removal of the solvent friction is not helpful – the implicit solvent advantage applies only to obtaining rapid equilibrium averages.

As for the quality of the energy landscape and its effect on solute dynamics, comparisons of PCA eigenvectors from simulations using either implicit or explicit solvation have been carried out for proteins (Cornell *et al.* 2001), DNA (Tsui and Case 2000) and RNA (Sherer and Cramer 2002) and have generally indicated high overlap between the two models. Nevertheless, some protein folding studies have identified serious deficiencies in GB landscapes that include overestimation of salt-bridge interaction energies (Zhou 2003) and a general tendency to overstabilize nucleation (Nymeyer and Garcia 2003). One alternative to propagating a trajectory using an implicit solvent model that has also been explored has been to take a trajectory generated with inclusion of *explicit* solvent and then post-process it to compute individual or average solvation free energies for various snapshots, whose computation would otherwise require more sophisticated simulation protocols as described in the next chapter.

Some work has also appeared describing MD with implicit solvation for solutes described at the DFT level. Fattebert and Gygi (2002) have proposed making the external dielectric constant a function of the electron density, thereby achieving a smooth transition from solute to solvent instead of adopting a sudden change in dielectric constant at a particular cavity surface. Non-electrostatic components of the solvation free energy have not been addressed in this model.

11.4.6 Equilibrium vs. Non-equilibrium Solvation

Most continuum models are properly referred to as 'equilibrium' solvation models. This appellation emphasizes that the design of the model is predicated on equilibrium properties of the solvent, such as the bulk dielectric constant, for instance. The amount of time required for a solvent to equilibrate to the sudden introduction of a solute (i.e., the solvent relaxation time) varies from one solvent to another, but typically is in the range of molecular vibrational and rotational timescales, which is to say on the order of picoseconds.

Processes that take place on longer timescales may thus be legitimately thought of as equilibrium processes with respect to solvation. However, the question arises of how applicable continuum models are to very fast processes. For instance, Figure 11.4 describes the relationship between gas-phase and solvated reaction coordinates for a reactive process, but the average amounts of time individual molecules spend at various positions on the reaction coordinate vary considerably. In the regions of the minima, equilibrium solvation seems assured, but transition state structures in principle live for only a single vibrational period. This suggests that the solvent may not have time to fully equilibrate to the TS structure, and a continuum model were it to be applied would overestimate the solvation free energy by assuming equilibration. In addition, considerable progress has been made in the extension of GB models to systems where an implicit membrane characterized by a dielectric constant different from the solvent is represented in both the electrostatic and non-electrostatic terms (Spassov, Yan, and Szalma 2002).

Still faster timescales are associated with phenomena like electron transfer (i.e., redox reactions) and photon absorption/emission and possible associated electronic excitation. Since these processes occur on the timescale of electronic motion, the surrounding solvent molecules may be regarded as frozen in place during the reaction, and clearly an equilibrium view of the instantaneous solvation is incorrect.

These issues will be addressed in more detail in Chapters 14 and 15. Here, we will simply note that successful extensions of continuum models to such ultrafast processes as electron transfer and photoexcitation requires that the response properties of the solvent be explicitly separated into slow and fast parts that interact with the solute over the appropriate timescales (see, for instance, Cossi and Barone 2000). This approach can also be used for dealing with non-equilibrium effects on transition states. However, unless there is a very sudden transfer of charge that takes place over a significant distance failure to account for non-equilibrium effects rarely has much consequence in estimating a free energy of activation – errors in the gas-phase potential energy difference between minima and the TS structure are typically larger in magnitude. Thus, unmodified continuum solvation models can still be quite useful in constructing diagrams like that shown in Figure 11.4 for the purpose of describing reactivity in solution.

11.5   Case Study: Aqueous Reductive Dechlorination of Hexachloroethane

Synopsis of Patterson, Cramer, and Truhlar (2001) 'Reductive Dechlorination of Hexachloroethane in the Environment. Mechanistic Studies via Computational Electrochemistry'.

Halogenated alkanes are very useful as solvents in a variety of industrial processes (at one time they were the solvents of choice for the dry cleaning of clothes, for example). The scale of their use is such that their accidental or deliberate discharge into the environment can lead to long-term contamination problems. As is true for many environmental contaminants, the molecule originally released may not be a particular danger from an environmental perspective, but some product into which it is transformed may be considerably more cause for concern.

An example is hexachloroethane (C_2Cl_6). In environmental aqueous phases, it typically undergoes reductive dechlorination relatively rapidly. One product of this dechlorination, produced in small amounts, is trichloroacetic acid (Cl_3CCO_2H), which is a regulated carcinogen in the United States. The authors studied the mechanistic aspects of C_2Cl_6 reductive dechlorination with both methodological and chemical goals. The methodological question involved identifying appropriate levels of theory for modeling the relevant reactions, while the chemical questions to be addressed were associated with identifying the relevant mechanistic pathways for reduction and any possible explanation for the generation of Cl_3CCO_2H as a product.

Various data were available for comparison in order to identify adequate theoretical levels for application. Table 11.5 illustrates the performance of two different levels of theory, CCSD(T)/aug-cc-pVDZ//BPW91/aug-cc-pVDZ and BPW91/aug-cc-pVDZ, with

Table 11.5 Reductive dechlorination benchmarks in the gas phase (eV) and aqueous solution (V)

Benchmark	Phase	Quantity	Experiment	CCSD(T)	BPW91
$C_2Cl_4 \rightarrow C_2Cl_4^{+\bullet} + e^-$	gas	ΔH_0 (IP)	9.33 (9.5)[a]	9.18 (9.42)[a]	8.81 (9.02)[a]
$Cl^\bullet + e^- \rightarrow Cl^-$	gas	$-\Delta H_0$ (EA)	3.61	3.40	3.64
	aqueous	$E_1^{\varnothing\prime}$	2.54	2.37[b]	2.62[b]
$C_2Cl_6 + 2e^- + H^+ \rightarrow$	gas	$\Delta G^\circ_{(g)}$	−18.54	−18.28	−18.55
$\quad C_2HCl_5 + Cl^-$	aqueous	$E_2^{\varnothing\prime}$	0.67	0.71[b]	0.85[b]
$C_2HCl_5 + 2e^- \rightarrow$	gas	$\Delta G^\circ_{(g)}$	−4.37	−3.90	−4.61
$\quad C_2Cl_4 + 2Cl^-$	aqueous	$E_2^{\varnothing\prime}$	1.15	1.09[b]	1.44[b]
$C_2HCl_5 + e^- \rightarrow$	gas	$\Delta G^\circ_{(g)}$	−1.16	−0.96	−1.45
$\quad C_2HCl_4^\bullet + Cl^-$	aqueous	$E_1^{\varnothing\prime}$	0.11	0.02[b]	0.52[b]

[a] Values in parentheses are for vertical process.
[b] Values differ from those originally reported by Patterson, Cramer, and Truhlar (2001) by 0.08 V per electron consumed. This difference reflects a more accurate measurement of the absolute potential of the normal hydrogen electrode as 4.36 V instead of 4.44 V since the time of that publication. See Lewis *et al.* (2004) and Section 11.4.1.

each level being supplemented by aqueous solvation energies computed from the SM5.42R/BPW91/DZVP//BPW91/aug-cc-pVDZ level of theory when appropriate. The reduction potentials in solution are in units of volts relative to the standard hydrogen electrode, and the authors provide a detailed appendix showing how to convert between the various standard states and conventions typically adopted in theoretical and experimental work. They note that the CCSD(T) level, combined with the continuum solvent model when needed, is a better choice than the DFT method; the mean unsigned error in predicted reduction potentials for the CCSD(T) model is 0.09 V while for the DFT model it is 0.24 V. As the DFT level does somewhat better for the gas-phase free energies of reaction than the CCSD(T) level, it appears that there is some modest cancellation of errors in the solvation free energies that improves the performance of the CCSD(T) model.

Having identified the optimal level of theory, the authors apply it to various structures, primarily stationary points on the gas-phase PES, to characterize the energetics associated with various postulated mechanistic pathways (Figure 11.13). They identify the first mechanistic step as electron transfer followed by barrierless chloride ion elimination to generate the pentachloroethyl radical ($C_2Cl_5^\bullet$). They discount the proposed heterolysis of this radical prior to a second electron transfer on the basis of the higher energy of the products $C_2Cl_4^{+\bullet}$ and Cl^-. Instead, they find that following a second electron transfer, there is again a barrierless elimination of a second chloride anion. However, this elimination is possible either from the same carbon as the first, in which case chloro(trichloromethyl)carbene is generated, or from the other carbon, in which case perchloroethylene is generated. While the former is much less exergonic than the latter, the barrierless nature of both reactions suggests that partitioning will be controlled by complex dynamic factors.

The importance of the former reaction is that it suggests a mechanism for the creation of Cl_3CCO_2H as a product. Based on analysis of the computed activation free energy for the rearrangement of chloro(trichloromethyl)carbene to perchloroethylene, the authors suggest that oxygen atom transfer to the carbene from some environmental source can

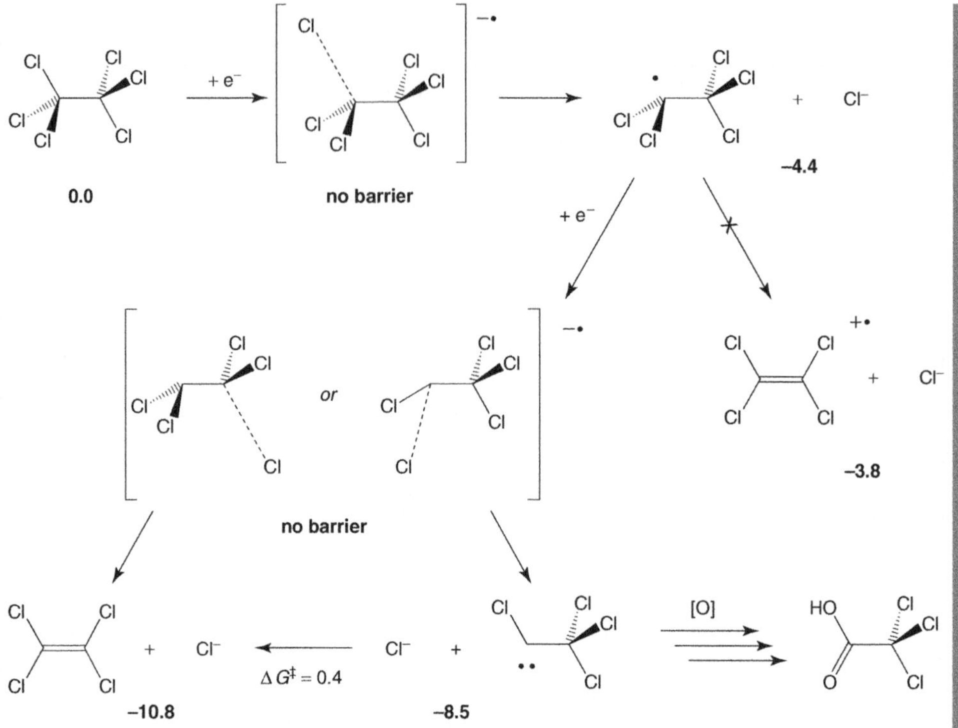

Figure 11.13   Relative aqueous free energies (eV) for various species on different reductive dechlorination paths of hexachloroethane as computed by Patterson and co-workers The relative energies are properly balanced, although for simplicity spectator species are not shown in every case. What is involved in achieving this balance for energy? What about free energy?

be kinetically competitive. Such a transfer would generate the acyl chloride equivalent of Cl_3CCO_2H, which would hydrolyze in short order to the carboxylic acid.

This paper provides an example of how accurate continuum models can open the door to the modeling of condensed-phase processes where solvation free energies have a very large influence on reaction energetics. It additionally offers a case study of how to first choose a model on the basis of experimental/theoretical comparisons over a relevant data set, and then apply that model with a greater expectation for its utility. The generality of this approach to other (equilibrium) electrochemical reactions seems promising.

Bibliography and Suggested Additional Reading

Bashford, D. and Case, D. A. 2000. 'Generalized Born Models of Macromolecular Solvation Effects' *Annu. Rev. Phys. Chem.*, **51**, 129.

Cramer, C. J. and Truhlar, D. G. 1999. 'Implicit Solvation Models: Equilibria, Structure, Spectra, and Dynamics', *Chem. Rev.*, **99**, 2161.

Cramer, C. J. and Truhlar, D. G., Eds., 1994. *Structure and Reactivity in Aqueous Solution*, ACS Symposium Series 568, American Chemical Society: Washington, DC.

Li, J. Cramer, and Truhlar, D. G. 1999. 'Application of a Universal Solvation Model to Nucleic Acid Bases. Comparison of Semiempirical Molecular Orbital Theory, *Ab Initio*, Hartree–Fock Theory, and Density Functional Theory', *Biophys. Chem.*, **78**, 147.

Llano, J. and Eriksson, L. A. 2002. 'First Principles Electrochemistry: Electrons and Protons Reacting as Independent Ions', *J. Chem. Phys.*, **117**, 10193.

Orozco, M. and Luque, F. J. 2000. 'Theoretical Methods for the Description of the Solvent Effect on Biomolecular Systems', *Chem. Rev.* **100**, 4187.

Reichardt, C. 1990. *Solvents and Solvent Effects in Organic Chemistry*, VCH: New York.

Roux, B. and Simonson, T., Eds. 1999. *Biophys. Chem.* **78**(1/2), [special issue devoted to implicit solvent models].

Schutz, C. N. and Warshel, A. 2001. 'What Are the Dielectric "Constants" of Proteins and How to Validate Electrostatic Models?', *Proteins*, **44**, 400.

Tapia, O. and Bertrán, J. 1996. *Solvent Effects on Chemical Reactivity*, Kluwer: Dordrecht.

Winget, P., Cramer, C. J., and Truhlar, D. G. 2004. "Computation of Equilibrium Oxidation and Reduction Potentials for Reversible and Dissociative Electron-Transfer Reactions in Solution", *Theor. Chem. Acc.*, **112**, 217.

References

Ángyán, J. G. 1992. *J. Math. Chem.*, **10**, 93.

Baik, M.-H. and Friesner, R. A. 2002. *J. Phys. Chem. A*, **106**, 7407.

Banavali, N. K. and Roux, B. 2002. *J. Phys. Chem. B*, **106**, 11026.

Banavali, N. K., Im, W., and Roux, B. 2002. *J. Chem. Phys.*, **117**, 7381.

Barone, V. and Cossi, M. 1998. *J. Phys. Chem. A*, **102**, 1995.

Beak, P. 1977. *Acc. Chem. Res.*, **10**, 186.

Best, S. A., Merz, K. M., and Reynolds, C. H. 1997. *J. Phys. Chem. B*, **101**, 10479.

Bordner, A. J., Cavasotto, C. N., and Abagyan, R. A. 2002. *J. Phys. Chem. B*, **106**, 11009.

Chambers, C. C., Giesen, D. J., Hawkins, G. D., Vaes, W. H. J., Cramer, C. J., and Truhlar, D. G. 1999. In: *Rational Drug Design*, Truhlar, D. G., Howe, W. J., Hopfinger, A. J., Blaney, J. M., and Dammkoehler, R. A., Eds., Springer: New York, 51.

Chandresekhar, J., Smith, S. F., and Jorgensen, W. L. 1985. *J. Am. Chem. Soc.*, **107**, 154.

Chen, I.-J. and MacKerell, A. D., Jr. 2000. *Theor. Chem. Acc.*, **103**, 483.

Chen, Y., Noodleman, L., Case, D. A., and Bashford, D. 1994. *J. Phys. Chem.*, **98**, 11059.

Cheng, A., Best, S. A., Merz, K. M., and Reynolds, C. H. 2000. *J. Mol. Graph. Model.*, **18**, 273.

Cheng, X. L., Hornak, V., and Simmerling, C. 2004. *J. Phys. Chem. B*, **108**, 426.

Chipman, D. M. 2002. *Theor. Chem. Acc.*, **107**, 80.

Chowdhury, S., Lee, M. C., Xiong, G., Duan, Y. 2003. *J. Mol. Biol.*, **327**, 711.

Christiansen, O. and Mikkelsen, K. V. 1999. *J. Chem. Phys.*, **110**, 1365.

Claverie, P. 1978. *Perspect. Quantum Chem. Biochem.*, **2**, 69.

Colominas, C., Luque, F. J., Teixidó, J., and Orozco, M. 1999. *Chem. Phys.*, **240**, 253.

Cornell, W., Abseher, R., Nilges, M., and Case, D. A. 2001. *J. Mol. Graph. Model.*, **19**, 136.

Cossi, M. 2004. *Chem. Phys. Lett.*, **384**, 179.

Cossi, M. and Barone, V. 2000. *J. Phys. Chem. A*, **104**, 10614.

Cossi, M., Barone, V., Cammi, R., and Tomasi, J. 1996. *Chem. Phys. Lett.*, **255**, 327.

Cossi, M., Scalmani, G., Rega, N., and Barone, V. 2002. *J. Chem. Phys.*, **117**, 43.

Cullis, P. M.. and Wolfenden, R. 1981. *Biochemistry,* **20**, 3024.

Curutchet, C., Bidon-Chanal, A., Orozco, M., and Luque, F. J. 2004. *Chem. Phys. Lett.*, **384**, 299.

Curutchet, C., Cramer, C. J., Truhlar, D. G., Ruiz López, M., Orozco, M., and Luque, F. J. 2003a. *J. Comput. Chem.*, **24**, 284.

Curutchet, C., Orozco, M., and Luque, F. J. 2001. *J. Comput. Chem.*, **22**, 1180.

Curutchet, C., Salichs, A., Barril, X., Orozco, M. and Luque, F. J. 2003b. *J. Comput. Chem.*, **24**, 32.

Dillet, V., Rinaldi, D., Ángyán, J. , G., and Rivail, J.-L. 1993. *Chem. Phys. Lett.*, **202**, 18.

Dolney, D. M., Hawkins, G. D., Winget, P., Liotard, D. A., Cramer, C. J., and Truhlar, D. G. 2000. *J. Comput. Chem.*, **21**, 340.

Dominy, B. N. and Brooks, C. L. 1999. *J. Phys. Chem. B,* **103**, 3765.

Edinger, S. R., Cortis, C., Shenkin, P. S., and Friesner, R. A. 1997. *J. Phys. Chem. B,* **101**, 1190.

Fattebert, J. L. and Gygi, F. 2002. *J. Comput. Chem.*, **23**, 662.

Feig, M., Onufriev, A., Lee, M. S., Im, W., Case, D. A., and Brooks, C. L. 2004. *J. Comput. Chem.*, **25**, 265.

Floris, F. M., Tomasi, I., and Pascual-Ahuir, J. L. 1991. *J. Comput. Chem.*, **12**, 784.

Foresman, J. B., Keith, T. A., Wiberg, K. B., Snoonian, J., and Frisch, M. J. 1996. *J. Phys. Chem.*, **100**, 16098.

Frediani, L., Cammi, R., Corni, S., and Tomasi, J. 2004. *J. Chem. Phys.*, **120**, 3893.

Fu, Y., Liu, L., Li, R. C., Liu, R., Guo, Q. X. 2004. *J. Am. Chem. Soc.*, **126**, 814.

Gallicchio, E. and Levy, R. M. 2004. *J. Comput. Chem.*, **25**, 479.

Giesen, D. J., Cramer, C. J., and Truhlar, D. G. 1994. *J. Phys. Chem.*, **98**, 4141.

Giesen, D. J., Chambers, C. C., Cramer, C. J., and Truhlar, D. G. 1997. *J. Phys. Chem. B,* **101**, 5084.

Gohlke, H. and Case, D. A. 2004. *J. Comput. Chem.*, **25**, 238.

Gouda, H, Kuntz, I. D., Case, D. A., and Kollman, P. A. 2003. *Biopolymers*, **68**, 16.

Guvench, O., Shenkin, P., Kolossvary, I., Still, W. C. 2002. *J. Comput. Chem.*, **23**, 214.

Hawkins, G. D., Cramer, C. J., and Truhlar, D. G. 1995. *Chem. Phys. Lett.*, **246**, 122.

Hawkins, G. D., Liotard, D. A., Cramer, C. J., and Truhlar, D. G. 1998. *J. Org. Chem.*, **63**, 4305.

Höfinger, S. and Zerbetto, F. 2003. *J. Phys. Chem. A,* **107**, 11253.

Hou, T. J., Qiao, X. B., Zhang, W., and Xu, X. J. 2002. *J. Phys. Chem. B,* **106**, 11295.

Jang, S., Shin, S., and Pak, Y. 2002. *J. Am. Chem. Soc.*, **124**, 4976.

Jayaram, B., Liu, Y., and Beveridge, D. L. 1998, *J. Chem. Phys.*, **109**, 1465.

Jayaram, B., Sprous, D., and Beveridge, D. L. 1998. *J. Phys. Chem. B,* **102**, 9571.

Klamt, A. 1995. *J. Phys. Chem.*, **99**, 2224.

Klamt, A. and Schüürmann, G. 1993. *J. Chem. Soc. Perkin Trans. 2*, 799.

Klamt, A., Jonas, V., Buerger, T., and Lohrenz, J. C. W. 1998. *J. Phys. Chem. A,* **102**, 5074.

Klicic, J. J., Friesner, R. A., Liu, S. Y., and Guida, W. C. 2002. *J. Phys. Chem. A,* **106**, 1327.

Lewis, A., Bumpus, J. A., Cramer, C. J., and Truhlar, D. G. 2004. *J. Chem. Educ.*, **81**, 596.

Li, J., Zhu, T., Hawkins, G. D., Winget, P., Liotard, D. A., Cramer, C. J., and Truhlar, D. G. 1999. *Theor. Chem. Acc,* **103**, 9.

Lipinski, C. A., Lombardo, F., Dominy, B. W., and Feeney, P. J. 1997. *Adv. Drug Delivery Rev.*, **23**, 3.

Lu, Q. and Luo, R. 2003. *J. Chem. Phys.*, **119**, 11035.

Luo, H. and Tucker, S. C. 1995. *J. Am. Chem. Soc.*, **117**, 11359.

Mennucci, B., Cossi, M., and Tomasi, J. 1996. *J. Phys. Chem.*, **100**, 1807.

No, K. T., Kim, S. G., Cho, K.-H., Scheraga, H. A. 1999. *Biophys. Chem.*, **78**, 127.

Nymeyer, H. and Garcia, A. E. 2003. *Proc. Natl. Acad. Sci. (USA)*, **100**, 13934.

Okur, A., Strockbine, B., Hornak, V., and Simmerling, C. 2003. *J. Comput. Chem.*, **24**, 21.

Onufriev, A., Bashford, D., and Case, D. A. 2000. *J. Phys. Chem. B,* **104**, 3712.

Onufriev, A., Case, D. A., and Bashford, D. 2002. *J. Comput. Chem.*, **23**, 1297.

Orozco, M., Colominas, C., and Luque, F. J. 1996. *Chem. Phys.*, **209**, 19.

Pascual-Ahuir, J. L., Silla, E., and Tuñon, I. 1994. *J. Comput. Chem.*, **15**, 1127.

Patterson, E. V., Cramer, C. J., and Truhlar, D. G. 2001. *J. Am. Chem. Soc.*, **123**, 2025.

Pierotti, R. A. 1976. *Chem. Rev.*, **76**, 717.

Pliego, J. R. and Riveros, J. M. 2002. *J. Phys. Chem. A*, **106**, 7434.

Qiu, D., Shenkin, P. S., Hollinger, F. P., and Still, W. C. 1997. *J. Phys. Chem. A*, **101**, 3005.

Rankin, K. N., Sulea, T., and Purisima, E. O. 2003. *J. Comput. Chem.*, **24**, 954.

Reinwald, G. and Zimmermann, I. 1998. *J. Pharm. Sci.*, **87**, 745.

Rinaldi, D. and Rivail, J.-L. 1973. *Theor. Chim. Acta*, **32**, 57.

Rinaldi, D., Bouchy, A., Rivail, J.-L., and Dillet, V. 2004. *J. Chem. Phys.*, **120**, 2343.

Rinaldi, D., Rivail, J.-L., and Rguini, N. 1992. *J. Comput. Chem.*, **13**, 675.

Sherer, E. C. and Cramer, C. J. 2002. *J. Phys. Chem. B*, **106**, 5075.

Spassov, V. Z., Yan, L., and Szalma, S. 2002. *J. Phys. Chem. B*, **106**, 8726.

Spuhl, O. and Arlt, W. 2004. *Ind. Eng. Chem. Res.*, **43**, 852.

Srinivasan, J., Trevathan, M. W., Beroza, P., and Case, D. A. 1999. *Theor. Chem. Acc*, **101**, 426.

Still, W. C., Tempczyk, A., Hawley, R. C., and Hendrickson, T. 1990. *J. Am. Chem. Soc.*, **112**, 6127.

Tannor, D. J., Marten, B., Murphy, R., Friesner, R. A., Sitkoff, D., Nicholls, A., Ringnalda, M., Goddard, W. A., III, and Honig, B. 1994. *J. Am. Chem. Soc.*, **116**, 11875.

Taylor, R. D., Jewsbury, P. J., and Essex, J. W. 2003. *J. Comput. Chem.*, **24**, 1637.

Thompson, J. D., Cramer, C. J., and Truhlar, D. G. 2003. *J. Chem. Phys.*, **119**, 1661.

Tissandier, M. D., Cowen, K. A., Feng, W. Y., Gundlach, E., Cohen, M. H., Earhart, A. D., Coe, J. V., and Tuttle, T. R., Jr. 1998. *J. Phys. Chem. A*, **102**, 7787.

Trasatti, S. 1986. *Pure Appl. Chem.*, **58**, 955.

Tsui, V. and Case, D. A. 2000. *J. Am. Chem. Soc.*, **122**, 2489.

Wang., J. M., Wang, W., Huo, S. H., Lee, M., and Kollman, P. A. 2001. *J. Phys. Chem. B*, **105**, 5055.

Winget, P., Cramer, C. J., and Truhlar, D. G. 2000. *Environ. Sci. Technol.*, **34**, 4733.

Winget P., Weber E. J., Cramer C. J., and Truhlar D. G. 2000 *Phys. Chem. Chem. Phys.*, **2**, 1231.

Young, P. E. and Hillier, I. H. 1993. *Chem. Phys. Lett.*, **215**, 405.

Young, P. E., Hillier, I. H., and Gould, I. R. 1994. *J. Chem. Soc., Perkin Trans. 2*, 1717.

Xie, L. and Liu, H. 2002. *J. Comput. Chem.*, **23**, 1404.

Zauhar, R. J. and Varnek, A. 1996. *J. Comput. Chem.*, **17**, 864.

Zhang, W., Hou, T. J., Qiao, X. B., and Xu, X. J. 2003. *J. Phys. Chem. B*, **107**, 9071.

Zhou, R. H. 2003. *Proteins*, **53**, 148.

Zoete, V., Michielin, O., and Karplus, M. 2003. *J. Comput.-aid. Mol. Des.*, **17**, 861.

12

Explicit Models for Condensed Phases

12.1 Motivation

At the heart of chemistry are atoms and molecules – they are the basis set in which chemical events are expressed. While the continuum models described in Chapter 11 can be very efficient and powerful in situations where the molecular nature of a surrounding condensed phase is superfluous to the question at hand, they are unsuitable when knowledge of the explicit behavior of the surroundings is deemed to be as important as its effect on some system embedded therein.

As noted in Chapter 11, the explicit representation of a condensed phase leads to a system characterized by an enormous number of degrees of freedom. This system thus has associated with it a phase space of high dimensionality, and typically one in which there are large volumes within a few $k_B T$ of one another in energy. Properties of such a system must be determined as statistical averages over phase space, as already discussed in some detail in Chapter 3. However, Chapter 3 was concerned primarily with observables other than thermodynamic properties, e.g., radial distribution functions, electrical moments, or vibrational frequencies. Here, the initial focus will be on carrying out simulations of condensed-phase systems specifically to extract thermodynamic information, including the free energy of solvation, the importance of which has already been amply discussed in Section 11.1.

12.2 Computing Free-energy Differences

As noted previously in Chapters 3 and 10, statistical thermodynamics relates all thermodynamic observables to the partition function Q. For ease of reference, the definition of Q and the equations defining various thermodynamic variables as a function of Q, some of which have appeared previously, are as follows

$$Q = \int\int e^{-E(\mathbf{q},\mathbf{p})/k_B T} d\mathbf{q} d\mathbf{p} \tag{12.1}$$

Essentials of Computational Chemistry, 2nd Edition Christopher J. Cramer
© 2004 John Wiley & Sons, Ltd   ISBNs: 0-470-09181-9 (cased); 0-470-09182-7 (pbk)

$$U = k_{\mathrm{B}} T^2 \left(\frac{\partial \ln Q}{\partial T} \right)_V \tag{12.2}$$

$$P = k_{\mathrm{B}} T \left(\frac{\partial \ln Q}{\partial V} \right)_T \tag{12.3}$$

$$H = U + PV \tag{12.4}$$

$$A = -k_{\mathrm{B}} T \ln Q \tag{12.5}$$

$$S = k_{\mathrm{B}} T \left(\frac{\partial \ln Q}{\partial T} \right)_V + k_{\mathrm{B}} \ln Q \tag{12.6}$$

$$G = H - TS \tag{12.7}$$

where U is the internal energy, P is the pressure, H is the enthalpy, A is the Helmholtz free energy, S is the entropy, and G is the Gibbs free energy (in subsequent discussion, the Gibbs free energy will be implied by the words 'free energy' unless the Helmholtz free energy is explicitly specified). Note that we have adopted the classical expression for Q by formulating it as a phase space integral over all spatial ($\mathbf{q}$) and momentum ($\mathbf{p}$) coordinates. This assumes that the energy levels, computed as the sum of kinetic and potential energy terms by the Hamiltonian H (not to be confused with the enthalpy), are sufficiently closely spaced that we may convert the sum-over-states formulation of Q (see Eq. (10.2)) into an integral.

12.2.1   Raw Differences

In chemistry, one is typically interested not in absolute values of thermodynamic functions but in their changes over the course of a chemical process. Consider, for instance, if one were to be interested in the difference in U for the proton shift reaction HCN → HNC in aqueous solution. Because U is a state function, the precise path over which the reaction occurs is not important – we need only evaluate U at the two endpoints to determine the difference. If we make use of the relationship

$$\frac{\partial \ln f}{\partial g} = \frac{1}{f} \frac{\partial f}{\partial g} \tag{12.8}$$

we may use Eqs. (12.1) and (12.2) to rewrite U as

$$U = \frac{\iint E(\mathbf{q}, \mathbf{p}) e^{-E(\mathbf{q}, \mathbf{p})/k_{\mathrm{B}} T} d\mathbf{q} d\mathbf{p}}{\iint e^{-E(\mathbf{q}, \mathbf{p})/k_{\mathrm{B}} T} d\mathbf{q} d\mathbf{p}}$$

$$= \iint E(\mathbf{q}, \mathbf{p}) P(\mathbf{q}, \mathbf{p}) d\mathbf{q} d\mathbf{p} \tag{12.9}$$

where Eq. (3.6) for the probability P of being at a particular point in phase space has been used. To evaluate Eq. (12.9) for both the aqueous HCN and HNC systems, we might carry out a Monte Carlo simulation of each, and determine U as an ensemble average of E over the probabilistically correct set of snapshots generated by the MC approach, i.e.,

$$
\langle U \rangle_B - \langle U \rangle_A = \frac{1}{M_B} \sum_i^{M_B} E_i - \frac{1}{M_A} \sum_i^{M_A} E_i
$$

$$
= \langle E \rangle_B - \langle E \rangle_A \tag{12.10}
$$

where B is aqueous HNC and A is aqueous HCN. To carry out the simulations in a rational way, one would want to employ the same conditions in each, e.g., number of solvent molecules, size of unit cell used within periodic boundary conditions, etc.

If one follows the procedure outlined above, the results are not very satisfying. The problem is that the total energies E are large numbers. Thus, even after sampling millions of configurations, the standard deviation in each ensemble average may still be on the order of, say, 10 kcal mol^{-1}. Taking the error in ΔU as the RMS of the two ensemble errors in E would then imply an error of 14 kcal mol^{-1}. Such a large error is not very useful in most instances, which is disappointing, particularly given the large investment of computational resources required to generate the ensemble averages. Note that, with ergodic trajectories, we could have taken time averages from MD simulations instead of ensemble averages from MC simulations, but the error problem would be the same.

Let us consider instead of ΔU the quantity ΔA. In order to engineer a probabilistic fashion to determine A we may rewrite Eq. (12.5) as

$$
A = k_B T \ln \frac{1}{Q}
$$

$$
= k_B T \ln \left[\frac{\iint e^{E(\mathbf{q},\mathbf{p})/k_B T} e^{-E(\mathbf{q},\mathbf{p})/k_B T} d\mathbf{q} d\mathbf{p}}{\iint e^{-E(\mathbf{q},\mathbf{p})/k_B T} d\mathbf{q} d\mathbf{p}} \right]
$$

$$
= k_B T \ln \left[\iint e^{E(\mathbf{q},\mathbf{p})/k_B T} P(\mathbf{q},\mathbf{p}) d\mathbf{q} d\mathbf{p} \right] \tag{12.11}
$$

in which case the Helmholtz free energy difference may be computed as

$$
\langle A \rangle_B - \langle A \rangle_A = k_B T \ln \left(\frac{1}{M_B} \sum_i^{M_B} e^{E_i/k_B T} \right) - k_B T \ln \left(\frac{1}{M_A} \sum_i^{M_A} e^{E_i/k_B T} \right)
$$

$$
= k_B T \ln \left\langle e^{E/k_B T} \right\rangle_B - k_B T \ln \left\langle e^{E/k_B T} \right\rangle_A
$$

$$
= k_B T \ln \left(\frac{\left\langle e^{E/k_B T} \right\rangle_B}{\left\langle e^{E/k_B T} \right\rangle_A} \right) \tag{12.12}
$$

At first glance, the situation looks if anything worse than was true for ΔU. Now the ensemble averages are not over the total energies (already large numbers), but over *exponentials* of the total energies expressed in multiples of $k_B T$! However, as long as the two systems A and B do not differ from one another by enormous amounts, the ratio of the two expectation values in the last line on the r.h.s. of Eq. (12.12) (which is the inverse of the equilibrium constant for the reaction A → B) is sufficiently close to unity that errors in the individual ensemble averages on the order of one or two percent have a much smaller impact on the error of the ratio than was the case for ΔU. Moreover, one takes the natural logarithm of the ratio to compute ΔA, so the absolute magnitude of the error is reduced still more.

Nevertheless, converging the individual expectation values to the level of a few percent error is a painfully slow task, since the reduced probabilities of high-energy points are balanced by their exponentially larger contributions to the partition function. However, one may take advantage of the relatively small difference between systems A and B to introduce a further approximation that is extraordinarily useful.

12.2.2 Free-energy Perturbation

If the ensembles in Eq. (12.2), over which the property averages for systems A and B are taken, were somehow to be the same, one would be able to take advantage of the properties of exponentials to write

$$\langle A \rangle_B - \langle A \rangle_A = k_B T \ln \left\langle e^{(E_B - E_A)/k_B T} \right\rangle_A \tag{12.13}$$

where we have arbitrarily chosen to label the ensemble average as having been selected based on system A. This formulation offers some enormous advantages over Eq. (12.12). One of the most important is that all contributions to the energy from solvent–solvent interactions (which are enormously dominant over solvent–solute interactions, since there are so many more solvent molecules) cancel out in the energy difference, since the ensembles are (somehow) identical.

What is meant by an identical ensemble for two different species? It is helpful to return to our specific example of HCN and HNC. To determine the proper identical ensemble for HNC based on one chosen in the usual fashion for HCN, we first stipulate that all particles that are common to the two systems, i.e., all solvent molecules, the carbon atom, and the nitrogen atom, have identical positions and momenta when we evaluate the energy in system B as when we evaluate it in A. Then, the only contribution to the energy *difference* in Eq. (12.13) would be the different interactions that the hydrogen atom has with all of the other atoms, based on whether it is attached to C or N (see Figure 12.1).

So, for each snapshot of the simulation that contributes to the ensemble (by either MC or MD evaluation), we compute the energy differential for all of the atoms interacting with H_B rather than H_A. In Figure 12.1, the particular case of one of the hydrogen atoms on a first-shell water molecule is illustrated. As this is a non-bonded interaction in each case, the contribution from H_D in a simple force field might be

$$\Delta E_{H_D} = \left(\frac{a_{HH}}{r_{H_B H_D}^{12}} - \frac{b_{HH}}{r_{H_B H_D}^6} + \frac{q_{H_B} q_{H_D}}{\varepsilon r_{H_B H_D}} \right) - \left(\frac{a_{HH}}{r_{H_A H_D}^{12}} - \frac{b_{HH}}{r_{H_A H_D}^6} + \frac{q_{H_A} q_{H_D}}{\varepsilon r_{H_A H_D}} \right) \tag{12.14}$$

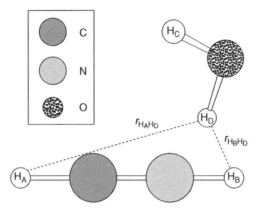

Figure 12.1 HCN(H) and one water molecule from the simulation box. In free energy perturbation, the simulation snapshots would be generated using the HCN system, and the energy difference for H_A being present compared to H_B would then be computed over that ensemble. For example, the differential interaction of the two with H_D would contribute to the full difference

where the functional forms of Eqs. (2.14) and (2.22) have been chosen to compute non-bonded interactions. This protocol defines free energy perturbation (FEP; Zwanzig 1954).

There are, however, potentially rather large problems involved in the scheme outlined thus far. In HCN, for instance, the nitrile lone pair is a fair hydrogen bond acceptor, and one may imagine that a water molecule will often be found hydrogen bonded to it, say at a distance of 2 Å. When a snapshot containing such a hydrogen bonded water is used to generate the HCN ensemble, H_B will be materialized with a normal bond distance to the nitrogen, say 1 Å, and this will create an HH non-bonded interaction of only 1 Å. Such a geometry will be extremely high in energy, so that it should contribute in only the most paltry way to any ensemble average. However, the nature of the HCN system is such that it might be expected to occur with great regularity. The HCN ensemble will therefore be a very poor source for a HNC ensemble, and the free-energy difference computed using Eq. (12.13) will be very bad.

In order to avoid this problem, the switching between the two molecules may be broken up into smaller steps using a coupling parameter λ that may take on values from 0 to 1. We then write the energy of the system as a general function of λ

$$E\left(\lambda\right) = \lambda E_B + (1-\lambda)\, E_A \tag{12.15}$$

which emphasizes that the endpoints are still the physically meaningful ones, but we are willing to consider, computationally at least, chimeric systems having partial HCN and partial HNC character. The utility of such systems is that we may now generalize Eq. (12.13) to

$$\langle A\rangle_B - \langle A\rangle_A = \sum_{\lambda=0}^{1} k_B T \ln \left\langle e^{(E_{\lambda+d\lambda}-E_\lambda)/k_B T}\right\rangle_\lambda \tag{12.16}$$

where λ is broken up into individual intervals of length $d\lambda$ (they need not all have identical widths, but in typical practice they do). Thus, for instance, in the HCN case we might decide to divide λ into 20 segments having a width of 0.05 each. In the first ensemble, generated for $\lambda = 0$, the energy difference associated with interaction with atom H_D would no longer be computed using Eq. (12.14), but instead according to

$$\Delta E_{H_D} = 0.05 \left(\frac{a_{HH}}{r_{H_B H_D}^{12}} - \frac{b_{HH}}{r_{H_B H_D}^{6}} + \frac{q_{H_B} q_{H_D}}{\varepsilon r_{H_B H_D}} \right) + 0.95 \left(\frac{a_{HH}}{r_{H_A H_D}^{12}} - \frac{b_{HH}}{r_{H_A H_D}^{6}} + \frac{q_{H_A} q_{H_D}}{\varepsilon r_{H_A H_D}} \right)$$

$$- \left(\frac{a_{HH}}{r_{H_A H_D}^{12}} - \frac{b_{HH}}{r_{H_A H_D}^{6}} + \frac{q_{H_A} q_{H_D}}{\varepsilon r_{H_A H_D}} \right)$$

$$= 0.05 \left[\left(\frac{a_{HH}}{r_{H_B H_D}^{12}} - \frac{b_{HH}}{r_{H_B H_D}^{6}} + \frac{q_{H_B} q_{H_D}}{\varepsilon r_{H_B H_D}} \right) - \left(\frac{a_{HH}}{r_{H_A H_D}^{12}} - \frac{b_{HH}}{r_{H_A H_D}^{6}} + \frac{q_{H_A} q_{H_D}}{\varepsilon r_{H_A H_D}} \right) \right] \quad (12.17)$$

The effect is to remove 95 percent of the unfavorable consequences of materializing the H_B atom, making the ensemble hopefully more relevant for the chimeric molecule than for 'full' HNC. Once sufficient statistics have been collected for this window, a fresh ensemble is generated using a simulation for the chimera with $\lambda = 0.05$, and evaluating the energy difference between $\lambda = 0.10$ and $\lambda = 0.05$. This process is repeated, interval by interval, until λ reaches 1, at which point all of the Helmholtz free-energy changes for each interval are summed together to give the total for HCN to HNC.

By creating new ensembles with each increase in λ, potentially offending water molecules in the region of the nitrogen atom like the one mentioned above are 'eased' out of the way, since in each new ensemble the presence of H_B becomes more manifest. The cost, however, is that now 20 simulations need to be undertaken instead of one (assuming an interval width of 0.05 as in the example).

When one is generating an ensemble for a fractional value of λ, it is equally easy to evaluate the energy change for $\lambda - d\lambda$ as it is for $\lambda + d\lambda$. The former is equivalent to imagining the reaction not as HCN $\rightarrow$ HNC but rather HNC $\rightarrow$ HCN. Evaluation in this fashion thus simultaneously determines the forward and reverse free-energy changes from the identical ensemble. In principle the free-energy change computed for the interval $[\lambda \rightarrow \lambda + d\lambda]$ should be exactly the opposite of that computed for the interval $[\lambda + d\lambda \rightarrow \lambda]$. In practice, however, this is rarely true, and the variations provide some indication of the potential error in the FEP process. For instance, in Figure 12.2 the reverse mutation predicts a negative free-energy change slightly larger in magnitude than the positive free-energy change for the forward mutation. This difference is sometimes reported as the error in the simulation. Because the free-energy change should be linear in λ if Eq. (12.15) is used (dotted line), the hysteresis of the FEP diagram is sometimes used as a more conservative estimate of the error.

An alternative procedure is known as 'double-wide sampling'. In this case, the ensemble is generated by MC or MD methods for the Hamiltonian corresponding to a given value of λ, but the evaluation of the free energy change is for the interval $[\lambda - 0.5d\lambda \rightarrow \lambda + 0.5d\lambda]$. Thus, the total interval width is still $d\lambda$, but the evaluation is over half-step changes left and right in the Hamiltonian parameters. In principle, this may lead to improved sampling

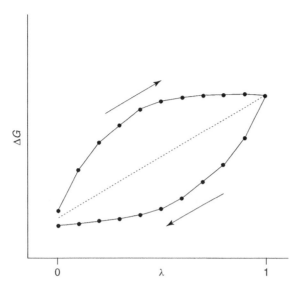

Figure 12.2   A typical FEP diagram showing the free-energy change in the forward (above) and reverse (below) directions for a λ-coupled mutation

because neither endpoint is evaluated using a Hamiltonian that is more than $0.5 d\lambda$ from the Hamiltonian used to generate the ensemble. Further discussion of technical points and error analysis is deferred to Section 12.2.6.

12.2.3   Slow Growth and Thermodynamic Integration

In Eq. (12.16), one may imagine taking λ intervals so small that ΔE on any given interval is arbitrarily close to zero. In that case, we may represent the exponential as a truncated power series, deriving

$$\langle A \rangle_B - \langle A \rangle_A = \lim_{d\lambda \to 0} \sum_{\lambda=0}^{1} k_B T \ln\left\langle 1 + \frac{(E_{\lambda+d\lambda} - E_\lambda)}{k_B T} \right\rangle_\lambda \qquad (12.18)$$

This expression may be further simplified by noting that $\ln(1 + x)$ is well approximated by x for sufficiently small values of x, so that we may write

$$\langle A \rangle_B - \langle A \rangle_A = \lim_{d\lambda \to 0} \sum_{\lambda=0}^{1} k_B T \left\langle \frac{(E_{\lambda+d\lambda} - E_\lambda)}{k_B T} \right\rangle_\lambda$$

$$= \lim_{d\lambda \to 0} \sum_{\lambda=0}^{1} \langle (E_{\lambda+d\lambda} - E_\lambda) \rangle_\lambda$$

$$= \lim_{d\lambda \to 0} \sum_{\lambda=0}^{1} (E_{\lambda+d\lambda} - E_\lambda) \qquad (12.19)$$

The removal of the ensemble average over the λ ensemble in the final line on the r.h.s. reflects the protocol of this technique, the so-called slow-growth method. It is assumed that if the Hamiltonian is infinitesimally perturbed at every step in the simulation, then the system will constantly be at equilibrium (following some initial period of equilibration), so separate ensemble averages need not be acquired.

In practice, then, the slow-growth technique is rather different from FEP when it comes to evaluating ΔE. Since each change in λ is also a step in the simulation, all of the intra-solvent energy terms change in addition to the solvent–solute interaction terms. With respect to the latter terms, however, the evaluation is similar to FEP in that chimeric molecules are involved.

A third simulation protocol for determining Helmholtz free-energy differences can be illustrated from further manipulation of Eq. (12.19). Thus we may write

$$\langle A \rangle_B - \langle A \rangle_A = \lim_{d\lambda \to 0} \sum_{\lambda=0}^{1} \langle (E_{\lambda+d\lambda} - E_\lambda) \rangle_\lambda$$

$$= \lim_{\Delta\lambda \to 0} \sum_{\lambda=0}^{1} \left\langle \frac{(E_{\lambda+\Delta\lambda} - E_\lambda)}{\Delta\lambda} \right\rangle_\lambda \Delta\lambda$$

$$= \int_0^1 \left\langle \frac{\partial E}{\partial \lambda} \right\rangle_\lambda d\lambda$$

$$\approx \sum_{\lambda=0}^{1} \left\langle \frac{\partial E}{\partial \lambda} \right\rangle_\lambda \Delta\lambda \tag{12.20}$$

where we first recognize the calculus relationship between the sum appearing on the r.h.s. in the second line and the definite integral in the third line (and simultaneously the definition of the partial derivative), and we then approximate the definite integral as a sum over small intervals. While the transformation from line 3 to line 4 may appear to simply reverse the transformation from line 2 to line 3, this is not the case, because the partial derivative remains in its analytic form; this is possible because most simulations evaluate the energy using $E(\lambda)$ functions that are trivially differentiated. Moreover, $\Delta\lambda$ in the final line is no longer infinitesimally small, i.e., this is a standard estimation of an integral by division of the integration range into discrete intervals with the function approximated over each interval by a single value, in this case the value at the start of the interval. This process defines the thermodynamic integration (TI) method. [TI can be derived in a much more rigorous and general way, and indeed, FEP may be regarded as a special case of TI; interested readers are referred to the bibliography at the end of the chapter.]

It is evident that TI and FEP are similar in that they involve multiple simulations over different windows $\Delta\lambda$, with accuracy expected to increase when more and smaller windows are employed. However, there are key differences as well. In TI, the ensemble average for one value of λ is not used to evaluate any energies involving a different value of λ; only the ensemble average of the energy derivative is accumulated. Moreover, different forms of $E(\lambda)$ may be conveniently evaluated, corresponding to different mutation paths from A to B. For example, one may choose the generalization of Eq. (12.15)

$$E(\lambda) = \lambda^n E_B + (1 - \lambda)^n E_A \qquad (12.21)$$

where n is an arbitrary exponent that may be freely chosen. [Note that for both FEP and TI, one may also couple λ more intimately into individual force-field terms; the only requirement is that the correct limits be maintained, i.e., $E(\lambda = 0) = E_A$ and $E(\lambda = 1) = E_B$.]

12.2.4 Free-energy Cycles

The discussion thus far has ignored certain rather tricky technical issues as well as certain very real practical difficulties that can arise in various types of simulations. Often, these problems can be avoided by the invocation of a free-energy cycle. For instance, Jorgensen and Ravimohan (1985) invoked such a cycle to study the difference in the free energies of aqueous solvation for methanol and ethane (Figure 12.3). The calculation of the absolute free energies of solvation for each of these two molecules would be subject to large errors, because the necessary perturbation would involve growing the molecules from 'nothing' both in the gas phase and in a box of water. While the former is a trivial exercise, the introduction of a solute into an equilibrated water box is a very difficult affair because no matter how small the first $d\lambda$ step is taken to be, there is a strong possibility of introducing the solute atoms into regions that result in unphysically high energies, thereby generating a poor sample. The difference between two solvation energies each with high associated errors would then have a still higher error, and might not be particularly useful as a result (for recent advances in addressing the challenge of computing absolute solvation free energies, see, for example, Åberg et al. 2004). Shirts et al. (2003) have demonstrated that the opposite process, i.e., disappearing a solute molecule from a water box, can be more useful for computing absolute solvation free energies, but a substantial commitment of computational resources is still required.

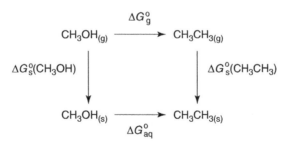

Figure 12.3 The vertical sides of this free-energy cycle correspond to free energies of aqueous solvation, while the horizontal sides correspond to chemical mutations that are not physically realistic but are accessible by FEP. The difference between the two vertical quantities must be equal to the difference between the two horizontal quantities. While the former difference is easier to measure, the latter is easier to compute

However, if one is indeed interested primarily in the *difference* in the solvation free energies, and not the absolute values, one can carry out the necessary two simulations in a completely different fashion. Instead of growing each solute molecule from nothing, one is transformed into the other using a chimeric approach (Jorgensen and Ravimohan used FEP). By the state-function nature of the free energy, the difference in the transmutation free energies in the gas phase and in aqueous solution must be equal to the difference in the absolute solvation free energies. A single transmutation in water is far simpler to carry out with good statistical accuracy than two separate 'creations' of solutes, and again the gas-phase mutation is trivial. Representing the solute molecules using the OPLS force field, Jorgensen and Ravimohan determined an aqueous solvation free-energy difference of 6.75 ± 0.2 kcal mol^{-1}, which is in good agreement with the experimental value of 6.93 kcal mol^{-1}.

Note that, in principle, one could use FEP to determine a 'web' of solvation free-energy differences between many different substrates, and then carry out a single calculation of an absolute free energy of solvation (i.e., growing one solute molecule from nothing) that would serve as an anchor to convert all of the relative free energies of solvation into absolute free energies of solvation (for the example of a set of substituted benzenes in water, and a comparison to predictions from the SM2 continuum model, see Jorgensen and Nguyen 1993).

Because they used a free energy cycle, Jorgensen and Ravimohan assumed that changes in the kinetic energy component of the mutation would cancel in the gas phase and in solution, so they did not compute them, i.e., they reduced the size of the phase-space problem for Monte Carlo sampling by a factor of 2 by removing all momentum degrees of freedom. This simplifying assumption remains standard in modern calculations. This is true for constant temperature MD simulations as well, since scaling the velocities to maintain temperature necessarily distorts the momentum sampling – modern simulations typically evaluate only the potential energy differences between mutated structures.

Free-energy cycles can be used to simplify simulations covering a wide variety of processes. For example, if we were interested not in the difference in free energies of solvation of methanol and ethane, but the difference in their partitioning between water and octanol, we can simply rewrite Figure 12.3 so the upper leg is in octanol instead of water, and carry out the same mutation of methanol to ethane described above, now once in octanol and once in water, to determine the free energy difference. The alternative procedures for analyzing the vertical legs would be very unpleasant indeed: we would either have to grow each solute from nothing in *two* different solvents, or we would have to mutate one *solvent* into another, which would be even worse.

A free-energy cycle finding particularly widespread use is one for evaluating differences in interactions between enzymes (or other molecular hosts) and alternative molecules in their active sites. By mutating one substrate into another, both in the presence of the enzyme and isolated in solution, differences in free energies of binding may be determined (Figure 12.4). An example is provided in Section 12.6 as a case study.

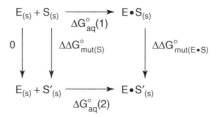

Figure 12.4 Differential binding free-energy cycle. The difference in binding free energies for two different substrates, S and S′, is equal to the difference in mutation free energies for changing S into S′ in solution, and E•S into E•S′ in solution. The leftmost vertical free-energy change is zero, since the free enzyme is a constant independent of substrate

12.2.5 Potentials of Mean Force

When free energy is expressed as a function of coordinate, it is referred to as a potential of mean force (PMF). The PMF W can be determined as

$$W(q) = -k_{\mathrm{B}}T \ln \pi(q) \tag{12.22}$$

where q is the coordinate, and π is the probability of the coordinate taking on a particular value, i.e.,

$$\pi(q) = Q^{-1} \int\!\!\int \delta[q'(\mathbf{q}) - q] e^{-E(\mathbf{q},\mathbf{p})/k_{\mathrm{B}}T} d\mathbf{q} d\mathbf{p} \tag{12.23}$$

where Q is the (normalizing) full partition function, δ is the Dirac delta function, and q' is the value of the PMF coordinate for any arbitrary point in phase space having positional coordinates $\mathbf{q}$.

In practice, one may evaluate these probabilities following a histogram approach like those outlined in Chapter 3. Over the course of a MC or MD simulation, the value of q' is collected and binned, and the probability of different ranges of values can be determined upon completion of the simulation based on the number of points in a bin compared to the total number of points. For example, we might be interested in the PMF for rotation about the C–O bond in fluoromethanol (see Figure 2.3). Over the course of a simulation, the torsional angle would be saved at every step, and with good sampling a probability histogram would permit conversion to a PMF accurately reflecting the true potential. In the case of fluoromethanol, the difference in energy between the lowest and highest points on the potential energy curve is about 3 kcal mol^{-1}. At 298 K, we would thus expect to sample points in the highest energy region about 100 times less frequently than points in the lowest energy region. Of course, if the width of a bin is, say, one degree, there are many other possibilities for bins to fill, and ultimately roughly one point in every 10 000 or so would be statistically expected to fall into the highest energy bin. To obtain reliable statistics, we might want this least populated bin to contain at least 100 points, so we would require a sample of some 1 000 000 snapshots. An ensemble of this size is accessible with current computational technology, but represents a reasonably significant investment of resources.

Now, consider if the highest energy point on the curve were to be 6 kcal mol^{-1} above the lowest at 298 K. Because the probability involves the exponential of the energy difference, doubling the difference squares the sampling ratio (i.e., the highest energy region is now sampled 10 000 times less frequently than the lowest energy region). Obtaining a statistically meaningful sample of low probability regions now becomes a significantly more difficult prospect, and statistically reliable PMFs cannot be obtained in this fashion.

The problem of low probability regions is even more severe when it comes to chemical reaction coordinates, where free energies of activation for chemically viable processes may range well above 20 kcal mol^{-1}. The probability of obtaining a snapshot in the region of a transition state structure having so high an energy (assuming for the moment that we have some Hamiltonian capable of describing bond-making/bond-breaking) is so remote that no brute force simulation can legitimately expect to capture even one relevant point, much less a statistically meaningful sample. This is the problem of sampling 'rare events'.

One approach to overcoming this problem is to apply a so-called 'umbrella potential' or biasing potential. This potential, a function of the coordinate of interest q, is added to the force-field energy with the aim of forcing q to be sampled heavily within a certain range of values that would not otherwise be statistically accessible. An ideal umbrella potential is one that is the exact negative of the PMF, since then the probability of sampling any value of q should be uniform. However, one rarely knows the PMF ahead of time (otherwise why would one be trying to calculate it?), so instead one typically applies rather simple biasing potentials (e.g., a quadratic potential) to force q to be sampled over some interval including a particular value q_0.

Consider, for instance, the S_N2 reaction of Br$^-$ with CH$_3$Br in aqueous solution, which has an activation free energy on the order of 20 kcal mol^{-1}. If we define our reaction coordinate as

$$q = r_{C-Br_A} - r_{C-Br_B} \qquad (12.24)$$

where A and B are the incoming and outgoing bromide ions, respectively, we see that the reactants correspond to large positive values of q, products to large negative values of q, and from our knowledge of bimolecular nucleophilic substitution reactions, we know that the transition state region will have values of q very near zero. Let us assume that we have a force field that provides an accurate potential energy curve in the gas phase for this S_N2 process – in spite of this, in a normal MC or MD simulation in a box of water we would be very unlikely to sample in regions anywhere near the TS because of the very low probabilities associated with such high-energy structures. However, if we apply biasing potentials of the form

$$U(q) = \frac{1}{2}k(q - q_0)^2 \qquad (12.25)$$

where q_0 is the particular value near which we want to sample, and we select the force constant k to be suitably large, we can ensure that the simulation will sample heavily within some distance of q_0, since structures having values of q significantly different from q_0 will be heavily penalized by addition of U.

When this procedure is followed, a different probability function $\pi^*(q)$ will be obtained over the sampled region. The *correct* PMF (i.e., for the *un*biased potential) is related to the new probability function according to

$$W(q) = -k_BT \ln \pi^*(q) - U(q) - k_BT \ln \langle e^{-U(q)/k_BT} \rangle_* \qquad (12.26)$$

where the ensemble average is accumulated with the biasing function added to the system Hamiltonian. This function is quite simple to evaluate for a typical selection of U. However, it is often the case for an unknown PMF that a single choice of functional form for U will not lead to a statistically useful sample over the entire range of interest for q. Instead, one carries out several simulations, with different choices for U (for instance, by varying choice of q_0 in Eq. (12.25)), and then patches together the relevant regions of the PMF to generate a single curve. This process is illustrated in Figure 12.5. Obtaining a good overlap of the individual pieces can be difficult in some instances, which contributes to error in the method when overlap is required. Indeed, Figure 12.5 is somewhat misleading, since each individual PMF fragment actually rises to infinitely positive free energy values at either end (that is, the probability of finding the system far to the right or left becomes so small that the corresponding free energy is very large). As these PMF walls have no physical meaning, but are artifacts of the umbrella function, they have been left out of Figure 12.5 for clarity, but in practice they can add to the difficulty associated with reliably overlapping different segments of the full reaction coordinate. The weighted histogram analysis method (WHAM;

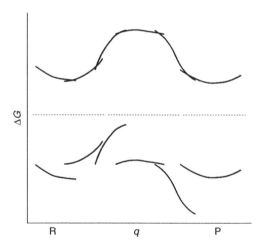

Figure 12.5 A reaction coordinate q constructed piecewise from reactants R to products P as a series of PMFs determined using different umbrella functions. The individual PMFs determined using Eq. (12.26), shown below the dashed line and taking each left endpoint as the relative zero, are held within their respective regions of the reaction coordinate by the umbrella function. Their overlap on a common energy scale generates the complete PMF shown above the dashed line

Kumar *et al.* 1992) is one of the more popular approaches for accomplishing this overlap; the details of WHAM, however, are beyond the scope of this text.

An alternative to extracting the proper PMF from one generated using a biasing potential is to employ the so-called constraint-force method. In this model, one or more degrees of freedom are held to a series of N fixed values (for simplicity we will continue to work with only one dimension q ranging then from q_1 to q_N). For a given fixed value q_i, with this value differing from q_{i+1} by a small amount Δq_i, the value of $\partial W / \partial q$ is evaluated. Once all average derivative values are in hand, it is a simple matter to reconstruct W by numerical integration, i.e.,

$$W(q_j) = \int_{q_{min}}^{q_j} \frac{\partial W}{\partial q} dq$$

$$\approx W(q_1) + \sum_{i=1}^{j-1} \frac{\partial W(q_i)}{\partial q} \Delta q_i \tag{12.27}$$

If readers fail to find inspection of Eqs. (12.22) and (12.23) particularly enlightening with respect to how precisely to evaluate $\partial W / \partial q$, they may consider themselves to be in good company. After substantial debate in the literature, the proper and rather complicated approaches to computing this quantity for the one- (den Otter and Briels 1998) and multidimensional (den Otter and Briels 2000) cases have been derived. In addition, Darve and Pohorille (2001) have described a generalization of this approach in which simulations may be run without the imposition of *any* constraints. In that case, a biasing potential still needs to be applied globally so that the system samples q fully, but the numerical integration of Eq. (12.27) avoids the problem of overlapping partial PMFs illustrated in Figure 12.5.

Rosso *et al.* (2002) have proposed an alternative sampling method in which the likelihood of rare event observation is enhanced by separating the reactive coordinate(s) from the remaining degrees of freedom and propagating the former components of the trajectory at high temperature with a fictitiously high mass. This combination permits the other degrees of freedom to respond adiabatically to the reactive coordinate(s), which are themselves able to generate a more complete unbiased free energy profile by virtue of the high temperature.

Irrespective of the protocol used for enhanced sampling, a key difficulty arises when the reaction mechanism is not well understood. In that case, even the definition of the reaction coordinate q can be problematic. This is a common problem in the simulation of enzyme active sites, where bond-forming or bond-breaking reactions may or may not occur with simultaneous proton transfer(s) between enzyme and substrate functional groups. In the event of multiple bond-making/bond-breaking events occurring simultaneously, it becomes quite difficult to construct suitable one-dimensional slices and biasing potentials through phase space that permit generation of useful PMFs.

In sum, the generation of accurate PMFs from probability distributions for processes with free energies of activation in excess of a few kilocalories per mole continues to be a significant challenge for modern simulation methods. Some alternative approaches, using both continuum and explicit solvation models, are discussed in Section 15.4.

12.2.6 Technical Issues and Error Analysis

Free-energy simulations are extremely demanding in a technical sense, and it is well beyond the scope of this book to fully prepare readers to apply the technology without further instruction. Nevertheless, there are a few technical issues that arise (on top of those already discussed for simulations in general in Section 3.6) that merit attention insofar as they affect many published free-energy simulation results. Much more authoritative treatments are available in the bibliography and suggested reading.

When perturbations from one molecule to another are carried out, there are two distinct approaches that may be taken. The 'single topology' approach involves a single solute species that is smoothly transformed from the first molecule to the second as a function of λ. In the HCN/HNC example above, the single topology approach would involve not only the steady disappearance of the carbon-bound hydrogen and the appearance of the nitrogen-bound hydrogen, but also any change in the C–N equilibrium length and force constant as it transforms from a nitrile to an isonitrile bond type. In addition, if the atomic partial charges on C and N were to be different for nitriles and isonitriles, these too would change as a function of λ. The solute molecule at intermediate values of λ is thus truly chimeric.

The 'dual topology' approach, on the other hand, involves having the distinct initial and final solutes simultaneously present, but no force-field interactions between the two are ever calculated. The interactions of both *are* calculated with the surrounding medium in the normal way, but at intermediate values of λ the total energy of the system will be derived as a λ-dependent function of the two. The dual topology approach is simpler in implementation but problems can arise if the two topologies drift away from one another during the course of the simulation (for instance, if one solute were to leave the active site of an enzyme while the other stayed in it, obviously the difference in binding free energies would not be calculable). Both single and dual topology calculations continue to see about equal use.

As already mentioned above, the sudden appearance of atoms at positions in space occupied by solvent molecules as the result of a mutation can lead to severe sampling problems. As a rule, changes in van der Waals interactions must be introduced much more slowly than changes in charge in order to maintain good equilibrium in ensemble averages. Since a free-energy change is independent of the mutation path (assuming perfect sampling), paths that carry out changes in charges more quickly than changes in van der Waals interactions are not uncommon.

The discussion in this chapter has focused almost exclusively on computing changes in the Helmholtz free energy A. However, most experimental measurements are carried out at constant pressure, not constant volume, so the majority of thermochemical data is in the form of Gibbs free energies G. As long as the total number of particles in a free-energy simulation remains constant, almost all simulations assume that ΔPV is zero, in which case the Gibbs and Helmholtz free energy changes are identical (this is readily derived from Eqs. (12.2)–(12.7)). When this is not the case, the additional contributions to G must be explicitly accounted for.

Of the three methods discussed above, FEP, TI, and slow growth, the first two see far more application than the third. The slow-growth condition, that the system is constantly at, or at least very, very near equilibrium, is quite hard to maintain over the course of a mutation. In principle, such an equilibrium can be obtained with sufficiently small changes in λ at each step, but in practice, numerical limitations in computing energy differences as a function of λ place a lower limit on that increment (as, of course, does the requirement that the total number of steps required not exceed available computational resources). Steps that are too large, on the other hand, lead to chaos, since high-energy interactions run the simulation into unrepresentative regions of phase space from which the system is unlikely ever to return.

In formalism, many aspects of free-energy simulations lend themselves more to implementation within a Monte Carlo sampling scheme than within a molecular dynamics scheme. Unfortunately, MC schemes applied to large flexible molecules (e.g., proteins) tend to be very inefficient, since most proposed moves of the large molecule are rejected as being too energetically unreasonable, so MD simulations remain the standard. Innovative attempts to combine some of the best features of both have been described, as already noted in Chapter 3.

Possibly the most vexing aspect of free energy calculations is that, as with most simulations of sufficiently complex systems, meaningful error analysis is almost impossible. The difficulty of demonstrating a properly converged sampling of phase space for any simulation has already been amply discussed in Section 3.6.4. Given that a free-energy simulation typically comprises some 10 or more individual simulations, whose errors may be expected to be highly correlated and the results from which are pieced together, uncertainty only grows. The cost of the simulations is such that they are rarely carried out more than once (using, for example, different starting conditions) to assess the statistical reliability of the free-energy change.

As a result, many free-energy simulations are carried out not necessarily to predict a specific value but rather to demonstrate agreement with experiment, after which interpretation of the simulation results can be carried out with enhanced confidence to understand *why* the free-energy change is what it is. This process in itself can be quite ambiguous, however. A typical analysis involves decomposing the free-energy change into constituent components, i.e., changes in electrostatic interactions, van der Waals interactions, bond torsion contributions, etc. However, while the total free-energy change is path independent, the changes in the components are *not*, so such a decomposition must be interpreted with caution.

12.3 Other Thermodynamic Properties

Properties other than free-energy changes are usually considerably more difficult to evaluate to an equivalent level of accuracy. One approach is simply to attempt a brute force calculation for different systems analogous to that outlined for U in Eqs. (12.9) and (12.10). However, this approach has little value in any but the simplest of systems owing to the large uncertainties in the absolute values of the thermodynamic quantities.

Another approach is to carry out free-energy simulations at several different temperatures, and then construct the equivalent of a van't Hoff plot to separate, say, the enthalpic and entropic contributions to the free energy. This approach is obviously extraordinarily demanding of resources, since every temperature point requires a new free-energy simulation, and unless there are many points, the error in the temperature dependence of the free energy determined by linear regression of the latter on the former may be rather large.

In some cases, it is possible to take advantage of various thermodynamic relationships to write some property as a fluctuation-dependent quantity. Thus, for example, the entropy change may be computed from

$$\langle S \rangle_B - \langle S \rangle_A = \int_0^1 \left(\langle E \rangle_\lambda \left\langle \frac{\partial E}{\partial \lambda} \right\rangle_\lambda - \left\langle E \frac{\partial E}{\partial \lambda} \right\rangle_\lambda \right) d\lambda \qquad (12.28)$$

evaluating the integral numerically, as is done for standard TI. Peter *et al.* (2004) have carried out a more detailed analysis of the accuracy and convergence of various approaches for computing entropy.

Similarly, the absolute constant-volume heat capacity may be computed as

$$\langle C_V \rangle = \frac{1}{k_B T^2} \left(\langle E^2 \rangle - \langle E \rangle^2 \right) \qquad (12.29)$$

As a rule, fluctuations are much slower to converge statistically than are the properties that are fluctuating, so analyses of Eqs. (12.28) and (12.29) require very long simulation times. Of course, Eq. (12.29) is simpler to evaluate than Eq. (12.28) since the latter does not involve a perturbation of one system into another.

12.4 Solvent Models

If a solvent is to be considered as 'explicitly' present in a simulation, obviously there must be some atomistic manner in which it is represented in the energy expression – this being the fundamental distinction from a continuum solvation model. However, since the solvent molecules greatly outnumber the solute molecule(s), there are advantages of efficiency that accrue from adopting as simple a representation as possible, and that is reflected in many of the solvent models in common use.

12.4.1 Classical Models

The simplest model for a solvent molecule is clearly one that is molecular-mechanics-like. That being said, various levels of complexity remain even within the choice of a classical representation. Of all possible solvents of interest to chemists, water is arguably the most important, and not surprisingly it has spawned the largest number of models. Besides differing in parameter values, the various classical models differ in the total number of interacting sites. Probably the simplest possible model for water is to treat it as a Lennard–Jones sphere, inside which two charges are embedded of equal magnitude and opposite sign to mimic water's dipole moment. A solute molecule thus sees three interaction sites: the center of the sphere characterized by characteristic ε and σ values (see Eqs. (2.16), (2.30), and (2.31)), and the two point charges. (An alternative model would be to put a point dipole at the center of the sphere, but the evaluation of dipole–dipole interactions is sufficiently more time-consuming than that of charge–charge interactions that there is no real simplification inherent in this approach.)

A slightly more complex representation is to put equal positive atomic charges on the hydrogen atoms and a negative charge on the symmetry axis, or equal negative charges in the lone pair regions, again to mimic water's dipole moment, but also to better represent its overall charge distribution. Such very simple models, with careful parameterization, do remarkably well in reproducing many properties of liquid water, e.g., bulk density, heat of vaporization, compressibility, heat capacity, etc. The most successful models along these lines, that are widely used in modern simulations, are the transferable-intermolecular-potentials-3- and -4-point-charge water models (TIP3P and TIP4P; Jorgensen *et al.* 1983). The similarly designed SPC (simple point charge) water model also continues to see modern use (Berendsen *et al.* 1981) including forms recently modified to improve its dielectric and diffusive properties (Glattli, Daura, and van Gunsteren 2003).

For non-aqueous solvents, the approximation of the solvent as a LJ sphere is usually less practical. However, substantial time savings can be realized by employing a united-atom approach for carbon atoms and their attached hydrogens. The most complete parameterization of organic solvents has been accomplished as part of the OPLS force field, including *inter alia* alkanes, aromatics, carbon tetrachloride, chloroform, furan, *n*-octanol, and pyrrole, many in both UA and AA representations (see Table 2.1 and also Jorgensen, Briggs, and Contreras 1990; Kaminski *et al.* 1994; McDonald and Jorgensen 1998). In these cases, as for water, solvent parameters were optimized based on comparison of bulk solvent properties to experimental measurements.

At the next level of complexity, the polarity of solvent models, as made manifest by their atomic partial charges, can be augmented with a polarizability. This allows the solvent molecule to respond to its surroundings in a fashion conceptually similar to the electronic component of the solvent polarization described in Section 11.1.1. Typically a polarizability tensor $\overline{\overline{\alpha}}$ is assigned either to the solvent molecule as a whole or to individual atoms. Then, the induced dipole moment at each polarizable position can be determined from

$$\mu^{\text{ind}} = \overline{\overline{\alpha}}\mathbf{E} \tag{12.30}$$

where $\mathbf{E}$ is the total electric field arising from all of the atomic point charges *and* all of the induced dipoles. Thus, μ^{ind} must be determined iteratively, with convergence potentially being problematic. Once converged, the additional contribution to the total electrostatic energy from the charge-induced dipole interactions can be computed according to

$$V = \frac{1}{2}\sum_i\sum_j \frac{q_i\mu_j^{\text{ind}}\cdot\mathbf{r}_{ij}}{r_{ij}^3} \tag{12.31}$$

where i runs over charge sites and j runs over polarizability sites and r is the intersite distance. In addition, induced-dipole–induced-dipole interactions contribute according to Eq. (2.23).

Owing to its particular importance, polarizable solvent models have largely been restricted to water, for which a sizable number have been developed (see, for example, Dang 1992; Rick, Stuart, and Berne 1994; Bernardo *et al.* 1994; Zhu and Wong 1994; Lefohn, Ovchinnikov, and Voth 2001). Because evaluating the terms deriving from solvent polarizability increases the amount of time required for a simulation by roughly an order of magnitude, the use of polarizable solvents has been primarily either for technical comparisons in model development, or for the simulation of particularly simple systems, where convergence for a given property of interest may be expected to occur quickly. Developers tend to focus on properties for which non-polarizable water models do poorly, e.g., the density anomaly in water where below 4 °C the liquid density begins to decrease with decreasing temperature. However, the failures of prior models to function well for such properties is not necessarily intrinsic, but may simply reflect a failure to have considered the property in the development of the non-polarizable model (Mahoney and Jorgensen 2000). More recent work with polarizable acetonitrile and acetone solvent models has indicated, not surprisingly, that polarizability critically improves the description of solvation structures and interaction energies associated with the solvation of monatomic ions in these solvents (Fischer *et al.* 2002).

A yet more complete but still formally classical solvent model has been developed for use when the solute is represented quantum mechanically. The electrostatic interactions between a classical solvent and a quantum mechanical solute are relatively simple to represent, and are discussed in detail in the next chapter on mixed QM/MM methods. The non-bonded interactions are somewhat more challenging. Gordon *et al.* (2001) have described an approach that they call the effective fragment potential (EFP) method that, by analogy to ECPs, replaces the direct computation of dispersion and exchange-repulsion interactions between solute and solvent electrons by an interaction between solute electrons and a solvent pseudopotential. The solvent pseudopotentials (and the representation of its electrostatic distribution and polarizability) are determined parametrically in order to create a transferable solvent model especially suitable for use in QM/MM calculations using HF theory as the QM component. The EFP model has since been extended to DFT as the QM component as well (Adamovic, Freitag, and Gordon 2003).

12.4.2 Quantal Models

When one refers to a quantum mechanical solvent model, the word 'model' reverts to its usual sense in the context of QM methods: it is the level of electronic structure theory used to describe the solvent. Thus, there is no real distinction between the solvent and the solute in terms of computational technology – the wave function for the complete supersystem (or the DFT equivalent) is computed without resort to methodological approximations beyond those inherent to the level of electronic structure theory. To avoid problems with basis-set imbalances, one might expect calculations representing the solvent in a fully QM fashion to employ a common level of theory for all particles, but this does not *have* to be the case.

At several points in this book, it has been emphasized that the prevalence of classical MC and MD simulations derives from the impracticality of carrying out fully QM dynamics. While this is largely true, for systems of only modest size where short trajectories may be profitably analyzed, fully QM MD simulations using the so-called Car–Parrinello technique are a viable option (Car and Parrinello 1985). In its most widely used formulation, the Car–Parrinello method employs DFT as the electronic-structure method of choice. In principle, every MD step should involve taking a phase point, computing the energy and the gradients for that point given the nuclear positions, and propagating a short time step prior to repeating this process. This formalism is extremely time-consuming. Car and Parrinello showed, however, that one does not need to fully converge the KS wave function at every step. Instead, the KS MO coefficients are treated as dynamical variables. That is, they are assigned a fictitious mass and have their 'coordinates' added to the usual $3N$ positional dimensions of phase space. By careful choice of the masses for the electronic degrees of freedom, and the time steps for the electronic and nuclear movements, it is possible to obtain a relevant MD sampling of phase space in favorable systems. To further increase speed, the method usually uses a plane-wave basis set, which is ideally suited to the periodic boundary conditions usually imposed in a condensed-phase simulation and allows fast Fourier transform methods to facilitate solution of the SCF equations.

The obvious advantage of a fully QM solvent representation is that intimate solvent participation in reactions, say as a proton donor or acceptor, or simply a charge-transfer partner with the solute, is handled entirely naturally. With improved DFT functionals and ever-increasing computer speeds, this method holds great promise for the future, although it is still sufficiently time-consuming that present day applications remain somewhat limited.

12.5 Relative Merits of Explicit and Implicit Solvent Models

The fundamental difference between the explicit and implicit solvent models is not that one has solvent and the other does not. Rather, the difference is that the implicit model employs a homogeneous medium to represent the solvent where the explicit model uses atomistically represented molecules. While the latter choice is clearly the more physically realistic, the practical limitations imposed by explicit representation dictate that it is not necessarily the best choice for a given problem of interest. This section compares and contrasts the relative strengths and weaknesses of the two models, including some illustrative applications.

12.5.1 Analysis of Solvation Shell Structure and Energetics

A reaction that has received a substantial amount of study using a variety of alternative solvent (and solute) models is the Claisen rearrangement, a [3,3] sigmatropic shift that converts an allyl vinyl ether into a γ,δ-unsaturated aldehyde (Figure 11.5). The motivation for its study has been two-fold. First, the conversion of chorismate to prephenate, which is the first committed step in the biosynthesis of aromatic amino acids in plants, involves an enzyme-catalyzed Claisen rearrangement. Secondly, although pericyclic reactions are conventionally thought of as being relatively insensitive to solvent effects, the rate acceleration for rearrangement of the parent allyl vinyl ether comparing the gas phase to aqueous solution has been estimated to be on the order of 1000-fold. How does aqueous solvation effect this large rate acceleration?

Storer et al. (1994) employed a SMx GB continuum solvent model to investigate this question. Because of the efficiency of the continuum model, they were able to examine various levels of electronic-structure theory in assessing the influence of aqueous solvation on the reaction coordinate. Their key findings were that (i) the TS structure was significantly better solvated than the reactant, primarily because of increased polarization free energy associated with increased polarity in the rearranging fragments and (ii) the solvation effect favored a change in the TS structure so that the rearranging fragments were separated by a larger distance, thus enhancing its polarity (Figure 11.5). These effects, combined with a very small hydrophobic acceleration associated with the two hydrocarbon termini coming together in the forming C–C bond, were sufficient to account for the full range of aqueous acceleration inferred experimentally.

By way of contrast, Severance and Jorgensen (1992) addressed the same problem using an explicit solvent model. In particular, they first generated the intrinsic reaction coordinate (a concept explained more fully in Chapter 15) for the Claisen rearrangement in the gas phase. They then selected specific structures along the reaction coordinate to serve as intermediates in a free-energy simulation using FEP methods. The rigid solute structures were treated classically using OPLS non-bonded parameters and ESP charges determined from gas-phase HF/6-31G(d) calculations, and the aqueous solvent was modeled with the TIP4P water model. By analyzing free energy changes as λ perturbed from one structure to the next along the reaction coordinate using MC simulations, they determined a rate acceleration in good agreement with that inferred experimentally. An investigation of the factors causing acceleration included an analysis of the radial distribution functions of water about the solute oxygen atom. The simulations indicated that the average number of hydrogen bonds to the reactant's ether-like oxygen atom was slightly in excess of one, while the number to the TS oxygen atom was closer to two. Moreover, the strengths of the solute–water interactions for hydrogen bonded waters were greater in the TS structure than in the reactant.

Thus, both studies came to similar conclusions with respect to the source of acceleration: greater polarity of the TS structure contributing to stronger aqueous solvation. However, the 'language' of the continuum model restricts the expression of that result to broader electrostatic terms while the explicit nature of the simulation permits a more fine-grained analysis that illustrates how improved hydrogen bonding is a part of the electrostatic component. In terms of describing the reaction path, the explicit model restricted itself to an analysis of the solvation of the *gas-phase* reaction coordinate. The continuum model, on the other hand, considered movement off the reaction coordinate and found that to be important in stabilizing the TS structure. Furthermore, the SCRF nature of the continuum model allowed for an analysis of the relative polarizability of the TS compared to the reactant, and the TS was found to be considerably more polarizable, again contributing to its improved solvation. (The use of HF/6-31G(d) ESP charges in the explicit simulation was motivated in part because the known systematic tendency for these charges to be too large in the gas phase may be taken as a compensating error for not including the effect of solvation on the electronic structure.)

One key issue, then, in deciding upon what type of solvation model to employ is the level of detail in solvent structure that is of interest to the researcher. An important point to make in this regard is that some solvent molecules really should not be thought of as solvent *per se*. For instance, various inhibitors of HIV-1 protease are known to bind strongly to the enzyme's active site only because there is an accompanying water molecule *also* bound in the site. Such a water cannot be considered simply to be a bulk region having a high dielectric constant; rather, it is more a component of a supermolecular solute. Such solvent molecules are something of a technical challenge for both kinds of solvent models. A continuum model may function perfectly well if the special solvent molecule(s) are included explicitly, but one needs to know ahead of time to include them. Similarly, in an explicit simulation, if they are not placed in the appropriate location, the timescale for entry from the bulk solution may be such that no solvent molecule ever occupies the correct position throughout the length of the simulation, in which case the results simply represent an unbalanced sampling of phase space and are not useful.

Interestingly, Morreale *et al.* (2003) have shown that when solvation free energies are decomposed into atom/group-specific contributions, there is in general good agreement between results obtained from continuum and explicit-solvent calculations. This observation suggests that future analyses of such fragment solvation free energies may assist in comparison of results from disparate solvation protocols.

12.5.2   Speed/Efficiency

For equivalent levels of theory used to represent the solute, continuum solvation models are inevitably several orders of magnitude faster for the analysis of solvation free energies than are explicit solvent models. As a rule, at the QM level the SCRF portion of a continuum solvation calculation usually adds no more than 15 percent or so to the total time of an SCF calculation; at the MM level a factor of 2 or so is involved. Obviously, however, there is no particular virtue in the speed with which a wrong answer (or one that fails to answer the question of interest) may be obtained. Thus, in those instances where an explicit model is called for, time must be invested in the calculation. The variety of system sizes that may be envisioned precludes generalizations about time requirements, but the current state of the art for simulations of solvated biomolecules having molecular weights in the range 10 to 100 kDa is typically a few nanoseconds for roughly one cpu-week of time on a modern processor. Advances in algorithms and hardware speeds are constantly improving on this.

12.5.3   Non-equilibrium Solvation

In Section 11.4.6, the limitations of continuum models in their ability to treat non-equilibrium solvation, at least in their simplest incarnations, were noted and discussed. In principle, explicit solvent models might be expected to be more appropriate for the study of chemical processes characterized by non-equilibrium solvation. In practice, however, the situation is not much better for the explicit models than for the implicit.

Consider a typical event that would be expected to exhibit non-equilibrium solvation, e.g., a chemical reaction with significant rearrangement of charge density in the region of the transition-state structure. The short lifetime of the TS and the corresponding 'sudden' change in charge distribution would be expected to limit the solvent's ability to solvate the reaction coordinate in a fully equilibrated fashion. In the abstract, it might seem that a molecular dynamics trajectory of the reaction would offer a useful tool for studying the non-equilibrium effects, since the time course of the reaction is realistically mimicked to within the accuracy of the employed force field. Unfortunately, the likelihood of any trajectory spontaneously following a reactive path is unacceptably small unless the barrier is quite low (and, if there is enough charge reorganization to give rise to significant non-equilibrium solvation effects, a low barrier is not expected).

In the absence of being able to observe spontaneous reactive events, one can attempt to take advantage of sampling methods like those outlined in Section 12.2.5 to force trajectories along the reaction coordinate. However, by changing the PES through addition of a biasing potential, or sampling over an extended period at constrained reaction coordinate values, one changes the length of time the trajectory spends in given regions of phase space, and the solvation is likely to become more equilibrium-like in character.

With Monte Carlo methods, the adoption of the Metropolis sampling scheme intrinsically assumes equilibrium Boltzmann statistics, so special modifications are required to extend MC methods to non-equilibrium solvation as well. Fortunately, for a wide variety of processes, ignoring non-equilibrium solvation effects seems to introduce errors no larger than those already inherent from other approximations in the model, and thus both implicit and explicit models remain useful tools for studying chemical reactivity.

12.5.4   Mixed Explicit/Implicit Models

Having identified the strongest points of the explicit and implicit solvent models, it seems an obvious step to try to combine them in a way that takes advantage of the strengths of each. For instance, to the extent first-solvation-shell effects are qualitatively different from those deriving from the bulk, one might choose to include the first solvation shell explicitly and model the remainder of the system with a continuum (see, for instance, Chalmet, Rinaldi, and Ruiz-López, 2001).

There are certain instances where this approach may be regarded as an attractive option. For example, Cossi and Crescenzi (2003) found that accurate computation of ^{17}O NMR chemical shifts for alcohols, ethers, and carbonyls in aqueous solution required at least one explicit solvent shell, but that beyond that shell a continuum could be used to replace what would otherwise be a need for a much larger cluster. However, just as the strengths of the two models are combined, so are the weaknesses. A typical first shell of solvent for a small molecule may be expected to be composed of a dozen or so solvent molecules. The resulting supermolecular cluster will inevitably be characterized by a large number of accessible structures that are local minima on the cluster PES, so that statistical sampling will have to be undertaken to obtain a proper equilibrium distribution. Thus, QM methods require a substantial investment of computational resources. In addition, certain technical points require attention, e.g., how does one keep the first solvent shell from 'exchanging' with the continuum since both, in principle, foster identical solvation interactions?

So, while there is growing interest in hybrid models of all sorts (as discussed in more detail in the next chapter), the choice of a mixed solvent model is not necessarily intrinsically better than a pure explicit or pure implicit model. In general, unless there is a strong suspicion that first-solvation-shell effects are drastically different from those more typically encountered, there is no particularly compelling reason to pursue a mixed modeling strategy. An example of such a situation might be the aqueous coordination sphere surrounding a highly charged metal cation. In that case, the electrostriction of the first shell makes the water molecules more ligand-like than solvent-like, and their explicit inclusion in the solute complex is entirely warranted.

12.6   Case Study: Binding of Biotin Analogs to Avidin

Synopsis of Kuhn and Kollman (2000) 'A Ligand That Is Predicted to Bind Better to Avidin than Biotin: Insights from Computational Fluorine Scanning'.

One of the strongest known interactions between a biopolymer and a small-molecule substrate is that between the protein avidin and D-(+)-biotin, the structure of which is shown in Figure 12.6. The binding energy for this complex has been measured to be -20.8 kcal mol^{-1}. While this represents an extraordinarily strong interaction, Kuhn and Kollman suggested that it might be possible to make it still stronger by replacing one or more hydrogen atoms on the biotin framework with fluorine atoms. Fluorine is roughly isosteric with hydrogen (i.e., the C–F and C–H bond lengths have roughly similar lengths and F and H have similar covalent radii), but is considerably more hydrophobic. Thus, if a region of the binding pocket interacts with biotin via non-polar interactions, and is adequately shaped to accommodate the very slightly larger fluorine atom, decreased aqueous solvation of the fluorinated analog would be expected to increase the binding free energy (note that the lower polarizability of fluorine compared to alkyl hydrogen also suggests the favorable dispersion interactions between the biotin analog and the protein will be reduced, but this is generally a smaller effect than enhanced hydrophobicity in the absence of steric constraints).

Kuhn and Kollman pursue several different algorithmic approaches to estimating the binding free energies of different fluorobiotins. The fastest approach, which they refer to as fluorine scanning, involves a combination of explicit and implicit solvation models to compute the horizontal legs of the free-energy cycle in Figure 12.6. First, an MD trajectory of the avidin–biotin complex is obtained under standard MD conditions, including explicit solvent and using periodic boundary conditions.

The trajectory is then 'post-processed' to determine absolute free energies in solution for biotin, avidin, and the avidin–biotin complex. This process begins by stripping the water from the trajectory, and then computing absolute free energy as

$$\langle G \rangle = \langle E_{\mathrm{MM}} + \Delta G_{\mathrm{solv}} \rangle - TS \tag{12.32}$$

where E_{MM} is the force-field energy, ΔG_{solv} is computed from a *continuum* solvation model (in this case a finite difference Poisson–Boltzmann (FDPB) model with hydrophobic atomic surface tensions), and the expectation value is taken over the snapshots of the MD trajectory. Evaluations of Eq. (12.32) for isolated biotin and avidin are carried out using the same snapshots as those for the complex, i.e., using those geometries found in the complex, but only the atoms of the individual component are retained. The solute entropies S are determined from the usual statistical mechanical formulae (Section 10.3) with the vibrational frequencies being determined from normal mode analysis of each solute optimized separately using a distance-dependent dielectric constant to mimic the effects of solvation.

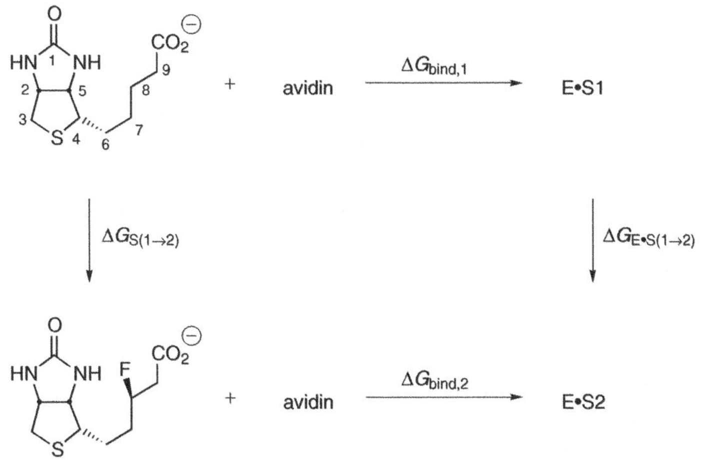

Figure 12.6 Free-energy cycle associated with the binding of biotin and fluorobiotin analogs to avidin. What issues arise in choosing a force field for explicit simulation of these systems? What methods are better suited to computing the vertical legs of the cycle and what methods the horizontal ones?

Note that since the free energies of the isolated components are computed using the same geometries as are employed in computing the free energy of the complex, the *internal* force-field energies cancel in computing a free energy of binding as

$$\langle G \rangle_{\text{bind}} = \langle G \rangle_{\text{complex}} - \langle G \rangle_{\text{biotin}} - \langle G \rangle_{\text{avidin}} \tag{12.33}$$

Only the force-field energy term associated with interactions between the biotin and avidin fragments remains. This is added to the differential solvation free energies and differential thermal terms to determine the full binding free energy.

To avoid the cost of multiple MD simulations, Eqs. (12.32) and (12.33) for fluoro-substituted biotin analogs are *also* evaluated using geometries from the original MD trajectory. The relevant hydrogen is simply replaced by a fluorine atom having the appropriate bond length oriented along the original C–H bond axis. Thus, there is no relaxation of the complex to relieve steric clashes if they are introduced. Again, the only force-field energy terms that survive in computing the free energy of binding are the interaction energies between the biotin analog and the avidin. It is further assumed that the entropy change computed for complexation with biotin remains the same for a fluorinated biotin.

The results of this rapid fluorine scanning are that substitution at positions *pro-R* 6 and 9 and *pro-S* 7, 8, and 9 are all predicted to *decrease* binding by more than 4 kcal mol^{-1}, substitution at position *pro-S* 6 is predicted to decrease binding by about 2 kcal mol^{-1}, substitution at position *pro-R* 7 is predicted to have only a small unfavorable effect, and substitution at position *pro-R* 8 is predicted to *increase* binding by a little less than 1 kcal mol^{-1}. The absolute free energy of binding for biotin itself with this method is computed to be -18.8 kcal mol^{-1}, although the fluctuations in the ensemble average and the possible error in the entropy calculation are sufficiently large to make the good agreement with the experimental value quoted above seem potentially slightly misleading.

Irrespective of the accuracy of the absolute binding free energies, the major goal of the scanning is to identify possible substitutions meriting further study by a more accurate methodology. First, as a check on the assumptions of the model, binding free energies for two substituted cases were computed from Eqs. (12.32) and (12.33) but using MD trajectories generated for the proper complexes. The results were sufficiently close to those obtained using the unsubstituted trajectory that no concerns were generated. Then, full FEP calculations using TI and explicit solvent were carried out mutating biotin into $8R$-fluoroavidin and $8S$-fluoroavidin, i.e., computing the vertical legs in Figure 12.6 (mutations were run in both the backward and forward directions). For the $8R$-fluoro analog, the binding free energy was computed to be 1.5 kcal mol^{-1} stronger than biotin, in reasonable agreement with the fluorine scanning value of 0.9 kcal mol^{-1}.

There are a few technical details in this paper that are rather more ill-defined than ideal for a 'canned' strategy – the description above of the fluorine scanning procedure glosses over some of the finer details associated with evaluating binding free energies for the substituted analogs. Nevertheless, this paper presents an interesting comparison of more and less time-consuming models for estimating differential binding free energies from explicit simulation. The joint application of explicit solvent and continuum solvent methodologies for biomolecular studies seems destined to increase in frequency.

Extensions of this case study are available for the interested reader. First, Dixon *et al.* (2002) have expanded the analysis presented above to include consideration of methylated biotin analogs and in the process developed a graphical approach for visualizing free energy changes. In addition, Lazaridis, Masunov, and Gandolfo (2002) have also considered the binding free energies of various ligands, including biotin and biotin analogs, to avidin and streptavidin. These authors decompose results from MD simulations with implicit solvation into ligand/enzyme interaction energies, reorganization energies, and entropy changes, and they conclude that the most difficult component to predict with acceptable precision is the reorganization energy of the macromolecule. The results from all three of these studies have important implications for docking models in general, and in particular for models that employ static ligand and/or receptor structures to improve efficiency, and thereby ignore relaxation energetics.

Bibliography and Suggested Additional Reading

Blondel, A. 2004. 'Ensemble Variance in Free Energy Calculations by Thermodynamic Integration: Theory, Optimal "Alchemical" Path, and Practical Solution', *J. Comput. Chem.*, **25**, 985.

Brooks, C. L., III and Case, D. A. 1993. 'Simulations of Peptide Conformational Dynamics and Thermodynamics' *Chem. Rev.* **93**, 2487.

Boresch, S. 2002. 'The Role of Bonded Energy Terms in Free Energy Simulations – Insights from Analytical Results', *Mol. Sim.*, **28**, 13.

Frenkel, D. and Smit, B. 1996. *Understanding Molecular Simulation*, Academic Press: New York.

Jensen, F. 1999. *Introduction to Computational Chemistry*, Wiley: Chichester.

Kollman, P. 1993. 'Free Energy Calculations: Applications to Chemical and Biochemical Phenomena', *Chem. Rev.*, **93**, 2395.

Levy, R. M. and Gallicchio, E. 1998. 'Computer Simulations with Explicit Solvent – Recent Progress in the Thermodynamic Decomposition of Free Energies and in Modeling Electrostatic Effects', *Annu. Rev. Phys. Chem.*, **49**, 531.

Lybrand, T. P. 1990. 'Computer Simulation of Biomolecular Systems Using Molecular Dynamics and Free Energy Perturbation Methods', in *Reviews in Computational Chemistry*, Vol. 1, Lipkowitz, K. B. and Boyd, D. B. Eds., VCH: New York, 295.

Orozco, M. and Luque, F. J. 2000. 'Theoretical Methods for the Description of the Solvent Effect on Biomolecular Systems', *Chem. Rev.*, **100**, 4187.

Sen, S. and Nilsson, L. 1999. 'Some Practical Aspects of Free Energy Calculations from Molecular Dynamics Simulation', *J. Comput. Chem.*, **20**, 877.

Straatsma, T. P. 1996. 'Free Energy by Molecular Simulation', in *Reviews in Computational Chemistry*, Vol. 9, Lipkowitz, K. B. and Boyd, D. B. Eds., VCH: New York, 81.

van Gunsteren, W. F., Luque, F. J., Timms, D., and Torda, A. E. 1994. 'Molecular Mechanics in Biology: From Structure to Function, Taking Account of Solvation', *Annu. Rev. Biophys. Biomol. Struct.*, **23**, 847.

References

Åberg, K. M., Lyubartsev, A. P., Jacobsson, S. P., and Laaksonen, A. 2004. *J. Chem. Phys.*, **120**, 3770.

Adamovic, I., Freitag, M. A., and Gordon, M. S. 2003. *J. Chem. Phys.*, **118**, 6725.

Berendsen, H. J. C., Postma, J. P. M., van Gunsteren, W. F., and Hermans, J. 1981. In: *Intermolecular Forces*, Pullman, B., Ed., Reidel: Dordrecht.

Bernardo, D. N., Ding, Y., Krogh-Jespersen, K., and Levy, R. M. 1994. *J. Phys. Chem.*, **98**, 4180.

Car, R. and Parrinello, M. 1985. *Phys. Rev. Lett.*, **55**, 2471.

Chalmet, S., Rinaldi, D., and Ruiz-López, M. F. 2001. *Int. J. Quantum Chem.*, **84**, 559.

Cossi, M. and Crescenzi, O. 2003. *J. Chem. Phys.*, **118**, 8863.

Dang, L. X. 1992. *J. Chem. Phys.*, **97**, 2659.

Darve, E. and Pohorille, A. 2001. *J. Chem. Phys.*, **115**, 9169.

den Otter, W. and Briels, W. 1998. *J. Chem. Phys.*, **109**, 4139.

den Otter, W. K. and Briels, W. J. 2000. *Mol. Phys.*, **98**, 773.

Dixon, R. W., Radmer, R. J., Kuhn, B., Kollman, P. A., Yang, J., Raposo, C., Wilcox, C. S., Klumb, L. A. Stayton, P. S., Behnke, C., Le Trong, I., Stenkamp, R. 2002. *J. Org. Chem.*, **67**, 1827.

Fischer, R., Richardi, J., Fries, P. H., and Krienke, H. 2002. *J. Chem. Phys.*, **117**, 8467.

Glattli, A, Daura, X., and van Gunsteren, W. F. 2003. *J. Comput. Chem.*, **24**, 1087.

Gordon, M. S., Freitag, M. A., Bandyopadhyay, P., Jensen, J. H., Kairys, V., and Stevens, W. J. 2001. *J. Phys. Chem. A*, **105**, 293.

Jorgensen, W. and Ravimohan, C. 1985. *J. Chem. Phys.*, **83**, 3050.

Jorgensen, W. L. and Nguyen, T. B. 1993. *J. Comput. Chem.*, **14**, 195.

Jorgensen, W. L., Briggs, J. M., and Contreras, M. L. 1990. *J. Phys. Chem.*, **94**, 1683.

Jorgensen, W. L., Chandrasekhar, J., Madura, J. D., Impey, R. W., and Klein, M. L. 1983. *J. Chem. Phys.*, **79**, 926.

Kaminski, G., Duffy, E. M., Matsui, T., and Jorgensen, W. L. 1994. *J. Phys. Chem.*, **98**, 13 077.

Kuhn, B. and Kollman, P. A. 2000. *J. Am. Chem. Soc.*, **122**, 3909.

Kumar, S., Bouzida, D., Swendsen, R. H., Kollman, P. A., and Rosenberg, J. M. 1992. *J. Comput. Chem.*, **13**, 1011.

Lazaridis, T., Masunov, A., Gandolfo, F. 2002. *Proteins*, **47**, 194.

Lefohn, A. E., Ovchinnikov, M, and Voth, G. A. 2001. *J. Phys. Chem. B*, **105**, 6628.

Mahoney, M. W. and Jorgensen, W. L. 2000. *J. Chem. Phys.*, **112**, 8910.

McDonald, N. A. and Jorgensen, W. L. 1998 *J. Phys. Chem. B*, **102**, 8049.

Morreale, A., Gelpi, J. L., Luque, F. J., and Orozco, M. 2003. *J. Comput. Chem.*, **24**, 1610.

Peter, C., Oostenbrink, C., van Dorp, A., and van Gunsteren, W. F. 2004. *J. Chem. Phys.*, **120**, 2652.

Rick, S. W., Stuart, S. J., and Berne, B. J. 1994. *J. Chem. Phys.*, **101**, 6141.

Rosso, L., Minary, P., Zhu, Z. W., Tuckerman, M. E. 2002. *J. Chem. Phys.*, **116**, 4389.

Severance, D. L. and Jorgensen, W. L. 1992. *J. Am. Chem. Soc.*, **114**, 10966.

Shirts, M. R., Pitera, J. W., Swope, W. C., and Pande, V. S. 2003. *J. Chem. Phys.*, **119**, 5740.

Storer, J. W., Giesen, D. J., Hawkins, G. D., Lynch, G. C., Cramer, C. J., Truhlar, D. G., and Liotard, D. A. 1994. In: *Structure and Reactivity in Aqueous Solution*, ACS Symposium Series, Vol. 568, Cramer, C. J. and Truhlar, D. G., Eds., American Chemical Society: Washington, DC, 24.

Zhu, S.-B. and Wong C. F. 1994. *J. Phys. Chem.*, **98**, 4695.

13

Hybrid Quantal/Classical Models

13.1 Motivation

An interest in understanding solvent structure represents one example of a situation that requires the explicit representation of a large system, as described in detail in the preceding chapter. For reasons of efficiency, such representation is most typically carried out at the molecular mechanics level. The chief drawback of the MM level of theory, however, is that it is almost never appropriate for the description of processes involving bond-making or bond-breaking, i.e., chemical reactions. To adequately model such processes, QM methods are required. However, the region of space within which significant changes in electronic structure occur along the course of a reaction coordinate is often relatively small compared to the size of the reacting system as a whole. For instance, a very large enzyme may catalyze the conversion of its substrate from one molecule to another, but the volume of space within which bonds are being made and broken is usually limited to the relatively small active site. The remainder of the enzyme may be important for maintaining its structure, recognizing other enzymes with which it works, folding, etc., but fails to exert any quantum mechanical influence on the catalytic active site.

Thus, from a modeling perspective, we may regard the situation in the abstract as described by Figure 13.1. Within a limited region, we wish to make use of the tools of quantum mechanics to accurately model an electronic-structure problem, while in the surrounding region the explicit representation of the supersystem is important, but the level of model applied can be reduced in complexity owing to the more simply understood influence of the outer region on the process as a whole. When the level applied to the outer system is MM, the complete Hamiltonian for the system must be some kind of hybrid of QM and MM methodologies, defining a so-called QM/MM technique. Put in a disarmingly simple fashion

$$H_{\text{complete}} = H_{\text{QM}} + H_{\text{MM}} + H_{\text{QM/MM}} \tag{13.1}$$

where H_{QM} accounts for the full interaction energy of all quantum mechanical particles with one other, H_{MM} accounts for the full interaction energy of all classical particles with one other, and $H_{\text{QM/MM}}$ accounts for the energy of all interactions between one quantum mechanical particle and one classical particle. Methods for the evaluation of the first two

Essentials of Computational Chemistry, 2nd Edition Christopher J. Cramer
© 2004 John Wiley & Sons, Ltd ISBNs: 0-470-09181-9 (cased); 0-470-09182-7 (pbk)

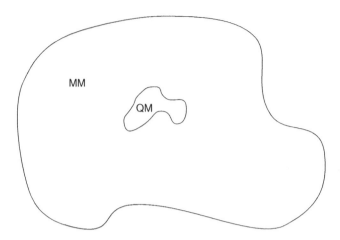

Figure 13.1 In large systems that require explicit representation, understanding bond-making/-bond-breaking processes can often be accomplished using a quantum mechanical representation of only a portion of the full system, with a molecular mechanics representation of the rest

terms on the r.h.s. of Eq. (13.1) have already been the subject of much discussion in preceding chapters – the devil for a hybrid method is in the details of the final term, and those are the subject of this chapter.

Many QM/MM modeling schemes have been described with varying levels of formalism. In terms of classification, perhaps the most fundamental distinction is whether or not the boundary separating the QM region from the MM region in Figure 13.1 cuts across any chemical bonds. If it does not, the coupling of the QM and MM regions can be represented with a reasonable degree of simplicity. If so clean a separation is *not* practical, however, e.g., the QM region consists of the substrate for a large enzyme *and* at least one atom from a side chain residue in the active site (that serves to accept a proton from the substrate, for example), then more complicated coupling schemes must be employed to stitch together the distinct subspaces.

13.2   Boundaries Through Space

In some sense, the simplest example of what might be called a QM/MM approach with a through-space boundary has already been alluded to in Section 12.2.5 and illustrated with the specific example of the Claisen rearrangement in Section 12.5.1. To evaluate the PMF for a reaction in solution, one useful approach is to compute the reaction coordinate using a QM method in the gas phase, and then determine changes in solvation free energy as the system is driven from one end of the coordinate to the other by the coupling parameter λ. For the FEP calculations themselves, the reacting system is represented classically (e.g., using fixed geometries, partial atomic point charges, and van der Waals parameters), but the gas-phase energies to which the solvation free energies are added, and also often the atomic partial charges, are taken from the antecedent QM calculations. As has already been emphasized, this approach ignores the effect solvent has on coordinates other than the reaction coordinate, and on the solute wave function, but it nevertheless may legitimately be referred to as a 'weakly coupled' QM/MM calculation. We now proceed to consider increasingly tightly coupled protocols for joining the two regions.

13.2.1 Unpolarized Interactions

A significant issue with modern force fields is that it can be difficult to simultaneously address both generality and suitability for use in condensed-phase simulations. For example, the MMFF94 force field is reasonably robust for gas-phase conformational analysis over a broad range of chemical functional groups, but erroneously fails to predict a periodic box of n-butane to be a liquid at $-0.5\,°C$ (Kaminski and Jorgensen 1996). The OPLS force field, on the other hand, is very accurate for condensed-phase simulations of molecules over which it is defined, but it is an example of a force field whose parameterization is limited primarily to functionality of particular relevance to biomolecules, so it is not obvious how to include arbitrary solutes in the modeling endeavor.

Kaminski and Jorgensen (1998) have proposed one particularly simple QM/MM approach to address this problem, which they refer to as AM1/OPLS/CM1 (AOC). In AOC, Monte Carlo calculations are carried out for solute molecules represented by the AM1 Hamiltonian embedded in periodic boxes of solvent molecules represented by the OPLS force field. Thus, H_{QM} in Eq. (13.1) is simply the AM1 energy for the solute, and H_{MM} is evaluated for all solvent–solvent interactions using the OPLS force field. The QM/MM interaction energy is computed in a fashion closely resembling the standard approach for MM non-bonded interactions

$$H_{QM/MM} = \sum_i^{solute} \sum_j^{solvent} \left[\frac{\alpha q_i^{CM1} q_j}{r_{ij}} + 4\varepsilon_{ij} \left(\frac{\sigma_{ij}^{12}}{r_{ij}^{12}} - \frac{\sigma_{ij}^6}{r_{ij}^6} \right) \right] \qquad (13.2)$$

where the Lennard–Jones parameters are determined from the usual combining rules (Eqs. (2.30) and (2.31)) assuming the solute atoms have ε and σ values characteristic for their atomic type in the OPLS force field. The single feature that is quantum mechanical is that the solute charges are determined from the CM1 charge model applied to the AM1 wave function (see Section 9.1.3.4). For charged molecules, the constant α is 1.0, while for neutral molecules, it is 1.2 to approximate the effect of solvent polarization on the gas-phase charge distribution.

The choice of AM1 as a particularly efficient level of electronic-structure theory is motivated by the large number of QM calculations potentially required in the MC sampling. In the standard AOC MC protocol, moves of solute internal coordinates are attempted every 50 MC steps, and accepted or rejected according to the standard Metropolis protocol as described in Section 3.4.2. If the move is accepted, the QM energy and CM1 charges are updated and used in Eq. (13.2) until the next accepted change of solute geometry. Note that QM calculations are *not* required unless a solute move is being attempted.

The AOC method successfully predicts the effects of polar solvents on rotameric equilibria for 1,2-dichloroethane and 2-furfural, as illustrated in Table 13.1. However, it is not very

Table 13.1 Differential solvation effects (kcal mol^{-1}) on rotameric equilibria

$\Delta G_S^o(\text{gauche}) - \Delta G_S^o(\text{trans})$

Solvent	AOC[a]	SM5.4/AM1[b]	Experiment
CCl$_4$	−0.10		−0.70[c]
C$_2$Cl$_4$		−0.31	−0.37[d]
CH$_3$CN	−1.19	−1.03	−1.01[c], −1.42[d]
H$_2$O	−1.48		

$\Delta G_S^o(\text{syn}) - \Delta G_S^o(\text{trans})$

	AOC[a]	Experiment[e]
CCl$_4$	0.10	−1.3 ± 0.5
CH$_3$OCH$_3$	−1.45	−1.7 ± 0.5
DMSO	−2.40	−2.3 ± 0.5

[a] Kaminski and Jorgensen (1998).
[b] Chambers *et al.* (1999).
[c] Wiberg *et al.* (1995).
[d] Depaepe and Ryckaert (1995).
[e] Abraham and Siverns (1972).

successful at predicting solvation effects on these equilibria in *non-polar* solvents, since the OPLS solvent molecules are not electronically polarizable. Such effects *are* included in continuum solvation models like SM5.4/AM1, being implicit in solvent dielectric constants on the order of 2, and data from that model may be compared for the 1,2-dichloroethane equilibrium in Table 13.1.

The AOC model has also proven efficient for modeling solvation free energy differences along a reaction coordinate generated from gas-phase calculations (as previously described in Section 12.5.1). Chandrasekhar, Shariffskul, and Jorgensen (2002) used this technique to obtain excellent agreement with experiment in predicting the aqueous acceleration of Diels–Alder cycloadditions of cyclopentadiene, supporting the conclusion from prior purely MM simulations that enhanced hydrogen bonding to hydrophilic functionality in the TS structures is responsible for the acceleration.

With respect to further developments of the AOC protocol, Udier-Blagovic *et al.* (2004) recently assessed the relative utility of scaled CM1 and CM3 charges from AM1 and PM3 calculations for use in computation of absolute solvation free energies via AOC. On an initial test set of 13 organic molecules, they found neither charge model based on PM3 to provide acceptable accuracy. However, scaling CM1 and CM3 charges derived from AM1 by factors of 1.14 and 1.15, respectively, gave average errors of only 1.0 and 1.1 kcal mol^{-1}, respectively, over a diverse test set of 25 organic molecules.

13.2.2 Polarized QM/Unpolarized MM

The next level of complexity involves accounting for environmentally induced relaxation of the QM wave function explicitly (as compared to, say, the implicit scaling factor α in Eq. (13.2)). The coupling Hamiltonian $H_{QM/MM}$ remains similar in spirit to that described by Eq. (13.2), in the sense that the interaction must be represented as a sum of electrostatic and other non-bonded interactions, but the next step is to determine the Fock operator (or analogous DFT operator) that is used to obtain orbitals minimizing the *complete* Hamiltonian. This is quite straightforward in practice given that we may write the coupling term as

$$H_{QM/MM} = \overset{\text{solute}}{\underset{i}{\sum}} \overset{\text{MM}}{\underset{m}{\sum}} \frac{q_m}{r_{im}} + \overset{\text{solute}}{\underset{k}{\sum}} \overset{\text{MM}}{\underset{m}{\sum}} \left[\frac{Z_k q_m}{r_{km}} + 4\varepsilon_{km} \left(\frac{\sigma_{km}^{12}}{r_{km}^{12}} - \frac{\sigma_{km}^{6}}{r_{km}^{6}} \right) \right] \tag{13.3}$$

Thus, the electrostatic interaction term of Eq. (13.2) has been separated into an operator acting on the QM electrons (the first term on the r.h.s. of Eq. (13.3)) and the classical term for the interaction of the MM atoms with the solute nuclei. The Lennard–Jones term is the same in Eqs. (13.2) and (13.3) (although the parameters may certainly be different from one model to another).

The next step is to find orbitals that minimize the expectation value of $H_{complete}$ in Eq. (13.1), given Eq. (13.3) for $H_{QM/MM}$. If we take as our wave function a standard normalized Slater determinant, we have

$$\langle \Psi | H_{complete} | \Psi \rangle = \langle \Psi | H_{QM} | \Psi \rangle + \langle \Psi | H_{MM} | \Psi \rangle + \langle \Psi | H_{QM/MM} | \Psi \rangle$$

$$= \left\langle \Psi \left| \sum_i^N -\frac{1}{2}\nabla_i^2 - \sum_i^N \sum_k^K \frac{Z_k}{r_{ik}} + \sum_{i<j} \frac{1}{r_{ij}} + \sum_{k<l} \frac{Z_k Z_l}{r_{kl}} \right| \Psi \right\rangle + \langle \Psi | \Psi \rangle H_{MM}$$

$$+ \left\langle \Psi \left| -\sum_i^N \sum_m^M \frac{q_m}{r_{im}} \right| \Psi \right\rangle + \langle \Psi | \Psi \rangle \sum_k^K \sum_m^M \left[\frac{Z_k q_m}{r_{km}} + 4\varepsilon_{km} \left(\frac{\sigma_{km}^{12}}{r_{km}^{12}} - \frac{\sigma_{km}^{6}}{r_{km}^{6}} \right) \right]$$

$$= \left\langle \Psi \left| \sum_i^N -\frac{1}{2}\nabla_i^2 - \sum_i^N \sum_k^K \frac{Z_k}{r_{ik}} - \sum_i^N \sum_m^M \frac{q_m}{r_{im}} + \sum_{i<j} \frac{1}{r_{ij}} + \sum_{k<l} \frac{Z_k Z_l}{r_{kl}} \right| \Psi \right\rangle$$

$$+ H_{MM} + \sum_k^K \sum_m^M \left[\frac{Z_k q_m}{r_{km}} + 4\varepsilon_{km} \left(\frac{\sigma_{km}^{12}}{r_{km}^{12}} - \frac{\sigma_{km}^{6}}{r_{km}^{6}} \right) \right] \tag{13.4}$$

where i and j run over N QM electrons, k and l run over the K nuclei in the QM fragment, and m runs over the M molecular mechanics atoms. The second equality in Eq. (13.4) simply expands the QM Hamiltonian into its usual individual terms and uses Eq. (13.3) to expand the QM/MM component of the Hamiltonian. The terms having no dependence on the electronic coordinates – H_{MM}, the QM-nuclei/MM-atom electrostatic interactions, and the LJ interactions – may be taken outside of expectation value integrals, which are then simply one by normalization of the wave function. The third equality simply collects terms together in a convenient fashion.

Note that the only operator acting on the electronic wave function for the QM/MM system that would not be present in the isolated QM system is that involving the charges of the MM atoms. In operator formalism, these atoms behave exactly like QM nuclei, except that they bear partial atomic charges instead of atomic-number-based charges. As such, they enter into the standard Fock operator just as nuclear charges do, i.e., as part of the one-electron operator. Elements of the QM/MM Fock matrix that minimize the energy computed from Eq. (13.4) are thus calculated from the generalization of Eq. (4.54) as

$$
F_{\mu\nu} = \left\langle \mu \left| -\frac{1}{2}\nabla^2 \right| \nu \right\rangle - \overset{\overset{\text{QM}}{\text{nuclei}}}{\sum_k} Z_k \left\langle \mu \left| \frac{1}{r_k} \right| \nu \right\rangle - \overset{\overset{\text{MM}}{\text{atoms}}}{\sum_m} q_m \left\langle \mu \left| \frac{1}{r_m} \right| \nu \right\rangle
$$
$$
+ \sum_{\lambda\sigma} P_{\lambda\sigma} \left[(\mu\nu|\lambda\sigma) - \frac{1}{2}(\mu\lambda|\nu\sigma) \right] \tag{13.5}
$$

where only the third term on the r.h.s. is different from the usual QM expression. The third term involves the computation of M one-electron integrals. Insofar as the bottlenecks in HF theory tend to be assembly of the two-electron integrals or diagonalization of the Fock matrix, the actual increase in computational time required for a QM/MM calculation compared to a purely QM calculation on the same fragment can be quite small.

DFT equations analogous to Eqs. (13.4) and (13.5) can be derived in a similarly straightforward way. Again, the ultimate influence of the MM system on the KS orbitals is made manifest only by the appearance of additional one-electron integrals associated with the MM atoms in the KS operator.

Of course, the simplicity of the QM/MM operator does not imply that it has only a small effect. Large atomic partial charges placed near the QM fragment would be expected to polarize the system strongly. Table 13.2 compares the dipole moments of the standard nucleic acid bases at the AM1 level evaluated in the gas phase and in a QM/MM calculation carried out modeling aqueous solvation with a periodic box of TIP3P water molecules. For comparison, results from the AM1-SM2 aqueous continuum solvation model are also provided.

It is important to recognize how a QM/MM calculation like that for the nucleic acid base solvated dipole moments is accomplished. We outline here a typical series of steps

1. Choose the particular QM and MM levels to be used.

2. Given those QM and MM levels, select a set of LJ parameters for the QM fragment. One option is to use the same parameters for atoms in the QM fragment as those that would

Table 13.2 Computed dipole moments (D) of the nucleic acid bases in the gas phase and in aqueous solution

	Gas	Aqueous solution	
Nucleic acid base	$\langle\Psi|\mu|\Psi\rangle_{AM1}$[a]	$\langle\langle\Psi|\mu|\Psi\rangle_{AM1/TIP3P}\rangle$[a]	$\langle\Psi|\mu|\Psi\rangle_{AM1-SM2}$[b]
Adenine	2.2	3.8	3.1
Cytosine	6.3	9.4	9.0
Guanine	6.2	9.4	8.5
Thymine	4.2	5.9	6.2
Uracil	4.3	6.2	6.4

[a]Gao 1994.
[b]Cramer and Truhlar 1992, 1993.

be applied to those atoms were they to be in the MM region (like the AOC model). Another option is to develop separate transferable LJ parameters to be used for the QM fragment whenever the particular QM/MM choice has been made (see, for example, Martin *et al.* 2002). The data in Table 13.2 were determined using such a procedure, with the parameters thus being part of the definition of the AM1/TIP3P model (Gao and Xia 1992).

3. Where necessary, determine system properties as ensemble expectation values (for the data in Table 13.2, a MC sampling scheme was employed, but MD methods are equally applicable). Every time the coordinates of any atom in the system change, i.e., at each time step in a MD trajectory or following an accepted MC move of either a QM or MM atom, recompute the QM wave function (since at least one term in the operator involving the relative positions of the QM electrons and the MM atoms must be different). Note the contrast with the AOC method, where only internal moves in the QM system's coordinates necessitate a recomputation of the wave function.

4. At each simulation step, the property or properties of interest are included in the ensemble average. For Table 13.2, the property is the evaluation of the dipole moment operator as an electronic expectation value over the QM subsystem. Thus, the QM/MM result for this case is an MC *ensemble* expectation value of a quantum mechanical *operator* expectation value.

A different application of the AM1/TIP3P model nicely illustrates the ability of QM/MM models to permit the analysis of quantities not typically simultaneously available to either pure QM or MM models. Gao (1994) employed the AM1/TIP3P model to determine the PMF for the Claisen rearrangement in water, a reaction already discussed in some detail in the context of pure continuum or explicit solvation models in Section 12.5.1 (see also Section 11.1.2). Similarly to the pure MM simulation, the computational protocol involved FEP along the gas-phase reaction coordinate using λ to drive the structure of the initial allyl vinyl ether through the TS to the unsaturated aldehyde product. At the AM1/TIP3P level, the same increase in hydrogen bonding to the ether oxygen noted in the pure MM study was observed. In the QM/MM model, however, the effect of this increased hydrogen bonding on

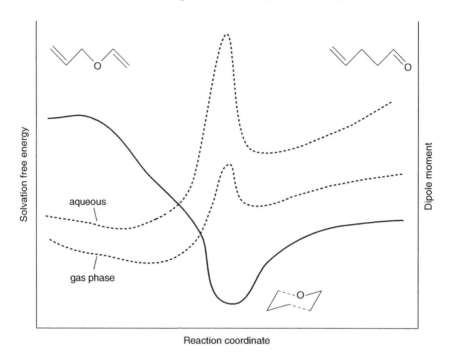

Figure 13.2 Schematic comparison of gas-phase and aqueous dipole moments (- - - - - -) to the free energy of solvation (———) along the reaction path for the Claisen rearrangement as computed at the QM/MM level. FEP technology allows access to the solvation free energy, while the QM treatment of the substrate permits evaluation of the dipole moment operator. Note the large increase in polarizability (as judged by the difference between gas-phase and aqueous dipole moments) in the region of the TS structure that contributes to a sharp increase in the magnitude of the favorable solvation free energy. Note also, however, that comparison of the relevant curves indicates that the total solvation free energy depends on more than just the dipole moment

the QM solute charge distribution could be directly analyzed. As illustrated in Figure 13.2, Gao found that the polarizability of the Claisen substrate was substantially larger in the region of the transition state (as judged by the induced dipole moment attributable to solvation), and that this contributed significantly to increasing the favorable relative solvation free energy of the TS structure compared to the reactants, thereby adding to rate acceleration. The same inference was made from analysis of pure QM continuum model results, but without an ability to correlate polarizability with hydrogen bonding propensity.

Of course, the rich information available from a QM/MM simulation does not come without cost. The QM/MM Claisen simulation required several million AM1 calculations to be carried out; while AM1 is a very efficient level of QM theory for a molecule as small as allyl vinyl ether, that still represents an enormous investment of computational resources. As a result, the application of QM/MM methodologies based on the formalism of Eqs. (13.4) and (13.5) has tended not to be especially systematic, i.e., choices of QM and MM models and necessary coupling parameters have tended to be made on an *ad hoc* basis, without regarding parameter transferability as being an issue of paramount concern.

With increasing use of such models, methods are likely to become more concisely defined in the near future. At present, the models for which protocols and parameters have been most clearly defined and where a fair number of applications have appeared applying those models in a consistent fashion include the already noted AM1/TIP3P model (more generally AM1/OPLS when solvents other than water are employed in the MM region) and a similarly fashioned HF/3-21G/OPLS model (Freindorf and Gao 1996). Implementations carrying the QM level as far as coupled-cluster theory have been reported (Kongsted *et al.* 2003).

A variation on the QM/MM theme that offers an increase in computational speed by sacrificing a certain level of microscopic detail has seen a moderate level of application when the MM system is simply a homogeneous solvent. In such cases, explicit atomistic representation of the solvent molecules is replaced with a set of integral equations governing their mutual interaction, for example, the reference interaction site model (RISM; Pratt and Chandler 1977). The equations are quite complex (and their solution can be a challenging numerical task) but in essence integral equation theories take as input the force-field parameters and interaction potentials associated with a molecular mechanics solvent molecule (e.g., TIP3P water or OPLS chloroform), and they output radial distribution functions describing the solvent's average structure (see Section 3.5). Alternatively, Freedman and Truong (2003) have described using RISM with solvent r.d.f.s computed from explicit simulations. In any case, integral equation models are in some sense intermediate between continuum models, which include no solvent structural information, and explicit models, where individual snapshots of a solvated system comprise the data from which averages are computed.

Several models combining QM solute representations with a RISM solvent treatment have been described recently. The QM solute is described as a sum of site potentials just as the MM solvent is, except that the electrostatic portion of the potentials derives from the QM wave function. Using the radial distribution functions for solvent charge sites about solute atoms permits the solvent electrostatic influence on the QM wave function to be determined. The RISM equations are then solved for the full set of site potentials until self-consistency is reached (Ten-no, Hirata, and Kato 1993). Given the final site–site distributions and the interaction potential between all sites, it is possible to compute the free energy of solvation (Lee and Maggiora 1993). In a QM/RISM hybrid model, experimental data for this quantity may be used to optimize LJ parameters for QM atoms, and this approach has been used to define the hybrid extended RISM and quantum mechanical solvation model XSOL (Shao, Yu, and Gao 1998) for use in modeling organic equilibria in aqueous solution.

Rather than treating the entire solvent via the RISM formalism, an alternative approach that has seen some study is to represent some solvent molecules explicitly, typically as MM species, and then embed the entire QM/MM cluster in a continuum dielectric medium according to the formalisms described in Chapter 11. Both Bandyopadhyay *et al.* (2002) and Cui (2002) adopted such an approach to study the neutral/zwitterionic equilibrium of glycine in water, the former group representing the explicit water molecules with the EFP model and Cui doing so with a modified TIP3P model. Both studies concluded that inclusion of some explicit solvent molecules gave critically improved accuracy over modeling the problem with exclusively continuum solvation. Note that inclusion of specific MM solvent molecules about a QM anion has the additional benefit of substantially reducing any likely instabilities associated with charge penetration (see Section 11.4.1.3).

13.2.3 Fully Polarized Interactions

Allowing the QM system to be polarized by the MM charges without at the same time accounting for polarization of the molecules comprising the MM system may be regarded as being possibly unbalanced. One approach for including polarizability in the MM system has already been described in Chapter 12, and its extension to a QM/MM system is algorithmically trivial. Thus, each MM molecule or atom is assigned a polarizability tensor $\overline{\overline{\alpha}}$, and the induced dipole at each polarizable center is determined from Eq. (12.30); in the QM/MM system, the electric field $\mathbf{E}$ has the same components from the MM partial charges and induced dipoles as in a fully classical system, and an additional component deriving from the nuclei and electronic wave function of the QM system that is straightforward to calculate. The interaction of the induced dipoles with the MM partial charges (Eq. (12.31)) and with one another (Eq. (2.23)) are added in the H_{MM} term of Eq. (13.1). In addition, the induced dipoles interact with the nuclei of the QM system according to Eq. (12.31), and with the electronic wave function as the expectation value of the operator equivalent of Eq. (12.31) (thereby adding additional one-electron integrals to the Fock operator, one for each induced dipole).

The evaluation of all of these terms must proceed iteratively until self-consistency is reached, since the induced dipoles and the relaxing QM wave function modify the electric field on which the induced dipoles are dependent. Thus, the increase in computational resources required to include MM polarizability can be quite significant – one order of magnitude is not uncommon. Comparisons between QM/MM systems modeled with and without MM polarizability have been largely equivocal on the utility of its inclusion (adding alternative three-body correction terms has also been examined for the hydrated manganous ion (Loeffler, Yague, and Rode 2002) and was similarly found to lead to no significant improvement in describing hydration structure). Given its very high cost of implementation, there seems to be little point in carrying the model to this degree of physicality. However, the lack of improvement in many cases may be attributable to the polarizability being added *post facto* to an already existing force field. By virtue of fitting to experiment, formally non-polarizable force fields must include polarization in some average way into their parameters, making it less likely that additional *explicit* accounting for polarization will show dramatic effects. It is likely that only ongoing efforts aimed at developing fully polarizable force fields from scratch will prove definitive in determining the level of additional physical insight that may be gained from having polarization present in explicit form (see, for instance, Banks *et al.* 1999).

Although complete, fully polarizable QM/MM schemes are computationally demanding, a simplified version of this formalism was arguably the first QM/MM approach to be described (Warshel and Levitt 1976), and the method still sees some use today. The simplification involves replacing explicit, polarizable MM molecules with a three-dimensional grid of fixed, polarizable dipoles – each a so-called Langevin dipole (LD) as it is required to obey the Langevin polarization law. Each dipole enters the Fock operator just as described above (Luzhkov and Warshel 1992).

Much like the RISM method, the LD approach is intermediate between a continuum model and an explicit model. In the limit of an infinite dipole density, the uniform continuum model is recovered, but with a density equivalent to, say, the density of water molecules in liquid water, some character of the explicit solvent is present as well, since the magnitude of the dipoles and their polarizability are chosen to mimic the particular solvent (Papazyan and Warshel 1997). Since the QM/MM interaction in this case is purely electrostatic, other non-bonded interaction terms must be included in order to compute, say, solvation free energies. When the same surface-tension approach as that used in many continuum models is adopted (Section 11.3.2), the resulting solvation free energies are as accurate as those from 'pure' continuum models (Florián and Warshel 1997). Unlike atomistic models, however, the use of a fixed grid does not permit any real information about solvent structure to be obtained, and indeed the fixed grid introduces issues of how best to place the solute into the grid, where to draw the solute boundary, etc. These latter limitations have curtailed the application of the LD model.

13.3 Boundaries Through Bonds

All of the QM/MM models discussed this far, much like continuum models, envision partitioning a chemical system into (at least) two distinct regions, where the boundary between these regions is everywhere characterized by a very low level of electron density. That is, no atoms on one side of the boundary are bonded to atoms on the other side. As a result, the $H_{QM/MM}$ term in the Hamiltonian of Eq. (13.1) is restricted to non-bonded interactions.

The situation is vastly more complicated when the boundary between the QM and MM regions passes across one or more chemical bonds. Somehow, the dangling valences from the two separate regions must be joined in a chemically (and computationally) sensible fashion. Developmental work is ongoing in this area; this section will focus on the current most widely used procedures.

13.3.1 Linear Combinations of Model Compounds

Many efforts in molecular design make use of sterically demanding groups, e.g., t-butyl groups, to enforce particular molecular geometries. Viewing the total molecule as some kind of sum of its functional groups, the intent is for the interaction between the large groups and the remainder of the molecule to be entirely steric in nature. In such a situation, the inclusion of the bulky group(s) in a fully QM calculation may be regarded as pointlessly expensive, since the size of the fragment(s) guarantees a large increase in the total number of QM basis functions, but the non-polarity of the fragments also indicates little likelihood of perturbing the electronic structure of the remainder of the molecule via electrostatic interactions (steric interactions are, of course, fundamentally electronic exchange-repulsion interactions, but for the moment we will ignore this level of detail and consider steric effects to be distinct from more classical electrostatic interactions). Thus, there is a clear motivation for passing a

QM/MM boundary through space in such a way that the sterically bulky groups fall on the MM side and the 'interesting' part of the molecule falls on the QM side. Finally, to avoid the question of how to deal with a cut bond, one may assume that the electronic structure of the QM region will be of similar quality with either the non-polar, bulky group as a cap, or with simply hydrogen atoms as caps. With such a philosophy, the energy of the system as a whole may be expressed as a linear combination of model compounds of different size and at different levels of theory. In simplest form

$$E_{\text{complete}} = E_{\text{QM}}^{\text{small}} + \left(E_{\text{MM}}^{\text{large}} - E_{\text{MM}}^{\text{small}} \right)$$
$$= E_{\text{MM}}^{\text{large}} + \left(E_{\text{QM}}^{\text{small}} - E_{\text{MM}}^{\text{small}} \right) \tag{13.6}$$

where the large system is the complete molecule, which is only treated at the MM level of theory, and the small system is the 'core' portion whose electronic structure is of primary interest, and it is computed at both the MM *and* QM levels. The two different term orderings on the r.h.s. of Eq. (13.6) are meant to emphasize the two primary motivations for pursuing this decomposition of the Hamiltonian.

The first motivation has already been emphasized above. There is some reason to believe that all of the important quantum effects are captured in the small system, and the steric energy associated with the bulky groups will be well captured as an 'embedding' energy, i.e., the difference between the MM energy of the small system and the large system. For example, Cramer and Pak (2001) modeled the reaction coordinate for intramolecular C–H bond cleavage from a benzyl position in $[(LCu)_2(\mu\text{-}O)_2]^{2+}$, L = 1,4,7-tribenzyl-1,4,7-triazacyclononane, by replacing the five non-reactive benzyl groups with H atoms in the small model system (Figure 13.3). As this QM system was treated at the density functional level

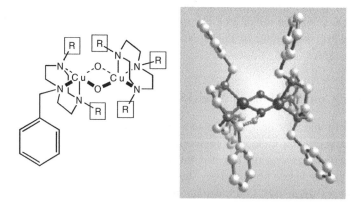

Figure 13.3 A bis(μ-oxo)dicopper complex represented using Eq. (13.6) where each boxed R group is H for the small QM system and benzyl for the large MM system. The structure on the right is a TS structure for H-atom transfer from C to O found by optimization at the hybrid level of theory. All other H atoms have been removed for clarity

of theory with a double-ζ basis set, reducing the system by 35 heavy atoms and 30 hydrogen atoms substantially reduced the total number of basis functions. The necessary MM energies were then computed with the UFF force field. Application of the model in this fashion has been especially attractive within the organometallic community, where large ligands can often be regarded as having a core portion that is electronically important, and remaining regions that are not. Thus, for example, Matsubara *et al.* (1996) have used combined DFT/MM models to study dihydrogen activation by platinum with different phosphine ligands, and Deng *et al.* (1997) have used other DFT/MM models to study the role of bulky substituents in Brookhart-type Ni(II) diimine-catalyzed olefin polymerization.

The alternative motivation for the second equality of Eq. (13.6) arises in cases where a force field may be regarded as being reasonably accurate except perhaps for some specific quantum mechanical effect(s) not well accounted for in the functional form of the force field. For example, French *et al.* (2000) constructed a (ϕ, ψ) potential energy surface for the torsions about the anomeric linkages in sucrose by adjusting an MM3 surface for the full molecule based on the difference between HF/6-31G(d) and MM3 surfaces for a tetrahydropyran-tetrahydrofuran ether model (i.e., sucrose without any hydroxyl groups, Figure 13.4). The MM3 force field exhibits a weakness in accounting for the so-called 'anomeric effect' in sugars (see Section 2.2.3). By correcting for this weakness using the QM results, French *et al.* were able to demonstrate that a sizable number of crystal structures containing sucrose moieties that had previously been assumed to be adopting abnormally high-energy conformations were instead in low-energy regions of the surface.

Note that the embedding philosophy of Eq. (13.6) may be applied more generally than simply in the context of QM/MM calculations. For example, one can imagine situations where the importance of a high-level accounting for electron correlation effects may be restricted to a small region of a large system, but the full system still requires an overall QM treatment. In such an instance, two different QM levels might be used in Eq. (13.6) instead of one QM and one MM level; obviously, the more efficient QM level is the one applied to the large system. For example, Sherer and Cramer (2001) studied the context dependence of the pK_a of the cytosine:2-aminopurine base pair in different double-helical RNA trimers by taking the base pair itself to be the small system and the trimer to be the large system, and choosing as the high and low levels of theory MP2/6-31G(d) and PM3, respectively, each augmented with an aqueous continuum solvation model (Figure 13.5).

Note that Eq. (13.6) is written in terms of energies and not Hamiltonian operators. That is because there is a certain ambiguity about how to define a wave function that would be simultaneously appropriate for all of the Hamiltonian operators that would otherwise appear on the r.h.s. This is not purely a notational issue, since it leaves open the question of the geometries used for the different energy terms. For instance, one approach would be to consider each energy on the r.h.s. to refer to complete geometry optimization at the appropriate level. This is clearly the simplest method, since every energy determination may be carried out completely independently of the others. However, if there are large differences between the corresponding regions of any pair of geometries, it calls into question the validity of the overall energy expression.

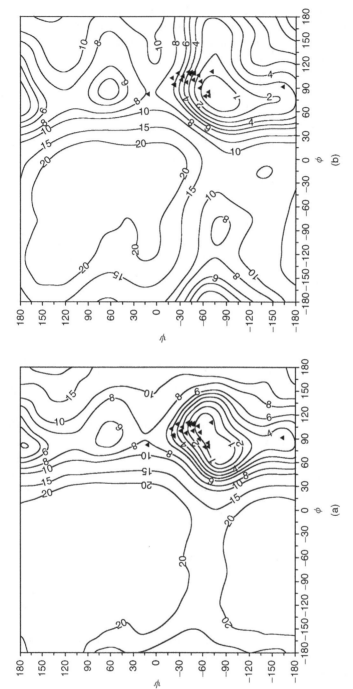

Figure 13.4 Torsional potential energy surfaces about the two C–O bonds linking the anomeric centers of sucrose at the MM3 level (a), 2-tetrahydrofuranyl-2-tetrahydropyranyl ether at the MM3 level (b), the same ether at the HF/6-31G(d) level (c), and the sum of the difference between the last two with the first (d). Thus, the last surface may be viewed either as the effect of the sucrose hydroxyl groups on the energy surface, evaluated at the MM3 level, added to the framework surface calculated at the *ab initio* level, or as an MM3 surface that has been partially 'corrected' quantum mechanically. Solid triangles represent anomeric torsions in sucrose units found in various X-ray crystal structures. Note that the hybrid surface is the only one that clusters the large majority of these triangles within low-energy contours

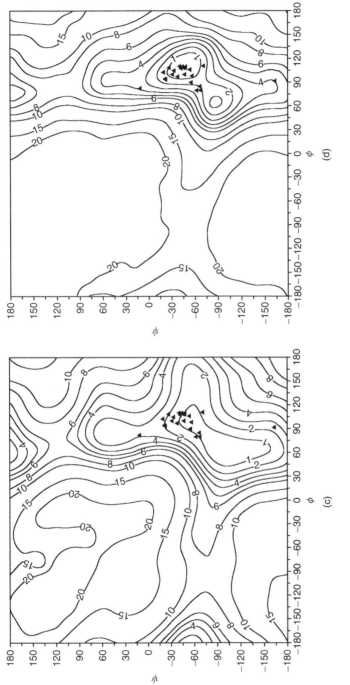

Figure 13.4   (continued)

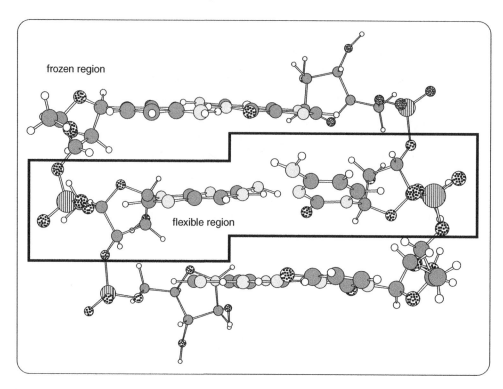

Figure 13.5   An application of a hybrid MO/MO philosophy to the indicated RNA trimer proceeds using correlated levels of electronic structure theory for various tautomers and protonation states of the central base pair, this pair then representing the small system in the MO/MO analog of Eq. (13.6), and semiempirical theory for both the small system and the frozen-geometry larger system

An alternative is to force those atoms common to the large and small models (i.e., all of the atoms in the small model except the capping hydrogens) to occupy the same coordinate locations in all three energy evaluations. Within this set of restraints, one may then write down fairly simple expressions for the gradients as sums of QM and MM gradients from the small and large systems, noting that there are some details associated with the capping atoms in the small system and the alignment of bonds to capping atoms with bonds in the full system (see, for example, Vreven *et al.* 2003 and Swart 2003). Maseras and Morokuma (1995) were the first to provide such gradient expressions, referring to the optimization approach as the integrated molecular orbital molecular mechanics method (IMOMM).

Subsequently, Humbel, Sieber, and Morokuma (1996) generalized the IMOMM optimization scheme to the case where two different levels of QM theory were used instead of a QM/MM approach, and Svensson, Humbel, and Morokuma (1996) examined the relative efficacy of different combinations of levels for prototype problems. Corchado and Truhlar (1998) later proposed a refinement of that methodology to improve computed vibrational frequencies and Rickard *et al.* (2003) showed that a combination of MP2 and HF theories permits the calculation of high-quality NMR chemical shifts within the high-level system.

Of course, Eq. (13.6) admits to further generalization. Rather than dividing a system into large and small models, there may be instances where a division into large, medium, and small models may be advantageous, with increasingly smaller regions treated with increasingly higher levels of theory. Svensson *et al.* (1996) generalized their geometry optimization scheme to this more general case, demonstrating the method for MO/MO/MM combinations, and refer to it as ONIOM, where the acronym, representing 'our own *n*-layered integrated molecular orbital molecular mechanics' scheme, is meant to emphasize the typical inward-to-outward, near-spherical layering of models that is typically chosen and is reminiscent of the almost eponymous lachrymatory bulb. To provide yet another layer of modeling when condensed-phase effects are of interest, the combination of ONIOM with the PCM continuum solvation model has been described (Vreven *et al.* 2001) as has a model for permitting explicit solvent molecules to morph from MM to QM while passing through a buffer region surrounding the QM subsystem (Kerdcharoen and Morokuma 2002).

13.3.2 Link Atoms

Situations arise where the influence of the MM region on the QM region to which it is bonded cannot be regarded simply as steric. In a large protein, for instance, polar and possibly charged residues in an MM region inevitably will polarize a QM region in the same protein. The only way to eliminate such QM/MM coupling is to include the entire protein in the QM region, and such an approach is extremely impractical for anything other than a possible single-point calculation at a fairly low level of electronic structure theory.

Of course, the strong coupling invoked here between the two regions is in no manner different than that dealt with in Section 13.2.2. What *is* different is that now there are interaction energy terms between the QM and MM regions that are *not* non-bonded terms, these new terms being associated with the bonds cut by the QM/MM boundary. In practice, coupled QM/MM calculations involving link atoms tend to adopt the following protocols for computation of the various terms.

1. H_{QM} is computed for the QM region capped with hydrogen atoms at every bond cut by the QM/MM boundary. The Fock operator may be like that defined in Eq. (13.5). However, since the capping hydrogen atom is not really a part of the system, the third term on the r.h.s. is not evaluated when μ or ν is a basis function on a capping hydrogen; similarly, no nuclear repulsion between the capping hydrogen nucleus and the MM atoms is computed.

2. The energies of bonds cut by the QM/MM boundary are evaluated using the standard MM bond-stretching term (i.e., as though the QM atom were an MM atom). In addition, a very large force constant is applied to the fictitious bond angle MM–atom–QM–atom–capping–H so that it remains essentially zero (note that this connectivity choice avoids the difficulty of working with bond angles near π radians).

3. Angle bending energies involving two MM atoms and one QM atom are computed using the standard force-field formulation. Angle bending terms involving one MM atom and

two QM atoms have been included in some studies and deliberately not included in others. There is no strong consensus on which, if either, approach is better.

4. Torsional energies involving two or three MM atoms and two or one QM atoms, respectively, are computed using the standard force-field formulation. Torsional energies involving one MM atom and three QM atoms have, like bond angles, been included in some studies and not in others.

5. Those MM point charges that are very close to the QM system, either coincidentally or because the capping hydrogen atoms bring electron density out to the MM boundary atoms, can have unphysically large influences on the electronic structure of the QM region, even when those portions of the Fock operator involving the basis functions of the capping atoms do not include their influence directly. As a result, some studies have zeroed charges on near-boundary atoms, others have scaled them, and still others have selectively kept and discarded particular interactions (Eurenius *et al.* 1996; Bakowies and Thiel 1996). There is increasing evidence that it is better to maintain all QM/MM charge interactions. Amara and Field (2003) have shown that these electrostatic interactions can be made considerably more stable by replacing MM point charges near the QM/MM boundary with spherical Gaussian charge distributions centered on the MM atoms in question.

6. All remaining terms associated with $H_{QM/MM}$ and H_{MM} are calculated in the usual way according to Eq. (13.4).

The use of a hydrogen atom as a capping atom is clearly motivated by simplicity. It is a reasonable choice based on other considerations as well, however. In general, the position of the QM/MM boundary is selected so that it will not cut across any particularly polar or polarizable bonds. This in principle allows the correct separation of the two electrons in the (single) bond to the one that will remain in the QM region and the one that will be eliminated in the MM region. In practice, then, the bonds that are inevitably cut in biomolecules, for instance, are C–C bonds between sp^3 carbon atoms. Hydrogen is then a reasonable choice for a capping atom because the electronegativities of H and C are not too different. Nevertheless, a potentially better choice is a pseudo-halogen having seven valence electrons and an electronegativity similar to that of carbon. The 'lone pairs' on such a capping atom will then resemble the electrons from the other bonding orbitals that would reside on the atom if the system were fully QM, which may offer a better representation of the system; Zhang, Lee, and Yang (1999) have provided an initial description of a method employing this protocol, using a pseudopotential for the core electrons that provides the appropriate electronegativity behavior.

To date, the use of link atoms has been associated with extra instability in MD simulations at the QM/MM level because of, *inter alia*, the stiff force constants maintaining linearity of bonds crossing the boundary and the large electrostatic interactions involving atoms near the boundary. Progress in this area, addressing the above and other issues, is expected to continue briskly.

13.3.3 Frozen Orbitals

In practice, the source of the greatest instability in the link atom approach is the strong interactions that can develop between the wave function in regions near the link atom and nearby MM atoms carrying partial atomic charges. As noted above, one can attempt to eliminate some of those interactions by fiat, but this tends to lead to other instabilities.

A significant contributor to this problem is the point charge nature of molecular mechanics charges. Smeared-out charge densities are more physically realistic and less prone to computational problems. This suggests that a worthwhile approach would be to have some sort of buffer layer between the polarizable QM region and the point-charge-represented MM region that would be itself represented by a continuous *unpolarizable* charge density, i.e., one that is not reoptimized as part of an SCF procedure.

While the details are somewhat mathematically tedious, the conceptual basis can be explained relatively simply. The full system is now partitioned into three regions, which may be called the MM region, the auxiliary region, and the QM region. The new auxiliary region is characterized by nuclei having their normal nuclear charges, and electron density expressed in some set of basis functions. Equation (13.1) may now be generalized to yield a Hamiltonian

$$H_{\text{complete}} = H_{\text{QM}} + H_{\text{aux}} + H_{\text{MM}} + H_{\text{QM/aux}} + H_{\text{QM/MM}} + H_{\text{aux/MM}} \qquad (13.7)$$

Compared to Eq. (13.1), there are three new terms, all involving the auxiliary region. However, two of these terms are entirely classical, H_{aux} and $H_{\text{aux/MM}}$. The first is simply the electrostatic interaction of the frozen density and its nuclei with themselves, while the second is the interaction of the frozen density and its nuclei with the MM point charges and non-bonded LJ terms between the two regions.

As for the $H_{\text{QM/aux}}$ term, it is in principle not much different than $H_{\text{QM/MM}}$, except that instead of adding one-electron integrals over atomic partial charges to the Fock operator it adds two-electron integrals with the orbitals for one electron being frozen. There is one additional subtle point, and that is that the MOs of the optimized QM wave function must be orthogonal to the MOs describing the frozen density.

Thus far, there have been two reasonably carefully described models implementing the broad philosophy outlined above. However, in reducing the above outlines to practical calculation, certain issues must be considered. First, it would be nice if the frozen density region would remain a constant throughout the course of a simulation (if it has to be recalculated constantly, it is really not much different from the fully QM region). One means to accomplish this is to associate the density with localized orbitals, e.g., sp^n hybrids on first-row atoms if the system is a protein. In addition, the generation of the frozen density is usually accomplished by a fully QM calculation on the sum of the QM and auxiliary regions, followed by the freezing of the auxiliary portion of the density. As such, the auxiliary region cannot be too large. To date, it has been limited to the atoms at the QM/MM boundary.

The first reported approach along these lines was the localized self-consistent-field (LSCF) method of Ferenczy *et al.* (1992), originally described for the NDDO level of theory. In this case, the auxiliary region consists of a single frozen orbital on each QM boundary atom, usually taken to be that atom's contribution to a localized orbital from a fully QM calculation on a slightly expanded region (that might itself have been capped at some boundary if necessary). The population of the orbital – as fully paired spin density – may either be treated as a free variable, or computed from the density matrix of the original QM calculation. At the NDDO level, once the spatial orientation of the orbital and its s to p ratio have been set, its orthogonality to all other orbitals may be very simply enforced in QM/MM calculations. To maintain an overall zero charge on the QM + auxiliary regions, it is necessary that the total number of electrons in the auxiliary orbitals be equal to the total number of such orbitals, so some care must be taken to ensure excess charge does not introduce problems. The LSCF method, then, looks very much like the link atom method, except that the orbitals describing QM-atom–capping-atom bonds are not optimized as part of the SCF, but are instead treated as frozen throughout. Extension of the LSCF formalism to *ab initio* HF and DFT levels of QM theory has been described by Philipp and Friesner (1999).

In comparison, a larger auxiliary region is employed in the generalized hybrid orbital (GHO) approach described by Gao *et al.* (1998). In this case, it is better to think of the QM/MM boundary as passing through an atom instead of through bonds, as certain carbon atoms are assigned both QM and MM character. On those atoms, three sp^3 orbitals are held frozen with paired-spin-density populations equal to one minus one-third of the partial atomic charge the atom would carry for MM purposes, i.e., there is an attempt to spread out the character of the boundary atom over its frozen orbitals. The remaining orbital, pointing to the QM region, is frozen in *shape* by orthogonality constraints, but its population and contribution to the various MOs is free to vary according to the SCF procedure. Thus, there is again a similarity to the link atom procedure, in that there is a fully optimized MO representing each bond at the QM/MM frontier, but in this case the orbital is surrounded by a much more realistic charge environment from the hybrid atom nucleus and its three frozen auxiliary orbitals. The three different approaches are compared schematically in Figure 13.6.

A subtle but key difference in the methodologies is that the orbital containing the two electrons in the C–X bond is frozen in the LSCF method, optimized in the context of an X–H bond in the link atom method, and optimized subject only to the constraint that atom C's contribution be a particular sp hybrid in the GHO method. In the link atom and LSCF methods, the MM partial charge on atom C interacts with some or all of the quantum system; in the GHO method, it is only used to set the population in the frozen orbitals.

The GHO approach has been designed in such a way that the QM/MM atoms at the boundary are intended to be transferable. Thus, hybrid atoms have modified semiempirical parameters and force-field parameters for use in computing the QM and MM portions of the QM/MM energy according to Eq. (13.4), supplemented by MM bond stretching, angle bending, and torsional terms whenever any one atom in the relevant linkage is a purely MM atom. The modifications have been made for the combination of the AM1 Hamiltonian and the CHARMM force field so as best to reproduce structural, energetic, and charge results from fully AM1 calculations for a spectrum of molecules bearing functional groups similar to those found in proteins. A CHARMM/PM3 implementation has also been reported (Garcia-Viloca and Gao 2004) as has the formalism for an *ab initio* Hartree–Fock GHO method (Pu, Gao, and Truhlar 2004).

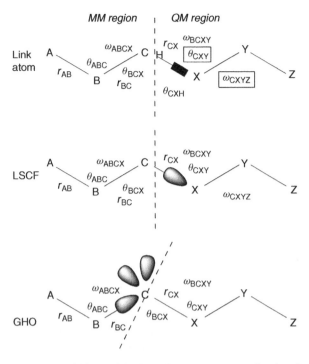

Figure 13.6 Comparison of QM/MM partitioning schemes across covalent bonds. Included MM bond stretch, angle bend, and torsion terms are indicated; those that are boxed are ignored by some authors. Frozen orbitals are in gray for the LSCF and GHO methods

Analytic derivatives have been reported for both the LSCF and GHO models, making them attractive options for MD simulations (Amara *et al.* 2000). Their generalization to *ab initio* levels of theory through the use of core pseudopotentials (along the lines of the pseudo-halogen capping atoms described above) ensures that they will see continued development.

13.4 Empirical Valence Bond Methods

A method that has certain connections with QM/MM techniques even if it does not usually involve simultaneous evaluation of QM and MM operators during a particular calculation is the empirical valence bond method (EVB; Warshel and Weiss 1980). At the heart of the EVB method is the notion that arbitrarily complex reactions may be modeled as the influence of a surrounding environment on a fundamental process that may be represented by some combination of valence bond resonance structures. For example, the proton transfer from one water molecule to another may, at any point along the reaction path, be envisaged as involving some admixture of the two VB wave functions corresponding formally to

$$\Phi_1 = HO^a{-}H^* + O^bH_2 \qquad (13.8)$$

$$\Phi_2 = HO^{a-} + H^*{-}O^bH_2{}^+ \qquad (13.9)$$

That is, these two wave functions are taken as the *basis* functions that are linearly combined to describe the system at an arbitrary point along the proton-transfer coordinate (the proton that is transferred has been labeled with an asterisk and the two oxygen atoms labeled a and b for ease of subsequent discussion).

Most chemists are quite comfortable thinking of chemical structure and reactivity in terms of valence bond notions – the resonance structures so often invoked in organic chemistry are one example of this phenomenon – so this approach has conceptual appeal. From a computational standpoint, the issue is how to derive a Hamiltonian operator that will act on VB wave functions so as to deliver useful energies.

13.4.1 Potential Energy Surfaces

VB wave functions like those in Eqs. (13.8) and (13.9) are in some sense MM-like representations of a chemical system. We insist, for instance, that the system described by Eq. (13.8) always has H^* bound to O^a, irrespective of the length of the bond at any given moment. Obviously, however, if the separation between those two atoms is large, it is absurd to imagine that there is a bond between them. Put more quantum mechanically, one would say that the contribution of that VB basis function to the 'correct' adiabatic ground state is very small. Thinking of the adiabatic wave function as a linear combination of the VB basis functions, we would say that the coefficient of Φ_1 in a CI-like expansion should be small, which is equivalent to saying that it must be at rather high energy relative to states making larger contributions.

All of these qualitative considerations suggest that a first step to designing a Hamiltonian for the VB system would be to use a simple force field where the making and breaking $O–H^*$ bonds are described by a Morse potential, the other OH bonds and bond angles in each of the two molecular fragments are described by harmonic potential functions, and interactions between the two fragments are modeled by standard electrostatic and LJ potential functions. That is, we would have for the uncharged VB Hamiltonian

$$H_1 = D_{OH} \left[1 - e^{-\alpha_{OH}(r_{O^aH^*} - r_{OH,eq})} \right]^2 + \sum_{H \neq H^*} \frac{1}{2} k_{OH} (r_{OH} - r_{OH,eq})^2$$

$$+ \frac{1}{2} k_{HOH} (\theta_{HO^aH^*} - \theta_{HOH,eq})^2 + \frac{1}{2} k_{HOH} (\theta_{HO^bH} - \theta_{HOH,eq})^2$$

$$+ \sum_{i \in a} \sum_{j \in b} \frac{q_i q_j}{\varepsilon r_{ij}} + \sum_{i \in a} \sum_{j \in b} 4\varepsilon_{ij} \left[\left(\frac{\sigma_{ij}}{r_{ij}} \right)^{12} - \left(\frac{\sigma_{ij}}{r_{ij}} \right)^6 \right] \qquad (13.10)$$

where the fragments a and b are the molecules containing the oxygen atoms having the same label. Although the bond involving H^* is unique in using a Morse potential, all MM terms are otherwise standard and assume a defined connectivity consistent with Eq. (13.8).

For the charged VB wave function, we take

$$H_1 = D_{OH} \left[1 - e^{-\alpha_{OH}(r_{O^bH^*} - r_{OH,eq})} \right]^2 + \sum_{H \neq H^*} \frac{1}{2} k_{OH} (r_{OH} - r_{OH,eq})^2$$

$$+ \sum_{H \in b} \frac{1}{2} k_{HOH} (\theta_{HO^bH} - \theta_{HOH,eq})^2$$

$$+ \sum_{i \in a} \sum_{j \in b} \frac{q_i q_j}{\varepsilon r_{ij}} + \sum_{i \in a} \sum_{j \in b} 4\varepsilon_{ij} \left[\left(\frac{\sigma_{ij}}{r_{ij}} \right)^{12} - \left(\frac{\sigma_{ij}}{r_{ij}} \right)^{6} \right] + \Gamma_2 \qquad (13.11)$$

where, with the exception of the final term, differences between Eqs. (13.10) and (13.11) simply reflect the differences in $O-H^*$ connectivity between Eqs. (13.8) and (13.9). The final term, Γ_2, is a parameter that is adjusted to make the relative energies of the two VB wave functions correct at their respective minima. In the current example Γ_2 may be determined as the energy of proton transfer from one water molecule to another in the gas phase (we are still considering only a two-molecule system), which energy is available from mass spectral measurements.

If we were to partially optimize the geometries of the systems having the two possible connectivities, holding fixed only the difference in the two bond lengths from O^a and O^b each to H^*, we would obtain the set of energy curves for the two VB Hamiltonians shown in Figure 13.7. A crude definition of the energy for the reaction coordinate would then be simply to take the minimum of H_1 or H_2 as one proceeds from left to right in the proton transfer process. Mathematically, that process is equivalent to taking as our energy the lowest eigenvalue of the 2×2 matrix

$$\mathbf{H} = \begin{bmatrix} H_{11} & H_{12} \\ H_{21} & H_{22} \end{bmatrix} \qquad (13.12)$$

where H_{11} is taken to be H_1, H_{22} is taken to be H_2, and H_{12} and H_{21} are taken to be zero. The matrix then being diagonal, the lowest eigenvalue is simply the lower of H_1 or H_2.

However, Eq. (13.12) is not simply an odd exercise in matrix algebra. Instead, it suggests a more chemical approach to obtaining the energy of the reacting system. Along the way from the VB structure of Eq. (13.8) to that of Eq. (13.9), the system obviously passes through a region where it is best described as a mixture of these two extreme resonances. The mathematical way in which this mixing can be accomplished is to allow the off-diagonal matrix elements in Eq. (13.12) to be non-zero. The most widely used approximation for these off-diagonal elements in a case like the one discussed thus far is

$$H_{12} = H_{21} = Ae^{-B(r_{OO} - r_{OO,'eq'})} \qquad (13.13)$$

where A, B, and $r_{OO,'eq'}$ are parameters to be optimized against experimental data. Thus, the coupling is designed to decrease exponentially as the two heavy atoms separate from some maximally coupled distance (for more complicated approaches to model H_{12} see, for example, Chang, Minichino, and Miller 1992 and Kim *et al.* 2000). With a non-zero coupling, diagonalization of $\mathbf{H}$ in Eq. (13.12) will give a curve for the lowest eigenvalue, H_0 shown in Figure 13.7, that smoothly connects the two VB extrema. In addition, the eigenvector

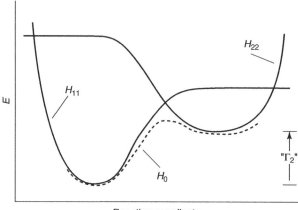

Reaction coordinate

Figure 13.7 Example of potential energy curves for separate VB Hamiltonians and the curve for the lowest energy eigenvalue when the separate Hamiltonians are coupled by an off-diagonal term in a 2×2 Hamiltonian matrix. Note that the difference between the minima for H_{11} and H_{22} is the term Γ_2 in the example given in the text only if all non-bonded and electrostatic interactions are identical in the two VB representations, and thus the quotes around the label

associated with that eigenvalue provides the coefficients for what amounts to a 2×2 CI wave function in the basis of the VB functions.

Note that the values of the various constants required, both in the 'normal' force field expressions and in the term(s) specific to the EVB formalism, may be determined from either experiment or high-level *ab initio* calculations or both.

The connection with QM/MM formalism arises when we consider immersing our gas-phase EVB system in an explicitly represented surrounding medium. In the proton transfer case, for instance, we might choose to immerse the system in a box of MM water molecules. The usual EVB assumption is that the interaction of the original VB system with the surroundings serves only to modify the diagonal terms in Eq. (13.12), and it does so only by adding the MM energy associated entirely with the solvent (which is the same whether added to H_{11} or H_{22}) and also adding non-bonded interactions that potentially *do* differentiate between the two VB basis functions (e.g., if the partial charge on H^* differs in Φ_1 and Φ_2). In the condensed phase, then, Eq. (13.12) becomes

$$\mathbf{H} = \begin{bmatrix} H_{11} + V_{1S} + V_{SS} & H_{12} \\ H_{21} & H_{22} + V_{2S} + V_{SS} \end{bmatrix} \tag{13.14}$$

where V_{SS} is the intrasolute energy, V_{iS} is the non-bonded energy between the surroundings and the VB system i, and all other terms are defined as previously.

13.4.2 Following Reaction Paths

The particularly attractive feature of EVB theory is that when it comes to modeling a reaction, one does not need to follow a reaction coordinate in some geometrical variable; rather, one follows a coordinate from one VB optimum to another. The full technical details are not examined here, but an outline is provided – further details are available in the additional reading at the end of the chapter.

To implement the reaction path following scheme in the EVB formalism, one defines a λ-dependent mapping

$$H_\lambda = (1 - \lambda)H_{11} + \lambda H_{22} \tag{13.15}$$

where λ is the usual coupling parameter for FEP that runs from 0 to 1 in discrete steps. Assuming sufficiently small steps in λ, one can define a free energy change using

$$\Delta G^* = \sum_\lambda -RT \ln\langle e^{-(H_{\lambda+1}-H_\lambda)/RT}\rangle_\lambda \tag{13.16}$$

which has the usual appearance for FEP when molar energies are used (cf. Eq. (12.16)).

However, for every value of λ Eq. (13.15) defines an energy surface that is different from the true PES, unless by coincidence at one or several points. This situation, of evaluating free energy changes from trajectories moving on adjusted free energy surfaces, is reminiscent of umbrella sampling (see Section 12.2.5), where energy modifications are introduced as a function of certain geometric coordinates in order to force trajectories into otherwise low probability regions of space. Just as is the case with umbrella sampling, the *correct* free energy change for the transition from one VB minimum to the other must be determined from an evaluation of each trajectory, using in particular the difference between the correct potential and the one used to generate the trajectory (see Eq. (12.26)). In the EVB case, that is the difference at any point between H_λ and the lowest eigenvalue of **H** as defined by Eq. (13.14).

The results of such simulations can be used to further refine the various parameters appearing in the VB energy expressions. In the case presented here, for instance, the parameters would be adjusted to provide the proper pK_a for water. The motivation for the extensive parameter validation is typically then to move the EVB system into an environment where experimental data are *not* well understood, e.g., an enzyme active site. Thus, the procedure outlined above was used by Åqvist and Warshel (1992) to model the initial deprotonation of water that is the rate-determining step in the hydration of carbon dioxide to bicarbonate catalyzed by carbonic anhydrase. After optimizing the EVB parameters on gas-phase and aqueous water–water proton transfers, the surrounding medium was changed to that of the solvated enzyme (represented by a force field and thus interacting with the VB Hamiltonian just as described above for surrounding water). Following this approach they obtained very close agreement with the experimentally measured catalytic effect of the enzyme. By inspecting simulation snapshots from the portion of the free-energy curve corresponding to the transition state region, they were able to gain structural insight into the catalytic mechanism of proton translocation.

13.4.3 Generalization to QM/MM

The EVB method as outlined above is not formally a QM/MM method during the course of any simulation. Instead, the connection to QM is that the parameters required for the EVB matrix elements may be determined in part from QM calculations. However, that could be said of almost any force-field parameter.

A closer relationship between EVB and QM/MM is apparent when some components of the matrix elements are computed 'on the fly' at a quantum mechanical level. Mo and Gao (2000), for instance, have described such a technique in the absence of an MM region where an EVB Hamiltonian that includes QM-computed terms is coupled with a surrounding QM region in a non-SCF fashion. The extension of this methodology to include an MM region follows naturally from the QM/MM couplings described above.

It should be apparent that when taken to its QM limit the EVB process simply becomes multi-state CI (see Chapter 14) for a QM system coupled to a classical environment. However, the enormous cost that would be associated with carrying out such a CI calculation at a level sufficiently accurate to compete with an empirically parameterized set of potential functions has inhibited any developments along these lines. Of course, there are interesting systems where data for empirical parameterization are lacking, but the cost of the multi-state CI treatment is still sufficiently expensive that it has not yet attracted any attention.

13.5 Case Study: Catalytic Mechanism of Yeast Enolase

Synopsis of Alhambra *et al.* (1999) 'Quantum Mechanical Dynamical Effects in an Enzyme-catalyzed Proton Transfer Reaction'.

The enzyme enolase catalyzes the dehydration of D-2-phosphoglycerate to phosphoenol-pyruvate, a crucial step in the synthesis of carbohydrates for energy storage. A remarkable feature of this reaction is that the rate determining step is removal of a proton from the C-2 position of the reactant that is expected to have a pK_a of no less than 32. The base used for this removal is a lysine ε-amino group; additional driving force for the reaction is stabilization of the intermediate tetraanion (the reactant is a trianion) by coordination to two magnesium ions (Figure 13.8).

Alhambra and co-workers adopted a QM/MM strategy to better understand quantum mechanical effects, and particularly the influence of tunneling, on the observed primary kinetic isotope effect of 3.3 in this system (that is, the reaction proceeds 3.3 times more slowly when the hydrogen isotope at C-2 is deuterium instead of protium). In order to carry out their analysis they combined fully classical MD trajectories with QM/MM modeling and analysis using variational transition-state theory. Kinetic isotope effects (KIEs), tunneling, and variational transition state theory are discussed in detail in Chapter 15 – we will not explore these topics in any particular depth in this case study, but will focus primarily on the QM/MM protocol.

The authors' approach does not actually use a trajectory to investigate the kinetics of the enolase-catalyzed reaction. Rather, the trajectory is designed to create a 'reasonable' configuration of the MM part, which is then held frozen as an environment surrounding the QM part, which is subjected to more detailed investigation. Thus, they began from a crystal structure of the enzyme with substrate bound, and propagated a classical trajectory (i.e., no portion was treated via QM) for 50 ps to equilibrate the system. The force field used in this case was CHARMM22. Then, $r_{breaking}$ and r_{making} in Figure 13.8 were held in the region of 1.58 and 1.21 Å, respectively, by adding strong harmonic constraints on these bond lengths, and the trajectory was propagated for an additional 10 ps. In principle, this

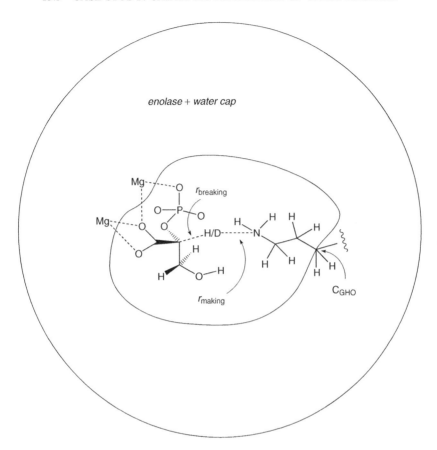

Figure 13.8 A 25-atom quantum subsystem embedded in an 8863-atom classical system to model the catalytic step in the conversion of D-2-phosphoglycerate to phosphoenolpyruvate by enolase. What factors influence the choice of where to set the boundary between the QM and MM regions? Alhambra and co-workers found, using variational transition-state theory with a frozen MM region that was selected from a classical trajectory so as to make the reaction barrier and thermochemistry reasonable, that the breaking and making bond lengths were 1.75 and 1.12 Å, respectively, for H, but 1.57 and 1.26 Å, respectively, for D

portion of the trajectory should be representative of the protein structure in the region of the TS for the reaction.

From this 10 ps region, random structures were selected, and QM/MM calculations were carried out at the AM1/CHARMM22 level, with the boundary carbon atom of the lysine side-chain modeled using the GHO approach. In these calculations, the geometry of the MM region was held frozen, but the QM region was optimized to find reactant, product, and transition-state structures. When the computed free energies of activation and of the overall reaction from these structures matched sufficiently closely to experiment, it was assumed that the MM structure was representative of a typical protein configuration in the vicinity of the TS, and a more detailed analysis of the kinetics was undertaken to better understand the experimentally observed isotope effects.

A key point made by the authors is that the neglect of quantum effects in purely classical simulations, particularly zero-point vibrational energy, can have very large effects on computed kinetic quantities. Thus, when the free energy of activation is computed using the QM/MM potential energies but a purely classical formalism for the vibrational partition function (i.e., by using a continuous integral in place of the discrete sum that is used for a QM harmonic oscillator; see Eq. (10.26)), the rate is underestimated by a factor of 34 compared to the complete QM/MM prediction. A still larger discrepancy is seen in the kinetic isotope effect: the purely classical treatment of vibrations predicts a KIE of 1.3, while the fully QM/MM protocol predicts 3.5 after accounting for a small tunneling effect. The latter method is in much better agreement with the experimental value of 3.3.

Having successfully matched the several experimental observables available for the enolase system, Alhambra and co-workers then examine the reaction coordinate to better understand the factors discriminating H from D reactivity. They discover that the TS for the reaction of H is much later than that for reaction of D, because the rapidly increasing zero-point energy of the N–H bond compared to the N–D bond offsets the drop in reaction coordinate potential energy and moves the free-energy bottleneck for H further towards products.

The authors finish by predicting a quantity that has not yet been measured, the secondary KIE that would be obtained with $-ND_2$ in place of $-NH_2$ as the reactive base. They also note the critical role of the Mg^{2+} counterions in facilitating the reaction. The binding to the two cations of the phosphoenolpyruvate tetraanion compared to the reactant trianion is predicted to be increased by some 240 kcal mol^{-1}; this offsets the highly unfavorable pK_a of the C-2 proton sufficiently to render the overall reaction thermochemistry only 2.8 kcal mol^{-1} endergonic.

One difficulty associated with the authors' methodology is that, in the absence of having substantial experimental data in hand, it is not in general obvious that a random selection of a frozen MM structure from an equilibrated trajectory will provide a useful configuration. In this instance, the authors were able to validate the quality of their framework geometry and go on to perform more in depth analyses of microscopic features of the reaction; an *a priori* prediction of the reaction's free-energy profile, however, would require a more complicated consideration of an ensemble of structures. A detailed protocol for such an endeavor was subsequently described (Alhambra *et al.* 2001), and has been applied to other systems.

Bibliography and Suggested Additional Reading

Åqvist, J. and Warshel, A. 1993. 'Simulation of Enzyme Reactions Using Valence Bond Force Fields and Hybrid Quantum/Classical Approaches', *Chem. Rev.*, **93**, 2523.

Elstner, M., Frauenheim, T., Kaxiras, E., Seifert, G., and Suhai, S. 2000. 'A Self-consistent Charge Density-functional Based Tight-binding Scheme for Large Biomolecules', *Phys. Stat. Sol. B*, **217**, 357.

Field, M. J. 2002. 'Simulating Enzyme Reactions: Challenges and Perspectives' *J. Comput. Chem.*, **23**, 48.

Gao, J. 1996. 'Methods and Applications of Combined Quantum Mechanical and Molecular Mechanical Potentials', in *Reviews in Computational Chemistry*, Vol. 7, Lipkowitz, K. B. and Boyd, D. B., Eds., VCH: New York, 119.

Gao, J., Amara, P., Alhambra, C., and Field, M. J. 1998. 'A Generalized Hybrid Orbital (GHO) Method for the Treatment of Boundary Atoms in Combined QM/MM Calculations', *J. Phys. Chem. A*, **102**, 4714.

Gao, J. and Thompson, M. A., Eds., 1998. *Methods and Applications of Hybrid Quantum Mechanical and Molecular Mechanical Methods*, ACS Symposium Series, Vol. 712, American Chemical Society: Washington, DC.

Orozco, M. and Luque, F. J. 2000. 'Theoretical Methods for the Description of the Solvent Effect on Biomolecular Systems', *Chem. Rev.*, **100**, 4187.

Reuter, N., Dejaegere, A., Maigret, B., and Karplus, M. 2000. 'Frontier Bonds in QM/MM Methods: A Comparison of Different Approaches', *J. Phys. Chem. A*, **104**, 1720.

Ryde, U. 2003. 'Combined Quantum and Molecular Mechanics Calculations on Metalloproteins' *Curr. Opin. Chem. Biol.*, **7**, 136.

Théry, V., Rinaldi, D., Rivail, J.-L., Maigret, B., and Ferenczy, G. G. 1994. 'Quantum Mechanical Computations on Very Large Molecular Systems: The Local Self-consistent Field Method', *J. Comput. Chem.*, **15**, 269.

Warshel, A. 1991. *Computer Modeling of Chemical Reactions in Enzymes and in Solutions*, Wiley: New York.

Zhang, Y. K., Lee, T. S., and Yang, W. T. 1999. 'A Pseudobond Approach to Combining Quantum Mechanical and Molecular Mechanical Methods', *J. Chem. Phys.*, **110**, 46.

References

Abraham, R. J. and Siverns, T. M. 1972. *Tetrahedron,* **28**, 3015.

Alhambra, C., Gao, J., Corchado, J. C., Villà, J., and Truhlar, D. G. 1999. *J. Am. Chem. Soc.*, **121**, 2253.

Alhambra, C., Corchado, J., Sánchez, M. L., Garcia-Viloca, M., Gao, J., and Truhlar, D. G. 2001. *J. Phys. Chem. B.*, **105**, 11326.

Amara, P. and Field, M. 2003. *Theor. Chem. Acc.*, **109**, 43.

Amara, P., Field, M. J., Alhambra, C., and Gao, J. 2000. *Theor. Chem. Acc.*, **104**, 336.

Åqvist, J. and Warshel, A., 1992. *J. Mol. Biol.*, **224**, 7.

Bakowies, D. and Thiel, W. 1996. *J. Phys. Chem.*, **100**, 10580.

Bandyopadhyay, P., Gordon, M. S., Mennucci, B., and Tomasi, J. 2002. *J. Chem. Phys.*, **116**, 5023.

Banks, J. L., Kaminski, G. A., Zhou, R. H., Mainz, D. T., Berne, B. J., and Friesner, R. A. 1999. *J. Chem. Phys.*, **110**, 741.

Chambers, C. C., Giesen, D. J., Hawkins, G. D., Vaes, W. H. J., Cramer, C. J., and Truhlar, D. G. 1999. In: *Rational Drug Design*, Truhlar, D. G., Howe, W. J., Hopfinger, A. J., Blaney, J. M., and Dammkoehler, R. A., Eds., Springer: New York, 51.

Chandrasekhar, J., Shariffskul, S., Jorgensen, W. L. 2002. *J. Phys. Chem. B*, **106**, 8078.

Chang, Y.-T., Minichino, C., and Miller, W. H. 1992. *J. Chem. Phys.*, **96**, 4341.

Corchado, J. C. and Truhlar, D. G. 1998. *J. Phys. Chem. A*, **102**, 1895.

Cramer, C. J., and Pak, Y. 2001. *Theor. Chem. Acc.*, **105**, 477.

Cramer, C. J. and Truhlar, D. G. 1992. *Chem. Phys. Lett.*, **198**, 74.

Cramer, C. J. and Truhlar, D. G. 1993. *Chem. Phys. Lett.*, **202**, 567 (erratum).

Cui, Q. 2002. *J. Chem. Phys.*, **117**, 4720.

Deng, L., Woo, T. K., Cavallo, L., Margl, P. M., and Ziegler, T. 1997. *J. Am. Chem. Soc.*, **119**, 6177.

Depaepe, J.-M. and Ryckaert, J.-P. 1995. *Chem. Phys. Lett.*, **245**, 653.

Eurenius, K. P., Chatfield, D. C., Brooks, B. R., and Hodoscek, M. 1996. *Int. J. Quantum Chem.*, **60**, 1189.

Ferenczy, G. G., Rivail, J.-L., Surján, P. R., and Náray-Szabó, G. 1992. *J. Comput. Chem.*, **13**, 830.

Florián, J. and Warshel, A. 1997. *J. Phys. Chem. B*, **101**, 5583.

Freedman, H. and Truong, T. N. 2003. *Chem. Phys. Lett.*, **381**, 362.

Freindorf, M. and Gao, J. 1996. *J. Comput. Chem.*, **17**, 386.

French, A. D., Kelterer, A.-M., Cramer, C. J., Johnson, G. P., and Dowd, M. K. 2000. *Carbohydr. Res.*, **326**, 305.

Gao, J. 1994. *J. Am. Chem. Soc.*, **116**, 1563.

Gao, J. and Xia, X. 1992. *Science*, **258**, 631.

Gao, J., Amara, P., Alhambra, C., and Field, M. J. 1998. *J. Phys. Chem. A*, **102**, 4714.

Garcia-Viloca, M. and Gao, J. 2004. *Theor. Chem. Acc.*, **111**, 280.

Humbel, S., Sieber, S., and Morokuma, K. 1996. *J. Chem. Phys.*, **105**, 1959.

Kaminski, G. and Jorgensen, W. L. 1996. *J. Phys. Chem.*, **100**, 18010.

Kaminski, G. and Jorgensen, W. L. 1998. *J. Phys. Chem. B*, **102**, 1787.

Kerdcharoen, T. and Morokuma, K. 2002. *Chem. Phys. Lett.*, **355**, 257.

Kim, Y., Corchado, J. C., Villà, J., Xing, J., and Truhlar, D. G. 2000. *J. Chem. Phys.*, **112**, 2718.

Kongsted, J., Osted, A., Mikkelsen, K. V., Christiansen, O. 2003. *J. Phys. Chem. A*, **107**, 2578.

Lee, P. H. and Maggiora, G. M. 1993. *J. Phys. Chem.*, **97**, 10175.

Loeffler, H. H., Yague, J. I., and Rode, B. M. 2002. *J. Phys. Chem. A*, **1106**, 9529.

Luzhkov, V. and Warshel, A. 1992. *J. Comput. Chem.*, **13**, 199.

Martin, M. E., Aguilar, M. A., Chalmet, S., Ruiz-López, M. F. 2002. *Chem. Phys.*, **284**, 607.

Maseras, F. and Morokuma, K. 1995. *J. Comput. Chem.*, **16**, 1170.

Matsubara, T., Maseras, F., Koga, N., and Morokuma, K. 1996. *J. Phys. Chem.*, **100**, 2573.

Mo, Y. and Gao, J. 2000. *J. Phys. Chem. A*, **104**, 3012.

Papazyan, A. and Warshel, A. 1997. *J. Phys. Chem. B*, **101**, 11254.

Philipp, D. M. and Friesner, R. A. 1999. *J. Comput. Chem.*, **20**, 1468.

Pratt, L. R. and Chandler, D. C. 1977. *J. Chem. Phys.*, **67**, 3683.

Pu, J., Gao, J., and Truhlar, D. G. 2004. *J. Phys. Chem. A*, **108**, 632.

Rickard, G. A., Karadakov, P. B., Webb, G. A., and Morokuma, K. 2003. *J. Phys. Chem. A*, **107**, 292.

Shao, L., Yu, H. A., and Gao, J. L. 1998. *J. Phys. Chem. A*, **102**, 10366.

Sherer, E. C. and Cramer, C. J. 2001. *J. Comput. Chem.*, **22**, 1167.

Svensson, M., Humbel, S., and Morokuma, K. 1996. *J. Chem. Phys.*, **105**, 3654.

Svensson, M., Humbel, S., Froese, R. D. J., Matsubara, T., Sieber, S., and Morokuma, K. 1996. *J. Phys. Chem.*, **100**, 19357.

Swart, M. 2003. *Int. J. Quant. Chem.*, **91**, 177.

Ten-no, S., Hirata, F., and Kato, S. 1993. *Chem. Phys. Lett.*, **214**, 391.

Udier-Blagovic, M., Morales de Tirado, P., Pearlman, S. A., and Jorgensen, W. L. 2004. *J. Comput. Chem.*, **25**, 1322.

Vreven, T., Mennucci, B., da Silva, C. O., Morokuma, K., and Tomasi, J. 2001. *J. Chem. Phys.*, **115**, 62.

Vreven, T., Morokuma, K., Farkas, Ö., Schlegel, H. B., Frisch, M. J. 2003. *J. Comput. Chem.*, **24**, 760.

Warshel, A. and Levitt, M. J. 1976. *J. Mol. Biol.*, **103**, 227.

Warshel, A. and Weiss, R. M. 1980. *J. Am. Chem. Soc.*, **102**, 6218.

Wiberg, K. B., Keith, T. A., Frisch, M. J., and Murcko, M. 1995. *J. Phys. Chem.*, **99**, 9072.

Zhang, Y., Lee, T.-S, and Yang, W. 1999. *J. Chem. Phys.*, **110**, 46.

14

Excited Electronic States

14.1 Determinantal/Configurational Representation of Excited States

An excited electronic state is one in which at least one electron is not in as low energy an orbital as it could be given the molecular geometry. Such a state may be generated by various processes, e.g., absorption of radiation by the ground state or as the product of a chemical reaction. Although unstable relative to collapse to the lower energy ground state, some excited states may have significant lifetimes owing to inefficient coupling with the ground state, as described in more detail below.

Given M doubly occupied molecular orbitals and N empty virtual orbitals (and possibly some number of singly occupied orbitals), the number of possible excited states that can be generated is enormous. For the moment, we will restrict our discussion to the case where the ground state is closed-shell and in the excited state only a single excited electron exists (implying it must have come from the HOMO). Extending the discussion to more general cases is intuitively straightforward, but notationally somewhat tedious.

The usual manner in which chemists think about an excited state is to take the ground state Ψ_0 as context. Thus, as shown in Figure 14.1, one considers our excited state to be generated by the removal of one electron from the HOMO of the ground state and its placement into some higher-energy orbital. Since the wave function for the ground state can be represented as a single Slater determinant, e.g.

$$^1\Psi_0 = \left| \psi_1^2 \psi_2^2 \psi_3^2 \cdots \psi_{N/2}^2 \right\rangle \tag{14.1}$$

(the Slater determinant appropriate for the RHF wave function of a singlet having N electrons), then the excited state might in general be written as

$$\Psi_{N/2}^a = \left| \psi_1^2 \psi_2^2 \psi_3^2 \cdots \psi_{N/2} \overline{\psi}_a \right\rangle \tag{14.2}$$

where we use the compact notation that if an orbital is not doubly occupied, it is multiplied by an α spin function unless there is a bar over it, in which case it is multiplied by a β

Essentials of Computational Chemistry, 2nd Edition Christopher J. Cramer
© 2004 John Wiley & Sons, Ltd ISBNs: 0-470-09181-9 (cased); 0-470-09182-7 (pbk)

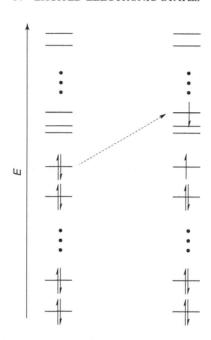

Figure 14.1 The singly excited state on the right may be qualitatively viewed as deriving from movement of an electron out of the ground-state HOMO into the indicated higher energy orbital (perhaps following absorption of a photon carrying the appropriate quantum of energy). Note, however, that the optimized orbitals of the ground state are at best approximations to those of the excited state

spin function. However, as described in more detail in Section 14.4 and Appendix C, the Slater determinant of Eq. (14.2) is not a pure spin state, but is instead an equal mixture of singlet and triplet states (the triplet being the so-called 'up-down' or '$S_z = 0$' triplet). It turns out that the corresponding pure spin states in this instance cannot be written as single Slater determinants, rather they require a linear combination of two. It is again somewhat cumbersome, however, constantly to specify the spin state rigorously in this manner, so for ease of presentation we will simply specify the spin state in a shorthand fashion, as a left superscript on the wave function, taking as implicit that the correct wave function uses the proper linear combination of determinants and is not a single determinant as written, i.e.,

$$^1\Psi_{N/2}^a = |\psi_1^2\psi_2^2\psi_3^2 \cdots \psi_{N/2}\overline{\psi}_a\rangle \tag{14.3}$$

is the singlet wave function generated from the appropriate combination of determinants having electrons of opposite spin singly occupying orbitals $N/2$ and a and

$$^3\Psi_{N/2}^a = |\psi_1^2\psi_2^2\psi_3^2 \cdots \psi_{N/2}\overline{\psi}_a\rangle \tag{14.4}$$

is the analogous triplet. Following the general notational scheme already introduced in Chapter 7, we will use the right sub- and superscripts in Ψ_i^a to indicate a state generated by removing one electron from occupied orbital i and placing it into previously empty orbital a.

Having dispensed with notational details, let us think more carefully about the chemical picture implied by Figure 14.1. By picturing an excited state as being a different occupation of the orbitals of the ground state, we are providing something of a privileged status to the ground-state orbitals. One must recall that these orbitals, in the HF procedure, are variationally optimized given an average electrostatic repulsion that depends on the shape *and occupation number* of all of the other MOs. When the excited state of Figure 14.1 is generated, the occupation number of the HOMO is reduced by 1 and the occupation number of the newly occupied orbital is increased by 1, and thus the HF field acting on *every* orbital has changed. This means that none of the occupied orbitals are optimal for the excited state, and as a result the energy of the excited state, e.g.,

$$E\left[{}^1\Psi_{N/2}^a\right] = \left\langle {}^1\Psi_{N/2}^a \,|H|\, {}^1\Psi_{N/2}^a \right\rangle \tag{14.5}$$

evaluated using Eq. (14.3) for the wave function (i.e., the Slater determinant formed from the optimized orbitals for the ground state), will be unphysically too high.

It is somewhat tempting to impose an incorrect dynamical view on the excited state when it is generated by ground-state absorption of a photon. One might imagine that the 'instantaneous' absorption process generates the wave function of Eq. (14.3), which then relaxes to the optimal wave function for the excited state by adjustment of all of the orbitals. This physical picture, however, ignores the common timescale of all electronic motion. Even as the electron is 'moving' from one orbital to the next, the orbitals whose occupation numbers are not changing are relaxing in response to changing electron–electron interactions.

To digress for a moment, a timescale separation that usually *is* valid is the Born–Oppenheimer separation of the nuclear and electronic motions. Thus, as illustrated in Figure 14.2, when the optimal geometry of the ground state is not the same as that for the excited state, we may view the absorption of radiation as taking place at the ground-state geometry, and the energy involved is called the 'vertical' excitation energy. On the timescale of nuclear dynamics, the geometry of the excited state ultimately relaxes to its own optimum. The energy difference ΔE between the two systems taking each to be at its own optimal geometry is referred to as the 'adiabatic' excitation energy. The same conceptual framework may be applied to the reverse process, emission. Note that excitation and emission energies are often expressed not in energy units but instead in terms of the wavelength of radiation corresponding to that energy according to

$$\Delta E = \frac{hc}{\lambda} \tag{14.6}$$

where h is Planck's constant, c the speed of light in a vacuum, and λ the radiation wavelength. A larger ΔE value is equivalent to a shorter wavelength, so one says that vertical excitations are blue-shifted (since blue light is on the short wavelength side of the visible spectrum) relative to adiabatic excitation energies, while vertical emissions with smaller ΔE values are red-shifted (red being at the opposite end of the visible spectrum) relative to the same adiabatic standard. The difference between the vertical excitation and emission energies (or wavelengths) is referred to as the Stokes shift.

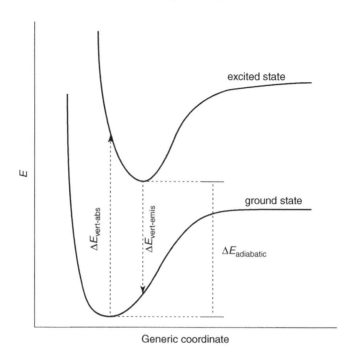

Figure 14.2   Schematic relationship between vertical absorption and emission energies and the adiabatic energy difference between the ground and excited states. [Note that a more rigorous treatment requires inclusion of ZPVE and thermal contributions in the adiabatic energy difference, and consideration of Franck–Condon overlap between quantized vibrational states for the vertical processes; some of these points are discussed in Section 14.5.]

In terms of computing adiabatic energy differences, if the Born–Oppenheimer PES for the excited state can be computed, geometry optimization of that state may be carried out using standard techniques. But, as we have been discussing above, we have not yet devised a scheme for computing the excited-state surface, since ground-state orbitals are not appropriate for minimum-determinantal excited-state wave functions. How then to obtain a better excited-state wave function?

The simplest approach, of course, is to maintain the minimum-determinantal description and reoptimize all of the orbitals. In practice, however, such an approach is practical only in instances where the ground-state and the excited-state wave functions belong to different irreducible representations of the molecular point group (cf. Section 6.3.3). Otherwise, the variational solution for the excited-state wave function is simply to collapse back to the ground-state wave function! And, even if the two states *do* differ in symmetry, the desired excited state may not be the *lowest energy* such state within its irrep, to which variational optimization will nearly always lead.

In certain favorable instances, one *can* coax the SCF equations to converge to different determinants of the same electronic state symmetry. For instance, phenylnitrenes have two different closed-shell singlet states, as re-illustrated in Figure 14.3 (cf. Section 8.5.3),

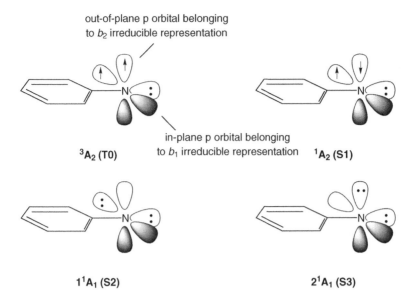

$^3A_2$ (T0) $^1A_2$ (S1)

1^1A_1 (S2) 2^1A_1 (S3)

Figure 14.3 Electronic configurations of phenylnitrene, differing in occupation of the nitrogen p orbitals, labeled according to their spin and spatial symmetries. Relative energy orderings for the four configurations are indicated in parentheses; T0 is the triplet ground state, and Sn represents the nth lowest singlet excited state

differing in whether the occupied nitrogen p orbital is in the plane of the aromatic ring ($S2$) or perpendicular to it and conjugated with the π system ($S3$). As both states are closed-shell, both belong to the totally symmetric irreducible representation (e.g., the two states are $^1A_1$ since phenylnitrene belongs to the C_{2v} point group), but the different symmetries of the alternative lone pair orbitals (one being b_1 and the other b_2) makes it relatively straightforward to converge variationally optimized wave functions that may be written as

$$^1\Psi_i = |\cdots_i b_1^2\rangle \tag{14.7}$$

and

$$^1\Psi_j = |\cdots_j b_2^2\rangle \tag{14.8}$$

(for HF example, see Kim, Hamilton, and Schaefer 1992; for corresponding DFT example, see Smith and Cramer 1996; Johnson and Cramer 2001). Note that in this case the ellipsis in each wave function carries a subscript to emphasize that while the orbitals thereby implied in Eqs. (14.7) and (14.8) are qualitatively similar, they are not *identical* (since each set was variationally optimized subject to different HOMOs). This point is not merely technical, but represents a critical problem associated with the approach, namely that there is no guarantee that the two wave functions of Eqs. (14.7) and (14.8) are orthogonal, as they should be.

The issue of orthogonality is an important one. Every excited state must be orthogonal to the ground state (as well as to all of the other excited states), and any technology for describing excited states that fails to enforce such orthogonality must be viewed with caution. One might, of course, be tempted to say that the failure of an excited state to be orthogonal to the ground state would be sufficiently damning to warrant no further use of the excited-state wave function. However, there is some room for ambiguity, insofar as the excited-state wave function must be orthogonal to the *exact* ground-state wave function, but we are almost never working with that wave function, only some approximation thereto. Thus, an *exact* excited-state wave function may very well fail to be orthogonal to, say, the HF *approximation* to the ground-state wave function.

In some cases, orthogonality is ensured by the individual natures of the two states. As already alluded to above, if the electronic states belong to two different irreps of the molecular point group, and the product of the two irreps fails to contain the totally symmetric representation, then the two states are necessarily orthogonal (see Appendix B). Taking again the phenylnitrene system in Figure 14.3 as an example, the lowest energy singlet is open-shell and has a single electron occupying *each* of the two nitrogen p orbitals. By analogy to Eqs. (14.3), (14.7), and (14.8), this formally two-determinantal wave function is

$$^1\Psi_k = \left| \cdots_k \, b_{1,k} \overline{b}_{2,k} \right\rangle \tag{14.9}$$

where the k subscripts on all orbitals emphasize their possible differences with those optimized for the wave functions of Eqs. (14.7) and (14.8). The electronic state symmetry of Eq. (14.9) is $^1A_2$. Since the product of A_2 with A_1 in the C_{2v} point group is A_2, which is not the totally symmetric representation, the orthogonality of the A_2 wave function of Eq. (14.9) with the A_1 wave functions of Eqs. (14.7) and (14.8) is ensured.

A different guarantee of orthogonality arises if the two states in question have different spin. Continuing with the phenylnitrene system, the *ground* state is the *triplet* version of Eq. (14.9), i.e.,

$$^3\Psi_0 = \left| \cdots_0 \, b_{1,0} \overline{b}_{2,0} \right\rangle \tag{14.10}$$

Orthogonality of the singlet and triplet spin coordinates ensures that the wave function of Eq. (14.10) is orthogonal to *all* of those in Eqs. (14.7)–(14.9).

Methods for generating excited-state wave functions and/or energies may be conveniently divided into methods typically limited to excited states that are well described as involving a single excitation, and other more general approaches, some of which carry a dose of empiricism. The next three sections examine these various methods separately. Subsequently, the remainder of the chapter focuses on additional spectroscopic aspects of excited-state calculations in both the gas and condensed phases.

14.2   Singly Excited States

For an average molecule, there are typically one or more low-energy excited states that may be reasonably well described as valence-MO-to-valence-MO single electronic excitations, and the language of spectroscopy reflects this point. Thus certain states are referred to as $n \rightarrow \pi^*$, $\pi \rightarrow \pi^*$, etc., indicating the orbital from which the electron is excited on the left and the orbital into which it is excited on the right. Finding wave functions for these states is sometimes facilitated by this relative simplicity of their character.

14.2.1 SCF Applicability

Ideally, one would like to study excited states and ground states using wave functions of equivalent quality. Ground-state wave functions can very often be expressed in terms of a single Slater determinant formed from variationally optimized MOs, with possible accounting for electron correlation effects taken thereafter (or, in the case of DFT, the optimized orbitals that intrinsically include electron correlation effects are use in the energy functional). Such orbitals are determined in the SCF procedure.

However, the problem of variational collapse typically prevents an equivalent SCF description for excited states. That is, any attempt to optimize the occupied MOs with respect to the energy will necessarily return the wave function to that of the ground state. Variational collapse can sometimes be avoided, however, when the nature of the ground and excited states prevents their mixing within the SCF formalism. This situation occurs most commonly in symmetric molecules, where electronic states belonging to different irreducible representations do not mix in the SCF, and also in any situation where the ground and excited states have different spin.

As an example of the former, consider the electronic states of fluorovinylidene illustrated in Figure 14.4. There are two different low lying triplet states, one having A′ electronic state symmetry and the other A″. Furthermore, within each respective irreducible representation, the states indicated are the *lowest energy* triplets. Thus, wave functions for each may be determined via an SCF approach. In this case, HF theory is not particularly attractive as an

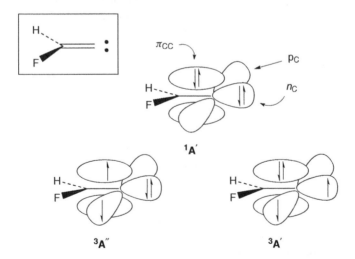

Figure 14.4 Valence MO occupations for three different electronic states of fluorovinylidene. The triplet states belong to different irreducible representations of the molecular point group because the singly occupied orbitals in which they differ belong alternatively to either the *a′* or *a″* irreps

option, owing to high spin contamination in the final wave functions. With DFT, however, this problem does not here arise (see Section 8.5), and at the BVWN5/cc-pVDZ and BLYP/cc-pVDZ levels of theory, the A″ triplet is predicted to be lower in energy than the A′ triplet by 2.3 and 1.6 kcal mol^{-1}, respectively (Worthington and Cramer 1997). This only slightly overestimates the experimental result of 0.9 kcal mol^{-1} (Gilles, Lineberger, and Ervin 1993).

Although the A″ triplet is the lowest energy electronic state of fluorovinylidene having triplet spin, the closed-shell singlet is lower in energy still, and is the ground state. Naturally, then, it too is amenable to an SCF description. Note that there can be no variational collapse of the A″ triplet to the A′ singlet, not only because the spatial symmetries of the two wave functions belong to different irreducible representations but also because the spin states are different. The predicted ΔSCF energy difference between the ^{1}A′ and ^{3}A″ states at the BVWN5/cc-pVDZ and BLYP/cc-pVDZ levels is 30.9 and 31.6 kcal mol^{-1}, respectively, which compares well with an experimental measurement of 30.4 kcal mol^{-1}.

In the case of the triplet of A′ symmetry, only the difference in spin states prevents variational collapse of the triplet to the singlet, but that is sufficient. Interestingly, DFT does a reasonably good job in predicting the singlet–triplet splitting between these two states, with BVWN5/cc-pVDZ and BLYP/cc-pVDZ both giving values of 33.2 kcal mol^{-1}, compared to an experimental measurement of 31.3 kcal mol^{-1}. If the fluorine is replaced by a t-butyl group, the same theoretical levels predict analogous splittings of 45.7 and 46.8 kcal mol^{-1}, respectively, compared to an experimental measurement (Gunion and Lineberger 1996) of 45.6 kcal mol^{-1}. These good agreements come in spite of the current formal status of DFT, where the Hohenberg–Kohn theorem has only been proven to apply to the lowest-energy state *irrespective of spin* in each irreducible representation of the molecular point group (see Section 8.2.1).

Many of the same considerations affecting these vinylidene examples arise in comparing the relative energies of the electronic states of phenylnitrene (Figure 14.3). In this system, there are many different theoretical data available to compare to experiment, which itself is available for the lowest two singlet states. Results from ΔSCF calculations at the HF and DFT levels of theory are listed in Table 14.1, as are results from many additional levels that will be discussed at appropriate points later in the chapter.

DFT levels of theory compare about as well with experiment for the splitting between the ^{3}A$_2$ and 1^1A$_1$ states of phenylnitrene as for the analogous states of the vinylidenes just discussed. Agreement is nearly quantitative at the BLYP level using a triple-ζ basis set. In general in Table 14.1, the energies of the excited states are predicted to be somewhat lower for equivalent levels of theory when triple-ζ basis sets are used in place of those of double-ζ quality. Such behavior is expected, insofar as ground states tend to have more electron density residing in the close-in valence region than do excited states, and thus the ground states are less demanding in terms of basis-set requirements. Moreover, most basis sets are optimized for ground-state atoms and molecules, so to the extent basis-set limitations affect the calculation, they should disproportionately affect excited states.

The DFT values for the ^{1}A$_2$ state derive from the sum method or projection techniques presented in Section 14.4, and discussion of those values is deferred to that point. As for the 2^1A$_1$ state, although no experimental measurement is available, comparison to other

Table 14.1 Energies (kcal mol^{-1}), where available, for lowest singlet excited states of phenylnitrene relative to the $^3A_2$ ground state[a]

Source	$^1A_2$	1^1A_1	2^1A_1
HF/6-31G(d)		64.6	80.1
BLYP/6-31G(d)	14.5[b], 22.8[c]	31.4	48.7
BPW91/cc-pVDZ (ΔSCF)[d]	14.3[b]	33.9	43.0
BLYP/cc-pVTZ (ΔSCF)[e]		29.5	41.0
CCSD(T)/cc-pVDZ[e]		35.2	47.2
CAS(8,8)/cc-pVDZ[d]	17.8	42.1	76.2
CASPT2/cc-pVDZ[d]	19.3	37.4	57.8
CASPT2/cc-pVTZ[e]	19.3	34.8	54.5
MRCISD/DZP[f]	21.0	39.8	(52.0)
Experiment[g]	18.	30.	n.a.

[a] Zero-point vibrational energies are not included in the theoretical energies, but ZPVE differences between alternative electronic states are predicted to be small at the few levels where they have been evaluated.
[b] Determined from sum method.
[c] Determined from spin projection.
[d] Johnson and Cramer (2001).
[e] Smith and Cramer (1996).
[f] Hrovat, Waali, and Borden (1992); Kim, Hamilton, and Schaefer (1992).
[g] Travers et al. (1992); Ellison, G. B., unpublished results.

levels of theory that would be expected to be reasonably accurate suggests that the DFT predictions are substantially too low. The DFT wave function for the 2^1A_1 state is that of Eq. (14.8) and, as discussed above, is found by fortuitous convergence of the SCF equations for this occupation scheme where variational collapse to the 1^1A_1 state of Eq. (14.7) would otherwise be expected. The apparently rather poor accuracy of the energy for the higher state suggests that this orthogonality issue cannot be ignored here, and the ΔSCF procedure must be regarded as unreliable.

As for the HF level, the ΔSCF approach for the closed-shell singlet states is identical to that in the DFT case (in this instance, the two-determinantal nature of the lower energy open-shell singlet requires an MCSCF description, so HF values are not reported for this state). However, both of the closed-shell singlets are subject to large non-dynamical correlation effects (as a consequence, in part, of being so close in energy to one another). Since HF theory is much more sensitive to such correlation than DFT, the energies of these two states are predicted to be *much* too high. This error is in some sense even worse than it appears, because severe spin contamination in the triplet, which exhibits an expectation value for S^2 in excess of 2.7, probably causes it too to be poorly represented at the HF level.

Of course, with HF wave functions in hand, it is possible to carry out post-HF calculations to partially correct for electron correlation effects. The poor quality of the HF wave functions, however, militate against any treatment much less sophisticated than coupled-cluster. At the CCSD(T)/cc-pVDZ level, the predicted energy of the lowest closed-shell singlet is in fair agreement with experiment (other data in the table suggest that use of a triple-ζ basis set would improve the CCSD(T) estimate). The energy of the second closed-shell singlet state looks to be somewhat low, however, again probably reflecting the non-orthogonal nature of the HF reference for this state compared to the lower energy one.

14.2.2   CI Singles

For many singly excited states, ΔSCF calculations are not an option under any circumstances. Within the context of using the orbitals of the ground state to describe the excited state, the simplest way to evaluate the energy of the excited state would be to evaluate the Hamiltonian for the determinant formed after promotion of the excited electron. Such an approach is rarely useful, however, and a significant drawback of these singly excited single-configuration wave functions is that, although each one will be orthogonal to the ground state (because of Brillouin's theorem, see Section 7.3.1), they are unlikely to be orthogonal to one another. However, as long as we limit our consideration to singly excited states, they can be made to be orthogonal to one another with fairly little computational effort, and in the process better descriptions of the states, and presumably better energies, may be determined.

This orthogonalization is the essence of the technique known as CI singles (CIS) because the CI matrix is formed restricting consideration to only the HF reference and all singly excited configurations (Figure 14.5). The matrix is essentially of size $M \times N$ where M is the number of occupied orbitals from which excitation is allowed, and N is the number of virtual orbitals into which excitation is considered. If excitation is allowed to occur with a spin-flip of the excited electron (e.g., permitting generation of triplet excited states from singlet ground states or vice versa; see, for example, Sears, Sherrill, and Krylov 2003) then the size increases, although none of the triplet states have matrix elements with any of the singlet states because of their different spins. Orthogonalization of the CIS matrix takes place only in the space(s) of the excited states, since they do not mix with the HF reference. The orthogonalization provides energy eigenvalues each of which has associated with it an eigenvector detailing the weight of every singly excited determinant in the state. That is, the CIS wave function for each excited state is written as

$$\Psi_k = \overset{\text{occupied}}{\underset{i}{\sum}} \overset{\text{virtual}}{\underset{a}{\sum}} c_{iak} \Psi_i^a \tag{14.11}$$

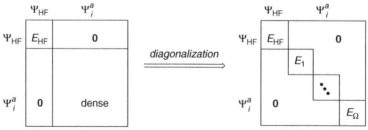

Figure 14.5   The CIS procedure diagonalizes the CI matrix formed only from the HF reference and all singly excited configurations. The diagonalization provides energy eigenvalues and associated eigenvectors that may be used to characterize individual states as linear combinations of single excitations

where the coefficients c are the components of the eigenvector for state k. With large enough basis sets, even the CIS matrix may grow cumbersomely large with which to work, and iterative methods designed to locate only lower energy roots are employed, just as in CI treatments considering higher excitations. Analytic gradients are available for CIS wave functions, so it is possible to optimize the geometry of a particular state, making CIS a useful method for obtaining either vertical excitation energies, or adiabatic excitation energies.

Note the difference in objectives between a ground-state CI calculation and a CIS calculation. In the former, the goal is to improve the description of the ground state, and excitations must be included at least through doubles (since singles do not mix with the ground state). In the CIS calculation, the ground state is important only to the extent it determines the orbitals, and the CI is carried out to orthogonalize the singly excited states.

Insofar as the latter process does not involve any orbital reoptimization for any particular state, it provides a wave function that is roughly equivalent in quality only to an HF wave function for the ground state. Of course, this may still be useful for a number of purposes. CIS results for six excited states of benzene are included in Table 14.2, as are results from other levels of theory that will be discussed later. The CIS results are qualitatively useful, insofar as the states are correctly ordered, and the error is fairly systematic – all states are predicted to be too high in energy by an average of 0.7 eV. The worst prediction is for the lowest excited state, which is known to have significant dynamical electron correlation, and is therefore challenging for the CIS method.

To improve CIS results beyond their roughly HF quality, various options may be considered. Particularly for spectroscopic predictions, semiempirical parameterization of the CIS matrix elements may be preferred over their direct evaluation in an *ab initio* sense using Eq. (7.12), and the most complete realization of this formalism is the INDO/S parameterization of Zerner and co-workers. A few examples of the excellent performance of this highly efficient model for the computation of excited-state energies have already been discussed (Table 5.1). Of additional interest, Hutchison, Ratner, and Marks (2002) found that CIS/INDO/S provided the highest accuracy of several methods (including *ab initio* CIS, RPA, and TDDFT; the last two are discussed later in this chapter) for predictions of first excited-state energies in 60 oligomers of various aromatic heterocycles. With increasing

Table 14.2 Energies (eV) for singlet excited states of benzene relative to the $^1A_{1g}$, ground state as predicted by various methods[a]

Excited state	CIS	RPA	TD-BPW91	TD-B3LYP	Expt.
$^1B_{2u}$	6.15	5.96	5.19	5.40	4.9
$^1B_{1u}$	6.31	6.01	5.93	6.06	6.2
$^1E_{1g}$	7.13	7.12	6.34	6.34	6.33
$^1A_{2u}$	7.45	7.43	6.87	6.84	6.93
$^1E_{2u}$	7.75	7.74	6.85	6.88	6.95
$^1E_{1u}$	7.94	7.52	6.84	6.96	7.0
Mean abs. error:	0.7	0.6	0.1	0.1	

[a] From Stratmann, Scuseria, and Frisch (1998). All calculations employed the 6-31+G(d) basis set.

length, these energies converge to the band gap for the one-dimensional solid, a property of considerable interest in solid-state chemistry and physics.

Improvement of *ab initio* CIS wave functions may in principle be accomplished with perturbation theory, but this tends to be very slowly convergent because the existence of a high-energy occupied orbital and a low-energy hole leads to denominators in perturbation theory expressions analogous to Eq. (7.48) that are very near zero, and thus the terms are very large. Head-Gordon *et al.* (1994) have proposed a more satisfactory approach where the effect of double excitations on the CIS state energy is evaluated; this method is referred to as CIS(D).

CIS technology has a particularly valuable application that is unrelated to an interest in excited states, *per se*. In systems where the exact orbital occupation of the ground state is not entirely certain, a CIS calculation allows a determination of whether the ground state has actually been found with respect to single excitations. If a CIS eigenvalue is found that is *below* the HF energy, then the HF reference either is not the ground state, or it may be, but within the accuracy of HF/CIS theory it is clear that a near degeneracy exists with another state or states. In highly symmetric systems, where orbital mixing is restricted by blocking of the Fock matrix, a check of the stability of any final wave function using this approach is a useful precaution.

A variation on the CIS scheme that is empirical in nature but has been demonstrated to offer surprisingly high accuracy in computational practice involves marrying some aspects of DFT with the CIS methodology (DFT-SCI; Grimme 1996). In particular, all HF orbital energies are replaced by KS equivalents, and the Coulomb integrals and diagonal elements are empirically scaled. Once this is done, the usual diagonalization process provides wave functions and state energies that compare nicely with more rigorous theoretical formulations. This technology has not yet seen widespread use.

14.2.3 Rydberg States

A particular type of singly excited state merits mention because of its unique characteristics. A so-called Rydberg state is one where the excited electron has an energy very near the level of the continuum, i.e., it is almost detached. Such states may conveniently be thought of as an electron attached to a molecular cation that acts as a central attractor, much as a nucleus acts as a central attractor in an atom. Thus, formal Rydberg orbitals may be of s, p, d, etc. character with the exact chemical nature of the underlying molecular system secondary in importance to its total charge. The Rydberg orbitals are by nature extremely diffuse compared to valence orbitals. As a result, any attempt to describe a Rydberg state requires an AO basis set that either includes diffuse functions on heavy atoms (and possibly H) or is supplemented by additional basis functions specifically tailored to Rydberg character that are not necessarily atom-centered, e.g., they may take the molecular center of mass as their origin (see, for example, Wiberg, de Oliveira, and Trucks 2002). The latter option is more efficient but introduces complications if gradients are desired for geometry optimization.

14.3 General Excited State Methods

Electronic states that cannot be well described as single excitations from the ground state require more general formalisms than those described thus far for the construction of their wave functions. Such formalisms, insofar as they are general, can certainly be used for states that *are* well characterized as single excitations as well; they simply tend to be somewhat more demanding in terms of computational resources, making methods like CIS economical alternatives when they can be applied. Some of the methods described below have already been discussed in Chapter 7 in the context of improving the description of the ground-state wave function, while others are specific to excited-state applications.

14.3.1 Higher Roots in MCSCF and CI Calculations

In the process of determining the expansion coefficients that define an MCSCF wave function along lines similar to Eq. (7.10), CI calculations are carried out in the space of the orbitals that are active in the MCSCF. Excited states that can be generated by electronic excitations within that active space then have a corresponding root in that limited CI window and, if one chooses, one can variationally optimize the orbitals for a root other than the one of lowest energy, i.e., other than the ground state.

In some instances, the root in question is dominated by a single CSF, allowing one to describe the state conveniently by reference to the ground state. In other instances, however, that will not be the case, and simple relationships between the excited state and the ground state cannot be easily formulated. This is, of course, purely a conceptual problem – the wave functions themselves are perfectly well defined and useful.

In any case, once a root is specified, the MCSCF process minimizes the energy for that root following the usual variational procedure. Problems can arise, however, along the way. Consider the situation illustrated in Figure 14.6, with two states having curves that cross along some geometrical coordinate. The point of crossing is a so-called 'conical intersection' in the corresponding PESs. In diatomics, such intersections are not permitted if the curves correspond to electronic states of the same symmetry (the 'non-crossing rule'), but in larger systems such restrictions are not in force, and conical intersections are ubiquitous. The state energies themselves are sensitive to which root is chosen for optimization. Obviously, the chosen root has a better representation since the orbitals are optimized for it, while the non-chosen root has a poorer representation, and thus its energy is erroneously too high when computed as a root of the MCSCF reduced CI matrix. In situations where the two begin close to one another in energy, e.g., near a conical intersection, it is possible that 'root switching' will occur during the optimization process. Thus, in Figure 14.6, if one is at the geometrical position indicated by the asterisk and one selects the second root as the state for optimization (assuming the initial HF orbitals resemble better the orbitals of State A), as the MCSCF proceeds, the energy of the second root will drop below that of the first, since the orbitals are dropping the energy of State B at the expense of State A. As a result, the MCSCF will suddenly switch from optimizing the orbitals for State B to optimizing them for State A, as it is *that* state that is now the second root. This situation is unstable.

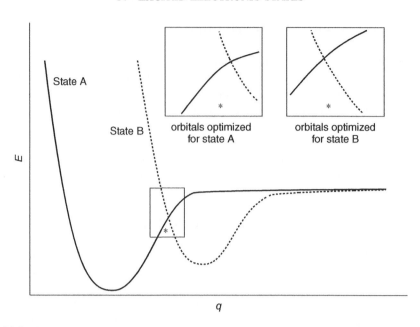

Figure 14.6 Two electronic states of an arbitrary system having a conical intersection. The inset region illustrates the effect on each curve of optimizing the orbitals for either State A or State B. At the coordinate position marked by an asterisk, the relative energies of the two states depend on which is chosen for orbital optimization, which can lead to root switching problems in an MCSCF calculation. Additionally, geometry optimization can cause root switching as well, if optimization passes through the conical intersection

To finesse this problem, it is possible to carry out a so-called 'state-averaged' MCSCF. In a state-averaged calculation, the orbitals are variationally optimized not for any one state energy, but rather for the average of the two (or more than two, if a larger number of states are of interest). A drawback to such a calculation is that the quality of any one state's wave function is lower than it would be were it to be the only state under consideration. On the other hand, a virtue of a state-averaged calculation is that all states are expressed using the same MOs, thereby ensuring orthogonality, which is critical if, say, transition dipoles between states are to be computed.

Nevertheless, root switching may still be problematic for geometrical reasons in the vicinity of conical intersections. Thus, for instance, any optimization of State B in Figure 14.6 that begins to the left of the asterisk in coordinate q will ultimately proceed to the right until State B falls below State A in energy, at which point it is the first root for *chemical* reasons, not technical reasons. The only remedy in this situation is careful analysis in the construction of state PESs.

MCSCF results for phenylnitrene using a complete active space formed from the six phenyl π orbitals and the two nitrogen p orbitals and the eight electrons contained therein are presented in Table 14.1. Note that, because of symmetry and spin restrictions, only the 2^1A_1 state must be determined as the second root of the MCSCF. The CAS results are quite good for the energy of the $^1A_2$ state (recall that MCSCF is designed to handle the two-determinantal character of that state in a fully rigorous fashion), but place the closed-shell singlets considerably too high in energy. Since MCSCF wave functions with small active spaces typically correct for non-dynamical correlation effects much more effectively than for dynamical correlation effects, it is not surprising that the closed-shell singlets, with their greater number of spin-paired electrons, are predicted to be unrealistically too high in energy.

To correct for dynamical electron correlation, the usual multireference methods may be applied. Table 14.1 lists CASPT2 and MRCI results for the phenylnitrene states. With a double-ζ basis, the two give roughly equivalent results in this system for the first two singlet states. The MRCI value for the 2^1A_1 state is actually not from a multireference calculation, but is instead a CISD result using the HF wave function of Eq. (14.8), so its somewhat lower energy is probably attributable to a failure to enforce orthogonality with the lower A_1 state. The CASPT2 predictions with a triple-ζ basis set (a basis set that is too large for practical use in an MRCISD calculation) are reasonably good, and probably represent the best estimate of any method listed in the table for the 2^1A_1 state. Note that the reasonably good agreement between the CASPT2/cc-pVTZ value for this state and that from the CISD calculation probably derives from the orthogonality error in the latter being fortuitously canceled by error associated with the limited size of the employed basis set.

On the whole, the CASPT2 method is the most robust general method for computing excited-state energies and wave functions spanning all manner of excitations. Design efforts with other methodologies tend to use CASPT2 values as developmental benchmarks. Grimme and Waletzke (2000) have proposed a multireference second-order perturbation scheme (MR-MP2) that is similar in formalism to CASPT2 but achieves substantially higher efficiency by truncation of the active space and the number of excitations allowed within it. For 22 excited-state energies in 14 very diverse molecules MR-MP2 achieved a mean unsigned error of 0.14 eV compared to experiment (Parac and Grimme 2002). Grimme and Waletzke (1999) have also proposed a DFT-MRCI analog to DFT-SCI that can be used for excited states where excitations beyond singles may need to be taken into account.

14.3.2 Propagator Methods and Time-dependent DFT

If a molecule is subjected to a linear electric field $\mathbf{E}$ that is fluctuating such that

$$\mathbf{E} = \mathbf{r}\cos(\omega t) \tag{14.12}$$

where $\mathbf{r}$ is the position vector in one dimension, t is time, and ω is the frequency of the fluctuation, it can be shown that the frequency-dependent polarizability is well approximated by

$$\langle\alpha\rangle_\omega = \sum_{i\neq 0}^{\text{states}} \frac{|\langle\Psi_0|\mathbf{r}|\Psi_i\rangle|^2}{\omega - (E_i - E_0)} \tag{14.13}$$

where the numerator of each term in the sum is a so-called transition dipole moment and the denominator involves the frequency and the energies of the excited states and the ground state. Note that, if the frequency corresponds exactly to the difference in energy between an excited state and the ground state, there is a pole in the frequency-dependent polarizability, i.e., it diverges since the denominator goes to zero.

Using propagator methodology (sometimes also called a Green's function approach or an equation-of-motion (EOM) method), the poles of the frequency-dependent polarizability can be determined without having to compute all of the necessary excited-state wave functions and their corresponding state energies. The necessary matrix equations are quite complex, and only a qualitative summary is provided here. Within the confines of the so-called random-phase approximation (RPA) the integrals that are required to compute the excitation energies are essentially those required to fill the CI matrix containing all single and double excitations and the transition dipole moments between the ground state and all singly excited configurations. Because the RPA method includes double excitations, it is usually more accurate than CIS for predicting excited-state energies. However, the method does not deliver a formal wave function, as CIS does. The RPA method may be applied to either HF or MCSCF wave functions. As with the CI formalisms they somewhat resemble, RPA solutions are most efficiently found by an iterative process that focuses only on a few lowest-energy excitations.

Table 14.2 includes RPA results for the six excited states of benzene already discussed in the context of CIS. The more complete RPA formalism does improve the results for those cases where CIS is most in error, but the net improvement in mean absolute error over all six states is only 0.1 eV in this case.

A DFT method that is strongly analogous to RPA is called time-dependent DFT (TDDFT). In this case, the KS orbital energies and various exchange integrals are used in place of matrix elements of the Hamiltonian. TDDFT is usually most successful for low-energy excitations, because the KS orbital energies for orbitals that are high up in the virtual manifold are typically quite poor. Casida, Casida, and Salahub (1998) have suggested that TDDFT results are most reliable if the following two criteria are met: (i) the excitation energy should be significantly smaller than the molecular ionization potential (note that excitations from occupied orbitals below the HOMO are allowed, so this is not a tautological condition) and (ii) promotion(s) should not take place into orbitals having positive KS eigenvalues.

Table 14.2 includes TDDFT results from the pure BPW91 and hybrid B3LYP functionals for the six excited states of benzene previously discussed for CIS and RPA methods. The pure functional is better for the most highly correlated $^1B_{2u}$ state, but the mean absolute error for the two methods over all six states is equivalent. The improved quality of the TDDFT results compared to CIS or RPA is substantial.

Many other comparisons between TDDFT and alternative methods have appeared. For example, Parac and Grimme (2002) found B3LYP TDDFT to give a mean unsigned error of 0.26 eV on the same 22 excited-state energies for 14 molecules discussed in the last section (cf. 0.14 eV for MR-MP2). In addition, Fabian (2001) compared B3LYP TDDFT results to CIS/INDO/S for various absorptions in 76 organosulfur compounds containing up to four sulfur atoms. The performance of TDDFT was again quite good: the mean unsigned error over all absorption maxima was 0.21 eV, which was superior to the CIS/INDO/S result of 0.35 eV. Interestingly the semiempirical PPP method, which by its nature is only applicable to excitations of the $\pi \rightarrow \pi^*$ variety, had an error of only 0.20 eV, illustrating that more expensive calculations are not always better calculations ...

A detailed comparison of several methods for local and charge-transfer excitation energies in benzenes substituted with donor and acceptor groups has been provided by Jamorski *et al.*

Table 14.3 Vertical excited-state energies (eV) of 4-dimethylaminobenzonitrile and 3,5-dimethyl-4-dimethylaminobenzonitrile relative to S_0 ground state

Method	S_1 (^{1}B, local)	S_2 (^{1}A, CT)	T_1 (^{3}A, CT)	T_2 (^{3}B, local)
CASPT2/DZP[b]	4.05	4.41	3.66	3.69
B3LYP-MRCI/TZVP[a]	4.33	4.62		
B3LYP-SCI/DZVP[c]	4.1	4.6	3.12	3.51
TD-LSDA//AM1[d]	3.89	4.17	3.10	3.43
LSDA//AM1 ΔSCF[d]			3.27	
TD-B3LYP//AM1[d]	4.38	4.54	3.10	3.73
B3LYP//AM1 ΔSCF[d]			3.32	
TD-PBE1PBE//AM1[d]	4.49	4.64		
TD-LSDA[d]	3.92	4.26	3.12	3.40
LSDA ΔSCF[d]			3.29	
TD-B3LYP[d]	4.38	4.62	3.14	3.68
B3LYP ΔSCF[d]			3.35	
Experiment[e]	4.25	4.56	3.36	3.50

Method	S_1 (CT)	S_3 (local)	T_1 (CT)	T_2 (local)
B3LYP-SCI/DZVP+[e]	4.17	4.95	3.77	4.03
TD-B3LYP[d]	3.91	4.98	3.08	3.94
B3LYP ΔSCF[d]			3.38	
TD-PBE1PBE[d]	4.09	5.16	3.02	4.01
PBE1PBE ΔSCF[d]			3.37	
TD-mPW1PW91[d]	4.10	5.16	3.00	4.04
mPWPW91 ΔSCF[d]			3.37	
Experiment[e]	4.27	5.00	3.48	4.31

[a] Parusel (2000).
[b] Serrano-Andrés et al. (1995).
[c] Parusel, Köhler, and Grimme (1998).
[d] Jamorski et al. (2002); 6-311+G(2d,p) basis set.
[e] Bulliard et al. (1999).

(2002) and some of their results are summarized in Table 14.3. An observation made in all of the above studies, and one that is particularly important for future developmental efforts, is that the TDDFT methodology performs relatively poorly for excitations characterized as charge-transfer (CT) or charge-resonance in weakly interacting composite chromophores (see also Casida et al. 2000 and Zyubin and Mebel 2003). Note, though, that the ΔSCF approach works well when it is possible to employ it.

Efforts to improve TDDFT for higher-energy excitations have shown some early success. Tozer and Handy (1998) have proposed a correction procedure to deliver functionals having asymptotically correct potentials, and Adamo and Barone (1999) have demonstrated that the hybrid PBE1PBE functional, for reasons that are not entirely clear, seems to be significantly less affected by high-energy errors than other hybrid functionals. Such ongoing developments together with efficient schemes for computing excited-state analytic derivatives (Furche and Ahlrichs 2002) make TDDFT the method of choice for molecules whose size precludes the use of very large multireference *ab initio* schemes. Within the area of inorganic chemistry, the accuracy of newly developed methods for computing transition-metal circular dichroism (CD) spectra with TDDFT is another very promising development (see, for example, Autschbach, Jorge, and Ziegler 2003). CD spectroscopy, together with optical rotatory dispersion (ORD), which is also open to computation (Beratan, Kondru, and Wipf 2002 and Rinderspacher and Schreiner 2004), is particularly useful in assigning absolute configuration for chiral molecules.

Finally, Shao, Head-Gordon, and Krylov (2003) have described a modification of TDDFT that permits formally multideterminantal target states to be described as spin–flip excitations from a single determinant reference state of different spin (SF-TDDFT). Thus, for example, the difficult singlet state(s) of trimethylenemethane, discussed in Chapter 7, may be generated from the single-determinantal triplet state by single spin–flip excitations (Slipchenko and Krylov 2003). This development substantially expands the range of excited states that may be addressed with TDDFT.

14.4 Sum and Projection Methods

The application of HF and KS-DFT is fundamentally limited to wave functions that can be expressed as single Slater determinants. This restricts their utility in dealing with states like the $^1A_2$ state of phenylnitrene, which is two-determinantal in character (Figure 14.3). One *can* apply HF or KS-DFT to the single determinant that, restricted to representation of the singly occupied orbitals and with normalization implicit, is written

$$^{50:50}\Psi = [b_1(1)\alpha(1)b_2(2)\beta(2) - b_1(2)\alpha(2)b_2(1)\beta(1)] \tag{14.14}$$

But, as described in more detail in Appendix C, this wave function, which configurationally corresponds to an α electron in the b_1 orbital and a β electron in the b_2 orbital, is neither a singlet nor a triplet, but a 50:50 mixture of the two, and this point is emphasized by the left superscript on Ψ in Eq. (14.14). While the wave function does not represent a pure spin state, we may take advantage of the prevailing situation by noting that we may write

$$\langle ^{50:50}\Psi|H|^{50:50}\Psi\rangle = \left\langle \left(\frac{1}{\sqrt{2}}{}^3\Psi + \frac{1}{\sqrt{2}}{}^1\Psi\right)|H|\left(\frac{1}{\sqrt{2}}{}^3\Psi + \frac{1}{\sqrt{2}}{}^1\Psi\right)\right\rangle$$

$$= \frac{1}{2}\langle ^3\Psi|H|^3\Psi\rangle + \frac{1}{2}\langle ^3\Psi|H|^1\Psi\rangle + \frac{1}{2}\langle ^1\Psi|H|^3\Psi\rangle + \frac{1}{2}\langle ^1\Psi|H|^1\Psi\rangle$$

$$= \frac{1}{2}\left(\langle ^3\Psi|H|^3\Psi\rangle + \langle ^1\Psi|H|^1\Psi\rangle\right) \tag{14.15}$$

where $^3\Psi$ and $^1\Psi$ represent the pure spin states (note that matrix elements of the Hamiltonian between different spin states are zero, leading to the simplification on going from the second to the third equality in Eq. (14.15)). The desired value in this process is the expectation value of the Hamiltonian for $^1\Psi$, which does *not* have a single-determinantal description. However, this expectation value is the only unknown in Eq. (14.15), because the expectation value for the triplet may be computed from the alternative triplet wave function

$$^3\Psi = [b_1(1)\alpha(1)b_2(2)\alpha(2) - b_1(2)\alpha(2)b_2(1)\alpha(1)] \tag{14.16}$$

which is the $S_z = 1$ single-determinantal triplet one usually works with in HF or KS-DFT electronic structure codes. Thus, rearrangement of Eq. (14.15) indicates that the energy of the open-shell singlet may be determined as

$$\left\langle {}^1\Psi | H | {}^1\Psi \right\rangle = 2 \left\langle {}^{50:50}\Psi | H | {}^{50:50}\Psi \right\rangle - \left\langle {}^3\Psi | H | {}^3\Psi \right\rangle \tag{14.17}$$

where the expectation values on the right are readily computed from single-determinantal HF or KS-DFT formalisms. This approach is known as the 'sum method'. Although presented here in the context of singlet and triplet states, it has been generalized to the construction of Heisenberg spin ladders in arbitrarily complex systems, such as the iron–sulfur clusters found in biological electron-transport systems (Noodleman *et al.* 1995).

To provide a specific example of the method, near UV experiments have led to assignments of the vertical and adiabatic excitation energies for the $1^1B_g \leftarrow 1^1A_g$ transition in *E*-diazene (HN=NH), where the ^1B_g state is open-shell. Table 14.4 compares sum-method predictions at the UHF and BLYP levels of theory to these experimental values, and also to published results at the MRCI level of theory. For this system, the HF results are systematically too high, and the DFT too low (cf. the sum method prediction for $^1A_2$ phenylnitrene in Table 14.1), but are competitive with the much more expensive MRCI results. Note that all three levels do quite well at predicting the *difference* in vertical and adiabatic excitation energies.

The sum method is simple and fast, and it is an SCF approach, which can make it an attractive option compared to other alternatives in some cases. However, it also has some serious possible drawbacks. First, geometry optimization is tedious, since, from Eq. (14.17), the gradients of the open-shell singlet will be the difference between twice the gradients of the 50:50 system and the gradients of the triplet system, all at the same geometry. Secondly, and more importantly, Eq. (14.15) is rigorously correct only when all three wave functions, singlet, triplet, and 50:50, are written using the same MOs, in practice those that are SCF

Table 14.4   Energies (kcal mol^{-1}) for the $1^1B_g \leftarrow 1^1A_g$ transition in *E*-diazene

Level of theory	Approach	Vertical	Adiabatic
UHF/cc-pVTZ[a]	sum method	75.3	65.3
	spin annihilation	55.4	44.9
BLYP/cc-pVTZ[a]	sum method	64.9	56.1
	spin annihilation	64.8	55.7
MRCI/DZP[b]		81.8	70.3
Experiment[c]		71.8 ± 2.3	61.8 ± 3.5

[a]Lim *et al.* 1996.
[b]Kim, Shavitt, and Del Bene 1992.
[c]Back, Willis, and Ramsay 1978.

optimized for the 50:50 determinant. In the limit of the two unpaired electrons being non-interacting (not only directly, but furthermore via polarization of the paired electrons) the use of identical MOs poses no problems, but that is the rather trivial case of the open-shell singlet being degenerate with the triplet (and thus also with the 50:50 wave function). When the unpaired electrons *do* interact, the two-determinantal $S_z = 0$ triplet energy that would be computed from the 50:50 MOs will be higher in energy than the $S_z = 1$ triplet energy computed with its orbitals fully optimized in the usual SCF way and employed in Eq. (14.17). This can contribute to significant error in unfavorable instances where the singlet and triplet orbitals are very different from one another. Finally, if the 50:50 wave function, which is an unrestricted wave function, shows significant spin contamination, then the use of the 50:50 orbitals to express the singlet and triplet wave functions in Eq. (14.15) will lead to their spin contamination as well, again causing possibly significant errors when the $S_z = 1$ triplet energy is used to replace the $S_z = 0$ triplet energy.

To reduce or eliminate spin contamination problems in unrestricted wave functions, spin projection methods have been developed that annihilate the contributions of certain spin states higher than the desired one. As derived in Appendix C, the PUHF energy for a wave function that has had contamination from the next higher spin state annihilated is computed as

$$E_{PUHF} = \langle \Psi^0 | H | \Psi^0 \rangle + \frac{\sum_i \langle \Psi^0 | H | \Psi^i \rangle \langle \Psi^i | A_{s+1} | \Psi^0 \rangle}{\langle \Psi^0 | A_{s+1} | \Psi^0 \rangle} \qquad (14.18)$$

where Ψ^0 is the original spin-contaminated wave function, i runs over all doubly excited determinants Ψ^i, and A_{s+1} is the spin annihilation operator

$$A_{s+1} = \frac{S^2 - \{(s+1)[(s+1)+1]\}}{[s(s+1)] - \{(s+1)[(s+1)+1]\}} \qquad (14.19)$$

where S^2 is the usual total spin operator, s is the desired spin state, and $s+1$ is the next higher spin state being annihilated.

In application to 'typical' UHF wave functions, the second term on the r.h.s. of Eq. (14.18) provides a small correction that improves the estimate of the state energy for slightly contaminated cases. In principle, however, there is no reason the formalism cannot be applied to a 50:50 wave function. The results of such an application to the already discussed diazene excitation are listed in Table 14.4.

At the UHF level, the excitation energies from spin annihilation represent a fairly severe underestimation of the excited-state energies, and disagree significantly with the sum method results from that same level of theory. UHF spin annihilation is generally not a worthwhile method to apply unless spin contamination effects are fairly small, which obviously is not the case here.

At the DFT level of theory, spin annihilation in principle has no analog, since the correct wave function for the KS density is not known (only the non-interacting KS wave function from which a portion of the kinetic energy is evaluated is known, see Section 8.3). However, Cramer *et al.* (1995) proposed a projected DFT (PDFT) procedure whereby the DFT energy of the mixed-state system is corrected by the same process employed in UHF theory, except with determinants formed from the KS orbitals, i.e.,

$$E_{PDFT} = {}^{50:50}E_{DFT} + \frac{\sum_i \langle \Psi_{KS}^0 | H | \Psi_{KS}^i \rangle \langle \Psi_{KS}^i | A_{s+1} | \Psi_{KS}^0 \rangle}{\langle \Psi_{KS}^0 | A_{s+1} | \Psi_{KS}^0 \rangle} \tag{14.20}$$

where i now runs over both single and double excitations since Brillouin's theorem no longer guarantees as zero the expectation values of the Hamiltonian appearing in the sum on the r.h.s. for singly excited determinants. This process applied to the diazene case of Table 14.4 gives energies almost identical to those determined from the sum method. In Table 14.1, the PDFT energy computed for the $^1A_2$ state of phenylnitrene overestimates this state's energy by about the same amount as the sum method underestimates it. Durant (1996) has also demonstrated success in using the PDFT model to evaluate transition-state properties for systems having substantial open-shell character.

The largest drawback of the spin annihilation procedure is similar to that of the sum method. That is, while the spin-annihilated wave function which results from the application of A_{s+1} to the 50:50 antecedent is in principle spin pure, it is expressed in the MOs that were optimized for the 50:50 case. These MOs minimize the energy of the contaminated state, but not that of the spin pure state, and errors can be significant. Nevertheless, the speed of the sum and projection methods, and their utility in many if not all instances, makes them useful for rough applications prior to resort to more expensive and sophisticated models.

14.5   Transition Probabilities

In electronic spectroscopy, one wants to know not only the energy difference between distinct electronic states but also the probability that a transition between them will take place under appropriate circumstances. Thus, in the recording of a classic UV/Vis spectrum for a molecule, the wavelengths of absorptions indicate the energetics of the transition, while the intensities of the absorptions indicate their 'allowedness', or probability.

The simplest approach to understanding the radiation- (light-) induced transition between electronic states is to invoke time-dependent perturbation theory. Thus, one starts from the time-dependent Schrödinger equation

$$-\frac{\hbar}{i} \frac{\partial \Psi}{\partial t} = H\Psi \tag{14.21}$$

where $\hbar$ is Planck's constant over 2π, i is the complex number $\sqrt{-1}$, and t is time. A complete set of eigenfunctions for Eq. (14.21) is given by

$$\Psi_j = e^{-(iE_j t/\hbar)} \Phi_j \tag{14.22}$$

where the wave functions Φ_j are the eigenfunctions of the time-*independent* Schrödinger Eq. (4.2) having eigenvalues E_j (it is a simple and worthwhile exercise to verify that

Eq. (14.22) is indeed an eigenfunction of Eq. (14.21)). Since the set of Ψ_j is complete, *any* wave function for the system may be expressed as

$$\Psi = \sum_k c_k e^{-(iE_k t/\hbar)} \Phi_k \qquad (14.23)$$

where the normalized expansion coefficients c run over all possible eigenstates k.

We may consider the presence of a radiation field as a perturbation on the otherwise time-independent H^0. Using the standard expression for the time-dependent electric field contribution to the Hamiltonian for radiation having a wavelength in the UV/Vis light region we have

$$H = H^0 + e_0 \mathbf{r} \sin(2\pi \nu t) \qquad (14.24)$$

where e_0 is the amplitude of the electric field associated with the light of frequency ν and $\mathbf{r}$ is the usual position operator (the sum of the $\mathbf{i}$, $\mathbf{j}$, and $\mathbf{k}$ operators in Cartesian space). With a time-dependent Hamiltonian, Eq. (14.23) is still valid for the description of any wave function for the system, except that the expansion coefficients c must also be considered to be functions of t.

A spectroscopic measurement, from a quantum mechanical perspective, may thus be envisioned as the following process. The system begins in some stationary state, in which case all values of c in Eq. (14.23) are 0, except for one, which is 1. For simplicity, we will consider the initial state to be the ground state, i.e., $c_0 = 1$. Beginning at time 0, the system is then exposed to radiation until time τ. During that time, the expansion coefficients will be in a constant state of change until, with the disappearance of the radiation, the Hamiltonian returns to being time-independent, at which point the expansion coefficients for Ψ cease to change. To the extent more than one coefficient is non-zero, the system exists in a superposition of states and the probability of any particular state k being observed by experiment, determined from evaluation of $\langle \Psi | \Psi \rangle$, is simply c_k^2.

To determine the latter probabilities, let us evaluate Eq. (14.21) for an arbitrary wave function expressed in the form of Eq. (14.23)

$$-\frac{\hbar}{i}\frac{\partial}{\partial t}\sum_k c_k e^{-(iE_k t/\hbar)} \Phi_k = [H^0 + e_0 \mathbf{r}\sin(2\pi \nu t)]\sum_k c_k e^{-(iE_k t/\hbar)} \Phi_k \qquad (14.25)$$

which may be expanded on both sides by explicitly taking the time derivative on the left and evaluating H^0 for the eigenfunctions on the right to

$$-\frac{\hbar}{i}\sum_k \frac{\partial c_k}{\partial t} e^{-(iE_k t/\hbar)} \Phi_k + \sum_k c_k E_k e^{-(iE_k t/\hbar)} \Phi_k$$

$$= \sum_k c_k E_k e^{-(iE_k t/\hbar)} \Phi_k + e_0 \mathbf{r}\sin(2\pi \nu t)\sum_k c_k e^{-(iE_k t/\hbar)} \Phi_k \qquad (14.26)$$

If we cancel the equivalent sums on the left and right we are left with

$$-\frac{\hbar}{i}\sum_k \frac{\partial c_k}{\partial t}e^{-(iE_kt/\hbar)}\Phi_k = e_0\mathbf{r}\sin(2\pi\nu t)\sum_k c_k e^{-(iE_kt/\hbar)}\Phi_k \qquad (14.27)$$

We now multiply on the left by Φ_m and integrate, where m indexes the stationary state Φ for which we are interested in measuring the probability of transition. This gives

$$-\frac{\hbar}{i}\sum_k \frac{\partial c_k}{\partial t}e^{-(iE_kt/\hbar)}\langle\Phi_m|\Phi_k\rangle = e_0\sin(2\pi\nu t)\sum_k c_k e^{-(iE_kt/\hbar)}\langle\Phi_m|\mathbf{r}|\Phi_k\rangle \qquad (14.28)$$

Note that the expectation value on the l.h.s. of Eq. (14.28) is simply δ_{mk}, because of the orthogonality of the stationary-state eigenfunctions. Thus, only the term $k = m$ survives, and we may rearrange the equation to

$$\frac{\partial c_m}{\partial t} = -\frac{i}{\hbar}e_0\sin(2\pi\nu t)\sum_k c_k e^{[i(E_m-E_k)t/\hbar]}\langle\Phi_m|\mathbf{r}|\Phi_k\rangle \qquad (14.29)$$

If we assume that our perturbation was small, and applied for only a short time, we may further assume that the expansion coefficients on the r.h.s. of Eq. (14.29) have their initial (ground-state) values. This leads to the further simplification

$$\frac{\partial c_m}{\partial t} = -\frac{i}{\hbar}e_0\sin(2\pi\nu t)e^{[i(E_m-E_0)t/\hbar]}\langle\Phi_m|\mathbf{r}|\Phi_0\rangle \qquad (14.30)$$

In order to determine c_m at (and after) time τ, we must integrate t from 0 to τ, giving

$$c_m(\tau) = -\frac{i}{\hbar}e_0\int_0^\tau \sin(2\pi\nu t)e^{[i(E_m-E_0)t/\hbar]}\langle\Phi_m|\mathbf{r}|\Phi_0\rangle dt$$

$$= \frac{1}{2i\hbar}e_0\left[\frac{e^{i(\omega_{m0}+\omega)\tau}-1}{\omega_{m0}+\omega} - \frac{e^{i(\omega_{m0}-\omega)\tau}-1}{\omega_{m0}-\omega}\right]\langle\Phi_m|\mathbf{r}|\Phi_0\rangle \qquad (14.31)$$

where

$$\omega = 2\pi\nu \qquad (14.32)$$

and

$$\omega_{m0} = \frac{E_m - E_0}{\hbar} \qquad (14.33)$$

We now ask the question, for what values of m is $|c_m|^2$ large? Given a particular frequency of radiation ω, the magnitude of c_m will be large if ω_{m0} is close to ω, thereby making the denominator in the second term in brackets very small (note that even when ω_{m0} is equal to ω, the expansion coefficient is well behaved because of the way the numerator approaches zero, cf. Section 10.5.2). This result is consistent with the notion that a photon of energy $h\nu$

is absorbed in the transition between the two states, although it takes a more sophisticated theoretical treatment to demonstrate this. However, this term fails to differentiate any one state m from another, all states being predicted to undergo transitions with equal probability at their respective frequencies.

It is the last term that accounts for differences in absorption probabilities. This term is the expectation value of the dipole moment operator (see Section 9.1.1) evaluated over different determinants. Its expectation value is referred to as the transition dipole moment.

The matrix elements r_{m0} are quite straightforward to evaluate. Before leaving them, however, it is worthwhile to make some qualitative observations about them. First, the Condon–Slater rules dictate that for the one-electron operator $\mathbf{r}$, the only matrix elements that survive are those between determinants differing by at most two electronic orbitals. Thus, only absorptions generating singly or doubly excited states are allowed.

In addition, group theory can be used to assess when transition dipole moments must be zero. The product of the irreducible representations of the two wave functions and the dipole moment operator within the molecular point group symmetry must contain the totally symmetric representation for the matrix element to be non-zero (note that, if the molecule does not contain an inversion center, the operator $\mathbf{r}$ does not belong to any *single* irrep, except for the trivial case of C_1 symmetry; see Appendix B for more details). A consequence of this consideration is that, for instance, electronic transitions between states of the same symmetry are forbidden in molecules possessing inversion centers.

The derivation above may be generalized to wave functions other than electronic ones. By evaluation of transition dipole matrix elements for rigid-rotor and harmonic-oscillator rotational and vibrational wave functions, respectively, one arrives at the well-known selection rules in those systems that absorptions and emissions can only occur to adjacent levels, as previously noted in Chapter 9. Of course, simplifications in the derivations lead to many 'forbidden' transitions being observable in the laboratory as weakly allowed, both in the electronic case and in the rotational and vibrational cases.

As a final point, let us consider the transition not simply between electronic states, but between wave functions described as products of (decoupled) electronic and vibrational states. That is, we consider wave functions Λ of the form

$$\Lambda = \Psi \Xi \tag{14.34}$$

where Ψ is the electronic wave function of Eq. (14.22) and Ξ is a vibrational wave function, e.g., as defined by Eq. (9.40). If we carry out the same analysis as above, for radiation of wavelengths that are far from regions associated with vibrational transitions (as UV/Vis is from IR), then we find that Eq. (14.31) generalizes to

$$c_{m,n}(\tau) = \frac{1}{2i\hbar} e_0 \left[\frac{e^{i(\omega_{m0}+\omega)\tau} - 1}{\omega_{m0} + \omega} - \frac{e^{i(\omega_{m0}-\omega)\tau} - 1}{\omega_{m0} - \omega} \right] \langle \Phi_m | \mathbf{r} | \Phi_0 \rangle \langle \Xi_n^m | \Xi_0^0 \rangle \tag{14.35}$$

where n indexes the vibrational wave functions of electronic state m, and we have assumed that the ground electronic state is also in its ground vibrational state. We now ask the question, when is the overlap between the vibrational wave functions (the so-called Franck–Condon overlap) large? The rough rule of thumb is that ground-state vibrational wave functions have their maximum amplitude near the equilibrium structure, while excited vibrational wave functions have their maximum amplitudes near their turning points, which is to say where the PES along a vibrational coordinate rises to roughly the vibrational energy (see also Figure 9.7).

This refines the situation depicted in Figure 14.2, which suggests that a vertical transition occurs from the ground PES to the excited PES. More realistically, it occurs from the ground vibrational level of the ground state (which does indeed usually have a wave function well centered about the equilibrium structure) to any number of vibrational levels of the excited state. However, the only vibrational wave function in the single coordinate depicted in Figure 14.2 that is likely to have significant amplitude at the ground-state's equilibrium geometry is the one with its turning point at that geometry. As the excited electronic state in this excited vibrational state has roughly the energy of the excited-state PES at the ground state's equilibrium geometry, the picture drawn in Figure 14.2 is quantitatively valid, if somewhat opaque in justification (similar arguments can be made even for dissociative excited states). The major remaining error in the figure is the one-half quantum of ZPVE that is being ignored in the ground state for every coordinate that changes significantly between the ground and excited states. However, two states rarely differ in more than a very small number of coordinates, and this remaining error is typically no worse than that associated with the computation of the state-energy difference. It can, in any case, be corrected for in cases where treatments of increased quantitative accuracy are desired.

14.6 Solvatochromism

A surrounding condensed phase can have enormous impacts on the electronic spectroscopy of a given molecule. Certain dye molecules are sufficiently sensitive to the nature of a surrounding solvent that the color of their solutions can vary across the entire visible spectrum depending on the particular solvent chosen. This solvent effect on spectroscopy is known as solvatochromism.

The influence of solvent on UV/Vis absorption spectra is in some ways analogous to its influence on reaction coordinates. In this case, it is not differential solvation of connected stationary points that is of interest, but rather differential solvation of the PESs for different electronic states (Figure 14.7). Solvatochromism, shown in Figure 14.7 as it affects vertical absorption, derives from the differential solvation of the ground- and excited-state potential-energy surfaces. The blue shift illustrated in this example results from the equilibrium free energy of solvation for the ground state being larger in magnitude than ΔG^* for the excited state. Note that ΔG^* is *not* the equilibrium free energy of solvation of the excited state (which cannot be determined from this diagram, since the PES for the excited state in *equilibrium* with solvent is not shown), nor even the non-equilibrium solvation free energy of the excited state at the ground-state geometry, since it also includes effects associated with the changing *ground-state* geometry at which the vertical excitation takes place.

The subtlety of the situation derives from the different timescales involved. If we restrict our discussion to absorption, for the moment, the timescale of the absorption has already been noted to be on the electronic scale – effectively infinitely fast from the point of view

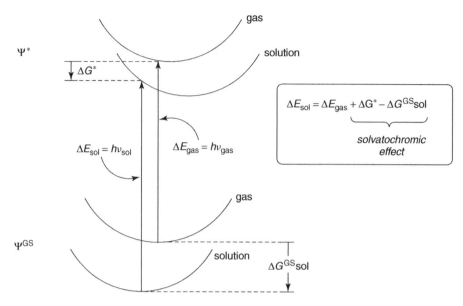

Figure 14.7 The dependence of solvatochromism on differential solvation of the ground- and excited-state PESs

of nuclear motion. Thus, when we speak of the solvation of the excited state, at the point of absorption the solvent is equilibrated to the charge distribution of the ground state, but its nuclear motion is frozen on the absorption timescale, so it cannot reorganize until after the fact. Thus, the solvation of the excited state is a non-equilibrium solvation.

Indeed, things are slightly more complicated, because the electrons of the solvent *can* respond on the timescale of the absorption. Thus, in discussing solvent effects, it is helpful to separate the bulk dielectric response of the solvent, which is a function of ε, into a fast component, depending on n^2 where n is the solvent index of refraction, and a slow component, which is the remainder after the fast component is removed from the bulk. The initially formed excited state interacts with the fast component in an equilibrium fashion, but with the slow component frozen in its ground-state-equilibrium polarization. The fast component accounts for almost the entire bulk dielectric response in very non-polar solvents, like alkanes, and about one-half of the response in highly polar solvents.

Because the solvent is fully equilibrated to the ground state but not to the excited state, it is often, but not always, the case that the ground state is better solvated than the excited state, and thus most absorptions are blue-shifted (moved to higher energy) in polar solvents. Applying identical arguments to the emission of a long-lived excited state, where the solvent has equilibrated to the excited state and thus will not be equilibrated to an instantaneously produced ground state, suggests that most spectroscopic emissions in polar solvents will be red-shifted.

While the above discussion has focused primarily on electrostatic interactions between solutes and polar solvents, experiment indicates that many absorptions in *non*-polar solutions are *red*-shifted. This appears not to be a polarization effect, but a manifestation of improved dispersion interactions between the excited state, which because of its more highly excited electron(s) tends to be more polarizable than the ground state, and the solvent. Differential hydrogen bonding interactions can also play an important role in situations where solute–solvent interactions of this type are manifest.

Given the disparate nature of the physical interactions between the different electronic states and the solvent, and the non-equilibrium nature of the solvation of at least one state in the vertical process, theoretical models require a fairly high degree of sophistication in their construction to be applicable to predicting spectroscopic properties in solution. This requirement, coupled with the rather poor utility of available experimental data (most solution spectra show very broad absorption peaks, making it difficult to assign vertical transitions accurately in the absence of a very complex dynamical analysis), has kept most theory in this area at the developers' level. A full discussion is beyond the scope of an introductory text, but we will briefly touch on a few of the key issues.

Continuum solvation models enjoy their usual advantage of efficiency, but the proper computation of the reaction field for the excited state requires that first the slow component is determined based on the ground-state charge distribution, and then the fast component based on the excited state, the latter process being iterative in the usual SCRF sense (Aguilar, Olivares del Valle, and Tomasi 1993; Mennucci, Cammi, and Tomasi 1998; Cossi and Barone 2000). In the absence of a surrounding solvent shell, however, differential dispersion and hydrogen bonding interactions must be accounted for in an *ad hoc* fashion after this accounting for polarization (Rauhut, Clark, and Steinke, 1993; Li, Cramer, and Truhlar 2000).

QM/MM approaches where the solute is QM and the solvent MM are in principle useful for computing the effect of the slow reaction field (represented by the solute point charges) but require a polarizable solvent model if electronic equilibration to the excited state is to be included (Gao 1994). With an MM solvent shell, it is no more possible to compute differential dispersion effects directly than for a continuum model. An option is to make the first solvent shell QM too, but computational costs for MC or MD simulations quickly expand with such a model. Large QM simulations with explicit solvent *have* appeared using the fast semiempirical INDO/S model to evaluate solvatochromic effects, and the results have been promising (Coutinho, Canuto, and Zerner 1997; Coutinho and Canuto 2003). Such simulations offer the potential to model solvent broadening accurately, since they can compute absorptions for an ensemble of solvent configurations.

14.7   Case Study: Organic Light Emitting Diode Alq3

Synopsis of Halls and Schlegel (2001) 'Molecular Orbital Study of the First Excited State of the OLED Material *tris*(8-hydroxyquinoline)aluminum(III)'.

Many modern display technologies make use of organic light emitting diodes. These devices typically include two layers, at least one of which is organic, through which electrons and holes propagate. When a hole meets an electron in a single layer or at an interface, the recombination leads to a singlet exciton that fluoresces in the light-generating event. One small organic molecule that has proven to be useful in this regard is *tris*-(8-hydroxyquinoline)aluminum(III), also called Alq3 (Figure 14.8).

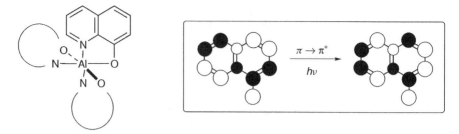

Figure 14.8 Structure of Alq3 with only one ligand drawn in full. The two states involved in the photoemission are computed to be highly localized to a single aryl ligand, and indeed to resemble the analogous HOMO and LUMO of 8-hydroxyquinoline (inset at right). How are issues of molecular geometry and absorption/emission spectroscopy related? What levels of theory would be expected to be most successful, and economical, for predicting geometry, or absorption/emission, or both? How does the inorganic aluminum atom complicate things if at all?

Halls and Schlegel approached Alq3 with an interest primarily in its excited-state properties. Prior studies on the ground state had provided some information about its molecular geometry, vibrational spectroscopy, and vertical absorption spectroscopy.

Geometry optimization of excited states can be tedious since analytic derivatives are available for a more limited range of theories than is true for ground states. When the excited state cannot be described at the single-configuration SCF level (i.e., it is not the lowest energy state of some irrep of the molecular point group), then the next simplest approach is CIS, for which analytic derivatives are indeed available. The size of Alq3 presents some basis set limitations, and Halls and Schlegel decided to employ 3-21+G**. [Note that it is not entirely clear what is meant by this basis set name, since usually the first '*' is really '(*)', and means polarization only on second row atoms (see Section 6.2.3), while the second '*' presumably implies p functions on H atoms; however, it would be very unbalanced to have polarization functions only on Al and H, so it seems likely that the authors used first-row polarization functions as well, perhaps borrowing them from the 6-31G(d) basis set, where such functions have been defined.] To check the likely utility of this basis set, they compared it to a large polarized valence triple-ζ basis set for the S_1 excited state of simple 8-hydroxyquinoline (which was small enough to permit geometry optimization to be carried out with the larger basis). Differences in the structures predicted using the two basis sets were sufficiently small that Halls and Schlegel were confident about using 3-21+G** for Alq3.

However, while the CIS level has been shown in many instances to be fairly good for optimizing excited-state geometries, it is not particularly good in this system for excitation energies, as shown in Table 14.5. The *ab initio* CIS results drastically overestimate the absorption energies in every case, irrespective of basis-set quality. Note, however, that the CIS/INDO/S results do very well indeed, particularly with the largest window of 15 occupied and 15 virtual orbitals used in generating the singly excited states. This illustrates how much one can achieve with careful parameterization.

The TDDFT results, in this case using the B3LYP functional, are the most satisfactory of the non-empirical results. Comparison of 3-21+G** results with previously published

Table 14.5 Measured and predicted absorption and emission energies (eV) for 8-hydroxyquinoline and Alq3

Process	Level of theory	8-hydroxyquinoline	Alq3
Absorption	CIS/3-21+G**	5.35	4.68
	CIS/pVTZ	5.80	
	CIS/INDO/S (12x12)		3.48
	CIS/INDO/S (15x15)	3.83	3.28
	TD-B3LYP/3-21+G**	3.76	3.00
	TD-B3LYP/6-31G(d)	3.72	2.90
	Experiment	3.84	3.20
Emission	CIS/3-21+G**		3.58
	TD-B3LYP/3-21+G**		2.30
	Experiment		2.40

6-31G(d) results again would seem to indicate that the former basis set may be regarded as adequate.

With respect to emission, the geometry computed for the excited-state Alq3 structure at the CIS/3-21+G** level was found to be significantly different from that for the ground state. This leads to a Stokes shift of 0.8 eV, which is not particularly well reproduced at the CIS level but is quite accurately predicted at the TD-B3LYP level.

Having obtained good agreement with experiment for the various spectroscopic data, Halls and Schlegel go on to analyze the MOs involved in the photoluminescence. They find that the orbitals involved are highly localized on a single one of the three aryl ligands in Alq3, and that these orbitals are quite similar to those involved in the $S_0 \rightarrow S_1$ absorption/emission of 8-hydroxyquinoline itself. They also express the difference between the excited-state geometry and the ground-state geometry of Alq3 in terms of the normal modes (this procedure is in essence a multilinear regression involving displacement vectors). They find that a particular normal mode having high intensity in the vibrational spectrum makes a significant contribution in this analysis, thereby rationalizing observations of vibrational structure in the absorption and emission spectra of Alq3 under matrix isolation conditions. They carry out their vibrational analysis using BLYP/6-31G(d) frequencies, as these were found to be in very good agreement with the experimental ground-state IR spectrum.

The work of Halls and Schlegel illustrates particularly effectively how different levels of theory may be used for studying different aspects of a complex chemical problem. Furthermore, repeated comparisons of theoretical predictions to experimental measurements in order to validate the chosen levels of theory provides solid support for the quality of further predictions using those levels.

Bibliography and Suggested Additional Reading

Aguilar, M. A. 2001. 'Separation of the Electric Polarization into Fast and Slow Components: A Comparison of Two Partition Schemes' *J. Phys. Chem. A*, **105**, 10 393.

Cave, R. J., Burke, K., and Castner, E. W., Jr. 2002. 'Theoretical Investigation of the Ground and Excited States of Coumarin 151 and Coumarin 120' *J. Phys. Chem. A*, **106**, 9294.

Ciofini, I. and Daul, C. A. 2003. 'DFT Calculations of Molecular Magnetic Properties of Coordination Compounds', *Coord. Chem. Rev.*, **238**, 187.

Foresman, J. B., Head-Gordon, M., Pople, J. A., and Frisch, M. 1992. 'Toward a Systematic Molecular Orbital Theory for Excited States' *J. Phys. Chem.*, **96**, 135.

Jensen, F. 1999. *Introduction to Computational Chemistry*, Wiley: Chichester.

Koch, W. and Holthausen, M. C. 2000. *A Chemist's Guide to Density Functional Theory*, Wiley-VCH: Weinheim.

Krylov, A. I., Slipchenko, L. V., and Levchenko, S. V. 'Breaking the Curse of the Non-dynamical Correlation Problem: the Spin–Flip Method', *ACS Symp. Ser.*, in press.

Levine, I. N. 2000. *Quantum Chemistry*, 5th Edn., Prentice Hall: New York.

Reichardt, C. 1990. *Solvents and Solvent Effects in Organic Chemistry*, VCH: New York.

Yarkony, D. R. 1998. 'Conical Intersections: Diabolical and Often Misunderstood', *Acc. Chem. Res.*, **31**, 511.

Zerner, M. C. 1996. 'Intermediate Neglect of Differential Overlap Calculations on the Electronic Spectra of Transition Metal Complexes', in *Metal–Ligand Interactions*, Russo, N. and Salahub, D. R., Eds., Kluwer: Dordrecht, 493.

Ziegler, T., Rauk, A., and Baerends, E. J. 1977. *Theor. Chim. Acta*, **43**, 261.

References

Adamo, C. and Barone, V. 1999. *Chem. Phys. Lett.*, **314**, 152.

Aguilar, M. A., Olivares del Valle, F. J., and Tomasi, J. 1993. *J. Chem. Phys.*, **98**, 7375.

Autschbach, J., Jorge, F. E., and Ziegler, T. 2003. *Inorg. Chem.*, **42**, 2867.

Back, R. A., Willis, C., and Ramsay, D. A. 1978. *Can. J. Chem.*, **56**, 1575.

Beratan, D., Kondru, R. K., and Wipf, P. 2002. *ACS Symp. Ser.*, **810**, 104.

Bulliard, C., Allan, M., Wirtz, G., Haselbach, E., Zachariasse, K. A., Detzer, N., and Grimme, S. 1999. *J. Phys. Chem. A*, **103**, 7766.

Casida, M. E., Casida, K. C., and Salahub, D. R. 1998. *Int. J. Quantum Chem.*, **70**, 933.

Casida, M. E., Gutierrez, F., Guan, J., Gadea, F.-X., Salahub, D. R., and Daudey, J. P. 2000. *J. Chem. Phys.*, **113**, 7062.

Cossi, M. and Barone, V. 2000. *J. Phys. Chem. A,* **104**, 10614.

Coutinho, K. and Canuto, S. 2003. *J. Mol. Struct. (Theochem)*, **632**, 235.

Coutinho, K., Canuto, S., and Zerner, M. 1997. *Int. J. Quantum Chem.*, **65**, 885.

Cramer, C. J., Dulles, F. G., Giesen, D. J., and Almlöf, J. 1995. *Chem. Phys. Lett.*, **245**, 165.

Durant, J. L. 1996. *Chem. Phys. Lett.*, **256**, 595.

Fabian, J. 2001. *Theor. Chem. Acc*, **106**, 199.

Furche, F. and Ahlrichs, R. 2002. *J. Chem. Phys.*, **117**, 7433.

Gao, J. 1994. *J. Am. Chem. Soc.*, **116**, 9324.

Gilles, M. K., Lineberger, W. C., and Ervin, K. M. 1993. *J. Am. Chem. Soc.*, , **115**, 1031.

Grimme, S. 1996. *Chem. Phys. Lett.*, **259**, 128.

Grimme, S. and Waletzke, M. 1999. *J. Chem. Phys.*, **111**, 5645.

Grimme, S. and Waletzke, M. 2000. *Phys. Chem. Chem. Phys.*, **2**, 2075.

Gunion, R. F. and Lineberger, W. C. 1996. *J. Phys. Chem.*, **100**, 4395.

Head-Gordon, M., Rico, R. J., Oumi, M., and Lee, T. J. 1994. *Chem. Phys. Lett.*, **219**, 21.

Hrovat, D. A., Waali, E. E., and Borden, W. T. 1992. *J. Am. Chem. Soc.*, **114**, 8698.

Hutchison, G. R., Ratner, M. A., and Marks, T. J. 2002. *J. Phys. Chem. A*, **106**, 10596.

Jamorski, C., Foresman, J. B., Thilgen, C., and Lüthi, H.-P. 2002. *J. Chem. Phys.*, **116**, 8761.

Johnson, W. T. G. and Cramer, C. J. 2001. *Int. J. Quantum Chem.*, **85**, 492.

Kim, S.-J., Hamilton, T. P., and Schaefer, H. F., III, 1992. *J. Am. Chem. Soc.*, **114**, 5349.

Kim, K., Shavitt, I, and Del Bene, J. E. 1992. *J. Chem. Phys.*, **96**, 7573.

Li, J., Cramer, C. J., and Truhlar, D. G. 2000. *Int. J. Quantum Chem.*, **77**, 264.

Lim, M. H., Worthington, S. E., Dulles, F. J., and Cramer, C. J. 1996. In: *Chemical Applications of Density-functional Theory*, ACS Symposium Series, Vol. 629, Laird, B. B., Ross, R. B., and Ziegler, T., Eds., American Chemical Society: Washington, DC, 402.

Mennucci, B., Cammi, R., and Tomasi, J. 1998. *J. Chem. Phys.*, **109**, 2798.

Noodleman, L., Peng, C. Y., Case, D. A., and Mouesca, J.-M. 1995. *Coord. Chem. Rev.*, **144**, 199.

Parac, M. and Grimme, S. 2002. *J. Phys. Chem. A*, **106**, 6844.

Parusel, A. B. J. 2000. *Phys. Chem. Chem. Phys.*, **2**, 5545.

Parusel, A. B. J., Köhler, G., and Grimme, S. 1998. *J. Phys. Chem. A*, **102**, 6297.

Rauhut, G., Clark, T., and Steinke, T. 1993. *J. Am. Chem. Soc.*, **115**, 9174.

Rinderspacher, B. C. and Schreiner, P. R. 2004. *J. Phys. Chem. A*, **108**, 2867.

Sears, J. S., Sherrill, C. D., Krylov, A. I. 2003 *J. Chem. Phys.*, **118**, 9084.

Serrano-Andrés, L., Merchán, M., Roos, B. J., Lindh, R. 1995. *J. Am. Chem. Soc.*, **117**, 3189.

Shao, Y., Head-Gordon, M., Krylov, A. I. 2003. *J. Chem. Phys.*, **118**, 4807.

Slipchenko, L. V. and Krylov, A. I. 2003. *J. Chem. Phys.*, **118**, 6874.

Smith, B. A. and Cramer, C. J. 1996. *J. Am. Chem. Soc.*, **118**, 5490.

Stratmann, R. E., Scuseria, G. E., and Frisch, M. J. 1998. *J. Chem. Phys.*, **109**, 8218.

Tozer, D. J. and Handy, N. C. 1998. *J. Chem. Phys.*, **109**, 10180.

Travers, M. J., Cowles, D. C., Clifford, E. P., and Ellison, G. B. 1992. *J. Am. Chem. Soc.*, **114**, 8699.

Wiberg, K. B., de Oliveira, A. E., Trucks, G. 2002. *J. Phys. Chem. A*, **106**, 4192.

Worthington, S. E. and Cramer, C. J. 1997. *J. Phys. Org. Chem.*, **10**, 755.

Zyubin, A. S. and Mebel, A. M. 2003. *J. Comput. Chem.*, **24**, 692.

15

Adiabatic Reaction Dynamics

15.1   Reaction Kinetics and Rate Constants

Consider an arbitrary equilibrium system

$$A + B + C + \cdots \rightleftharpoons \cdots + X + Y + Z \qquad (15.1)$$

where no particular stoichiometry is implied. When the system is displaced from equilibrium, by addition of more of a particular species, by a change in temperature and/or pressure, or by any other influence, empirical observation has shown that the rate at which equilibrium is reestablished may be expressed as

$$\text{rate}(t) = k_\phi(t) \frac{[A]^a [B]^b [C]^c \cdots}{\cdots [X]^x [Y]^y [Z]^z} \qquad (15.2)$$

where k_ϕ is a phenomenological rate constant (distinguished from an elementary rate constant as defined later on), [W] represents the concentration of species W (usually expressed in units of molarity or partial pressure), and each concentration term has associated with it an exponent that is sometimes referred to as the 'molecularity' of the species. Often, but not always, molecularities have integral values, including zero. Note that since we are measuring a return to equilibrium, all concentration terms are functions of time t, as are k_ϕ and the rate itself.

The *a priori* prediction of all of the variables appearing on the r.h.s. of Eq. (15.2) is a challenging task, to say the least. This is particularly true because the equilibrium of Eq. (15.1) may involve the simultaneous operation of a large number of individual chemical reactions, with some possibly involving very low concentrations of reactive intermediates, the presence of which may be difficult to establish experimentally. In order to make progress, a critical simplification is to break the overall process down into so-called elementary steps. To simplify matters a bit, we will consider only adiabatic reaction steps, that is, reactions taking place on a single PES without any change in electronic state (the topic of non-adiabatic dynamics is discussed briefly in Section 15.5). For practical purposes, there are only two kinds of elementary reactions: unimolecular and bimolecular.

Essentials of Computational Chemistry, 2nd Edition Christopher J. Cramer
© 2004 John Wiley & Sons, Ltd   ISBNs: 0-470-09181-9 (cased); 0-470-09182-7 (pbk)

15.1.1   Unimolecular Reactions

The simplest unimolecular reaction may be expressed in equilibrium form as

$$A \underset{k_{-1}}{\overset{k_1}{\rightleftharpoons}} B \qquad (15.3)$$

where A and B are isomeric, e.g., conformationally or constitutionally. The unimolecular rate constants above and below the equilibrium arrows are associated with the forward and reverse steps in the equilibrium process. Thus, the rate at which A is converted into B is $k_1[A]$ while the rate at which B is converted into A is $k_{-1}[B]$. These rate constants truly are 'constants', i.e., they are independent of time.

Note that at equilibrium, the rate at which A is converted into B must be exactly equal to the rate at which B is converted into A, i.e., the system is stationary with respect to reactant and product concentrations. Thus

$$k_1[A]_{eq} = k_{-1}[B]_{eq} \qquad (15.4)$$

This may be rearranged to yield

$$\frac{k_1}{k_{-1}} = \frac{[B]_{eq}}{[A]_{eq}}$$
$$= K_{eq} \qquad (15.5)$$

where K_{eq} is the equilibrium constant for Eq. (15.3). So, it is a straightforward task to measure the ratio of the elementary rate constants, but how is any one measured individually?

If we consider the system perturbed from equilibrium – let us suppose that there is an excess of A – then the rate of return to equilibrium may be expressed either as the rate of disappearance of A, i.e., $-d[A]/dt$, or as the rate of appearance of B, i.e., $d[B]/dt$. Using the first choice, we may write

$$-\frac{d[A]}{dt} = k_1[A] - k_{-1}[B] \qquad (15.6)$$

If the second term on the r.h.s. can be ignored, either because $k_{-1} \ll k_1$ (a so-called 'irreversible' reaction), or because we start with $[A] \gg [B]$ and only observe the system over a time frame where that relationship continues to hold, then we may rearrange Eq. (15.6) to give the first-order rate expression

$$-\frac{d[A]}{[A]} = k_1 dt \qquad (15.7)$$

where 'first order' implies that the sum of the exponents for concentration terms on the r.h.s. of the general rate expression written in the form of Eq. (15.2) is one. Integration of both sides from time 0 to time τ leads to

$$\ln\left(\frac{[A]_0}{[A]_\tau}\right) = k_1\tau \tag{15.8}$$

Thus, experimentally, one plots the logarithm of the concentration ratio against time under conditions where Eq. (15.7) holds in order to determine k_1. The reverse rate constant k_{-1} may either be determined analogously, or from Eq. (15.5) once k_1 is known.

Note that for a first-order reaction, the time required for the reactant concentration to drop by some constant factor is a simple function of the rate constant. Thus, for instance, the half-life $\tau_{1/2}$, which is the time required such that $[A]_{\tau_{1/2}} = \frac{1}{2}[A]_0$ for any starting concentration, may be determined from Eq. (15.8) to be $k_1^{-1}\ln 2$.

Fragmentation is another possible unimolecular reaction. A fragmentation reaction may be expressed as

$$A \underset{k_{-1}}{\overset{k_1}{\rightleftharpoons}} B + C \tag{15.9}$$

(in principle, fragmentations involving more than two products all of which are produced simultaneously are possible, but examples are very rare). The rate for disappearance of A when in excess of its equilibrium value is

$$-\frac{d[A]}{dt} = k_1[A] - k_{-1}[B][C] \tag{15.10}$$

The only difference from an experimental viewpoint between Eqs. (15.6) and (15.10) is that Eqs. (15.7) and (15.8) can now be made to apply by ensuring that either one (or both) of B and C have vanishingly small concentrations over the course of the rate measurement.

15.1.2 Bimolecular Reactions

The opposite of a fragmentation reaction is a condensation reaction, i.e.,

$$A + B \underset{k_{-1}}{\overset{k_1}{\rightleftharpoons}} C \tag{15.11}$$

Note that in practice the abstract species A and B may themselves already be molecules or supermolecules formed from prior condensations, but simple probability arguments make condensation reactions simultaneously involving more than two species impossible under most sets of experimental conditions. The rate law associated with eq. 15.11 is

$$-\frac{d[A]}{dt} = k_1[A][B] - k_{-1}[C] \tag{15.12}$$

where k_1 is a second-order rate constant because it multiplies a set of concentrations whose exponents sum to 2. The simplest evaluation of k_1 proceeds by arranging for a vanishingly small concentration of C over the course of the measurement (or choosing an irreversible reaction) and a vast excess of B. In that case, the rate expression becomes

$$-\frac{d[A]}{[A]} = k_1^* dt \tag{15.13}$$

which is identical to Eq. (15.7) except that k_1^*, the so-called pseudo-first-order rate constant which may be measured in exactly the fashion already described above for a normal first-order rate constant, is the product of the second-order rate constant k_1 and the effectively constant $[B]_0$ when B is in excess.

Bimolecular reactions having more products than the single species produced from a condensation are also possible, and their rate laws are constructed and measured in a fashion analogous to Eqs. (15.12) and (15.13). Note that the special case of a bimolecular reaction involving two molecules of the same reactant has a rate law that is particularly simple to integrate and work with.

15.2   Reaction Paths and Transition States

Theory may play two particularly important roles in rationalizing and predicting chemical reaction dynamics. As noted in the last section, the first step to understanding the dynamical behavior of a complex chemical system is breaking down the overall system into its constituent elementary processes. From a theoretical standpoint, the likely importance of various processes may be qualitatively assessed from the potential energy surfaces of putative reactions. Reactions with very high barriers will be less likely to play an important role, while low-barrier reactions will be more likely to do so.

Moreover, the PES helps to define the scope of each elementary reaction. Thus, for instance, a bimolecular condensation that involves the formation of two new bonds between the reacting species may either proceed in a concerted fashion, with only a single predicted TS structure, or it may proceed as a stepwise process with two TS structures; the stepwise process is really two elementary reactions – first a condensation and then a unimolecular rearrangement.

To say that an overall process involves two different TS structures presupposes, however, some sort of trajectory that the reacting system maps out on the PES. In general, when chemists think of a system moving on a PES, they tend to think about a particular path called the minimum-energy path (MEP) or sometimes the intrinsic reaction coordinate (IRC). The MEP is the path downwards from a saddle point to a minimum that would be followed by a ball rolling on a surface if its velocity were infinitely damped at every point; an example of such a path is given in Figure 1.4. When the potential energy surface is expressed in mass-weighted coordinates, the MEP is also the path that follows the steepest gradient at every point. The mass-weighted Cartesian coordinates for an atom are simply the Cartesian coordinates scaled by the square root of the atomic mass. Mass-weighted internal coordinates can be generated by diagonalization of the mass-weighted Cartesian coordinate force constant matrix. This coordinate system is a very convenient one in which to work since the gradients for many electronic structure methods are available to facilitate the following of the MEP.

It is worth digressing for a moment to note that following an MEP is often crucial to understanding the nature of a TS structure. Sometimes, when a molecule has a single imaginary frequency, visualization of the corresponding normal mode does not necessarily make it obvious what the reaction coordinate is. It can often happen that the TS structure that has been located corresponds to some process other than the one of interest, e.g., a TS structure for the internal rotation of a methyl group may be found when the desired TS structure was for some bond-making or bond-breaking process. In such a case, following the MEP will lead, in each direction, to the ultimate minimum energy structures connected by the TS structure. On complex potential energy surfaces, such connections can be critical to understanding the overall topology of the PES (see, for instance, Gustafson and Cramer 1995).

Although the MEP and its connection to TS structure(s) is tremendously useful as a conceptual tool, it can also be somewhat misleading to the extent that it focuses analysis on the PES itself. It should always be kept in mind that the equilibria and kinetics of reacting systems are nearly always governed by the free energy of *populations* of molecules, and not the potential energy of single molecules. To the extent the free energy describes a thermal distribution of particles composing the reacting system, one may think of the system as a cloud hovering over the PES, with the density of the cloud thinning as it rises according to Boltzmann statistics. Within the cloud, individual molecules may be exchanging energy with one another to rise and fall relative to the PES, but the net distribution remains dictated by temperature. A reacting system may be thought of as a cloud over the PES headed towards a mountain pass whose saddle point is the TS structure. However, the passage of the cloud over the pass need by no means take place directly *over* the TS structure. Depending on how wide the pass is and how tall the cloud is, many cloud particles may be able to pass arbitrarily far to the left and right of the TS structure (when the pass is very narrow one says that the reaction has an entropic bottleneck, meaning that little variation in degrees of freedom other than the reaction coordinate is permitted).

So, while the TS structure, by virtue of being a stationary point on the PES, can be informative about the height of the pass, and local topology (by Taylor expansion of the surface about the stationary point), it is only one representative of the *population* of molecules passing from reactants to products. As such, one should be rather careful not to confuse the TS *structure*, which is the stationary point, with the transition state, which may be somewhat more rigorously defined for an N-atom system as a *surface* having $3N - 7$ degrees of freedom (i.e., one less than the reactants) through which the reactive flux is maximized. That is, the ratio of the number of molecules crossing the surface in the direction reactants $\rightarrow$ products to the number crossing in the opposite direction in a given time interval is maximal. To make the distinction between the TS structure and the transition state more clear, it is helpful to return to a somewhat older term for the latter, namely, the 'activated complex'. The remainder of this chapter will hew to this distinction as closely as possible.

Returning to kinetics, while theory can be advantageously used to decompose a complex system into its constituent series of elementary reactions, we have not yet described any relationship between a theoretical quantity associated with the individual elementary reactions and their forward and reverse rate constants. It is axiomatic that reactions with high-energy

TS structures must proceed more slowly than reactions with low-energy TS structures, but a more quantitative analysis requires that we invoke more sophisticated models describing the relationship between the properties of the activated complex and kinetics. Of such models, the most versatile is transition-state theory (TST).

15.3   Transition-state Theory

15.3.1   Canonical Equation

The fundamental equations of transition-state theory may be derived in a number of different ways. Presented here is a somewhat less rigorous derivation that has the benefit of being pleasantly intuitive. Other derivations may be found in the sources listed in the bibliography at the end of the chapter, or in references therein.

Consider the simple unimolecular reaction of Eq. (15.3), where the objective is to compute the forward rate constant k_1. Transition-state theory supposes that the nature of the activated complex, $A^{\ddagger}$, is such that it represents a population of molecules in equilibrium with one another, and also in equilibrium with the reactant, A. That population partitions between an irreversible forward reaction to produce B, with an associated rate constant $k_{\ddagger}$, and deactivation back to A, with a (reverse) rate constant of k_{deact}. The rate at which molecules of A are activated to $A^{\ddagger}$ is k_{act}. This situation is illustrated schematically in Figure 15.1. Using the usual first-order kinetic equations for the rate at which B is produced, we see that

$$k_1[A] = k_{\ddagger}[A^{\ddagger}] \tag{15.14}$$

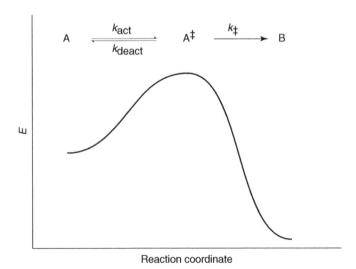

Reaction coordinate

Figure 15.1   The nature of a unimolecular reaction within the framework of transition-state theory

As we seek an expression for k_1, we may rearrange this to

$$k_1 = \frac{k_{\ddagger}[A^{\ddagger}]}{[A]}$$

$$= k_{\ddagger} K^{\ddagger} \tag{15.15}$$

where $K^{\ddagger}$ is the equilibrium constant between the activated complex and the reactants. Using the usual relationship between free energy and the equilibrium constant, we may write

$$K^{\ddagger} = e^{-(G_{\ddagger} - G_A)/k_B T} \tag{15.16}$$

where the difference in free energy between the activated complex and the reactants is referred to as the activation free energy, $\Delta G^{\ddagger}$. Using the thermodynamic concepts of Chapter 10 (see Eqs. (10.1) and (10.3)–(10.6)), we may write the free energy of a species as

$$G = U_0 + PV - k_B T \ln Q \tag{15.17}$$

where Q is the partition function. Combining Eqs. (15.16) and (15.17) we may write.

$$K^{\ddagger} = e^{-[(U_{\ddagger,0} + PV_{\ddagger} - k_B T \ln Q_{\ddagger}) - (U_{A,0} + PV_A - k_B T \ln Q_A)]/k_B T}$$

$$= \frac{Q_{\ddagger}}{Q_A} e^{-(U_{\ddagger,0} - U_{A,0})/k_B T} e^{-(PV_{\ddagger} - PV_A)/k_B T}$$

$$\approx \frac{Q_{\ddagger}}{Q_A} e^{-(U_{\ddagger,0} - U_{A,0})/k_B T} \tag{15.18}$$

Assuming that PV changes are negligible in activation of A leads to the final line of Eq. (15.18), and this assumption is usually quite reasonable.

If we combine Eqs. (15.15) and (15.18) we have

$$k_1 = k_{\ddagger} \frac{Q_{\ddagger}}{Q_A} e^{-(U_{\ddagger,0} - U_{A,0})/k_B T} \tag{15.19}$$

Note that the zero-point-energy-including difference in internal energies between A and $A^{\ddagger}$ in the exponential term is easily computable from an electronic structure calculation (for the electronic energy) and a frequency calculation (to determine the ZPVE) for the minimum energy and TS structures corresponding to A and $A^{\ddagger}$, respectively. In addition, the availability of frequencies for A permits ready computation of Q_A, as described in Chapter 10. Some attention needs to be paid, however, to the nature of the partition function for the activated complex, $Q_{\ddagger}$.

Following up on the discussion in Section 15.2 about the nature of the activated complex, the TS structure should be recognized as a species that is a minimum in $3N - 7$ degrees of freedom – the 'missing' degree of freedom is the reaction coordinate. Thus, we may readily define the electronic, translational, and rotational components of the partition function associated with the TS structure in the usual way. For the vibrational component, we will separate out the partition function for the reaction coordinate degree of freedom (see Eq. (10.28)) and write

$$k_1 = \frac{k_{\ddagger}}{1 - e^{-\hbar\omega_{\ddagger}/k_B T}} \frac{Q^{\ddagger}}{Q_A} e^{-(U_{\ddagger,0} - U_{A,0})/k_B T} \tag{15.20}$$

where $Q^{\ddagger}$ is the reduced partition function over the $3N - 7$ bound degrees of freedom and $\omega_{\ddagger}$ is the 'vibrational frequency' associated with the reaction coordinate. If we use a power series expansion for the exponential function of $\omega_{\ddagger}$ on the r.h.s. of Eq. (15.20), truncating after the first two terms, we have

$$k_1 = \frac{k_{\ddagger}}{1 - \left(1 - \dfrac{\hbar\omega_{\ddagger}}{k_B T}\right)} \frac{Q^{\ddagger}}{Q_A} e^{-(U_{\ddagger,0} - U_{A,0})/k_B T}$$

$$= \frac{k_{\ddagger} k_B T}{\hbar\omega_{\ddagger}} \frac{Q^{\ddagger}}{Q_A} e^{-(U_{\ddagger,0} - U_{A,0})/k_B T} \tag{15.21}$$

Notice that the only two unknowns remaining are $k_{\ddagger}$ and $\omega_{\ddagger}$. In this case, the vibrational frequency $\omega_{\ddagger}$ should not be thought of as the imaginary frequency that derives from the standard harmonic oscillator analysis, but rather the real inverse time constant associated with motion along the reaction coordinate. However, it is exactly motion along the reaction coordinate that converts the activated complex into product B. That is, $k_{\ddagger} = \omega_{\ddagger}$. Eliminating their ratio of unity from Eq. (15.21) leads to the canonical TST expression

$$k_1 = \frac{k_B T}{h} \frac{Q^{\ddagger}}{Q_A} e^{-(U_{\ddagger,0} - U_{A,0})/k_B T} \tag{15.22}$$

For the bimolecular reaction case involving reactants A and B, the derivation above generalizes to

$$k_1 = \frac{k_B T}{h} \frac{Q^{\ddagger}}{Q_A Q_B} e^{-(U_{\ddagger,0} - U_{A,0} - U_{B,0})/k_B T} \tag{15.23}$$

A point of occasional confusion arises with respect to units. In Eq. (15.22), all portions are unitless except for $k_B T/h$, which has units of sec^{-1}, entirely consistent with the units expected for a unimolecular rate constant. In Eq. (15.23), the same is true with respect to the r.h.s., but a bimolecular rate constant has units of concentration^{-1} sec^{-1}, which seems paradoxical. The point is that, as with any thermodynamic quantity, one must pay close attention to standard-state conventions. Recall that the magnitude of the translational partition function depends on specification of a standard-state volume (or pressure, under ideal gas conditions). Thus, a more complete way to write Eq. (15.23) is

$$k_1 = \frac{k_B T}{h} \frac{Q^{\ddagger}}{Q_A Q_B} \frac{Q_A^\circ Q_B^\circ}{Q^{\ddagger,\circ}} e^{-(U_{\ddagger,0} - U_{A,0} - U_{B,0})/k_B T} \tag{15.24}$$

where the various Q° terms have values of one and carry the standard-state volume units used for the translational partition function (the same generalization applies to Eq. (15.22), but it is sufficiently rare for a unimolecular reaction to have different standard-state volumes for the activated complex and the reactant that one rarely gives thought to this point). Care must be taken then such that if the molecular translational partition function is computed for a volume of, say, 24.5 L (the volume occupied by one mole of an ideal gas at 298 K and 1 atm pressure), and the rate constant is in, say, molecules cm^{-3} sec^{-1}, the appropriate conversion in standard states is made.

In a very general form, then, we have the canonical expression

$$k = \frac{k_B T}{h} \frac{Q^{\ddagger}}{Q_R} \frac{Q_R^o}{Q^{\ddagger,o}} e^{-\Delta V^{\ddagger}/k_B T} \tag{15.25}$$

where R refers generically to either unimolecular or bimolecular reactants, and $\Delta V^{\ddagger}$ is the difference in zero-point-including potential energies of the reactants and TS structure. When working in molar quantities, Eq. (15.25) becomes

$$k = \frac{k_B T}{h} \frac{Q^{\ddagger}}{Q_R} \frac{Q_R^o}{Q^{\ddagger,o}} e^{-\Delta V^{\ddagger}/RT} \tag{15.26}$$

in which case one often absorbs the standard-state partition functions back into the exponential to write

$$k = \frac{k_B T}{h} e^{-\Delta G^{o,\ddagger}/RT} \tag{15.27}$$

where $\Delta G^{o,\ddagger}$ is referred to as the free energy of activation. Note that using Eq. (10.6) we may also write

$$k = \frac{k_B T}{h} e^{-\Delta H^{o,\ddagger}/RT} e^{\Delta S^{o,\ddagger}/R} \tag{15.28}$$

15.3.1.1 Relation between theory and experiment

Operationally, the theoretical computation of a rate constant using TST typically employs Eq. (15.26). One locates all necessary stationary points – one TS structure and one or two minima – and evaluates their energies and their partition functions under the rigid-rotor-harmonic-oscillator approximation. Experiment, on the other hand, measures rate constants according to the methodologies outlined in Section 15.1, typically with the goal of deriving such quantities as the free energy of activation. However, the experimental data may be analyzed in a variety of ways, and it is critically important to ensure experimental/theoretical comparisons are made under consistent conditions.

One analysis of experimental data involves carrying out rate constant measurements at a series of temperatures, and then plotting $\ln(k/T)$ against $1/T$ (a so-called Eyring plot). We may rearrange Eq. (15.28) to

$$\ln\left(\frac{k}{T}\right) = -\frac{\Delta H^{o,\ddagger}}{RT} + \frac{\Delta S^{o,\ddagger}}{R} + \ln\left(\frac{k_B}{h}\right) \tag{15.29}$$

which shows that the slope of such a plot should be $-\Delta H^{o,\ddagger}/R$ and the intercept is a function of $\Delta S^{o,\ddagger}/R$. With these quantities in hand, the activation free energy may be easily computed for any temperature within the range of the data points and compared directly to a theoretical computation of this quantity (extrapolation outside the range of the data points can be dangerous because enthalpy and entropy are themselves both dependent on temperature, so it represents an approximation to assume their constancy over a given measurement range).

An alternative analysis having a long history, however, is to simply plot $\ln k$ vs. $1/T$, this procedure being motivated by the empirically derived Arrhenius expression

$$k = Ae^{-E_a/RT} \tag{15.30}$$

where A is the so-called pre-exponential factor and E_a is the Arrhenius activation energy. Rearranging Eq. (15.30) readily illustrates that a plot of $\ln k$ vs. $1/T$ will have slope $-E_a/R$ and intercept $\ln A$. Simple algebra allows us to express the relationships between the Arrhenius quantities and the thermodynamic quantities as

$$E_a = \Delta H^{o,\ddagger} + RT \tag{15.31}$$

and

$$A = \frac{k_B T}{h} e^{(1+\Delta S^{o,\ddagger}/R)} \tag{15.32}$$

Because these two different conventions exist (as well as other conventions, e.g., one based on collision theory, that will not be discussed here), when the term 'activation energy' is used without qualification, it is critical for accurate comparisons that it be established whether this term refers to an Arrhenius activation energy, a TST activation free energy, a difference in stationary-point potential energies, a difference in zero-point-including stationary-point potential energies, etc. The term 'barrier' is also often used ambiguously, and care should be taken to establish its meaning in a given situation.

One point of interest deriving from the equations of TST (and Arrhenius theory) is that the upper limit for the 298 K rate constant of a unimolecular reaction that takes place with zero activation energy (of whatever sort) is roughly 10^{13} sec^{-1}. This is, in some sense, a conceptually obvious result since that is on the order of a molecular vibrational frequency, which is thought of as the 'mechanism' by which a transition state goes to its products.

15.3.1.2 *Kinetic isotope effects*

As noted in Chapter 10, the zero-point energy, and the translational, rotational, and vibrational partition functions all depend on the isotopic masses of the atoms. Thus, so too does the rate constant for a given reaction. A difference in rates observed for two different isotopically substituted reactants is referred to as a kinetic isotope effect (KIE), usually expressed as a ratio of rates. Isotope effects are divided into two classes: primary isotope effects refer to situations where the isotopic substitution involves one of the two atoms involved in a breaking (or making) bond, while secondary isotope effects cover all other possibilities. In general, then, a KIE may be computed as

$$
\frac{k_{light}}{k_{heavy}} = \frac{\dfrac{Q^{\ddagger}_{light}}{Q_{R,light}} e^{-\Delta V^{\ddagger}_{light}/k_B T}}{\dfrac{Q^{\ddagger}_{heavy}}{Q_{R,heavy}} e^{-\Delta V^{\ddagger}_{heavy}/k_B T}}
$$

$$
= \frac{Q^{\ddagger}_{light}}{Q^{\ddagger}_{heavy}} \frac{Q_{R,heavy}}{Q_{R,light}} e^{-(\Delta ZPVE^{\ddagger}_{light} - \Delta ZPVE^{\ddagger}_{heavy})/k_B T} \tag{15.33}
$$

From a theoretical perspective, isotope effects are fairly trivially computed. The stationary points on the PES and their electronic energies are independent of atomic mass, as are the molecular force constants. Thus, one simply needs to compute the isotopically dependent zero-point energies and translational, rotational, and vibrational partition functions, and evaluate Eq. (15.33).

Primary isotope effects tend to be dominated by the difference in zero-point energies, as illustrated in Figure 15.2. Because the reaction coordinate is the breaking bond, and because there is little or no ZPVE associated with this mode in the TS structure, the full difference in reactant ZPVEs enters into the difference in zero-point-including potential energy barriers.

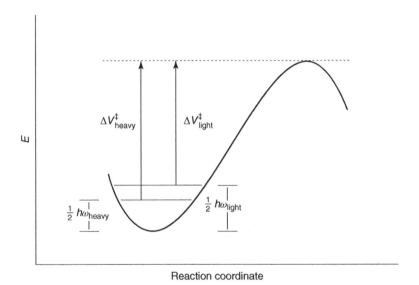

Reaction coordinate

Figure 15.2   The dominant contribution to a primary KIE is the differential loss of zero-point vibrational energy in the reaction coordinate when an isotopically substituted bond is broken. Because the light isotopomer has a higher vibrational frequency, it has more ZPVE, and a lower potential energy of activation (thus primary isotope effects expressed as k_{light}/k_{heavy} are essentially always greater than 1). Effects from other normal modes are ignored in this diagram (cf. Figure 15.3)

To a rough approximation, then, in the limit of a fully broken bond in the TS structure, the primary KIE is

$$\frac{k_{\text{light}}}{k_{\text{heavy}}} \approx e^{-\frac{1}{2}h(\omega_{\text{heavy}}^{\text{R},\ddagger} - \omega_{\text{light}}^{\text{R},\ddagger})/k_B T} \tag{15.34}$$

where $\omega^{\text{R},\ddagger}$ refers to the frequency in the reactants of the bond being broken in the TS structure. One of the largest possible differences in isotopic frequencies involving elements occurs for hydrogen/deuterium substitutions, with X–H bonds typically having stretching frequencies about 35% larger than X–D bonds. Using this relationship and a light isotope frequency of 3100 cm^{-1}, Eq. (15.34) suggests that the maximum primary KIE for a hydrogen/deuterium-substituted system at 298 K is about 7. Of course, if the bond is not fully broken in the TS structure, smaller values may be observed/computed. Larger values may also be observed, owing to quantum mechanical tunneling, as described in Section 15.3.3. When heavier elements are used, isotope effects become smaller, but a number of experimental techniques have proven to be sufficiently accurate to measure very small differences (see, for instance, Keating *et al.* 1999).

Secondary KIEs are also typically much smaller than primary KIEs, because the isotopically substituted modes are not lost in the TS structure (see Figure 15.3). In addition,

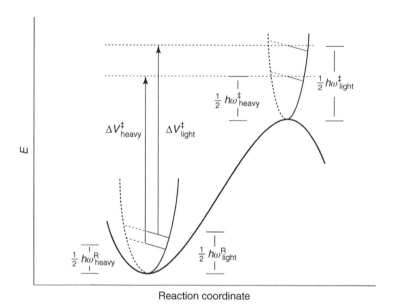

Reaction coordinate

Figure 15.3 Secondary KIEs are associated with normal modes other than the reaction coordinate, one of which is shown in this diagram. The heavy and light vibrational frequencies both change on going from the reactant (R) to the TS structure ($\ddagger$); because in this example the mode is 'tighter' in the TS structure, the difference between the heavy and light ZPVEs increases, and this causes the potential energy of activation to be larger for the light isotopomer than the heavy one (an example of an inverse secondary KIE). In a real many-atom system there are potentially a large number of modes that will contribute to the secondary KIE, some in a normal fashion and some in an inverse fashion

secondary KIEs can be 'inverse', which is to say that the light atom rate over the heavy atom rate can be less than one. In this case, no particular simplifications of Eq. (15.33) are general, and each partition function may play a role in addition to those of the ZPVEs. This is particularly true because different vibrational modes may cancel one another in the secondary KIE. That is, one mode may lead to a large normal KIE but be canceled by another mode that leads to a large inverse KIE, such that more subtle effects associated with, say, rotational motion, may be made manifest.

One caveat that must be observed when comparing computed and experimental isotope effects is that experimental measurements can sometimes be for multistep reactions. When a particular elementary reaction is not rate-determining, that is, it is not the bottleneck in the overall process, then it does not matter whether or not that reaction has associated with it a large KIE; it will not influence the observed overall rate. A separate caveat with light atoms at low to moderate temperatures is that tunneling effects may play a significant role.

15.3.2 Variational Transition-state Theory

Canonical TST defines the free energy of the activated complex based on the TS structure. This is convenient because, as it is a stationary point, we can use the machinery of the rigid-rotor-harmonic-oscillator approximation to compute the necessary partition functions to define its (reduced-dimensionality) free energy. However, it is by no means guaranteed that the free energy associated with the TS structure really is the *highest* free energy of any point along the MEP – it is only guaranteed that it is the highest point of *potential* energy along the MEP. As a simple example, it might be the case that the potential energy wells associated with some normal modes tighten up after the TS structure is reached, even though the bottoms of those wells are at a point on the MEP slightly below the energy of the TS structure. The increase in ZPVE resulting from those tighter potentials may exceed the drop in bottom-of-the-well energy such that the free energy of the non-stationary point is higher than that of the TS structure.

Variational transition-state theory (VTST), as its name implies, variationally moves the reference position along the MEP that is employed for the computation of the activated complex free energy, either backwards or forwards from the TS structure, until the rate constant is minimized. Notationally

$$k^{\text{VTST}}(T, s) = \min_s \frac{k_{\text{B}} T}{h} \frac{Q^{\ddagger}(T, s)}{Q_{\text{R}}} \frac{Q_{\text{R}}^{\text{o}}}{Q^{\ddagger,\text{o}}} e^{-\Delta V^{\ddagger}(s)/k_{\text{B}} T} \tag{15.35}$$

where s is a position on the MEP at which k^{VTST} is evaluated. By convention, $s = 0$ refers to the saddle point, and negative and positive values are displaced to the reactant and product sides of the saddle point, respectively.

To compute the r.h.s. of Eq. (15.35), we need to define how we compute the partition function (and the ZPVE) for the non-stationary point s. In this case, we simply continue to take advantage of our decision to treat the activated complex as a species having $3N - 7$ bound degrees of freedom. In order to define this space for an arbitrary point on the MEP, we simply project out motion along the MEP at point s and evaluate the partition functions (and the ZPVE) for the remaining degrees of freedom, all of which are bound and at their local minima for that choice of s, in the usual fashion. In practice, minimization of the rate constant with respect to s is accomplished by standard search techniques for situations where analytic gradients of the function to be minimized are not available. Since the partition functions depend on both s and T, changes in T may lead to changes in the value of s minimizing Eq. (15.35). In other words, the variational transition state can move with changes in temperature.

Such detailed surveys of the potential energy surface are computationally fairly demanding since not only are energies sought at every point but also gradients (to assess the degree to which one is still on the MEP) and force constants for all bound degrees of freedom (to compute the vibrational partition functions). If the potential energy surface for the reacting system can be described by an analytic function, then the computational complexity is not much worse than for any typical propagation of a molecular mechanics MD trajectory. However, while molecular mechanics can accurately describe regions of the PES in the close vicinity of minima, it is much more difficult to develop analytic functions for regions within which bond breaking (or making) are taking place and, moreover, to connect any such functions to the 'usual' ones about minima so that the entire PES can be described in a smooth, differentiable way.

One alternative to having a global analytic function is simply to compute, on the fly, energies, derivatives, and force constants whenever they are required. Such an approach, usually called 'direct dynamics' (Truhlar and Gordon 1990), in principle permits the use of sophisticated quantum mechanical methods at every point. While this will typically improve on the quality of the computed quantities compared to a global analytic function, this improvement comes at the expense of potentially having to do many, many such QM calculations (this latter point is especially telling when one is actually propagating a trajectory, as opposed to simply looking for the variational TS). These constraints have provided much of the motivation for the development of rapid MM and QM/MM models, like the MCMM and VB methods discussed in Chapters 2 and 13, that can accurately reproduce features of the PES in the vicinity of TS structures, where VTST optimization takes place.

Note that the conventional TST expression is simply the special case of VTST where evaluation is done exclusively for $s = 0$. As such, the VTST rate constant will always be less than or equal to the conventional TST rate constant (equal in the event that $s = 0$ minimizes Eq. (15.35)). Put differently, when very accurate potential energy surfaces are available, the conventional TST rate constant is typically an overestimate of the exact classical rate constant. (Note that it is possible, however, for a compensating or even offsetting error to arise from overestimation of the barrier height if the potential energy surface is *not* very accurate.)

Allison and Truhlar have compared TST and VTST to accurate solution of the time-dependent Schrödinger equation for a number of three-atom chemical reactions (it is only for such small systems that the accurate solution of the time-dependent Schrödinger equation is practical) and those results are listed in Table 15.1. On the high-quality surfaces available for this comparison, VTST is typically accurate to within 50% at temperatures above 600 K.

Table 15.1 Logarithmically averaged percent errors in TST and VTST compared to accurate quantal rate constants for a series of 3-atom reactions[a]

T (K)	Number of reactions	LAPE (%)	
		TST	VTST
200	37	1480	1952
250	40	452	569
300	48	283	296
400	49	131	148
600	37	65	51
1000	34	53	24
1500	26	63	18
2400	8	139	21

[a]Allison and Truhlar 1998. Logarithmically averaged percent errors treat each factor of 2, whether an overestimate or an underestimate, as a 100% error.

At the highest temperatures, VTST is about five times more accurate than canonical TST, but even canonical TST is still accurate to within about a factor of 2. Note that improved performance of VTST as temperature increases is to be expected since entropic effects increasingly dominate under those conditions, and it is primarily entropy that moves the optimal dividing surface away from the potential energy saddle point. At low temperatures, TST appears to outperform VTST, but that is an artifact of not considering tunneling contributions to the rate constant (see Section 15.3.3). Tunneling effects *are* included in the accurate quantal rate constants; since TST usually overestimates the *classical* rate, it is accidentally in better agreement with the *quantal* rate, which is always increased over the exact classical rate by tunneling.

Note that, with the minimized rate constant in hand, a generalized activation free energy can be defined as the difference between the free energy of the reactants and that for the point s_{min}. Note also that for the computation of isotope effects, VTST proceeds exactly like conventional TST, except that there is no requirement at a given temperature that the value of s that minimizes the rate constant for the light-atom-substituted system will be the same value of s that minimizes the rate constant for the heavy-atom-substituted system. Each must be determined separately, at which point the ratio of rate constants for that temperature may be expressed.

15.3.3 Quantum Effects on the Rate Constant

The metaphor invoked in Section 15.2 of a reacting system as a cloud wandering through a mountain pass is, by virtue of being macroscopic, necessarily a classical metaphor. In visualizing that situation, we accept as a given that those portions of the cloud below the level of the pass (i.e., at too low an energy) fail to go through and portions above the pass always do. Like the cloud in the mountain pass, the probability of transmission from reactants to products in a classical system is a Heaviside function of the energy, as illustrated in Figure 15.4. In a quantum system, however, the transmission probability is sigmoidal in shape, reflecting the phenomena of tunneling and non-classical reflection. Tunneling refers to the ability of a quantum system having an energy below the saddle point to 'tunnel' through the barrier to the products side, while non-classical reflection refers to the possibility that a quantum system above the saddle-point energy will suffer from destructive interference in a way that prevents it from crossing to products. This situation is compared to the classical one in Figure 15.4.

Insofar as tunneling increases the rate constant by allowing lower-energy systems to be reactive and non-classical reflection decreases the rate constant by reducing the reactivity of higher-energy systems, one might imagine that the two could safely be assumed to cancel. However, a thermally equilibrated Boltzmann population has a much larger percentage of

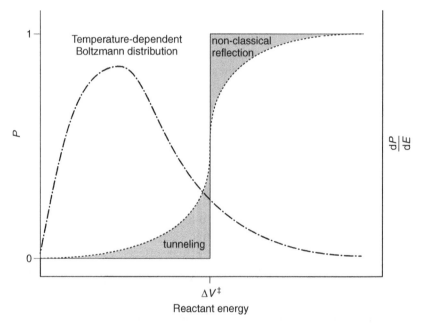

Figure 15.4 Probabilities of reaction (P) for systems moving towards a parabolic barrier for a reaction with a zero-point-including potential energy of activation $\Delta V^{\ddagger}$. Classical systems (———) below the barrier height have zero probability of reaction and above the barrier height have unit probability (i.e., the 'curve' describes a Heaviside function). Quantum systems (------), on the other hand, have increasingly non-zero probabilities as the barrier energy is approached from below because of tunneling and increasingly less than unit probabilities as the barrier energy is approached from above because of non-classical reflection. Note that because of the Boltzmann distribution of energies in a thermalized population of reacting systems (--·--·-- referenced to the right ordinate), typically many more molecules have energies in the region where tunneling can increase the reaction rate than have energies in the region where non-classical reflection can reduce the reaction rate. As a result, the former is the more quantitatively important of the two quantum phenomena

low-energy systems than high-energy ones, so tunneling effects tend to predominate over non-classical reflection, and the inclusion of quantum mechanical tunneling can be critical to predicting accurate rate constants. Within a TST (or VTST) framework, one writes

$$k = \kappa(T)\frac{k_B T}{h}\frac{Q^{\ddagger}}{Q_R}\frac{Q_R^{o}}{Q^{\ddagger,o}}e^{-\Delta V^{\ddagger}/k_B T} \tag{15.36}$$

where κ is called the transmission coefficient. The transmission coefficient is a function of temperature and, importantly, of the shape of the PES in the region about the activated complex. In the classical limit $\kappa = 1$ but, particularly at low temperatures, κ can become arbitrarily large.

Qualitatively, κ depends on the shape of the barrier (both height and width), the mass of the particle (the lighter the particle, the greater the probability of tunneling), and the temperature, the latter because T dictates the Boltzmann population of the reactant and activated-complex energies. In the context of a many-atom system, tunneling through the barrier may occur along any one or more coordinates, and the mass in question for each case may be considered to be the reduced mass of the normal mode. Thus, tunneling effects can be present even when the reaction coordinate itself is dominated only by heavy-atom motion.

Highly accurate prediction of transmission coefficients including many degrees of freedom is a very difficult quantum mechanical problem. A simplifying approximation is to consider tunneling only in the degree of freedom corresponding to the reaction coordinate. Within this one-dimensional formalism, various levels of approximation are available.

The simplest approximation is that of Wigner (1932), which takes

$$\kappa(T) = 1 + \frac{1}{24}\left[\frac{h\,\text{Im}(\nu^{\ddagger})}{k_B T}\right]^2 \tag{15.37}$$

where $\nu^{\ddagger}$ is the imaginary frequency associated with the reaction coordinate (the notation $\text{Im}(x)$ means that we take only the imaginary part of the frequency, which is to say that we treat it as though it is a real number rather than a complex one). The Wigner correction works well provided that $h\,\text{Im}(\nu^{\ddagger}) \ll k_B T$.

A more robust approximation to κ has been provided by Skodje and Truhlar (1981), generalizing earlier work by Bell (1959) for parabolic barriers. For notational convenience we take

$$\alpha = \frac{2\pi}{h\,\text{Im}(\nu^{\ddagger})} \tag{15.38}$$

and

$$\beta = \frac{1}{k_B T} \tag{15.39}$$

In the Skodje and Truhlar approximation, one takes for $\beta \le \alpha$

$$\kappa(T) = \frac{\beta\pi/\alpha}{\sin(\beta\pi/\alpha)} - \frac{\beta}{\alpha - \beta}e^{[(\beta-\alpha)(\Delta V^{\ddagger}-V)]} \tag{15.40}$$

where $\Delta V^{\ddagger}$ is the zero-point-including potential energy difference between the TS structure and the reactants, and V is 0 for an exoergic reaction and the (positive) zero-point-including potential energy difference between reactants and products for an endoergic reaction. In the case where $\alpha \leq \beta$, the corresponding expression is

$$\kappa(T) = \frac{\beta}{\beta - \alpha} \left\{ e^{[(\beta - \alpha)(\Delta V^{\ddagger} - V)]} - 1 \right\} \tag{15.41}$$

An inspection of the power series expansion for the exponential in Eqs. (15.40) and (15.41) indicates that neither expression diverges as α and β become arbitrarily close to equal (an analogous consideration of the power series expansion for the sine function in Eq. (15.40) indicates the first term on the r.h.s. to be similarly free from singularities).

A still more sophisticated approach involves fitting the reaction coordinate to a so-called Eckart potential (Eckart 1930). The Eckart potential permits an exact, analytic solution of the probability of tunneling through the barrier (and of non-classical reflection) from the time-independent Schrödinger equation for systems of fixed energy E. When that result is numerically integrated over all energies, weighted by the Boltzmann probability of the reacting system having a particular energy at a given temperature T, a very good estimate of κ in the limit of tunneling along a single dimension is obtained. [Note that when transition state theory is formulated for a system of constant energy, as opposed to constant temperature, it is called microcanonical TST (μTST) or Rice–Ramsperger–Kassel–Marcus (RRKM) theory for the unimolecular case; a microcanonical variational TST (μVTST) can be applied in a fashion analogous to VTST, with the choice of dividing surface location s now potentially different at each energy E.]

It should be noted, however, that even the best one-dimensional tunneling estimate is still likely to underestimate the full tunneling contribution, since tunneling may occur through dimensions of the PES *other* than the reaction coordinate. Multi-dimensional tunneling approximations are sufficiently complex, however, that they will not be further discussed here.

Another important point that must be borne in mind is that failure to account for tunneling, or to recognize its contribution in the first place, can lead to significant errors in the interpretation of experimental data. For example, Watson (1990) analyzed an Eyring plot of apparent rate constants for methane metathesis by methyllutetiocene (Figure 15.5) to infer

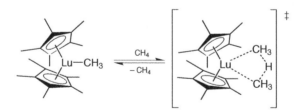

Figure 15.5 Transition-state structure for rate-determining hydrogen atom transfer in the methane metathesis reaction of methyllutetiocene. Note that the kinetics for this narcissistic reaction may be followed by using a ^{13}C label either in the reacting methane or in the methyl group of the starting organometallic

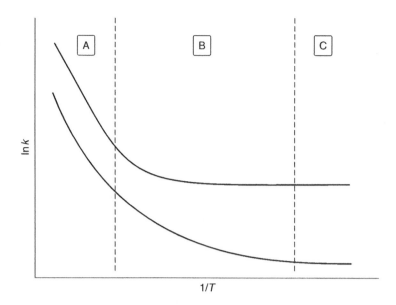

Figure 15.6 Typical changes in rate constants as a function of temperature for light-isotope (above) and heavy-isotope (below) substituted systems. In the high-temperature regime A, Arrhenius or TST plots will be essentially linear and yield good estimates of the activation parameters. In the intermediate region B, such plots may still be linear if the sampled temperature range is too small, but any activation parameters inferred therefrom will be of little utility. Additionally, KIE values in this region are very sensitive to the temperature. In the low-temperature regime C, the rate constant is almost entirely a result of tunneling, and little information about the PES can be gleaned from kinetic analysis

an activation enthalpy of 11.6 kcal mol^{-1}. However, Sherer and Cramer (2003) found for the hydrogen-atom-transfer rate-determining step that Eqs. (15.40) and (15.41) predict the tunneling transmission coefficient κ to drop from 93 to 4 over the experimental temperature range of 300 to 400 K. When the apparent rate constants were divided by their corresponding κ values, an Eyring plot of the corrected 'true' semiclassical rate constants provided an activation enthalpy of 19.2 kcal mol^{-1}. This latter result agreed well with a value of 20.3 kcal mol^{-1} computed from DFT, and illustrates the magnitude of the quantitative difference that may arise when tunneling is ignored in experimental Eyring analyses.

Figure 15.6 forms the basis for a more general discussion of tunneling, reaction rates, and kinetic isotope effects. The rate constant for an exergonic chemical reaction does not actually go to zero as the temperature goes to zero. Instead, after the temperature drops sufficiently, all reactant systems will be in their lowest energy state, that state will have some rate of tunneling through the barrier, and that rate is the non-zero asymptote that will be approached. Of course, an analysis that *neglects* tunneling would interpret a rate that is independent of temperature as corresponding to an activation enthalpy of zero (see Section 15.3.1.1) which may be very far from correct.

Moreover, because tunneling is less efficient for heavy isotopes, the transition to tunneling-dominated kinetics occurs at lower temperatures for heavier isotopes (Figure 15.6, region B), leading to regions where KIEs will change rapidly as a function of temperature, which may also confuse interpretation. These effects are not limited to esoterically low temperatures: enzymes catalyzing proton, hydride, and hydrogen atom transfers can exhibit large rate contributions from tunneling at biological temperatures (Kohen and Klinman 1998).

15.4   Condensed-phase Dynamics

Solvent effects on reaction coordinates have already been discussed in a general fashion in Section 11.1.2. In terms of estimating condensed-phase rate constants, we consider three levels of approximation. In the simplest model, referred to as separable equilibrium solvation (SES), we assume that the effects of a surrounding condensed phase are limited simply to changing the free energy along the MEP. In that case, the condensed-phase free energy of activation is simply the sum of the gas-phase free energy of activation and the free energies of solvation of the activated complex and the reactant(s) (see Figure 11.4), and we may use Eq. (15.27) to compute the rate constant directly from the condensed-phase $\Delta G^{o,\ddagger}$. The free energy of the activated complex in solution may either be evaluated for the gas-phase TS structure, or may be taken as the maximum free energy along the solvated MEP, which would be a variational-like treatment. Operationally, we may most readily compute the solvation free energy for each point on the MEP by assuming the solvent to be fully equilibrated to that point and using any convenient solvation model (a continuum model being the most efficient choice).

At the next level of approximation, we continue to imagine the solvent to be fully equilibrated to the reacting system at every point, but instead of working with the solvated MEP from the gas-phase surface, we find the equilibrium solvation path (ESP) which is the MEP on the fully solvated surface (see Figure 11.1). While both the gas-phase and solvated surfaces are defined entirely in terms of solute coordinates, the ESP may be quite different from the gas-phase MEP because solvation effects may 'push' the path in directions orthogonal to the gas-phase reaction coordinate (see Figure 11.5). With the ESP in hand, TST (or VTST) analysis may be carried out in the usual way to obtain a condensed-phase rate constant.

The beauty of the prior approximations is that by assuming a mean-field influence of solvation we can continue to work in a phase space having the same dimensionality as that for the gas phase; that being the case, analysis using the tools of TST is mechanically identical for the two phases. When the solvent is *not* fully equilibrated with the complete reaction path, however, the reacting system can no longer legitimately be described exclusively in terms of solute coordinates.

Note that the region where solvent is least well equilibrated to the solute is expected to be in the vicinity of the activated complex, since it has so short a lifetime. Since non-equilibrium solvation is less favorable than equilibrium solvation, the non-equilibrium free energy of the activated complex is higher than the equilibrium free energy, and the non-equilibrium lag in solvent response thus slows the reaction. This effect is sometimes referred to as solvent 'friction' and can be accounted for by inclusion in the transmission factor κ.

Explicit inclusion of all solvent degrees of freedom, e.g., in an MD simulation, is not a very effective approach to modeling the non-equilibrium solvent influence, however. One issue that should be apparent is that one can no longer really define a TS structure under such conditions – on the enormously high-dimensional PES constructed from all solute and solvent coordinates there will be a huge number of saddle points having similar energies in regions between reactants and products, and the related problem of running trajectories through this potentially high-energy volume of phase space to estimate rate constants has already been noted in Chapter 12.

To simplify matters, it is usually assumed that the influence of the solvent can be modeled with so-called effective solvent coordinates. A typical choice is to treat the solvent coordinate as having a harmonic potential that is linearly coupled to the solute. If one extends this approach to use an infinite number of solvent harmonic oscillators one obtains the so-called generalized Langevin equation for solute dynamics (Zwanzig 1973). In the other limit of reducing consideration of the *solute* to a *single* coordinate, one obtains Kramers–Grote–Hynes theory (Kramers 1940; Grote and Hynes 1980). The development of more sophisticated treatments for the solvent coordinates in non-equilibrium solvation models remains an active area of research.

15.5   Non-adiabatic Dynamics

15.5.1   General Surface Crossings

When two (or more) potential energy surfaces corresponding to different electronic states of a chemical system are close to one another in energy, the electronic wave function should really be written as a linear combination of the different adiabatic wave functions. For simplicity, let us consider the case of only two states, in which case we would write

$$\Psi(\mathbf{Q}, \mathbf{q}) = c_1(\mathbf{Q})\psi_1(\mathbf{Q}, \mathbf{q}) + c_2(\mathbf{Q})\psi_2(\mathbf{Q}, \mathbf{q}) \tag{15.42}$$

where ψ_1 and ψ_2 are the two adiabatic states that depend on the electronic coordinates $\mathbf{q}$ and the nuclear coordinates $\mathbf{Q}$ and the coefficients also depend on the nuclear coordinates because the mixing of the states will vary with different geometries. The situation is illustrated for a single internal coordinate in Figure 15.7. Note that since the coefficients c_1 and c_2 depend only on nuclear coordinates, each is a nuclear wave function.

In this case, the Schrödinger equation becomes

$$H\Psi(\mathbf{Q}, \mathbf{q}) = E_{full}[c_1(\mathbf{Q})\psi_1(\mathbf{q}; \mathbf{Q}) + c_2(\mathbf{Q})\psi_2(\mathbf{q}; \mathbf{Q})] \tag{15.43}$$

where the Hamiltonian operator now includes nuclear kinetic energy as well as the nuclear repulsion, i.e.,

$$H = \sum_{k}^{nuclei} -\frac{1}{2m_k}\nabla_k^2 + H_{el} + V_N \tag{15.44}$$

where m_k is the mass of nucleus k in atomic units, ∇^2 is defined as in Eq. (4.4), and H_{el} and V_N are defined in Eq. (4.16) and the following discussion. To determine a given c as a function of nuclear coordinates $\mathbf{Q}$, we can multiply both sides of Eq. (15.43) on the left

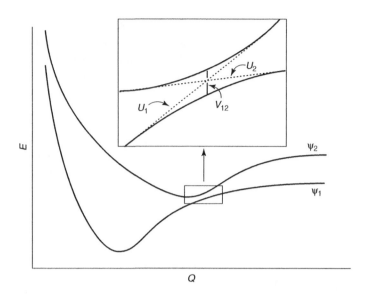

Figure 15.7   Near approach (or avoided crossing) of two electronic states as a function of nuclear coordinate Q. The inset expands the region of the avoided crossing to facilitate the definition of quantities appearing in the Landau–Zener surface-hopping-probability model

by the adiabatic state corresponding to that c and integrate. After taking advantage of the orthonormality of the adiabatic states, and noting that each adiabatic electronic wave function is an eigenfunction of the electronic Hamiltonian with an associated energy eigenvalue, we derive

$$\sum_k^{\text{nuclei}} -\frac{1}{2m_k}\left[\nabla_k^2 + \sum_{j=1}^{2}(2\langle\psi_i|\nabla_k|\psi_j\rangle\cdot\nabla_k + \langle\psi_i|\nabla_k^2|\psi_j\rangle)\right]c_i = (E_{\text{full}} - E_i)c_i \qquad (15.45)$$

where E_i is the energy eigenvalue for the ith electronic state ψ_i and the vector operator ∇ is defined as

$$\nabla_k = \left(\frac{\partial}{\partial x_k}, \frac{\partial}{\partial y_k}, \frac{\partial}{\partial z_k}\right) \qquad (15.46)$$

Note that Eq. (15.45) is itself a Schrödinger equation for nuclear eigenfunction c_i. The Born–Oppenheimer approximation, previously discussed in Section 4.2.3, involves assuming a value of zero for all of the integrals in Eq. (15.45) involving the nuclear ∇ or ∇^2 operator acting on electronic wave functions (cf. Eq. (9.37)). Under that assumption, the nuclear and electronic wave functions are separable, but spontaneous changes in electronic states, i.e., surface-to-surface crossings, are not permitted. Any model addressing such state–state interconversions must instead start from Eq. (15.45) (possibly generalized to a larger number of electronic states).

Unfortunately, Eq. (15.45) does not admit to simple analytic solutions under realistic sets of chemical conditions. Moreover, if we now try to extend Eq. (15.43) to its time-dependent analog, which is time-dependent simultaneously in the nuclear and the electronic wave functions, things get very messy indeed. Fully quantum mechanical solutions to non-adiabatic problems are still limited to the simplest of systems.

A popular alternative in these instances is to propagate classical trajectories on the mix of surfaces, where the probability of 'hopping' from one surface to another is periodically evaluated over the time course of the trajectory. By following the time course of a very large number of trajectories, the rate constant can be determined from a plot of reaction probability as a function of time. (A large number of trajectories is required in order to generate this plot reliably as a histogram of reaction times.) Thus, for instance, the system might be started with a certain amount of kinetic energy moving inward on the upper PES of Figure 15.7. The 'wave packet' will increase its kinetic energy as the potential energy drops, then climb the repulsive wall of the surface until it reaches a potential energy equal to the sum of its original potential and kinetic energies, and then will reverse direction outward. In a single internal coordinate, this could correspond to two atoms colliding and rebounding. If the trajectory hops from one surface to another during the course of the simulation, and then fails to hop back (or hops an even number of times after the initial hop) then the outgoing trajectory will correspond to the lower energy electronic state and be counted as a reactive event.

Various models to compute the probability of hopping exist. One of the simplest is the Landau–Zener model for avoided crossings in a single coordinate. The probability of the hop is determined as

$$P = e^{-(\pi V_{12}^2/h\upsilon|\dot{\psi}_1-\dot{\psi}_2|)} \tag{15.47}$$

where V_{12} is the energy gap between the surfaces at the point of the avoided crossing, υ is the velocity at which the wave packet is traveling in coordinate Q, and

$$\dot{\psi}_i = \frac{dU_i}{dQ} \tag{15.48}$$

where U represents a continuation of the diabatic states across the adiabatic avoided crossing as illustrated in Figure 15.7. Qualitatively, the Landau–Zener model says that the probability of hopping increases when (i) the adiabatic states approach one another increasingly closely, (ii) the wave packet passes the region of avoided crossing quickly, and (iii) the shapes of the adiabatic surfaces change suddenly in the region of the avoided crossing. Note, of course, that if a trajectory is to be reactive, it must cross only *once*, either on the way in or the way out, or an odd total number of times.

More sophisticated hopping schemes have been proposed for multi-dimensional surfaces and for more general situations than avoided crossings (see, for example, Tully 1976; Hack *et al.* 2001; Heller, Segev, and Sergeev 2002). However, further discussion on this topic is not undertaken here.

15.5.2 Marcus Theory

A special case of a non-adiabatic reaction is electron transfer. The dynamics of electron-transfer processes have been studied extensively, and the most robust model used to describe them is Marcus theory (Marcus 1964). The full scope of Marcus theory is very broad, and we consider here only the simplest application of the model. We will take the generic electron transfer reaction

$$A^- + B \rightarrow A + B^- \tag{15.49}$$

For this simple case, Marcus theory predicts the rate constant for electron transfer to be

$$k_{ET} = Z_{AB}e^{-(\Delta G^\circ_{AB}+\lambda)^2/4\lambda RT} \tag{15.50}$$

where Z_{AB} is the collision frequency for the reactants (typically in the range of 10^9 to 10^{10} sec^{-1} for reactions in non-viscous liquids at ambient temperatures), ΔG°_{AB} is the free energy change for the electron transfer, λ is the so-called reorganization energy, R is the universal gas constant, and T is the temperature.

The reorganization energy term derives from the solvent being unable to reorient on the same timescale as the electron transfer takes place. Thus, at the instant of transfer, the bulk dielectric portion of the solvent reaction field is oriented to solvate charge on species A, and not B, and over the course of the electron transfer only the optical part of the solvent reaction field can relax to the change in the position of the charge (see Section 14.6). If the Born formula (Eq. (11.12)) is used to compute the solvation free energies of the various equilibrium and non-equilibrium species involved, one finds that

$$\lambda = (\Delta q)^2 \left(\frac{1}{\varepsilon_\infty} - \frac{1}{\varepsilon_0} \right) \left(\frac{1}{2r_A} + \frac{1}{2r_A} - \frac{1}{r_{AB}} \right) \tag{15.51}$$

where Δq is the amount of charge transferred (1 for the reaction of Eq. (15.49)), ε_∞ is the fast dielectric constant (sometimes called the optical dielectric constant, equal to the square of the index of refraction – around 2 for typical solvents), ε_0 is the slow, or bulk, dielectric constant, r_A and r_B are the radii of species A and B, respectively, and r_{AB} is the distance between them at reaction. The quantity in Eq. (15.51) is sometimes called λ_o because it considers only 'outer-sphere', which is to say solvent, reorganization. More sophisticated approaches can be used when inner-sphere reorganization is also important, e.g., for ligated metal systems where the metal–ligand bond lengths might vary significantly as a function of charge. In such instances, inner-sphere reorganization energies can often be estimated from calculations of relaxation energies when the geometry of the species for the initial charge state is allowed to relax to the final charge state.

The exact form of Eq. (15.50) is made more intuitive by considering the simple reaction coordinate diagrams of Figure 15.8. In these cases, we consider two parabolic potential energy surfaces corresponding to the two sides of Eq. (15.49). The reaction coordinate may, in particularly simple instances, be thought of as a generalized solvent coordinate. Thus, when the solvent is optimally configured for $A^- + B$, the energy of the curve for state $A + B^-$ is quite high. If the free energies of the left and right sides of Eq. (15.49) are the same (which would happen if A and B were different isotopes of the same metal, for instance), the separation of the two curves at either minimum is exactly λ. From the mathematics of parabolae, this requires the intersection of the two curves to take place at $\lambda/4$ energy units

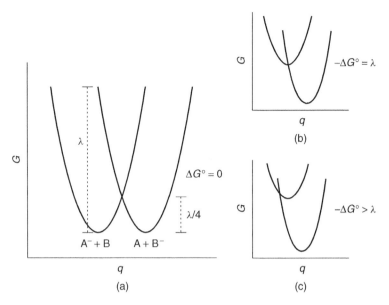

Figure 15.8 Electron-transfer reaction coordinate diagrams used in Marcus theory. Diagram (a) refers to a case with no net free energy of reaction, in which case the intersection of the two curves occurs at $\lambda/4$ above the minima and is taken as the barrier to the electron transfer (a barrier associated with solvent reorganization in the simplest limit). When the overall driving force is equal in magnitude to λ (b), the two curves cross at the equilibrium solvent configuration of the first state, and reaction is barrierless. However, when the driving force becomes still greater (c), the crossing of the two curves proceeds to the *left* on the reaction coordinate, and occurs at higher energy than the minimum of the reactant curve. This situation creates the inverted region where rate decreases with increasing exergonicity

above the two equal minima. This situation in illustrated in the first reaction coordinate diagram of Figure 15.8, and rationalizes the denominator of the exponential in Eq. (15.50): if ΔG°_{AB} is zero, then the argument of the exponential is $\lambda/4RT$ which is indeed the 'barrier' for reaction in the system with no thermochemical driving force in either direction.

Note that Marcus theory in the form of Eq. (15.50) makes a rather surprising prediction. If ΔG°_{AB} is equal to λ in magnitude but of opposite sign, which is to say the exergonicity of the electron transfer exactly cancels the reorganization energy, than the argument of the exponential is zero and the rate is predicted to be diffusion-controlled. However, if the driving force becomes greater still, then the argument of the exponential returns to positive, and the rate is predicted to *decrease* (Figure 15.8). This corresponds to the so-called inverted region of Marcus theory. That is, as one of a pair of reactants in an electron-transfer reaction is varied so that the reaction becomes more and more favorable in a free-energy sense, the rate is predicted to reach a maximum and then decrease. Experimental verification of this prediction did not occur until many years after the initial publication of the theory, in part because the required driving force is so high and in part because of the technical challenges associated with measuring very large rate constants. Nevertheless, an inverted region has now been demonstrated in several instances, and this validation of Marcus theory no doubt contributed to it being the subject of the Nobel Prize in 1992.

A key point that must be made is that quantum mechanical tunneling through the Marcus-theory barrier when it is non-zero can increase the rate for electron transfer just as is true for any other activated process. Because the electron is so light a particle, tunneling can be a major contributor to the overall rate. Models for electron tunneling will not, however, be presented here.

15.6 Case Study: Isomerization of Propylene Oxide

Synopsis of Dubnikova and Lifshitz (2000) 'Isomerization of Propylene Oxide. Quantum Chemical Calculations and Kinetic Modeling'.

When the commodity chemical propylene oxide is heated to high temperature in the gas phase in a shock tube, unimolecular rearrangement reactions occur that generate the C_3H_6O isomers allyl alcohol, methyl vinyl ether, propanal, and acetone (Figure 15.9). Dubnikova and Lifshitz carried out a series of calculations to determine the mechanistic pathway(s) for each isomerization, with comparison of activation parameters to those determined from Arrhenius fits to experimental rate data to validate the theoretical protocol.

Because of the complexity of the molecular hypersurface, the authors chose the B3LYP level in combination with the cc-pVDZ basis set for their search for TS structures. Based on judgment and intuition, they chose initial geometries for TS structures and optimized them subject to the constraint of there being a single imaginary frequency in the product stationary

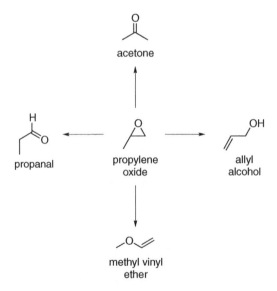

Figure 15.9 Isomerizations of propylene oxide. What plausible geometries might be proposed as starting guesses for TS structures? Once a TS structure is found, will it be obvious for which process it is the TS? What features of the TS may have an impact on the level of theory used to determine its energy?

point. They then carried out IRC calculations to verify that the TS structure connected in one direction to the desired product and in the other direction to propylene oxide.

For the isomerizations to allyl alcohol, propanal, and acetone, they found concerted TS structures that represented the only barrier between reactant and product, and these structures were predicted to have stable, closed-shell singlet wave functions. However, for the isomerization to methyl vinyl ether, a pathway involving three TS structures and two intermediates was identified, with several stationary points having a high degree of biradical character. To deal with this problem, they used a broken-symmetry SCF procedure (see Section 8.5.3). Another multistep pathway involving a carbene intermediate was also found for the isomerization of propylene oxide to methyl vinyl ether, but Dubnikova and Lifshitz assigned it as being kinetically unimportant based on significantly higher TS energies than those found for the first pathway.

While the B3LYP/cc-pVDZ level was judged to be a good choice for locating stationary points, it was not expected to be quantitatively useful in computing activation enthalpies. For this purpose, single-point CCSD(T)/cc-pVDZ calculations were carried out. Dubnikova and Lifshitz are not clear on what, if any, special precautions were taken with the biradical species (i.e., were single-reference HF wave functions somehow generated, or were mixed-state UHF reference wave functions used?) The potential energies were combined with the ZPVEs and thermal enthalpic contributions calculated from scaled B3LYP frequency calculations to determine absolute H values for all species. Absolute entropies S were computed from the B3LYP geometries and scaled vibrational frequencies. The energies for several of the stationary points relative to propylene oxide varied by as much as 4 kcal mol^{-1} comparing CCSD(T) to B3LYP. Although it is not *a priori* obvious which might be expected to do better, the general rule that B3LYP somewhat underestimates barrier heights compared to CCSD(T) suggests the latter will be of greatest utility.

With all activation parameters in hand, Dubnikova and Lifshitz convert them to A and E_a of the Arrhenius equation (Eqs. (15.30)–(15.32)) to compare to measured values; the data are provided in Table 15.2. In the case of the rearrangement to methyl vinyl ether, the data for the highest energy TS structure along the path were used. It is interesting to note that the comparison of the rate constants derived from the activation parameters at some particular temperature – 1000 K is shown in Table 15.2 – appears more favorable than a direct comparison of the activation parameters themselves. This occurs because in every case the error in activation energy is compensated for by an error in the pre-exponential factor. That is, if the activation energy is predicted to be too high, which would predict too

Table 15.2   Predicted and measured activation parameters for unimolecular rearrangements of propylene oxide

Product	Source	A (sec^{-1})	E_a (kcal mol^{-1})	k_{1000} (sec^{-1})
Allyl alcohol	Experiment	7.9×10^{12}	57.1	2.7
	Theory	2.2×10^{13}	60.2	1.6
Methyl vinyl ether	Experiment	3.2×10^{13}	58.8	4.7
	Theory	1.3×10^{14}	59.3	14.9
Propanal	Experiment	2.5×10^{14}	58.5	42.8
	Theory	3.5×10^{13}	54.4	47.0
Acetone	Experiment	1.7×10^{14}	60.7	9.6
	Theory	1.1×10^{14}	54.2	163.3

small a rate constant, the pre-exponential is predicted to be too large, which returns the rate constant to a reasonable value, and vice versa. In spite of such compensating errors, in the case of acetone the final error in the rate is almost a factor of 20. At lower temperatures, this error would increase dramatically.

Nevertheless, the agreement that is obtained – which is probably the best one should expect given the small size of the basis set used in the CCSD(T) calculations and the possible problems associated with biradical character in the methyl vinyl ether pathway – suggests that the theoretically predicted TS structures are accurate representations of the actual transition states. This establishes the concerted nature of three of the rearrangements and the stepwise nature of the fourth.

Bibliography and Suggested Additional Reading

Chuang, Y.-Y., Cramer, C. J., and Truhlar, D. G. 1998. 'The Interface of Electronic Structure and Dynamics for Reactions in Solution', *Int. J. Quantum Chem.*, **70**, 887.

Chuang, Y.-Y., Radhakrishnan, M. L., Fast, P. L., Cramer, C. J., and Truhlar, D. G. 1999. 'Direct Dynamics for Free Radical Kinetics in Solution: Solvent Effect on the Rate Constant for the Reaction of Methanol with Atomic Hydrogen', *J. Phys. Chem. A*, **103**, 4893.

Espenson, J. H. 1995. *Chemical Kinetics and Reaction Mechanisms*, 2nd Edn., McGraw-Hill: New York.

Garrett, B. C. and Truhlar, D. G. 1979. 'Semiclassical Tunneling Calculations', *J. Phys. Chem.*, **83**, 2921.

Hynes, J. T. 1996. 'Crossing the Transition State in Solution', in *Solvent Effects and Chemical Reactivity*, Tapia, O. and Bertrán, J. Eds., Kluwer: Dordrecht, 231.

Jensen, F. 1999. *Introduction to Computational Chemistry*, Wiley: Chichester.

Jensen, F. and Norrby, P.-O. 2003. 'Transition States from Empirical Force Fields', *Theor. Chem. Acc.*, **109**, 1.

Johnston, H. S. 1966. *Gas Phase Reaction Rate Theory*, Ronald Press: New York.

Lowry, T. H. and Richardson, K. S. 1981. *Mechanism and Theory in Organic Chemistry*, 2nd Edn., Harper & Row: New York.

Steinfeld, J. I., Francisco, J. S., and Hase, W. L. 1999. *Chemical Kinetics and Dynamics*, 2nd Edn., Prentice Hall: Upper Saddle River, NJ.

Truhlar, D. G., Garrett, B. C., and Klippenstein, S. J. 1996. 'Current Status of Transition-state Theory', *J. Phys. Chem.*, **100**, 12771.

Tucker, S. C. and Truhlar, D. G. 1989. 'Dynamical Formulation of Transition State Theory: Variational Transition States and Semiclassical Tunneling', in *New Theoretical Concepts for Understanding Organic Reactions*, Bertrán, J. and Czismadia, I. G., Eds., Kluwer: Berlin, 291.

Worth, G. A. and Robb, M. A. 2002. 'Applying Direct Molecular Dynamics to Non-adiabatic Systems', *Adv. Chem. Phys.*, **124**, 355.

References

Allison, T. C. and Truhlar, D. G. 1998. In: *Modern Methods, for Multidimensional Dynamics Computations in Chemistry*, Thompson, D. L., Ed., World Scientific: Singapore, 618.

Bell, R. P. 1959. *Trans. Faraday Soc.*, **55**, 1.

Dubnikova F. and Lifshitz, A. 2000. *J. Phys. Chem. A*, **104**, 4489.

Eckart, C. 1930. *Phys. Rev.*, **35**, 1303.

Grote, R. G. and Hynes, J. T. 1980. *J. Chem. Phys.*, **73**, 2715.

Gustafson, S. M. and Cramer, C. J. 1995. *J. Phys. Chem.*, **99**, 2267.

Hack, M. D., Wensmann, A. M., Truhlar, D. G., Ben-Nun, M., and Martinez, T. J. 2001. *J. Chem. Phys.*, **115**, 1172.

Heller, E. J., Segev, B., and Sergeev, A. V. 2002. *J. Phys. Chem. B*, **106**, 8471.

Keating, A. E., Merrigan, S. R., Singleton, D. A., and Houk, K. N. 1999. *J. Am. Chem. Soc.*, **121**, 3933.

Kohen, A. and Klinman, J. P. 1998. *Acc. Chem. Res.*, **31**, 397.

Kramers, H. A. 1940. *Physica*, **7**, 284.

Marcus, R. A. 1964. *Annu, Rev. Phys. Chem.*, **15**, 155.

Sherer, E. C. and Cramer, C. J. 2003. *Organometallics*, **22**, 1682.

Skodje, R. T. and Truhlar, D. G. 1981. *J. Phys. Chem.*, **85**, 624.

Truhlar, D. G. and Gordon, M. S. 1990. *Science*, **249**, 491.

Tully, J. 1976. In: *Dynamics of Molecular Collisions*, Part B, Miller, W. H., Ed., Plenum: New York, 217.

Watson, P. L. 1990. In *Selective Hydrocarbon Activation: Principles and Progress*, Davies, J. A., Watson, P. L., Liebman, J. F., and Greenberg, A., Eds., VCH: New York, 79.

Wigner, E. Z. 1932. *Z. Phys. Chem. B*, **19**, 203.

Zwanzig, R. 1973. *J. Stat. Phys.*, **9**, 215.

Appendix A

Acronym Glossary

Note: Basis set abbreviations are detailed in Chapter 6 and are, for the most part, not included here. Only the most common *combinations* of exchange and correlation functionals are included as separate acronyms. Unit abbreviations are not listed.

6-12	The inverse power dependence of Lennard–Jones terms
AA	All-atom (as opposed to united-atom)
ACM	Adiabatic connection method
ADF	Amsterdam density functional code
AIM	Atoms in molecules
AM1	Austin Model 1
AMBER	Assisted model building with energy refinement
AO	Atomic orbital
AOC	AM1/OPLS/CM1
B	Becke (1988) exchange functional
B1B95	ACM one-parameter functional
B3LYP	ACM using B exchange and LYP correlation functionals
B3PW91	ACM using B exchange and PW91 correlation functionals
B86	Becke (1986) exchange functional
B95	Becke correlation functional
B97	ACM functional of Becke
B97-1	ACM functional of Becke reparameterized by Hamprecht *et al.*
B98	ACM MGGA exchange-correlation functional of Becke
BAC	Bond-additivity correction
BB1K	B1B95 optimized for kinetics
BD	CCD using Brueckner orbitals
BH&H	Becke half-and-half exchange functional
BKO	Born–Kirkwood–Onsager
BLYP	B exchange and LYP correlation functionals
Bm	Modification of Becke exchange functional for use with $\tau 1$
BPW91	B exchange and PW91 correlation functionals

Essentials of Computational Chemistry, 2nd Edition Christopher J. Cramer
© 2004 John Wiley & Sons, Ltd   ISBNs: 0-470-09181-9 (cased); 0-470-09182-7 (pbk)

BR	MGGA exchange functional of Becke and Roussel
BSSE	Basis set superposition error
CAM	Cambridge GGA exchange functional
CAS	Complete active space
CASPT2	Complete active space second-order perturbation theory
CASSCF	Complete active space self-consistent field
CBS	Complete basis set
CCD	Coupled cluster with double substitution operator
CCSD	Coupled cluster with single and double substitution operators
CCSD(T)	CCSD with perturbative estimate for connected triples
CCSDT	Coupled cluster with single, double, and triple substitution operators
CCSDTQ	Coupled cluster including single through quadruple excitations
CD	Circular dichroism
CFF	Consistent force field
CHARMM	Chemistry at Harvard molecular mechanics
CHELP	Charges from electrostatic potentials
CI	Configuration interaction
CID	CI including only double electronic excitations
CIS	CI including only single electronic excitations
CISD	CI including single and double electronic excitations
CIS(D)	CIS including a correction for double excitations
CISDT	CI including single, double, and triple electronic excitations
CISDTQ	CI including single through quadruple electronic excitations
CISD(Q)	CISD with Langhoff–Davidson estimate for quadruples
CMn	Charge model n (where n is a version number)
CNDO	Complete neglect of differential overlap
CoMFA	Comparative molecular field analysis
COSMIC	Computation and structural manipulation in chemistry
COSMO	Conductor-like screening model
CP	Counterpoise; Car–Parrinello
C-PCM	Conductor formulation of PCM
CS	Correlation functional of Colle and Salvetti
CSF	Configuration state function
CT	Charge transfer
CVFF	Consistent valence force field
DFT	Density functional theory
DFTB	Density functional tight-binding theory
DFT-SCI	Density functional theory singles configuration interaction
D-PCM	Dielectric formulation of PCM
DZ	Double zeta (basis set)
DZP	Double zeta polarized (basis set)
EA	Electron affinity
ECEPP	Empirical conformational energy program for peptides

ECP	Effective core potential
EDF1	Empirical density functional 1
EFP	Effective fragment potential
EHT	Extended Hückel theory
EOM	Equation of motion
EPR	Electron paramagnetic resonance
ESFF	Extensible systematic force field
ESP	Electrostatic potential; Equilibrium solvation path
ESR	Electron spin resonance
EVB	Empirical valence bond
FDPB	Finite difference Poisson–Boltzmann
FEP	Free energy perturbation
FLOGO	Floating Gaussian orbitals
FT97	Filatov and Thiel (1997) density functional
Gn	Gaussian-n theory ($n = 1$, 2, or 3)
G3S	Scaled G3 theory
G96	GGA functional of Gill
GAPT	Generalized atomic polar tensor
GB	Generalized Born
GDAC	Geometry-dependent atomic charge
GGA	Generalized gradient approximation
GHO	Generalized hybrid orbital
GIAO	Gauge-including atomic orbital
GROMOS	Gröningen molecular simulation
GTO	Gaussian-type orbital
GUI	Graphical user interface
GVB	Generalized valence bond
H&H	Half-and-half adiabatic connection formula
HCTH	GGA exchange-correlation functional of Hamprecht, Cohen, Tozer, and Handy
HF	Hartree–Fock
h.f.s.	Hyperfine splitting
HOMO	Highest occupied molecular orbital
IEF	Integral equation formalism
IGLO	Individual gauge for localized orbitals
IMOMM	Integrated molecular orbital molecular mechanics
IMOMO	Integrated molecular orbital molecular orbital
INDO	Intermediate neglect of differential overlap
INDO/S	INDO parameterized for spectroscopy
IP	Ionization potential
IPCM	PCM with a gas-phase isodensity surface as the cavity surface
IR	Infrared
IRC	Intrinsic reaction coordinate

ISM	MGGA correlation functional of Imamura, Scuseria, and Martin
IUPAC	International Union of Pure and Applied Chemistry
KCIS	MGGA correlation functional of Kriger, Chen, Iafrate, and Savin
KIE	Kinetic isotope effect
KMLYP	Kang and Musgrave ACM functional including LYP
KS	Kohn–Sham
LANL	Los Alamos National Laboratory
Lap	MGGA correlation functionals
LCAO	Linear combination of atomic orbitals
LD	Langevin dipole
LDA	Local density approximation
LG	Lacks-Gordon density functional
LJ	Lennard–Jones
LMP2	Localized MP2
LSCF	Localized self-consistent field
LSDA	Local spin density approximation
LYP	Lee-Yang-Parr correlation functions
LUMO	Lowest unoccupied molecular orbital
MBPTn	Many-body perturbation theory of order n
MC	Monte Carlo
MC	Multicoefficient (as a prefix to a level of theory being scaled)
MCMM	Multiconfiguration molecular mechanics
MCPF	Modified coupled-pair functional
MCSCF	Multiconfiguration self-consistent field
MD	Molecular dynamics
MEP	Minimum energy path; Molecular electrostatic potential
MGGA	Meta-generalized gradient approximation
MINDO/3	Modified intermediate neglect of differential overlap (version 3)
MKS	Multiplicative Kohn–Sham (NMR model)
MM	Molecular mechanics
MMFF	Merck molecular force field
MNDO	Modified neglect of differential overlap
MNDOC	MNDO including electron correlation effects
MNDO/d	MNDO augmented with d functions for some atoms
MO	Molecular orbital
MP4SDQ	MP4 including single, double and quadruple excitations
MPn	Møller–Plesset perturbation theory of order n
mPBE	Modified PBE functional
MPEOE	Modified partial equalization of orbital electronegativity
mPW	Modified Perdew–Wang density functional
MPW1K	mPW1PW91 optimized for kinetics
mPW1N	mPW1PW91 modified for halide/alkyl-halide nucleophilic substitutions
mPW1PW91	One-parameter ACM using PW91 functionals

*m*PW1S	*m*PW1PW91 modified for sugar conformational analysis
MRCI	Multireference CI
MRCISD	Multireference CI including single and double excitations
MR-MP2	Multireference second-order perturbation theory
MST	Miertus–Scrocco–Tomasi (polarized continuum) model
MST-ST	MST model augmented with atomic surface tensions
μTST	Microcanonical transition state theory
μVTST	Microcanonical variational transition state theory
NAO	Natural atomic orbital
NBO	Natural bond orbital
NDDO	Neglect of diatomic differential overlap
NHE	Normal hydrogen electrode
NIST	National Institute of Standards and Technology (U.S.)
NMR	Nuclear magnetic resonance
nOe	Nuclear Overhauser effect
NPA	Natural population analysis
O	OPTX exchange functional
O3LYP	ACM using O exchange and LYP correlation functionals
OLYP	O exchange and LYP correlation functionals
OM1	Orthogonalization method 1
OM2	Orthogonalization method 2
ONIOM	Our own *n*-layered integrated molecular orbital molecular mechanics
o.o.p.	Out-of-plane
OPLS	Optimized potentials for liquid simulations
ORD	Optical rotatory dispersion
P	Perdew exchange functional
P86	Perdew correlation functional
PA	Proton affinity
PB	Poisson–Boltzmann
PBC	Periodic boundary condition
PBE	Perdew, Burke, and Enzerhof functional
PBE1PBE	ACM functional derived from PBE
PCA	Principal components analysis
PCM	Polarized continuum model
pc-*n*	Polarization consistent n-ζ basis sets of Jensen
PD	Pairwise descreening
PDDG	Pairwise distance directed Gaussian
PDFT	Projected density functional theory
PEG	Polyethyleneglycol
PEOE	Partial equalization of orbital electronegativity
PES	Potential energy surface
PKZB	MGGA exchange-correlation functional of Perdew, Kurth, Zupan, and Blaha

PM3	Parameterized (NDDO) model 3
PM3(tm)	PM3 with a d orbital extension to transition metals
PME	Particle-mesh Ewald
PMF	Potential of mean force
PMPn	Projected Møller–Plesset theory of order n
POS	Points on a sphere
PP	Perfect pairing
PPP	Pariser–Parr–Pople
PUHF	Projected UHF
PW	Perdew–Wang (1991) exchange functional
PW91	Perdew–Wang (1991) correlation functional
QCISD	Quadratic configuration interaction including singles and doubles
QCISD(T)	QCISD with perturbative estimate for connected triples
QEq	Charge equilibration
QM	Quantum mechanics
QMHO	Quantum mechanical harmonic oscillator
QM/MM	Quantum mechanics/molecular mechanics hybrid
QSPR	Quantitative structure–property relationship
RAS	Restricted active space
r.d.f.	Radial distribution function
RESP	Restrained ESP
RHF	Restricted Hartree–Fock
RISM	Reference interaction site model
RMS	Root mean square
RMSD	Root-mean-square deviation
ROHF	Restricted open-shell Hartree–Fock
ROKS	Restricted open-shell Kohn–Sham theory
ROSS	Restricted open-shell singlet density functional theory
RPA	Random-phase approximation
RRKM	Rice–Ramsperger–Kassel–Marcus
S	Slater exchange functional
SAC	Scaling all correction
SAM1	Semi-*ab initio* method 1
SAM1D	SAM1 with d orbitals
SAR	Structure–activity relationship
SASA	Solvent-accessible surface area
SCC-DFTB	Self-consistent charge density functional tight-binding theory
SCF	Self-consistent field
SCIPCM	PCM with a liquid-solution-phase isodensity surface as the cavity surface
SCRF	Self-consistent reaction field
SCS-MP3	Spin-component-scaled MP3
SES	Separable equilibrium solvation

SF-CISD	Spin–flip CISD
SF-CIS(D)	Spin–flip CIS(D)
SF-TDDFT	Spin–flip TDDFT
SINDO1	Symmetric orthogonalized INDO model
SMx	Solvation model x (using Cramer–Truhlar GB formalism)
SOMO	Singly occupied molecular orbital
SPC	Simple point charge
SRP	Specific reaction (or range) parameters
S–T	Singlet–triplet
STO	Slater-type orbital
$\tau 1$	MGGA correlation functional
τHCTH	MGGA modification of HCTH
TCSCF	Two-configuration self-consistent field
TDDFT	Time-dependent density functional theory
TI	Thermodynamic integration
TIPnP	Transferable intermolecular potentials n point charge water model
TMM	Trimethylenemethane
TPSS	MGGA exchange-correlation functional of Tao, Perdew, Staroverov, and Scuseria
TPSSh	ACM functional derived from TPSS
TraPPE	Transferable potentials for phase equilibria
TS	Transition state
TST	Transition-state theory
TZ	Triple zeta (basis set)
TZP	Triple zeta polarized (basis set)
UA	United-atom (as opposed to all-atom)
UFF	Universal force field
UHF	Unrestricted Hartree–Fock
UV	Ultraviolet
UV/Vis	Ultraviolet/visible
VB	Valence bond
VDD	Voronoi deformation density
VSEPR	Valence-shell electron-pair repulsion
VSIP	Valence-shell ionization potential
VSXC	Exchange-correlation functional of van Voorhis and Scuseria
VTST	Variational transition-state theory
VWN	A Vosko, Wilk, Nusair correlation functional
VWN5	A Vosko, Wilk, Nusair correlation functional
WHAM	Weighted histogram analysis method
Wn	Weizmann-n theory ($n = 1, 2, 3,$ or 4)
XSOL	Extended RISM and quantum mechanical solvation model
ZPVE	Zero-point vibrational energy

Appendix B

Symmetry and Group Theory

B.1 Symmetry Elements

To say that something is symmetric, or that it possesses symmetry, usually is to say that an internal motif is repeated in some fashion. In the context of chemistry, that motif is a spatial arrangement of atoms. We may classify the nature of an object's symmetry based on the fashion in which its repeated motifs are made manifest. In describing the symmetric positioning of atoms in a molecule, there are only a few different operations that are relevant for chemical systems, and these operations are referred to as 'symmetry elements'. When a particular symmetry element is present, the molecule is said to 'possess' that symmetry element. The four symmetry elements that may be used to characterize a molecular structure are:

Plane of symmetry. If a plane can be placed in space such that for every atom of the molecule not in the plane there is an identical atom (which is to say, the same atomic number and isotope) on the other side of the plane at equal distance from it (i.e., a 'mirror image'), the molecule is said to possess a plane of symmetry. The Greek letter σ is often used to represent both the plane of symmetry and the 'operation' of mirror reflection that it performs. An example of a molecule possessing a plane of symmetry is methylcyclobutane, as illustrated in Figure B.1. Note that a planar molecule *always* has at least one σ, since the plane of the molecule satisfies the above symmetry criterion in a trivial way (the set of reflected atoms is the empty set). Note also that if we choose a Cartesian coordinate system in such a way that two of the Cartesian axes lie in the symmetry plane, say x and y, then for every atom found at position (x,y,z) where $z \neq 0$ there must be an identical atom at position $(x,y,-z)$.

Proper rotation axis. If a molecule can be rotated about some axis so that the positions originally occupied by every atom are subsequently occupied by identical atoms, the molecule is said to possess a proper rotation axis. The axis and the rotation operation performed about it are typically represented by the notation C_n, where n is the order of the rotation. The order is the largest value of n for which it is true that a rotation of $2\pi/n$ radians about the axis reproduces the original structure; this is also referred to as a n-fold rotation axis.

Essentials of Computational Chemistry, 2nd Edition Christopher J. Cramer
© 2004 John Wiley & Sons, Ltd ISBNs: 0-470-09181-9 (cased); 0-470-09182-7 (pbk)

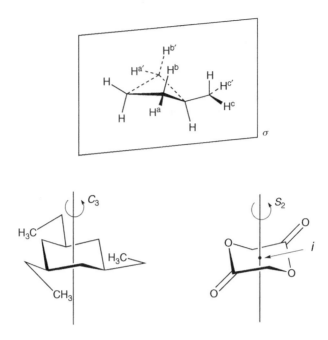

Figure B.1 Symmetry elements possessed by various molecular conformations. The molecules illustrated here have no elements beyond those indicated, but arbitrary molecules may be characterized by many different symmetry operations

An example of a molecule with a three-fold rotation axis is the conformation of *sym*-1,3,5-triethylcyclohexane shown in Figure B.1. Note that all molecules possess a trivial C_1 axis (indeed, an infinite number of them). Note also that if we choose a Cartesian coordinate system where the proper rotation axis is the z axis, and if the rotation axis is two-fold, then for every atom found at position (x,y,z) where x and y are not simultaneously equal to 0 (i.e., not on the z axis itself) there will be an identical atom at position $(-x,-y,z)$. If the rotation axis is four-fold, there will be an identical atom at the *three* positions $(-x,y,z)$, $(x,-y,z)$, and $(-x,-y,z)$. Note finally that for linear molecules the axis of the molecule is a proper symmetry axis of *infinite* order, i.e., C_∞.

Improper rotation axis. Rotation about an improper axis is analogous to rotation about a proper symmetry axis, except that upon completion of the rotation operation, the molecule is mirror reflected through a symmetry plane perpendicular to the improper rotation axis. These axes and their associated rotation/reflection operations are usually abbreviated S_n, where n is the order of the axis as defined above for proper rotational axes. Note that an S_1 axis is equivalent to a σ plane of symmetry, since the initial rotation operation simply returns every atom to its original location. Note also that the presence of an S_2 axis (or indeed any S axis of even order n) implies that for every atom at a position (x,y,z) that is not the origin, there will be an identical atom at position $(-x,-y,-z)$; the origin in such a system is called a 'point of inversion', since one may regard every atom as having an identical partner related by inversion through the origin, and the inversion operation itself is usually denoted i. An example of a molecule containing an S_2 axis is the chair conformation of 2,5-dioxo-1,4-tetrahydropyran (Figure B.1). Lastly, note that the presence of higher order improper axes implies the simultaneous presence of one or more proper rotation axes. In particular, improper axes S_n where n is odd imply a coincident C_n axis and n perpendicular C_2 axes, and improper axes S_n where n is even imply a coincident $C_{n/2}$ axis.

Point of inversion. The action of a point of inversion is described above in the context of improper rotation axes. Note that planes of symmetry and points of inversion are somewhat redundant symmetry elements, since they are already implicit in improper rotation axes. However, they are somewhat more intuitive as separate phenomena than are S_n axes, and thus most texts treat them separately.

B.2 Molecular Point Groups and Irreducible Representations

An individual molecular structure may possess no symmetry elements at all, or a single symmetry element, or some combination of multiple symmetry elements. It turns out that there are a finite number of possible combinations, and each such combination defines what is referred to as a point group. The names of the various molecular point groups together with a flow chart indicating how to assign a molecule to a point group are provided in Figure B.2. [In crystallography, solids can be characterized by space groups, which are analogous to point groups but more numerous as additional symmetry elements relating different molecules in the crystal must also be considered. No further discussion of space groups is provided here.]

There is a special algebra associated with the different point groups, and the mathematical field of group theory is devoted to this topic. Group theory is a fascinating topic, but only its most basic aspects are addressed here. To begin, all point groups other than the non-symmetric C_1 group are characterized by two or more so-called irreducible representations, or irreps for short. Operationally, an irreducible representation defines how a signed or phased fragment (e.g., an orbital) of the symmetric structure 'transforms' under the various possible symmetry operations that compose the point group.

For example, the C_s point group contains two irreps, usually called a' and a''. Fragments belonging to the a' irrep are unchanged upon reflection through the symmetry plane of the molecule. Irreps leaving fragments unchanged under all symmetry operations of the point group are referred to as 'totally symmetric' irreps. The a'' irrep of the C_s point group, on the other hand, reverses the phase of a fragment on reflection through the mirror plane.

A much more detailed example is provided in Figure 6.7, where the C_{2v} point group to which the water molecule belongs is characterized by four irreps. The totally symmetric a_1 irrep includes orbitals unchanged by rotation about the C_2 axis or reflection through either of the two vertical mirror planes. The a_2 irrep includes orbitals that are unchanged by rotation about the C_2 axis but that are inverted by reflection through either of the two vertical mirror planes (Figure 6.7 does not list any such orbitals since none exist in

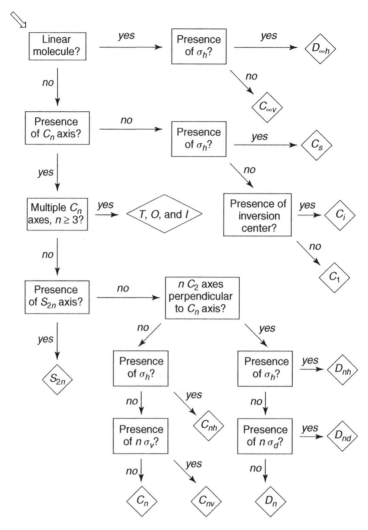

Figure B.2 Flow chart for point-group assignment. A symmetry plane that is perpendicular to a proper axis of rotation is a σ_h plane, one that includes the unique proper axis of rotation is a σ_v plane, and one that includes the highest order proper axis of rotation and bisects the remaining two-fold axes of rotation is a σ_d plane

a minimal basis set representation, but the d_{xy} orbital on oxygen or the antisymmetric combination of two p_y orbitals on the H atoms would belong to this irrep in a polarized basis set representation). The b_1 irrep includes orbitals that are inverted by rotation and reflection through the σ_{xz} symmetry plane, but left unchanged by reflection through the σ_{yz} symmetry plane, and the b_2 irrep includes orbitals that are inverted by rotation and reflection through the σ_{yz} symmetry plane, but left unchanged by reflection through the σ_{xz} symmetry plane.

B.3   Assigning Electronic State Symmetries

Individual molecular orbitals, which in symmetric systems may be expressed as symmetry-adapted combinations of atomic orbital basis functions, may be assigned to individual irreps. The many-electron wave function is an antisymmetrized *product* of these orbitals, and thus the assignment of the wave function to an irrep requires us to have defined mathematics for taking the product between two irreps, e.g., $a' \otimes a''$ in the C_s point group. These product relationships may be determined from so-called character tables found in standard textbooks on group theory. Tables B.1 through B.5 list the product rules for the simple point groups C_s, C_i, C_2, C_{2h}, and C_{2v}, respectively.

Assignment of an electronic wave function to an irrep is typically straightforward. All doubly filled orbitals are ignored, as the product of all of them with one another is the totally symmetric irrep, which is the multiplicative 'one' in all point groups. Thus, we need only take the product of all of the singly occupied orbitals to determine the irrep of the wave function. For a doublet, there is only one singly occupied orbital, so the irrep to which it belongs determines the irrep of the wave function. Figure 6.9 illustrates this point for H_2NO. Note that, to distinguish it from orbital irreps, the wave-function state symmetry is usually written with a capital letter. In triplets (and open-shell singlets), there are two singly occupied

Table B.1   Product rules for the C_s point group[a]

$\otimes$	a'	a''
a'	a'	a''
a''	a''	a'

[a] See text for irrep definitions.

Table B.2   Product rules for the C_i point group[a]

$\otimes$	a_g	a_u
a_g	a_g	a_u
a_u	a_u	a_g

[a] Objects unchanged by inversion belong to the a_g irrep; objects that change phase on inversion belong to the a_u irrep.

Table B.3   Product rules for the C_2 point group[a]

$\otimes$	a	b
a	a	b
b	b	a

[a] Objects unchanged by rotation belong to the a irrep; objects that change phase on rotation belong to the b irrep.

Table B.5 Product rules for the C_{2h} point group[a]

$\otimes$	a_g	a_u	b_g	b_u
a_g	a_g	a_u	b_g	b_u
a_u	a_u	a_g	b_u	b_g
b_g	b_g	b_u	a_g	a_u
b_u	b_u	b_g	a_u	a_g

[a]Objects unchanged by rotation belong to a type irreps, while objects changing phase on rotation belong to b type irreps; objects unchanged by reflection through the horizontal σ belong to $-_g$ type irreps, while objects changing phase on reflection belong to $-_u$ type irreps.

Table B.4 Product rules for the C_{2v} point group[a]

$\otimes$	a_1	a_2	b_1	b_2
a_1	a_1	a_2	b_1	b_2
a_2	a_2	a_1	b_2	b_1
b_1	b_1	b_2	a_1	a_2
b_2	b_2	b_1	a_2	a_1

[a]See text for irrep definitions.

orbitals, so their product must be taken to determine the state symmetry. Figure 14.4 provides examples of this process. In more complex open-shell systems, the sequential product of all of the singly occupied orbitals determines the electronic state symmetry.

Some complications can arise. Although in many point groups the product of any two irreps is another irrep (as is true for the examples in Tables B.1 through B.5), in some cases the product of two irreps can only be expressed as a linear combination of two or more different irreps. A determinant that does not belong to a single irrep is not a true wave function, but must be combined with other determinants to construct a wave function having a pure state symmetry. Such situations are beyond the scope of this text.

B.4 Symmetry in the Evaluation of Integrals and Partition Functions

Mathematical functions and operators can be assigned to irreps just as orbitals can be. This has enormous implications for practical computations because the integral over all space of any product that does not contain the totally symmetric representation vanishes, i.e., there is no point evaluating it. The Fock and Hamiltonian operators both belong to the totally symmetric irrep of *any* point group, because they depend only on interparticle distances and the ∇^2 operator, and these quantities are unaffected by changes in the coordinate system brought about by rotations, reflections, etc. Thus, in evaluating Fock matrix elements of the form

$$F_{\mu\nu} = \int\int \phi_\mu(1)F\phi_\nu(2)d\mathbf{r}(1)d\mathbf{r}(2) \tag{B.1}$$

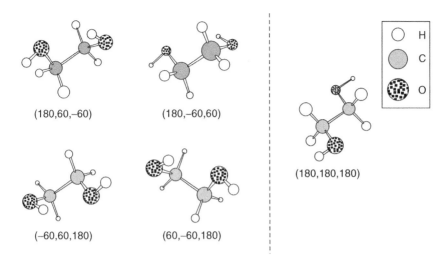

(180,60,−60) (180,−60,60)

(180,180,180)

(−60,60,180) (60,−60,180)

Figure B.3 The conformer of 1,2-ethanediol shown on the left may be generated, to within enantiomeric equivalence, by four different combinations of the left-to-right dihedral angles (ω_{HOCC}, ω_{OCCO}, ω_{CCOH}). The conformer on the right, on the other hand, is unique. To account for the greater phase-space volume associated with the four-fold degenerate conformer, computation of its free energy must include a term of $-RT\ln 4$

if the product of the irreps to which ϕ_μ and ϕ_ν belong does not contain the totally symmetric representation, the integral need not be evaluated, as already discussed in the context of Figure 6.7. This point is further discussed in the context of transition dipole moments in Section 14.5. Note that we use the language 'contains' the totally symmetric representation to account for cases where the product of the irreps is a linear combination of irreps, any *one* of which is the totally symmetric one.

Finally, recall that the presence of certain symmetry elements reduces the rotational partition function as described in Section 10.3.5 and Table 10.1. In addition, if a molecule can adopt the same conformation (to within enantiomerism) using different values for internal degrees of freedom, then this 'structural degeneracy' contributes to the free energy a term of $-RT\ln n$ where n is the number of otherwise identical conformations that employ different values for the internal degrees of freedom. For example, the two conformations of 1,2-ethanediol shown in Figure B.3 belong to the C_1 and C_{2h} point groups, respectively. The former is 4-fold degenerate, while the latter has no degeneracy. This effect can be important in the evaluation of Boltzmann-averaged conformational populations in potentially symmetric molecules.

Appendix C

Spin Algebra

C.1 Spin Operators

Electrons (and many other particles) have associated with them an intrinsic angular momentum that has come to be called 'spin'. One of the greatest successes of relativistic quantum mechanics is that spin is seen to arise naturally within the relativistic formalism, and does not need to be added *post facto* as it is in non-relativistic treatments. As with orbital angular momentum, spin angular momentum has x, y, and z components, and the operators S_x, S_y, and S_z, together with orthonormal eigenfunctions α and β of electron spin, are defined from

$$S_x\alpha = \tfrac{1}{2}\hbar\beta \tag{C.1}$$

$$S_x\beta = \tfrac{1}{2}\hbar\alpha \tag{C.2}$$

$$S_y\alpha = \tfrac{1}{2}i\hbar\beta \tag{C.3}$$

$$S_y\beta = -\tfrac{1}{2}i\hbar\alpha \tag{C.4}$$

$$S_z\alpha = \tfrac{1}{2}\hbar\alpha \tag{C.5}$$

$$S_z\beta = -\tfrac{1}{2}\hbar\beta \tag{C.6}$$

where $i = \sqrt{-1}$.

Thus, α and β are eigenfunctions of the operator S_z, with eigenvalues of $1/2$ and $-1/2$, respectively, in atomic units (recall that the value of $\hbar$ is 1 in atomic units, see Table 1.1). The spin operator S is defined by

$$S = S_x + S_y + S_z \tag{C.7}$$

and repeated application of Eqs. (C.1) through (C.6) reveals that

$$S^2\alpha = \tfrac{1}{2}(\tfrac{1}{2} + 1)\hbar^2\alpha \tag{C.8}$$

Essentials of Computational Chemistry, 2nd Edition Christopher J. Cramer
© 2004 John Wiley & Sons, Ltd ISBNs: 0-470-09181-9 (cased); 0-470-09182-7 (pbk)

and

$$S^2\beta = \tfrac{1}{2}(\tfrac{1}{2}+1)\hbar^2\beta \tag{C.9}$$

That is, α and β are also eigenfunctions of the operator S^2 with eigenvalues $s(s+1)$ where, for a single electron, s is $\tfrac{1}{2}$.

For an N-electron spin function, the total spin angular momentum is additive, i.e.,

$$S = \sum_{i=1}^{N} S(i) \tag{C.10}$$

where $S(i)$ is the operator of Eq. (C.7) applied to electron i. The individual Cartesian components of the spin angular momentum are also additive. Thus, for a normalized N-electron spin function Ψ, Eqs. (C.5) and (C.6) imply that

$$S_z\Psi = \sum_{i=1}^{N} s_z(i)\Psi \tag{C.11}$$

where $s_z(i)$ is the eigenvalue $\pm\tfrac{1}{2}$ of the S_z operator for electron i. Thus Ψ is an eigenfunction of the z component of the total spin angular momentum with an eigenvalue equal to the sum of the eigenvalues of the individual electrons.

Consider the operator S^2 for a many-electron spin function. From Eq. (C.10) and also Eqs. (C.1) to (C.6) it follows that

$$S^2 = \sum_{i=1}^{N} S^2(i) + 2\sum_{i<j}^{N} \left[S_x(i)S_x(j) + S_y(i)S_y(j) + S_z(i)S_z(j) \right] \tag{C.12}$$

One may ask under what circumstances are many-electron wave functions eigenfunctions of S^2, and that question is addressed next.

C.2 Pure- and Mixed-spin Wave Functions

For ease of notation only the two-electron case is considered here. Generalization to more electrons is entirely straightforward, if algebraically tedious. In the simplest case, both electrons are spin-paired in the same orbital, in which event we have

$$^{\mathrm{CSS}}\Psi = \frac{1}{\sqrt{2}} \begin{vmatrix} a(1)\alpha(1) & a(1)\beta(1) \\ a(2)\alpha(2) & a(2)\beta(2) \end{vmatrix}$$

$$= \frac{1}{\sqrt{2}} a(1)a(2)[\alpha(1)\beta(2) - \alpha(2)\beta(1)] \tag{C.13}$$

where the superscript CSS emphasizes that Ψ is a closed-shell state and a is the normalized spatial part of the doubly occupied molecular orbital. If we evaluate S^2 for this wave function we have

$$\langle {}^{CSS}\Psi|S^2|{}^{CSS}\Psi\rangle = \tfrac{1}{2}\langle a^2(1)a^2(2)\rangle\langle[\alpha(1)\beta(2) - \alpha(2)\beta(1)]|S^2|[\alpha(1)\beta(2) - \alpha(2)\beta(1)]\rangle$$

$$= \tfrac{1}{2}[\langle\alpha(1)\beta(2)|S^2|\alpha(1)\beta(2)\rangle - \langle\alpha(1)\beta(2)|S^2|\alpha(2)\beta(1)\rangle$$

$$- \langle\alpha(2)\beta(1)|S^2|\alpha(1)\beta(2)\rangle + \langle\alpha(2)\beta(1)|S^2|\alpha(2)\beta(1)\rangle] \tag{C.14}$$

Note that since the spatial part of the molecular orbital is independent of spin, it may be integrated out (to 1). As for the remaining expectation values, if we evaluate Eq. (C.12) for the spin product function $\alpha(1)\beta(2)$ we find

$$S^2\alpha(1)\beta(2) = S^2(1)\alpha(1)\beta(2) + S^2(2)\alpha(1)\beta(2) + 2S_x(1)S_x(2)\alpha(1)\beta(2)$$

$$+ 2S_y(1)S_y(2)\alpha(1)\beta(2) + 2S_z(1)S_z(2)\alpha(1)\beta(2)$$

$$= \tfrac{1}{2}(\tfrac{1}{2} + 1)\alpha(1)\beta(2) + \tfrac{1}{2}(\tfrac{1}{2} + 1)\alpha(1)\beta(2) + \tfrac{1}{2}\beta(1)\alpha(2) + \tfrac{1}{2}\beta(1)\alpha(2)$$

$$- \tfrac{1}{2}\alpha(1)\beta(2) \tag{C.15}$$

where we employ atomic units to avoid writing $\hbar^2$ repeatedly and evaluate the one-electron spin operators using Eqs. (C.1) through (C.6) and (C.8) and (C.9.) Similarly, we have

$$S^2\alpha(2)\beta(1) = \tfrac{1}{2}(\tfrac{1}{2} + 1)\alpha(2)\beta(1) + \tfrac{1}{2}(\tfrac{1}{2} + 1)\alpha(2)\beta(1) + \tfrac{1}{2}\beta(2)\alpha(1) + \tfrac{1}{2}\beta(2)\alpha(1)$$

$$- \tfrac{1}{2}\alpha(2)\beta(1) \tag{C.16}$$

Using Eq. (C.15) allows us to evaluate the first integral on the r.h.s. of the last equality in Eq. (C.14) as

$$\langle\alpha(1)\beta(2)|S^2|\alpha(1)\beta(2)\rangle = \iint \alpha(1)\beta(2)\tfrac{1}{2}(\tfrac{1}{2} + 1)\alpha(1)\beta(2)d\omega(1)d\omega(2)$$

$$+ \iint \alpha(1)\beta(2)\tfrac{1}{2}(\tfrac{1}{2} + 1)\alpha(1)\beta(2)d\omega(1)d\omega(2)$$

$$+ \iint \alpha(1)\beta(2)\tfrac{1}{2}\beta(1)\alpha(2)d\omega(1)d\omega(2)$$

$$+ \iint \alpha(1)\beta(2)\tfrac{1}{2}\beta(1)\alpha(2)d\omega(1)d\omega(2)$$

$$- \iint \alpha(1)\beta(2)\tfrac{1}{2}\alpha(1)\beta(2)d\omega(1)d\omega(2)$$

$$= \tfrac{1}{2}(\tfrac{1}{2} + 1) + \tfrac{1}{2}(\tfrac{1}{2} + 1) + 0 + 0 - \tfrac{1}{2}$$

$$= 1 \tag{C.17}$$

where the orthonormality of the α and β spin functions for each electronic spin coordinate ω permits the trivial evaluation of the individual integrals. Since the fourth integral on the r.h.s. of the last equality in Eq. (C.14) differs only by assignment of the electron labels 1 and 2, it also must have a value of 1. By the same symmetry argument, the second and third integrals must be equal to one another. Evaluating the second using Eq. (C.16) gives

$$
\langle\alpha(1)\beta(2)|S^2|\alpha(2)\beta(1)\rangle = \int\int \alpha(1)\beta(2)\tfrac{1}{2}(\tfrac{1}{2}+1)\alpha(2)\beta(1)d\omega(1)d\omega(2)
$$

$$
+ \int\int \alpha(1)\beta(2)\tfrac{1}{2}(\tfrac{1}{2}+1)\alpha(2)\beta(1)d\omega(1)d\omega(2)
$$

$$
+ \int\int \alpha(1)\beta(2)\tfrac{1}{2}\beta(2)\alpha(1)d\omega(1)d\omega(2)
$$

$$
+ \int\int \alpha(1)\beta(2)\tfrac{1}{2}\beta(2)\alpha(1)d\omega(1)d\omega(2)
$$

$$
- \int \alpha(1)\beta(2)\tfrac{1}{2}\alpha(2)\beta(1)d\omega(1)d\omega(2)
$$

$$
= 0 + 0 + \tfrac{1}{2} + \tfrac{1}{2} - 0
$$

$$
= 1 \tag{C.18}
$$

Thus, the expectation value of S^2 from Eq. (C.14) for the closed-shell state is simply $\tfrac{1}{2}(1 - 1 - 1 + 1) = 0$.

Another wave function of interest is the one formed from two α-spin electrons in two different spatial orbitals a and b. This $S_z = 1$ (see Eq. (C.11)) wave function is written as

$$
{}^3_1\Psi = \frac{1}{\sqrt{2}} \begin{vmatrix} a(1)\alpha(1) & b(1)\alpha(1) \\ a(2)\alpha(2) & b(2)\alpha(2) \end{vmatrix}
$$

$$
= \frac{1}{\sqrt{2}}\alpha(1)\alpha(2)\left[a(1)b(2) - a(2)b(1)\right] \tag{C.19}
$$

In this case, integration over the normalized non-spin-dependent spatial portion of the wave function leaves only a fairly simple integral to evaluate for the expectation value of S^2, namely

$$
\langle\alpha(1)\alpha(2)|S^2|\alpha(1)\alpha(2)\rangle = \int\int \alpha(1)\alpha(2)\tfrac{1}{2}(\tfrac{1}{2}+1)\alpha(1)\alpha(2)d\omega(1)d\omega(2)
$$

$$
+ \int\int \alpha(1)\alpha(2)\tfrac{1}{2}(\tfrac{1}{2}+1)\alpha(1)\alpha(2)d\omega(1)d\omega(2)
$$

$$
+ \int\int \alpha(1)\alpha(2)\tfrac{1}{2}\beta(1)\beta(2)d\omega(1)d\omega(2)
$$

$$
- \int\int \alpha(1)\alpha(2)\tfrac{1}{2}\beta(1)\beta(2)d\omega(1)d\omega(2)
$$

$$+ \iint \alpha(1)\alpha(2)\tfrac{1}{2}\alpha(1)\alpha(2)d\omega(1)d\omega(2)$$

$$= \tfrac{1}{2}(\tfrac{1}{2}+1) + \tfrac{1}{2}(\tfrac{1}{2}+1) + 0 - 0 + \tfrac{1}{2}$$

$$= 2 \tag{C.20}$$

In the case of the $S_z = -1$ state (i.e., two β electrons instead of α), it is straightforward to show that the expectation value may be evaluated by the analog of Eq. (C.20) with all spin functions permuted α to β and vice versa. The resulting expectation value is still 2.

Thus far, we have described wave functions for which

$$S^2\Psi = s(s+1)\hbar^2\Psi \tag{C.21}$$

where $s = 0$ for the singlet and $s = 1$ for the triplet. (We have not formally proven that the wave functions are eigenfunctions of S^2, but inspection of the 'right-hand portions' of the integrals in the expectation values of Eqs. (C.14) and (C.20) makes this a simple exercise.) This situation *defines* what is meant by a singlet ($s = 0$) or triplet ($s = 1$) wave function.

Let us now consider the wave function with an α electron in spatial orbital a and a β electron in spatial orbital b, i.e.,

$$^{50:50}_{\pm}\Psi = \frac{1}{\sqrt{2}} \begin{vmatrix} a(1)\alpha(1) & b(1)\beta(1) \\ a(2)\alpha(2) & b(2)\beta(2) \end{vmatrix}$$

$$= \frac{1}{\sqrt{2}}[a(1)\alpha(1)b(2)\beta(2) - a(2)\alpha(2)b(1)\beta(1)] \tag{C.22}$$

which has been superscripted 50:50 and subscripted $\pm$ for reasons that will be apparent later. Evaluation of the expectation value for S^2 involves

$$\langle^{50:50}\Psi|S^2|^{50:50}\Psi\rangle = \tfrac{1}{2}\big[\langle a(1)\alpha(1)b(2)\beta(2)|S^2|a(1)\alpha(1)b(2)\beta(2)\rangle$$
$$- \langle a(1)\alpha(1)b(2)\beta(2)|S^2|a(2)\alpha(2)b(1)\beta(1)\rangle$$
$$- \langle a(2)\alpha(2)b(1)\beta(1)|S^2|a(1)\alpha(1)b(2)\beta(2)\rangle$$
$$+ \langle a(2)\alpha(2)b(1)\beta(1)|S^2|a(2)\alpha(2)b(1)\beta(1)\rangle\big] \tag{C.23}$$

Note that the second and third integrals on the r.h.s. are zero because of the orthonormality of the spatial orbitals a and b, whose products appear over the same electronic coordinate in those integrals. The spatial functions integrate to one in the first and fourth integrals, and the remaining spin expectation values are just those of Eq. (C.17). Thus, the expectation value of Eq. (C.23) is $\tfrac{1}{2}(1 - 0 - 0 + 1) = 1$. With additional work, it can be shown that $^{50:50}\Psi$ is *not* an eigenfunction of S^2.

By permutational symmetry, it is easy to show that the expectation value of S^2 for the other possible 50:50 wave function, i.e.,

$$^{50:50}_{\mp}\Psi = \frac{1}{\sqrt{2}}\begin{vmatrix} a(1)\beta(1) & b(1)\alpha(1) \\ a(2)\beta(2) & b(2)\alpha(2) \end{vmatrix}$$

$$= \frac{1}{\sqrt{2}}[a(1)\beta(1)b(2)\alpha(2) - a(2)\beta(2)b(1)\alpha(1)] \tag{C.24}$$

is also 1, and it too fails to be an eigenfunction of S^2. In order to construct proper eigenfunctions, we must take linear combinations of the two 50:50 functions, thereby creating two-determinantal wave functions. In particular, we can construct

$$^{OSS}\Psi = \frac{1}{\sqrt{2}}\left(^{50:50}_{\pm}\Psi - ^{50:50}_{\mp}\Psi\right)$$

$$= \tfrac{1}{2}[a(1)b(2) + a(2)b(1)][\alpha(1)\beta(2) - \alpha(2)\beta(1)] \tag{C.25}$$

where the spin function is identical to that appearing in the closed-shell singlet wave function of Eq. (C.13). Since S^2 operates only on the spin part of the wave function, $^{OSS}\Psi$ must be an eigenfunction of S^2 with eigenvalue $s = 0$ just as is true for $^{CSS}\Psi$. Thus, $^{OSS}\Psi$ is a singlet, and in particular it is an open-shell singlet (hence the 'OSS' superscript), i.e., at least two electrons are in singly occupied orbitals.

The other linear combination of the 50:50 wave functions is

$$^{3}_{0}\Psi = \frac{1}{\sqrt{2}}\left(^{50:50}_{\pm}\Psi + ^{50:50}_{\mp}\Psi\right)$$

$$= \tfrac{1}{2}[a(1)b(2) - a(2)b(1)][\alpha(1)\beta(2) + \alpha(2)\beta(1)] \tag{C.26}$$

After integration over the spatial coordinates, we may evaluate the expectation value of S^2 as

$$\langle^{3}_{0}\Psi|S^2|^{3}_{0}\Psi\rangle = \tfrac{1}{2}\big[\langle\alpha(1)\beta(2)|S^2|\alpha(1)\beta(2)\rangle + \langle\alpha(1)\beta(2)|S^2|\alpha(2)\beta(1)\rangle$$

$$+ \langle\alpha(2)\beta(1)|S^2|\alpha(1)\beta(2)\rangle + \langle\alpha(2)\beta(1)|S^2|\alpha(2)\beta(1)\rangle\big]$$

$$= \tfrac{1}{2}(1 + 1 + 1 + 1)$$

$$= 2 \tag{C.27}$$

where the integrals on the r.h.s. were simplified using Eqs. (C.17) and (C.18). Thus, the wave function of Eq. (C.26) is a triplet wave function, and it is the so-called $S_z = 0$ triplet.

Equation (C.25) and (C.26) make it apparent that we may also write

$$^{50:50}_{\pm}\Psi = \frac{1}{\sqrt{2}}\left(^{OSS}\Psi + ^{3}_{0}\Psi\right) \tag{C.28}$$

if a and b are spatially identical in the singlet and triplet wave functions. Equation (C.28) is the foundation of the sum method, described in Section 14.4.

C.3 UHF Wave Functions

In evaluating Eq. (C.23), we invoked orthonormality between the spatial orbitals a and b, each of which contains an electron of different spin. However, in a UHF wave function, the α and β orbitals are not necessarily orthogonal to one another (only within each set, either α or β, are all of the orbitals mutually orthogonal to one another). In that case, the second and third terms on the r.h.s of Eq. (C.23) survive as $-\langle a|b\rangle^2$. In general, one can show that for a UHF wave function where the number of α electrons is greater than or equal to the number of β electrons, the expectation value of S^2 may be computed as

$$\langle{}^{\mathrm{UHF}}\Psi|S^2|{}^{\mathrm{UHF}}\Psi\rangle = \frac{n_\alpha - n_\beta}{2}\left(\frac{n_\alpha - n_\beta}{2}+1\right)+n_\beta - \sum_{i\in\alpha, j\in\beta}^{\mathrm{occupied}}\langle\phi_i|\phi_j\rangle^2 \qquad (\mathrm{C.29})$$

where n_ζ is the number of electrons of spin ζ and $\{\phi\}$ is the set of UHF molecular orbitals.

Consider the behavior of Eq. (C.29) in certain idealized limits. If there are no β electrons, then Eq. (C.29) reduces to the correct eigenvalue for a system of all parallel spins (cf. Eq. (C.21)). If there *are* β electrons, and for every occupied β MO there is a spatially identical occupied α MO, then the sum on the r.h.s. of Eq. (C.29) is equal to n_β (there is one overlap integral value of unity for each occupied β MO with its partner α MO, and all other overlap integrals must be zero because since other α MOs must be orthogonal to the partner α MO, so too they must be orthogonal to the spatially identical β MO), and the expectation value is again the correct eigenvalue for a high-spin system with excess α electrons. Note, however, that this expectation value can also be achieved to within arbitrary accuracy *without* requiring every occupied β MO to have a spatially identical occupied α MO: all that is required is that the sum of the squares of the overlap integrals approach its limiting value, n_β.

Finally, consider the case where the overlap between the α and β orbitals is exactly zero (which could happen, for instance, if all α MOs were on one atom and all β MOs on another atom with the two atoms infinitely far apart). In that case, the expectation value will be larger than the pure spin state where only the excess α electrons are unpaired, but smaller than the value expected for the pure spin state where all electrons are unpaired (i.e., a low-spin $(n+1)$-multiplet where n is the *total* number of electrons, for which the expectation value would be computed using $s = (n_\alpha + n_\beta)/2$ instead of $s = (n_\alpha - n_\beta)/2$). Such a system is said to be 'spin-contaminated' because it is a mixture of the lowest spin state and varying contributions from states of higher spin multiplicity. Obviously, such wave functions are of limited utility, since expectation values of other properties will also represent an admixture of the properties of the different states.

C.4 Spin Projection/Annihilation

When a spin-contaminated wave function is obtained from a UHF calculation, the desired spin state is inevitably the one of lower spin (otherwise one would have constructed the high-S_z component of the higher spin state). The contaminated wave function can be improved by removing the undesirable higher spin states through a process known as projection or annihilation. Consider the case where the UHF wave function for the desired state $^s\Psi$ is contaminated by the next higher possible spin state $^{(s+1)}\Psi$, i.e.,

$$^{\text{UHF}}\Psi = c_s\,^s\Psi + c_{(s+1)}\,^{(s+1)}\Psi \tag{C.30}$$

where each pure spin wave function is normalized and the sum of the squares of the coefficients c is 1 for normalization of $^{\text{UHF}}\Psi$. When the annihilation operator of Eq. (14.19) is applied to the spin-contaminated wave function we have

$$A_{s+1}\,^{\text{UHF}}\Psi = c_s \frac{S^2 - \{(s+1)[(s+1)+1]\}}{[s(s+1)] - \{(s+1)[(s+1)+1]\}}\,^s\Psi$$

$$+ c_{(s+1)} \frac{S^2 - \{(s+1)[(s+1)+1]\}}{[s(s+1)] - \{(s+1)[(s+1)+1]\}}\,^{(s+1)}\Psi$$

$$= c_s \frac{[s(s+1)] - \{(s+1)[(s+1)+1]\}}{[s(s+1)] - \{(s+1)[(s+1)+1]\}}\,^s\Psi$$

$$+ c_{(s+1)} \frac{\{(s+1)[(s+1)+1]\} - \{(s+1)[(s+1)+1]\}}{[s(s+1)] - \{(s+1)[(s+1)+1]\}}\,^{(s+1)}\Psi$$

$$= c_s \cdot 1 \cdot\,^s\Psi + c_{(s+1)} \cdot 0 \cdot\,^{(s+1)}\Psi$$

$$= c_s\,^s\Psi \tag{C.31}$$

Thus, the annihilation operator completely removes the next higher spin state and delivers a wave function that is a pure s spin state. Note, however, that it is not a normalized wave function, since $c_s < 1$ (otherwise the original wave function would not have been spin contaminated). Normalization is simple in this case, since we have

$$\langle^{\text{UHF}}\Psi|A_{s+1}|^{\text{UHF}}\Psi\rangle = \langle c_s\,^s\Psi + c_{(s+1)}\,^{(s+1)}\Psi|c_s\,^s\Psi\rangle$$

$$= \langle c_s\,^s\Psi|c_s\,^s\Psi\rangle + \langle c_{(s+1)}\,^{(s+1)}\Psi|c_s\,^s\Psi\rangle$$

$$= c_s^2\langle^s\Psi|^s\Psi\rangle + c_s c_{(s+1)}\langle^{(s+1)}\Psi|^s\Psi\rangle$$

$$= c_s^2 \tag{C.32}$$

With the annihilated wave function in hand, any property may be computed in the usual fashion as an expectation value of the appropriate operator. The Hamiltonian operator is a particularly simple operator to work with because we can make good use of the original UHF wave function in evaluating the expectation value. Thus

$$E_{\text{PUHF}} = \frac{\langle^{\text{UHF}}\Psi|H A_{s+1}|^{\text{UHF}}\Psi\rangle}{\langle^{\text{UHF}}\Psi|A_{s+1}|^{\text{UHF}}\Psi\rangle}$$

$$= \frac{\langle c_s\,^s\Psi + c_{(s+1)}\,^{(s+1)}\Psi|H|c_s\,^s\Psi\rangle}{\langle^{\text{UHF}}\Psi|A_{s+1}|^{\text{UHF}}\Psi\rangle}$$

$$= \frac{\langle c_s{}^s \Psi | H | c_s{}^s \Psi \rangle + \langle c_{(s+1)}{}^{(s+1)} \Psi | H | c_s{}^s \Psi \rangle}{\langle {}^{\mathrm{UHF}} \Psi | A_{s+1} | {}^{\mathrm{UHF}} \Psi \rangle}$$

$$= \frac{\langle c_s{}^s \Psi | H | c_s{}^s \Psi \rangle}{\langle {}^{\mathrm{UHF}} \Psi | A_{s+1} | {}^{\mathrm{UHF}} \Psi \rangle}$$

$$= \langle {}^s \Psi | H | {}^s \Psi \rangle \tag{C.33}$$

where the final line is the desired result. To solve for the projected UHF (PUHF) energy, one employs the so-called 'resolution of the identity' technique in the first line of Eq. (C.33), giving

$$E_{\mathrm{PUHF}} = \frac{\displaystyle\sum_i^{\mathrm{states}} \langle {}^{\mathrm{UHF}} \Psi | H | \Psi^i \rangle \langle \Psi^i | A_{s+1} | {}^{\mathrm{UHF}} \Psi \rangle}{\langle {}^{\mathrm{UHF}} \Psi | A_{s+1} | {}^{\mathrm{UHF}} \Psi \rangle}$$

$$= \langle {}^{\mathrm{UHF}} \Psi | H | {}^{\mathrm{UHF}} \Psi \rangle + \frac{\displaystyle\sum_i^{\substack{\mathrm{excited} \\ \mathrm{states}}} \langle {}^{\mathrm{UHF}} \Psi | H | \Psi^i \rangle \langle \Psi^i | A_{s+1} | {}^{UHF} \Psi \rangle}{\langle {}^{\mathrm{UHF}} \Psi | A_{s+1} | {}^{\mathrm{UHF}} \Psi \rangle} \tag{C.34}$$

which is the result expressed in Eq. (14.18). Note that the first term on the r.h.s. of the last equality of Eq. (C.34) is simply the energy of the spin-contaminated wave function, so the second term may be considered the 'correction' associated with spin annihilation. Note also that the only Hamiltonian matrix elements that will be non-zero will be for excited states differing from the ground state by double excitations (the singles are zero by Brillouin's theorem, and triples and higher are zero because the Hamiltonian is a two-electron operator). The Hamiltonian matrix elements required to compute the correction term are exactly those needed for an MP2 calculation (cf. Eq. (7.47)), so the computational effort required for a PUHF calculation is essentially that for an MP2 calculation.

While the PUHF energy is better than the more highly contaminated UHF energy, there are many potential problems associated with it. To begin, the molecular orbitals were optimized for the contaminated wave function, not the annihilated wave function, and as such they may be considerably less than ideal for the single Slater determinant describing the pure spin state. In addition, the geometry of the contaminated state may not be a good representation of the geometry of the pure state. As analytic derivatives are not available for the PUHF energy, reoptimization is tedious.

Another potential problem is that the wave function may be contaminated not only by state $s + 1$, but also states $s + 2$, $s + 3$, etc. The $s + 1$ annihilation operator will reduce the weights of these states in the annihilated wave function, but it will not eliminate them. Inspection of Eq. (C.33) should make clear that the higher states will contribute to the PUHF energy if they appear on both the left and right sides of the Hamiltonian expectation values with non-zero coefficients. When such contamination is important, recourse to a more complete projection operator, that annihilates an arbitrary number of spin states is available, but the computational cost increases to essentially that of an MP4 calculation. Note that the problems of the orbitals being non-ideal for the pure lowest spin state persist in this instance.

In order to account for electron correlation, projection operator methods within the MPn perturbation theory formalism have also been described (Schlegel 1988). Such PMPn methods can be valuable in assessing convergence of projected pure-spin-state energies, but it should be recalled that perturbation theory is most successful when the true wave function differs from the HF determinant by only a small amount. Since the HF determinant *starts* with contamination, it is a given that it is potentially too far removed from the true wave function to be of much use in perturbation theory, and some caution should be exercised. Krylov (2000) has demonstrated that coupled-cluster theory can be more robust than perturbation theory in relying on spin-contaminated UHF wave functions so long as the reference orbitals are chosen properly, but this method too ultimately breaks down as the spin contamination becomes especially severe. As a final resort, MCSCF methods can always be used to construct multideterminantal, pure spin states, although this may not be the most convenient choice for a particular problem.

References

Krylov, A. I. 2000. *J. Chem. Phys.*, **113**, 6052.
Schlegel, H. B. 1988. *J. Phys. Chem.*, **92**, 3075.

Appendix D

Orbital Localization

D.1 Orbitals as Empirical Constructs

It is both flattering and vexing to quantum chemists how ubiquitous it has become to rationalize chemical behavior using molecular-orbital-based arguments. It is flattering to the extent
that it indicates how large an impact quantum mechanics has had on modern chemistry, but
it is vexing because orbitals themselves, unlike the wave function, are not really a rigorous
part of quantum mechanics.

Indeed, there are other formulations of quantum mechanics, all of which have been shown
to be entirely equivalent in a formal sense to the matrix-algebraic-molecular-orbital version,
that do not in any way require an invocation of orbitals. However, the matrix-algebraic
method lends itself most readily to implementation on the architecture of a digital computer,
and thus it has come to overwhelmingly dominate modern computational chemistry. As a
result, the orbitals that are part of the computational machinery for approximately solving the
matrix algebraic equations have taken on the character of unassailable parts of the quantum
mechanical formalism, but that status is undeserved.

Consider, for example, two orbitals which might be obtained from a HF calculation on
ethylene, namely the orthonormal σ and π bonding orbitals, both of which are doubly occupied (Figure D.1). If we restrict our consideration to only these two orbitals, and moreover
we use restricted HF theory so that we can ignore the details of spin orbitals, we can write
the properly antisymmetric HF wave function for this system of two orbitals as

$$\Psi = \tfrac{1}{\sqrt{2}}\sigma(1)\pi(2) - \tfrac{1}{\sqrt{2}}\sigma(2)\pi(1) \tag{D.1}$$

The energy of the system is calculated as the expectation value of the Hamiltonian

$$\langle\Psi|H|\Psi\rangle = \tfrac{1}{2}\langle\sigma(1)\pi(2)|H|\sigma(1)\pi(2)\rangle - \tfrac{1}{2}\langle\sigma(1)\pi(2)|H|\sigma(2)\pi(1)\rangle$$
$$- \tfrac{1}{2}\langle\sigma(2)\pi(1)|H|\sigma(1)\pi(2)\rangle + \tfrac{1}{2}\langle\sigma(2)\pi(1)|H|\sigma(2)\pi(1)\rangle \tag{D.2}$$

Now consider a different wave function, formed from different orbitals. In particular, we
will take the positive and negative linear combinations of the σ and π orbitals shown in

Essentials of Computational Chemistry, 2nd Edition Christopher J. Cramer
© 2004 John Wiley & Sons, Ltd   ISBNs: 0-470-09181-9 (cased); 0-470-09182-7 (pbk)

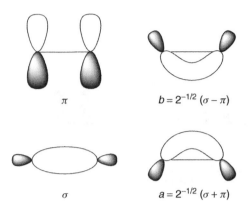

Figure D.1 The usual σ and π orbitals of a doubly bonded system (left) and the banana bonds formed by their linear combination (right)

Figure D.1 to create so-called 'banana-bond' orbitals. We define these as

$$a = \tfrac{1}{\sqrt{2}}\sigma + \tfrac{1}{\sqrt{2}}\pi \qquad\qquad (D.3)$$

and

$$b = \tfrac{1}{\sqrt{2}}\sigma - \tfrac{1}{\sqrt{2}}\pi \qquad\qquad (D.4)$$

It is a simple exercise to show that if σ and π are orthonormal then a and b are too. Let us now consider the antisymmetric wave function

$$\Phi = \tfrac{1}{\sqrt{2}}a(1)b(2) - \tfrac{1}{\sqrt{2}}a(2)b(1) \qquad\qquad (D.5)$$

The energy of Φ is

$$\langle\Phi|H|\Phi\rangle = \tfrac{1}{2}\langle a(1)b(2)|H|a(1)b(2)\rangle - \tfrac{1}{2}\langle a(1)b(2)|H|a(2)b(1)\rangle$$
$$- \tfrac{1}{2}\langle a(2)b(1)|H|a(1)b(2)\rangle + \tfrac{1}{2}\langle a(2)b(1)|H|a(2)b(1)\rangle \qquad (D.6)$$

which may be rewritten using Eqs. (D.3) and (D.4) as

$$\langle\Phi|H|\Phi\rangle = \tfrac{1}{8}\langle[\sigma(1)+\pi(1)][\sigma(2)-\pi(2)]|H|[\sigma(1)+\pi(1)][\sigma(2)-\pi(2)]\rangle$$
$$- \tfrac{1}{8}\langle[\sigma(1)+\pi(1)][\sigma(2)-\pi(2)]|H|[\sigma(2)+\pi(2)][\sigma(1)-\pi(1)]\rangle$$
$$- \tfrac{1}{8}\langle[\sigma(2)+\pi(2)][\sigma(1)-\pi(1)]|H|[\sigma(1)+\pi(1)][\sigma(2)-\pi(2)]\rangle$$
$$+ \tfrac{1}{8}\langle[\sigma(2)+\pi(2)][\sigma(1)-\pi(1)]|H|[\sigma(2)+\pi(2)][\sigma(1)-\pi(1)]\rangle \quad (D.7)$$

A somewhat tedious separation and collection of the 64 individual integrals in Eq. (D.7) (an exercise left for the industrious reader) leads to

$$\langle \Phi | H | \Phi \rangle = \tfrac{1}{2} \langle \sigma(1)\pi(2) | H | \sigma(1)\pi(2) \rangle - \tfrac{1}{2} \langle \sigma(1)\pi(2) | H | \sigma(2)\pi(1) \rangle$$

$$- \tfrac{1}{2} \langle \sigma(2)\pi(1) | H | \sigma(1)\pi(2) \rangle + \tfrac{1}{2} \langle \sigma(2)\pi(1) | H | \sigma(2)\pi(1) \rangle \qquad \text{(D.8)}$$

which has the same r.h.s. as Eq. (D.2). That is, the expectation value of the energy for the wave function Φ is the same as that for the wave function Ψ. Since the HF process was variational, this implies that Φ is an 'equally optimal' wave function as Ψ, and there is no obvious reason to prefer one over the other.

This particular example illustrates what can be shown more formally to be true in general: the energy of the wave function is invariant to expressing the wave function using *any* normalized linear combination of the occupied HF orbitals, as are the expectation values of all other quantum mechanical operators. Since all such choices of linear combinations of orbitals satisfy the variational criterion, one may legitimately ask why the HF orbitals should be assigned any privileged status of their own as chemical entities.

The answer to that question is that, empirically, the HF orbitals have proven to be useful models for rationalizing certain chemical phenomenon. To that extent, like many other chemical models that do not have rigorous quantum mechanical foundations, they are useful tools that now find widespread use in the chemical community. It is worthwhile to consider briefly the advantages and disadvantages of various schemes for formulating MOs that have appeared in the chemical literature.

The persistence of the HF orbitals as objects for chemical discussion stems from three features. First, they are the fundamental products of the HF SCF process, and thus immediately at hand following an HF calculation. Secondly, and most importantly, they have associated with them specific orbital energies. The HF MOs are precisely those MOs that diagonalize the Fock operator, and so they have associated energy eigenvalues. It is known from photoelectron spectroscopy that electrons do reside at distinct energy levels, and from an experimental standpoint, a molecular orbital is 'defined' by the energy of an electron and the instantaneous difference in electron density between the pre- and post-ionized state (of course, measuring the density difference on a timescale faster than electronic relaxation is impossible, but the relaxed difference may in some cases be reasonably close to the pre-relaxed difference). Thus, the HF orbitals are natural objects for the discussion of spectroscopic quantities. Finally, the HF orbitals can be assigned to individual irreps of the molecular point group; this will be true for other orbitals only if linear combinations of the HF orbitals are restricted to be exclusively within their individual irreps.

Other orbital localization schemes create MOs that do *not* diagonalize the Fock operator, and thus it is more difficult to assign orbital energies to them. However, the canonical HF orbitals have certain features that are occasionally regarded as undesirable, and this has motivated the development of alternative localization methods. Thus, for example, many of the HF orbitals in large systems tend to be highly delocalized, but most chemical reactivity concepts are local in nature. For the discussion of such concepts, it is desirable to have highly localized MOs. In general, localization schemes impose some sort of critical constraint on the final orbitals. For instance, one constraint might be that the repulsion between electrons in the same orbitals be maximized – this results in very compact orbitals. An alternative is that interactions between electrons in *different* orbitals might be *minimized* – this results in maximally 'distinct' orbitals. It should be obvious that an infinite number of choices might be made for how best to combine the HF orbitals to make new orbitals, with goodness being determined only by how useful the final orbitals prove to be in a given qualitative or semi-quantitative empirical chemical model.

D.2 Natural Bond Orbital Analysis

A localization algorithm that hews particularly closely to intuitive chemical concepts is the natural bond orbital (NBO) method of Weinhold and co-workers (Reed, Curtiss and Weinhold 1988; for latest generation code and additional technical discussion see www.chem.wisc.edu/~nbo5/). NBO localization is a multistep process the details of which are sufficiently complicated that they do not warrant specific identification here – however, a general overview of the procedure can be described fairly simply. In an initial step, orbitals that are associated almost entirely with a single atom, e.g., core orbitals and lone pairs, are localized as so-called natural atomic orbitals (NAOs). Next, orbitals involving bonding (or antibonding) between pairs of atoms are localized by using only the basis set AOs of those atoms. Finally, the remaining Rydberg-like orbitals are identified, and all orbitals are made orthogonal to one another. The result is that, except for very small contributions from other AOs to ensure orthogonality, all NAOs and Rydberg orbitals are described using the basis-set AOs of a single atom and all NBOs are described using the basis-set AOs of two atoms (in cases where resonance or other delocalization effects require orbitals delocalized over more than two atoms, additional work is required). Thus, NBO analysis provides an orbital picture that is as close as possible to a classical Lewis structure for a molecule.

This localization scheme permits the assignment of hybridization both to the atomic lone pairs and to each atom's contributions to its bond orbitals. Hybridization is a widely employed and generally useful chemical concept even though it has no formal basis in the absence of high-symmetry constraints. With NBO analysis, the percent s and p character (and d, f, etc.) is immediately evident from the coefficients of the AO basis functions from which the NAO or NBO is formed. In addition, population analysis can be carried out using the NBOs to derive partial atomic charges (NPA, see Section 9.1.3.2).

Another useful chemical concept is hyperconjugation, which rationalizes certain chemical phenomena in terms of filled-orbital–empty-orbital interactions (see Section 2.2.3). Consider for instance the torsional potential about the $C(3)$–$O(4)$ bond in 2,4-dioxaheptane (Figure D.2). Delocalization of high-energy oxygen lone pair density into the low-energy σ^* $C(3)$–$O(2)$ antibonding orbital is expected to stabilize torsional conformations that maximize the overlap between such orbitals (in an oxahydrocarbon case like this, the phenomenon is referred to as the anomeric effect).

NBO analysis can be used to quantify this phenomenon. Since the NBOs do not diagonalize the Fock operator (or the Kohn–Sham operator, if the analysis is carried out for DFT instead of HF), when the Fock matrix is formed in the NBO basis, off-diagonal elements will in general be non-zero. Second-order perturbation theory indicates that these off-diagonal elements between filled and empty NBOs can be interpreted as the stabilization energies

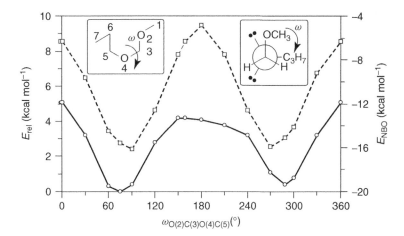

Figure D.2 The torsional potential for 2,4-dioxaheptane (———, left ordinate) exhibits a two-fold periodicity suggesting a large influence from hyperconjugative effects. The sum of the NBO interaction energies for the two O(4) lone pairs delocalizing into the C(3)–O(2) σ^* orbital (- - - - - -, right ordinate) shows the same periodicity, with an amplitude of about 11 kcal mol^{-1}. The smaller amplitude of the full torsional potential reflects primarily the presence of other influences as well as the approximate nature of the NBO analysis. See Cramer, Kelterer, and French 2001

deriving from hyperconjugation. Figure D.2 illustrates how the minima in the torsional potential energy curve for 2,4-dioxaheptane are influenced by the changes in hyperconjugation between the O(4) lone pairs and the C(3)–O(2) antibonding σ^* orbital as quantified by NBO analysis (a similar analysis may be carried out for filled-orbital–filled-orbital repulsive interactions, which are the electronic component of steric interactions). This approach is a nice energetic complement to other analyses focusing on structural changes or changes in partial atomic charges (cf. Figure 9.3) in the investigation of hyperconjugative effects. Nevertheless, it must be stressed that the NBO procedure is only a conceptual model, since it is based on orbitals, and limitations in its utility may be expected in cases where chemical species are poorly represented as Lewis structures.

References

Cramer, C. J., Kelterer, A.-M., and French, A. D. 2001. *J. Comput. Chem.*, **22**, 1194.
Reed, A. E., Curtiss, L. A., and Weinhold, F. 1988. *Chem. Rev.*, **88**, 899.

Index

Ab initio, (*see* Molecular orbital theory)
Acidity, 11, 176, 194, 410–413, 415, 469, 481
Activated complex, 523–527, 531, 535, 538
Activation energy, 62, 132, 149–150, 267,
    285–288, 300, 349, 378, 390, 422,
    440–442, 483–484, 527–539, 543–545
Active space, (*see* Orbital)
Adiabatic connection, 264–268, 273, 278–300,
    340, 371
Adiabatic process, 331, 489–490, 497, 505,
    519
Alkane solvation, 388, 407
Allyl system, 116–119, 234
Alq3, 513–515
AM1, 145–156, 193, 281, 287, 313, 319–323,
    338, 340, 375, 381–382, 459, 465, 476
AM1*, 154
AM1/d, 154
AM1/OPLS and AM1/TIP3P, (*see* QM/MM)
AMBER, 51, 59, 99
AMBER*, 51, 60
Amine basicity, 91
Amsterdam Density Functional, 273
Analytic derivatives, (*see* Gradient)
Anions, 119, 148, 176, 182, 244–246, 414
Anomeric effect, 23, 469, 578
Antisymmetry, 122–126, 190, 265
AOC, (*see* QM/MM)
Aqueous solvation, (*see* Water, as solvent)
Arrhenius equation, 528, 545
Atom types, 31, 38, 40, 48–49, 310, 356, 404,
    408
Atomic partial charge, (*see also* Population
    analysis) 31–32, 100, 135, 151–152, 171,
        270, 309–324, 402–404, 411, 443, 446,
        458, 462, 474–476, 480, 579
    atoms-in-molecules, 309, 315–318
    classes, 310–324
    CMn, ($n = 1$–3), 319–324, 404, 459–461
    conformational dependence, 313, 319
    discretization, 399
    electronegativity equalization, 54, 310–312
    ESP, 318–319, 322–323, 449
    GAPT, 315
    Löwdin, 314–315, 320
    Mulliken, 312–314, 320, 322–323, 404
    NPA, 314–315, 322–323, 578
    SCRF calculations, 404
Atomic radii, 27, 403
Atomic units, 15
Atomization energy, 111, 192, 267, 280–284,
    367, 375, 381–382
Atoms-in-molecules, 315–318
Autocorrelation function, 86–88
Avidin, 452–454
Avogadro's number, 359
Avoided crossing, 499, 540–541

B exchange, 263, 266–267, 295
B1B95, 267–268, 287, 290, 295
B1LYP, 267, 283, 292, 295, 339
B1PW91, 267, 283, 292, 295, 339
B3LYP, 241, 267–268, 271, 278–279,
    284–286, 290–292, 294–295, 298–299,
    330, 339–340, 347, 350, 381–382, 414,
    544–545
B3LYP*, 268, 295
B3P86, 284, 290–291, 330

Essentials of Computational Chemistry, 2nd Edition Christopher J. Cramer
© 2004 John Wiley & Sons, Ltd ISBNs: 0-470-09181-9 (cased); 0-470-09182-7 (pbk)

B3PW91, 266–267, 284, 288, 290–291, 293–295, 330, 339–340
B86 exchange, 263, 295
B88 correlation, 263, 295
B95 correlation, 264, 295
B97, 285, 287, 295–296, 347
B98, 264, 285, 287, 295
BAC, (*see* Bond additivity correction)
Band structure, 192, 498
Basis set, (*see also* Orbital; note that individual basis sets below are indexed by name *not distinguishing* between number or type of polarization or diffuse functions), 114–115, 117, 128–129, 139, 143, 155, 158, 166–180, 220, 227–230, 256, 273–274, 448, 578
3-21G, 172, 175–176, 192–194, 197, 340
4-31G, 172
6-21G, 172
6-311G, 172, 174, 176–177, 329, 340
6-31G, 172, 174–177, 192–193, 195, 340
additivity principle, 177–178
auxiliary, 261–262, 273
cc-pCVnZ, ($n = $ D, T, Q,...), 171, 176, 228–229
cc-pVnZ, ($n = $ D, T, Q,...), 171–172, 176–177, 192–193, 197, 228–229, 235, 274, 340
correlation-consistent, 171, 173, 179, 228
d functions, 173–174
density functional, 260–262
diffuse functions, 176, 180, 194, 279, 331, 412, 414
effective core potential, 178–179, 192, 224, 345, 447
EPR-III, 328
excited-state demands, 494
f functions, 174–175, 228
g functions, 174–175
IGLO-III, 328
linear dependence, 182
MAXI-n, ($n = $ 1, 2, ...), 172
MIDI!, 175–176, 199, 321
MIDI-n, ($n = $ 1, 2, ...), 172
MIDIY, 175–176
MINI-n, ($n = $ 1, 2, ...), 171
minimal, 170–172, 181–182, 214, 313
pc-n, ($n = $ 1, 2, ...), 175–176, 274
plane waves, 273, 448

polarization functions, 173–175, 197, 228, 514
Rydberg-state demands, 498
sp functions, 170–172, 180
splitting, 170–173, 313
STO-3G, 155, 169–171, 184–185, 192–193, 214
STO-MG, 169–170, 172
superposition error, 195–196, 279, 293
VB, 478–480
BB1K, 268, 295
Becke exchange, (*see* B exchange)
Benzene, 183, 497, 502
Benzyne, (*see* Didehydrobenzene)
Berendsen coupling, 91
Bergman cyclization, 349
BH&HLYP, 266, 283, 286, 290, 292, 296, 339
Biasing potential, (*see* Umbrella potential)
Biotin, 452–454
BLYP, 263, 272, 282, 285, 287, 289, 291–294, 330, 339–340, 347, 494–495, 505
Bm, 285, 296
Bohr magneton, 15, 327
Boltzmann distribution, 160, 377, 451, 523, 534–535
Boltzmann's constant, 71, 358
Bond additivity correction, 243, 371
Bond critical point, 316
Bond dipole moment, 33
Bond dissociation energy, 20–21, 148, 156, 216, 243, 279, 419
Bond length, 3, 6–8, 11, 17, 22, 26–28, 34, 42, 44, 145, 160, 183, 197, 235, 243, 291, 293, 453, 479, 483, 542
equilibrium, 18–19, 40, 337, 342
Bond order, 38, 320–321
Bond separation reaction, 373
Bond stretching, (*see* Potential energy functions)
Bonding, 5, 28, 34, 38, 49, 112, 118, 153, 171–172, 193–194, 216, 275, 311, 324, 381, 575, 578
dative, 197, 279
overemphasis at RHF level, 188, 197
Born equation, 396–397, 402, 542
effective radius, 402–403
generalized, 402–404, 408–409, 415, 420–421, 448

Born-Oppenheimer approximation, 110–111, 331, 489–490, 540
Boundary-element method, 400, 404
BP86, 277, 282, 285–289, 291–292, 340, 347–348
BPW91, 233, 257, 266, 282, 285, 288–289, 291–292, 294, 321, 339, 422–423, 495, 502
BR exchange, 264, 296
Brillouin's theorem, 213–214, 221, 496, 507, 573
Broken symmetry, (*see* Density functional theory)
Bromide/methyl bromide, 440
Brownian dynamics, 80
Brueckner orbital, 226, 231–234
1,3-Butadiene, 207–208, 293
t-Butylvinylidene, 494
BVWN, 282, 330
BVWN5, 494

Cage critical point, 316
CAM exchange, 263, 296
Canonical ensemble, 357–358
Car-Parrinello, 447–448
Carbon monoxide, 244, 294, 299–300, 347–348
Carbon tetrachloride solution, 409, 446, 460
Carbonic anhydrase, 481
CASPT2, (*see* Perturbation theory, multireference)
CASSCF, (*see* Self-consistent field, multiconfiguration)
Cavitation, 387–388, 406–407, 417
Cavity, 388, 394–406, 410–411, 415, 419–421
  charge penetration, 415
  general solute, 395, 398, 410
  spherical radius, 395–398, 406
CBS, (*see* Multilevel methods)
Centrifugal distortion, 334
CFF, 50, 53
Charge transfer, 196, 269, 279, 293, 415, 422, 448, 503
Charge-charge interaction, (*see also* Electrostatic and Nonbonded interactions), 48, 90, 157, 307, 309, 400, 404, 445
Charge-dipole interaction, 307, 446
CHARMM, 52, 60, 99, 408, 476, 482–483
CHELPG, (*see* Atomic partial charge, ESP)

Chem-3D, 50, 55
Chem-X, 53, 60
Chemical shift, 344–349, 451, 472
  scaled, 345–346
Chloride/allyl chloride, 198, 293
Chloride/methyl chloride, 185–186, 390
Chloroform solution, 387, 409, 411, 416, 446, 465
Circular dichroism, 504
CIS, 140, 187, 214, 496–499, 502, 514–515
CISD, (*see* Configuration interaction)
Claisen rearrangement, 392, 448–449, 463–464
CM*n*, (*n* = 1–3; *see* Atomic partial charge)
CNDO, 136–139
Coarse-grained models, 35, 98
Code, (*see* Software)
Collective coordinates, 35, 98
Collision theory, 528, 542
Comparative molecular field analysis, 308–310
Complete basis set, (*see* Multilevel methods)
Compressibility, 418, 446
Condensed-phase effects, (*see also* Solvation and Solvatochromism), 379, 385–393, 538–539
Condon-Slater rules, 212–213, 221, 510
Configuration interaction, (*see also* CIS), 211–216, 224–227, 244–246, 328, 336, 401, 496–502
  full, 211, 224, 278
  matrix elements, 212–213
  multireference, 216, 495, 501, 505
  quadratic, 226–227, 281, 286–287, 292, 336, 340
  single-reference, 211–216, 496–497
  spin-flip, 215–216, 234, 496
  VB, 477–481
Configuration state function, 206–212, 220, 499
Conformational analysis, 19, 97, 150, 313, 459
Conformational averaging, 64–66, 97, 193, 288, 377–378, 563
Conical intersection, 499–500
Continuum solvation, (*see* Solvation)
Contraction, (*see* Orbital)
Convergence
  binding energy, 196
  correlation energy, 228–229, 236
  DFT with respect to basis set size, 274, 288

Convergence (*continued*)
  finite-field calculation, 327
  geometry optimization, 40–50, 141, 183,
    191
  HF SCF, 121, 128–129, 166, 181–182, 207,
    262, 491
  induced dipole moment, 446
  KS SCF, 274, 491, 495
  multipole-multipole interactions, 307
  quadratic, 181
  SCRF, 396–397
  simulation, 93–96, 444
  solvation free energy, 401
Core electrons, 134, 178–179, 195, 228, 240,
    345, 474
Core potential, (*see* Basis set)
Correlation, (*see also* Electron correlation), 110
Correlation energy density, 259, 263
COSMO, (*see* Solvation)
Coulomb integral, (*see* Two-electron integral)
Coulomb radius, 403–404
Coulomb's law, 2, 14, 37, 402
Counterpoise correction, 195
Coupled-cluster theory, 224–227, 229–237,
    242, 244–246, 336, 401, 465, 574
  predictions from, 244–246, 281, 292, 339,
    375, 381–382, 422–423, 495, 544–546
  scaled, 229–230
  spin-flip, 227
Coupling parameter, 265, 433–437, 443–444,
    449, 458, 464, 481
Cross terms, 34–36
CS correlation, 296
Curtin-Hammett principle, 300
Cutoff distance, 47, 88–90
CVFF, 53, 60
Cyclobutene, 207–208

d orbitals, 153–155, 167, 285, 291
Darwin relativistic correction, 223–224
Davidson correction, 215
Debye-Hückel parameter, 395, 403
Decay time, 87, 95
Degeneracy, 204–206, 231, 244, 324, 333,
    350, 359–364, 498, 563
  structural, 364, 563
Degrees of freedom, 6, 20, 34, 42–43, 69, 75,
    78–80, 88, 92, 183–186, 338, 342, 394,
    429, 442, 523, 525–526, 531–532, 535,
    538, 563

Delta SCF, 194–195, 288, 330–331, 494–496,
    503
Density functional theory, 249–300, 371
  broken symmetry, 275–276, 545
  multideterminantal, 276
  overdelocalization, 279–280, 294, 330–331
  predictions from, (*see* individual functional
    names)
  projected, 494, 506–507
  SCI and MRCI, 498, 501
  tight-binding, 268–271, 321–322, 404
  time-dependent, 497, 501–504, 514–515
Density matrix, 127–128, 181, 188–189, 196,
    261–262, 308, 328, 396, 404, 476
  spin, 189, 328, 330
  spin-difference, 328
Deprotonation, (*see* Acidity)
Diazene, 505–507
1,2-Dichloroethane, 398, 459–460
Didehydrobenzene
  1,3-, 197
  1,4-, 231–234, 254, 275, 277, 349–350,
    374–375
2,5-Didehydropyridinium cation, 231–234, 275
Dielectric constant, 2, 32–33, 98, 101, 394–397,
    403, 405, 417, 421, 452, 460, 512, 542
Diels-Alder reaction, 285, 460
Diffusion, 88, 543
Dipole moment, 32–33, 37, 82–84, 143, 152,
    198, 294, 306–307, 310, 315, 320–326,
    332, 342, 387–388, 397, 411, 445,
    463–464
  induced, 33, 325, 387–388, 446, 463–464
Dipole-dipole interaction, (*see also*
    Electrostatic and Nonbonded interactions),
    23–28, 32, 47, 90, 307, 445
Dirac δ, 84–85, 224, 439
Direct dynamics, 532
Direct methods, 13, 191
Dispersion, (*see also* Nonbonded interaction),
    28–29, 149, 155, 192, 195, 198, 271, 293,
    371, 388, 406–408, 447, 513
Divide-and-conquer formalism, 274
DNA, (*see* Nucleic acids)
Docking, 62–64, 404, 420, 454
Double-wide sampling, 434
DREIDING, 38, 53
Drug design, 62, 152–153, 309–310
Dry cleaning, 422

Dual topology, 443

Dynamics,
  molecular, 72–80, 91–96, 273–274, 399,
    420–421, 431–434, 438–440, 444,
    447–454, 463, 474–477, 482–484,
    513, 538
  non-adiabatic, 539–544
  reaction, 357, 423, 482–484, 519–546

ECEPP, 54
Eckart potential, 536
EDF1, 268, 282, 296
Effective core potential, (*see* Basis set)
Effective fragment potential, 447, 465
Eigenfunction, 106, 111, 120–122, 126, 166,
    173, 182, 188, 190, 212, 216–218, 220,
    255–256, 324–325, 328, 332, 336,
    507–509, 565–570
Eigenvalue, 95, 106–107, 110–111, 121–122,
    149, 190, 206, 214, 216, 219–220, 250,
    253, 255, 272, 325, 330–333, 335–337,
    362, 479, 496, 502, 507, 540, 565–570
Electric multipole moment, (*see* Multipole
    moment, electric)
Electrochemistry, 410, 413–415, 422–424,
    541–544
Electron affinity, 137, 176, 195, 270, 285,
    288–290, 311, 330–331, 414, 423
Electron correlation, 111, 128–129, 132–133,
    149, 165, 173, 178, 192–195, 203–246,
    251, 280, 330, 388, 493, 574
  angular, 228
  core, 228, 242–243
  core-valence, 228, 240–241
  dynamical, 203–205, 211, 216, 223, 233,
    497, 501
  effect on geometries, 197–198, 235
  effect on solvation free energy, 401, 406
  effect on vibrational intensities, 341
  energy, 129, 132–133, 149, 165, 178, 214,
    224, 242, 370–372
  exchange, 128, 189, 251, 265–267, 274, 278
  non-dynamical, 182, 203–205, 209, 212,
    216, 223, 246, 275–277, 285, 291, 351,
    495, 501
  radial, 228
  scaled energies, 238–239
Electron density, (*see also* Density functional
    theory and Gradient), 61, 112, 249–280,
    314–318, 421, 475–476, 577

Electron spin resonance, 189, 305, 327–330
Electron transfer, 422–424, 541–544
Electronegativity, 23, 31, 152, 171, 270, 307,
    310, 313, 318, 474
Electronic energy, 110–111, 121, 148, 154,
    203, 206, 220, 238, 332–333, 366, 375,
    412, 525
Electronic excited state, 140–141, 176,
    186–187, 254, 273, 360–361, 487–513
Electronic *g* value, 327, 330
Electrostatic interaction, (*see also*
    Charge-charge, Dipole-dipole, and
    Nonbonded interactions), 30–34, 88, 90,
    100, 195, 198, 387–388, 393–406, 444,
    447, 461–462, 467, 474, 478
Electrostatic potential, (*see also* Atomic partial
    charge), 199, 308–309, 318–319,
    394–395, 399–400, 405
Electrostriction, 452
Elementary reaction, 519–523, 531
Empirical valence bond, 477–482
Enantiomeric excess, 160
Enediyne, 349–350
Enolase, 482–484
Ensemble, 69, 82, 91–93, 99, 355–366,
    432–434, 440, 463
Ensemble average, (*see also* Expectation
    value), 70, 83–88, 429, 431–437, 441,
    443, 452–454, 463
Enthalpy, (*see also* Heat of formation), 10, 92,
    355–356, 358, 366–378, 381–383, 412,
    430, 444, 527–528, 537, 545
Entropy, 355, 358–366, 376–378, 386, 430,
    445, 452–453, 527–528, 545
  bottleneck, 523, 533
Enzyme-substrate binding, 62–63, 400,
    438–439, 442, 452–454, 457–458,
    482–484
Equation-of-motion method, (*see* Propagator
    method)
Equilibration, 92–93, 96, 311
Equilibrium constant, 11, 41, 62, 132,
    379–380, 386, 389, 416, 432, 520,
    524–525
Equilibrium fraction, 377
Ergodicity, 72, 93, 431
ESFF, 54, 60
Essential dynamics, (*see* Principal components
    analysis)

Euler's approximation, 75, 77
Even function, 342–343
Ewald sum, 47, 89–90, 101
Exchange, (*see* Electron correlation)
Exchange energy density, 258–259
Exchange functional, 257, 263–264
Exchange integral, (*see* Two-electron integral)
Exchange repulsion, 24, 28, 407, 447, 467
Exchange-correlation energy, 256, 260, 262, 265–266, 271, 338
Excited state, (*see* Electronic excited state)
Expectation value, (*see also* Ensemble average), 61, 71, 83, 94, 142, 190, 203, 209, 218, 220, 223, 244, 253–256, 262, 265–266, 271–275, 324–327, 342, 387, 397, 432, 452, 461–463, 466, 495, 505–510, 567–573
Extended Hückel theory, (*see* Hückel theory)
Extrapolation, 176–177, 230, 239–244, 528

Fast multipole methods, 48, 191, 274
Fenske-Hall theory, 135
Fermi contact integral, 327–328
Fermi hole, 125, 189, 251
Fermi resonance, 341
Ferromagnetism, 136
Finite-field calculation, 327
First derivative, (*see* Gradient)
Fluorine scanning, 452–454
Fluoromethane, 333, 347–348
Fluoromethyl radical, 343–344
Fluorovinylidene, 493–494
Fock operator, 126–129, 189, 203, 219, 462, 562, 577, 578
Force constant, 18–21, 35, 37, 39, 192, 338–339, 341, 443, 473, 522, 529
  scaled, 341
Force field, 17–66, 75, 98–99, 156, 312, 318–319, 404, 411, 452–454, 459, 465
Force-biased Monte Carlo, 82
Four-index integral, (*see* Two-electron integral)
Fourier decomposition, 23–24
Fourier transform, 88, 101, 448
Franck-Condon overlap, 490, 511
Free energy, 355, 389, 429–444
  activation, (*see* Activation energy)
  cycle, 391–393, 412, 418, 437–439, 452–454
  Gibbs, 358, 368–369, 376–380, 430, 437–443, 452–454, 531, 542–543

Helmholtz, 430–436, 443
  perturbation, 432–444, 449, 453, 458, 463–464, 481
  solvation, (*see* Solvation, free energy)
  transfer, 386–389
Free rotor, 376–377
Frequency, (*see also* Vibrational spectroscopy), absorption, 333, 335, 341, 489, 507–511
  imaginary, 338, 523, 526, 535
  intensity, 341, 507–511
  isotopic dependence, 357, 530
  reaction coordinate, 526, 535
  scale factor, 339–341, 345, 349–350, 356, 545
  vibrational, 228, 245, 305, 315, 334–342, 345, 349–351, 356–357, 365–366, 370, 375–376, 429, 452, 472, 526, 528–529, 545
Friction, 73, 80, 334, 420, 421, 538
FT97 exchange, 263, 296
Functional derivative, 256
2-Furfural, 459–460

Gn, ($n = 1, 2, \ldots$; *see* Multilevel methods)
G96, 296
Gas constant, (*see* Universal gas constant)
Gauge origin, 344–345
Gaussian function, 168–169
Gaussian-type orbital, (*see* Orbital)
GB/SA, (*see* Solvation)
Gear predictor-corrector algorithms, 78
Generalized Born equation, (*see* Born equation)
Generalized gradient approximation, 263–264, 266–268, 278–279, 285, 291
Geometry
  comparing theory to experiment, 61, 69
  effect on SCF convergence, 181
  optimization, 40–50, 182–187, 190–192, 196–198, 238, 334, 472–473, 490, 497, 499–500, 505, 514
  quality of predicted, 150–151, 196–198, 235–236, 291–294, 329–330, 573
  relaxation, 489, 542
  rotational spectroscopy dependence, 332, 364
  UV/VIS spectroscopy dependence, 140, 489–490
  vibrational spectroscopy dependence, 338
GHO, (*see* Orbital, generalized hybrid)
GIAO, 345

Global minimum, 23, 46, 97, 146, 383
Glucose, 60, 150–151, 193, 235, 240, 385
Gradient,
    corrected density functionals, (see
        Generalized gradient approximation)
    electron density, 263–264
    potential energy surface, 43–45, 133, 144,
        196–198, 221, 234–235, 238, 243, 260,
        291, 319, 401, 472, 477, 497, 505, 522,
        532, 573
Green's function, (see Propagator method)
Grid, 62–64, 260, 308, 318, 338, 399–401,
    466–467
    docking, 62–64
    for ESP charge, 318
    integration over, 260, 338, 399–400
GROMOS, 54, 60, 99
Ground state, 109, 115, 360, 487–504,
    507–508, 511, 513–515
Group theory, (see Point group and Symmetry)
GVB, 209

H&H, 266, 296
Half-electron method, (see Hartree-Fock
    theory)
Half-life, 521
Hamiltonian, 72, 106–111, 119–122, 154, 157,
    166, 179, 203, 212, 215, 219–220, 223,
    249–250, 252–255, 262, 321, 325–327,
    387, 396–397, 434, 436, 457, 459, 461,
    478, 496, 508, 562, 572–575
    determination from electron density,
        249–250, 252–254, 475
    EVB, 477–482
    including radiation field, 508
    non-interacting, 122, 219–220, 255–256,
        265
    QM/MM, 457, 459–462, 467–469
Hardness, 270
Harmonic oscillator, 61, 72–74, 336–342, 356,
    364–365, 376, 484, 527, 531, 539
Harris functional, 269
Hartree product, 120–122
Hartree-Fock theory, 126–129
    ab initio, 126–129, 165–199, 203–205, 327
    and DFT, 258, 267
    half-electron, 148
    instability, 234
    limit, 128–129, 165–166, 173, 176–178,
        228, 230

periodic, 192
predictions from, 192–199, 281, 287–289,
    292–294, 322–323, 330–331,
    338–340, 346–348
projected, 506, 571–574
QM/MM modifications, 462
restricted, 126–128, 190, 197, 205, 234, 487
restricted open-shell, 188–190, 206, 325,
    328–329
semiempirical, 128, 131–147
TS structures, 197–198
unrestricted, 148, 188–190, 234, 244, 272,
    324–325, 328, 506, 545, 571–574
Hartree-Fock-Slater method, 252
HCTH, 264, 274, 283, 285, 287, 289, 292, 296
Heat capacity, 366, 445–446
Heat of formation, 37, 40–41, 142, 147–148,
    155, 192, 240–244, 356, 366–375, 378,
    381–383
Heaviside function, 534
Heisenberg spin ladder, 505
Hellmann-Feynman theorem, 264, 326
Hessian, 44–46, 185, 191, 221, 260, 336–338,
    365–366
Hexachloroethane, 422–424
HF/3–21G/OPLS, (see QM/MM)
Hindered rotor, 376–377
Histogram, 83–86, 439, 541
Hohenberg-Kohn theorems, 252–254, 273, 494
Hole function, 251, 257, 278
Hooke's law, 18
Hückel theory, 115–119, 269
    extended, 134–136, 181
Hund's rule, 204
Hybrid DFT, (see Adiabatic connection)
Hydrazine, 138–139, 151
Hydrogen bonding, 33, 50, 112, 145, 149–151,
    156, 158, 193, 195, 198, 279, 293, 309,
    386, 407, 433, 449, 460, 463, 513
Hydrogen cyanide, 316, 322, 347, 430–435
Hydrogen electrode, 410, 414, 423
Hydrogen fluoride, 176, 236, 294, 347–349
Hydrogen-atom transfer, 267, 286, 537–538
Hydrophobicity, 152, 388, 407–408, 449, 452
Hydroxylamine, 381–383
8-Hydroxyquinoline, 513–515
Hyperconjugation, 24–25, 313, 578–579
Hyperfine coupling, 10, 189, 305, 327–330,
    343–344

Hyperpolarizability, 325–327
Hypervalency, 143, 148, 153–155, 174–175, 197
Hysteresis, 434

Ideal gas assumption, 358–359, 361–362, 379, 527
IGLO, 345
IMOMM, (*see* QM/MM)
Implicit solvation, (*see* Solvation)
Improper torsions, 27
Index of refraction, 409, 512, 542
INDO, 139–143, 153, 181
INDO/S, 139–141, 153, 497, 502, 514–515
Infrared spectroscopy, (*see* Vibrational spectroscopy)
Intensity, (*see* Frequency)
Internal coordinates, 6–7, 29, 34, 36, 46–48, 82, 336, 459, 522, 539, 541
Internal energy, 92, 356, 358–366, 376–377, 430–432, 444, 453, 525
Intrinsic reaction coordinate, (*see* Reaction coordinate)
Ion convention, 378
Ionic strength, 394–395
Ionization potential, 116, 135, 137, 194–195, 270, 272, 285, 311, 330–331, 414, 502
    predicted values, 141, 143, 149, 194–195, 288–290, 423
    valence-shell, 135
IPCM, (*see* Solvation)
IRC, (*see* Minimum-energy path)
Irreversible reaction, 520, 522, 524
ISM, 264, 296
Isodesmic equation, 166, 372–375, 381–382, 413
Isogyric, 373–374
Isotope effect, 357, 528–531
    kinetic, 482–484, 528–531, 533, 537–538

Jahn-Teller distortion, 206
Jellium, 250

KCIS, 264, 296
Kinetic-energy density, 264
Kinetic-energy functional, 250, 255–258, 262, 264, 274–275
Kinetic-energy operator, 107, 266, 269, 274, 332, 344

Kinetic isotope effect, (*see* Isotope effect)
Kinetics, 199, 267, 334, 344, 390, 393, 421, 482–483, 519, 523, 537
Kirkwood-Onsager equation, 396–397
KMLYP, 286–288, 296
Kohn-Sham theory, (*see also* Self-consistent field), 255–257, 274–278, 397, 448, 578
    QM/MM modifications, 461–462
Koopmans' theorem, 149, 194–195, 272, 330–331
Kramers-Grote-Hynes theory, 539
Kronecker δ, 107, 224

Landau-Zener model, 541
Langevin dipole, 466–467
Langevin dynamics, 80, 539
Lap correlation, 264, 297
Laplacian operator, 107, 127, 562
LCAO approach, 111–113
Leapfrog algorithm, 77–78, 101
Lennard-Jones potential, 29–30, 33, 47, 155, 271, 461–463, 478
Lewis structure, 209, 578–579
LG exchange, 263, 297
Line search, 44
Linear response theory, 387
Linear scaling, (*see* Scaling behavior)
Link atom, 473–477
Liquid crystal, 417
Local density approximation, 258–263, 266
Local minimum, 6, 41–46, 61, 69, 183, 185–186, 235, 291, 337–338, 377–378, 419, 522–523
Local spin density approximation, 259–267, 278, 282, 285–286, 288–289, 291–292, 294, 345–348
Locally enhanced sampling, 98
London forces, (*see* Dispersion)
LT2A, 297
LYP correlation, 263, 266, 291, 297

MacroModel, 50–51
Magnetic multipole moment, (*see* Multipole moment, magnetic)
Marcus theory, 541–544
Markov chain, 82
Mass-velocity relativistic correction, 223
Mataga-Nishimoto integral, 137

Matrix diagonalization, (*see also* Scaling behavior), 14, 212–214, 262, 274, 462, 480, 496, 522

Matrix elements, (*see also* Configuration interaction), 114, 116, 119, 127–128, 138, 184–185, 213, 222, 257, 269, 272, 345, 462, 496–497, 502, 510, 562, 573

Matrix isolation, 349–351

Mayer bond order, 320–321

MBPT*n*, (*see* Perturbation theory)

MCG3, (*see* Multilevel methods, multicoefficient models)

MCSCF, (*see* Self-consistent field, multiconfiguration)

Membrane, 418, 421

Memory, 13, 191

Menschutkin reaction, 393

meta-Generalized gradient approximation, 264, 268, 278, 285

Metal, (*see also* Solid and Transition metal), 26, 38, 61, 141, 179, 275, 286, 291, 299, 328, 452, 542

Metastability, 96

Methanol, 100, 156, 208, 211, 299–300, 319, 347, 437–438

Methyl radical, 116, 188–189, 330, 343–344, 374

Methyldiazonium cation, 317

Metropolis sampling, 81–82, 451, 459

Microcanonical ensemble, 91

MINDO/3, 141–145

Minimum, (*see* Global and Local minima)

Minimum-energy path, (*see also* Reaction coordinate), 7, 449, 522–523, 531–532, 538, 545

Missing parameters, 39–41

MM2, 38, 40, 50, 55, 59–60

MM2*, 50–51

MM3, 14, 20–21, 38, 50, 55, 59–60, 64–65, 341, 469–471

MM3*, 50–51

MM4, 55

MMFF, 38, 50, 56, 60, 341, 459

MMX, 56

MNDO, 143–156, 158, 193, 281, 287, 346–347

MNDO/d, 145, 153–155

MNDOC, 145

Model, (definition), 2

Model chemistry, 180, 240

Mole, 355

Molecular dynamics, (*see* Dynamics)

Molecular electrostatic potential, (*see* Electrostatic potential)

Molecular mechanics, 17–66, 150, 264, 341, 445, 457–484, 532

Molecular orbital, (*see* Orbital)

Molecular orbital theory, 105–129, 203–244, 271–280, 285, 575–579

   ab initio, 129, 131, 133, 143, 165–246

   semiempirical, 128, 131–162, 237, 260, 375

Molecular rotational constant, 333

Molecular weight, 100, 152, 362, 380, 450

Molecularity, 519

MOMEC, 56

Moment of inertia, 6, 94, 332–334, 362–364, 377

Monte Carlo, 64, 80–82, 90, 92–93, 259, 431–434, 438–440, 444, 447, 449, 451, 459, 463, 513

Morse potential, 20–21, 30, 311, 478

*m*PBE exchange, 263, 279, 297

MP*n*, ($n = 2, 3, 4, \ldots$; *see* Perturbation theory)

*m*PW exchange, 263, 297

MPW1K, 267–268, 279, 283, 285–288, 297

*m*PW1N, 268, 297

MPW1S, 268, 297

*m*PW1PW91, 268, 283, 287, 290, 292, 297, 339

*m*PW3PW91, 284, 293, 339

*m*PWPW91, 283, 285, 287, 290, 292, 339, 503

MST, (*see* Solvation, PCM)

Mulliken, (*see* Population analysis)

Multiconfigurational molecular mechanics, 50, 532

Multilevel methods, 239–244, 260

   CBS, 239, 242, 281, 286–289, 371, 382

   G*n*, ($n = 1–3$), 240–243, 281, 284, 289, 291, 371, 382

   multicoefficient models, 242–243, 281, 287

   W*n*, ($n = 1–4$), 242, 282, 289

Multiple-minima problem, 96–98, 451

Multipole expansion, 154, 307–308, 397–398, 401, 416

Multipole moment,
  electric, 30–33, 144, 154, 305–308,
    317–318, 387–388, 397–398
  magnetic, 326, 344

NBO, (*see* Orbital, natural)
NDDO, 143–158, 371, 476
Newton's equations of motion, 74
Newton's second law, 74
Newton's third law, 79
Newton-Raphson minimization, 44–46
NHE, (*see* Hydrogen electrode)
Non-classical reflection, 534–535
Non-crossing rule, 499
Non-local DFT, (*see* Generalized gradient
  approximation and Gradient, electron
  density)
Nonbonded interaction, (*see also*
  Charge-charge, Dipole-dipole, and Steric
  interactions and Dispersion), 19, 27, 30,
  33–35, 40, 47, 62–64, 88, 100, 149, 151,
  157, 195, 271, 278–279, 293, 311, 371,
  432–434, 443, 459, 461, 467, 480
Normal mode, 46, 337, 341–342, 356,
  364–366, 376, 391, 452, 514–515, 523,
  530–531, 535
Normalization, 71, 84–87, 107, 124, 168, 203,
  206, 218, 321, 361, 418, 439, 462, 504,
  572
Nosé-Hoover coupling, 92
Nuclear magnetic resonance, 10, 36, 64–66,
  305, 344–349, 411
Nuclear motion spectroscopy, (*see also*
  Rotational and Vibrational
  spectroscopies), 331–344
Nuclear Overhauser effect, 35–36, 99
Nuclear repulsion energy, 106, 110–111, 145,
  158
Nucleic acids, 35, 38, 47, 57, 99, 156, 279,
  387, 408, 411, 416–417, 421, 462, 469,
  472

O exchange, 263, 267, 297
O3LYP, 267, 284, 287–288, 290, 297
Occupation number, 206, 210, 489
Octanol solution, 409, 446
Octanol-water partitioning, 152, 416, 438
Odd function, 342
OMn, (n = 1, 2), 158

One-electron integral, 127, 137–138, 153, 184,
  262
  external potential, 262
  symmetry, 184
Open-shell singlet, 136, 190, 208, 276–277,
  505, 570
Operator, 4, 106, 216–221, 249, 271, 396, 463,
  565–566
  cluster, 224–226
  Coulomb and exchange, 127, 220, 265,
    273–274
  Fock, (*see* Hartree-Fock theory and Reaction
    field)
  Hamiltonian, (*see* Hamiltonian)
  Kirkwood-Onsager, 396
  Kohn-Sham, (*see* Kohn-Sham theory)
  one-electron, 120–122, 126–129, 203, 223,
    253, 255–256, 462, 466, 475, 510
  permutation, 123–124
  S^2, 188, 324–325, 328, 506, 565–572
  spin-annihilation, 506, 571–573
  S_z, 122, 272, 324–325, 327, 565–571
OPLS, 38, 56, 60, 98–99, 438, 446, 449,
  459–460
OPLSA*, 51
Optical rotatory dispersion, 504
Orbital, (*see also* Basis set)
  active space, 207–210, 233–234, 239,
    499–501
  banana-bond, 576
  complex, 234
  contracted, 168–173, 180
  core, 134, 171–172, 209, 215, 224, 328, 345
  energy, 111, 115–119, 128, 195, 219,
    230–231, 269, 498, 502, 577
  floating, 173
  frozen, 215, 475–476
  gauge-including, 345
  Gaussian-type, 155, 167–180, 273–274, 328
  generalized hybrid, 476–477, 483
  hydrogenic, 112, 128, 134, 167–168
  Kohn-Sham, 256–257, 264, 272, 306, 448,
    502
  localized, 209, 221, 345, 475–476, 575–579
  molecular, 39, 100, 105–129, 149, 158,
    165–199, 203–244, 306, 324, 328, 343,
    475, 487–500, 506, 559–562, 571, 573,
    575–577
  natural, 578

primitive, 168–173
Slater-type, 134, 136–138, 141, 148, 155, 158, 167–172, 181–182, 270, 273, 328, 347–348
spin, 124–126, 188–190, 234, 324, 575
valence, 134, 140, 145, 155, 171–173, 328, 345, 498
virtual, 195, 205–212, 216, 221, 228–229, 272, 487–496
Orthogonality, 107, 126, 158, 177, 182, 206, 221, 314, 475–476, 491–492, 495–496, 501, 509
Orthonormality, 107, 218, 312, 314, 540, 565, 567, 569
Oscillator strength, 141
Overlap integral, 114, 134–138, 262, 312–314, 321, 325, 328, 571
Overtone, 341
Oxygen atom, 227–228

P exchange, 263, 297
P86 correlation, 263, 290, 297
Pairlist, 47, 90
Pairwise descreening, 404, 420
Parameterization, 36–39, 140–142, 145–147, 154, 172, 237–244, 266–271, 341–342, 410–411, 418, 445–447, 459–463, 477–482, 497, 514
Parameterized correlation methods, 237–244, 370–371
Pariser-Parr approximation, 137
Pariser-Parr-Pople model, 138, 502
Partial charge, (see Atomic partial charge)
Particle in a box, 361
Particle-mesh Ewald, (see Ewald sum)
Partition coefficient, 411, 416–419
Partition function, 71, 357–366, 429–432, 525–531, 562–563
    activated complex, 525–532, 535
    electronic, 359–361
    molecular, 359–366
    rotational, 359, 362–364, 528–529, 563
    translational, 359, 361–362, 526–529
    vibrational, 359, 364–366, 484, 528–529
Pauli exclusion principle, 123–124, 179, 251
PBE, 263, 283, 285, 287, 289, 292, 297
PBE1PBE, 267, 283, 285, 287, 290, 292, 298, 330, 347, 503–504
PCI-80, 238, 286, 370

PCM, (see Solvation)
PDDG, 158
PEF95SAC, 54
Penalty function, 36–37, 41, 146, 154
Perchloroethylene, 423–424, 460
Periodic boundary conditions, 86–89, 101, 273, 431, 448, 452, 459
Permutation operator, 123–124
Perturbation theory, 216–224, 226, 228–243, 336, 401, 469, 498, 573–574
    convergence, 228–229
    localized MP2, 222
    multireference, 223, 233–234, 375, 495, 501
    predictions from, 281, 287–289, 291–292, 294, 321–323, 327–330, 338–340, 343, 346–347, 366
    spin-component-scaled, 239
    spin-projected, 574
    time-dependent, 507
Phase angle, 23, 25–26, 35, 40
Phase point, 70–72, 80–83, 448
Phase space, 70–82, 92–98, 420, 429–430, 439, 443–444, 448, 450–451, 538
Phenylnitrene, 276, 490–492, 494–496, 500–501, 504–505
Phenylnitrenium cation, 272
Pierotti's formula, 406
pK_a, (see Acidity)
PKZB, 264, 283, 285, 289, 292, 298
Planck's constant, 15, 107, 356, 489, 507
PM3, 146–156, 158, 160–162, 193, 281, 320–323, 338, 340, 469
$PM3_{BP}$, 156
PM3(tm), 154
PM4 and PM5, 151, 155
Point group, 183–186, 232, 234, 254, 362–363, 490–492, 494, 514, 559–563, 577
Poisson equation, 394–402
Poisson-Boltzmann equation, 394–395, 399–400, 453
Polarizability, 33–34, 90–91, 155, 294, 325–327, 387, 407, 446–447, 451, 464, 466–467, 513
    frequency-dependent, 502
Polarization, 90–91, 156, 173–175, 387, 411, 459, 462, 466–467, 512
    free energy, 394–405, 449

Population analysis, (*see also* Atomic partial charge), 315, 476, 578
  bond order, 320
  Löwdin, 314–315, 320
  Mulliken, 199, 270, 312–315, 319–323
  natural, 314–315, 321–323, 578
Potential energy functions
  bond stretching, 18–21, 473
  torsions, 22–27, 377, 474
  valence angle bending, 21–22, 26, 473
  van der Waals, 27–30, 46
Potential energy surface, 6–10, 17, 99, 111, 183–187, 216, 221, 238, 244–246, 331, 334–338, 356, 469, 511–512, 522–523, 535–536
  electron-transfer, 542–543
  EVB, 478–481
  excited-state, 490, 500, 511–512, 539–544
  solvated, 389–393, 409, 415, 380, 423, 511–512, 538–539
Potential of mean force, 146, 419–420, 439–442, 463–464
PPP, (*see* Pariser-Parr-Pople)
Pressure, 92–93, 358–359, 380, 385, 430, 443
  standard-state, 361–362
Principal components analysis, 95, 421
Probability, (*see also* Transition probability), 4, 12, 69, 81, 83, 85, 92–93, 98, 106, 112, 121, 125, 166, 262, 377, 430–432, 439–441, 541
Projection, (*see* Spin, projection)
Prolate top, 333, 362
Propagation, 74–79
Propagator method, 497, 501–504
Propylene oxide, 544–546-497
Protein, (*see also* Enzyme-substrate binding), 31, 35, 38, 54, 56–59, 77, 96–97, 99, 157, 408, 421, 444, 452, 473, 475–476, 481–484, 538
Proton affinity, 148, 194, 291
Proton transfer, 442, 477–481
Pseudobond/pseudoangle, 35–36
Pseudohalogen, 474, 477
PW exchange, 263, 267, 298, 330
PW91 correlation, 263, 266, 291, 298
PWPW91, 283, 285, 287, 289, 339

QCISD, (*see* Configuration interaction, quadratic)

QM/MM, 61, 157, 447, 457–484, 513, 532
  AM1/OPLS, 465
  AM1/TIP3P, 463–465
  AOC, 459–460, 463
  boundary, 458, 467–468, 473–476
  HF/3–21G/OPLS, 465
  IMOMM, 472
  ONIOM, 473
  XSOL, 465
QSAR and QSPR, (*see* Structure-activity relationship)
Quadrature, 260, 338
Quadrupole, 31, 144, 154, 307, 398
Quantization, 105
Quantum mechanics, 1, 4–5, 28, 105–129, 309, 326, 357, 457–484, 575
Quantum numbers, 122–123, 134, 167, 229, 324, 333, 336, 361

R12 method, 229–230, 249
Radial distribution function, 84–86, 449, 465
Radical, 105, 117, 119, 148, 179, 185–190, 194–195, 199–200, 234, 244, 279, 329, 331, 349
Radical polymerization, 199–200
Radius, (*see* Born equation, Cavity, and van der Waals)
Raman spectroscopy, 341
Random number, 82
Rare event, 440, 442
RASSCF, (*see* Self-consistent field, multiconfiguration)
Rate constant, 12, 519–522, 532–538, 541–546
Rate-determining step, 481–482, 537
Reaction coordinate, 48–49, 207, 211, 241, 277, 391–392, 409, 419, 439–442, 449, 451, 457–458, 463–464, 479–484, 511, 522–526, 536–538, 542–546
  intrinsic, (*see* Minimum-energy path)
Reaction dynamics, (*see* Dynamics)
Reaction field, (*see also* Solvation), 387, 394, 396, 400–401, 404, 513, 542
  charge transfer, 415, 448
  self-consistent, 393–406, 410–424, 449–450, 513
Reaction quotient, 379
Redox potential, (*see* Electrochemistry)
Reduced mass, 17, 332, 336–337, 357, 535

Reductive dechlorination, 422–424
Reference interaction site model, 465–467
Relativistic effects, 107, 140, 179, 223–224, 345, 368, 565
Relativity, 123, 129, 565
Renner-Teller, 244
Reorganization energy, 454, 542–543
Resolution of the identity, 573
Resonance energy, 118–119
Resonance integral, 114, 134, 144
Rice-Ramsperger-Kassel-Marcus theory, 536
Rigid-rotor approximation, 332–334, 362, 527, 531
Ring critical point, 316
RNA, (see Nucleic acids)
Root switching, 499–500
Roothaan, 126–128
Root-mean-square deviation, 94
Rotation, 23, 97, 119, 150, 194, 280, 332–335, 363, 370, 439, 557–558
  barriers about bonds, 150, 158, 194
  spectroscopy, 332–334
RPA, (see Propagator method)
RRKM theory, (see Rice-Ramsperger-Kassel-Marcus theory)
Runge-Kutta integration, 78
Rydberg states, 113, 141, 498

Saddle point, (see also Transition state), 6, 9, 522–523, 531, 533–534, 539
SAM1, 154–155
Sampling, (see also Metropolis sampling), 72, 75, 82, 93, 95, 98, 415, 420, 432–445, 448, 450–451, 463
Scaling behavior, 46–48, 128, 190
  CCSD, 225, 237
  CCSD(T), 237
  CCSDT, 225, 237
  CISD, 214, 237
  DFT, 262, 273–274
  Ewald summation, 47
  fast multipole methods, 48
  Hartree-Fock theory, 128, 178, 190, 237, 262
  linear, 14, 48, 157, 191, 222, 237, 274
  matrix inversion/diagonalization, 14, 45, 185, 262, 274
  molecular mechanics, 29, 47–48
  MP2, 221, 237

MP4, 222, 237
QCISD, 237
Schrödinger equation, 106, 539
  electronic, 110, 120, 122, 128–129, 134, 165, 167, 170, 211, 225, 228, 237, 246, 250, 254–255, 278, 332–333, 507
  non-linear, 225, 396–397, 400
  nuclear, 331, 540
  one-dimensional, 334–335, 536
  rigid-rotor, 332–333, 362–363
  time-dependent, 507, 532, 536
SCIPCM, (see Solvation)
Second derivatives, (see Hessian)
Secular equation, 113–115, 134–136, 212, 256
Selection rules, 332–333, 336, 341, 510
Self-consistent field, (see also Density functional, Hartree-Fock, and Kohn-Sham theories), 121–122, 126–129, 448, 475, 482, 493–496, 505, 514, 577
  localized, 475–477
  multiconfiguration, 205–211, 233–235, 246, 336, 349–350, 495, 499–501, 574
  reaction field, (see Reaction field, self-consistent)
Self-interaction error, 251, 263, 278, 280
Semiempirical, (see Molecular orbital theory)
SHAKE, 79
SHAPES, 26, 58
Shear viscosity, 88
Simulated annealing, 97
Simulation, 38, 69–99, 319–320, 357, 429–454, 459, 462–465, 475–477, 513
SINDO1, 141–143, 153
Single-point calculation, 129, 192, 345, 473
Single topology, 443
Singlet-triplet splitting, 138, 216, 232–234, 275–277, 350, 493–495
Size-consistency, 215, 221, 225–226, 241, 257
Slater determinant, 124–126, 166, 203, 255–256, 265, 272, 274, 328, 396, 400, 461, 487–488, 493, 504, 573
Slater exchange, 252, 258, 260
Slater-type orbitals, (see Orbital)
Slow growth, 435–437, 443
SMx model, (see Solvation)
S_N2 reaction, 185–186, 198, 293, 390, 440
Software, 12–15

Soil, 418
Solid, (*see also* Metal), 9, 49, 89, 99–101,
    136, 186, 192, 252, 269, 273, 366, 418,
    470, 498, 559
Solubility, 418
Solvation, (*see also* Condensed-phase effects
    and Reaction Field),
  continuum, 385–386, 393–424, 448–454,
      460, 462, 467, 469, 473, 513, 538
  convergence, 401
  COSMO and COSMO-RS, 404–406
  effect on UV/Vis spectroscopy, (*see*
      Solvatochromism)
  effective coordinate, 539, 542–544
  electrostatic component, 397–410, 449
  explicit modeling, 91, 429–454, 459–460,
      462
  explicit/implicit hybrid models, 421,
      451–452
  free energy, 386–424, 429, 437–439, 458,
      463–466
  generalized Born, (*see* Born equation)
  GB/SA, 408–409
  IEF-PCM, 401
  ions, 90, 447
  IPCM, 401
  MST-ST, 408–410, 416–417
  non-electrostatic component, (*see also*
      Cavitation and Dispersion), 406–410,
      421
  non-equilibrium, 421–422, 450–451, 538,
      542–543
  non-isotropic, 418
  PCM, 400–401, 405, 417, 473
  proton, 410, 412
  SCIPCM, 401
  separable equilibrium, 538
  shell, 86, 385–386, 410, 420, 449–452,
      512–513
  SM*x*, 387–388, 404, 409, 411, 416–418,
      422–424, 438, 448, 460, 462
  supermolecule, 415–416, 450–451
  time scale, 421–422
Solvatochromism, 393, 511–5133
Solvent-accessible surface area, (*see* Surface
    area)
Solvent model, 400–403
Solvent-separated pair, 419–420

Solvent-solvent interaction, 432, 436
Space group, 559
SPC, (*see* Water)
Spherical harmonics, 134, 332–333
Spin, 122–124, 136, 258, 324–325, 360,
    487–488, 492, 565–574
  contamination, 190, 234, 272–273,
      275–276, 324–325, 328–330,
      494–495, 506–507, 571–574
  degeneracy, 360
  density, 189, 199, 330
  polarization, 188–190, 199, 258–259, 328
  projection, 234, 325, 328–329, 506–507,
      571–574
Spin-orbit coupling, 107, 129, 242, 361,
    368–369
Spin-orbital, (*see* Orbital)
Spin-spin coupling constant, 305, 345–349
SRP, 155–156
Standard state, 361–362, 366–375, 386, 526
  conversion, 378–379, 386, 423
  elemental, 366–372
State-averaging, 500
Stationary point, 24–25, 45–46, 133, 135, 183,
    185, 192, 262, 300, 338, 356, 389–393,
    423, 511, 523, 527, 529, 531, 545
Statistical mechanics, 357–366, 377, 386
Steepest descent minimization, 44
Steric interaction, 19, 24, 27–30, 452–454,
    467, 579
Stirling's approximation, 359, 362
STO-3G, (*see* Basis set)
Stochastic dynamics, 79–80
Stokes shift, 489, 515
Strain energy, 21, 40–41, 355–356
Structure-activity relationship, 152–153, 418
Structure prediction, 19, 61
Sublimation, 418
Sucrose, 469–472
Sum method, 494–495, 504–507, 570
Supercritical fluid, 417
Surface area, 47, 152, 386, 400, 407
  solvent-accessible, 407–409
Surface hopping, 540–541
Surface tension, 407–410, 452, 467
SVWN, 260, 282, 286, 289, 291, 294, 339
Switching function, 47, 100, 394
Switching parameter, (*see* Coupling parameter)

SYBYL, 58

Symmetry, 182–188, 209, 215, 275–276, 279, 335, 490–491, 498–500, 510, 557–562
  number, 362–364, 377

$\tau 1$ correlation, 264, 298

τHCTH, 264, 285, 298

T_1 diagnostic, 226

TDDFT, (see Density Functional Theory)

Temperature, 28, 61, 69–71, 75, 77–78, 82, 91–92, 97, 336, 356–357, 360, 362–363, 369–370, 372, 377, 380, 438, 442, 444, 447, 519, 523, 527–528, 531–533, 535–538, 542, 544, 546

Tetramethylsilane, 346

Theory, (definition), 1

Thermal electron convention, 378

Thermodynamic integration, 435–437, 443, 454

Thermodynamics, 11, 39–40, 355–380, 429–445

Thomas-Fermi functional, 251

Thomas-Fermi-Dirac functional, 252

TIPnP, (n = 3, 4; see Water)

Torsions, (see also Potential Energy Functions), 22, 27, 356, 376–377, 469, 578–579

TPSS, 264, 279, 283, 290, 292, 298

TPSSh, 283, 290, 292, 298

Trajectory, (see also Propagation), 70–80, 482–484, 522, 541

Transition dipole moment, 501–502, 510, 563

Transition metal, (see also Metal), 58, 140–141, 154, 179, 268, 275, 285–286, 291, 299, 345, 504

Transition probability, 272, 305, 507–511

Transition state, (see also Geometry), 7, 9, 11, 46, 48–50, 62, 133, 159–162, 185–186, 208, 235, 279, 293, 300, 338, 378, 390–93, 421–422, 440, 450–452, 463–464, 468, 482–484, 507, 522–533, 536, 538–539, 544–546

Transition-state theory, 160, 524–537
  microcanonical, 536
  variational, 531–533, 536, 538

Translation, 6, 327, 361–362, 378, 386

Transmission coefficient, 535–537

Trichloroacetic acid, 422–424

Trimethylenemethane, 204–209, 277, 504

Tunneling, 482, 530, 533–538, 544

Turning point, 511

Two-electron integral, 127–129, 132, 136–140, 143–144, 154, 158, 166, 169, 191, 221, 251, 257, 273, 462, 475
  Coulomb, 122, 125–127, 138, 189–191, 273–274, 402
  exchange, 125–127, 189, 191, 214, 261
  symmetry, 186

UFF, 38, 58, 60, 469

Umbrella potential, 99, 440–441, 451, 481

Uncertainty principle, 334

Uniform electron gas, 250–252, 259–260, 262

United-atom model, 38, 51, 446

Universal gas constant, 359, 542

Urey-Bradley term, 26, 100

UV-Vis spectroscopy, 135–136, 139–141, 393, 411, 507–508, 511–515
  effect of solvation, (see Solvatochromism)

VALBOND, 38, 58

Valence angle bending, (see Potential Energy Functions)

Valence bond, (see also Empirical valence bond), 49

van der Waals,
  interaction, (see Potential Energy Functions)
  radius, 311, 399, 401, 408
  surface, 308, 318, 400, 408

van't Hoff plot, 444

Vapor pressure, 152, 405

Variational principle, 108–110, 172, 203, 212, 215, 222, 251, 253–254, 257, 490–491, 499

Verlet algorithm, 77–79

Vertical process, 331, 423, 489–490, 505–506, 511–513

Vibrational averaging, 61, 342–344

Vibrational frequency, (see Frequency)

Vibrational motion, 76, 334–336, 356

Vibrational spectroscopy, 17, 21, 35, 88, 101, 332, 334–342, 349–351, 411, 514

Voronoi cell, 318

VSXC, 264, 283, 285, 287, 290, 292, 298, 340

VTST, (see Transition-state theory)

VWN, 259–260, 298

VX, 177–178

W*n*, (*n* = 1–4; *see* Multilevel methods)
Water, 2, 82–83, 85, 91, 151, 172–173, 184,
  230, 236, 291, 294, 306–307, 309, 315,
  347–348, 559
  acidity, 477–481
  as solvent, 385–387, 390, 392, 407–408,
    422–424, 429–433, 437–438, 445,
    449, 452, 460, 462–464, 480–481
  density anomaly, 447
  polarizable models, 90, 446–447
  SPC, 446
  TIP*n*P, (*n* = 3, 4), 446, 449, 462
Water dimer, 149–151
Wave function, 4, 105–129, 166, 181,
  186–190, 230, 244, 249, 312, 324, 326,
  411, 459–466, 561–562, 575–577
  adiabatic, 478, 539–540
  CIS, 214, 496–498
  determination from electron density,
    253–254
  excited-state, 487–494, 497–501, 504–507
  harmonic oscillator, 335–337, 510–511
  Kohn-Sham, 271–273, 448, 506
  mixed-spin state, 504–507, 566–570
  multideterminantal, 190, 203–216, 230–235,
    274–277, 324, 487–488, 562

  nuclear, 111, 331–333, 335–337, 342–343,
    539–541
  rigid-rotor, 332–333, 510
  single-determinantal, 203–205, 222, 226,
    230–232, 271–273, 487–488, 493
  stability, 187, 192, 234, 275, 475, 498
  VB, 477–479
  vibrational, 61, 334–337, 342–343,
    510–511
Weighted histogram analysis method, 441–442
Wigner correction, 535

X exchange, 263, 298
X*α* method, 252, 259
X-ray crystallography, 3, 61–62, 84, 315,
  470–471
X3LYP, 267, 279, 284, 290, 298
XLYP, 283, 290
XSOL, (*see* QM/MM)

Zeolite, 99–101, 312
Zero-flux surface, 316–317
Zero-point vibrational energy, 62, 69,
  356–357, 364–365, 368–369, 375, 484,
  490, 495, 511, 525, 529–532, 545
Zwitterion, 392, 465

Printed and bound by CPI Group (UK) Ltd, Croydon, CR0 4YY

27/10/2024

14580201-0004